Acoustical Imaging

Volume 10

Acoustical Imaging

Volume 1 Proceedings of the First International Symposium, December 1967, edited by A. F. Metherell, H. M. A. El-Sum, and Lewis Larmore

Volume 2 Proceedings of the Second International Symposium, March 1969, edited by A. F. Metherell and Lewis Larmore

Volume 3 Proceedings of the Third International Symposium, July 1970, edited by A. F. Metherell

Volume 4 Proceedings of the Fourth International Symposium, April 1972, edited by Glen Wade

Volume 5 Proceedings of the Fifth International Symposium, July 1973, edited by Philip S. Green

Volume 6 Proceedings of the Sixth International Symposium, February 1976, edited by Newell Booth

Volume 7 Proceedings of the Seventh International Symposium, August 1976, edited by Lawrence W. Kessler

Volume 8 Proceedings of the Eighth International Symposium, May 29–June 2, 1978, edited by A. F. Metherell

Volume 9 Proceedings of the Ninth International Symposium, December 3–6, 1979, edited by Keith Y. Wang

Volume 10 Proceedings of the Tenth International Symposium, October 12–16, 1980, edited by Pierre Alais and Alexander F. Metherell

A Continuation Order Plan is available for this series. A continuation order will bring delivery of each new volume immediately upon publication. Volumes are billed only upon actual shipment. For further information please contact the publisher.

Acoustical Imaging

Volume 10

Edited by
Pierre Alais
Henri Beghin Laboratory
of Physical Mechanics
University of Paris VI
Saint-Cyr-l'Ecole, France

and
Alexander F. Metherell
Department of Radiology
South Bay Hospital
Redondo Beach, California

PLENUM PRESS · NEW YORK AND LONDON

The Library of Congress cataloged the first volume of this series as follows:

International Symposium on Acoustical Holography.

Acoustical holography; proceedings. v. 1-
New York, Plenum Press, 1967-

v. illus. (part col.), ports. 24 cm.

Editors: 1967- A. F. Metherell and L. Larmore (1967 with H. M. A. el-Sum)
Symposiums for 1967- held at the Douglas Advanced Research Laboratories, Huntington Beach, Calif.

1. Acoustic holography–Congresses–Collected works. I. Metherell. Alexander A., ed. II. Larmore, Lewis, ed. III. el-Sum, Hussein Mohammed Amin, ed. IV. Douglas Advanced Research Laboratories. v. Title.
QC244.5.I 5 69-12533

ISBN 978-1-4684-3946-5 ISBN 978-1-4684-3944-1 (eBook)
DOI 10.1007/978-1-4684-3944-1

Proceedings of the Tenth International Symposium on Acoustical Imaging, held October 12–16, 1980, in Cannes, France

Softcover reprint of the hardcover 1st edition 1982
A Division of Plenum Publishing Corporation
233 Spring Street, New York, N.Y. 10013

PREFACE

This volume contains the proceedings of the Tenth International Symposium on Acoustical Imaging held in Cannes, France, October 12th through the 16th, 1980.

Fifty-seven papers were presented over the course of the four day meeting. Fifty-two manuscripts were received in time for publication of the proceedings. There was representation from 14 nations, including England, France, U.S.A., West Germany, Canada, Italy, Japan, Poland, The Netherlands and Norway among the authors and in addition, Switzerland, Spain, Belgium, and Denmark were represented.

The following papers were presented at the meeting for which manuscripts were not received in time for publication: "Improved Phased Array Imaging and Medical Diagnosis" by F.L. Thurstone; "Scanning Acoustic Microscope Operating in the Reflection Mode" by H. Kanda, I. Ishikawa, T. Kondo, and K. Katakura; "Empirical Determination of Flaw Characteristics Using the Scanning Laser Acoustic Microscope - SLAM" by D. Yuhes, C.L. Forres, and L.W. Kessler; "A Wide Angled Fraction Limited Holographic Lens System for Acoustical Imaging" by H. Heier; "Progress in the Development of Sonographic Contrast Agents" by J. Ophir, and F. Maklad, A. Gobuty and R.E. McWhirt.

This meeting marks the first occasion that one of these symposia has been held outside of the United States. All of the nine previous symposia were held in California; Chicago, Illinois; Key Biscayne, Florida; and Houston, Texas. It was extremely gratifying to see the attendance maintained at such a high level. A total of 134 attendees were present during the conference and their names and addresses are listed at the end of the book. Since this meeting took place in Europe, it was attended by a higher proportion of international attendees. We hope that the trend of moving the symposium to various places in the world will continue to make these meetings available to a larger number of scientists and engineers and further add to tne international character of the meeting.

As editors and symposium chairmen, we would like to express our appreciation to the International Committee who helped in organizing the meeting including Mahfuz Ahmed, U.S.A.; C.B. Burckhardt, Switzerland; B.P. Hildebrand, U.S.A.; C.R. Hill, England; Larry W. Kessler, U.S.A.; C.T. Lancee, Netherlands; Rolf Mueller, U.S.A.; John P. Powers, U.S.A.; Jerry L. Sutton, U.S.A.; F.L. Thurstone, U.S.A.; Robert C. Waag, U.S.A.; Glen Wade, U.S.A.; and Keith Wang, U.S.A. The National Committee was responsible for most of the local arrangements and these include P. Alais,Paris; M. Auphan, Limeil; C. Gazanhes, Marseille; G. Grau, Rueil-Malmaison;A. Lansiart, Saclay; H. Mermoz, Toulon; P. Peronneau, Paris; J. Perrin, Paris; M. Perulli, Compiegne; L. Pourcelot, Tours; G. Quentin, Paris; J. Roux, Bordeaux; R. Torquet, Valenciennes; P. Tournois, Cagnes; and A. Zarembovitch, Paris. In addition, the editors also wish to express their thanks to Marrie- Theresa Larmande and the other members of the staff in the Laboratoire de Mecanique who helped run the registration and take care of all of the detailed organization during the meeting. The editors also wish to express their thanks to the Session Chairman, C. B. Burckhardt, D.E. Huyas, P. Tournois, Glen Wade, C.R. Hill, Pierre Peronneau, J. Perrin, F.L. Thurstone, M. Ahmed, R. Torguet, and E.J. Pisa. Finally, the help and assistance of P.J. Aviani in helping prepare the manuscript is appreciated.

The Eleventh International Symposium on Acoustical Imaging will be held May 4 through 7th, May, 1981 at the Naval Post-Graduate School, Monterey, California. Information may be obtained from the Chairman: Professor J.P. Powers, Department of Electrical Engineering, Naval Post-Graduate School, Monterey, California, 93940.

The Twelfth International Symposium on Acoustical Imaging will be held the week prior to the World Federation of Ultrasound in Medicine and Biology in July, 1982 at the Institute of Electrical Engineers Headquarters, London, England. Further information may be obtained from the Symposium Chairman: Dr. C.R. Hill, Institute of Cancer Research, Sutton, Surrey, England.

Pierre Alais
Alexander F. Metherell

CONTENTS

ARRAYS

METHODS

TISSUE CHARACTERIZATION

COMPUTER TOMOGRAPHY

DIFRACTION EFFECTS AND IMAGE FORMATION

UNDERWATER IMAGING AND NON-DESTRUCTIVE EVALUATION

REAL-TIME CONSTANT DEPTH SCANNING WITH PHASED ARRAYS

T.A. Whittingham

Regional Medical Physics Department
Newcastle General Hospital
Newcastle upon Tyne, England

INTRODUCTION

Constant depth scanning (C-scanning) differs from B-scanning in that the imaged cross-section lies in a plane which is perpendicular to the scanning beam rather than in a plane which contains it. The potential merits of C-scanning in medicine include the ability to image cross-sections which lie parallel or nearly parallel to the body surface, e.g. coronal cross-sections of the orbit or breast, and the ability to take advantage of strong focussing since large beamwidths before or after the focal zone are irrelevant. Since the high lateral resolution, high sensitivity and large angular aperture of a strongly focussed beam are qualities which are well suited to the detection and mapping of scattered ultrasound, C -scanning is a promising mode for tissue differentiation.

The water baths and slow mechanical movements associated with early C-scanners have done little to stimulate widespread interest in the technique, although Hill and Carpenter (1976) reduced both problems with a contact scanner employing a fast spiral scan.

The purpose of this paper is to examine some of the possibilities for real-time C-scanning by electronically controlled arrays, since this approach would overcome the present problems of bulky mechanics or water baths and of relatively slow image forming times.

The fundamental problem with any form of raster scanning with a transducer operating in the conventional pulse-echo mode is that most of the scanning time is wasted waiting for the echo to return from the required depth. In medical applications this means that for each image point the transducer must wait typically 200 μs (range 15cm) for an echo to return. Even a coarse 100 x 100 image matrix would thus take 2 seconds to compile, and in practice there would be further delays to allow for reverberations and echoes from beyond the target range. One solution is to detect the echo from all points along one line of the image at the same time, using hundreds of signal detection and processing channels working in parallel, one for each point in the line. The two dimensional image can thus be built up by repeating the process for a hundred or more closely spaced lines at the same range. This is the approach used, for example, by Green et al., (1974) in the S.R.I. Transmission Camera.

This 'parallel processing' approach involves a considerable amount of electronics and is likely to be expensive. An alternative solution is to transmit a pulse which, as before, excites all target points in one line of the object plane at the same time and to rapidly sweep the receiving beam along the line to interrogate each point in turn. This 'within-pulse' scanning approach has been used successfully in marine applications (e.g. Welsby et al., 1973) where the long pulse length necessary to give sufficient time (typically 100 μs) to scan across a 30° sector is not a serious disadvantage. The corresponding range resolution of up to 150mm clearly would be a considerable problem for medical applications.

A third approach is presented here. The idea is illustrated in Fig. 1 for both a linear and a sector scanning format. In the linear case (upper diagram) an active group of elements (T) in a linear array advances steadily along the array 'spraying' a narrow beam of continuous waves across the field of view. After a fixed time delay (τ) a similar receiving group (R) advances across the array receiving the echoes from a fixed range ($\frac{1}{2}$ c τ). Similarly in the sector scanning format (lower diagram) a narrow beam of continuous waves from a 'phased array' is sprayed across a sector, in the manner of a hosepipe which is rapidly swung in an arc, while a receiving beam is again arranged to follow at a fixed time interval behind, thus receiving echoes from the arc A B. In both cases A represents a target point from which an echo has just returned to the receiver and B represents a target point at the same range from which an echo has just started its return journey. Notice that although continuous waves are used, a given target point is irradiated only for a brief interval, and thus effectively 'experiences' a pulse (Fig. 2).

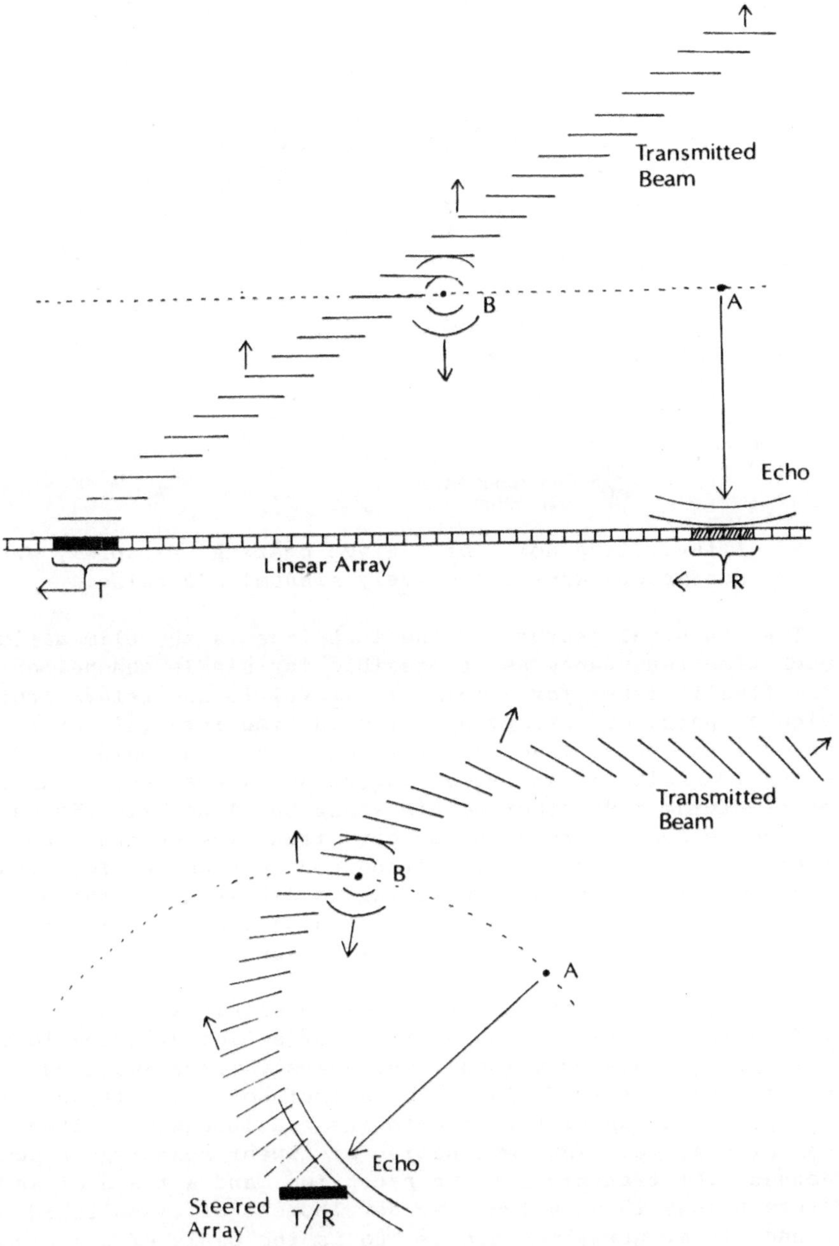

Fig. 1. Illustrating how a continuous wave transmitting array (T) and a tracking, but delayed, receiving array (R) may scan a line (AB) at constant range, for both a linear (upper) and sector (lower) scan format.

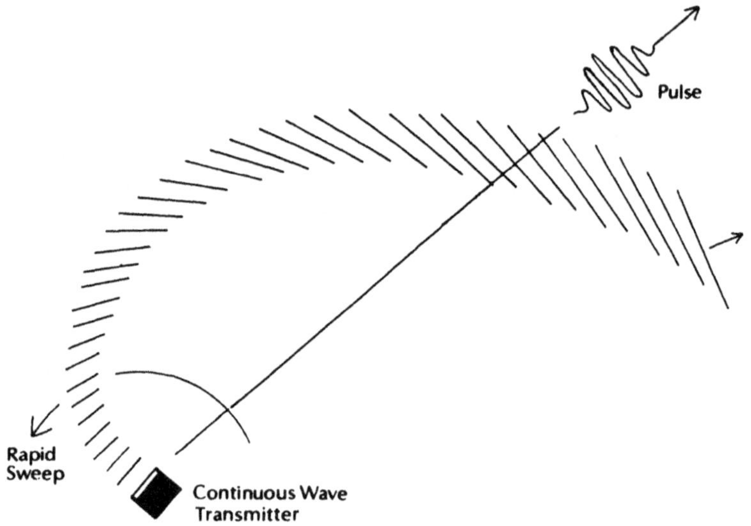

Fig. 2. Illustrating how, for a given bearing, a rapidly swept continuous wave effectively transmits a pulse.

The essential feature of the technique is the elimination of as much time redundancy as is possible for single channel operation. In the time it takes for a pulse to travel to and return from a particular point on the scanned line (a time interval for which a conventional pulse echo raster scanning system would be idle), both the transmitting and receiving channels are busy transmitting to or receiving from other points along the line. If the same transducer elements are used for both transmission and reception then some time redundancy is introduced in exchange for economy and system simplification, resulting in a halving of the maximum possible frame rate and the restriction of the minimum target range to that with a go and return time equal to the sweep period.

A number of variations are possible within the basic concept, including the choice between a linear or sector scanning format, the use of separate or common transducers for transmission and reception, and the transmission of a train of discrete but closely spaced pulses as an alternative to the continuous wave transmission illustrated above. The combination of linear scanning, a common transducer for transmission and reception, and a train of short discrete pulses is an attractive development of established pulse-echo and linear array techniques and is the basis of a system being developed at the present time in this laboratory. A description of this system will be published in due course.

For the remainder of this paper, however, the possibility of a more radical system using continuous waves in a sector scanning mode will be considered.

SYSTEM DESCRIPTION

The operating principle of a system employing separate focussed transmitter and receiver arrays is shown in Fig. 3. The arrays may be accurately described as 'phased arrays' since a single frequency is involved and beam steering is achieved by adjusting the phase shift between adjacent elements. It is shown below that a sweep time of 200 μs is practicable for a 60° arc, so that a slow orthogonal sweep, achieved by mechanical deflection of the beams, could build up a two dimensional image containing, say 200 lines in 40ms. The receiving beam follows the transmitting beam with a time delay τ, which could have any value up to the sweep duration T. If longer values were to be used then the receiver could not discriminate between echoes from the intended range ($\frac{1}{2}$ c τ) and those from a closer range $\frac{1}{2}$ c (τ-T). This restriction would not apply if the closer range corresponded to a water bath. Thus for a sweep time of 200 μs, ranges of up to 150mm could be accommodated. If a common transmission and reception array were to be used the reception sweep could not commence until the end of the transmission sweep and the minimum value of delay τ would equal the duration of the sweep T. The period between transmission sweeps would therefore be the sum of the sweep time and the delay, i.e. T + τ

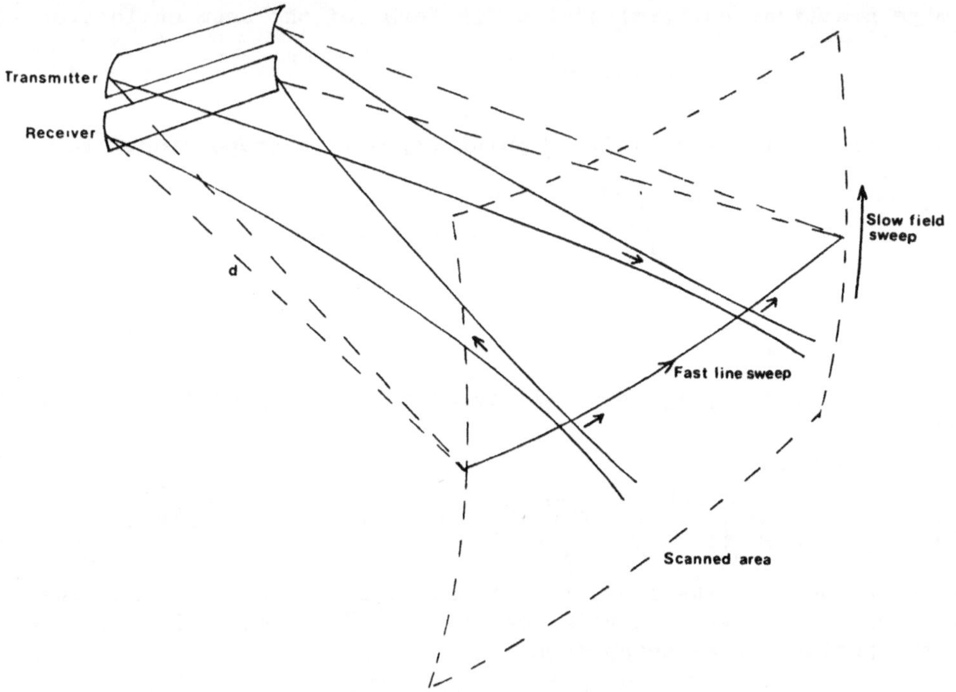

Fig. 3. A possible C-scan system. The relatively slow field sweep is achieved by mechanical deflection.

Thus the depicted system could scan an approximately square surface with sides of over 100mm at a depth of 100mm or so at real time rates. It will be seem that the range resolution is relatively coarse, essentially due to the narrow-band nature of the technique, so that the scanned zone would take the form of a curved slab, concentric with the centre point of the arrays.

THEORY FOR MODERATE SWEEP RATES, FAR-RANGE TARGETS, NO FOCUSSING

Transmitted amplitude distribution

The array is represented by a continuous line distribution of infinitessimal sources. A range (r) which is large compared to the length of the array (2a) is considered. With reference to Fig. 4 (a) each point on the array, situated a distance ℓ from the centre is considered to be a generator of amplitude:

$$u(\ell,t) = A.e^{j\,(wt + k\ell t)} \qquad 1$$

where k ℓ defines the rate of change with time of the phase difference between a point on the array a distance ℓ from the centre and the centre itself. k is a constant for all points on the array. A more practical definition of k, in terms of the beam deflection (Θ) is:

$$k = \frac{2\pi}{\lambda} \cdot \frac{d}{dt}(\sin\Theta) \qquad 2$$

The total amplitude at a field point (r, Θ) is proportional to

$$U\,(r,\Theta,t) = \int_{-a}^{+a} \frac{1}{d} \cdot u\,(\ell,\, t - \frac{d}{c})\; d\ell \qquad 3$$

where the term (t - d/c) makes allowance for the time necessary for sound to cover the distance (d) between each point of the array and the field point.

Since r >> ℓ then, ignoring all terms in ℓ/r higher than the first order

$$d = r + \ell \sin\Theta \qquad 4$$

Thus $$U\,(r,\Theta,t) \simeq \frac{A}{r} \cdot \int_{-a}^{+a} e^{j(w + k\ell)\,(t - \frac{r}{c} - \frac{\ell \sin\Theta}{c})}\, d\ell \qquad 5$$

If we assume that the sweep rate is sufficiently slow so that the time needed to cover the path length ℓ sin Θ is negligible compared to the period of the sweep then

$$U\,(r,\Theta,t) = \frac{2Aa}{r} \cdot e^{jw(t - \frac{r}{c})} \cdot \mathrm{sinc}\left[ka(t - \frac{w \sin\Theta}{kc} - \frac{r}{c})\right] \qquad 6$$

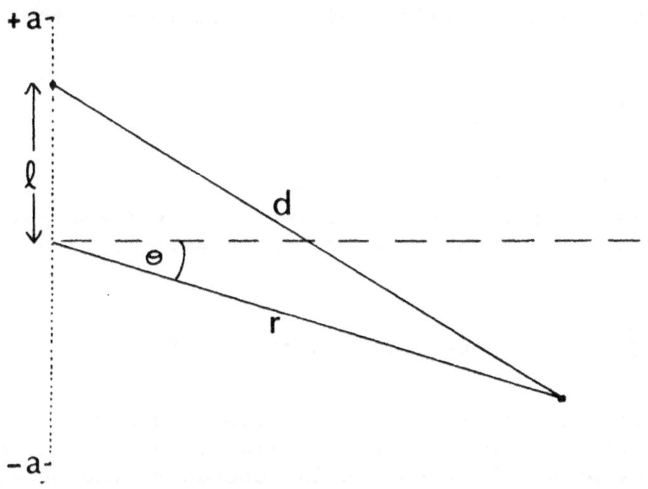

Fig. 4(a). Representation of the array as a line source.

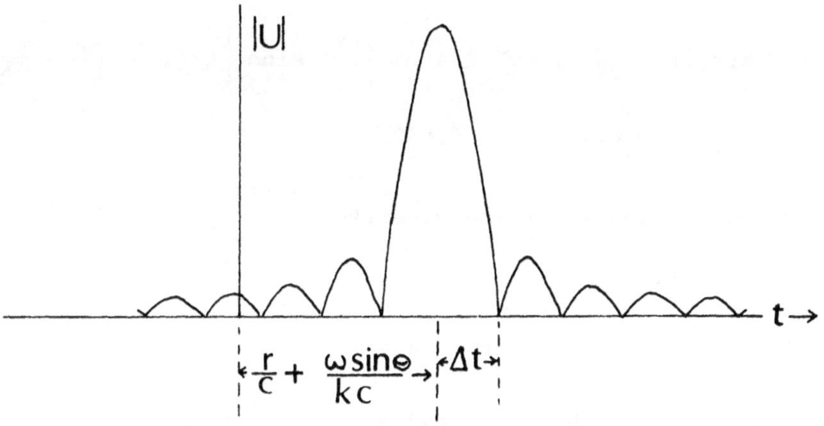

Fig. 4(b). Variation of transmitted amplitude U with time, for a given r and Θ

The variation of $|U|$ with t is indicated in Fig. 4(b). It has the form of a sinc function in common with the expression for the far field of a plane transducer. The time taken for the main lobe to pass a field point (r, Θ) is 2 Δt, where Δt is the change in t necessary to change the variable of the sinc function by π, i.e.

$$\Delta t = \frac{\pi}{ka} = \frac{\lambda}{2a} \cdot \frac{1}{\frac{d}{dt}(\sin\Theta)} \qquad 7$$

The angular half-width Δ (sinΘ) of the main lobe is the same as for a plane transducer, i.e. :

$$\Delta\,(\sin\Theta) = \frac{\lambda}{2a} \qquad 8$$

Combined transmit - receive response

If the back scattering coefficient at the field point (r,θ) is $S(r,\Theta)$, the integrated response of the receiver will be :

$$B(r,\Theta,t) = \int_{-a}^{+a} S(r,\Theta)\,.\,\frac{U(r,\Theta,t - d/c)}{d}\; e^{jk\ell(t-\tau)}\,d\ell \qquad 9$$

where again the term $(t - d/c)$ makes allowance for the time taken for each contribution to cover a path length d, and the phase factor $k\ell\,(t - \tau)$ is the phase distribution introduced by the receiver array processing, deliberately delayed by a time τ with respect to the transmitter phase distribution. Again assuming that $a \sin\Theta / c$ is small compared to the sweep period:

$$B(r,\Theta,t) = S(r.\Theta)\,.\,\frac{4Aa}{r^2}\,.\,e^{jw(t - 2r/c)}\,.\,\mathrm{sinc}\left[ka\left(t - \frac{2r}{c} - \frac{w \sin\Theta}{kc}\right)\right]$$

$$\times\ \mathrm{sinc}\left[ka\left(t - \tau - \frac{w \sin\Theta}{kc}\right)\right] \qquad 10$$

Maximum response occurs for the condition :

$$\tau = \frac{2r}{c} \qquad 11$$

An overall maximum also requires that

$$t = \tau + \frac{w \sin\Theta}{kc} = \tau + \frac{\sin\Theta}{\frac{d}{dt}(\sin\Theta)} \qquad 12$$

Thus if a single point scatterer is scanned by the proposed transmit-receive arrays the maximum response will only be obtained if the value of τ is chosen to correspond to the range (r) of the target. Also the time (t) in the sweep at which the peak response is obtained will be linearly proportional to sinΘ.

Range Resolution

A measure of the range resolution of the system may be derived by considering the response, for given values of τ and t, produced by a point scatterer at the same bearing (Θ) as the point (r,Θ) giving a peak response, but at a range $r + \Delta r$. As r is increased from zero the first zero in the response will occur when the respective variables of the two sinc terms differ by π. Since subsiduary maxima are more than 14dB below the principal maximum of a sinc function the response must be reduced by more than 14 dB for values of Δr greater than:

$$\Delta r_{14dB} = \frac{\pi c}{2ka} = \frac{c}{4} \cdot \frac{\lambda}{a} \cdot \frac{1}{\frac{d}{dt}(\sin\Theta)} \qquad 13$$

If a 20dB criterion is chosen instead, then we must increase the difference between the sinc function variables to just over 5 $\pi/2$, i.e. increase Δr by a factor of 5/2. For a 20mm long array (a = 10mm) operating at 3 MHz, a 60° sweep in 200 μs would produce a Δr_{14dB} value of approximately 4mm or a Δr_{20dB} value of approximately 10 mm.

Lateral resolution

The lateral resolution may be estimated similarly by calculating the change in sin Θ necessary to reduce the peak response by a given amount. Since sin Θ appears equally in both sinc terms of equation 10, there will be a more rapid variation in response, according to a $(\text{sinc})^2$ function. Thus a change of 0.6 π or more in the sinc variable will reduce the response by more than 14dB so that

$$\Delta (\sin \Theta)_{14dB} = \frac{0.6\ \pi}{a\ w/c} = \frac{0.3\ \lambda}{a} \qquad 14$$

A 20 dB criterion would involve a change of sinc variable of slightly more than 0.7 π so that only a marginally greater value for Δ (sin Θ) would result. If the same example as was used above for range resolution is considered, and a target range of 100mm is assumed (τ = 130μs), the half beam width would be approximately 1.5mm.

Simulation of a B-scan

In order to gain a visual impression of the range and lateral resolution of the system a computer simulation of a hypothetical B-scan was performed. The program evaluated $|B|^2$, as defined in equation 10, for a particular field point as a function of time, for a number of evenly spaced values of τ. This represented building up a B-scan from a number of constant-depth line sweeps, each sweep being responsive to a slightly greater depth than the previous one.

Fig. 5 shows the response $|B|^2$ expressed in dB relative to the peak value, presented as it would be in a sector B-scan display. The data used was as follows : point target range (r) = 75 mm, bearing (Θ) = -15°; array length (2a) = 30mm; frequency = 1.9 MHz; peripheral rate of phase change (ka) = 6.10^5 radians s^{-1} (60° in 200μs)

The response was calculated at 5 μs time (i.e. bearing) intervals from t = 5 μs to t = 75 μs and the delay τ (i.e. range) was incremented in 10 μs steps from 10 μs to 310 μs

A prominent feature of the response is the orientation of the regions of high response along two preferred axes: one along a line of constant bearing, the other along an oblique line. This may be explained by reference to equation 10, since one of the sinc factors is unaltered if the variable t remains constant whilst the other remains constant if (t - τ) remains constant. The fine structure of maxima and minima which would be expected from the sinc factors is not apparent, owing to the coarse 5 μs sampling of t used here. This structure was seen in other plots made for 2 μs steps of t. The relatively poor axial resolution predicted by equation 13 is clearly demonstrated.

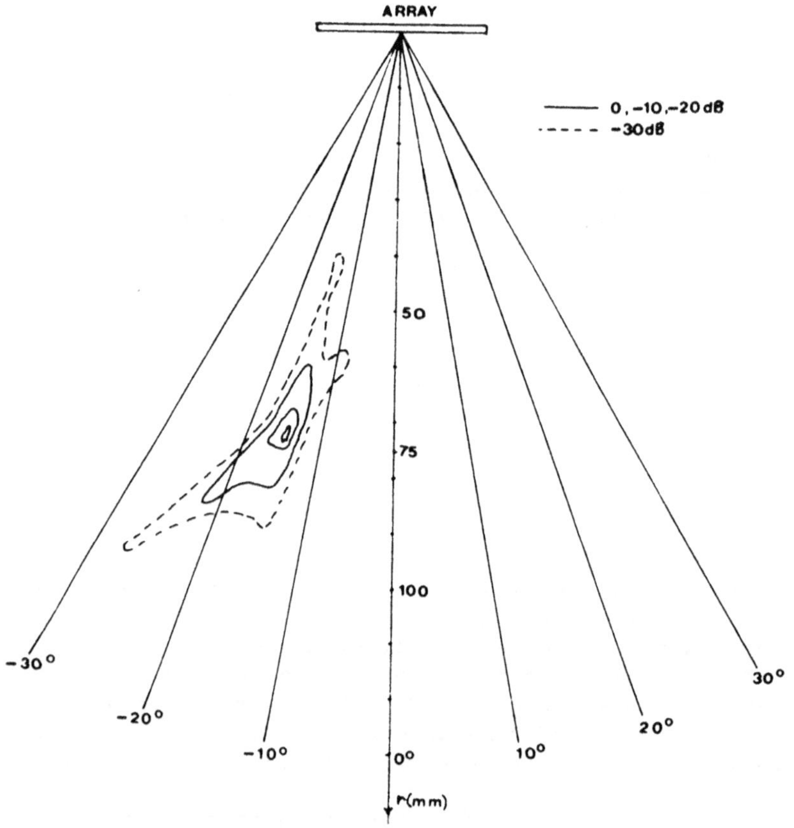

Fig. 5. Simulated B-scan of a point target, to illustrate resolution. Contours at 10 dB intervals. 1.9 MHz, unfocussed 30mm array.

TRANSMITTED AMPLITUDE DISTRIBUTION FOR HIGHER SWEEP RATES AND FOCUSSING.

Effect of an imposed quadratic phase distribution

As before, the transmission array is described by Fig. 4 (a) but we now assign to each point generator along the array an amplitude u (ℓ, t) where:

$$u(\ell,t) = A.\ e^{j(wt + k\ell t + k'\ \ell^2)} \qquad 15$$

Here, $k\ell$ describes the rate of change of phase, measured with respect to the centre of the array, at an array point at ℓ, and $k'\ell^2$ describes a fixed quadratic phase distribution imposed on the array elements to achieve focussing. The total amplitude $U(r,\Theta,t)$ at a field point (r, Θ) is given by equation 3 as before. However since terms involving $(\ell/r)^2$ are now included we replace equation 4 by:

$$d = r + \ell \sin\Theta + \frac{\ell^2}{2r} \cos^2\Theta \qquad 16$$

Thus

$$U(r,\Theta,t) = \frac{A}{r}.\ e^{jw(t - \frac{r}{c})} \cdot \int_{-a}^{+a} e^{jk\ell(t - \frac{r}{c} - \frac{w \sin\Theta}{kc})} \times e^{j\ell^2(k' - \frac{w \cos^2\Theta}{2rc} - \frac{k \sin\Theta}{c})} d\ell \qquad 17$$

The term in ℓ^2 is responsible for the focussing action of the beam. The first contribution $(k'\ell^2)$ is the imposed quadratic phase distribution, intended to cancel out the second contribution $(w\ell^2\cos^2\Theta/2rc)$ due to the path length differences between the array and a point target. The third contribution $(k\ell^2\sin\Theta/c)$ arises from the retardation term $(t - d/c)$ introduced into equation 3; it is discussed further below. At a range equal to the focal length (R_Θ) this quadratic term should cancel to zero, i.e.

$$k' - \frac{w \cos^2\Theta}{2 R_\Theta c} - \frac{k \sin\Theta}{c} = 0$$

i.e.
$$\frac{1}{R_\Theta} = \frac{1.}{\cos^2\Theta} \cdot \left[\frac{1}{R_o} - \frac{2 \sin\Theta}{c} \cdot \frac{d(\sin\Theta)}{dt} \right] \qquad 18$$

where R_o is the focal length for the on-axis case (Θ =0) i.e.

$$R_o = \frac{\pi}{\lambda k'} \qquad 19$$

This is the expected focal length for the imposed phase distribution, in the absence of beam sweeping. The consequence of the bracketed term of equation 18 is a focal length which increases progressively through the sweep. The $\cos^2\Theta$ term reduces the focal length symmetrically for deflections on either side of the principal axis. The two effects oppose each other to produce a substantially constant focal length (R_o) in the second half of the sweep, but combine to produce a focal length which decreases with off-axis bearing in the first half. In the complete transmit-receive system the resolution and sensitivity will be degraded for all bearings where the focal length differs significantly from the preferred range ($\frac{1}{2}$ c τ). It may be shown by a graphical solution of equation 17 that these effects can be reduced by using moderate sweep times (200 to 400 μs) and moderate sweep angles (40° to 60°).

Discussion of the bearing dependence of focal length

The variation of the focal length with bearing may be understood by considering the difference in path length to a given field point (r,Θ) between contributions from the periphery of the array and those from the centre. In the theory for moderate sweep rates the propagation time (a sin Θ/c) corresponding to this difference in path length was assumed to be negligible with respect to the sweep period, but it has been taken into account here. Its effect is illustrated in the upper half of Fig. 6. Without any deliberate focussing ($k' = 0$) waves from the nearer end of the array take less time to reach a given field point than do those from the centre. Consequently, since both contributions arrive together, that from the nearer end of the array must have left the array slightly later in the sweep than that from the centre. Similarly the contribution from the further end must have left the array slightly earlier in the sweep. Thus the contribution from the nearer end is transmitted on a bearing which is slightly ahead of that from the centre, whilst that from the further end is transmitted on a bearing which is slightly behind. The result is a naturally converging beam in the first half of the sweep and a naturally diverging beam in the second half. When combined with a deliberate focussing mechanism the resultant focal length progressively increases from less to more than its nominal value as the sweep progresses, being equal to its nominal value at $\Theta = 0$.

Although the previous analysis has considered only the transmitted wave, the same physical reasoning would predict that the focal length of the received beam would progressively reduce during the sweep. With reference to the lower half of Fig. 6 a wavefront scattered from the field point (r,Θ) will reach the nearer end of the array before it reaches the centre. Thus it arrives when the receiving array is 'pointing' at a slightly earlier bearing in the sweep than that of the scattering field point. Similarly the same wavefront will reach the further end of the array when it is 'pointing' to a later bearing than that of the field point. The result is a progressive decrease in resultant focal length during the sweep.

The significance of the $\cos^2\Theta$ term is a consequence of the fact that the effective aperture coordinate of a particular point on the array, as 'seen' from a bearing Θ, is $\ell \cos\Theta$, whilst the focussing phase correction ($k' \ell^2$) is calculated for the real aperture coordinate ℓ. This factor is not a consequence of the fast sweep but is associated with all electronically focussed and deflected beams.

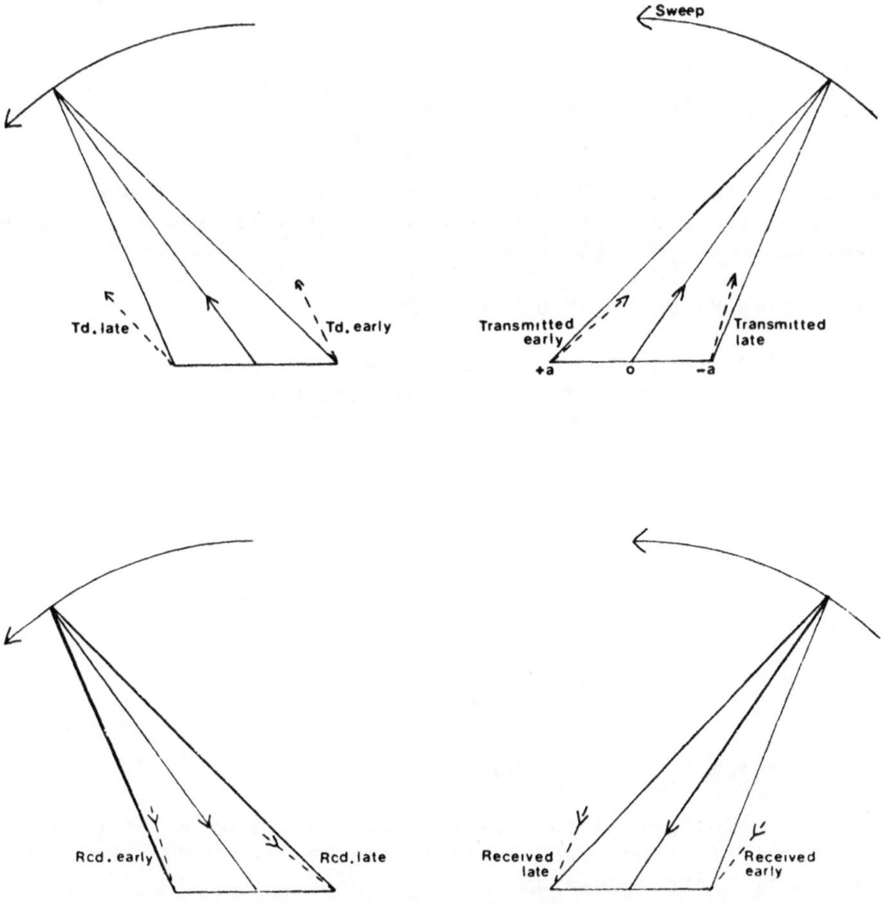

Fig. 6. Illustrating how differences in path length lead to natural beam convergence or divergence, for a rapidly swept unfocussed array.

Compensation for the effect of sweep-rate on focal length

If the phase distribution imposed to produce focussing is modified from the simple quadratic function previously considered it should be possible to compensate for the effects just discussed.

Thus, instead of the element excitation described by equation 15, we might excite each array element according to

$$u(\ell, t) = A.\, e^{j\left(wt + k\ell\left(t + \frac{\ell\sin\Theta}{c}\right) + k'\ell^2\right)} \qquad 20$$

When this expression is substituted in equation 3 the troublesome term ($k\ell^2\sin\Theta/c$) is cancelled out. Since $\sin\Theta$ is proportional to time (t) the required phase distribution is a constant focussing term ($k'\ell^2$) plus a time varying term which varies quadratically with ℓ:

$$u(\ell, t) = e^{j\left(wt + k'\ell^2 + \left(k\ell + \frac{k^2\ell^2}{w}\right)t\right)} \qquad 21$$

Correction of the $\cos^2(\Theta)$ dependency could also be achieved by further modification of the phase distribution if this was considered worthwhile.

IMPLEMENTATION BY MEANS OF DELAY LINES

A simple method of producing a phase difference which increases linearly with time is to drive a delay line (delay T) with a frequency ($w_o + mt$) which increases linearly with time. The phase shift across the delay line is

$$\Delta\phi = - w_o T + \tfrac{1}{2} m T^2 - mTt \qquad 22$$

Comparison with equation 21 suggests that if we choose a delay (T_ℓ) for each element (coordinate ℓ) in the transmitter array such that:

$$k\ell + \frac{k^2\ell^2}{w} = - m T_\ell \qquad 23$$

$$\text{and} \quad k'\ell^2 = \tfrac{1}{2} m T_\ell^2 - w_o T_\ell \pm 2n\pi \qquad 24$$

where n is an integer, then driving each element with the phase derived from the delay line should automatically produce both the sweeping action and focussing.

The frequency of the ultrasonic elements (w) need be neither swept nor close to the delay line frequency sweep ($w_o + mt$). The phase shifts generated across the delay lines could be transferred to the elements of the transmitter array by mixing the delay line outputs with a swept local oscillator signal whose frequency would be maintained a constant amount (w) above or below the delay line frequency. This difference frequency could be detected, filtered-out and used to drive the array elements. Similar delay lines and frequency shifting techniques could be used to shift the phase of the receiving array signals to form a swept and focussed receiving beam. An extra delay (τ) would have to be inserted in series with the phase-shifting delay lines in the receiving system in order to provide the necessary lag of the receiving beam behind the transmitting beam. If the same array were to be used to transmit and receive then the same set of delay lines could provide the phase shifts for both transmitting and receiving beams.

Thus the phase shifts necessary to deflect the beams and provide sweep-rate compensated focussing may be produced relatively easily by tapping a swept-frequency delay line at appropriate points. For moderate sweep rates the term ($k^2 \ell^2/w$) in equation 23 may be omitted and the T_ℓ values are directly proportional to ℓ. This allows the simplication of regularly spaced tapping points.

CONCLUSION

A novel method for scaning a line at constant depth has been introduced and theoretically explored. A number of different options are possible depending whether a linear or sector scanning format is required, whether separate transmission and reception arrays are employed or a single array fulfills both functions, and whether the continuous transmission, which is characteristic of the concept, takes the form of a true continuous wave or a train of closely spaced but discrete pulses. The continuous wave sector scanning arrangement discussed in this paper has been shown to offer good lateral resolution but relatively poor axial resolution. Very high scanning rates cause the focal length to vary within the sweep, but this effect is not serious at moderate sweep rates and sector widths. In any event it may be corrected by a simple modification of the phase distribution across the arrays. The necessary distribution to sweep and focus the arrays may be conveniently derived from tapping points on a delay line. These tapping points are regularly spaced for moderate sweep rates but are dependent on element position if the effects of high sweep rate must be corrected.

The relatively poor range resolution is a consequence of the narrow frequency bandwidth of the system. The bandwidth may be increased by transmitting a train of closely spaced pulses instead of a continuous wave, and this is the basis of a linear scan system currently under development in this laboratory. This approach, however involves sacrificing the simplicity of the form considered in this paper.

ACKNOWLEDGEMENTS

I would like to thank Professor F.T. Farmer and Professor K. Boddy for supporting this work.

REFERENCES

Green, P.S., Schaefer, L.F., Jones, E.D. and Suarez, J.R., 1974. A new high-performance ultrasonic camera, in: "Acoustical Holography and Imaging, Vol. 5", P.S. Green, ed., Plenum Press.

Hill, C.R. and Carpenter, D.A., 1976, Ultrasonic echo imaging of tissues: instrumentation, Brit. J. Radiol., 49: 238-243.

Welsby, V.G., Creasey, D.J. and Barnickle, N., 1973, Narrow beam focussed array for electronically scannered sonar: some experimental results, J. Sound and Vibration, 30: 237-248.

ULTRAFAST ACOUSTICAL IMAGING

WITH SURFACE ACOUSTIC WAVE COMPONENTS

Pierre Cauvard and Pierre Hartemann

THOMSON-CSF, Laboratoire Central de Recherches

Domaine de Corbeville, 91401 Orsay, France

ABSTRACT

A new technique for processing the electric signals resulting from an ultrasound multielement probe has been used to achieve a high rate B-mode imaging system. The images are formed with surface-acoustic-waves (SAW) propagating on a PZT substrate and read out with SAW transducer arrays. The imaging system operation is explained and the configuration of specific SAW components is described. For showing the feasibility of this processing method, experimental results about a 20 parallel channel system are reported. A rate of 800 frames/sec. has been reached for an image made of 1500 uncorrelated points. However more detailed images have been obtained with a frame rate of 400/sec. and a point number of 3000 In this case the resolution is 1 mm in depth and 2.5mm in width.

INTRODUCTION

Generally echo-tomographic acoustical imaging systems operate at a slow frame rate (below on hundred per second) and do not allow an accurate analysis of motion of different heart parts. As reported in this paper, a very fast frame rate has been carried out by processing the signals related to a multielement transducer array with new surface-acoustic-wave (SAW) components : a scaled down image of the probed tissue slice is formed on a SAW propagating surface and read out with a SAW transducer matrix. For showing the feasibility, this method has been applied to the processing of signals supplied by a 20 transducer linear array, the step of which is equal to 2.5 mm. However this technique is usable for ultrasound probes with other features. The first section of this paper is devoted to a description of the imaging system operation, the second to the frame rate and resolution. Some characteristics of SAW components are given in the third part.

IMAGING SYSTEM OPERATION

The electric signals resulting from a linear array transducer are processed according to the arrangement schematically illustrated in fig.1. The image forming is achieved with a SAW component which consists of a launching transducer array and a receiving transducer matrix deposited on a PZT substrate. For SAW, the wavelength is equal to 140 µm at 15.7 MHz; whereas the acoustic wavelength in human tissues is close to 500 µm, the operating frequency of the probe being 3 MHz. The launching SAW transducer array has been designed with dimensions (step and aperture) close to those of the probe array divided by the ratio of wavelengths in both propagating media; that is 3.57. Every receiving transducer of the probe is connected to a SAW transducer through a frequency changing mixer. The signal at the difference frequency is selected to get the phase reserval necessary to focusing. In these conditions, the device operates as an acousto-electronic lens. The product of frequency and delay time due to acoustic paths between a point object in the patient body and the corresponding point image on the PZT surface is identical for both propagating media. The only limiting condition is that at a SAW point image the waves at least partially overlap in time. A scaled down image of the probed tissue slice is formed with SAW and a tissue depth of 17 cm is reduced to a length of 4.76 cm on the SAW substrate.

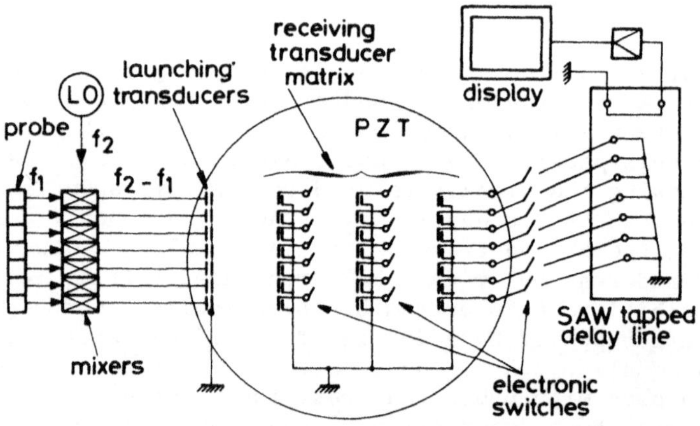

Fig. 1 – Schematic representation of the imaging system structure.

The SAW image is read out with the transducer matrix which has been designed for observing the tissue located between 2 and 17 cm from the patient skin. The transducers of each matrix column are simultaneously switched on during a short time (300 ns) to the inputs of a SAW tapped delay line according to an appropriate timing. The number of matrix columns must be sufficient to visualize the whole probed tissue slice. It is dependent on the field depth of the acousto-electronic lens. For the experimental set-up, the step of the 20 element probe array and the SAW launching transducer array is respectively equal to 2.5 mm and 0.7 mm. The angular aperture of an elementary transducer is close to 12 degrees and the field depth of the lens is larger than 5 cm in biological tissues. Then a transducer matrix with 3 columns can be used.

A parallel-to-serial conversion of electrical signals is performed with a SAW tapped delay line. That is necessary to observe B-mode images by sweeping the screen of a display unit with vertical lines.

A photograph of the main part of the system is presented in fig.2.

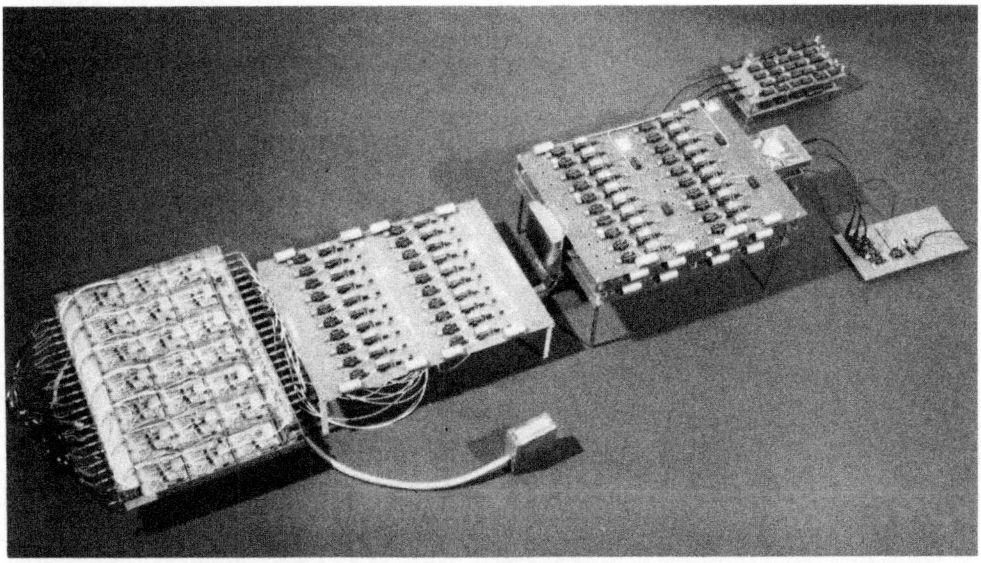

Fig. 2 - Photograph of the imaging system without power supplies, signal generators and display.

FRAME RATE AND RESOLUTION

The repetition rate of pulses applied to the ultrasound probe is close to 4000 cycles per second and only a part of the whole image is visualized during one cycle. All transducer matrix columns are switched to the inputs of the tapped delay line with a rate of 5 per cycle and N cycles are necessary to reconstruct the frame part related to one column. Then the whole image is obtained with a frame rate of 4000/N by interlacing the lines. At first, N was equal to 5 and the frame rate was 800/sec. with 75 lines made of 20 points (number of probe array elements = 20) that is 1500 uncorrelated points for a probed tissue slice 5 cm wide and 15 cm deep. The resolution in depth was 2 mm. That one in width is given by the spacing between two adjacent elementary transducers of the probe; that is 2.5 mm. But the image appeared as a dotted pattern. Then finer images have been achieved for a value of N equal to 10. In this case the frame rate is 400/sec. with 150 lines and the number of points is equal to 3000 for the same probed tissue slice. The limit of resolution in depth given by the quality of the acousto-electronic lens is reached. It is close to 1 mm.

Triggering of the ultrasound probe and electronic switches is performed in TTL logic.

SAW COMPONENTS

- image forming device

The main element of this signal processing method is the SAW image forming device. The SAW propagates on a hot pressed piezoelectric PZT disc the diameter of which is equal to 55 mm (see fig.3). Each array is made of 46 transducers with a step of 700 μm. The dummy spacing between two adjacent launching transducers is equal to 150 μm. It is impossible to realize a dummy spacing equal to that one of the probe transducers divided by the wavelength ratio. Then the angular aperture of the SAW launching transducers is slightly larger than that one of the probe transducers. The insertion loss is close to 25 dB for the first receiving array, 30 dB for the second and 40 dB for the third one. Only 20 channels have been bound to 4 printed circuit connectors. The size of this component in its package is 8x7x2 cm (see fig. 4).

-tapped delay line

The parallel-to-serial conversion of signals is performed with a SAW tapped delay line and the 20 instantaneous signals supplied by a matrix column are expanded in about 11 μs. A particular delay line configuration has been designed for reducing the parasitic reflections of SAW to a level better than - 40 dB.

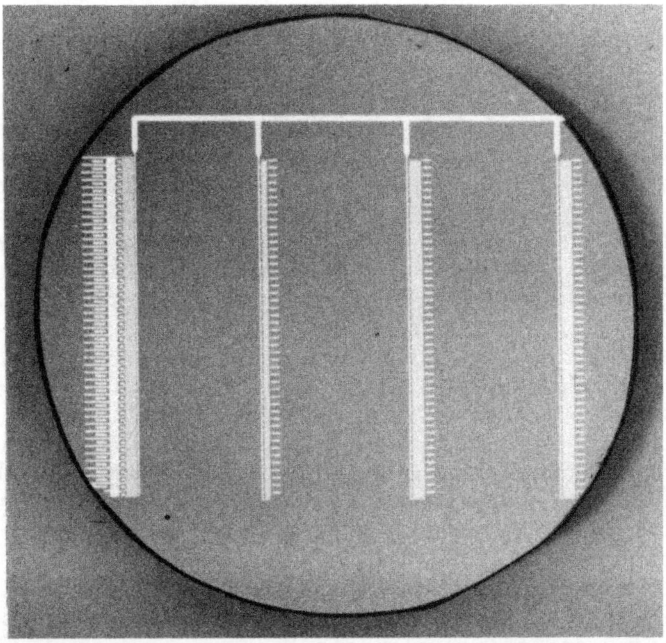

Fig.3- SAW image forming component before mounting. The launching transducer array is located on the left side of the figure and the 3 other transducer arrays form the receiving matrix. The spacing between two adjacent transducer columns is close to 14 mm.

Fig.4- SAW image forming device in its package (8x7x2cm).

CONCLUSION

The feasibility of an ultrafast acoustic imaging system by means of surface-acoustic-wave components has been shown. A B-mode image of 3000 points for a tissue slice of 15 x 5 cm has been achieved with a rate of 400/sec., the resolution in depth being 1mm. Real-time images of a living human heart have been visualized on the screen of a display unit. However it is impossible by looking at the screen to take completely advantage of this very high rate. Then the electric signals after the SAW image forming component can be memorized during a sequence and read out with a slow rate to observe with accuracy the complicated motion of different parts of a heart. For improving the quality of images, the number of channels can be increased. However this number will be inferior to 100 because of the difficulty to bind a great number of connections on the PZT surface. Moreover the step of the probe array can be decreased to reduce the relative level of orders due to the sampling. But the number of the SAW transducer matrix columns which depends upon the field depth of the acousto-electronic lens must remain weak for limiting the SAW propagation perturbations and the number of connections.

The relative phase of ultrasound signals is not lost during this processing and theoretically a Doppler imaging system can be performed. The implementation of this processing using SAW is quite simple in comparison with that of a delay line network processing and frame rate as high as 1000 per second could be obtained with very simple and compact electronic circuits. However a rate of 400/sec. seems sufficient for the future development of the medical research.

ACKNOWLEDGMENT

This work was supported in part by the "Direction Générale à la Recherche Scientifique et Technique" (France).

ULTRASOUND SIMULATOR

J. Souquet, P. Stonestrom, M. Nassi

Varian Associates, Inc.
Systems & Techniques Laboratory
Palo Alto, California, 94303, U.S.A.

ABSTRACT

An ultrasound simulator has been created to calculate the beam pattern of various array configurations. This simulator works either in the transmit mode or in the transmit/receive mode. It takes advantage of the computing and imaging display capabilities of the V-76 computer.

This simulator primarily addresses the problem of suppression of unwanted parasitic lobes. This can be achieved in various ways. This paper presents some simulated and experimental results of various array configurations while varying some parameters: element spacing, pulse length, effect of time quantization, effect of nonlinear processing of received signals, etc.

INTRODUCTION

In the field of medical ultrasound imaging, transducer arrays, either linear[1] or phased[2], are now widely used. Each element of the array is driven separately in order to achieve beam steering and focusing. In the transmit mode the focus is fixed, while in the receive mode dynamic focusing can be achieved by varying the delays on each of the transducer elements to allow the focal range to track down the incoming signal.

The imaging quality of ultrasound systems greatly depends upon the acoustic beam pattern generated from the transducer; i.e., lateral resolution, transverse resolution, sidelobe level, grating lobe level. All of these factors are a function of the transducer geometry: resolution depends upon the transducer aperture (for a given frequency), sidelobe and grating lobe depend upon the element spacing in the transducer array. Therefore, we have developed a tool, the ultrasound simulator, which helps us optimize a transducer array given some constraints; eg., resolution, unwanted lobe, cost of system, etc. The ultrasound simulator is a computer program which runs on the V-76 minicomputer. Various kinds of displays are available as output and they will be described in a following section. Let us first describe the simulator capabilities.

ULTRASOUND SIMULATOR CAPABILITIES

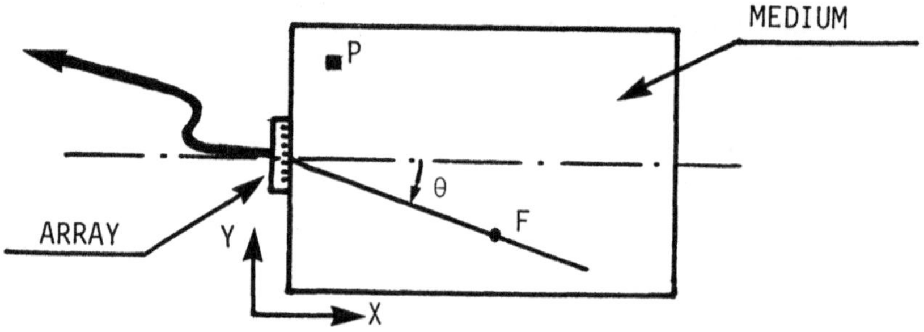

Fig. 1. Experiment Setup

The transducer array is affixed to a propagation medium (usually water); through its electronics, steering angle (θ) and focus (F) are held constant. A sensing probe (P) mechanically scans the propagation medium from pulse to pulse. The probe may either transmit, receive, or simply act as a passive reflector.

Therefore the types of beam patterns we can simulate are:

- transmit beam pattern (one way)
- receive beam pattern (one way)
- round trip patterns (echo off probe)

In order to properly match the physical world, we have tried to carefully simulate each subgroup of the ultrasound system: acoustic section, transducer section, RF section. Let us review each of these groups and describe the current features and limitations of the present simulator.

Acoustic Section

We only have at the present time considered a homogeneous propagation medium with frequency independent acoustic attenuation. We have done so because it was easier to compare the simulated beam pattern in water with the experimental one in the same propagation medium.

The future extension of the simulator could try to more carefully model the tissues by taking into account the frequency dependent absorption, the nonuniform absorption, and the nonuniform speed of sound.

Transducer Section

The simulator is able to handle any arbitrary 1-dimensional, 2-dimensional, or 3-dimensional array geometry. We also have taken into account the radiation pattern of individual elements in the array. We have used for the radiation pattern of a narrow-strip acoustic transducer (ω) the formulation derived by Selfridge, Kino and Khuri-Yakub[3] which we have verified experimentally and gives us satisfaction:

$$R(\theta) = \frac{\sin\,(\pi\omega/\lambda\,\sin\,\theta)}{(\pi\omega/\lambda\,\sin\,\theta)} \cdot \cos\,\theta$$

It differs from the usual far-field radiation pattern of a single element by the cos θ term, which might be considered as the obliquity factor.

Any kind of pulse shape and length can be simulated. We can either digitize a real pulse, either use a theoretical pulse or a mathematical pulse. The pulse we usually use is a three-cycle pulse with a cosine envelope. Amplitude apodization (or shading) is also included in the simulator.

Some other features might eventually be incorporated in the simulator: noise, element-to-element bandwidth variation, interelement coupling (acoustical and electrical).

RF Section

This section includes delay errors and quantization. In the receive or transmit/receive mode it also incorporates the capability of performing nonlinear RF processing on the receive echoes. Such a processing can be a compression law (log or square root) followed after summation by the corresponding expansion law.

ULTRASOUND SIMULATOR DISPLAYS

As we have mentioned previously, various kinds of diplays are available for the simulation of the array beam pattern.

One-Dimensional Display

This kind of display is the one which is usually shown when presenting a beam pattern. It represents the radiation pattern of a transducer array at a constant depth in front of the array versus normalized amplitude. In our case, since we will be mainly interested in phased array, the displayed beam pattern will be the one computed for a constant radius in depth in front of the array. This kind of display is very useful to accurately measure the FWHM (full width to half maximum) of the beam pattern and the various lobe amplitudes.

Two-Dimensional Display

This display can be called "the time-compressed pulse-mode beam pattern." It represents the total beam pattern issued from the array and seen by the probe scanning the medium. At each particular point in the medium, the probe senses a pulse which has a given time duration. For this kind of display the probe only retains the peak value of this pulse. Therefore, the display shown on the Varian CT display monitor is a pessimistic estimate of the real world, but it gives us a good grasp of the actual beam pattern.

In this display, various levels of gray are representations of the normalized amplitude of the beam pattern. In order to enhance some features of the beam pattern, we can use all of the capabilities of the CT display: gamma curves, zoom, thresholding, distance measurement, etc.

Pseudo Three-Dimensional Display

This display is similar to the previous one with the exception that the third dimension now represents the normalized amplitude (cf. one dimensional display).

In the following section of this paper we will use, as illustration, any of the displays previously described. They will be referred to as 1D, 2D, and 3D.

RESULTS

Comparison with Experiment

The first step in checking the simulator is to compare its results with some actual measurements. The experimental values we have used were obtained with the first generation Varian sector scanner; namely the V-3000. The system operates at 2.25 MHz, and the transducer array is composed of 32 elements, 0.6λ apart. On the front face of the array there is a lens which has a fixed focus of 7.5 cm.

Figure 2 shows a comparison between experiment and simulation for a one-way experiment. The beam pattern is sensed by the probe

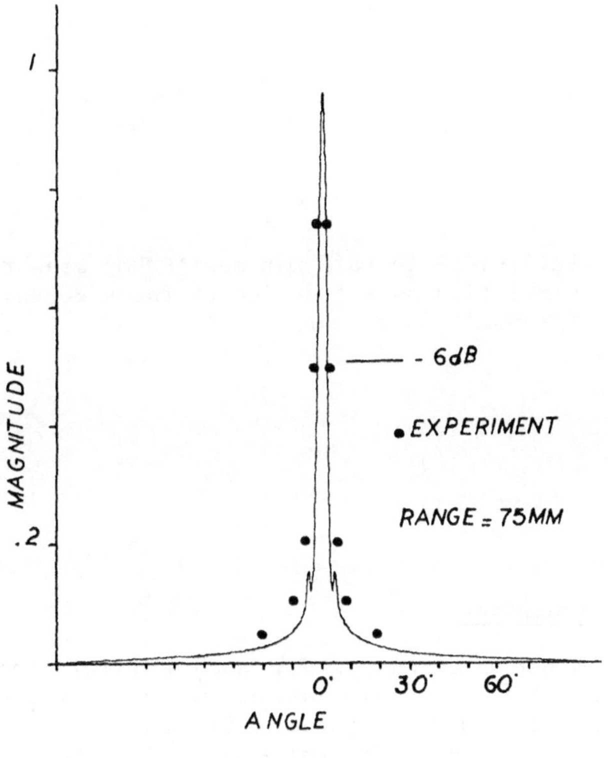

Fig. 2. Comparison between simulation and experiment on a 32-element array.

(real or simulated) placed 7 cm in front of the array straight ahead. The agreement is good. Figure 3 shows measurements of FWHM at various depths in front of the transducer. The agreement is acceptable.

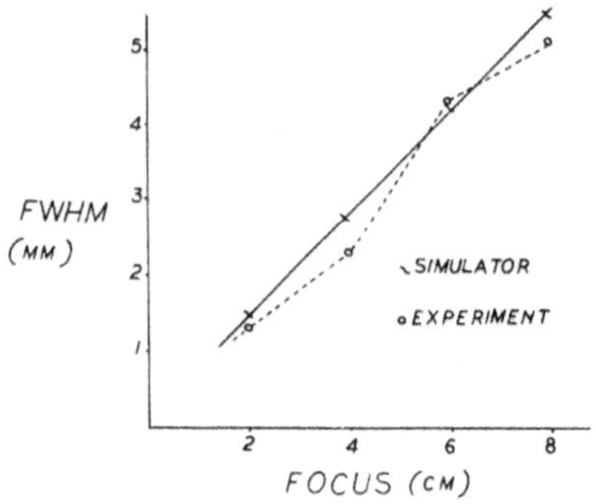

Fig. 3. Full width to half maximum (FWHM) measurements and simulation as a function of focus depth. (Transmit focus only)

Round-Trip Simulation

As explained previously, for such a simulation the probe P acts as a reflector. The simulations are performed for the V-3000 system: 32 transmitters spaced 0.6λ apart, and the 16 center elements as receivers. The pulse used for the simulation is a 3-cycle pulse centered at 2.25 MHz cosine weighted. Figures 4(a), (b) and (c) show the three different outputs of the simulator. Figure 4(a) is the 1-D output at 75 mm in depth. Figure 4(b) is the 2-D output and (c) the 3-D output. In each case the simulator has been asked to perform the following task: transmit beam steered at 45° and focussed at 7.5 cm in depth (simulating the fixed lens affixed to the array). The receiver was also steered at 45° and focussed at 7.5 cm.

As can be seen from the simulation, the worst lobe (grating lobe) lies 22 dB down from the main lobe 105° away, which is not in the field of view of the sector scanner. The 3 dB beamwidth at 7.5 cm is equal to 5.3 mm when steered at 45° and 4.6 mm when steering straight ahead.

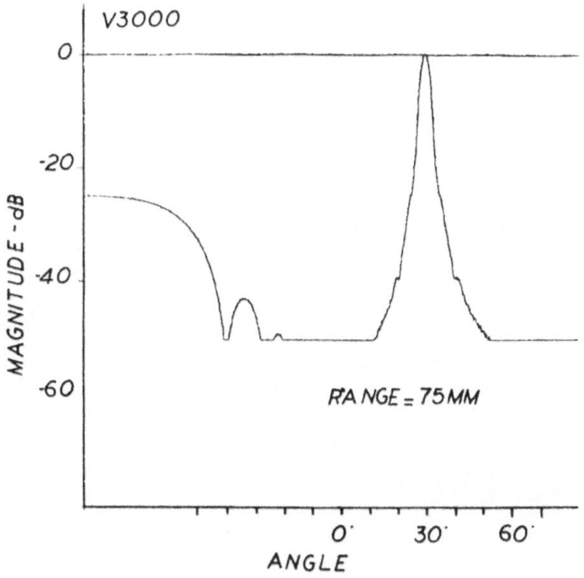

Fig. 4(a). Round trip simulation of V-3000 array. 1D display

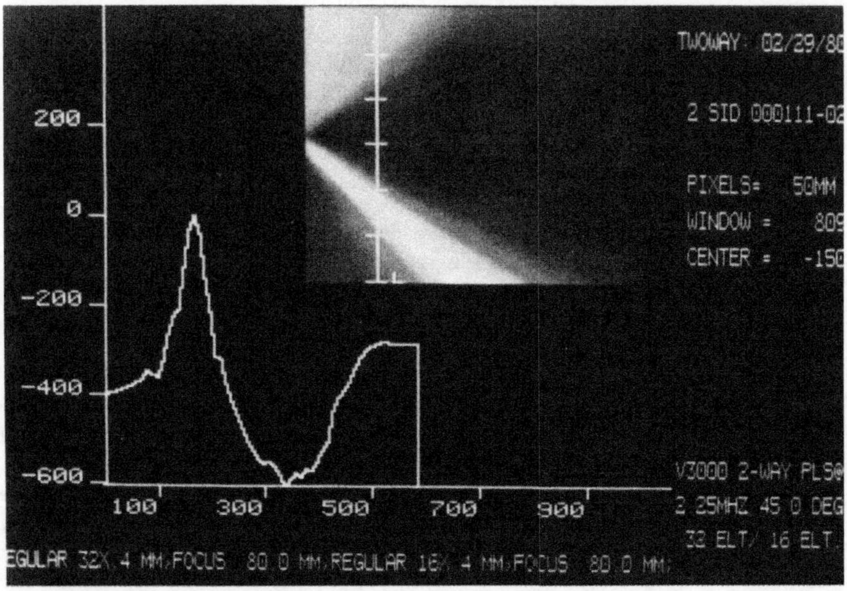

Fig. 4(b). Round trip simulation of V-3000 array. 2D display.

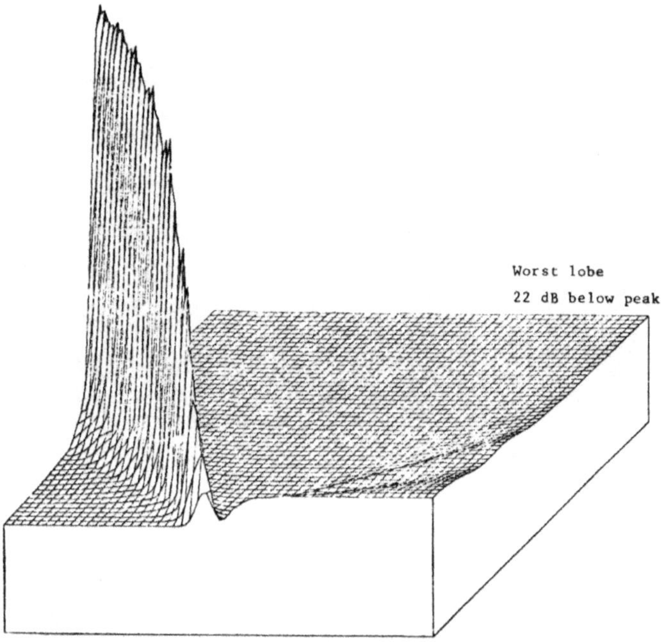

Fig. 4(c). Round trip simulation of V-3000 array. 3D display.

Simulated Experiments

Resolution Improvement

Resolution is one factor among many others which is important to improve in order to enhance image quality. One obvious way in the previous example would be to use 32 elements for receive thus doubling the number of receivers, thereby substantially increasing the cost of the system.

Various other alternative can be considered. One of them is to use the 32 elements for transmitting and 16 elements for receiving. The 32 transmitters are placed 0.6λ apart and the 16 receivers 1.2λ apart. Therefore, all the elements of the array are transmitting and every other element is receiving. One then benefits on receive from a wider aperture, thereby improving resolution. The combined transmit-receive beam pattern is of great importance. In this example, the grating lobes which could be expected from the 1.2λ periodicity of the receive elements are pulled down because they occur at a minimum of the transmit pattern. Figures 5(a) and (b) are an illustration of this case. In Fig. 5(a) the simulation is performed using alternate elements for transmit (16 elements spaced 1.2λ apart) and alternate elements for receive (16 elements spaced 1.2λ apart). As before, the beam is steered 45° on transmit and focused 7.5 cm in front of the array. The same procedure is used for receive. The worst lobe is 15 dB below peak and is located in the field of view 50° away from the main lobe. In Fig. 5(b) the simulation is done for the kind of array described previously (32 transmitters 0.6λ and 16 receivers 1.2λ). In this case the worst lobe is 25 dB below peak. We have gained 3 dB compared to Fig. 4. In Fig. 4 two first order grating lobes were reinforcing each other because the transmit and receive array had the same periodicity. In Fig. 5 the second order grating lobe of the receiver is reinforcing the first order grating lobe of the transmitter giving a total lobe of a slightly lower amplitude. The FWHM of Fig. 5(b) is 3.7 mm, which is an improvement over Fig. 4.

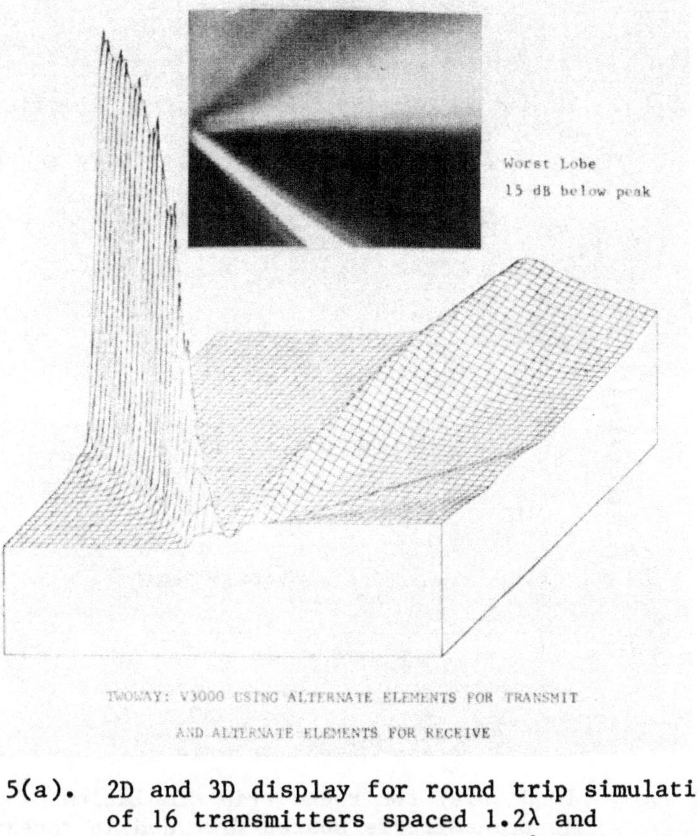

Fig. 5(a). 2D and 3D display for round trip simulation of 16 transmitters spaced 1.2λ and 16 receivers spaced 1.2λ. 45° beam steering.

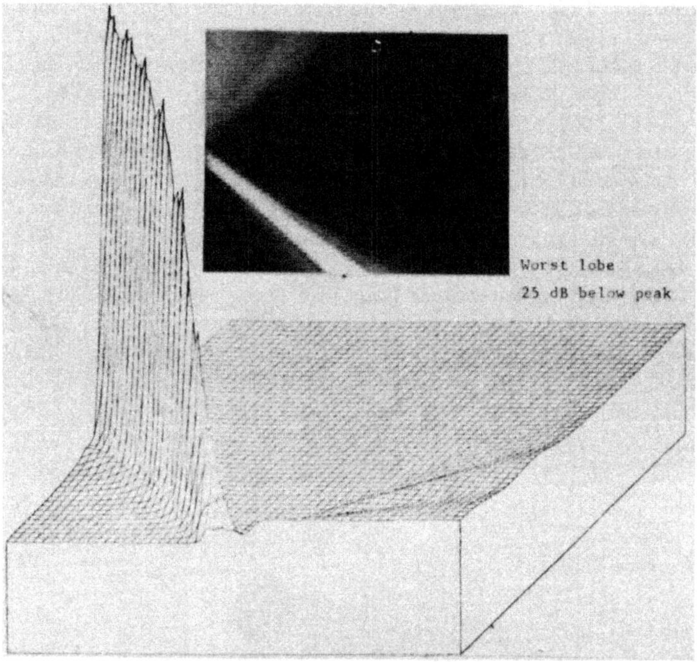

Fig. 5(b). 2D and 3D display for round trip simulation of 32 transmitters spaced 0.6λ and 16 receivers spaced 1.2λ. 45° beam steering.

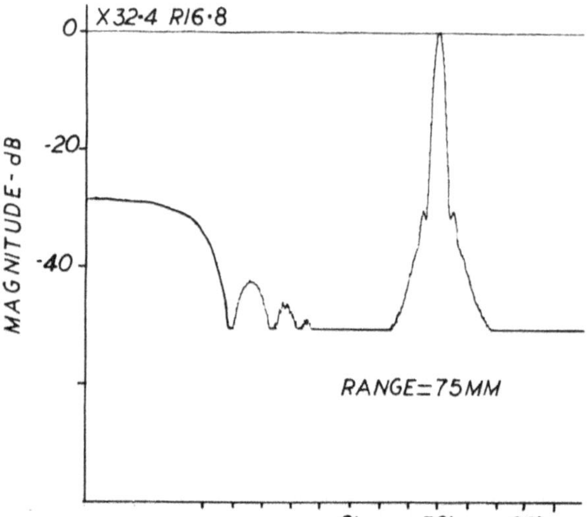

Fig. 5(c). 1D display for round trip simulation of 32 transmitters spaced 0.6λ and 16 receivers 1.2λ. Beam steered 30°.

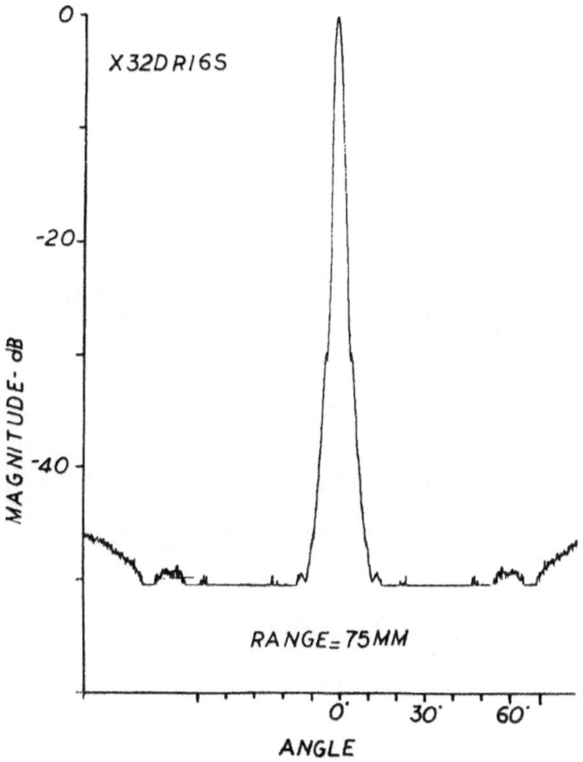

Fig. 5(d). Same as Fig. 5(c) but steered straight ahead.

We also have investigated the case where the 16 elements of the receiver are randomly spread over the total aperture of the array. Such a configuration should be helpful for sidelobe reduction without increasing the width of the main lobe. Figure 6(a) shows the round trip simulation of an array with 32 dense transmitters and 16 random sparse receivers when the electronics is steered straight ahead. When compared to Fig. 5(c) one sees that the sidelobe located at -30 dB tends to vanish

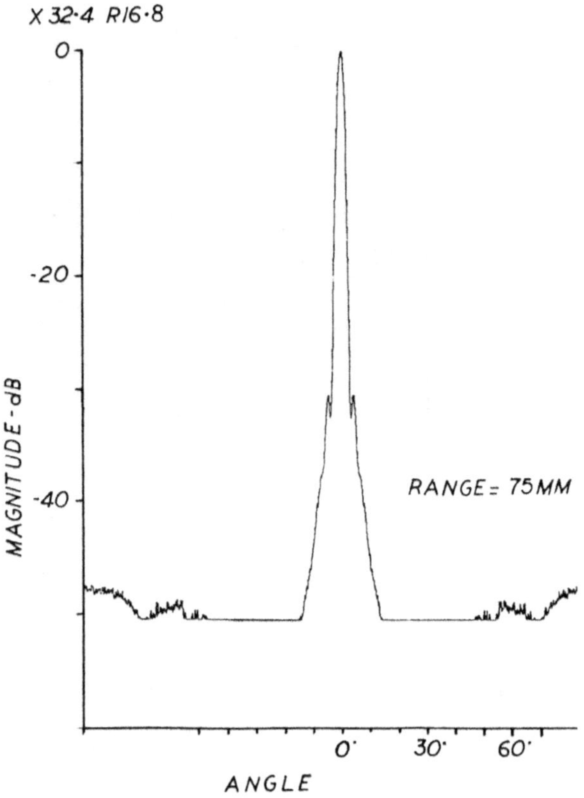

Fig. 6(a). Same as Fig. 5(d) but sparse 16 receivers.

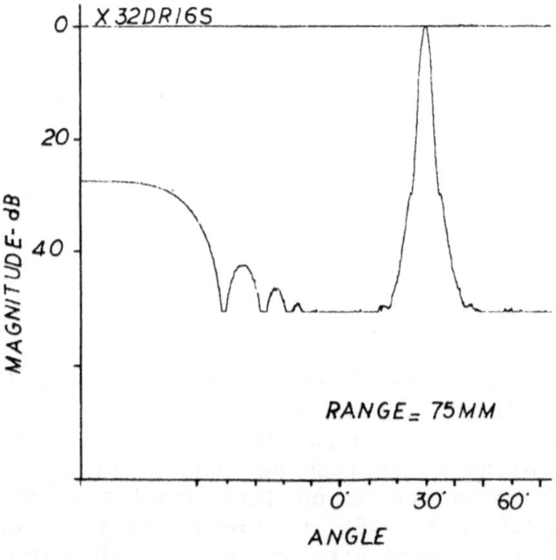

Fig. 6(b). Same as Fig. 5(c) but sparse 16 receivers.

without any deterioration of the main lobe. The same comparison can be made between Figs. 5(d) and 6(b) where the beam is now steered 30° away from the perpendicular to the array.

Nonlinear Processing

As described previously, the simulator is capable of handling some processing which can be performed in the RF section of the actual system. We have tried to simulate the effect of nonlinear processing on a single target. The nonlinear processing we simulated was a logarithmic compression of each receive channel followed by an expansion after the summation. The theoretical effect of such processing should be the decrease of any unwanted lobes, either sidelobe or grating lobe, without any deterioration of the main lobe. It should be mentioned we only have considered a single target, thus avoiding any problems which can be encountered with cross-product terms for more than one target. Figure 7 shows the results obtained for the last array described and should be compared with Fig. 6. There is an improvement in sidelobe and grating lobe. The sidelobe located at -30 dB on Fig. 6(a) has now completely vanished on Fig. 7. A similar impovement has been achieved with grating lobe.

Effect of Pulse Length

All the simulations performed in the previous examples used the same kind of pulse, namely a 3-cycle pulse amplitude weighted with a cosine law. Reducing the pulse length (i.e., making wider band transducers) improves the grating lobe level. If one uses a 1.5-cycle pulse instead of 3, the improvement on the grating lobe should be 6 dB. This can be checked on Figs 8(a) and (b) where, for a given array, we have simulated the transmitted beam pattern with a 3-cycle pulse (Fig. 8(a)) and a 1.5-cycle pulse (Fig. 8(b)). The improvement achieved is 5.7 dB.

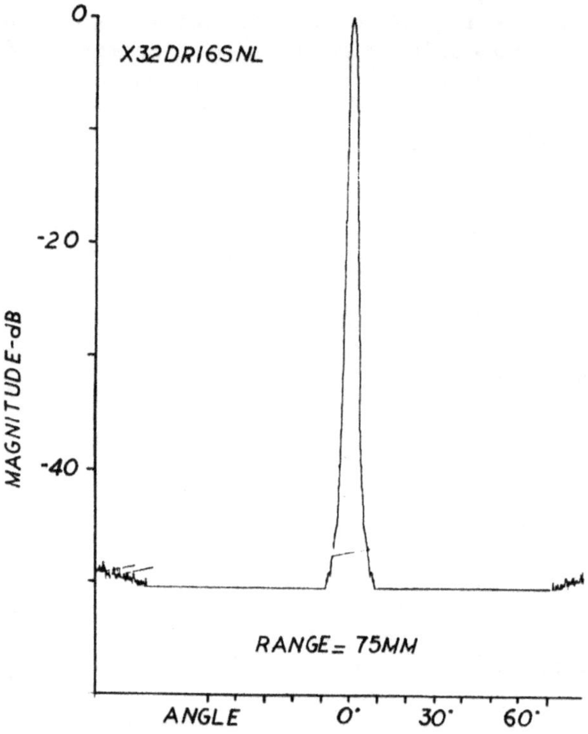

Fig. 7. Effect of nonlinear processing on array described in Fig. 6(a).

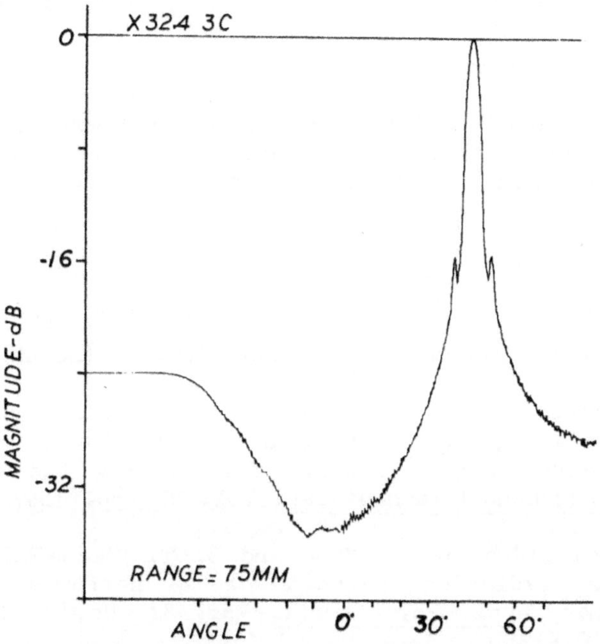

Fig. 8(a). One way simulation of V-3000 array with a 3-cycle pulse.

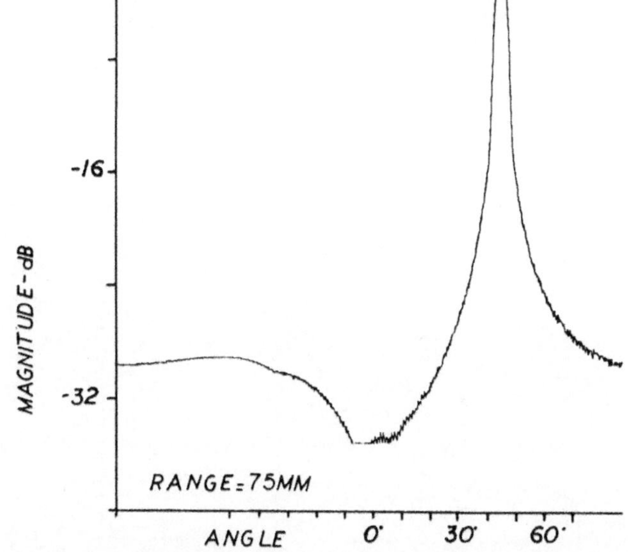

Fig. 8(b). One way simulation of V-3000 array with a 1.5 cycle pulse.

CONCLUSIONS

We have in this paper demonstrated the usefulness of our ultrasound simulator. It helps us in designing new types of arrays with improved performances. A natural extension of such a tool is the understanding of what can happen when the beam propagates in tissue, evolving from a beam simulator to an image simulator.

REFERENCES

1. N. Bom, Dutch Patent Application #7104271 (March 31, 1971).

2. W. A. Anderson, J. T. Arnold, L. D. Clark, W. T. Davids, W. J. Hillard, W. J. Lehr, L. T. Zitelli, A new real-time phased array sector scanner for imaging the entire human heart, Ultrasound in Medicine, Vol. 3B, pp. 1547-1557 (1977).

3. A. R. Selfridge, G. S. Kino and B. T. Khuri-Yakub, A theory for the radiation pattern of a narrow-strip acoustic transducer, Appl. Phys. Lett., Vol. 37, No. 2, July 1, 1980, pp. 35-36 (1980).

DYNAMIC FOCUSING AND COMPOUND SCANNING IN LINEAR ARRAY OF TRANSDUCERS

Marceau Berson*, Alain Roncin, L. Pourcelot

Service de Médecine Nucléaire in vivo et Ultrasons
C.H.R. Bretonneau, 37044-TOURS CEDEX, France

INTRODUCTION

Over the last few years, improvements in electronic focusing techniques and electronic steering of the acoustic beam have resulted in the very rapid development of apparatus using arrays of piezo-electric transducers. Such instruments, whether they use linear or sectorial electronic scanning, make real-time exploration of organs possible, supplying images with excellent resolution. One can therefore by using linear scanning with 5 MHz frequency probes, visualize surface vessels with a lateral resolution at 6 dB better than 1 mm.

Each type of linear and sectorial scanning process has its own advantages and disadvantages. The use of these two types of scanning on the same array might prove to be very worthwhile since it would mean that the following could be used :

- classical electronic scanning,
- compound with inclined parallel beams at an angle selected by the operator,
- a T.M. or Doppler exploration by means of an inclined beam issuing from a chosen area of the array. The angle of deflexion and the area explored are,in that case, chosen in relation to the image of the structures supplied in real-time, by the whole of the array.

*Attaché de Recherche I.N.S.E.R.M..

OUTLINE OF THE PROBLEM

As we know, given a group of N rectangular transducers of width a, on an array of step d, the distribution of acoustic pressure around a focal point is given by :

$$P = k \frac{\sin X}{X} \frac{\sin Ny}{N\sin y}$$

with $$X = \frac{\pi a \sin\theta}{\lambda}$$

$$y = \frac{\pi d(\sin\theta - \sin\phi)}{\lambda}$$

θ = angle which gives the direction in relation to the normal at the array

ϕ = angle of deviation of the beam.

The term $\frac{\sin X}{X}$ stands for the caracteristics of each element, and we get the term $\frac{\sin Ny}{N\sin y}$ from the network of N elements : which gives the distribution of acoustic pressure between the principal lobe and secondary lobes and grating lobes. It can be shown that when step d of the array (distance between the centres of the two neighbouring elements) is greater than the ultrasonic wave lenght λ in the propagation area, a grating lobe which may be fairly large appears. In order to avoid the grating lobe supplying incorrect information, the step of the array should be smaller than the wavelength i.e. always be less than 0.75 mm, since ultrasonic frequencies used in medicine are higher than 2 MHz. This goes to show the difficulties we came up against.

We know besides that the relative amplitude of pressure in the different lobes is modulated by the elementary function $\frac{\sin X}{X}$. This function should therefore decrease gradually so that the fall in pressure should not be too great as the inclination of the beam increases. This slight decrease in relation to ϕ is observed if each element is omnidirectional.

It is therefore necessary that the elementary width should be very small and the elements be well insulated. This is achieved by cutting very deeply into the transducer and this considerably weakens the probes.

All that has been said so far shows the technical difficulties involved in the making of probes and also accounts for the present limitation in the use of electronic steering for high ultrasonic frequencies.

COMPOUND SCANNING

Principle and application

It is well known that with compound scanning which leads to the exploration of one single structure under several incidences, better definition and overall knowledge of the structure can be obtained. This scanning process in often used in B manual echography and we have tried to transfer it onto an array in order to enlarge the interrogated area, to get a better knowledge of rounded structures and to have access to areas otherwise difficult to reach. Figure 1 shows the principle of this scanning. The resulting image is made up by the superposition of three images obtained successively with incidences 0, +ϕ and -ϕ and shown successively on the same visualization screen. The optical superposition is realized by retinal presistence or integration on a film.

Figure 2 shows the signal processing which takes place during scanning. Two series of delays (T and T') bring about the electronic focusing (T) as well as the orientation of the beam (T'). A commutation permits the use (2 and 3) or non use (1) of delays T'. Between images 2 and 3, the sequence is reversed so as to switch from incidence +ϕ to incidence -ϕ.

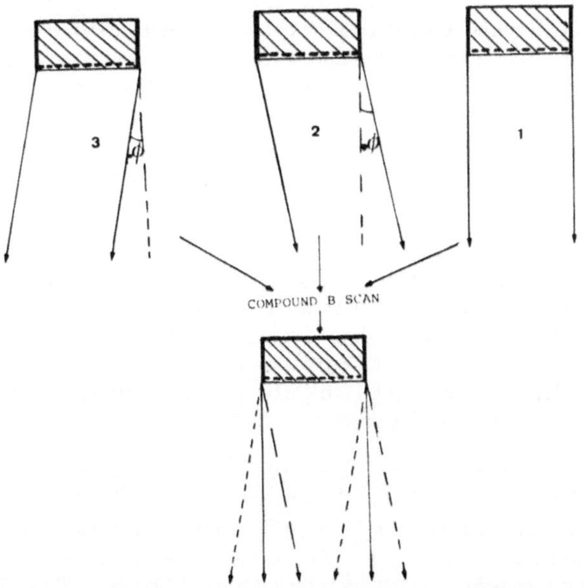

Figure 1 : Principle of compound scanning

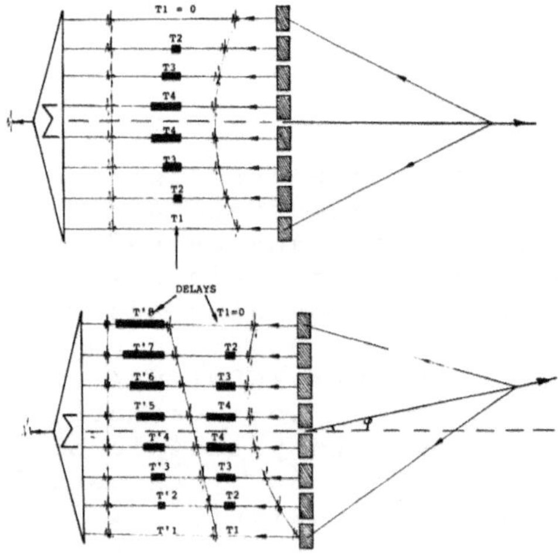

Figure 2:
Signal processing during compound scanning

Compound scanning has been achieved on our linear electronic scanning apparatus (S.S.A.B.E.L.) with an classical probe, that does not match the characteristics described in the theoretical introduction and which would have made it possible to obtain a suitable deflexion of the beam.

Indeed, the step between each transducer is 0.5 mm for a resonance frequency of 3.5 MHz (λ = 0.42 mm), the elements being electrically connected two by two. In addition, the insulation between the elements is inadequate. Hence the appearance of a grating lobe and a loss in sensitivity. These problems compelled us to limit the angle of deflection ϕ to 12 degrees approximately, but we were determined to test the principle and the electronic equipment.

Results

The system of correction of scans X and Y was set up by means of a phantom placed in water, made up of several wires. Figure 3 shows the images of these wires (a) without and (b) with correction of the scans: the good superposition of the information obtained under three different incidences can be observed.

Figure 4 shows a cross-section of liver (L), right kidney (R.K.), renal vein (R.V.) and vena cava (V.C.). Note once again the excellent superposition of the echoes in the area common to all three images. Some areas are slightly blurred due to the presence of the grating lobe. The interrogated area is noticably increased by the deflexions + and $-\phi$ (25^0).

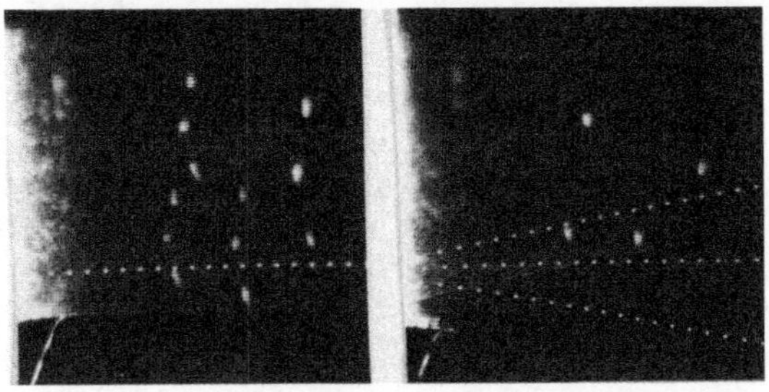

Figure 3:
Images of four wires with compound scanning

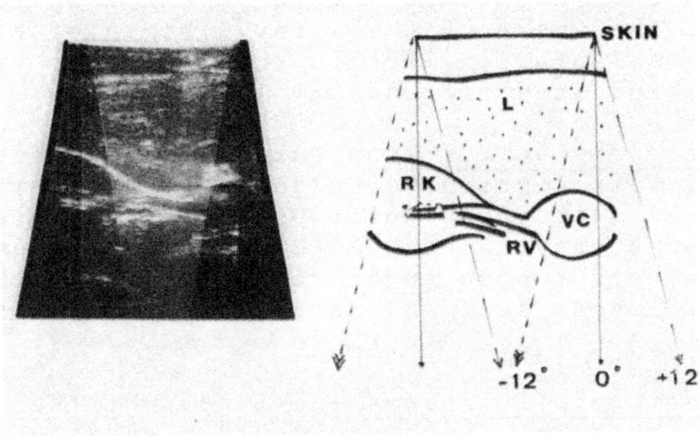

Figure 4: An in vivo cross-section

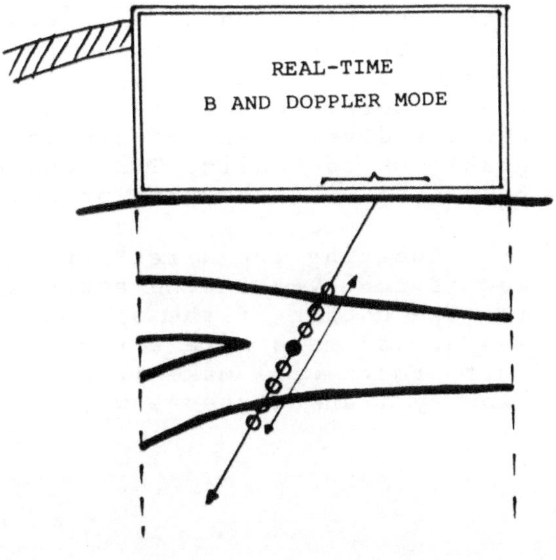

Figure 5:
Principle of real-time B and Doppler mode

DISCUSSION

The problems involved in the use of an unsuitable probe will be greatly reduced by the design of a captor which meets the theoretical criteria i.e. with a width of each element less than λ and with better insulation between the elements. In addition, the simultaneous use of 32 elements will make it possible to maintain an aperture large enough for the electronic lens.

Experiments carried out and the results obtained have shown that it was possible to orientate satisfactorily the ultrasonic beams and to obtain a uniform response over the whole of the array. The interest of such electronic orientation lies in the fact that we can then use the captor in different ways :

1° An electronic sector scan taken from any section of the array, this section being chosen in relation to the interrogated area which forms part of the overall image. Although less easy to handle than the probes usually used for a sector scan such a captor has all the advantages of a linear scan and most of those of a sector scan.

2° A scan of parallel beams inclined over the whole length of the array. Such use would make it possible to explore by means of parallel lines those structures which are not easily accessible and to achieve excellent steering for biopsy puncture needles.

3° A Doppler (or T.M.) exploration from one area of the probe. Figure 5 shows this process which is truly remarkable since it corresponds to the indispensable combination of B-scan, and Doppler using the same array of transducers. We are presently working on this project and the development of electronic orientations should enable us to realize T.M. and Doppler explorations, and B imaging with the same probe.

Locating the direction in which the Doppler is aimed is achieved in the echographic image of the structure. The choice of the optimum incidence will mean a much easier and much more effective Doppler exploration, and furthermore will make it possible to calculate the blood flow by transcutaneous way.

4° Doppler imaging : real-time Doppler imaging has already been achieved in our laboratory using parallel beams perpendicular to the plane of the array of transducers. This technique could become routine procedure thanks to the electronic orientations of the Doppler beams (figure 6) and because of definite improvements as regards the digitalisation and the memorisation of the signal. A very high number of Doppler points can in fact be obtained,and it thereby becomes possible to get an image of the blood flow without the equipment being too complex.

CONCLUSION

The work and the results outlined in this paper show the importance of the electronic steering of the beams on a multielement array. This technique greatly increases the potentialities of such an array. Combined B-mode and Doppler explorations which will eventually lead to an overall image of structures and blood flow are one of the extremely wide range of uses that this technique makes possible.

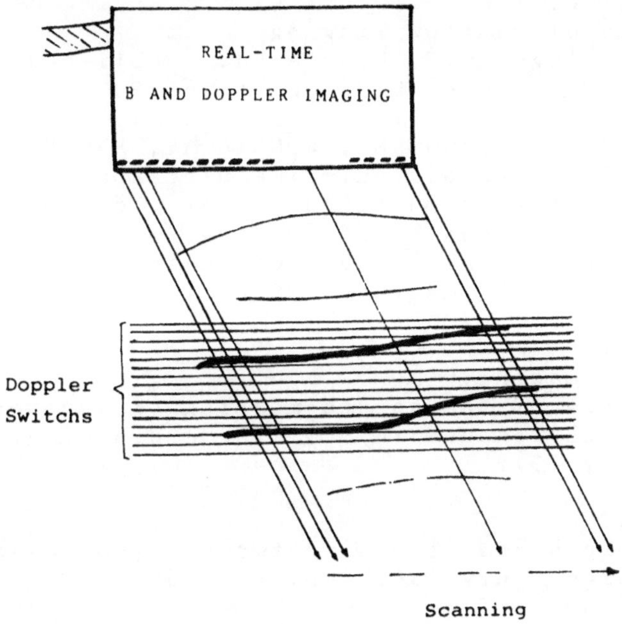

Figure 6 : Principle of real-time B and Doppler imaging

REFERENCES

- BOM N., ROELANDT J., KLOSTER F.E., LANCEE C.T., and HUGENHOLTZ P.G.
  Multielement system and its applications in cardiology. Proceedings of the second world congress on Ultrasonics in Medicine. Rotterdam, 4-8 june 1973.

- CARPENTER D.A., DADD M.J., and KOSSOF G.
  A multimode real-time scanner.
  Ultrasound in Medicine and Biology, 1980, vol. 6, n°3, pp. 279-284.

- DEFRANOULD and SOUQUET.
  Ultrasonic array design and performance.
  Echocardiology, C.T. Lancée Ed., published by Martinus Nijhoff, The Hague 1979, pp. 395-412.

- LIGVOET C.M., RIDDEER J., LANCEE C.T.
  A dynamically focused multiscan system.
  Echocardiology, N. Bom Ed., published by Martinus Nijhoff, The Hague, 1977, pp. 313-332.

- POURCELOT L.
  Real-time blood flow imaging.
  Echocardiology, C.T. Lancée Ed., published by Martinus Nijhoff, The Hague 1979, pp. 421-429.

- POURCELOT L., BERSON M., RONCIN A., BESSE D., PEJOT C.
  New developments of real-time ultrasonic tomography using an array of multitransducer : dynamic focusing, compound scan, and blood imaging.
  IVth international conference on information processing in medical imaging. INSERM 1979, vol. 88, 49-68.

- POURCELOT L.
  Vascular imaging.
  Proceedings of the second meeting of the world federation for ultrasound in medicine and biology. Miyazaki, 22-27 july 1979.

- SOMER J.C.
  Electronic sector scanning for ultrasonic diagnosis.
  Ultrasonics, july 1968, pp. 153-159.

- THURSTONE F.L., VON RAMM O.T.
  A new ultrasound technique employing two dimensional electronic beam steering.
  Acoustical Holography, vol. 5, P.S. Green Ed., Plenum Press.

A PHASED ARRAY ACOUSTIC IMAGING SYSTEM FOR MEDICAL USE

H. Edward Karrer, J. Fleming Dias, John D. Larson, Richard D. Pering
Hewlett-Packard Laboratories
1501 Page Mill Road
Palo Alto, California 94304
Samuel H. Maslak, Acuson, 900 N. San Antonio Rd.
Los Altos, California 94022
David A. Wilson, SRI International, 333 Ravenswood Avenue
Menlo Park, California 94025

I. ABSTRACT

An experimental linear phased array acoustic imaging system has been developed for cardiac, obstetric, and radiological use. This is a real time, pulsed B scan system using either a 2.5 MHz or a 3.5 MHz hand held transducer. The display format is a sector scan with dimensions of 90° x 24 cm and the system is dynamically focussed. The system uses an analog heterodyne architecture to greatly simplify the electronics for a densely packed array and to allow use with transducers of different frequencies. With a 20 mm aperture segmented into 64 elements, at 2.5 MHz, the system achieves a 6 dB resolution at 7 cm in water of 2.5 mm in azimuth, 1 mm in range, and 3.5 mm in elevation. Emphasis was put on a system with wide dynamic range; side lobe levels of the array beam plot less than -65 dB have been achieved over the sector. A digital scan converter is used to convert the sector scan to a raster display and for image processing. Images will be shown to demonstrate system performance.

II. INTRODUCTION AND BACKGROUND

Linear phased array imaging systems differ primarily in array aperture*, element density and means of focussing. These parameters

* The apertures discussed are rectangular with long, narrow elements. The azimuth direction is along the row of elements and the elevation direction is parallel to the length of the elements.

have major effects on the required transducer technology and the system architecture. Simply stated, the three major trade offs are as follows:

(1) The azimuthal resolution in the far field and in the focussed near field is proportional to λ/d and improves with larger apertures (d) and higher frequency as predicted from simple diffraction theory. The maximum aperture size for medical use is limited by the need for a small body contact transducer and various anatomical limits. For instance, for cardiac use, the intercostal space limits the maximum aperture dimension to about 25 mm. The maximum frequency is limited by the body attenuation to 5 MHz for normal adults (range 2-24 cm).

(2) The azimuthal aperture also determines the requirement for focussing. For typical apertures of 10-30 mm and frequencies of less than 5 MHz, much of the imaging takes place in the near field ($< d^2/4\lambda$) of the array and focussing is required. This can be either fixed depth focussing with an acoustic lens, or variable electronic or dynamic focussing, the latter providing superior resolution over the entire depth.

(3) Another limit comes because the azimuthal aperture is sampled with a finite number of elements. It can be shown that the element spacing must be $\approx \lambda/2$ in order to eliminate grating lobe responses in the imaging field. The grating lobes cause image clutter and severely limit the acoustic dynamic range of the image. This restriction requires element densities of at least 32 elements/cm of aperture at 2.5 MHz.

The linear phased arrays described in the literature to date have various blendings of these three trade offs. Somer's (1) original 21 element system achieved the $\lambda/2$ element spacing, but at the low frequency of 1.3 MHz where λ was 1.2 mm. With an 11 mm azimuthal aperture, he had a wide beam width ($\sin^{-1} \lambda/d$) of 6°. The near-far field transition was 2.6 cm, and therefore azimuthal focussing was not required and only beam steering was used. The elevation aperture was 10 mm and not focussed. Thurstone's (2) system used a 25 mm aperture at 2.25 MHz. This gave a respectable beam width of 1.5°, but with only 16 elements and an element spacing of 2.3 λ, there were grating lobes in the imaging field. The near-far field point was at 23 cm and azimuthal dynamic focussing was successfully demonstrated throughout the near field. The elevation aperture was 14 mm and was not focussed. The system of Anderson et. al., (3) was between the former two with an azimuthal aperture of 12 mm at 2.25 MHz and a beam width of 3.1°. The near-far field transition was at 6 cm and a fixed focus spherical lens was employed for azimuthal and elevation focus.

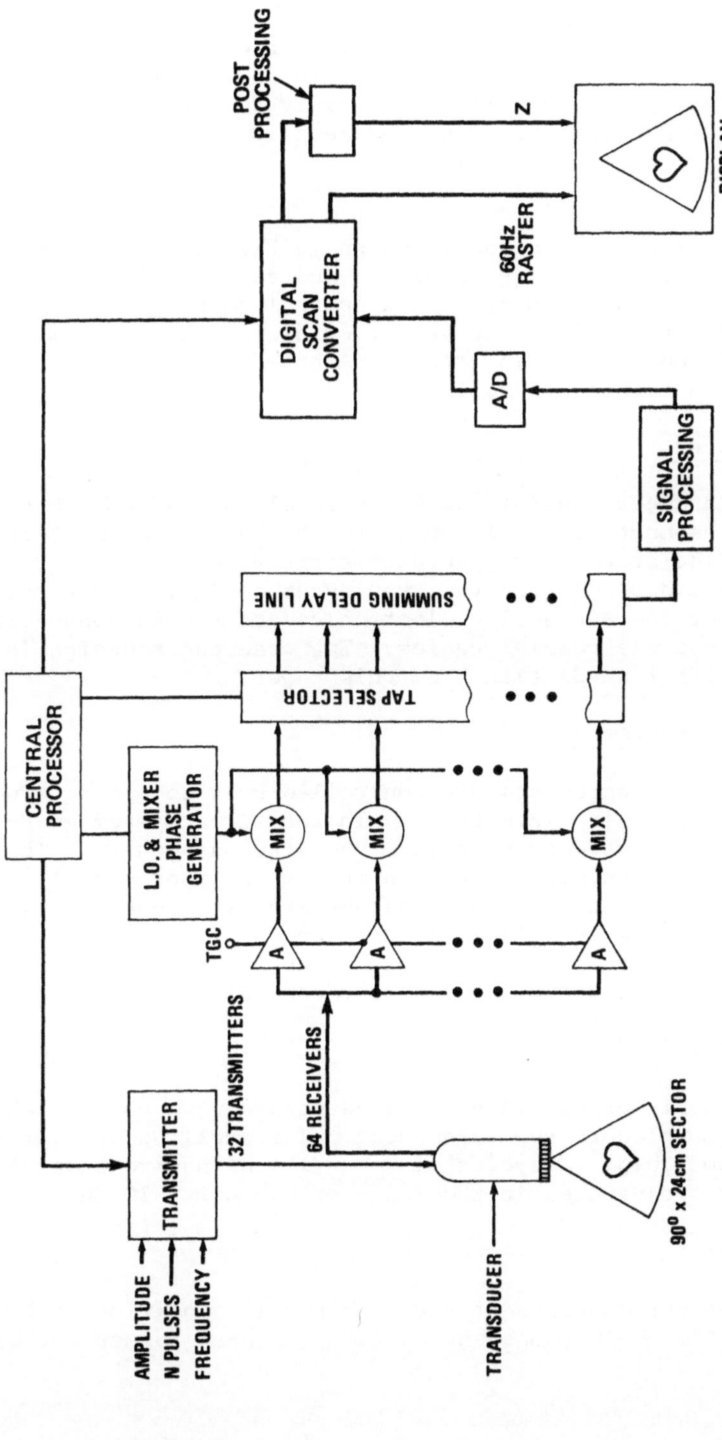

Figure 1. Acoustic Imaging System Block Diagram

The approach taken in this work was to select an aperture as large as possible consistent with anatomical considerations and then to pack it with elements of sufficient density to eliminate grating lobes. The resulting aperture is 20 mm x 14 mm with 64 elements. The system is fully dynamically focussed in aziumth and uses a fixed focus acoustic lens in elevation.

III. SYSTEM DESCRIPTION

The system block diagram is shown in Figure 1. It consists of a transducer, scanner (transmitter and receiver), scan converter, and display, all under control of a central computer. This experimental system was built to be very flexible so that the imaging process could be studied in detail. A brief description of the various system components follows.

A) Transducer

The transducer consists of 64 uniformly spaced PZT-5H ceramic elements. The central 32 elements are shared as transmitters and receivers. The transducer generates short pulses of energy in the 1-5 MHz range, depending on the transducer design. The transducer probe does not contain active electronics and the 64 connections are brought out via coaxial cables. The transducer design is discussed further in Section V of this paper.

B) Central Processor

The various components are controlled by a computer. Two versions of the system were built. In the first system an HP 2100 computer was used to calculate the scan parameters. These parameters were then stored in core memory and subsequently read out during system use. The scan parameters were under software control for optimum scan flexibility. In the final version of the system, the scan parameters are stored in ROM and the components of the system are under microprocessor control.

C) Transmitter

The transmitter generates a gated square wave voltage which is transformer coupled to the transducer. The amplitude is variable from 0 to 400 V p-p. N cycles ($1 \leq N \leq 15$) of square wave can be generated at frequencies in the range of $0.5 \leq f \leq 10$ MHz.

D) Receiver

The receiver consists of a variable gain amplifier in each channel for time gain compensation (TGC), a beam former for beam steering and focussing, and an RF detector. The receiver front end measures the short circuit current generated by the transducer in order to minimize cable capacitance effects. The TGC has manual control with eight slide pots to vary the gain to compensate for body loss over the 24 cm range. The loss consists of attenuation (1-2 dB/cm-MHz) and reflection coefficient losses. Each TGC control has 60 dB of adjustable gain, while the TGC dynamic range at any gain setting is 60 dB. The TGC linearity is $\pm$ 0.5 dB. The variation in channel to channel electronic gain is $\pm$ 0.5 dB. The receiver noise figure (the ratio of signal to noise at the input and output) is 5 dB and is limited by the transducer.

The beam is steered $\pm$ 45° from the normal and focussed from 2 to 24 cm using a combination of time delay and phase adjustment (4, 5, 6). A tapped L C summing delay line is used for beam steering while vernier phase adjustment in each channel is used for focussing. The equation for the time delays required for each channel makes no paraxial approximations and only assumes a constant sound velocity. Each radial line of the sector scan is divided into a number of focal zones during which the element delays are constant. Each receive channel is heterodyned with a local oscillator to create an IF frequency of 1.7 MHz. The phase of this LO is adjustable in small increments, and the mixer output has this phase variation. This variable phase, under Central Processing Unit control, provides a means by which the pulses from the various elements can be brought into alignment. For the coarse time delay adjustments for steering, a tapped summing delay line and tap selector (switch matrix) are employed. An L-C delay line was chosen over other possible types to achieve wide dynamic range. The tap selector switch allows any receiver channel to be switched into any tap and is changed once per line. The output of the summing delay line is then IF filtered and envelope detected.

The heterodyning method, when used with a summing delay line, has the following advantages:

1) It provides adjustable time delays for each channel without a separate string of switchable delays in each channel. The latter is economically prohibitive in systems with many channels.

2) The bandwidth required for the delay lines is reduced since the IF frequency is lower than the carrier. This is a consequence of the fact that the information bandwidth is less than the carrier frequency.

3) Transducers of different frequencies can be easily used by changing the local oscillator frequency.

The heterodyning method, in conjunction with phase shifting to achieve focussing, requires a restricted information bandwidth since the phasing is accurate only at one frequency. In practice, this is not a severe limitation.

E) Signal Processing, Digital Scan Converter, and Display

The summed output of the receiver is IF filtered, compressed, and analog-to-digital converted.

This digitized sector line information is passed to a digital scan converter which converts the sector information to a rectangular format, and stores the data for subsequent display or display refresh. This technique a) eliminates flicker in the displayed picture regardless of data rate, b) allows inter-sector line fill-in, c) provides still pictures, d) allows post processing on the picture such as light-dark reversal, grey scale enhancement, etc. and e) maintains uniform picture brightness at all angles and ranges.

There is a fundamental limit to the rate at which data may be gathered. This is set by the velocity of sound in tissue (1500 m/s) and the maximum range (24 cm one way), which together give a round trip time of flight for one sector line of 320 μs. In order to gather data at a "real time" (>25 Hz) rate, we may have a maximum of 120 sector lines in the picture. The digital scan converter eases limitations placed on the displayed picture by the maximum data rate.

The final image, along with appropriate alphanumeric information, is displayed on a CRT monitor in a television raster format. There are 480 lines by 640 pixels per line, each with 32 possible levels of grey. This may be interlaced or not, in American or European TV format. Postprocessing may be done at the output of the digital scan converter.

IV. DISCUSSION OF ACOUSTIC DYNAMIC RANGE

The acoustic dynamic range of the imaging system is limited by the side lobe level of the azimuthal beam pattern expressed in dB. (Here the term side lobes also includes any grating lobes.) The dynamic range limits the ability of the system to differentiate between target echoes of differing strength in the same region of space. Using the human heart as a target, we made many dynamic range measurements and showed, for instance, that the echoes from the inside heart muscle (endocardium) are at least 45 dB below echoes from the rear heart wall from which they must be distinguished. During the course of this work seven different methods of controlling side lobes were theoretically and experimentally evaluated. The summary results are given:

1) The theoretical results of Tancrell (7) and our experimental results show a slower rise time and longer time duration for the grating lobe signal as compared to the main beam. This suggests that derivative processing could be used to reject grating lobes. Some improvement in rejection is possible using this technique.

2) Thurstone (2) used non-linear processing by taking the log of each receiver signal before summing in order to reject side lobes and improve resolution. This is essentially multiplicative as opposed to additive processing and has been used in sonar work (8). Our experimental and theoretical results showed that this log processing does reject side lobes better than additive processing but that it only works well for single targets. With multiple targets that have differing reflectivities at the same range, it causes target dropouts and results in an image with a granular texture.

3) The final array beam pattern is the product of the transmit and receive array patterns. Thus if the array is configured so that the receiver grating lobes are not illuminated by the transmit grating lobes, then the grating lobe responses will be reduced. This could be done for instance by using different receive and transmit element spacing.

4) Another technique is to use non-uniform spacing (9) of the elements in order to widen the aperture and thus achieve better resolution without increasing the number of elements. The grating lobe, which would appear due to spatial undersampling of the aperture, is smeared out due to the non-uniform spacing.

5) Apodization or element weighting is a technique commonly used on CW systems for achieving side lobe rejection at the expense of beam width. Theoretically, it is not as effective in the pulsed case since the entire aperture is not illuminated at one time and the proper phase cancellations to create deep nulls in the array response do not occur. Tancrell (7) has shown theoretically that apodization can lower the close-in side lobes with negligible effect on grating lobe rejection. Experimentally, we have shown that apodization emproves the beam response when transducers with several cycles of pulse response are used.

6) The pulse shape can affect the side lobe levels. As the number of cycles in the acoustic pulse increases, the response approaches the CW case with larger grating lobes. With pulses of one cycle or less, close-in side lobes come up since deep nulls cannot occur. Somewhere in between is best. The pulse shape will depend on the transmit excitation and the transducer response and is only partially controllable.

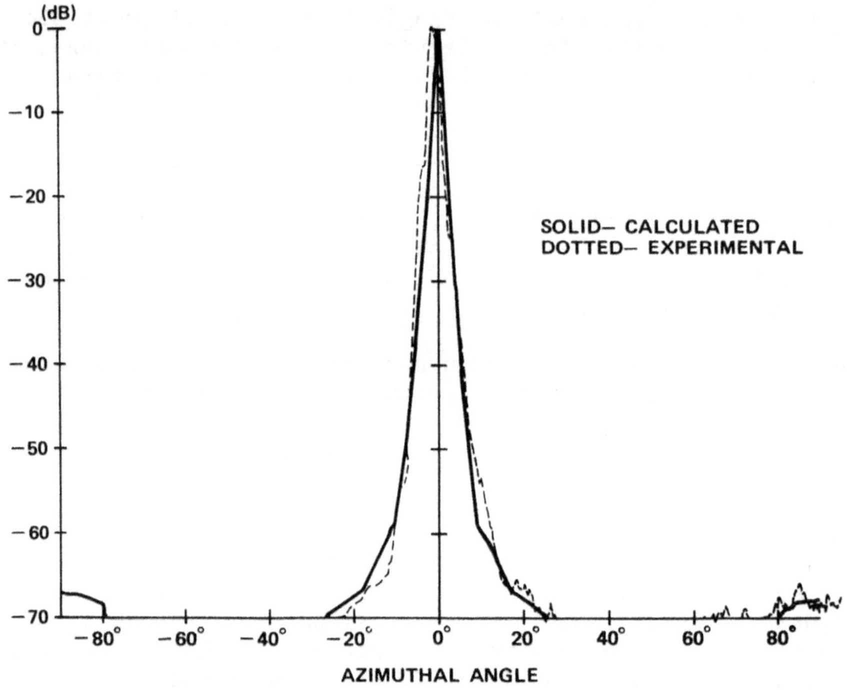

Figure 2. Azimuthal Beam Pattern

7) The most effective way to reduce the grating lobes is to increase the element density. This requires narrow elements which also is necessary to achieve wide sector scan angles. Our approach was to use 64 receivers uniformly distributed across 20 mm with 32 shared transmitters in the central region. The transmitter is fixed focussed at 7 cm.

Although all the above methods were evaluated, only items 5, 6, and 7 were effective enough to be incorporated in the final imaging system. The system results are shown in Figure 2 which shows the agreement between experimental and theoretical beam plots. (The transducer array and beam former were modeled using an extension of Stephanishen's (10) method). It should be noted that in an ideal system with no quantization and with 64 receive elements, the far out side lobes should be at a level of -66 dB round trip. A prototype system is shown in Figure 3.

V. TRANSDUCER

The transducer probe is shown in Figure 3. The acoustic portion of the transducer consists of a "stack" of various layers which are epoxy bonded together. The base of the stack is a backing material which is acoustically lossy and impedance matched to the rest of the stack. Next in the stack is PZT-5H piezoelectric ceramic. The ceramic is lapped and polished, then vacuum deposited electrodes are applied prior to bonding. At this point 64 long narrow elements are defined by sawing. Each PZT element is 0.24 mm wide, 14 mm long, and 0.48 mm thick. The next layer is a 0.04 mm brass foil which protects the elements, provides a common electrical connection to the tops of all the elements, and forms an RFI shield. Finally a convex cylindrical acoustic lens made of silicone rubber completes the stack. The elements are connected to the receiver passively over individual leads.

In order to meet the imaging system requirements, the array elements must have the following properties:

- Close element-to-element spacing to eliminate grating lobes
- Narrow elements to illuminate a full 90° sector
- Small element-to-element coupling
- Wide bandwidth to achieve short pulses
- Sufficient sensitivity to get images free of thermal noise at reasonable transmit voltages
- Small element-to-element gain variations.

These items are discussed in detail next.

IMAGING SYSTEM

TRANSDUCER

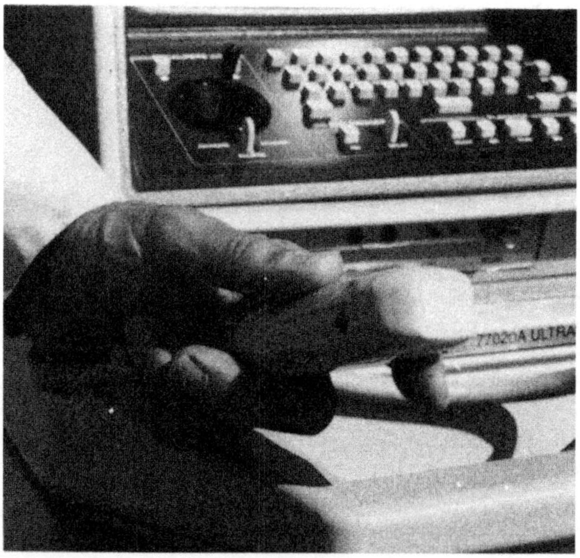

Figure 3. Prototype Equipment

(1) Element Spacing - The element spacing is picked to be about $\lambda/2$ at 2.5 MHz (.32 mm) so that there are no grating lobes in the imaging region.

(2) Element Width - The selection of element width is a compromise among three factors. The element must be narrow in order to illuminate the entire 90° sector. A 4-6 dB roll-off is achieved for a ± 45° sector. As the element becomes narrower, its capacitance decreases and sensitivity is lost since each element must drive the cable capacity. Finally, the element width selection is critical in controlling the modes of vibration of the element. When the thickness and width of the element are equal, the desired thickness mode couples to the dilatational or width-expansion mode of the element. This coupling can lengthen the acoustic pulse since the lateral damping of the element is small. The solution is to make the thickness several times greater than the width in order to separate the modes. For the elements in question the thickness mode is at 2.5 MHz and the dilatational mode is at 8.0 MHz.

(3) Cross Coupling - The element-to-element cross coupling is increased as the elements are brought closer together. Acoustic coupling can occur because of lateral wave propagation along the surface of the backing or through the protective metal foil. Measurements of elements with a small piezoelectric probe show that this coupling is about -15 dB between elements. This coupling has at least two detrimental effects on the final image. When elements are coupled, the effective width of the element is increased and the radiation roll-off with angle will be more rapid. This limits the sector size. This also reduces the effective number of elements in the aperture. A more serious problem is termed "ring-down".

(4) Ring-down - When a simple transducer is placed in a large water tank with no targets, and the transmitters are excited, the receivers will pick up spurious signals during a "ring-down" period of ~50μs. (This is the time for the receiver outputs to fall to 50 dB below their peak value). A fixed pattern clutter in the image occurs out to a range of about 7 cm. The cause is energy which travels out laterally from the transmitters via surface waves on the backing surface and plate waves in the metal foil, and which excites the adjacent receivers. This coupling energy has a spectral distribution which is centered about 1 MHz and is excited by a mode of motion of an element which has been called the mass-spring mode (11). The mode is excited because of the motion of the center of mass of the element on a compliant backing, while its frequency is determined by the element mass and backing spring constant. The mode has a high Q because of the coherent reflection of laterally traveling waves from adjacent elements, and the low loss per wavelength in the backing.

The cross coupling and ring-down can be reduced by several methods:

(a) by extending the saw cuts between the elements deep into the backing (at least by λ) to stop surface wave propagation

(b) by selecting the metal foil thickness so that the plate wave reflections from adjacent elements interfere destructively with the surface wave reflections (12)

For the final arrays, ring-down times of 10-12 μs are achieved.

(5) Bandwidth - A wide bandwidth is achieved by forming the elements directly on a lossy, matched backing. The backing material is a tungsten-vinyl composite formed at high pressure and temperature (13,14). The typical measured impedance is 25-30 x 10^6 rayls and the attenuation is 20 dB/cm for a 1.2 μs, 2.5 MHz pulse. The epoxy bonding was done using carefully polished surfaces under clean room conditions. Bonds of < 4 μm thickness were achieved.

The single element 3 dB bandwidth, as measured from an input impedance plot, is 50%. Figure 4 illustrates the transmit/receive pulse response of the phased array with a thin rod as a target. The trace shows the received IF waveform.

(6) Sensitivity - The PZT-5H provides sufficient sensitivity when mounted directly on a matched, lossy backing. There is a 40-50 dB round trip transduction loss due to the mismatch between tissue ($Z = 1.5 \times 10^6$ rayls) and PZT (34×10^6 rayls). Thus with a peak transmit voltage of 200V, the maximum receiver signal will be 1V, If the receiver thermal noise is 3 μV, there is a 110 dB dynamic range which is sufficient for good imaging. (For instance 50 dB of body attenuation and 60 dB of range between specular and diffuse echoes at any depth). It should be noted that the receiver noise is limited by the thermal noise in the backing.

(7) Gain Variations - The element-to-element gain variations on the transducer alone are about ± 2 dB. Amplitude variations of this level have little effect on the side lobe levels down to the 60 dB level (15).

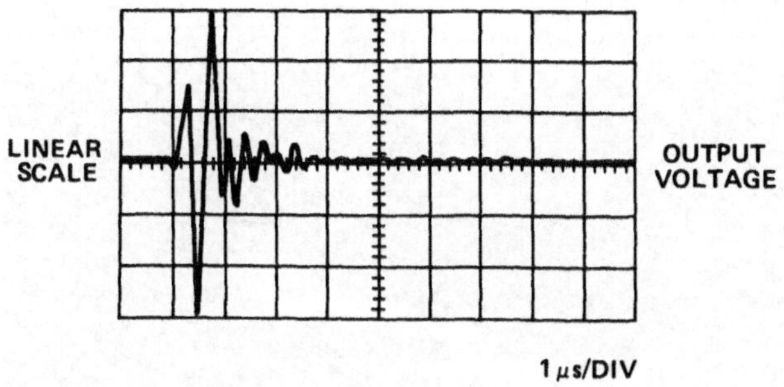

Figure 4. Transducer Pulse Response

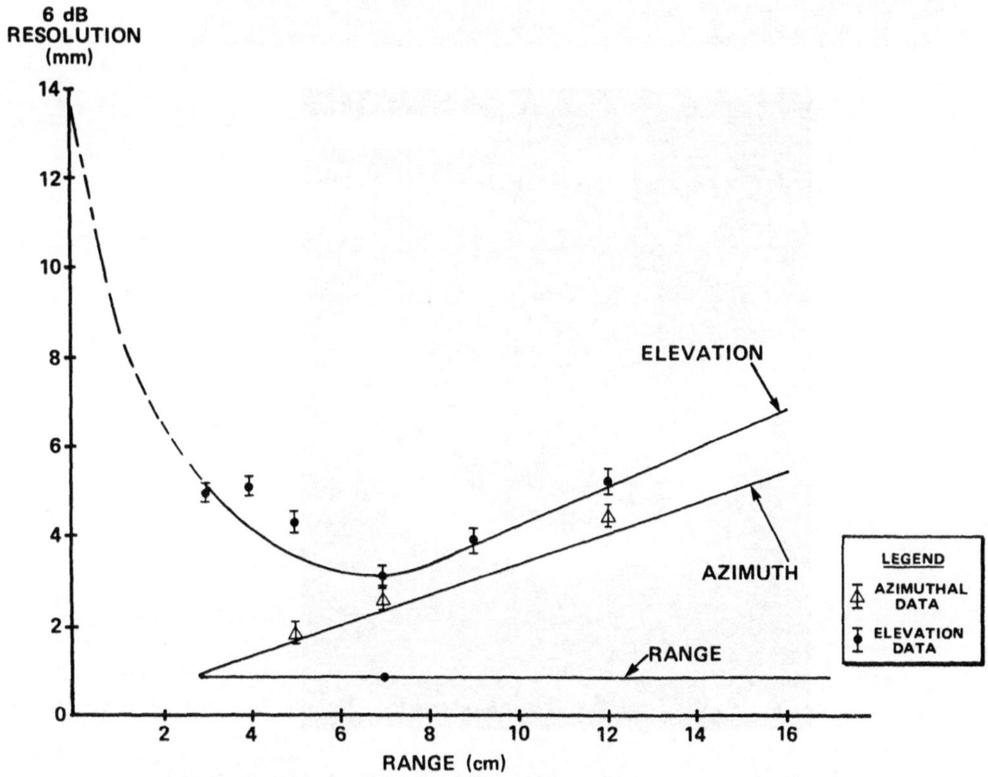

Figure 5. System Resolution

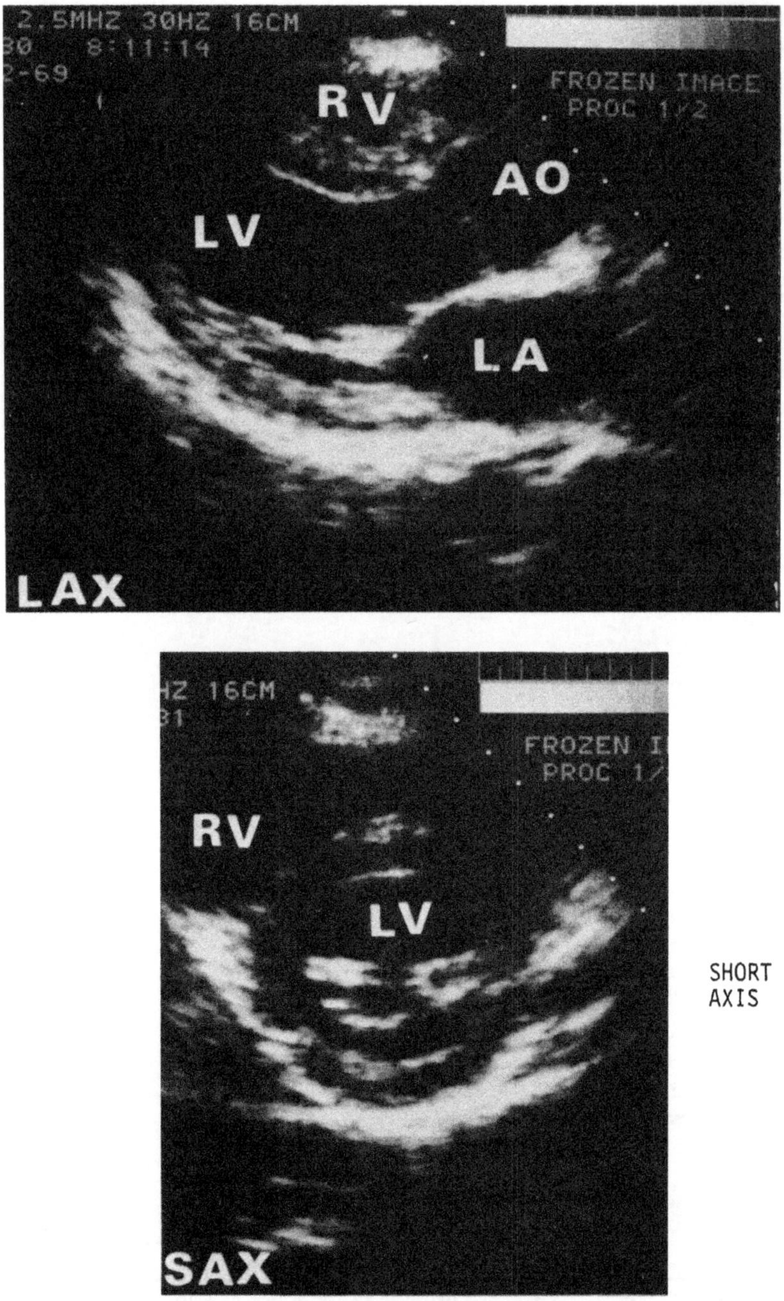

LONG AXIS

SHORT AXIS

Figure 6. Long and Short Axis Views of Adult Heart

VI. RESULTS

A) Resolution - The azimuthal, elevation and range resolutions of the system are discussed below:

1) Azimuth - Azimuthal resolution is shown in Figure 5. This applies over the range of dynamic focus of 2-24 cm. The transmitter was apodized.

2) Elevation - The elevation resolution is determined by the properties of the cylindrical lens which is fixed focussed. The 6 dB elevation beam width as a function of range is also shown in Figure 5.

3) Range - The primary limitations on range resolution are the length of the acoustic pulse generated by the transducer and the system bandwidth. The system bandwidth is 0.9 MHz centered at 2.5 MHz. The 6 dB acoustic pulse length from the transducer (excited with a single cycle square wave at 2.5 MHz) and through the receiver is 1.25 μs. The range resolution is then 1 mm. Other smaller contributions to the range resolution error include the pulse spreading due to frequency dependent attenuation in the body and diffraction errors caused by the transmitter being focussed only at one range. The range resolution is summarized in Figure 5.

B) Acoustic Power - The acoustic power of the system was measured with a calibrated piezoelectric microprobe as well as a thermocouple calorimeter. At full transmit voltage the spatial and temporal average intensity at the transducer face is 23 mW/cm^2.

C) Images - Polaroid photos showing long and short axis views of an adult human heart are presented in Figure 6.

VII. ACKNOWLEDGEMENTS

The authors would like to acknowledge the contributions that many Hewlett-Packard people made to this imaging system. It was developed originally at HP Laboratories and experimental prototypes were built there. The effort was then transferred to the HP Andover Division where the system was refined and put into its final configuration.

At the HP Andover Division, Dave Perozek and John Hart provided project leadership; Arthur Dickey, Ron Gatzke, and Larry Banks led the scanner design; Ray O'Connell did the scan converter. Hugh Larson did the signal processing algorithms. Amin Hanafy, Jim Fearnside, Dave Miller, Jerry Leach and Dan Latina contributed greatly to the transducer design.

At HP Laboratories the project was encouraged by Don Hammond. John Dukes contributed significant improvements to the video processing Henry Yoshida, George Nelson and Johnny Ratcliff gave skilled fabrication support. Dr. Richard Popp from the Stanford Hospital gave enlightened consultation and clinical evaluation during the entire project.

To all these people, and many others, we gratefully acknowledge the spirit which allowed such a system to be built.

VIII. REFERENCES

1. J.C. Somer, "Electronic Sector Scanning for Ultrasonic Diagnosis," Ultrasonics, July 1968, pp. 153-159.

2. F.L. Thurstone, and O.T. Von Ramm, "Electronic Beam Steering for Ultrasonic Imaging," 2nd World Congress on Ultrasonics in Medicine, June 4-8, 1973, Rotterdam, Netherlands, Excerpta Medica, Amsterdam, Netherlands, 1974, pp. 43-48.

3. W.A. Anderson, et. al., "A New Real Time Phased Array Sector Scanner for Imaging the Entire Adult Human Heart," Ultrasound in Medicine, 3B, 1977, pp. 1547-1558.

4. S.H. Maslak, "Acoustic Imaging Apparatus," U.S. Patent #4,140,022, Feb. 20, 1979.

5. C.B. Burckhardt, et. al., "A Simplified Ultrasound Phased Array Sector Scanner," Echocardiography, Third Symposium, Martinus Nijhoff Publishers, The Hague, pp. 385-393.

6. G. Manes, et. al., "A Single Channel Reduced Bandwidth SAW Based Processor for Ultrasound Imaging," 1979 Ultrasonics Symposium Proceedings, pp. 179-183, IEEE 79CH1482-9.

7. R.H. Tancrell et. al., "Near Field Transient Acoustic Beam Forming with Arrays," 1978 Ultrasonics Symposium Proceedings, pp. 339-343, IEEE 78CH1344-1 SU.

8. V.G. Welsby, "Multiplicative Receiving Arrays," Journal Brit. IRE., 22, July 1961, pp. 5-12.

9. J.L. Allen, "Array Antennas," IEEE Spectrum, 1, November 1964, pp. 115-130.

10. P.R. Stephanishen, "Transient Radiation from Pistons in an Infinite Baffle," J. Acoust. Soc. Am., 49, 1971, pp. 1629-1638.

11. J.D. Larson, "A New Vibration Mode in Tall, Narrow Piezoelectric Elements," 1979 Ultrasonics Symposium Proceedings, pp. 108-113, IEEE 79CH1482-9.

12. Amin Hanafy, private communication.

13. S.L. Lees et. al., "Acoustic Properties of Tungsten-Vinyl Composites," IEEE Transactions on Sonics and Ultrasonics, SU-20, January 1973, pp. 1-2.

14. J.D. Larson, J.G. Leach, "Tungsten-Polyvinyl Chloride Composite Materials - Fabrication and Performance," 1979 Ultrasonics Symposium Proceedings, pp. 342-345, IEEE 79CH1482-9.

15. K.N. Bates, "Tolerance Analysis for Phased Arrays," 9th Acoustical Imaging Symposium, Houston, Texas, Dec. 3-6, 1979, K.Y. Wang, Editor, Plenum Publishers , pp. 239-262.

ORIGIN OF ANOMALOUS BEHAVIOR IN TRANSDUCERS USED IN ACOUSTICAL IMAGING

B. Delannoy and C. Bruneel

Laboratoire D'Opto-Acousto-Electronique- ERA CNRS
N° 593 - Universite de Valenciennes
59326 Valenciennes Cedex - France

INTRODUCTION

Good design of the transducer arrays frequently used in acoustic imaging devices like those reported by a number of authors[1-8] is of major importance for obtaining good images after reconstruction. This proves even more crucial if one tries to get very good resolution, with a correspondingly small spatial period, say of the order of one acoustic wavelength, of the array. The major drawback of the coupling which then arises between neighbouring transducers is a narrowing of their radiation pattern. This leads in turn to a decrease of their angular aperture together with a degradation of the lateral spatial resolution of the image, or a narrowing of the scanning angle in sector scan devices like those designed by some authors[9-10].

In order to find some criteria leading to an optimum design for these transducers, a systematic study of their radiation has been performed for various boundary conditions and dimensions.

First, in a preceding paper [11] we have shown theoretically and experimentally that for the particular case of scalar diffraction of longitudinal acoustic waves in a fluid medium, the nature of the medium surrounding the ultrasonic transducer determines its directivity factor. This allowed us, to set up precisely the criteria for choosing between the different mathematical formulas given by Kirchoff, Rayleigh and Sommerfeld, for the radiation from acoustic sources.

On the other hand, for extended sources, say one or several wavelengths in width, the experimental measurements showed severe deviations from the theoretical predictions, consisting mainly in the occurence of parasitic sidelobes inclined by a 50 degree angle to the transducer normal. In another paper (12-13) it has been shown that these lobes are due to parasitic zeroeth order Lamb waves propagating in the transducer material and radiating into the fluid medium. For single transducers, which may resonate under Lamb wave propagation, the parasitic lobe amplitude is very sensitive to the transducer width, however the resonance effects become weaker as the width increases.

For transducers grouped in linear arrays and machined simply by cutting the metallization over one face of the piezoelectric platelet, the parasitic lobe amplitude is nearly width-independant. Moreover, the contribution of higher order Lamb waves to the main lobe remains negligible with respect to that arising from electrical coupling between neighbouring transducers. As we have shown in another paper (14) this effect, constitutes the major cause of the narrowing of the main lobe for each individual source.

We report the main results of the study, together with some suggestions for reducing the parasitic coupling phenomena.

I - Radiation of a single transducer.

The study of the radiation pattern of an isolated transducer surrounded by a fluid medium brings two distinct phenomena to evidence, with a relative strength depending on the transducer dimensions. Thus, for sources which are very narrow at the acoustic wavelength scale, the directivity factor will be given by the boundary conditions of the problem, but for larger sources, the resonance effect for Lamb waves propagating inside the transducer arise.

A - Narrow transducers - Boundary Conditions.

The theoretical description of the radiation of acoustic sources is based upon a choice between the different mathematical formulas suggested by Kirchhoff, Rayleigh and Sommerfeld. Precise criterions have been defined and verified experimentally by us for the application of these formulas in the particular case of scalar diffraction of longitudinal acoustic waves in a fluid medium.
The mathematical description is then given by a Helmholtz-Kirchhoff integral :

$$p(\vec{r}) = \int_S \left(g \frac{\partial p}{\partial n} - p \frac{\partial g}{\partial n}\right) dS \qquad (1)$$

where the closed integration surface S surrounds the observation point P and p stands for the complex amplitude of the local acoustic pressure and g is a Green's function for harmonic waves, which must be determined. Mathematically speaking, several Green's functions may be used. The choice may be made her unambiguously when thinking to their physical meanings. In fact, the Green's function gives (in all space) the amplitude distribution which would be induced by a fictitious source lying at the observation point P, with position vector $\vec{r}$, owing to the reflections onto the closed surface S.

Let us consider a first limiting case where the source radiation occurs in a free space (i.e. $Z_1=Z_2$, fig. 1a). No reflection occurs, so that only the direct wave from point P is observed at the current point M with position vector $\vec{r}_o$ and the Green's function becomes a free space propagation function. Eq. 1 may so be written as :

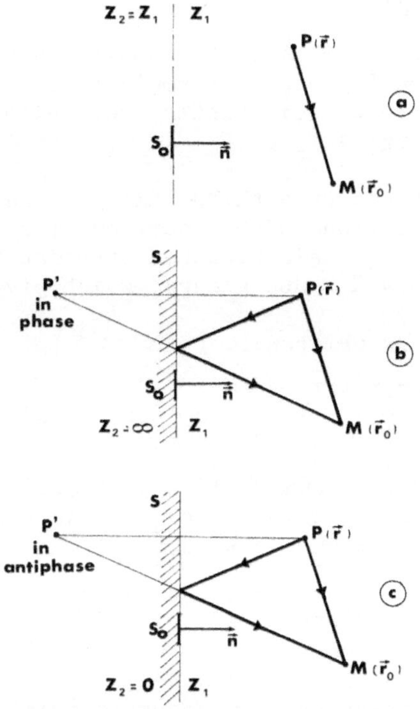

Fig. 1 - Boundary conditions. (a) source in a free space; (b) source in a rigid plane; (c) source in a soft space.

$$p(\vec{r}) = j\,\frac{p_o}{\lambda}\int_{S_o} \frac{\exp(-jk|\vec{r}-\vec{r}_o|)}{|\vec{r}-\vec{r}_o|}\cdot\frac{1+\cos(\vec{n},\vec{r}-\vec{r}_o)}{2}\,dS \qquad (2)$$

This particular form of Kirchhoff's formula describes accurately the case of a source radiating in a free space where the $\frac{1+\cos(\vec{n},\vec{r}-\vec{r}_o)}{2}$ term is the directivity factor (α_K).

Let us consider now the second limiting case, where the source is imbedded in a half space with acoustically hard boundary ($Z_2 = \infty$, fig. 1b).

Two waves are then received at point M : the direct one and that which has been reflected without phase change on the rigid plane. The Green's function (g_+) is then the sum of the contributions of the source at P and that of its image P' through the rigid plane.

The relation (1) may then be written as :

$$p(\vec{r}) = \frac{jp_o}{\lambda} \int_{S_o} \frac{\exp(-jk|\vec{r}-\vec{r}_o|)}{|\vec{r}-\vec{r}_o|} \, dS \qquad (3)$$

Where S_o is the active surface of the source in the plane boundary. This relation, known as Rayleigh's formula describes the radiation of a source imbedded in a rigid baffle, for which the directivity factor equals unity (α_R=1).

Let us finally consider a third limiting case where the source is imbedded in an half plane with acoustically soft boundary (Z_2=0, fig. 1c). Two waves are again received at point M, but the reflected wave suffers now from a Π phase change, which gives the Green's function (g_-).

The relation (1) is now written :

$$p(\vec{r}) = \frac{jp_o}{\lambda} \int_{S_o} \frac{\exp(-jk|\vec{r}-\vec{r}_o|)}{|\vec{r}-\vec{r}_o|} \cos(\vec{n},\vec{r}-\vec{r}_o) \, dS \qquad (4)$$

This relation is known as Sommerfeld's diffraction formula and describes the radiation of a rigid source in a soft baffle, where the directivity factor is now $\alpha_S = \cos(\vec{n},\vec{r}-\vec{r}_o)$.

To summarize, it appears that the nature of the boundary conditions around the source plays a very significant role on its acoustic pressure field distribution.

For transducer arrays used in acoustic imaging devices each element of the array is long enough to allow the assumption of a bidimensional cylindrical spreading. As we have shown[11] the directivity factors remain, in this case, the same as for the spherical tridimensional spreading.

Then, in the far-fied approximation, Eq (2),(3) and (4) takes a simple analytical form and may be expressed as the product of two terms : the directivity factor $\alpha(\theta)$ taking the boundary conditions into account (i.e, Z_2/Z_1) and a second term of the form $[\sin X(\Theta)]/X(\Theta)$ depending on the source width 2b, where $X(\Theta)$ equals :

$X(\Theta) = (2\pi b \sin \Theta)/\lambda$

$\Theta = (\vec{n}, \vec{r} - \vec{r}_o)$ and λ is the wavelength.

Then the variation, versus the angle Θ, of the characteristic parameter A, i.e, the electrical voltage in the following experiment, will follow a law of the form :

$$A = A_o \frac{\sin X(\Theta)}{X(\Theta)} \cdot \alpha(\Theta) \qquad (5)$$

The far-field directivity patterns of ultrasonic transducers in the three different boundary conditions have been measured in a pulsed quasicontinuous mode.

In figs. 2-4 the realization of three different boundary conditions is illustrated, and the measured directivity patterns are shown, together with the theoretical predictions according to Eq (5). The patterns are normalized versus the on-axis (Θ=0) values.

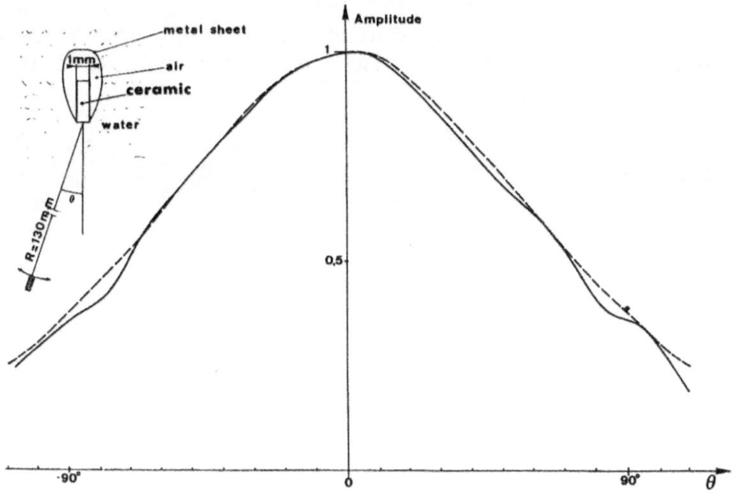

Fig.2 - Calculated (---) and measured (—) directivity pattern of a small transducer in a free space (f=570KHz)

B - Wide transducers - resonance effects induced by Lamb waves.

Anomalous behavior was evidenced during an experimental study of the radiation pattern of single transducers in water as a function of their physical dimensions.

The discrepancies with respect to the theoretically expected results, as shown in fig. 5 for an extreme case, consist mainly in the occurence of parasitic lobes, symmetrical about the main lobe, but inclined with respect to it.

The directions of the parasitic lobes being nearly frequency

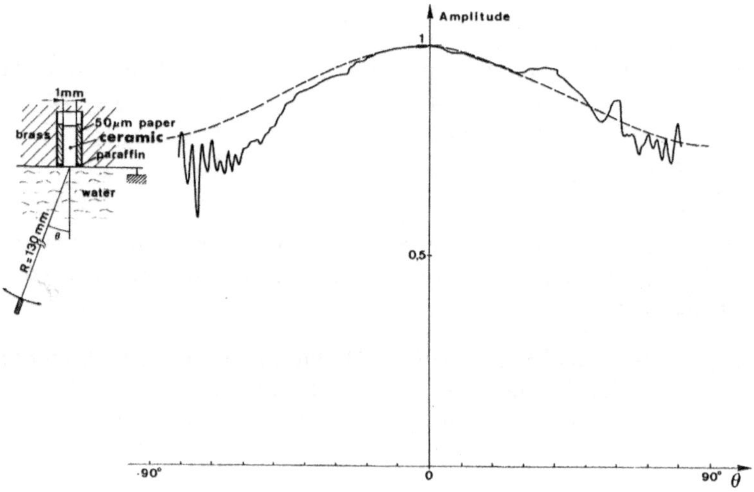

Fig. 3 - Calculated (---) and measured (___) directivity pattern of a small transducer in a rigid planar baffle (f=570 KHz).

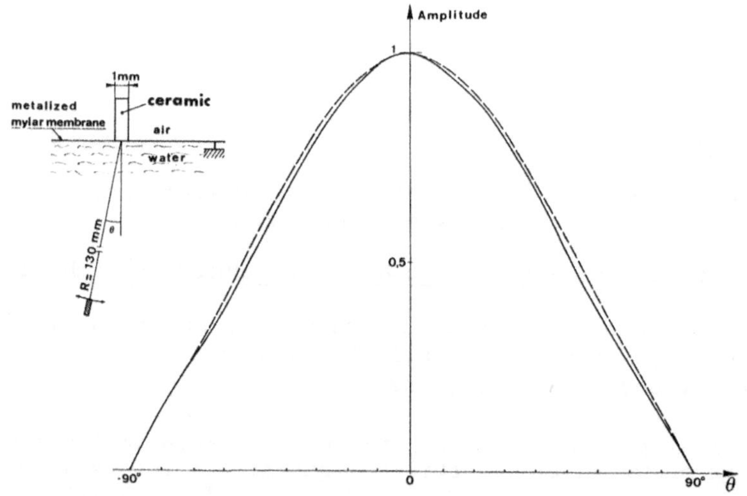

Fig. 4 - Calculated (---) and measured (___) directivity pattern of a small transducer in a soft planar baffle (f=570 KHz).

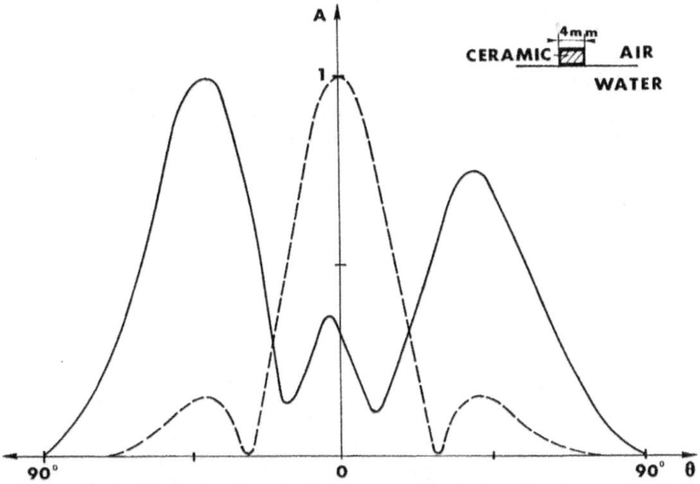

Fig. 5 - Characteristic radiation pattern. Transducer width is $2b=4$ mm - $f=f_o=$ 800 KHz - $2b/\lambda \simeq 2.1$.

independant, they may be thought to be caused by parasitic vibration modes propagating along the transducer surface. These modes radiate energy into the water in the form of longitudinal acoustic waves along the symmetrical directions given by the wavevector conservation law :

$$\Theta = \pm \sin^{-1} (K_p/K_w) \qquad (6)$$

Where the parasitic K_p and water K_w wavevectors are given in terms of the corresponding wavelengths

Λ and λ by : $K_p = 2\pi/\Lambda$ (7)

$$K_w = 2\pi/\lambda \qquad (8)$$

The values measured for the angle Θ are nearly equal to ± 50 degrees, so the parasitic waves must travel with a velocity nearly equal to 2000 m/s and, as we have shown (12-13) these speeds are very close to those of Lamb zeroeth order modes.

The characteristic radiation patterns given in Fig. 6a and 6b illustrate the behavior for the same transducer radiation at two different frequencies. These transducers are disposed against a watter boundary, so as to meet the Sommerfeld boundary conditions. The radiation patterns are normalized to the axis ($\Theta=0$) values for the theoretical predictions (dashed curves) and to the highest value observed for the experimental findings (solid curves). A pulsed, but quasicontinuous, mode of operation was used.

By considering the width 2b of the transducer and the values of the velocity in these two cases, it appears that in the first case the transducers width is nearly equal to $3\Lambda/2$ for the Lamb mode S_o waves, hence the important perturbation in the radiation pattern. In the second case, antiresonance occurs for the same S_o mode and the perturbation becomes very weak.

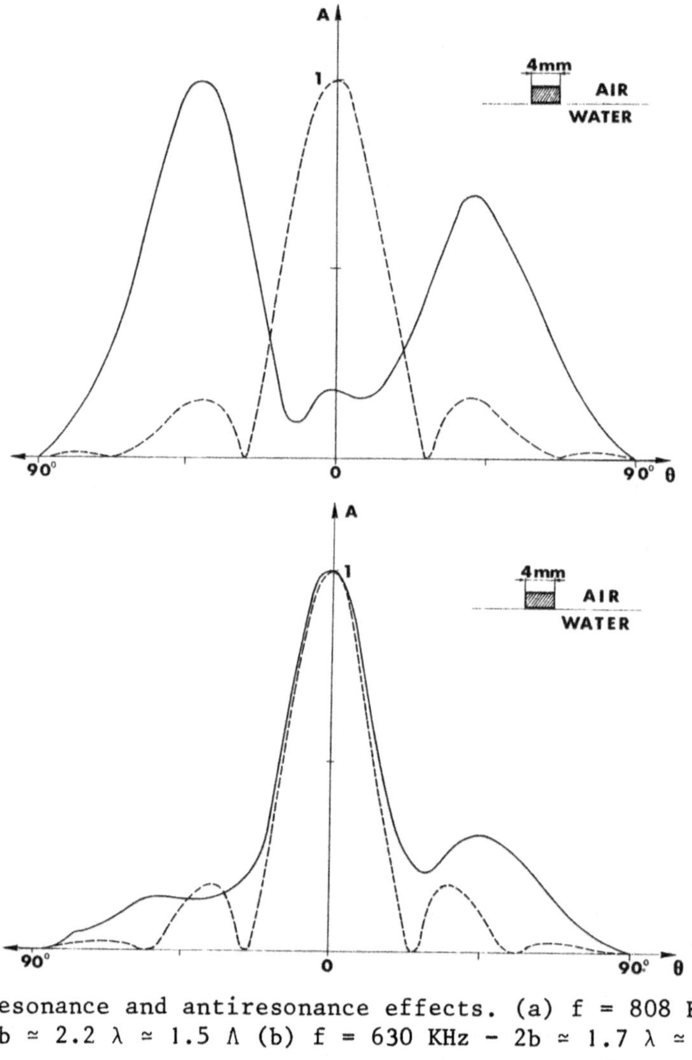

Fig. 6 - Resonance and antiresonance effects. (a) f = 808 KHz - $2b \simeq 2.2\ \lambda \simeq 1.5\ \Lambda$ (b) f = 630 KHz - $2b \simeq 1.7\ \lambda \simeq \Lambda$

However, for wider transducers, the perturbation associated with zeroeth order Lamb waves decreases and the main lobe contains the major part of the energy (fig. 7). Moreover it is seen that higher order Lamb waves that have higher phase velocities contribute to the broadening of the main lobe, but this contribution remains

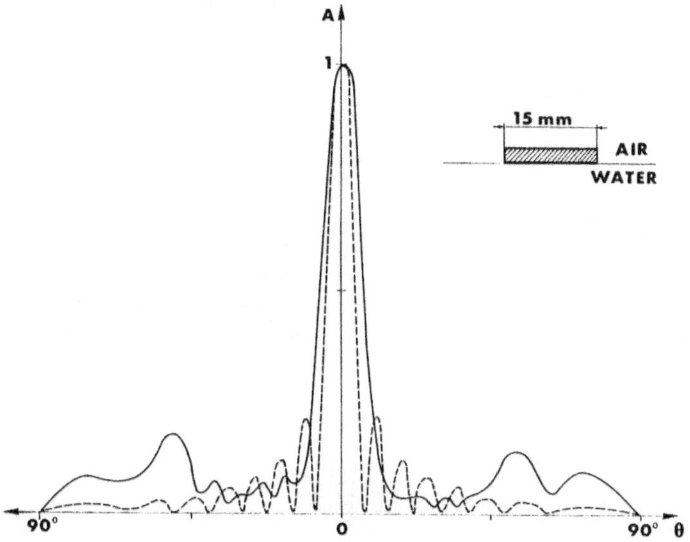

Fig. 7 - Radiation from wide sources - $f = f_o = 800$ KHz - $2b \simeq 8\ \lambda$

negligible with respect to that arising from electrical coupling between transducers in the case of transducers arrays (see II-B).

For transducers narrower than one wavelength λ, the main and parasitic radiation patterns have very large angular widths and their respective contribution may hardly be resolved in the experimental result.

II - Radiation of array transducers.

For the transducers integrated in a array structure, an electric coupling phenomenon between neighbouring transducers is added to the acoustic coupling via Lamb waves progagating inside the ceramic platelet.

Both phenomena have been studied theoretically assuming a negligible interaction between them. The effects derived from this study for the radiation pattern have been verified experimentally.

A - Mechanical coupling.

The mechanical coupling via Lamb waves is evidenced by parasitic lobes whose amplitude doesn't depend of the transducer width, unlike the isolated transducer case. As we have shown experimentaly [12-13] the Lamb Waves are launched at the transducer edges, that is in the transition zone of the driving electric field, where a gradient of piezoelectric stress exists, and these Lamb waves propagate out of the transducer inside the ceramic plate, producing therefore a non negligible mechanical coupling between neighbouring transducers.

It is seen in fig. 8 that, even though the transducer width is equal to the critical width of Fig. 5, the parasitic lobe level related to zeroeth order Lamb modes is greatly decreased. There is no resonance here because the Lamb waves are not reflected at the transducer boundaries.

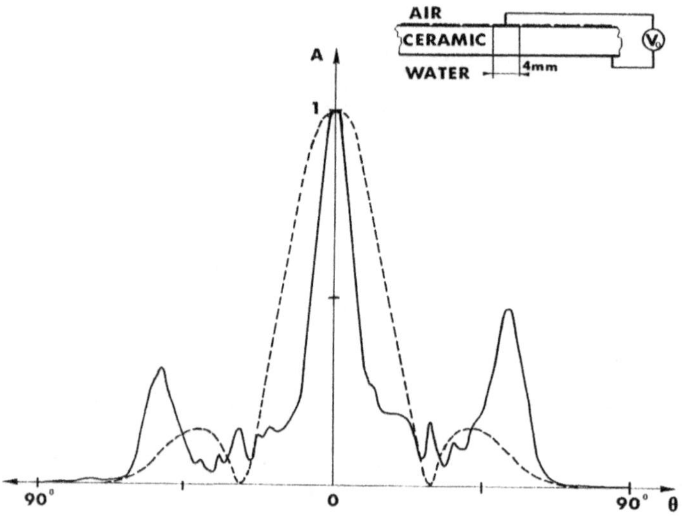

Fig. 8 - Radiation from a transducer in a piezoelectric array.
$f = f_o = 800$ KHz $2b/\lambda \simeq 2.1$.

The exponential decay law of the zeroeth order Lamb wave has been measured, and the value of the attenuation coefficient α for zeroeth order Lamb waves coupled to water may be derived as about 2 dB/λ where λ is the wavelength in the water. The higher order Lamb wave radiating along directions weakly slanted with respect to the transducer normal, give a significant contribution to the narrowing of the radiation pattern. But, this kind of coupling doesn't explain the experimental findings of fig. 8. As we shall see, the electrical coupling between neighbouring transducers also play a very important role in array transducers.

B - Electrical coupling.

The transducer array may be viewed as a series of capacities with more or less coupling between them via fringing fields.

The spatial period a of the array must be chosen to correctly sample the acoustic field, an optimum value being one half a wavelength in the acoustic propagation medium i.e, $\lambda/2$, according to Shannon's sampling criterion. For example, in biomedical imaging, the velocity in tissues, which are water-like media, is nearly equal to 1500 ms^{-1}, that is, three times lower than in the piezoelectric ceramic (for which velocity is typically 4200 ms^{-1}). For a piezoelectric platelet, typically one half an acoustic wavelength in tickness, say W_o, the a/W_o ratio may therefore be equal to about 1/3. In this case, the electric field is not very well confined to the theoretically active area of the individual transducers. The piezoelectric effect is therefore responsible for the launching of ultrasonic waves over a wider-than-wished active area. For large transducer, this effect may be of no importance, but it increases drastically as the ratio a/W_o is reduced.

Rheographic simulation.

As we have shown[13-14] it is possible to assume that only electrical coupling exists, which is the case for an unpolarized ceramic.

As the physical dimensions considered are much lower than the electromagnetic wavelength at the operating frequency, say some megahertz, study of the electric field map may be performed using the static approximation. The problem is then reduced to the solution of Laplace's equation in two dimensions.

An analogical solution has been preferred, using the rheological analogy where the precision is sastisfactory.

From this simulation method, a plot of the electrical potential distribution on the surface along the array axis x has been obtained, from which accurate mean values of the potentials V_n of the nth order electrodes with width 2b and gap width has been obtained. These values agree closely with those experimentally observed on unpolarized ceramics. Knowing the potentials on the electrodes, it is now possible to deduce the perturbations induced by electrical coupling on the radiation pattern.

Influence of the coupling on the radiation pattern.

In the first approximation, the amplitude A_n of the corresponding generated acoustic wave is then proportional to V_n and the radiation pattern of a transducer with width a equal to spatial period, neglecting any coupling, is given by:

$$A(\Theta) = \frac{\gamma}{a}\int_{-a/2}^{a/2} A_o \exp\left[\frac{(j2\pi x \sin\Theta}{\lambda}\right] dx = \gamma A_o \frac{\sin X}{X} \qquad (9)$$

where $X = \pi a \sin\Theta/\lambda$, A_o is the amplitude of the acoustic waves at the transducer surface, λ is the wavelength in the propagating medium, γ stands for a proportionality factor depending on the observation distance from the transducer and Θ is an angle measured from the transducer normal.

The coupling leads to the generation of a wave, not only on a single transducer, but also on its neighbours. The amplitude of the generated wave may therefore now be written as :

$$A(\Theta) = \frac{\gamma}{a} \int_{-\infty}^{+\infty} A(x) \exp\left[j2\pi x \sin\Theta)/\lambda\right] dx \qquad (10)$$

where if 2b is the width of the transducer and n the transducer rank

$$A(x) = \begin{cases} A_o \text{ for } |x| < b \\ A_n \text{ for } |x - na| < b \\ 0 \text{ otherwise} \end{cases} \qquad (11)$$

Assuming that 2b is very close in value to a, then Eq. 10 becomes :

$$A(\Theta) = \gamma \sum_{n=0}^{\infty} (2n+1)(A_n - A_{n+1}) \frac{\sin(2n+1)X}{(2n+1)X} \qquad (12)$$

The radiation pattern, normalized to the value in the direction $\Theta=0$ is shown in Fig (9) assuming that the A_n are proportional to the V_n and all parameter values being those used for the simulation.

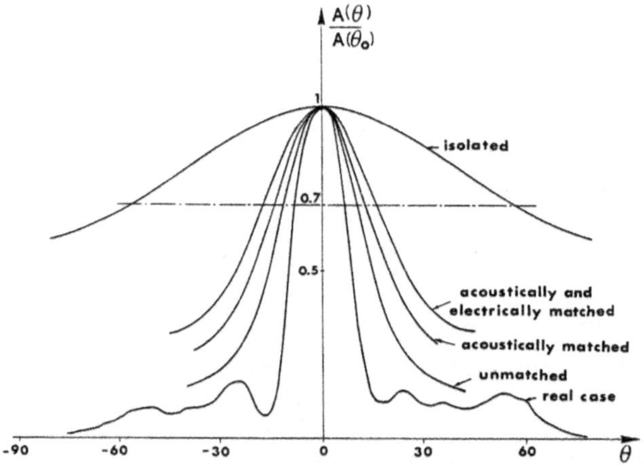

Fig. 9 - Effect of coupling and matching on the radiation pattern.

The effective angular aperture is decreased by a factor nearly equal to 5 for a particular transducer in a array obtained by cutting only the metallization with respect to an isolated transducer. The experimentally obtained diagram is even more perturbed, because of mechanical coupling via higher order Lamb waves which have been neglected here (fig. 9).

In the preceding calculations, it was assumed that the potential distribution is not affected by the ultrasound generation process. Such an approximation is certainly valid when a severe acoustic impedance mismatch exist between the ceramic and the propagation medium. ($Z_{ceramic}/Z_{water} \simeq 20$). But if the propagation medium is nearly acoustically matched to the ceramic, either naturally or by using a multilayer structure (as GOLL and al [15]), the electrical energy dissipated in the electromechanical conversion can no longer be neglected.

This conversion leads to a faster decrease of the potentiels V_n on the neighbouring transducers, so that the coupling is significantly reduced and the radiation pattern is broadened, as shown in fig. 9.

If, moreover, all the transducers of the array are electrically loaded and matched, the decay will be even faster . (Fig. 9).

However, when using an acoustically matching multilayer, one may not be surprised to see an increase in the coupling, since acoustic energy leakage occurs then via propagating Lamb waves guided in the multilayer structure.

III - Reduction of parasitic coupling.

As it may be thought, cutting the ceramics reduce coupling effects of both origins. Separate studies have been made, for cuts with several depths between neighbouring transducers, on the reduction in electrical coupling and on the mechanical coupling for an isolated transducer or a transducer integrated in on array.

A - Effect on the electric coupling.

Owing to the very high dielectric constant of the ceramic, ($\varepsilon_r \simeq 1500$), it may be felt that cutting it deeply between the transducers (not only the metallization, but the materiel itself), will result in a drastic reduction of the electrical coupling. This effect has been studied for several cut depths W ($W-W_0$ corresponds to the isolated transducer) by BRUNEEL and al [14-13]. A strong decrease of the potential is found when the cut depth W is increased and the effect of coupling on radiation pattern is shown Fig. 10, for the unmatched transducers.

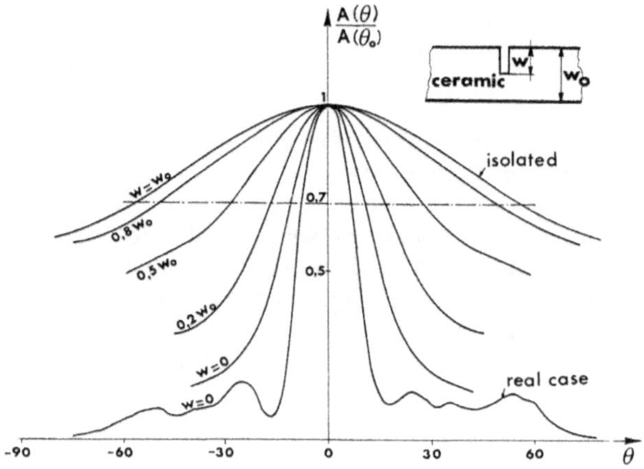

Fig. 10 - Effect of cutting on the radiation pattern.

Finally, for limiting the technologic problem, the electric coupling may be reduced by cutting slots into the array to a minimum depth limited by the rated performance and the mechanical strength of the array.

B - Effect on the mechanical coupling.

The parasitic effects may be strongly reduced by suppressing the Lamb wave propagation inside the transducer as we have shown by a theoretical study(12-13).

That high Lamb wave attenuation results if the transducer of width 2b is cut apart into N elementary sources, each narrower than the half Lamb wavelength $\Lambda_{L/2}$. This can be expressed as a condition for N :

$$N > 4b/\Lambda_L \quad (13)$$

This criterion has been experimentally verified for an initial transducer width 2b = 4 mm, corresponding to the maximum parasitic effects previously recorded Fig. 5. The diagrams given here (Fig. 11a-b) are for two different cuts. The first with N=2 (fig. 11a) shows an improvement over the initial transducer (fig. 5). The second with N=4 (fig. 11b) shows a satisfying correction to the radiation pattern, which becomes comparable to the theoretical one without Lamb waves (the criterion gives N ≃ 3).

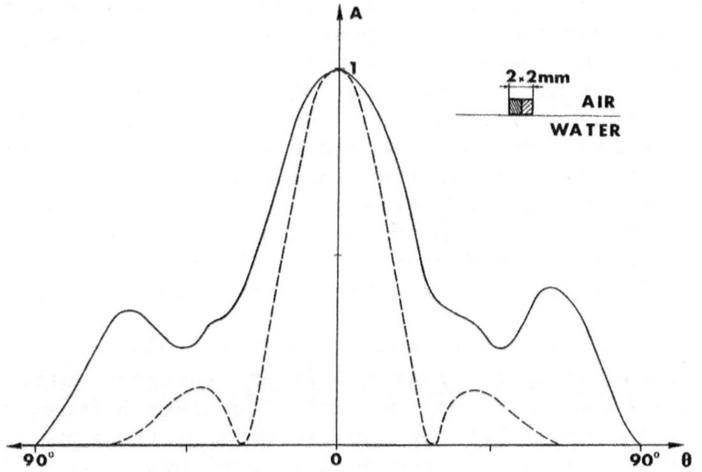

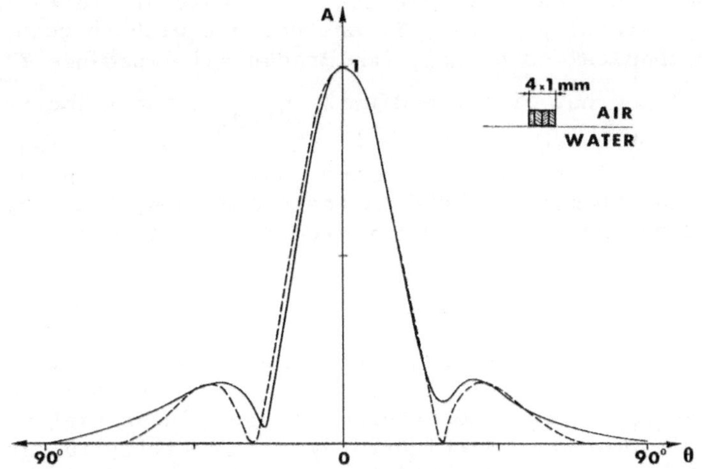

Fig. 11 - Effect of cutting the transducer apart. (a) N = 2, $f = f_o$ = 800 KHz, 2b≃ 2.1λ, (b) N = 4, $f = f_o$ = 800 KHz, 2b≃2.1

The suppression of the parasitic modes by subdividing the transducer into several separate, electrically interconnected elements, is important for all acoustic sources with dimensions of the order of the wavelength.

C - Technological problems for realization.

In order to both improve the response time constant of transducer (which enables the use of acoustic pulse with durations close to one time period) and to keep the electroacoustic conversion efficiency for longitudinal waves as high as possible, it is necessary to perform an acoustic match between the propagation medium and the piezoelectric transducer, using quarter-wavelengths antireflecting layers bonded on the transducer front face. The isolation of transducer by cutting the platelets doesn't therefore solve longer the problem of parasitic Lamb wave propagation inside the impedance matching materials. Finally, for the most severe case where the sampling period is very close to the half acoustic wavelength, the only solution is to cut also the matching layers, which leads to a very crucial problem of mechanical strength for the acoustic probe.

It proves also useful to put a protective platelet over the front face of the array in order to prevent the penetration of the coupling paste (used to insure mechanical contact between the probe and the studied medium) between the elements of the array, which would again induce parasitic mechanical coupling.
However, the mechanical characteristics of this protective membrane must be very carefully chosen, if one doesn't wish to reintroduce parasitic mechanical coupling : its mechanical impedance $Z_{membrane}$ must be equal to that of the medium Z_{medium} or the velocity of zeroeth order Lamb waves must be lower than that of longitudinal waves in the membrane. In the first case the whole acoustic energy is transmitted through the membrane and in the second one, the coupling between Lamb waves and the external medium is not allowed.

Conclusion.

It appears that the boundary conditions have a very significant influence to the pressure fied distribution of radiating acoustical sources. Pratically, for the transducers used in acoustical imaging, the diffraction formula is weighted by a directivity factor which depends on the acoustic properties of the medium surrounding the source.

It has been shown that Lamb waves which propagate inside the piezoelectric ceramic induce parasitic radiation in water around a frequency-independant direction, characteristic of each mode. For single transducers, the amplitude of this radiation varies with frequency because of resonance effects, but it remains low for large transducer widths. This is not the case for array transducers, where the Lamb waves induce mechanical coupling between neighbouring transducers for array transducers, in addition there is electrical coupling as well as coupling due to higher order Lamb waves. This, in fact, constitues the major cause of the narrowing of the main lobe for each individual source.

It is possible to reduce these parasitic couplings by at least partial cutting of the transducers, but this introduces probe strength problems.

The technological problems become very hard in the realization of array with a spatial period very close to the half acoustic wavelength, where a perfect isolation between neighbouring elements is needed. Moreover, the choice of the material used for the protective membrane is critical, if one doesn't wish to reintroduce mechanical coupling there.

REFERENCES.

1) N. BOM, CT LANCEE and J. HONKOOP.
"Ultrasonic viewer for cross sectional analyses of moving cardiac structures". Bio-med. Eng. 6,500 (1971).
2) FL. THURSTONE, OT. Von RAMM.
"A new ultrasound imaging technique two dimensional electronic beam steering."
Acoustical Holography Vol. 5, Proceedings of the Fifth International Symposium, Ed. Newell Booth, Publ. Plenum Press. New-York and London, p. 249-259 (1973).
3) P. ALAIS, M. FINK.
"Fresnel zone focussing of linear arrays applied to B and C echography" Acoustical Holography and Imaging. Vol. 7 Chicago (1976).
4) M. BERSON, A. RONCIN, L. POURCELOT.
"Appareils multitransducteurs à focalisation dynamique".
Société Française pour les Applications des Ultrasons en Médecine et en Biologie. (S.F.A.U.M.B.) Paris (1977).
5) C. BRUNEEL, B. NONGAILLARD, R. TORGUET, E. BRIDOUX, JM. ROUVAEN.
"Reconstruction of an acoustical image using an "acousto electronic lens device". Ultrasonics Vol. 15 N° 6, pp. 263-264 Nov.(1977).
6) C. BRUNEEL, R. TORGUET, JM ROUVAEN, E. BRIDOUX and B. NONGAILLARD.
"Ultrafast echotomographic system using optical processing of ultrasonic signals".
Appl. Phys. lett., 30,371 (1977).
7) JC. SOMER.
Symposium on Echocardiology - Rotterdam - June 1977.
Echocardiology N. BOM Editor 1977, pp. 325-334.
8) B. DELANNOY, R. TORGUET, C. BRUNEEL, E. BRIDOUX and JM ROUVAEN.
"Acoustical image reconstruction in parallel-processing analog electronic system."
J. Appl. Phys. Vol. 50, n° 5, pp. 3153-3159, May 1979.
9) JC. SOMER.
"Electronic Sector scanning for ultrasonic diagnosis."
Ultrasonics, Vol. 6, Nr 3, pp. 153-159 (1968).

10) P. CAUVARD, P. HARTEMAN, C. BRUNEEL and F. HAINE.
"Ultrasound beam scanning driven by surface acoustic-waves."
Ultrasonics Symposium Proceedings.
I.E.E.E. Ultrasonics Symposium - September 1978.

11) B. DELANNOY, H. LASOTA, C. BRUNEEL, R. TORGUET and E. BRIDOUX.
"The Infinite Planar Baffles problem in acoustic radiation and its experimental verification." J. Appl. Phys. vol. 50, n° 8, pp. 5189-5195 - August 1979.

12) B. DELANNOY, C. BRUNEEL, F. HAINE and R. TORGUET.
"Anomalous behavior in the radiation pattern of piezoelectric transducers induced by parasitic Lamb wave generation."
J. Appl. Phys. Vol. 51, n° 7 - July 1980.

13) B. DELANNOY, Thèse -"Reconstruction d'images acoustiques par échantillonnage et correction de phase" Université de Valenciennes (1979).

14) C. BRUNEEL, B. DELANNOY, R. TORGUET, E. BRIDOUX, H. LASOTA.
"Electrical coupling effects in an ultrasonic transducer array."
Ultrasonics Nov. Vol. 17 n° 6, pp. 255-260. November 1979.

15) JH. GOLL and B.A. AULD.
"Multilayer impedance matching schemes for broadbanding of water loaded piezoelectric transducers and high Q electric resonators."
I.E.E.E. trans on Sonics and Ultrasonics.
Vol. SU-22 n° 1 (1975), 52.

A STATISTICAL ESTIMATION APPROACH TO MEDICAL IMAGE RECONSTRUCTION FROM ULTRASONIC ARRAY DATA

E. J. Pisa and C. W. Barnes*

Rohe Scientific Corp., Philips Medical Systems
Santa Ana, California 92704 USA
*School of Engineering, University of California, Irvine
Irvine, California 92717 USA

INTRODUCTION

Images that purport to represent the distribution of scattering objects in living tissue can be generated from ultrasonic pulse-echo data by a number of means. In the simplest form, pulses are transmitted and received on a single fixed-focus transducer that is scanned mechanically or by hand to generate a conventional B-scan image.

Tracking focus [1-4] has been proposed as a method of improving the lateral resolution of B-scan images. In one form, tracking focus can be implemented with a concentric array of annular transducer elements. During reception, the focus of the composite transducer is varied, to follow the pulse in the medium, by summing the outputs of the annular elements with varying relative time delays.

The use of arrays of transducers opens up many additional possibilites for producing images. Angular beam scanning can be accomplished through conventional phased array techniques. Lateral beam translation can be obtained through sequenced switching of array elements. Both of these techniques can be combined with tracking focus.

Kino's group at Stanford has implemented a synthetic aperture approach for arrays that generates an image by the "time-delay-and-sum" method based on simple geometric considerations [5].

Another approach to image generation, proposed by Norton [6-8], uses an array of point transducers, each of which radiates a spherical wave. Each time-sample of the signal resulting from transmission on one point transducer and reception on another point transducer, represents an ellipsoidal projection (with the two point transducers at the foci) of the medium scattering amplitude. An approximation to the scattering amplitude distribution is then computed from a set of these projections. The general approach is analagous to that used in x-ray computer aided tomography, although the proposed reconstruction algorithm is different.

It is not clear that these imaging methods make the best use of the available data, nor is it clear that they are optimal for imaging in living tissue. In none of these methods is a criterion of optimality explicit. In addition, they all ignore at least one of the following effects that are potentially significant in living tissue:

1) frequency response of the transducer(s)
2) finite aperture of the transducer(s)
3) dispersive attenuation in the propagation medium
4) space-variant frequency dependence of the scattering amplitude
5) measurement error.

We propose here a general statistical estimation approach to image generation that takes account of all of the effects listed above, and is optimal in the minimum mean square error (MMSE) sense. In the remainder of this paper the general approach is defined, the explicit estimator equations to be solved are given, and a number of algorithmic approaches to solving these equations are discussed. In addition, a computer model for generating the required system response functions is described.

The major problem remaining to be solved is finding an efficient algorithm for implementing the MMSE estimator.

THE DATA ACQUISITION MODEL

We consider an array of transducers in a linear wave-propagating medium. A single data record is generated by applying a signal waveform to a transmitting transducer (or group of transducers) and subsequently receiving a scattered waveform (the data record) on a receiving transducer (or group of transducers). A set of data records is generated by varying the transmitting and receiving transducer configurations and, possibly, the waveforms applied to the transmitting transducers. We denote the kth data record by $a_k(t)$.

Our signal processing approach is based upon the following mathematical model for the generation of the data records (see Appendix A):

$$A_k(\nu) = \int \underline{dx}\ H_k(\underline{x},\nu)\ U(\underline{x},\nu),$$

where

$$A_k(\nu) = \int dt\ e^{2\pi i\nu t}\ a_k(t),$$

= the temporal Fourier transform of a data record.

$H_k(\underline{x},\nu)$ = frequency-domain round-trip system response, associated with the kth record, to a unit amplitude point scatterer located at coordinate $\underline{x}$. This response function includes the effects of the waveform applied to the transmitting transducer(s), the frequency responses of both transmitting and receiving transducer(s), the radiation pattern(s) associated with the finite aperture(s) of transmitting and receiving transducer(s), and the effects of attenuation and dispersion in the propagating medium.

$U(\underline{x},\nu)$ = complex scattering amplitude per unit volume at coordinate $\underline{x}$ and at frequency ν.

ν = frequency (Hertz).

$\underline{x}$ = (x,y,z) = spatial coordinates.

$\underline{dx}$ = dxdydz = volume element.

We prefer to state our fundamental assumptions regarding the data acquisition model in terms of an equation rather than in terms of a specific physical model for the scattering mechanism, since a specific physical model may be more restrictive than necessary. The equation that we use to describe the generation of a data record may be compatible with more than one physical model.

In effect, our model contains two fundamental assumptions:

1) The scattering in the medium (whatever the physical mechanism) can be represented by a single complex scalar function of $\underline{x}$ and ν.
2) The pulse-echo response is a linear function of this complex scalar scattering function.

These assumptions clearly exclude anisotropic scattering and multiple scattering. However, this model should correctly describe weak specular reflections at a surface if the surface is one of "local reaction" [9, p. 260]. We conjecture that these assumptions represent a reasonable model for the weak scattering associated with the internal structure of soft tissue.

The time-domain data records can be expressed directly in terms of the scattering amplitude function by

$$a_k(t) = \int d\nu \int \underline{dx}\, e^{-2\pi i \nu t} H_k(\underline{x},\nu)\, U(\underline{x},\nu).$$

We shall consider the case where the raw data consists of discrete-time samples of the pulse-echo responses. In addition, we complete the model by adding a term for measurement error. The final form for our data acquisition model is

$$a_k(nT) = \int d\nu \int \underline{dx}\, e^{-2\pi i \nu nT} H_k(\underline{x},\nu)\, U(\underline{x},\nu) + e_k(nT), \tag{1}$$

where

T = sampling period (for simplicity we consider here the case of uniform sampling; the model is easily generalized to non-uniform sampling)

$e_k(nT)$ = error associated with the nth sample of the kth record.

Our problem is to reconstruct $U(\underline{x},\nu)$, as well as we can, from the data set $\{a_k(nT)\}$.

MINIMUM MEAN SQUARE ERROR (MMSE) ESTIMATION

The Approach

Our approach to the problem of generating an image from the pulse-echo data is a statistical one in which we model the unknown scattering function and the measurement error as random processes.

We construct an estimate, $\hat{U}(\underline{x},\nu)$, of the true scattering function, $U(\underline{x},\nu)$, based on the data set $\{a_k(nT)\}$ such that for each value of $\underline{x}$ and ν

$$E\left\{|U(\underline{x},\nu) - \hat{U}(\underline{x},\nu)|^2\right\}$$

is minimized*. $E\{\}$ denotes statistical expectation.

We shall limit ourselves to estimates that are linear functions of the data; i.e., we shall construct the linear MMSE estimate of $U(\underline{x},\nu)$. For the special case where the scattering amplitude and measurement error can be modeled as Gaussian random processes, the linear MMSE estimate is the best MMSE estimate.

Vector Space Formulation

This problem can be conveniently formulated in terms of operators and vector spaces.

We let $\mathcal{V}_u$ denote the infinite-dimensional Hilbert space of square integrable scattering functions; i.e.,

$$\mathcal{V}_u = \left\{U(\underline{x},\nu) : \int d\nu \int \underline{dx}\ |U(\underline{x},\nu)|^2 < \infty\right\}.$$

The standard scalar product in this space is defined by

$$\langle U,V\rangle = \int d\nu \int \underline{dx}\ U^*(\underline{x},\nu)\ V(\underline{x},\nu)\ .$$

We let $\mathcal{V}_a$ denote the finite-dimensional Hilbert space of data sets; i.e.,

$$\mathcal{V}_a = \left\{a_k(nT)\right\}\ .$$

*Another statistical approach to image reconstruction from array data, in the ocean environment, has been investigated by Duckworth [10], who proposed the use of a data adaptive maximum likelihood technique for estimating the frequency-wave number spectrum of the received data.

The standard scalar product in this space is defined by

$$\langle \underline{a}, \underline{b} \rangle = \sum_{n,k} a_k^*(nT)\, b_k(nT) \; .$$

The data acquisition model, as expressed by (1), can now be written as an operator equation

$$\underline{a} = \mathbb{H}\, U + \underline{e} \; ,$$

where $\mathbb{H}$ is the measurement operator,

$$\mathbb{H} : \mathcal{H}_u \longrightarrow \mathcal{H}_a ,$$

defined by (1), $\underline{a}$ is the data vector of measurements, $\underline{e}$ is the measurement error vector, and U is the complex scattering function.

The construction of a MMSE estimate only requires knowledge of second order statistics of $U(\underline{x},\nu)$ and $e_k(nT)$. We assume that $U(\underline{x},\nu)$ and $e_k(nT)$ have zero means (if the means are not zero, their effects can be easily subtracted from the measurements), and covariances given respectively by

$$R_u(\underline{x},\underline{x}' \; ; \nu,\nu') = E\left\{ U(\underline{x},\nu)\; U^*(\underline{x}',\nu') \right\}$$

and

$$R_e(k,n \; ; k',n') = E\left\{ e_k(nT)\; e_{k'}^*(n'T) \right\} .$$

In addition, we assume that $U(\underline{x},\nu)$ and $e_k(nT)$ are uncorrelated.

The covariance functions can be considered to be the kernels of operators

$$\mathbb{R}_u : \mathcal{H}_u \longrightarrow \mathcal{H}_u$$

$$\mathbb{R}_e : \mathcal{H}_a \longrightarrow \mathcal{H}_a$$

defined as follows:

$$V = \mathbb{R}_u U$$

means

$$V(\underline{x},\nu) = \int d\nu' \int \underline{dx}' \; R_u(\underline{x},\underline{x}' \; ; \nu,\nu') \; U(\underline{x}',\nu')$$

and

$$\underline{b} = \mathbb{R}_e \underline{a}$$

means

$$b_k(nT) = \sum_{n',k'} R_e \; (k,k' \; ; n,n') \; a_{k'}(n'T).$$

Before writing out the expressions for the MMSE estimate, one additional mathematical device is required. The adjoint $\mathbb{H}^+$ of the measurement operator $\mathbb{H}$ is a linear mapping from $\mathscr{V}_a$ into $\mathscr{V}_u$, defined by

$$\left\langle \mathbb{H}^+ \underline{a}, U \right\rangle = \left\langle \underline{a}, \mathbb{H} U \right\rangle .$$

Writing out the expressions for these two scalar products, we find that

$$W = \mathbb{H}^\dagger \underline{a}$$

means

$$W(\underline{x},\nu) = \sum_{n,k} H_k^*(\underline{x},\nu) \; e^{2\pi i n \nu T} a_k(nT).$$

The Estimator Equations

With this background we can now write expressions for the linear MMSE estimate in operator notation. Two forms for the MMSE estimate are well-known (see, for example, [12, Ch. 5] or [15, Ch. 6]. Although these two forms are equivalent in the sense that they lead to the same estimate, they represent quite different algorithms.

The first form is expressed in terms of the covariance operators and is sometimes referred to as the "covariance form." For this form the MMSE estimate is given by

$$\hat{U}(\underline{x},\nu) = \mathbb{R}_u \, \mathbb{H}^\dagger \, (\mathbb{H} \mathbb{R}_u \mathbb{H}^\dagger + \mathbb{R}_e)^{-1} \, \underline{a}.$$

The associated covariance function for the estimate error is given by

$$\Lambda(\underline{x},\underline{x}' ; \nu,\nu') = E\left\{[U(\underline{x},\nu) - \hat{U}(\underline{x},\nu)]\,[U(\underline{x}',\nu') - \hat{U}(\underline{x}',\nu')]^*\right\} \quad (2)$$

If this function is interpreted as the kernel of an integral operator

$$\mathbb{\Lambda} : \mathcal{H}_u \longrightarrow \mathcal{H}_u ,$$

then estimation theory tells us that

$$\mathbb{\Lambda} = \mathbb{R}_u - \mathbb{R}_u \mathbb{H}^\dagger (\mathbb{H} \mathbb{R}_u \mathbb{H}^\dagger + \mathbb{R}_e)^{-1} \mathbb{H} \mathbb{R}_u.$$

The second form for the MMSE estimate is expressed in terms of the inverses of the covariance operators and is sometimes referred to as the "information form." For this form the MMSE estimate is given by

$$\hat{U}(\underline{x},\nu) = \left(\mathbb{H}^\dagger \mathbb{R}_e^{-1} \mathbb{H} + \mathbb{R}_u^{-1}\right)^{-1} \mathbb{H}^\dagger \mathbb{R}_e^{-1} \underline{a} ,$$

and the covariance operator for the estimate error can be computed from

$$\mathbb{\Lambda} = \left(\mathbb{H}^\dagger \mathbb{R}_e^{-1} \mathbb{H} + \mathbb{R}_u^{-1}\right)^{-1} .$$

In the following section we offer interpretations of these two estimator forms.

INTERPRETATION OF THE ESTIMATOR EQUATIONS

The Covariance Form

The covariance form for the scattering image estimator can be represented in block diagram form as shown in Figure 1. The

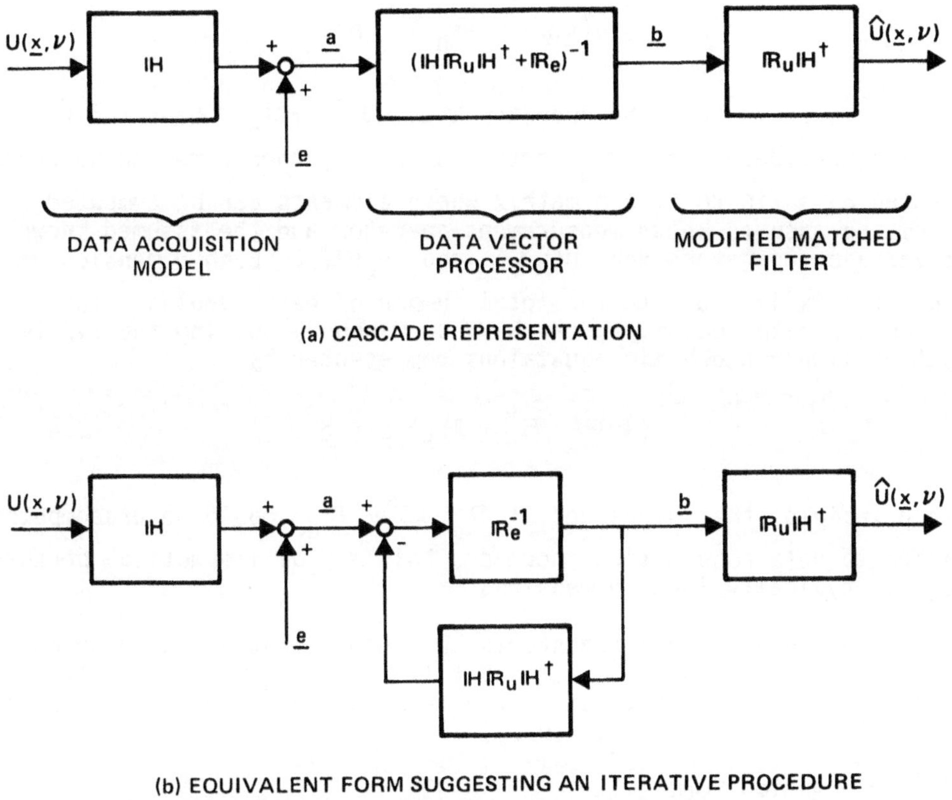

Figure 1. Block Diagram of Covariance Form of Scattering Image Estimator

estimator is a cascade of two components. The first component is a data vector processor that transforms the input data vector, $\underline{a}$, into a processed data vector, $\underline{b}$, according to

$$\underline{b} = (\mathbb{H}\,\mathbb{R}_u\,\mathbb{H}^{\dagger} + \mathbb{R}_e)^{-1}\,\underline{a}\,.$$

The processed data vector, $\underline{b}$, is then transformed into the scattering image according to

$$\hat{U}(\underline{x},\nu) = \mathbb{R}_u\,\mathbb{H}^{\dagger}\,\underline{b}\,.$$

We observe that the operator $\mathbb{H}\,\mathbb{R}_u\,\mathbb{H}^{\dagger} + \mathbb{R}_e$ maps the finite dimensional data space $\mathcal{V}_a$ into itself and, therefore can be represented by a finite square matrix whose elements can be computed from the assumed known measurement operator and the assumed known covariance operators for $U(\underline{x},\nu)$ and $e_k(nT)$. Each dimension of this matrix is equal to the total number of data samples. The processed data vector, $\underline{b}$, can be determined by solving the finite set of linear algebraic equataions represented by

$$(\mathbb{H}\mathbb{R}_u\,\mathbb{H}^{\dagger} + \mathbb{R}_e)\,\underline{b} = \underline{a}\,.$$

In principle, the inverse of $\mathbb{H}\,\mathbb{R}_u\,\mathbb{H}^{\dagger} + \mathbb{R}_e$ could be precomputed prior to data acquisition; however, this may be impractical because of its typically large dimensions.

The operator that transforms $\underline{b}$ into a scattering image can be regarded as a modified matched filter since the operator $\mathbb{H}^{\dagger}$ represents a filter matched to the response to an ideal monochromatic point source, $\delta(\underline{x}-\underline{x}_0)\,\delta(\nu-\nu_0)$.

The Information Form

The information form for the scattering image estimator can be represented in block diagram form as shown in Figure 2. This estimator can also be regarded as the cascade of two components. This first component is a modified matched filter that transforms the input data vector into a zero order approximation to the scattering image according to

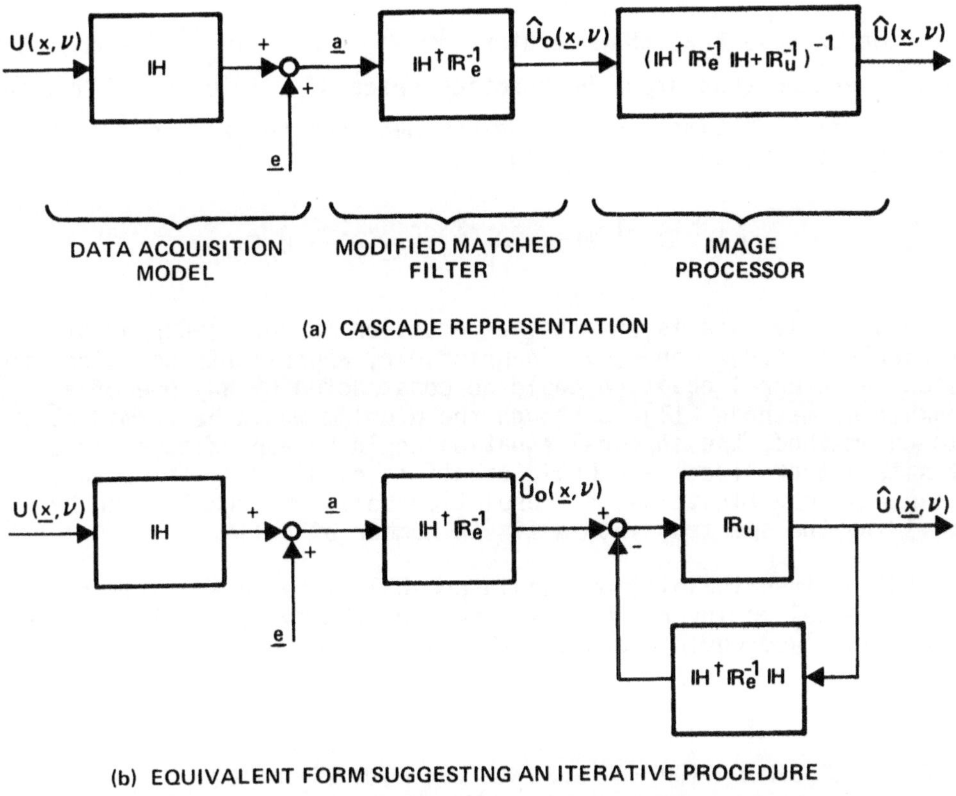

Figure 2. Block Diagram of Information Form of Scattering Image Estimator

$$\hat{U}_o(\underline{x},\nu) = \mathbb{H}\,\mathbb{R}_e^{-1}\,\underline{a}\,. \tag{3}$$

The zero order approximation is then transformed into the linear MMSE estimate according to

$$\hat{U}(\underline{x},\nu) = \left(\mathbb{H}^{\dagger}\,\mathbb{R}_e^{-1}\,\mathbb{H} + \mathbb{R}_u^{-1}\right)^{-1}\,\hat{U}_o(\underline{x},\nu).$$

We observe that the operator $\mathbb{H}^{\dagger}\mathbb{R}_e^{-1}\,\mathbb{H} + \mathbb{R}_u^{-1}$ is an integral operator that maps the function space $\mathscr{U}_u$ into $\mathscr{U}_u$. Thus the linear MMSE estimate can be determined from the zero order estimate by solving the integral equation

$$\left(\mathbb{H}^{\dagger}\,\mathbb{R}_e^{-1}\,\mathbb{H} + \mathbb{R}_u^{-1}\right)\hat{U}(\underline{x},\nu) = \hat{U}_o(\underline{x},\nu)\,.$$

Unfortunately this is an integral equation in four independent variables: x,y,z and ν. In principle, approximate solutions to such an integral equation could be constructed by any one of a number of methods [13], although the problem would be formidable. In one method, the integral equation could be approximated by a finite set of linear algebraic equations by dividing the scanned volume into a finite set of resolution cells or "voxels" and by dividing the spectrum into a finite number of cells.

Also, in principle, one could pre-calculate an approximate inverse kernel or inverse matrix; however, the typically large dimensions involved would probably make this approach unattractive.

ZERO ORDER SIMULATION FOR INFORMATION FORM ESTIMATOR

To gain insight into the estimator approach in the information form, we computed the zero order response, as described by (3), for a linear array of 11 rectangular elements. Explicitly we computed $\hat{U}(\underline{x},\nu) = \mathbb{H}^{\dagger}\mathbb{R}_e^{-1}\,\mathbb{H}\,U(\underline{x},\nu)$, where the scattering object was a single non-dispersive point scatterer and the measurement error was assumed to be white noise ($\mathbb{R}_e = \sigma^2\,\mathbb{I}$). Data records were generated by sequentially transmitting and receiving on one element at a time. The resulting estimate can be viewed as a point spread function of the zero order estimator. Plots of this response for an on-axis point and an off-axis point are shown in Figure 3 and 4 respectively.

From Figure 3 we see that the on-axis point spread function is localized on the point scatterer and is fairly symmetrical. Its lateral dimensions are on the order of what one would expect from the Rayleigh resolution criterion; however, the sidelobes are unacceptably high. The off-axis point spread function, Figure 4, is badly skewed and is clearly unacceptable for imaging purposes.

These results clearly indicate the need for the higher order corrections that should be supplied by the image processor. The localized nature of the zero order estimate suggests that an iterative approach, as suggested by Figure 2b, for generating higher order estimates may be feasible.

RECURSIVE IMPLEMENTATION OF ESTIMATOR

General Discussion

An alternate procedure for constructing a MMSE estimate of the scattering function consists of dividing the entire data set into a sequence of smaller vector blocks of data and then recursively computing the MMSE estimate taking one data vector at a time. In this approach the MMSE estimate is obtained by solving a sequence of small problems rather than by solving one large problem. In addition, this approach opens up the possibility of on-line processing: i.e., one need not wait for the entire data set to be acquired before beginning processing; instead, the data vectors can be processed sequentially as they arrive.

We let $\underline{a}_m$ denote the mth data vector and $\mathbb{H}_m$ denote the measurement operator associated with this data vector. Our measurement equations then become

$$\underline{a}_m = \mathbb{H}_m U(\underline{x},\nu) + \underline{e}_m, \qquad m = 0,1,2, \ldots ,$$

where $\underline{e}_m$ is the error vector associated with the mth data vector. Let $\hat{U}_{m-1}(\underline{x},\nu)$ denote the MMSE estimate of $U(\underline{x},\nu)$ based on the data set $\{\underline{a}_0,\underline{a}_1, \ldots , \underline{a}_{m-1}\}$.

The problem is: given the next data vector, $\underline{a}_m$, how do we update our estimate and compute $\hat{U}_m(\underline{x},\nu)$? This is a standard problem in recursive estimation theory, and for a general discussion the reader is referred to [11, Ch. 4], [12, Ch. 6], [14], or any other standard reference on recursive estimation. This

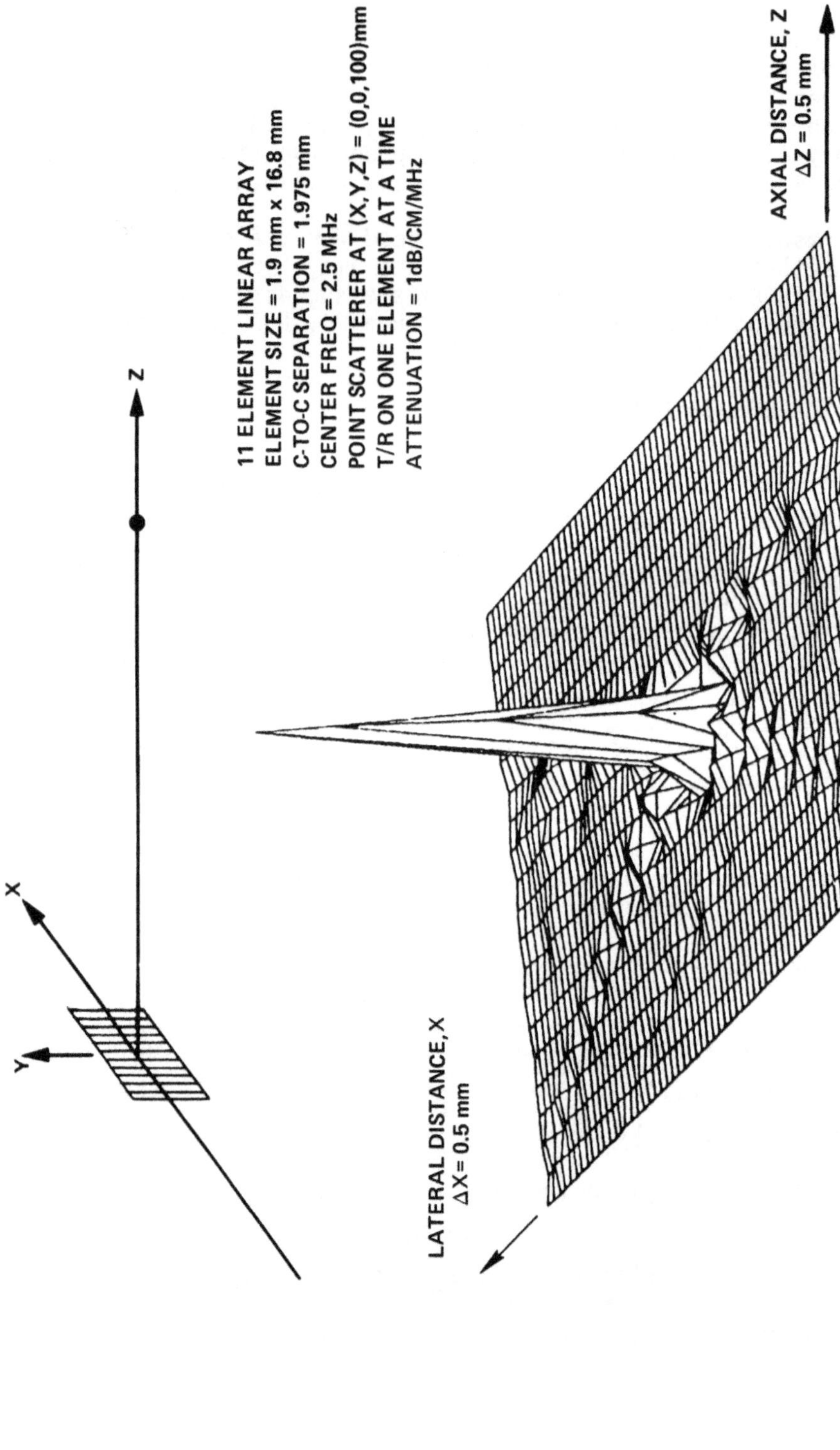

Figure 3. On-Axis Point Spread Function of Zero Order Estimator

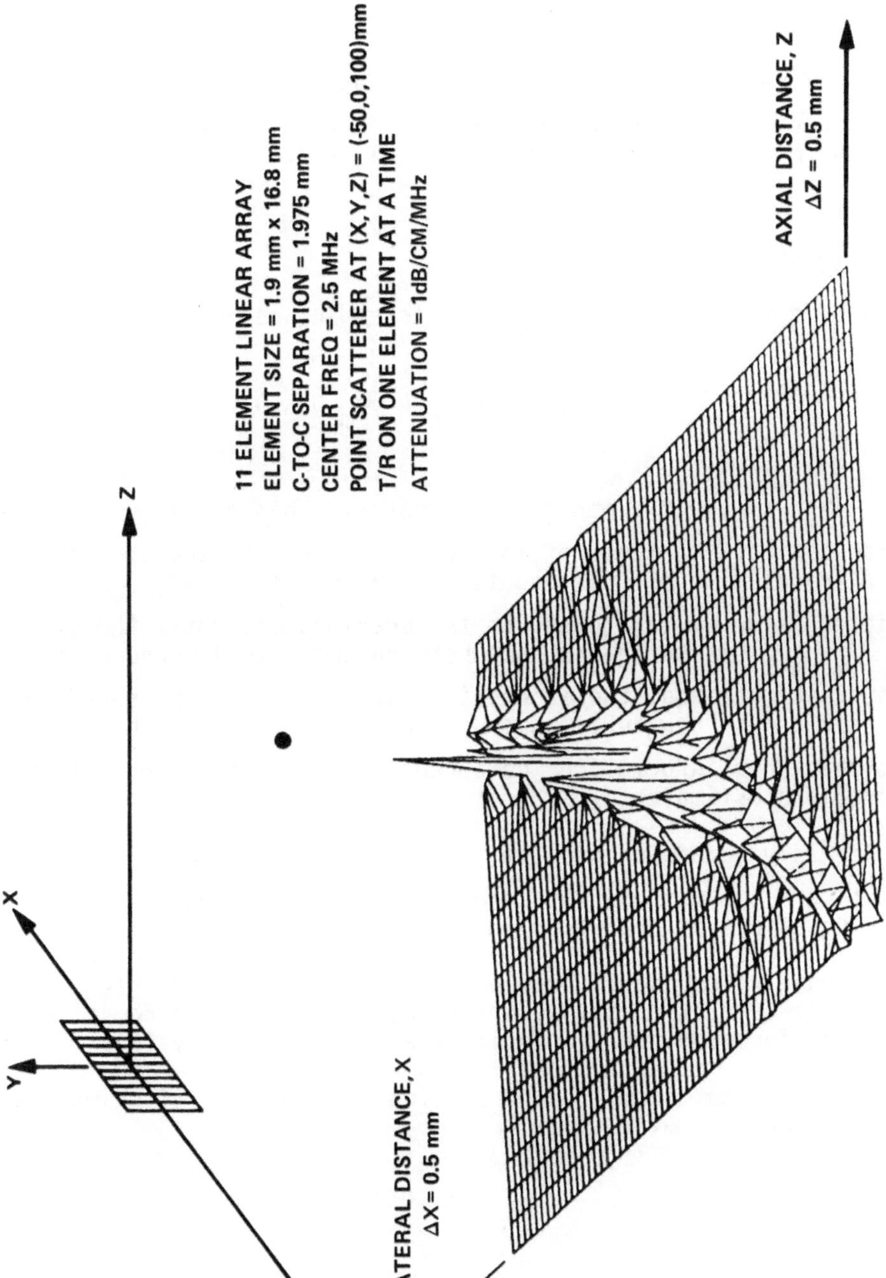

Figure 4. Off-Axis Point Spread Function of Zero Order Estimator

problem can also be viewed as a degenerate case of a Kalman filter applied to a static system with time-varying measurements, as discussed in Appendix B.

As before, we shall assume that U and $\underline{e}_m$ have zero means and are uncorrelated with each other. In addition we assume that for each m, $\underline{e}_m$ is uncorrelated with all prior data.

Recursive Estimator Equations

The update equation for the linear MMSE estimate is [12,14,15]

$$\hat{U}_m = \hat{U}_{m-1} + \mathbb{K}_m(\underline{a}_m - \mathbb{H}_m \hat{U}_{m-1}) , \tag{4}$$

where $\mathbb{K}_m$ is the "Kalman gain" operator. This equation expresses the new estimate in terms of the previous estimate and the new data. The interpretation of this equation is as follows: $\mathbb{H}_m \hat{U}_{m-1}$ is the predicted value of $\underline{a}_m$ based on the previous estimate; thus, $\underline{a}_m - \mathbb{H}_m \hat{U}_{m-1}$ is the difference between the actual measured data and its estimate and, therefore, is that part of the measured data that represents "new information." The operator $\mathbb{K}_m$ then transforms this "new information" component of $\underline{a}_m$ into a correction to the previous estimate.

The only remaining problem, then, is how to compute $\mathbb{K}_m$. We present here two methods that are based on the standard Kalman filter equations as discussed in Appendix B. We shall simply present the equations; for a derivation, the reader is referred to any standard reference on Kalman filters, such as [12], [14], or [15]. A by-product of the recursive estimation procedure is the covariance function for the estimate error, defined by (2).

Two equivalent forms for generating the sequence of Kalman gain operators are:

Covariance Form

$$\mathbb{K}_m = \Lambda_{m-1} \mathbb{H}_m^{\dagger} \mathbb{C}_m^{-1} \tag{5}$$

$$\mathbb{C}_m \triangleq \mathbb{H}_m \Lambda_{m-1} \mathbb{H}_m^{\dagger} + \mathbb{R}_{\underline{e}_m}$$

$$\mathbb{R}_{\underline{e}_m} = E\left\{\underline{e}_m \underline{e}_m^{\dagger}\right\}$$

$$\Lambda_m = \Lambda_{m-1} - \mathbb{K}_m \mathbb{C}_m \mathbb{K}_m^{\dagger} .$$

Note: $\underline{e}_m^{\dagger}$ is the conjugate transpose of the column matrix $\underline{e}_m$.

Information Form

$$\mathbb{K}_m = \Lambda_m \mathbb{H}_m^{\dagger} \mathbb{R}_{\underline{e}_m}^{-1} \tag{6}$$

$$\Lambda_m^{-1} = \Lambda_{m-1}^{-1} + \mathbb{H}_m^{\dagger} \mathbb{R}_{\underline{e}_m}^{-1} \mathbb{H}_m .$$

In the covariance form, the error covariance matrix is computed recursively. At each stage of the recursion it is necessary to invert the operator $\mathbb{C}_m$, which can be represented by a square matrix with each dimension equal to the dimension of the data vector $\underline{a}_m$.

In the information form, the inverse of the error covariance matrix is computed recursively, and at each stage of the recursion it is necessary to determine Λ_m from its inverse Λ_m^{-1}; i.e., we must invert an integral operator.

To start the recursion, initial values are needed for $\hat{U}_0$ and Λ_0. If no prior information about the scattering function is available, other than the a priori statistics

$$E\{U\} = 0$$

$$E\left\{U(\underline{x},\nu)\ U^*(\underline{x}',\nu')\right\} = R_u(\underline{x},\underline{x}';\nu,\nu') \ ,$$

then an appropriate set of initial conditions is given by

$$\hat{U}_o = 0$$

$$\Lambda_o = \mathbb{R}_u$$

$$\Lambda_o^{-1} = \mathbb{R}_u^{-1} \ .$$

If prior information is available (say from a conventional B-scan), then it could be used as a starting point.

In the following two sections we consider two ways in which the sequence of data vectors can be chosen for recursion. In the first method, the data vectors are chosen to be single scalar samples from the data records; i.e., the data vectors are one-dimensional and the scattering function estimate is updated with each scalar sample. This method has been proposed by Wood, Macovski and Morf [16] for recursive processing of x-ray projections to compute MMSE estimates of images in computer aided tomography. In the second method, each data vector is the set of samples from a single record generated by the transmission and reception of a single pulse. In this case the scattering function estimate is updated after each data record.

Recursion on Single Data Data Samples

It is convenient to revert to a two-index notation and let

$a_k(nT) = n^{th}$ sample of kth record,

$\hat{U}_{k,n}(\underline{x},\nu)$ = MMSE estimate of $U(x,\nu)$ based on data up to and including $a_k(nT)$,

$$\Lambda_{k,n}(\underline{x},\underline{x}'\ ;\nu\ ,\nu')$$
$$= E\left\{[U(\underline{x},\nu) - \hat{U}_{k,n}(\underline{x},\nu)]\ [U(\underline{x}',\nu) - \hat{U}_{k',n'}(\underline{x}',\nu')]^*\right\}.$$

Using (4), the recursion on single data samples can be explicitly expressed as

$$\hat{U}_{k,n}(\underline{x},\nu) = \hat{U}_{k,n-1}(\underline{x},\nu)$$

$$+ K_{k,n}(\underline{x},\nu) \left[a_k(nT) - \int \underline{dx} \int d\nu \, H_k(\underline{x},\nu) \, e^{-2\pi i n\nu T} \, \hat{U}_{k,n-1}(\underline{x},\nu)\right].$$

Notice that the Kalman gain is a scalar function of $\underline{x}$ and ν.

The recursion equations for computing the Kalman gain function in the covariance form are

$$K_{k,n}(\underline{x},\nu)$$

$$= C_{k,n}^{-1} \int \underline{dx}' \int d\nu' \Lambda_{k,n-1} (\underline{x},\underline{x}' ;\nu,\nu') \, H_k^*(\underline{x}',\nu') \, e^{-2\pi i n\nu' T},$$

$$C_{k,n} = \int \underline{dx} \int d\nu \int \underline{dx}' \int d\nu' \, H_k(\underline{x},\nu) \, \Lambda_{k,n-1}(\underline{x},\underline{x}' ;\nu,\nu') \cdot$$

$$H_k^*(\underline{x}',\nu') \, e^{2\pi i n(\nu-\nu')T} + \sigma_{e_k(n)}^2$$

and

$$\Lambda_{k,n}(\underline{x},\underline{x}' ;\nu,\nu')$$

$$= \Lambda_{k,n-1}(\underline{x},\underline{x}' ;\nu,\nu') - C_{k,n} K_{k,n}(\underline{x},\nu) \, K_{k,n}^*(\underline{x}',\nu').$$

In the information form, the Kalman gain is computed from

$$K_{k,n}(\underline{x},\nu)$$

$$= \sigma_{e_k(n)}^{-2} \int \underline{dx}' \int d\nu' \, \Lambda_{k,n}(\underline{x},\underline{x}' ;\nu,\nu') \, H_k^*(\underline{x}',\nu') \, e^{2\pi i n\nu' T}$$

and

$$\Lambda_{k,n}^{-1}(\underline{x},\underline{x}';\nu,\nu') = \Lambda_{k,n-1}^{-1}(\underline{x},\underline{x}';\nu,\nu')$$

$$+ \sigma_{e_k(n)}^{-2} H_k^*(\underline{x},\nu) H_k(\underline{x}',\nu') e^{2\pi i n(\nu-\nu')T},$$

where $\Lambda_{k,n}^{-1}(\underline{x},\underline{x}';\nu,\nu')$ is the kernel of an inverse integral operator; i.e.,

$$\int \underline{dx}' \int d\nu' \Lambda_{k,n}^{-1}(\underline{x},\underline{x}';\nu,\nu') \Lambda_{k,n}(\underline{x}',\underline{x}'';\nu',\nu'')$$

$$= \delta(\underline{x}-\underline{x}'') \delta(\nu-\nu'') .$$

These recursions have been expressed in terms of sequential samples of a single record. In going from the end of one record to the beginning of the next record, we begin the recursion again using the estimates obtained at the end of the previous record as initial values.

We observe that for the case of recursion on each scalar data sample, no operator inversion is required in the covariance form since $C_{k,n}$ is a scalar. However, the information form still requires inversion of an integral operator.

Recursion on Single Data Records

We assume that each record contains N samples, and let $\underline{a}_k$ denote a column matrix of N samples from the kth record. In this case the Kalman gain operator, $\mathbb{K}_k$, as determined by (5) or (6), is a row matrix of length N whose elements are functions of $\underline{x}$ and ν; $\mathbb{C}_k$ is an N-by-N matrix of numbers, and Λ_k is an integral operator. Thus, if the covariance form is used, then inversion of an N-by-N matrix is required to compute the Kalman gain matrix at each step of the recursion.

General Comments on Recursive Method

It is important to recognize that the Kalman gain operators and the error covariance functions are not data dependent and can be precomputed and stored. However, the storage requirements might be quite high. The storage requirements for the covariance functions will be very demanding since they are functions of eight real scalar variables.

Also the evaluation of $\mathbb{K}_m$ and $\mathbb{C}_m$ requires the evaluation of integrals throughout the volume under observation and throughout the frequency band spanned by the signals. These integrals must be performed numerically and, therefore, will require finite approximations.

CONCLUSIONS

We have proposed a statistical estimation approach to image generation which is optimal in the MMSE sense. A computer model for generating the required system response functions was described. Explicit equations were given for the linear MMSE estimate of the complex scattering amplitude density, in both the covariance form and the information form. The estimator equations were interpreted, and the zero order response for the information form was simulated.

By viewing the estimation problem as a degenerate case of a Kalman filter applied to a static system with time varying measurements, the estimator may be obtained recursively. Recursive estimator equations were given, in both covariance and information form. Two possible recursion schemes were considered: recursion on single data samples and recursion on single data records.

In both the recursive and the non-recursive versions of the estimator, a significant number of terms are data-independent and can be precomputed and stored. However, the storage requirements can be quite high.

The transmit-receive configuration used to obtain the data records is totally unconstrained by the statistical estimation approach to image generation. The estimate error covariance, which is available from the estimator equations, may be helpful in comparing transmit-receive configurations. In any case, the full synthetic aperture advantage should be achievable with the statistical estimation approach.

The major problem remaining to be solved is finding a practical algorithm for implementing the MMSE estimator.

ACKNOWLEDGMENTS

We would like to thank Dr. James Chan of the Magnavox Research Laboratory, Torrance, California for providing the simulation results. This work was supported by Philips Laboratories, Briarcliff Manor, New York.

REFERENCES

1. Hubelbank, M., and O. Tretiak. "Focused Ultrasonic Transducer Design," M.I.T. Electronics Res. Lab Report No. 98, 1970, 169-177.
2. Walker, J. T., and J. D. Meindl. "A Digitally Controlled CCD Dynamically Focused Phased Array," Proc. IEEE Ultrasonics Symposium, 1975, 80-83.
3. Norton, S. J. "Theory of Acoustic Imaging," Ph.D. Thesis, Stanford University, Stanford Electronic Lab. Tech Report No. 4956-2, 1976.
4. McKeighen, R. E., and M. P. Buchin. "New Techniques for Dynamically Variable Electronic Delays for Real Time Ultrasonic Imaging," Proc. IEEE Ultrasonics Symposium, 1977, 250-254.
5. Corl, P. D., and G. S. Kino. "A Real-Time Synthetic Aperture Imaging System," in Acoustical Imaging, Vol. 9, K. Wang editor. New York: Plenum Press, 1980, 341-355.
6. Norton, S. J., and M. Linzer. "Ultrasonic Reflectivity Tomography: Reconstruction with Circular Transducer Arrays," Ultrasonic Imaging, 1, 1979, 154-184.
7. Norton, S. J. "Reconstruction of a Reflectivity Field from Line Integrals Over Circular Paths," J. Acoust. Soc. Am., 67 (3), March 1980, 853-863.
8. Norton, S. J. "Reconstruction of a Two-Dimensional Reflecting Medium Over a Circular Domain: Exact Solution," J. Acoust. Soc. Am., 67 (4), April 1980, 1266-1273.
9. Morse, P. M., and K. U. Ingard. Theoretical Acoustics. New York: McGraw-Hill, 1968.
10. Duckworth, G. L. "Adaptive Array Processing for Acoustic Imaging," in Acoustical Imaging, Vol. 9, K. Wang editor. New York: Plenum Press, 1980, 177-201.
11. Luenberger, D. G. Optimization by Vector Space Methods. New York: Wiley, 1969.
12. Schweppe, F. C. Uncertain Dynamic Systems. Englewood Cliffs, NJ: Prentice-Hall, 1973.
13. Atkinson, K. E. A Survey of Numerical Methods for the Solution of Fredholm Integral Equations of the Second Kind. Philadelphia, PA: SIAM, 1976.
14. Meditch, J. S. Stochastic Optimal Linear Estimation and Control. New York: McGraw-Hill, 1969.
15. Anderson, B.D.O., and J. B. Moore. Optimal Filtering. Englewood Cliffs, NJ: Prentice-Hall, 1979.
16. Wood, S. L., A. Macovski, and M. Morf. "Reconstructions with Limited Data Using Estimation Theory," in Computer Aided Tomography and Ultrasonics in Medicine, Raviv et.al. editors, IFIP, North Holland Pub. Co., 1979, 219-233.
17. Mason, W. P. Electromechanical Transducrs and Wave Filters. Princeton, NJ: Van Nostrand, 1948.

18. Auld, B. A. Acoustic Fields and Waves in Solids, Vol. 1, Ch. 8. New York: Wiley, 1973,
19. Sachse, W., and N. N. Hsu. "Ultrasonic Transducers for Materials Testing and Their Characterization," in Physical Acoustics, XIV, Mason and Thurston editors. New York: Academic Press, 1979.
20. Stepanishen, P. R. "Transient Radiation from Pistons in a Finite Planar Baffle," J. Acoust. Soc. Am., 49, 5 (part 2), 1971, 1629-1638.
21. Stepanishen, P. R. "The Time-Dependent Force and Radiation Impedance on a Piston in a Rigid Infinite Planar Baffle," J. Acoust. Soc. Am., 49, 3(Part 2), 1971, 841-849.
22. Barnes C. W. Mathematical Modeling of Ultrasonic Array Imaging Systems, Tech. Rpt. 79-1, Rohe Scientific Corp., Santa Ana, CA., March 1979.
23. Lockwood, J. C., and J. G. Willette. "High-Speed Method for Computing the Exact Solution for the Pressure Variations in the Nearfield of a Baffled Piston," J. Acoust. Soc. Am., 53, 3, 1973, 735-741; Erratum: 54, 6, 1973, p. 1762.
24. O'Donnell, M., E. T. Jaynes, and J. G. Miller. "Mechanisms: Relationship Between Ultrasonic Attenuation and Dispersion," Third Int. Symp. Ultrasonic Imaging and Tissue Characterization, June 5-7, 1978; N.B.S., Gaithersburg, MD, p. 36 of program and abstracts.

APPENDIX A: DATA ACQUISITION MODEL

Introduction

This appendix provides a brief overview of the mathematical model for the generation of data records. The reader will find a more complete description in the cited references.

A functional block diagram of the system model appears in Figure 5. We have broken the basic pulse-echo system into three subsystems: (1) transducer electro-mechanical equivalent circuit; (2) transducer radiative impulse response; and (3) medium. We use a linear system approach based on an input-output description for each subsystem shown. That is, each subsystem is described by a time domain impulse response or an equivalent frequency domain transfer function. The total system response can be obtained by temporal convolution of subsystem impulse responses or by multiplication of subsystem transfer functions. This approach is well suited for digital computer simulation.

Transducer Equivalent Circuit

In our model we assume that the (flat) piezoelectric transducer face moves as a rigid piston, so that higher order transverse

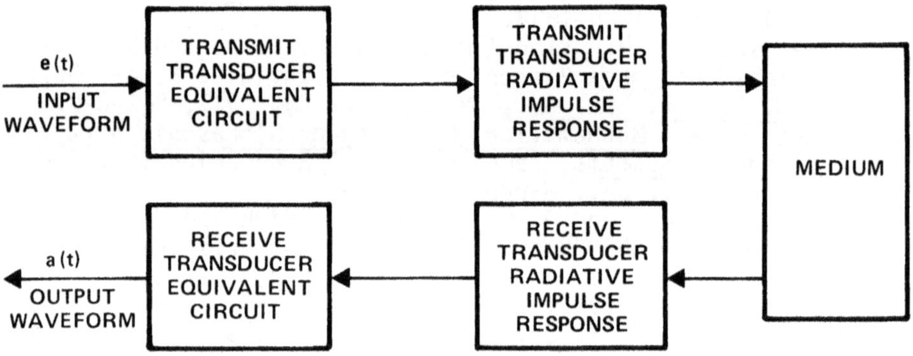

Figure 5. Functional Block Diagram of System Model

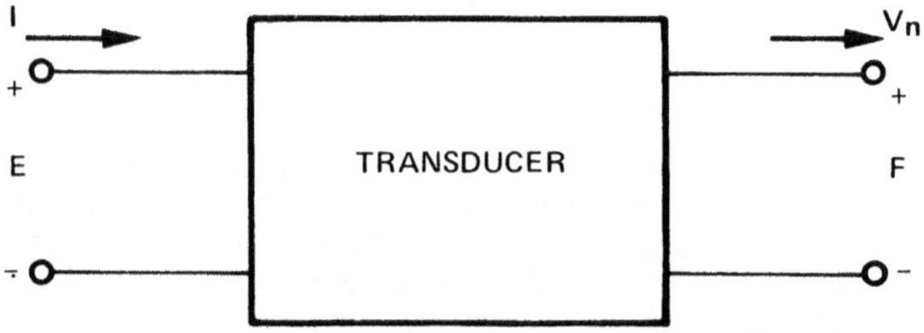

Figure 6. Transducer as a Two-Port Network

modes of oscillation are not present. Equivalent circuits for such thickness mode expander piezoelectric transducers have been derived by Mason [17] and others [18,19]. The Mason equivalent circuit for a thickness mode expander is a lossless, one-dimensional, linear, three-port, distributed network, with one electrical port and two mechanical ports. The two mechanical ports correspond to the two faces of the transducer. This equivalent circuit can be converted to a linear, bilateral, two-port network as shown in Figure 6 by internally including the effects of loading on the back face of the transducer. The properties of this two-port network can be described in the frequency domain by the complex "ABCD" transmission parameters as follows

$$E(\nu) = A(\nu)\ F(\nu) + B(\nu)\ V_n(\nu)$$

$$I(\nu) = C(\nu)\ F(\nu) + D(\nu)\ V_n(\nu)\ ,$$

where E = voltage at electrical terminals
I = current at electrical terminals
F = force on transducer face
V_n = normal velocity of transducer face

and $AD - BC = 1.$

If the transducer (with backing) is operated as a transmitter into a medium with radiation load impedance Z_{rad}, then the resulting normal velocity of the front face due to an input voltage $E(\nu)$ is

$$V_n(\nu) = \frac{E(\nu)}{A(\nu)\ Z_{rad}(\nu) + B(\nu)}\ .$$

If the transducer is operated as a receiver with open circuit electrical terminals, then

$$E_{oc}(\nu) = -\frac{V_n(\nu)}{C(\nu)} = \frac{F(\nu)}{D(\nu)}\ .$$

By using simple two-port pi and tee networks, we can include dielectric loss, electrical matching and tuning networks, and mechanical matching layers ("quarter-wave" windows) in the transducer equivalent circuit model.

If the lateral dimensions of the transducer are large compared to a wavelength, then Z_{rad} is independent of frequency and

$$Z_{rad} \approx \rho ca,$$

where ρ = density of medium
c = speed of sound in medium
a = area of transducer face.

More generally, we write

$$Z_{rad}(\nu) = \rho ca(\theta(\nu) - i\chi(\nu)) .$$

For the case of rectangular or circular apertures, analytical expressions for $\theta(\nu)$ and $\chi(\nu)$ may be found in [9, pp. 385, 394].

Radiative Impulse Response

We assume a planar piston transducer in an infinitely rigid baffle (all in the $z = 0$ plane). The transducer is assumed to move as a rigid piston, normal to the $z = 0$ plane. Using a Green's function approach [20,21] or a plane wave angular spectrum approach [22], the exact solution to this boundary value problem can be shown to be

$$\phi(\underline{x},t) = g(\underline{x},t) * v_n(t),$$

where

$\phi(\underline{x},t)$ = velocity potential at field point $\underline{x}$

$g(\underline{x},t)$ = radiative impulse response from transducer face to field point $\underline{x}$

$v_n(t)$ = normal velocity of transducer face.

Using a temporal Fourier transform, this solution can be conveniently expressed in the frequency domain as

$$\Phi(\underline{x},\nu) = G(\underline{x},\nu)\, V_n(\nu),$$

where $G(\underline{x},\nu)$ = radiative transfer function.

Stepanishen [20] has shown that the radiative impulse response $g(\underline{x},t)$ is a function of geometry only and can be expressed as

$$g(\underline{x},t) = \begin{cases} \dfrac{L(ct)}{2\pi t \sin\theta(ct)} , & \dfrac{R_{min}}{c} \le t \le \dfrac{R_{max}}{c} , \\ 0 , & \text{otherwise} \end{cases}$$

where $L(ct)$ = arc length on transducer face

$\theta(ct)$ = angle between the path ct and the normal from the fieldpoint, $\underline{x}$, to the plane of the transducer face

R_{min}, R_{max} = respectively, shortest and longest distance from $\underline{x}$ to the transducer face.

Stepanishen [20] has derived an explicit analytical expression for the radiative impulse response of a circular piston, which we use in our computer model. The radiative impulse response of an annular piston can be obtained by subtracting the two circular responses. Lockwood and Willette [23] have derived an explicit analytical expression for the radiative impulse response of a rectangular piston. We employ a general numerical technique, utilizing a triangular net, to obtain the radiative impulse response of rectangular or hexagonal pistons. In all cases the radiative impulse response is calculated in the time domain, and the radiative transfer function is obtained by Fourier transformation using an FFT algorithm.

Three additional points should be noted. First, because of the reciprocity property of the Green's function, the radiative impulse response of a transmit transducer is the same as that of a receive transducer of the same characteristics. Second, the concept of radiative impulse response of a single transducer can be extended to an array of transducers by the principle of linear superposition. Third, the raditive impulse response is valid throughout the half space (nearfield and farfield) and applies for narrow- or wideband operation.

<u>Medium</u>

In order to solve the piston radiator boundary value problem to obtain a radiative impulse response which can be computed, we must assume a homogeneous medium. The effects of attenuation and dispersion in this medium are included as follows.

The effects of attenuation can be modeled by letting the wave number, k, be complex; i.e.,

$$k = \frac{\nu}{c} + i\,\frac{\alpha(\nu)}{2\pi} ,$$

where $\alpha(\nu)$ = attenuation coefficient (nepers/meter).

Causality, as reflected in the Kramers-Kronig relations, requires that frequency-dependent attenuation be accompanied by velocity dispersion. A local relation that applies for weak dispersion, as derived by O'Donnell, Jaynes and Miller [24], is given by

$$\alpha(\nu) = \frac{\pi^2 \nu^2}{c_o^2}\,\frac{dc}{d\nu} ,$$

where c_o is the phase velocity at some reference frequency ν_o. Thus

$$c(\nu) = c_o + \frac{c_o^2}{\pi^2}\int_{\nu_o}^{\nu} d\nu\,\frac{\alpha(\nu)}{\nu^2} .$$

In biological tissue, for frequencies in the range of one to ten megahertz, the attenuation is approximately proportional to frequency. Thus, if we let

$$\alpha(\nu) = \kappa\,\nu ,$$

then

$$c(\nu) = \frac{c_o^2\,\kappa}{\pi^2}\,\ln(\nu/\nu_o) + c_o .$$

Thus

$$k(\nu) = \frac{\nu}{c_0} + \frac{\beta(\nu)}{2\pi} + i\,\frac{\alpha(\nu)}{2\pi} ,$$

where

$$\beta(\nu) = \frac{2\pi\nu}{c_0}\left\{\left[\frac{c_0\kappa}{\pi^2}\,\ell n(\nu/\nu_0) + 1\right]^{-1} - 1\right\}$$

$$c_0 = c(\nu_0) .$$

Since the radiative transfer function for a lossless medium can be efficiently computed by Stepanishen's method, it would be desirable if we could obtain the transfer function with loss directly from the lossless transfer function. This can be done approximately if all points on the radiating aperture are approximately equidistant from the observation point $\underline{x}$. Thus, if we define

$$\overline{R}(\underline{x}) = \text{average distance from radiating aperture to } \underline{x} ,$$

and if

$$\exp\left\{(-\alpha + i\beta)\, R\right\} \approx \exp\left\{(-\alpha + i\beta)\, \overline{R}\right\}$$

for all points on the radiating aperture, then [22]

$$G(\underline{x};\nu) \approx W(\overline{R},\nu)\; G_0(\underline{x};\nu) ,$$

where $G_0(\underline{x};\nu)$ is the radiative transfer function in a lossless medium and

$$W(\overline{R},\nu) = \exp\left\{[-\alpha(\nu) + i\beta(\nu)]\,\overline{R}\right\} .$$

Under this approximation, the radiative impulse response is given by

$$g(\underline{x},t) \approx w(\overline{R},t) * g_0(\underline{x},t) ,$$

where $g_0(\underline{x},t)$ is the radiative impulse response for a lossless medium (which can be computed by Stepanishen's method), and

$$w(\overline{R},t) = \int d\nu\; e^{-2\pi i\nu t}\; W(\overline{R},\nu)$$

characterizes the time-spreading of an impulse caused by attenuation and dispersion in the medium.

Total System Response

We can obtain the frequency-domain round-trip system response to a point scatterer at $\underline{x}$ by using the "ABCD" parameters of the transducer to directly relate the radiation field to the electrical input and output of the transducer.

The pressure at $\underline{x}$ due to radiation from the transducer is given in the frequency domain by [22]

$$P_{inc}(\underline{x};\nu) = S^T(\nu)\, G^T(\underline{x};\nu)\, E(\nu),$$

where

$$S^T(\nu) = \text{transfer function of transducer in transmission}$$

$$= \frac{-2\pi i \rho\nu}{A(\nu)\, Z_{rad}(\nu) + B(\nu)}$$

$$\rho = \text{density of medium}$$

$$G^T(\underline{x};\nu) = \text{transmit radiative transfer function}$$

$$E(\nu) = \text{transmit voltage waveform} \; .$$

We assume that the point scatterer generates a spherical wave with amplitude proportional to the incident pressure; i.e.,

$$P_{scat}(\underline{x};\nu) = U(\nu)\, P_{inc}(\underline{x};\nu)\, \frac{e^{2\pi i k|\underline{r}-\underline{r}_o|}}{|\underline{r}-\underline{r}_o|},$$

where $U(\nu)$ is the frequency-dependent scattering amplitude for the point scatterer.

The open circuit voltage produced in the transducer by this scattered spherical wave can be shown [22] to be

$$E_{oc}(\nu) = S^R(\nu)\, G^R(\underline{x};\nu)\, U(\nu)\, P_{inc}(\underline{x};\nu),$$

where

$$S^R(\nu) = \text{transfer function of transducer in reception}$$

$$= \frac{4\pi}{C(\nu)\, Z_{rad}(\nu) + D(\nu)}$$

$G^R(\underline{x};\nu)$ = receive radiative transfer function.

Thus the overall round-trip transfer function is given by

$$\frac{E_{oc}(\nu)}{E(\nu)} = S^T(\nu)\, G^T(\underline{x};\nu)\, U(\nu)\, G^R(\underline{x};\nu)\, S^R(\nu)$$

or

$$E_{oc}(\nu) = H(\underline{x};\nu)\, U(\nu)\,,$$

where $H(\underline{x};\nu)$ is the frequency-domain round-trip system response used in the data acquisition model:

$$H(\underline{x};\nu) = E(\nu)\, S^T(\nu)\, G^T(\underline{x};\nu)\, G^R(\underline{x};\nu)\, S^R(\nu) \quad .$$

If we transmit and receive on the same transducer, then $G^T(\underline{x};\nu) = G^R(\underline{x};\nu)$. Again, linear superposition can be used to handle transmit-receive configurations involving groups of transducers.

Also, if the point scatterer is approximately equidistant from all points on the transducer face, then the effects of attenuation and dispersion in the propagating medium can be approximated by

$$G(\underline{x};\nu) \approx W(\overline{R};\nu)\, G_0(\underline{x};\nu)\,,$$

as discussed above.

Thus the response to a medium with scattering described by a complex scattering amplitude per unit volume given by $U(\underline{x};\nu)$ becomes

$$E_{oc}(\nu) = \int \underline{dx}\, H(\underline{x};\nu)\, U(\underline{x};\nu)\;,$$

where the integration is carried out over the entire scattering volume.

APPENDIX B: THE KALMAN FILTER APPROACH

The Kalman filter is a recursive MMSE estimator for the state vector $\underline{x}(n)$ of a system described by the state equation

$$\underline{x}(n+1) = \mathbf{\Phi}(n)\,\underline{x}(n) + \mathbb{G}(n)\,\underline{w}(n)\ ,$$

where $\underline{w}(n)$ is a white noise forcing vector. The estimates are based on the observations, $\underline{z}(n)$, which are related to the state vector by the measurement equation

$$\underline{z}(n) = \mathbb{H}(n)\,\underline{x}(n) + \underline{v}(n)\ ,$$

where $\mathbb{H}(n)$ is the measurement operator and $\underline{v}(n)$ is white observation noise.

We shall not repeat here the Kalman filter equations since they are available in any standard text on recursive estimation (e.g., [12], [14], or [15]). We only wish to show how the image estimation problem can be viewed in this context.

The scattering function, $U(\underline{x},\nu)$, can be viewed as the state vector of the system under observation. If we assume that the scattering function is not changing with time, then the operators in the state equation are given by

$$\mathbf{\Phi}(n) = \mathbb{I} \qquad \text{(identity operator)}$$

and

$$\mathbb{G}(n) = \mathbb{O} \qquad \text{(zero operator)}.$$

Thus, we have time-varying measurements on a system with no dynamics. In this case, the Kalman filter equations reduce to those given in the section on recursive estimator equations.

A COMPUTER MODEL FOR SPECKLE IN ULTRASOUND IMAGES: THEORY AND APPLICATION

R.J. Dickinson

GEC Hirst Research Centre
Wembley, Middlesex
England

1 INTRODUCTION

The ultrasonic B-scan of regions of parenchymal tissue has a number of properties which are similar to laser speckle, and distinguish it from imaging techniques using incoherent radiation such as radiography. The special properties of the image mean that conventional techniques of image processing and analysis are inadequate, and to develop more appropriate techniques and relationship between the final image and the tissue properties and the ultrasonic field must be understood in more detail. This paper describes the use of a theoretical model and a computer simulation to examine this relationship, and its possible application to the evaluation of image processing routines.

2 A MODEL FOR SPECKLE IN B-SCAN

2.1 Theory

The B-scan image of regions of tissue without large interfaces is caused by the backscattering of an ultrasonic pulse by fluctuation in density and compressibility (Gore and Leeman 1977). The scattering is governed by the following inhomogeneous wave equation.

$$\nabla^2 p_1 - \frac{1}{c^2}\frac{\partial^2 p_1}{\partial t^2} = \underline{\nabla} p_0 . \underline{\nabla} \rho_1 / \rho_0 + \frac{1}{c^2} \left[\frac{\rho_1}{\rho_0} + \frac{\beta_1}{\beta_0}\right] \frac{\partial^2}{\partial t^2} p_0 \qquad (1)$$

where $p_0(\underline{r},t)$ is the incident pulse

$p_1(\underline{r},t)$ is the backscattered (received) pulse
β_1,ρ_1 are the (small) deviations of compressibility and density about their means β_0 and ρ_0. A solution of this equation for the pulse-echo case where the same transducer is used for transmit and receive is given by Dickinson (1980).

$$p_r(y,z,t) = \int H(2x'-ct,y'-y,z'-z)T(x',t',z')d^3\underline{r}' \quad (2)$$

where $p_r(y,z,t)$ is the A-scan produced when the transducer is positioned at (o,y,z).

$H(\underline{r})$ is the pulse echo response of the transducer from a point scatterer at $\underline{r}$.

$T(\underline{r})$ is the tissue impulse response which is the following differential function of density and compressibility.

$$T(\underline{r}) = \frac{1}{4}\frac{\partial^2}{\partial x^2}\left[\frac{\rho_1}{\rho_0} - \frac{\beta_1}{\beta_0}\right] - \left[\frac{\partial^2}{\partial y^2} + \frac{\partial^2}{\partial y^2}\right]\frac{\beta_1}{\beta_0} \quad (3)$$

To obtain the B-scan image, each A-scan is demodulated and smoothed, and the transducer is scanned in a line. For the work reported here, a linear scanning mode is modelled. A computer simulation of the process of formation of a B-scan image from a particular transducer pulse and tissue impulse response has been reported on previously (Bamber and Dickinson 1980). The tissue impulse response cannot be exactly defined as this would presuppose a knowledge of the density and compressibility on a scale less than the acoustic wavelength. Although acoustic microscopy may provide this in the future, a stochastic model of the tissue impulse response can be used to examine some properties of the B-scan image. Two such stochastic models are common, and will now be examined.

2 2 The discrete scatterer model

This models the tissue as an ensemble of point scatters randomly positioned in a homogeneous matrix. Hence

$$T(\underline{r}) = \sum_{n=1}^{N} a_n \delta(\underline{r}-\underline{r}_n) \quad (4)$$

The demodulated B-scan of this model is given by

$$I(\underline{r}) = \left|\sum_{n=0}^{N} a_n H(\underline{r}-\underline{r}_n)\right| \quad (5)$$

For a quasi-monochromatic pulse the incident pulse can be expressed as $H(\underline{r}) = A(\underline{r})e^{i\underline{k}_o\cdot\underline{r}}$ (6)

where k_o is the fundamental frequency and A is pulse envelope.

At any point in the A-scan, the echo amplitude is a random function of position.

$$I = \left| \sum_{n=0}^{M} a_n e^{ik_o \cdot r_n} \right|$$

M is the number of scatterers in a resolution cell.

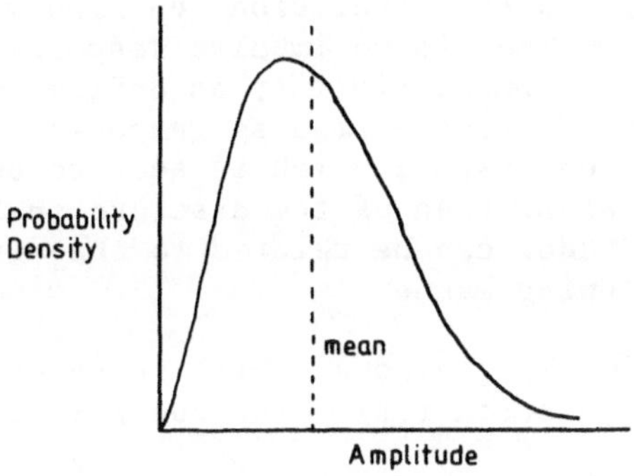

Fig 1 The Rayleigh distribution

Thus the echo amplitude at any point is governed by an expression identical to that governing laser speckle and has the following statistics.

The first-order statistics are given by the Rayleigh probability function (figure 1, Beecham 1966)

$$p(A) = \frac{A}{\langle A \rangle} e^{-A/\langle A \rangle} \tag{8}$$

A discrete scatterer model, and a simulated B-scan of the model is shown in figure 2. The amplitude histogram of the scan is shown in figure 3.

When considering soft tissue as a scattering matrix, it is not obvious which components can be identified as point scatterers, and which the homogeneous substrate. A more realistic model is the homogeneous continuum model, in which the density and compressibility are continuous, and exhibit small fluctuations about their respective means according to a set of predefined statistics.

2 3 The inhomogeneous continuum model

In the model to be used here, the density is assumed to be constant, and the compressibility a random function of position with a specified auto-correlation function. The tissue model used here has a Gaussian auto-correlation function with a correlation length $\underline{a}$; $e^{-r^2/\underline{a}^2}$. The tissue impulse response is then a double differential of the compressibility as defined by eq (3). Figure 4 shows such a random tissue impulse response, with its associated simulated B-scan, which is seen to be similar in appearance to the equivalent scan of the discrete scatterer model (fig 2). This tissue model can be related to the discrete scatterer model in the following manner.

The tissue impulse response, $T(\underline{r})$ is constrained to have an auto-correlation junction $\tilde{T}(\underline{r})$. The spatial power spectrum is given by

$$|T(\underline{k})|^2 = F(\tilde{T})) \tag{10}$$

F is the Fourier operator so

$$T(\underline{k}) = [F(\tilde{T}(\underline{r})]^{\frac{1}{2}} e^{i\phi(\underline{k})} \tag{11}$$

where $\phi(k)$ is a random variable.

Using the shift property of Fourier transforms

$$T(\underline{r}) = \sum_{n=0}^{N} t(r-\underline{r}_n) \tag{12}$$

where

$$t(r) = F^{-1}|T(\underline{k})| \tag{13}$$

So the tissue impulse response can be considered to be composed of a series of randomly positioned blobs. Each blob has a shape and size determined by the auto-correlation function of the tissue

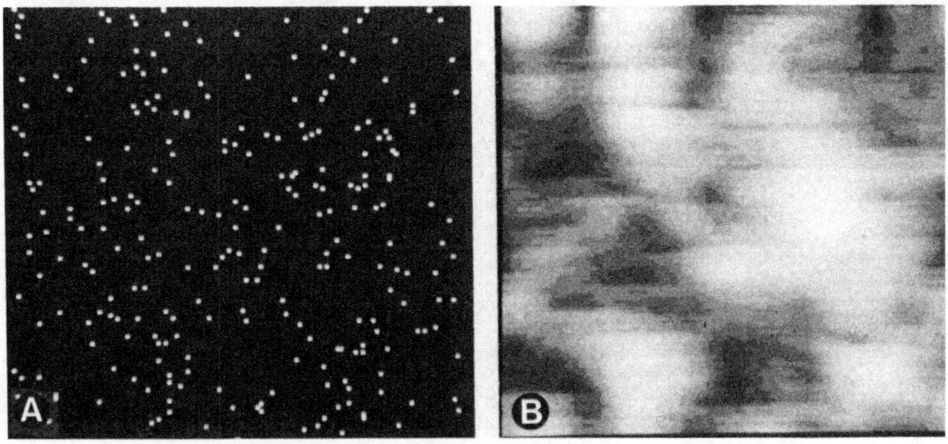

Fig 2 (a) The discrete scatterer model, mean spacing = 8 pixels
(b) Simulated B-scan of (a) wavelength = 6, pulse length = 8, beamwidth = 12. Image area = 128 × 128 pixels.

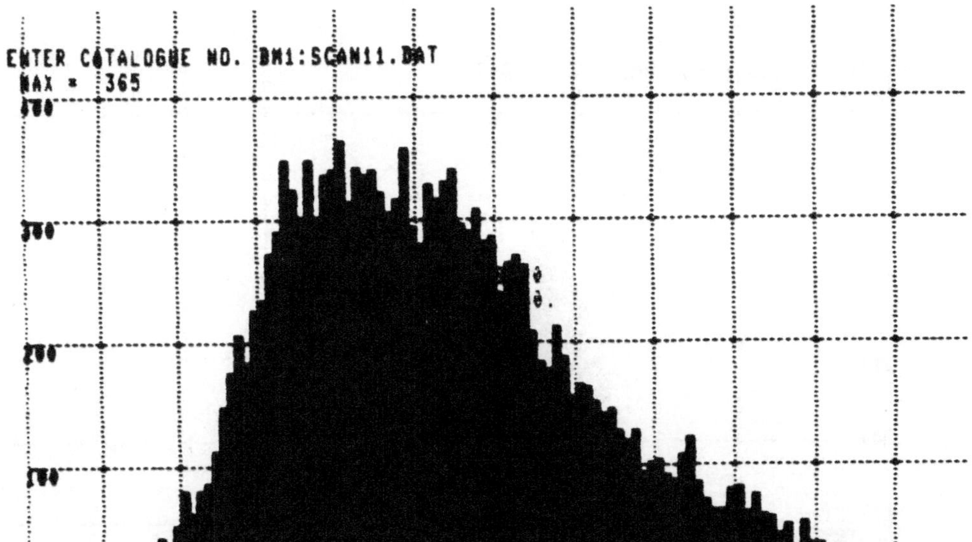

Fig 3 Amplitude histogram of figure 2(b)

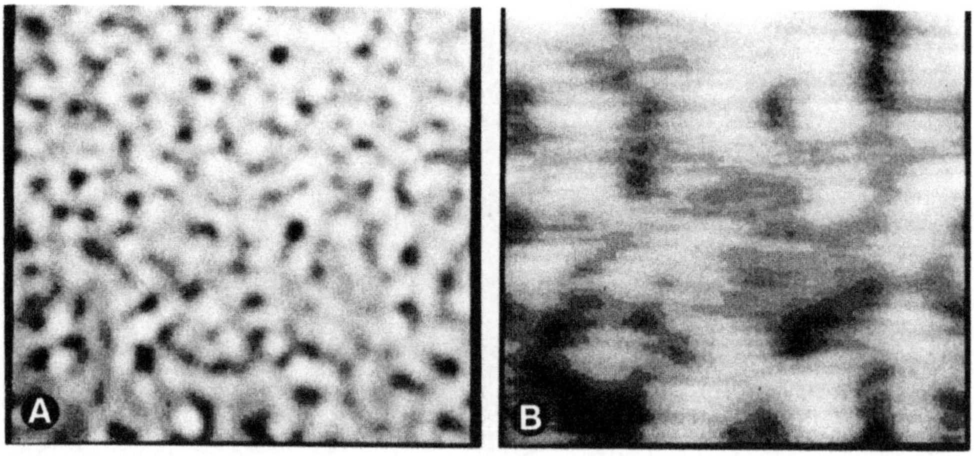

Fig 4 (a) Tissue impulse response for the inhomogeneous continuum model $\bar{a}$ = 6.0
(b) Simulated B-scan of (a) pulse parameters as in fig 2(b)

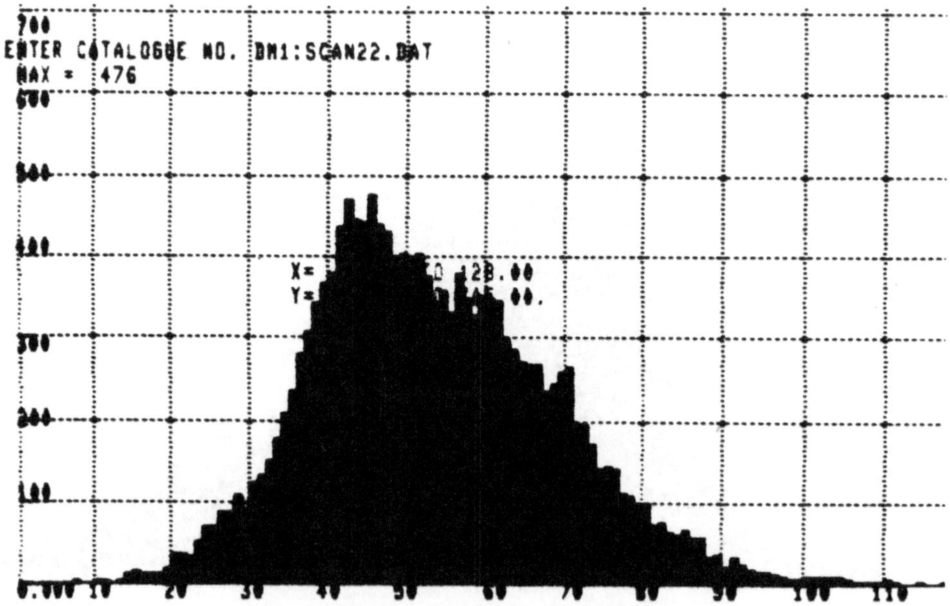

Fig 5 Amplitude historgram of figure 4(b)

impulse response.

The scan of such a model has properties similar to those of the discrete scatterer model.

Again, by looking at the power spectrum

$$I(\underline{k}) = |T(\underline{k})|e^{i\phi(k)}A(k-k_o) \tag{14}$$

$$|I(k)|^2 = |T(k)|^2|A(\underline{k}-\underline{k}_o)|^2 \tag{15}$$

Comparison of eqs 4 and 15 show that if $|T(\underline{k})|^2$ is constant over the range of $\underline{k}$ where $A(\underline{k}-\underline{k}_o)$ is significant, or alternatively, that the width of the auto-correlation function is small compared to the ultrasonic wavelength and pulse size, then the spatial dependence of the two scans is identical, and this is borne out by figures 2 and 4. The inhomogeneous continuum model thus has similar properties to the discrete scatterer model under certain conditions, and because it can also model a range of different conditions it will be examined in more depth.

3 PROPERTIES OF SIMULATED SCANS

3 1 First order statistics

The amplitude histogram shows the typical Rayleigh curve for tissue models with a range of correlation coefficients $\bar{a}$=2.0-8.0. An example is given in figure 5.

3 2 Second order statistics

These are examined by calculating the auto-correlation of the scans in two directions, parallel to and perpendicular to the transducer axis. The half-width of this function, d(0.5), as defined by figure 6 gives a simple estimate of the size of the spatial fluctuations of the scan. If this parameter, calculated for axial separation, is plotted as a function of the correlation length of the tissue model (fig 7), it can be seen that for $\bar{a}$ less than 12, there is no systematic dependence. The axial correlation dependence is more dependent on the beam width than the pulse length. If the same parameter is examined for undemodulated scans, this effect is not observed, and there is a systematic dependence on $\bar{a}$ (fig 8). Thus the property of B-scans that their spatial dependence is independent of the tissue structure is a direct consequence of the nonlinear processing introduced by demodulation, and hence disregarding phase.

The equivalent dependence of the lateral correlation distance on $\bar{a}$ is shown in figure 9. For small values of $\bar{a}$ there

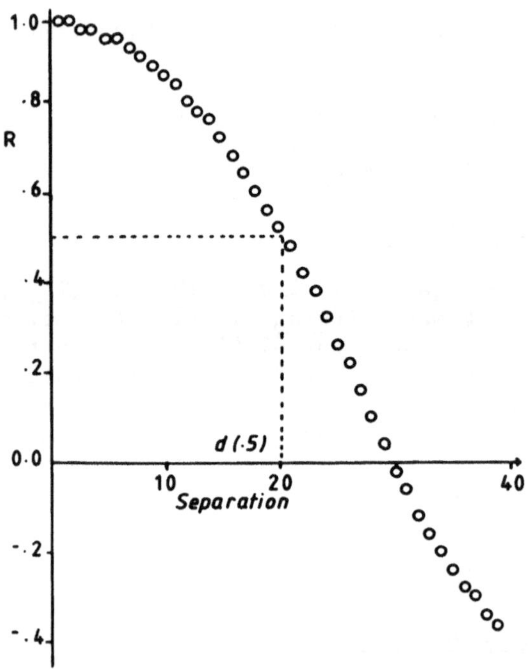

Fig 6 Lateral autocorrelation function of a simulated B-scan showing correlation distance d(0.5) a = 10, beamwidth = 30, pulse length = 14

is a plateau region where the correlation distance is independent of a. Again the equivalent curve for undemodulated A-scans does not exhibit this plateau (fig 10).

The dependence of the structure on the transducer properties can also be examined using the simulation. The results are as follows. The axial correlation distance has a slight dependence on the wavelength of the pulse, but no dependence on the pulse length. There is a dependence on the beamwidth as seen in figure 7. The lateral correlation distance depends only on the beam-width (fig 11).

So the inhomogeneous continuum model can be used to investigate the properties of the B-scan image. It is also proposed to use it to evaluate image processing methods. These are described in the next section together with preliminary results.

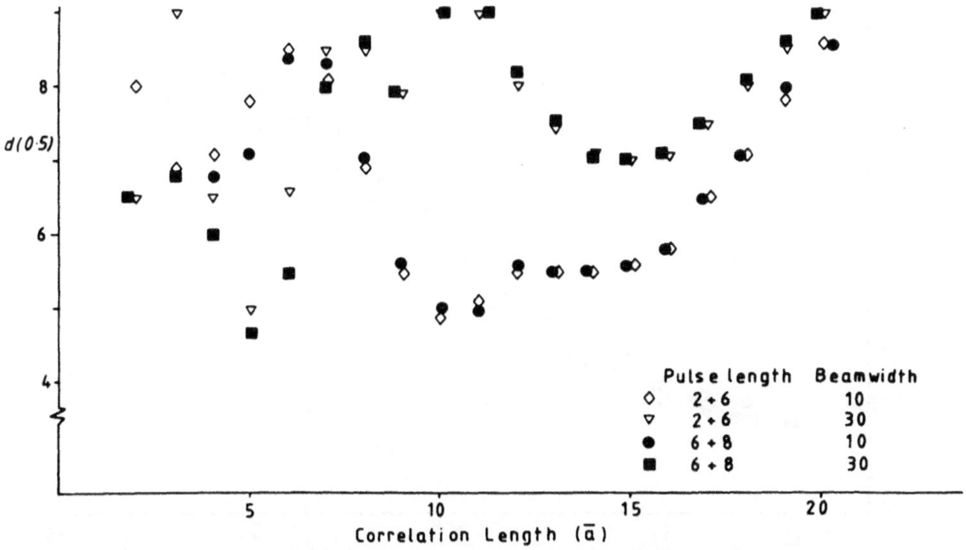

Fig 7 Axial correlation distance d (0.5) between demodulated A-scans as a function of the correlation length of the tissue model

4 IMAGE PROCESSING

The B-scan of an inhomogeneous tissue devoid of major interfaces has been shown to have first and second order statistics which are independent of the tissue model. The human visual system can perceive changes in the first and second order statistics only. The one characteristic of the tissue that is reflected in the scan is the mean scattering level and it is difficult to readily perceive small changes in a mean level because of the high contrast speckle modulating the picture. An example is shown in figure 12, where a circular region of increased mean level is easy to perceive in one case but hard in the second case. Thus the properties of the B-scan image are not suited for visual display, and some processing of the echoes is really required to extract the useful information for display, and suppress the speckle, which contains no information and is confusing. This work on image processing is under way in this laboratory. A suitable example to test such processing is shown in figure 13, which is a scan of region of zero mean scattering level. This region gives rise to an anechoic (black) region in

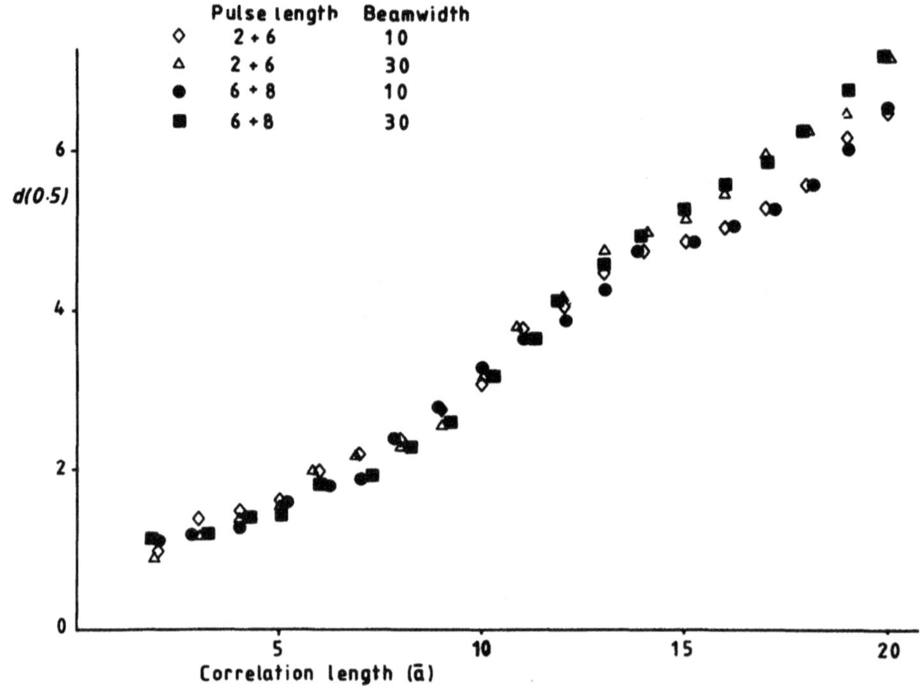

Fig 8 Axial correlation distance d (0.5) between demodulated A-scans as a function of the correlation length of the tissue model

the centre, but there are other anechoic regions, albeit smaller, distributed throughout the scan. A suitable image processing routine should reliably distinguish the two.

One possible technique of image processing is deconvolution of the beamprofile, in particular of the sidelobes. The previous tissue model with a region of zero mean scattering level will be useful in assessing the usefulness of this deconvolution. Figure 14 shows two scans, one with a pulse whose beam profile is given by a sinx/x function appropriate to a rectangular aperture, and the other with a Gaussian beam profile of equivalent 6 dB beamwidth. It can be seen that the Gaussian beamprofile gives a better image of the circular hole, and so it seems that it will be worth performing such a deconvolution.

5 CONCLUSIONS

The properties of the B-scan image of a region of inhomogeneous tissue have been shown to be similar to laser speckle for two statistical models of the tissue, using both a

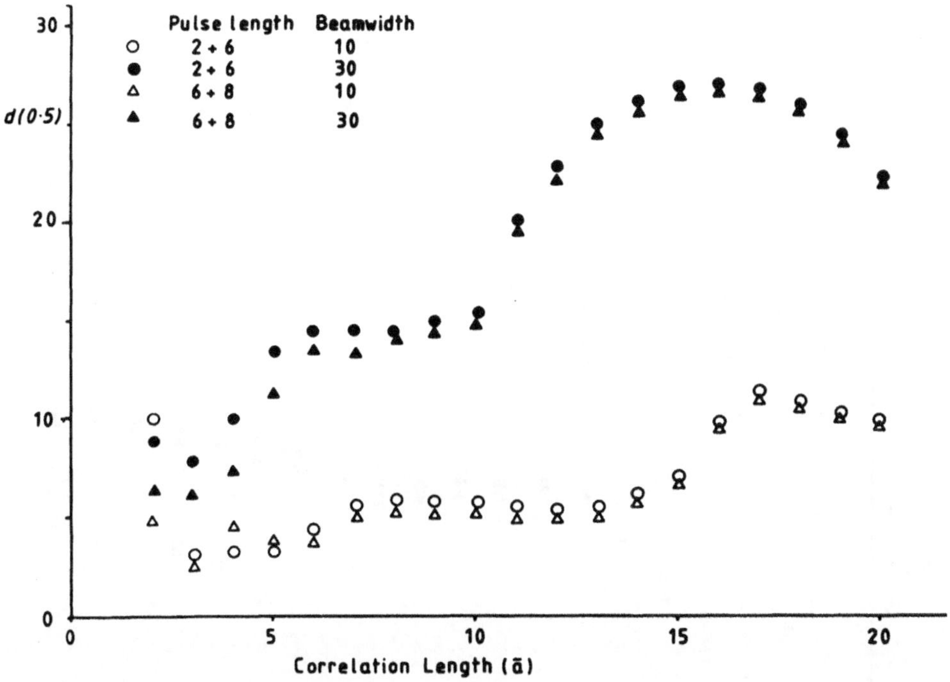

Fig 9 Lateral correlation distance d (0.5) between demodulated A-scans as a function of the correlation length of the tissue model

theoretical analysis and a computer simulation of the scanning process. If the tissue is modelled as having random compressibility fluctuation with a specified auto-correlation function, then for small correlation lengths the first and second order statistics are largely independent of the correlation length. The second order statistics are dependent on the parameters of the interrogating pulse, especially the beamwidth. The fact that the fluctuations in the axial direction are influenced by the beamwidth is an interesting result and would benefit from further work.

One conclusion that can be made from these properties is that the B-scan image is not very suitable for visual analysis, and it is intended to use the simulation to develop and evaluate image processing routines. One method of image processing that looks promising is deconvolution of the sidelobes, but many other techniques are possible and it is hoped to report on these at a later date.

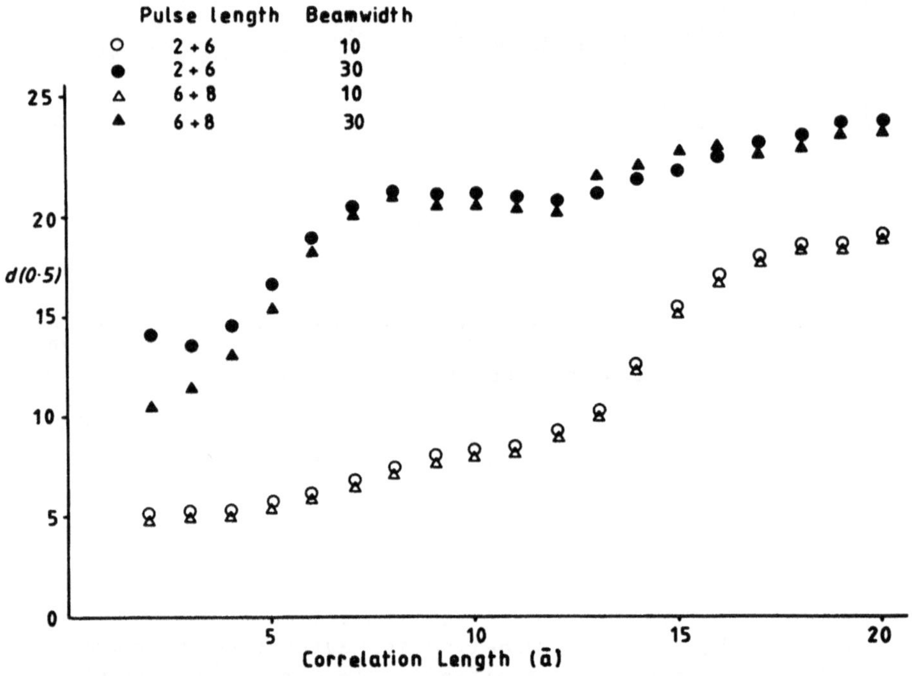

Fig 10 Lateral correlation distance d (0.5) between undemodulated A-scans as a function of the correlation length of the tissue model

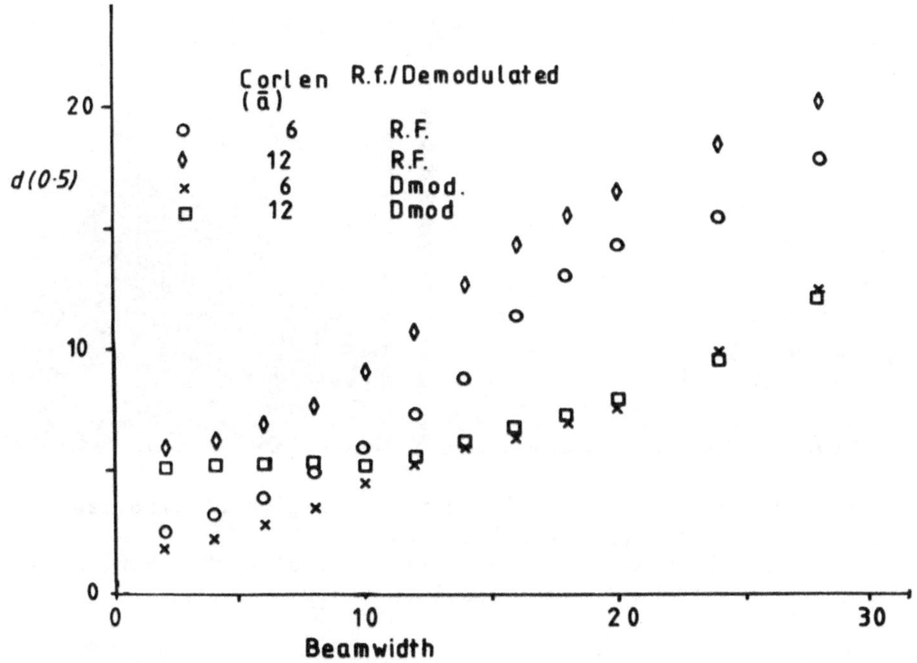

Fig 11 Lateral correlation distance d (0.5) between A-scans as a function of the pulse beamwidth

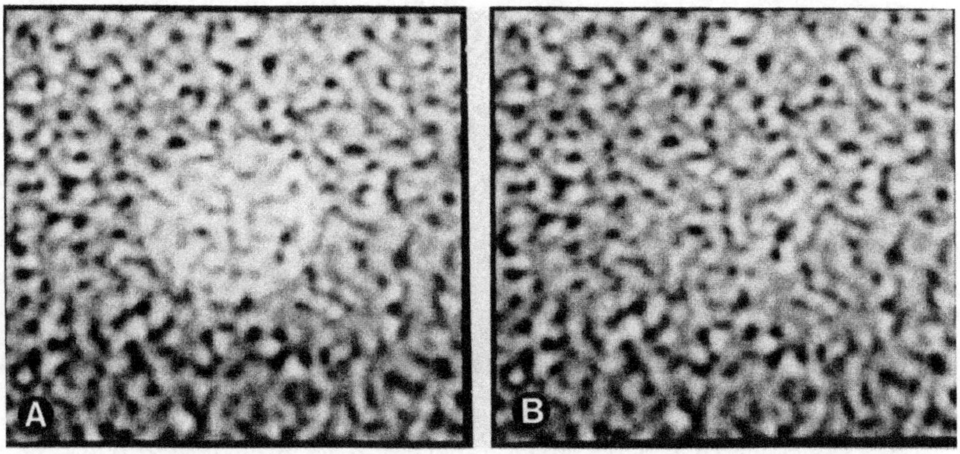

Fig 12 Tissue models with circular regions of higher mean level
(a) Difference in mean level = 50, standard deviation = 100, correlation length = 4.0
(b) Difference in mean level = 10, standard deviation = 100, correlation length = 4.

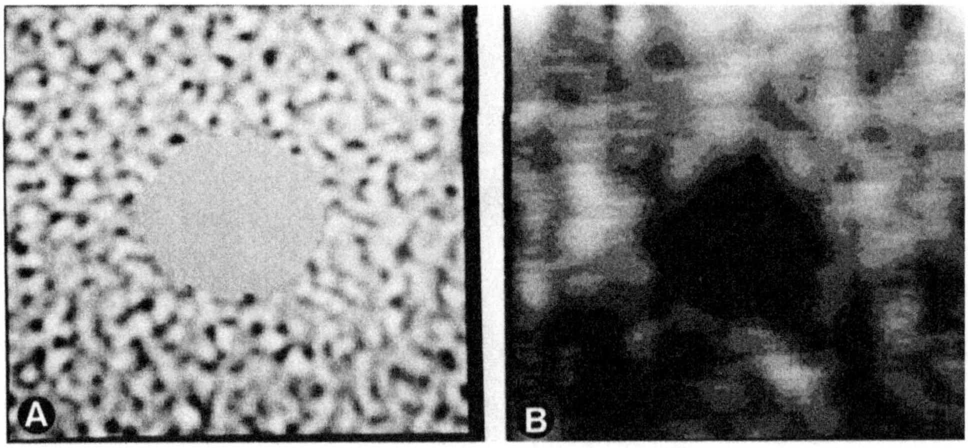

Fig 13 (a) Tissue model with circular region of zero mean scattering level $\bar{a}$ = 2.0
(b) Simulated B-scan of (a) : wavelength = 6.0, pulse length = 4.0, beamwidth = 6.0

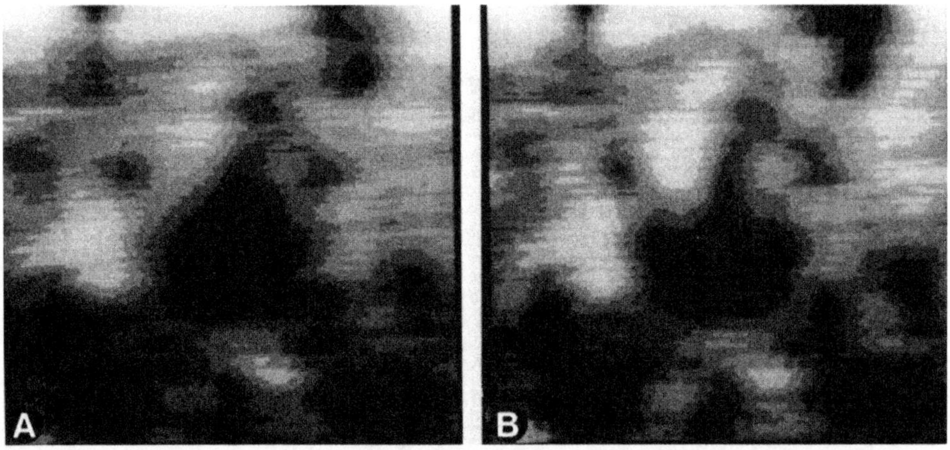

Fig 14 Two scans of figure 13(a) with different beam profiles. Wavelength = 6.0, pulse length = 8.0, 6 dB beamwidth = 12
(a) with Gaussian beam profile (no sidelobes)
(b) with a sinc function beam profile (sidelobes)

6 REFERENCES

Bamber, J.C. and Dickinson R.J.: 1980, Ultrasonic B-scanning: a computer simulation, Phys.Med.Biol., 25:463-479

Beecham, D.: 1966, Ultrasonic scatterer in metals, its properties and relation to grain size determination, Ultrasonics 4:67-76

Dickinson, R.J.: 1980, Ultrasonic echo analysis in the investigation of tissue motion, Ph.D. thesis submitted to the University of London

Gore, J.C. and Leeman, S.: 1977, Ultrasonic backscattering from human tissue : A realistic model, Phys.Med.Biol., 22:317-326

COHERENCE AND NOISE IN ULTRASONIC TRANSMISSION IMAGING

U. Röder, C. Scherg and H. Brettel

Gesellschaft für Strahlen- und Umweltforschung mbH
Ingolstädter Landstr. 1, D-8042 Neuherberg/München
Germany

INTRODUCTION

Ultrasonic transmission imaging[1] has improved so far that practical applications in medical diagnosis can be expected for the near future. Especially diffuse, spatially incoherent ultrasound, as described by Havlice et al[2], has made image content more reliable. Images of hand or foot displaying tendons and blood vessels are readily available and, at the moment, a first practical application will probably be in orthopedics.

Transmission images of liver and kidneys show internal struture only in vitro. An actual problem is to improve the in vivo image quality to the level of the orthopedics examples. It should be noted that scattering along the way through the patient's body is one reason for the information loss when organs are imaged in vivo.

The present paper is concerned with the imaging properties of spatially incoherent ultrasound. We discuss the influence of elementary sources which create the incoherent wavefield, especially with regard to noise arising from object planes outside the in-focus plane. We also show first experimental results on a new type of diffuser providing a large number of independent elementary sources.

SPATIAL INCOHERENCE

An extended source for spatially incoherent ultrasound has to be composed of many small ultrasonic transducers with statistically varying phase changes among their emitted wave fields. A suitable description in this case is based on intensities rather than on amplitude and phase, normally used for the characterization of an ultrasonic wave train. In fully incoherent ultrasound, the resultant intensity is the sum of the component intensities. The smallest size of the elementary source for such a fully incoherent wave field is defined in such a way that subdiving the area in two or more smaller sources with mutually changing phases among their wave fields makes no further contribution to the incoherence of the total field.

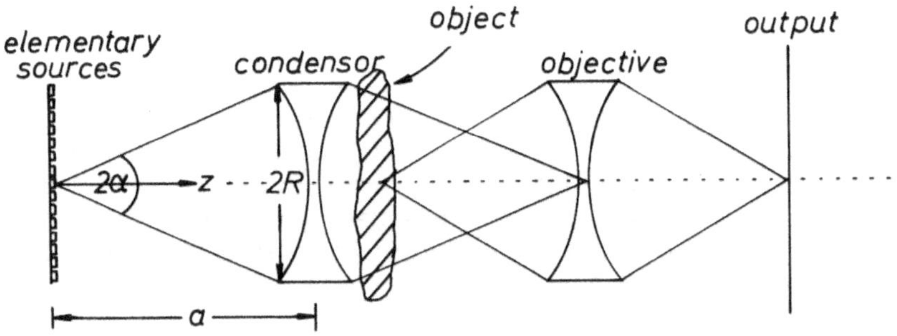

Fig. 1 Configuration for ultrasonic transmission imaging.

A lower limit for the size of the coherent elementary area can be derived from the coherence conditions of a wave field. We assume an arrangement like that of Fig. 1. Because each mode interferes only with itself, the ranges in coordinates (x,y,z) and momentum (p_x,p_y,p_z) are given by

$$\Delta x \Delta p_x = h \quad (1\ a)$$

$$\Delta y \Delta p_y = h \quad (1\ b)$$

$$\Delta z \Delta p_z = h \quad (1\ c)$$

h : Planck's constant.

For the condition of spatial coherence eq. 1a and eq. 1b can be rewritten in terms of the dimensions of the ultrasonic elementary source Δx and Δy and the angular aperture $2\alpha_x$ and $2\alpha_y$ of the wave.

With $\sin\alpha_x = \Delta p_x/2p$ (2 a)

$\sin\alpha_y = \Delta p_y/2p$ (2 b)

and $p = \hbar \cdot k = h/\lambda$ (3)

it follows

$2\Delta x \sin\alpha_x = \lambda$ (4 a)

$2\Delta y \sin\alpha_y = \lambda$ (4 b)

or with $\sin\alpha \approx \mathrm{tg}\,\alpha = R/a$

$$A_{el} \approx \frac{\lambda^2 a^2}{A_{co}} \quad (5)$$

A_{el} is the area of the elementary source and A_{co} is the aperture area of the condensor lens. It is assumed that the object behind the condensor fills out the whole aperture of this lens.

If the extended source has an area A_{so}, the maximum number N of incoherent elementary sources is

$$N = \frac{A_{so}}{A_{el}} \approx \frac{A_{co}\, A_{so}}{\lambda^2 a^2} \quad (6)$$

The intensity I in the output plane is the sum of the intensities of the N independent elementary wave fields i_k in this plane

$$I = \sum_{k=1}^{N} i_k \quad (7)$$

This equation is important for the calculation of the signal-to-noise ratio in the final image. As Goodman[3] has shown for optical wave fields, the signal-to-noise ratio is

$$SNR = \frac{\langle I \rangle}{\sigma_I} \quad (8)$$

with the statistical expectation $\langle I(r) \rangle$ of the intensity at one point and its variance $\sigma_I^2(r)$ in the presence of noise. Assuming that the elementary contributions are random variables and identically distributed, one gets[4]

$$SNR = \sqrt{N} \quad (9)$$

In a transmission system like that of Fig. 1, N has an upper limit determined by the system's geometrical parameters. If the source image formed by the condensor exceeds the aperture of the objective lens, no further gain in signal-to-noise ratio is possible by using a larger source.

The result in eq. 9 can be interpreted in terms of information transmission: The transfer of information from the input (object) to the output (detector plane) is carried out by N channels. Each elementary source corresponds to one channel and creates an image at the output. Besides the signal information the image formed by each channel exhibits noise due to diffraction of ultrasound within the 3-dimensional object, or from reflections within the object and the acoustic components. The spatial distribution of the noise in the image, however, depends on the position of the elementary source which is used for the information transfer: the noise is inherent to each channel. The total output is the incoherent superposition of the different outputs of the N channels. As the noise is averaged by using the input from many channels, signal-to-noise ratio is improved.

COHERENCE TIME

Spatially incoherent ultrasound is created by many mutually random phase changes among their wave fields, as described above. Consequently the intensity fluctuates in space and time. Just like in optics one can define a coherence time τ which can be derived from eq. 1c. The coherence time τ corresponds to the average half-width of the intensity maxima as a function of time[5]. If the observation time T is short compared to τ, no fluctuations in time are registrated, but the intensity will be randomly distributed in space. This granulation or speckle pattern can be reduced by time averaging if $T \gg \tau$.

So, on the one hand, one gains in signal-to-noise ratio by using many information carrying channels, but on the other hand, the superposition of the wave fields creates an inherent granulation pattern which makes it necessary to use a sufficient observation time $T \gg \tau$ for its suppression. For the residual speckles which appear during the observation time, the signal-to-noise ratio is proportional to the square root of the number of uncorrelated granulation patterns which are time averaged[6].

EXPERIMENTS WITH HIGHLY INCOHERENT ULTRASOUND

On the basis of the results which were derived in the foregoing, an optimal spatial incoherence of an ultrasonic wave field is obtained, if

a) the image of the source fills the pupil area of the objective lens,

b) the size of an elementary source is equal to the coherence area A_{el},

c) the statistical phase changes of the elementary wave fields are such that many uncorrelated granulation patterns are averaged during the observation time.

An estimation for an ultrasound transmission system with usual values

ν = 2 MHz $\quad$ $2\,R_{co} = 2\,R_{so}$ = 25 cm
(λ= 0.75 mm) $\quad$ a = 50 cm

gives as maximal number N of elementary sources for an optimal gain in signal-to-noise ratio

$$N_{max} = 1.7 \cdot 10^4$$

Till now, the realization of spatially incoherent ultrasound has received a variety of treatments. Havlice et al[2] used 25 ultrasonic sources which were turned on independently, and the final image was the summation of the 25 single exposures. Such a system requires considerable electronic outlay and an extension of this concept to $10^3 - 10^4$ sources seems to be doubtful, especially with a one-dimensional detector line.

Another approach is to use a moving groundglass for diffuse, incoherent ultrasound, which was experimentally verfied by Havlice[2] and Alphonse et al[7]. This method has the advantage of many elementary wave fields forming a highly incoherent insonification. But to move a large ultrasonic scattering plate with high velocity through water or another fluid cause construction problems.

A simpler method is to scatter ultrasound at moving particles where the difference in impedance of the particles relative to the surrounding fluid provides scattering of the coherent wave. We used a chamber filled with water and polystyrene cylinders with dimensions of about the wavelength of the ultrasound for efficient scattering. To create a statistical movement of the scatterers inside the chamber, we made a turbulent flow in it by a special arrangement of the water supply and draining nozzles. The chamber was closed on each side with 2 mm thick lucite windows. The thickness of the layer of moving particles which the incoming coherent ultrasound has to penetrate, was about 25 mm. The velocity of the scatterers can be varied by regulating the water flow.

In this way the requirements of many independent elementary sources are fulfilled to a high degree, as every polystyrene particle is origin of an elementary wave and the final incoherent wave field is the intensity summation of all elementary waves.

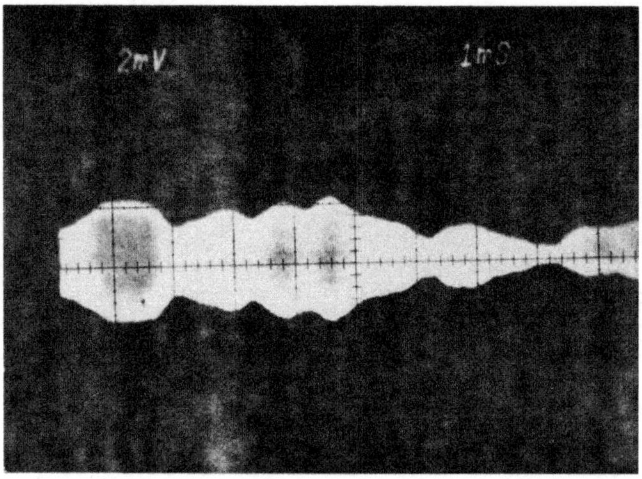

Fig. 2 Time dependence of ultrasonic amplitude from the hydrophone.

The coherence time of this device was measured with a small hydrophone. Fig. 2 shows the ultrasound amplitude as a function of time, when a constant amplitude transducer of 2 MHz was set in front of the turbulent flow chamber and the hydrophone at a distance of about 50 cm from behind. The amplitude modulation in Fig.2 comes from the ultrasonic granulation pattern moving over the detector. Due to shielding problems of the hydrophone, the curve is not fully modulated. With a maximal velocity of the polystyrol particles in our construction of a turbulent flow chamber, a coherence time of about 1 msec is possible.

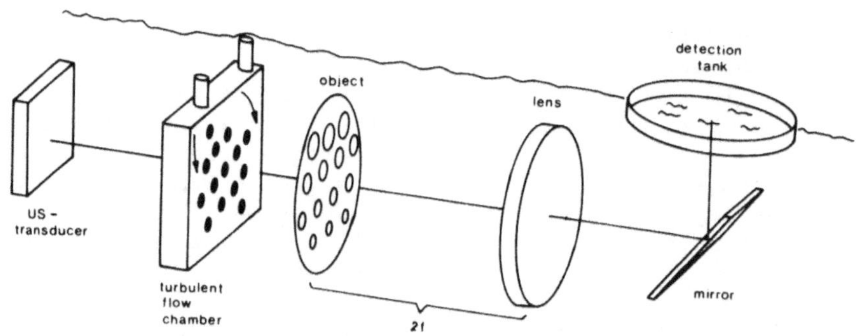

Fig. 3 Transmission imaging system

To test the effectiveness of our diffuser for incoherent transmission imaging, we made several preliminary experiments with a simple acoustic transmission system[8]. Transducer, test object, imaging lens, and a reflecting mirror to image onto the water surface were immersed into a water tank (Fig. 3). For spatially incoherent imaging the turbulent flow chamber was inserted between transducer and object. The transducer was an air-backed PZT transducer of 100 x 100 mm size which delivered continuous ultrasound of 1.6 MHz. The 1:1 imaging lens was a spherical polystyrene lens of 90 mm diameter and 100 mm focal length. A small detection tank with a thin polyethylene foil at its base, filled with water, made the detection system more insensitive against water undulations of the larger object tank. A Schlieren optical system was used to make the ultrasonic image at the surface visible. With this simple transmission system we could image test objects with about 50 mm diameter and a resolution of about 4 mm by coherent insonification (Fig. 4a). The test object in Fig. 4 was a 1 mm thick brass disk with holes of different diameters and distances between the holes in different rows. The holes in the upper row were of 2.5 mm diameter being at a interhole distance of 2.5 mm. This corresponds to a main spatial frequency of 1/5 mm^{-1} in horizontal direction. The subsequent rows had 1/4 mm^{-1}, 1/3 mm^{-1} and 1/2 mm^{-1} main spatial frequency (2, 1.5 and 1 mm hole diameter with a distance equal to the diameter between the holes).

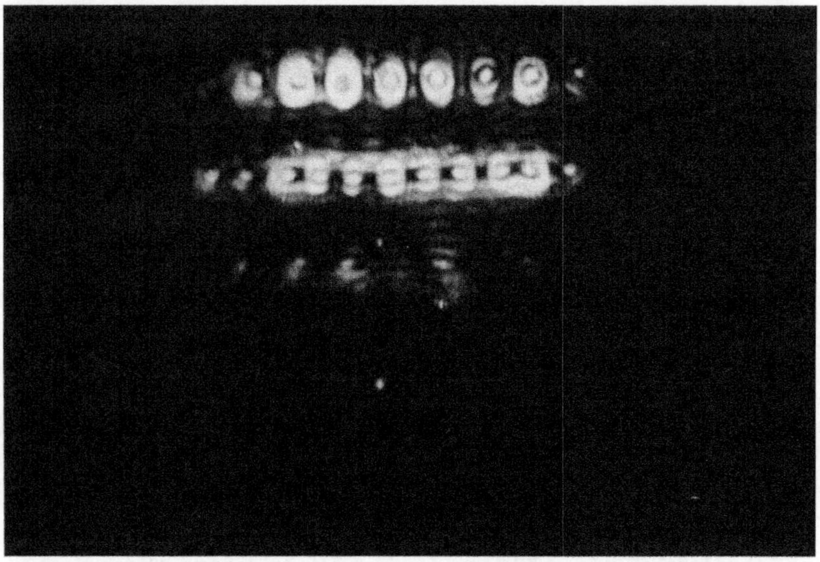

Fig. 4 Effect of spatial incoherence in imaging.
a) coherent insonification

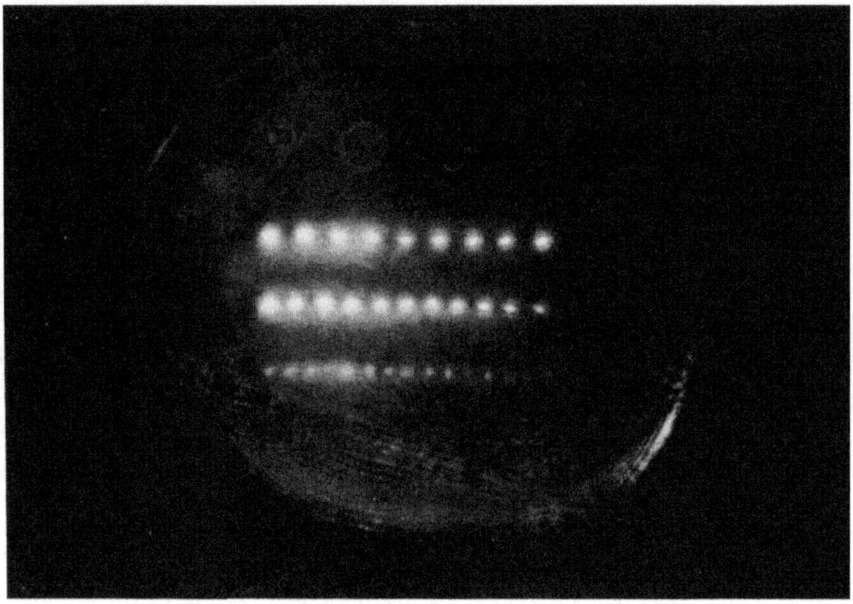

Fig. 4 Effect of spatial incoherence in imaging.
b) incoherent insonification

The coherent image in Fig. 4a shows typical artefacts like rings and other interference structures filling the whole field of view. As the object is in the near field of the source, the insonification is rather inhomogeneous. An uniform insonification in the far-field of the ultrasonic source would require a distance of greater than 250 cm between object and source. The first two rows of the test object are imaged quite well, the third row is hard to see and barely resolved and the lowest row could not be detected because of its weak intensity compared with the upper rows.

For spatial incoherent insonification the turbulent flow chamber was inserted between transducer and object. With the incoherent system (Fig. 4b) the illumination of the object is more uniform, most of the interferences are smeared out and the lateral resolution has increased to at least 3 mm (1/3 mm^{-1} spatial frequency). It should be noted that some structures in both images are due to imperfections in the foil of the detection tank.

An example for the effect of observation time T on the quality of the image is shown in Fig. 5. The object (a small wrench) is imaged with an incoherent system of $\tau \approx 1$ sec and with an obser-

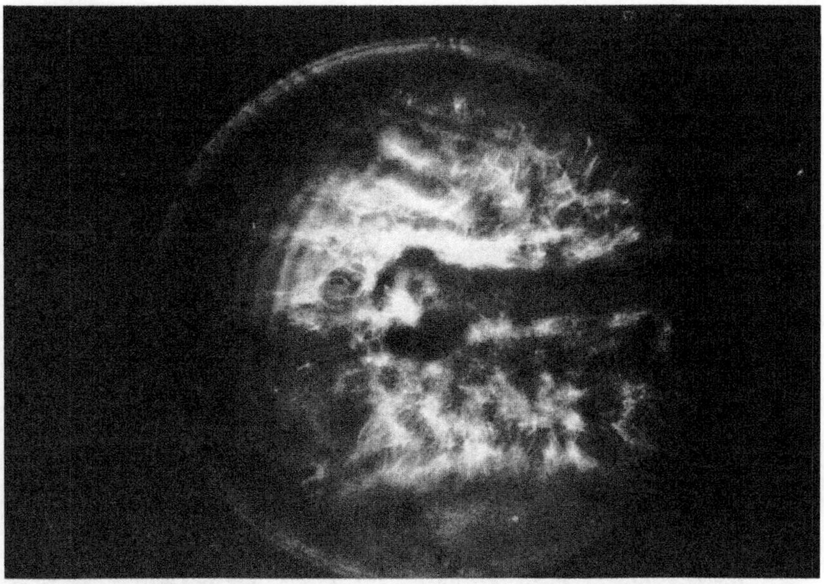

Fig. 5 Effect of coherence time τ in imaging
a) observation time $T \approx \tau$

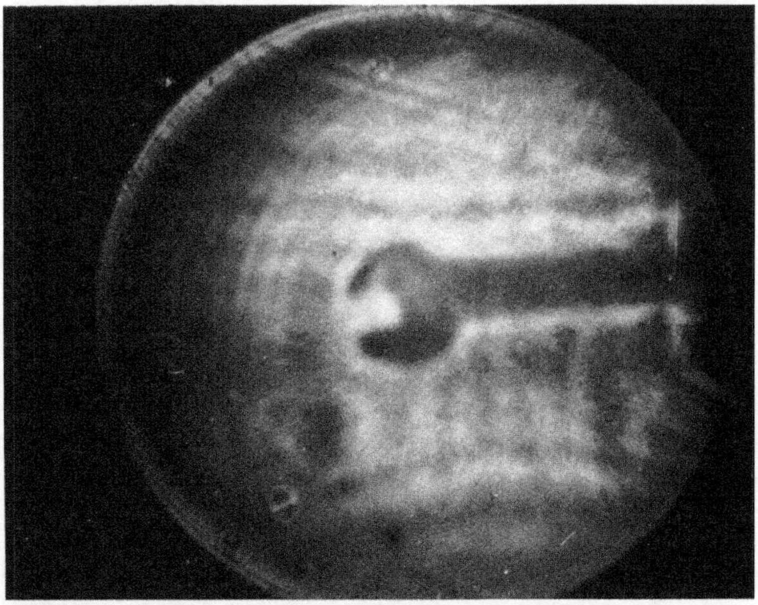

Fig. 5 b) observation time $T > \tau$

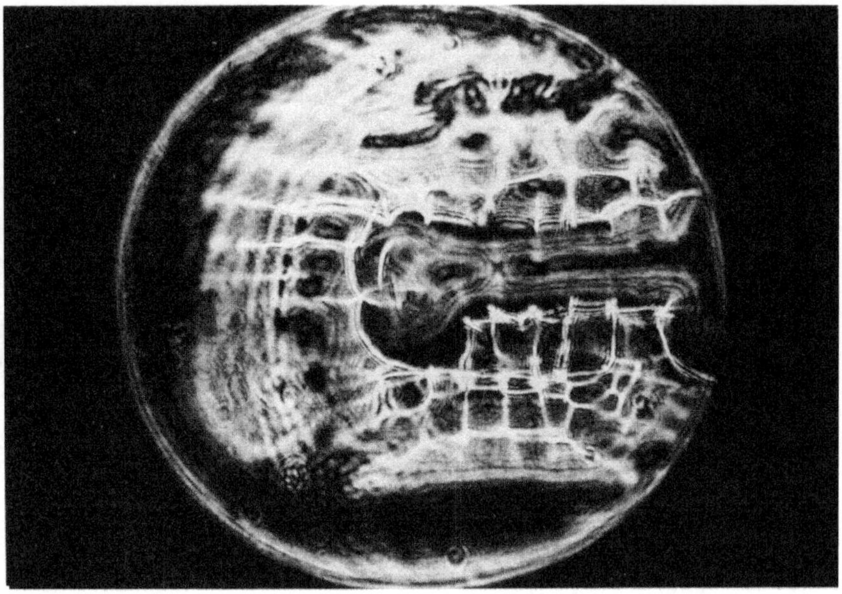

Fig. 5 Effect of coherence time τ in imaging
c) coherent insonification

vation time of 1 sec. in Fig. 5a. After this time, the influence of a speckling pattern, arising from the independent scatterers of the source, is still to be seen. Only longer exposure times yield a satisfactory image. The observation time in Fig. 5b is 10 sec. The speckles have disappeared and the image qualitiy is excellent. For comparison, Fig. 5c shows the ultrasonic image with coherent insonification. In this case, it is especially difficult to recognize the object, as it exhibits sharp edges surrounded with a bright field, and the advantage of incoherent insonification is striking.

The relatively long observation times in these experiments are due to resonance properties of the water surface detector. Only at slow velocities of the scatterers, the detection system worked well[8].

CONCLUSIONS

The scattering of ultrasound at randomly moved particles provides an insonification with a very high degree of spatial incoherence. Using this principle, we realized an incoherent ultrasound source, and showed that image quality is clearly improved. The intensity fluctuations in time at a given point in the wave field were measured to be as short as 1 msec.

Medical diagnostic applications require the high sensitivity of piezoceramic detectors. The most promising system to date[9] sweeps the image over a 1-dimensional piezoelectric detector line by means of counterrotating prisms. Another possible detector system would mechanically scan the 1-dimensional detector line over the image. In order to profit from the incoherence properties temporal integration over at least 100 individual speckle patterns is necessary[10]. In both types of scanning systems mentioned above the coherence time has to be shortened by 1-2 orders of magnitude. A possible solution is the combination of many scattering devices[6,10].

ACKNOWLEDGEMENTS

The authors would like to thank Prof. Dr. W. Waidelich for helpful suggestions and discussions.

REFERENCES

1. Green, P.S., Orthographic Transmission Imaging, to appear in "Clinics in Diagnostic Ultrasound", Vol. 5, New Techniques and Instrumentation, K.J.W. Taylor, ed., Churchill Livingstone/Medical Division of Longman Inc.

2. Havlice, J.F., Green, P.S., Taenzer, J.C., and Mullen, W.F., Spatially and Temporally Varying Insonification for the Elimination of Spurious Detail, in "Acoustical Holography" Vol.7, L.W. Kessler, ed., Plenum Press, New York (1977).

3. Goodman, J.W., Film-Grain, Noise in Wavefront-Reconstruction Imaging, J.Opt.Soc.Am., 57, : 493 (1967).

4. Chavel, P., and Lowenthal, S., Noise and Coherence in Optical Image Processing. II. Noise Fluctuations, J.Opt.Soc.Am. 68,6: 721 (1978).

5. Martienssen, W., and Spiller, E., Coherence and Fluctuations in Light Beams, Am. Journ. Phys., 32: 919 (1964).

6. Lowenthal, S., and Joyeux, D., Speckle Removal by a Slowly Moving Diffuser Associated with a Motionless Diffuser, J.Opt.Soc.Am., 61,7: 847 (1971).

7. Alphonse, G.A., and Vilkomerson, D., Broadband Random Phase Diffuser for Ultrasonic Imaging, Ultrasonic Imaging, 1: 325 (1979).

8. Röder, U.,and Scherg, C., Scattered Ultrasound for Incoherent Insonification in Transmission Imaging, to appear in Ultrasonics.

9. Green, P.S., Schaefer, L.F., Jones, E.D., and Suarez, J.R., A New High Performance Ultrasonic Camera, in "Acoustical Holography", Vol. 5, P.S. Green, ed., Plenum Press, New York (1974).

10. Schröder, E., Elimination of Granulation in Laser Beam Projections by means of Moving Diffusers, Opt. Commun., 3, 1: 68 (1971).

CONTOUR PLOTTING BY A SONIC PHASE LOCK LOOP

R.J. Redding

Director, Design Automation (London) Limited
September House, Cox Green Lane
Maidenhead, Berks

Introduction

The use of a sound wave as a measurement tool is increasing for a number of reasons. Medically it is a benign ray. Scientifically it is a means of measuring with the minimum of disturbance and in an engineering sense it provides digital information which is readily processed by a computer. However, sonic measurements have largely used a pulse-echo technique which has been refined to a high degree by advances in electronic circuitry and components. Effort has been concentrated in the ultrasound region for practical reasons.

Concurrently radio communication has made immense strides and as a result of mass markets a number of electronic components have become available and found use in other areas. The present performance of pulse gating systems is directly attributable to the components for radio purposes.

If we consider radio and sound as wave motion but with a speed difference of about 1 million to 1 then the components and techniques developed for radio will obviously work well on mechanical sound vibrations. This paper describes first attempts at using the highly developed techniques known in hi-fi radio to sound paths for measurement and imaging purposes.

Pulse versus Continuous Wave Operation

Almost all ultrasound measurements have employed the pulse-echo technique in which the resolution is a faction of the band width and has been largely determined by the sharpness of the pulse and the gating circuits in order to distinguish the main echo from subsidiary ones. The higher the frequency of the ultrasound pulse the greater the resolution, but at the expense of the range because the attenuation of the sound wave increases roughly as the square of the frequency.

If one attempts to use a continuous wave then the result is a pattern of standing waves which is unstable in the event of movement due to interaction between the outgoing wave and returning reflections. This effect is well known in paractice. e.g. in speech amplification equipment the feedback from the loudspeaker to the microphone results in a "howl" whose frequency is determined by the path length and circuit characteristics.

If, however, we raise the continuous frequency to an ultrasonic one we can use directional transducers and receivers and thus get a communication channel between two transducers, either directly or from a reflecting surface. Such a channel is of little use for measurement purposes because attenuation and random reflections will cause changes in amplitude. However, if we frequency modulate this carrier channel with a low frequency, we can compare the phase of the sent and received signals, which modulation can be virtually unaffected by changes in the amplitude of the carrier. By locking the phase difference in a phase-locked loop, the modulation frequency is a measure of the transit time of the wave between the transducers and thus a measure of the distance. In effect we have a frequency modulated radio transmitter system where the microphone and loud speaker are connected together to "howl" around the loop. By changing the antennas to mechanical wave transducers, the velocity of the wave is reduced from the speed of light to that of sound so that the frequency is determined by the path length and fluid since any other delays in the electronics being about 1 million times faster, are not significant.

The Advantages of Frequency Modulation

The superiority of frequency modulation over other forms are well known in radio communication and telemetry, and seem equally applicable to mechanical wave motion. In some aspects they seem to work better on sound waves because the attenuation increases roughly with the frequency. Thus we can choose the carrier frequency and power to give just sufficient signal at the receiver for limiting to occur. Thus very little energy is imparted to disturb or damage the specimen and echoes and standing waves are kept to a minimum. Further the well known interference free "single signal" performance of FM systems can be invoked. The strongest signal captures the detector so that any echoes or spurious paths are ignored unless they are stronger than the desired one. This effect is the basis of the code name for the system - COHMOD - being short for Coherently Modulated Ultrasound.

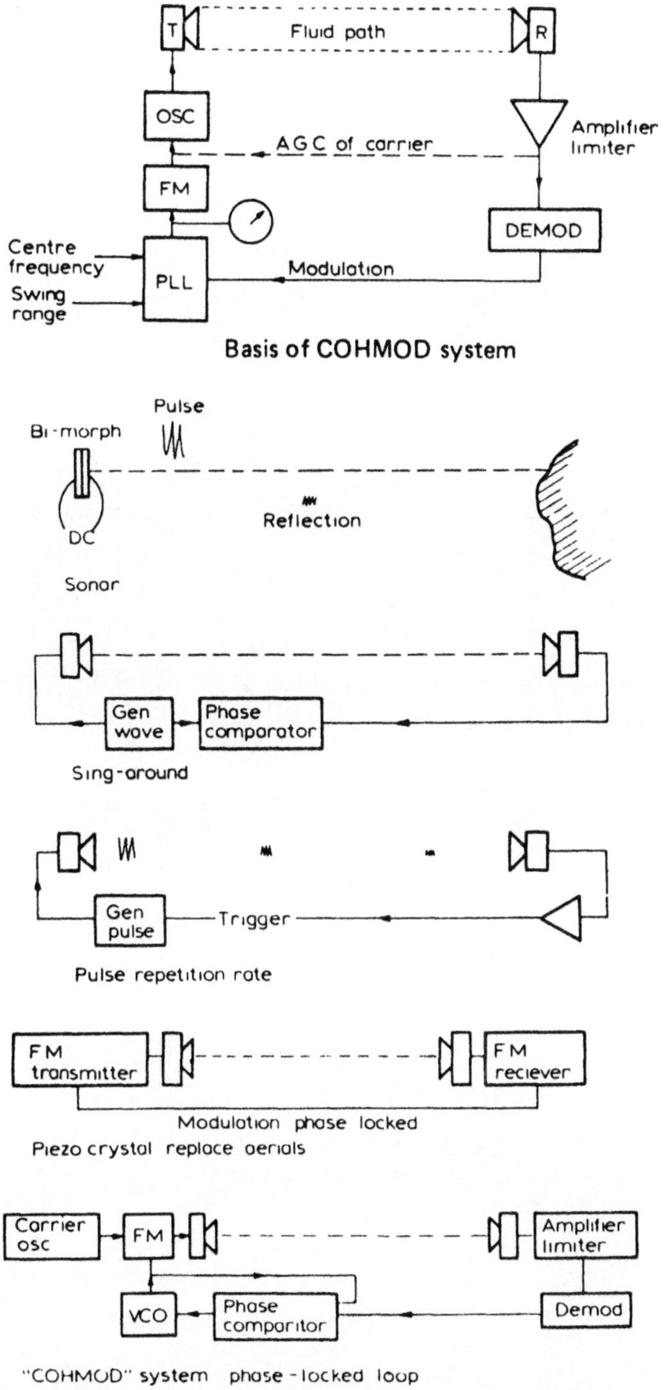

Fig. 1. Various ways of using ultrasound

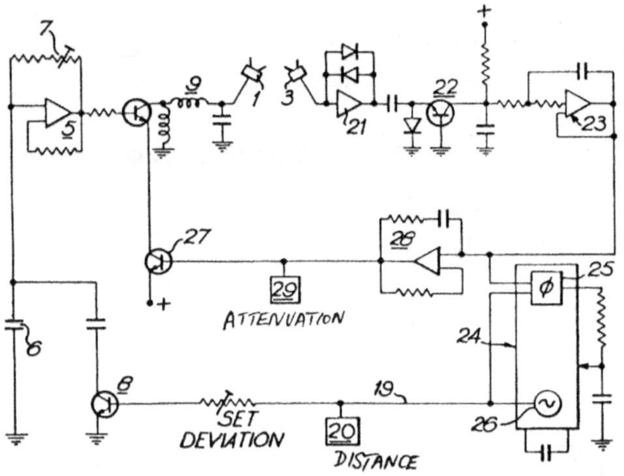
7
5
9
1
3
21
22
23
27
28
29
ATTENUATION
25
24
6
8
SET DEVIATION
20
DISTANCE
19
26

Fig. 2

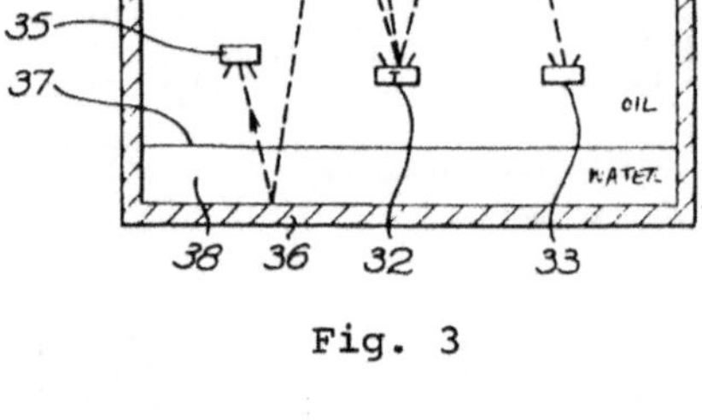
31
30
34
35
37
X
Y
Z
OIL
WATER
38
36
32
33

Fig. 3

Further advantages are that the system is not responsive to amplitude modulation and therefore noise and naturally occurring ultrasound have little effect on the performance. It is also a "narrow band" system allowing highly resonant transducers, the deviation being chosen to suit the band width of the transducer. Typically the power input is 5v. at 20mA for the drive and the resolution, the output being both a digital frequency and an analogue voltage (being the drive to the voltage controlled oscillator of the PLL). By metering the automatic gain control (which corresponds to AVC of radio but is applied to the transmitter), a continuous signal representing the attenuation of the path is also provided as an analogue voltage.

Experimental Experience

The COHMOD system originated from attempts at detecting flammable gas, e.g. methane and hydrogen, in the atmosphere of industrial plants. In the absence of site trials there is no progress to report. The system has been examined in relation to specific problems where the expected contamination is known, e.g. the accumulation of CO_2 in cellars since the system is not a gas detector but depends on changes in the atmospheric density for its discrimination. The use of the system for fire detection is described in Reference (1) the system has considerable advantages over both mechancial and earlier ultrasonic methods for measuring physical parameters such as distance and perhaps flow and these are discussed in paper (2) which suggests that there is a new genus of transducer based on sonic resonance in a fluid path, and bearing comparison with the superiority of the quartz crystal clock over earlier mechanical ones. The performance in terms of speed of response, discrimination and convenience compared with current hardware is considerable but because it depends on sonic velocity the reading is not absolute unless compensated for ambient conditions. A very elegant way of doing such compensation is to use a dual comparator system in which an unknown fluid path is compared with a reference one, both being subjected to the same conditions. Although contrary to traditional measurement practice the scope for such built-in reference systems seems bright, particularly since a large number of readings can be processed as a pattern to elucidate trends and give early warning of departures from normal. This feature of pattern recognition can be extended to allow comparison with the build up of previous events particularly for the early warning of catastrophes (Ref. 2).

Acoustic Imaging in Air

Work has been mainly conducted in the atmosphere using 25 and 40 kHz carrier frequencies because transducers are readily and cheaply available. Lower frequencies using a "tweeter" loud speaker as the transmitter and electret microphones as the receiver have been shown to detect passing vehicles at a distance of at least 10m. whilst ignoring intervening pedestrians for road traffic control. For anti-collision of vehicles and rail-mounted machines e.g. cranes and locomotives, a distance of 100m. seems feasible using a bi directional system.

A practical philosophy of design consists of choosing the carrier frequency as high as possible consistent with a reliable communication path; the modulation frequency is then chosen separately according to the output information required. There can be a fraction or many wave lengths of the modulation in the path although the former is preferable since any confusion due to a false lock is easily avoided by the circuit design. In practice the availability of a transducer in a suitable housing has been the main limitation. The deviation can be chosen to suit the band width of the unit. It is not critical and preferably narrow, e.g.5%

Use in Liquids and Solids

In some respects use of the system in liquids and solids is easier than gases because the attenuation and speed is higher. However, the mounting of the transducer becomes more pertinent since additional sound paths are provided via the structure. These effects may be minimised in various ways, mechanically and electrically. In particular the PLL can be set to work instead over a specific frequency range, clear of unwanted transit times.

The lower attenuation can mean that the system is blinded by excess energy in a confined space and hence the "ACG" system becomes an important feature since it reduces the energy input until there is a 'just adequate' communication path in the preferred direction, thus reducing reflection effects. This can be used to good effect when combined with a variable "lock range" facility by tuning the PLL. Thus, having settled at one interface the system can be tuned to a shorter range and the power increased until a shorter range reflection occurs if some other inhomogenity is present. Thus both the spacial position and the degree of opaqueness appears to be available as convenient electrical signals, suitable for computation.

The transmitter, receiver and the electronic circuit can be combined in one package which may be an epoxy resin encapsulation. Thus it is possible to measure the distance to an air-liquid interface such as a storage vessel from above looking down, or from below the surface and looking up. The latter case is superior since the sonic paths are more likely to be stable with temperature below the surface.

Figure 3 shows a development being pursued to determine accurately the amount of liquid fuel in tanks. One transmitter 32 operates successively with three transducer receivers, 33 & 34 looking at the surface, and 35 which looks at the bottom. From the known spacings X, Y and Z and the transit times, a simple computation gives the oil and water levels, and information about the temperature gradients in the tank. Such temperatures are of vital importance in the storage of liquified petroleum gases under pressure in order to warn against a phenomenon known as "Roll-over" due to cooling on draw off of gas.

Comparison with other Techniques

The COHMOD system is akin to "sing-around" but continuous in operation and employing separate carrier and measuring frequencies. It does not utilise the Doppler effect and it gives a discrete measurement or reading rather than a pictorial image. Thus it seems to have nothing to offer for the conventional "photographic" type of imaging, but it can draw "contour lines" if the transducers are moved or caused to scan a reflecting surface.

The purpose of this paper is to invite consideration of whether the COHMOD technique has merit for medical purposes since the sound reflecting properties of animate, mineral or liquid surfaces would appear similar. Possible uses are the monitoring of movement of the chest wall, and perhaps heart valves - directly on a chart recorder.

Similarly, the use of multiple beams for fire and gas detection described in Ref. 4 might have application to the recognition of malignant growths and the science of tomography since the signals provided are ideal for computing. However, the potentially useful application needs to be specified and provided with background knowhow in order that the electronic assistance can be adopted effectively.

References:

1. British Patent Specification No. 1,523,231 et seq. U.S. Patent No. 4,119,950 etc.
2. A New Genus of Transducer? R.J. Redding, "Measurement and Control", Vol.13, pp 27-30, January 1980
3. Sound Beams for Fire Alarm and Gas Detection Systems. "Fire", March 1980
4. Paper to I.E.E.E. Ultrasonics Symposium, Boston, November, 1980

PSEUDO-HOLOGRAPHIC ACOUSTICAL IMAGING
WITH A LIQUID CRYSTAL CONVERTOR

J.L. Dion, A. LeBlanc and A.D. Jacob

Département d'Ingénierie
Université du Québec à Trois-Rivières
C.P. 500, Trois-Rivières, Québec G9A 5H7

ABSTRACT

A new acoustical holography technique is presented, based on the acousto-optical conversion properties of a specially designed nematic liquid crystal cell. In particular conditions, an ultrasonic hologram can be "written" on a thin L.C. layer as a spatial modulation of the molecular orientation. Direct and real-time conversion to a visible hologram or image of the object can be achieved by means of incoherent light. The system has inherent intensity or phase imaging properties. Application to interferential acoustical holography is also considered. The images presented show that the theoretical limit of resolution can be practically achieved at frequencies of a few megahertz.

1. INTRODUCTION

We have previously demonstrated that an ultrasonic compressional wave passing through an oriented layer of nematic liquid crystal induces a molecular reorientation perpendicular to the acoustic displacement [1-3]. This was shown to be related to the anisotropy of ultrasonic attenuation in the medium [3] and, more recently, to the theorem of minimum entropy production [4]. Our theory predicts [3] that transmission of polarized light through the LC cell varies as the fourth power of the acoustic intensity and attenuation anisotropy, and also as the tenth power of the layer thickness, for small distortions. This is well experimentally supported [3, 5].

We have established that this new phenomenon makes possible an original acousto-optical hologram convertor. This convertor could be used in various acoustical holography systems operating at frequencies between 1 and 10 MHz. A first working system has been built. Basically, the ultrasonic hologram is formed on the LC convertor by superposing the image given by acoustic lenses with a reference plane wave. The resulting vibrations in the LC layer vary in direction and amplitude from point to point, so producing a LC structure distortion pattern corresponding to the image hologram. Then, by means of an appropriate optical system, this pattern is converted into a fringed image which can be directly observed or photographed. The corresponding signal from a video camera can be simply processed to give an un-fringed image, where brightness changes according to acoustic intensity. We present various images obtained by these means, which particularly reveal the high resolving power of the technique. We are planning to eventually use this acousto-optical convertor to experiment with phase imaging and acoustical interferential holography systems to which it is naturally applicable. This technique may be called "pseudo-holographic" since image reconstruction is achieved without a laser, using an incandescent light source. Contrarily to others [6, 7, 8], we have succeeded in this application of liquid crystal to acoustical imaging because of the unique properties of the cell walls used. Without these, the basic phenomenon is nearly impossible to observe and use.

2. PRINCIPLES OF OPERATION

2.1 THE ACOUSTO-OPTICAL EFFECT IN LIQUID CRYSTALS

When an ultrasonic wave propagates obliquely through an oriented nematic liquid crystal layer (Fig. 1), its molecular structure is distorted under the action of a torque [1] which was shown [3, 5] to be expressed as

$$C_u = \frac{2Ia\Delta\alpha}{v} \sin 2\phi \qquad \text{N m/m}^3 \tag{1}$$

where C_u is the torque per unit volume, I is the acoustic intensity, a is the thickness of the layer, v is the velocity of sound in the L.C., $\Delta\alpha = \alpha_{\parallel} - \alpha_{\perp}$ is the anisotropy of acoustic attenuation and $\phi = \beta - \theta$ is the angle between the wave propagation direction and the molecular director. We have explained this torque by introducing a minimization principle for acoustical losses [1, 3]: "In a medium with acoustical anisotropy, the molecules tend to reorient such as to minimize propagation losses". Later, this has been related to the theorem of minimum entropy production of irreversible thermodynamics [4].

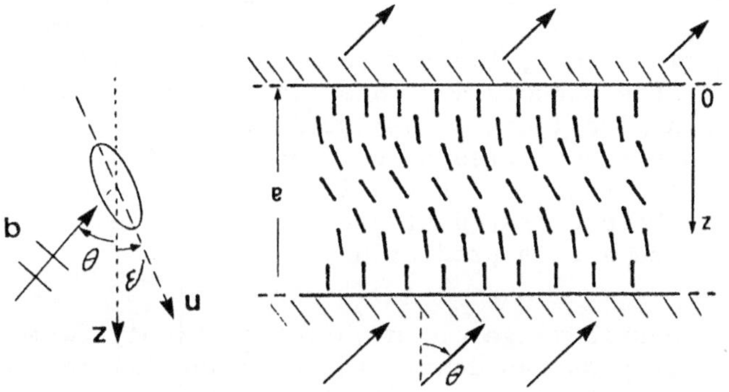

Fig. 1 Liquid crystal distortion produced by an ultrasonic wave

This layer being birefringent, if it is placed between crossed polarizers with axes at 45° from the plane of the figure, the transmission of light τ will depend on β (z), and consequently on the acoustic intensity. Our theory predicts [3] that for small distortion β, with $\theta \simeq 40^o$,

$$\tau = k_1 \, I^4 \tag{2}$$

where k_1 is a constant depending particularly on the type of liquid crystal, thickness and temperature. This has been verified for various nematic compounds of different thickness from 100 to 250 microns [5]. Other researchers, using fairly different experimental conditions, found a similar fourth power law [8] and even a law with an exponent depending strongly on temperature [7]. Their explanation of the phenomenon is very different from ours. Furthermore, theory and experiment show that, as a function of thickness $\underline{a}$ for a given acoustic intensity:

$$\tau = k_2 \, a^{10} \tag{3}$$

where k_2 is a constant. We have also observed [5] that light transmission response to a constant ultrasonic excitation may be described by

$$\tau = \sin^2 A(1 - e^{-Dt}) \quad (4)$$

where A and D are constants which particularly depend on the type of L.C., thickness, acoustic intensity and temperature. When irradiation stops, light transmission variation due to relaxation of the structure obey a similar law. The effect (Eq. 2) should also increase as the fourth power of the ultrasonic frequency since attenuation anisotropy (Eq. 1) is nearly proportional to frequency in the 1 to 10 MHz range [3, 9, 10]. Until now, we have operated at frequencies around 3.5 MHz, but we intend to check this prediction.

If the acoustic image of an object is simply formed on a L.C. cell as Greguss has done [11], equation (1) indicates that the effect should be quite small since ϕ is nearly zero, as we have experimentally observed. Furthermore, if acoustic transmission is high at normal incidence through a L.C. cell made with half-wave-thick simple glass windows, it decreases rapidly as the angle of incidence increases; multiple reflections between the windows then become important and troublesome. In these conditions, a fairly high acoustic intensity (several mW/cm^2) is required to produce visible effects which are mainly due to acoustic streaming [3, 7]. The resulting acoustic images are of very poor quality and have a very narrow dynamic range. In view of our theory, these factors particularly explain the previous unsuccessfull attempts to develop a working nematic liquid crystal acoustical image convertor.

2.2 ACOUSTICAL HOLOGRAPHY WITH A LIQUID CRYSTAL CONVERTOR

We will now see how this acoustical torque phenomenon in liquid crystals can be used in ultrasonic holography. Figure 2 shows a nematic L.C. layer submitted to two plane acoustic waves 1, 2, which pass freely through the layer. Oblique wave 1 is a reference wave, or pre-orientation wave because it produces a uniform distortion in the plane of the layer. The superposition of the object wave 2 to the first causes an acoustical displacement $\vec{d}$ as shown by the arrows, which amplitude and direction periodically varies from left to right. This in turn produces a periodically varying torque (Eq. 1), and distortion as indicated by the dotted lines. The acoustical displacement is generally elliptical as shown, and the long axis of the molecules tend to orient along the short axis of the ellipse. We may therefore say that the uniform re-orientation produced by reference wave 1 is spatially modulated by "object" wave 2, according to its phase and intensity. Consequently, because of bire-

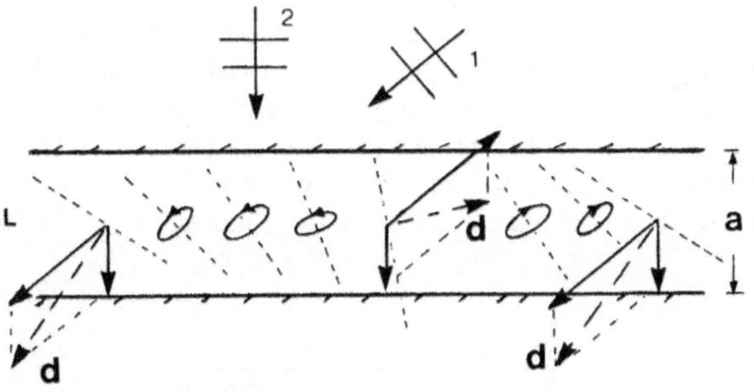

Fig. 2 Acoustic displacement produced by two waves, and resulting L.C. orientation modulation

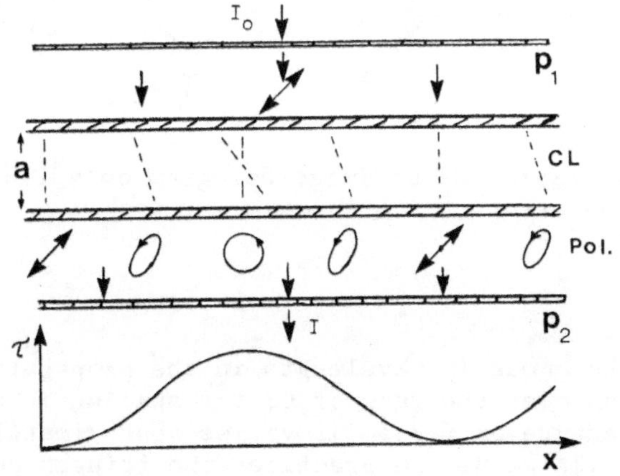

Fig. 3 Transmission of light through an orientation modulated liquid crystal

fringence, the polarization state of light simultaneously crossing the layer will be accordingly modulated. If the cell is placed (Fig. 3) between crossed polarizers p_1, p_2, the transmission of light τ will also vary periodically and one may observe directly visible fringes corresponding to the acoustical interference fringes of the hologram. An essential condition for this to be possible is high acoustical transmission through the cell for waves of widely varying angles of incidence. This is realized using layered cell windows [5].

In figure 4, the image of an object Ob irradiated by plane acoustic waves is formed on liquid crystal convertor C by an acoustic lens L_a. A plane reference wave produced by transducer T_r is then superposed to the object wave on C, so producing the image-hologram of Ob. If there is no object, this hologram is simply a system of hyperbolic fringes on C, symmetrical relative to the plane of the figure. In this plane, the fringe spacing is given by

$$b = \frac{\Lambda}{\sin \theta_r - \sin \theta_o} \qquad (5)$$

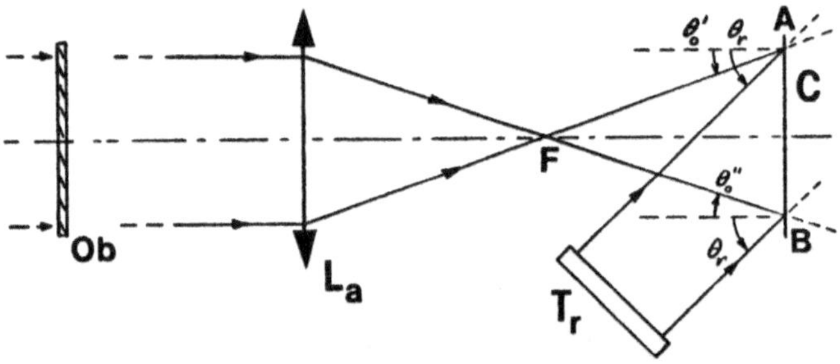

Fig. 4 Formation of an image-hologram on a liquid crystal convertor (L.C.C.).

where Λ is the acoustic wavelength in the propagation medium. Since θ_o varies over the face of C, the spacing is not uniform. For a lens aperture of f/3.5, it varies approximately from 1.8Λ at A to 1.2Λ at B. In practice, the fringes curvature is not very large and they can be considered as parallel for all

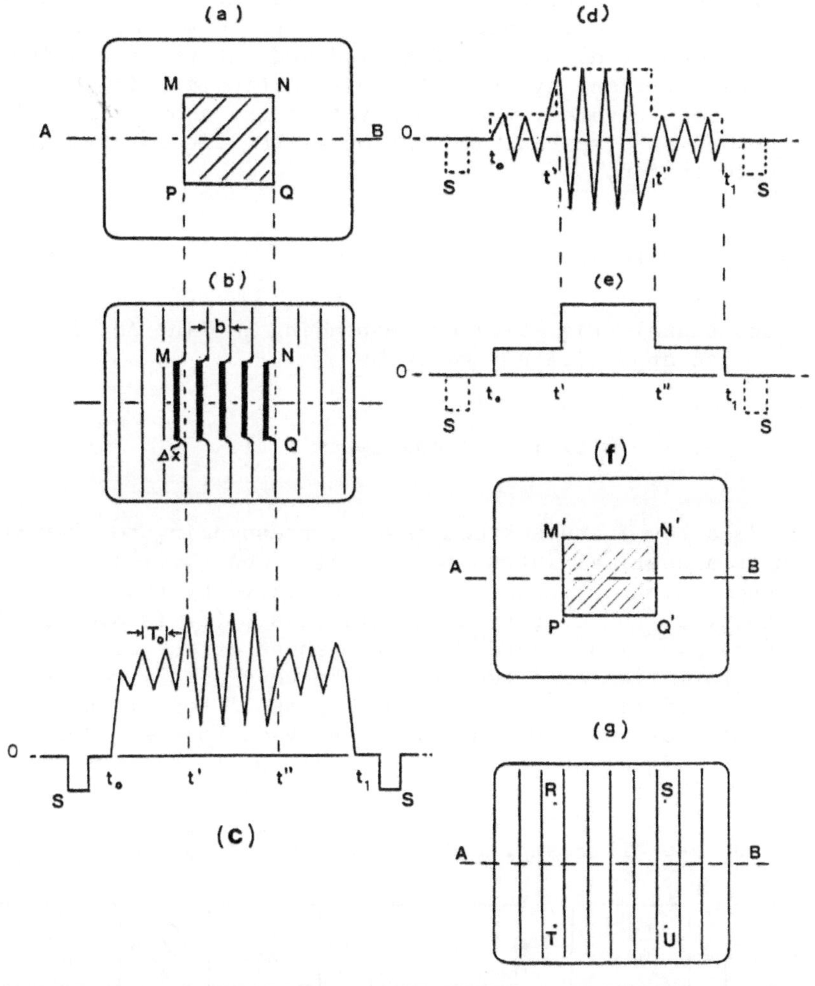

Fig. 5 Principle of conversion of LCC image-hologram into an intensity image.

practical purposes. Therefore, the acoustical hologram of a simple object such as the clear area MNPQ of fig. 5-a, as seen through a suitable optical system, should sensibly be as in fig. 5-b. The visible fringes of image area MNPQ have higher contrast than the surrounding ones corresponding to higher acoustical intensity. Phase differences of the object wave in the plane of the L.C. convertor are converted into a shift Δx of the fringes relative to the background. It follows that this visual pattern can be directly visualized and interpreted using a simple acousto-optical system with this particular liquid crystal convertor. Overall, we may say that the acoustical hologram is recorded as a spatial modulation of the liquid crystal molecules orientation.

2.3 INTENSITY AND PHASE IMAGING

The video signal (Fig. 5-c) corresponding to line AB of fig. 5-b is then approximately given by

$$S(t) = C(t) + A(t) \cos [\omega t + \psi(t)] \qquad (6)$$

where C(t) is a low frequency component corresponding to average brightness or acoustical intensity, and A(t) cos (...) is a carrier signal which amplitude A(t) is modulated by the acoustical intensity of the object wave. Its phase $\psi(t)$ is modulated by the fringes position according to the phase of the object wave. The information on intensity can be very simply recovered with the system of fig. 6 where the high-pass filter blocks C(t). The detector and low-pass filter provide to the video monitor a video signal proportional to A(t), while the synchronization pulses lost in the filtering process are re-injected at the output. Fig. 5-d shows the signal after high-pass filtering, and 5-e after amplitude detection and low-pass

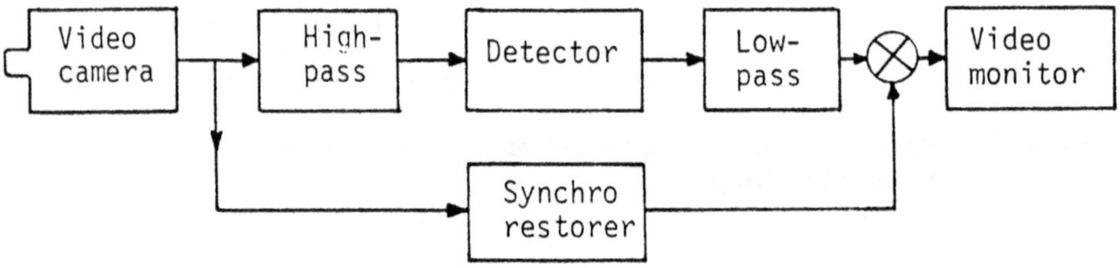

Fig. 6 Image-hologram processing system to provide an intensity image.

filtering. Finally, the picture shown on the video monitor (fig. 5-f) is a bright area M'N'P'Q' representing acoustical object MNPQ.

To obtain a phase image corresponding to the fringe shift Δx (fig. 5-b), a reference signal is required. This can be provided by recording a hologram of the background, without an object, such as in fig. 5-g. Along line AB of this hologram, the corresponding video signal can be approximated by

$$S_o(t) = C_o(t) + A_o(t) \cos \omega t \qquad (7)$$

If the object hologram (Eq. 6) is now superposed to this last one, a Moiré figure is then obtained which is the interferogram or interferometric hologram. Along line AB, the signal is

$$S'(t) = S_o(t) + S(t)$$

$$S'(t) = C'(t) + A'(t) \cos[\omega t + \psi'(t)] \qquad (8)$$

where $$A'(t) = [A_o^2 + 2A_oA \cos\psi(t) + A^2]^{\frac{1}{2}} \qquad (9)$$

Therefore, the carrier amplitude $A'(t)$ will follow the phase $\psi(t)$ of the object wave, and simple processing by system of fig. 6 will provide a phase image of the acoustical object.

2.4 INTERFERENTIAL ACOUSTICAL HOLOGRAPHY

From what has just been said it is easily seen that this acousto-optic liquid crystal convertor makes possible fairly simply implementation, not only of phase imaging, but also of interferential acoustical holography or acoustical holographic interferometry. $S_o(t)$ and $S(t)$ of eq. (8) only have to be two successive holograms of the same object. Until now, this technique [12, 13] has known a rather slow development because of the lack of suitable detectors which would allow imaging in reasonable time. We think that the present liquid crystal convertor (LCC) and the associated imaging process should provide an interesting answer to the problem. In a second phase of the development, we are particularly considering using it to realize electro-optical processing of acoustical information as described by Pasteur and Seyzeriat [14].

2.5 LENS-LESS ACOUSTICAL HOLOGRAPHY

If the acoustic lens of fig. 4 is removed, an ordinary acoustical hologram is recorded by convertor C in the form of nearly parallel interference fringes ("off-axis holography"). Therefore, with an appropriate optical system, a beam of coherent light can be diffracted by this hologram, and a visible image recovered after spatial filtering, as in liquid surface acoustical holography [15].

3. THE EXPERIMENTAL SYSTEM

The experimental system is schematically represented in fig. 7. Its main parts are: 1- The electro-acoustical and acousto-optical elements; 2- The optical elements; 3- The electro-optical elements, processing and display.

3.1 ELECTRO-ACOUSTICAL AND ACOUSTO-OPTICAL ELEMENTS

An open tank was used, with a water-antifreeze mixture as a propagation medium. Object transducer T_o is a 25-mm quartz disc naturally resonant at 3.5 MHz, while reference transducer T_r is made from a 12 X 12 cm^2, 130-micron thick, polyvinylidene fluoride sheet (PVDF, a piezoelectric polymer). This last transducer can give nearly plane reference waves at frequencies about 3.5 MHz with good uniformity and sufficient intensity ($\sim$1 mW/cm^2). The transducers are powered by separate RF amplifiers driven by a common generator. Acoustic energy transmitted by object Ob is refracted by inflatable acoustic lens L_a and reflected by glass plate M_1 toward convertor C where it forms an acoustical image of Ob. Energy transmitted by C is reflected by glass plate M_2 to an acoustical absorber A. The reference wave from T_r is incident on C at about 45° where it interferes with the object wave to form the image-hologram. Transmitted energy is also absorbed by A. Liquid crystal convertor C has a usable area of 8 X 8 cm^2 and is formed by a 250-micron nematic liquid crystal in the homotropic structure between multi-layered walls [5]. Its acoustical amplitude transmission is over 90% for angles of incidence between 0° and 12°, and between 37° and 50°, at a frequency of 3.75 MHz.

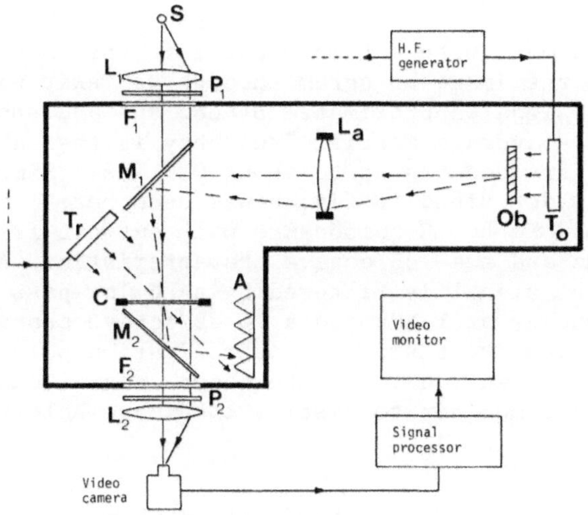

Fig. 7 Acoustical holography system using a liquid crystal convertor.

3.2 OPTICAL ELEMENTS

The optical sub-system provides a nearly cylindrical polarized beam of incoherent light passing through the tank and L.C.C. toward a detector such as a video camera. Light source S is a small light bulb (12 V, 50 W) at the focus of lens L_1. The polarizer axis is set at 45° over the horizontal plane. For better contrast, a color filter may be placed between S and L_1. The light beam enters through window F_1 and exits through F_2 where it is partially blocked by crossed polarizer P_2. The image of C formed by lens L_2 may then be seen at its focus.

3.3 ELECTRO-ACOUSTICAL ELEMENTS, PROCESSING AND DISPLAY

A video camera with entrance pupil at the focus of F_2 is used to view the image-hologram through the exit window and give a video signal approximated by eq. (6) and shown in fig. 5-c. The average carrier frequency is then about 2 MHz, modulated at frequencies up to about 0.9 MHz. Simple signal processing as described in fig. 6 was used here. High-pass filtering is done by LC componants with cut-off frequency at about 500 kHz and a 40 dB/octave characteristic. After diode detection, the signal is filtered by a LC low-pass filter with cut-off frequency at 1 MHz and a 40 dB/octave characteristic. The signal, with the restored synchronization pulses added, is then fed to a video monitor. The camera signal can also be sent directly to the monitor to display the image-hologram.

4. EXPERIMENTAL RESULTS

The two test objects used in this particular experiment are shown in fig. 8. The first one is an opening in form of a cross, with 20 mm-long arms, 5 mm wide, cut-out from a 0.45 mm-thick aluminum sheet. The second one is an acrylic plastic block, 6 mm-thick, with holes of 1.5 and 3 mm diameters drilled in its middle plane.

The image-holograms of the two objects are shown in fig. 9-a and 9-b, as photographed on the video screen. The acoustical lens was set to give a magnification of about 1.7. They clearly show that the directly observed acoustical image-hologram may be considered as a spatial carrier (the fringes) which intensity (or fringe contrast) is modulated by the intensity of the object acoustic wave. The estimated average acoustic intensity of the object wave was about 500 $\mu W/cm^2$. We have indications that this convertor can usefully respond to a few μW/cm^2. As such, it can therefore be directly interpreted without further processing. However, the overall contrast which is due to the non-linear

Fig. 8 Test objects.

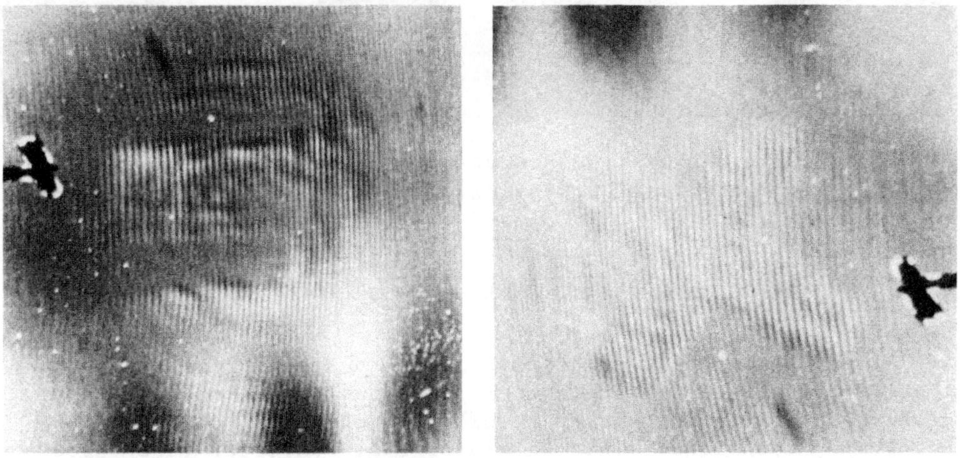

Fig. 9 Image holograms of the two test objects of Fig. 8.

response of the LCC is not very high: if the response were linear, the image would hardly be visible. Since the information on acoustic intensity distribution is carried by the fringes contrast, amplitude detection processing gives a highly contrasted picture as shown in fig. 10-a and 10-b. The visible artifacts have two main causes: orientation defects in the liquid crystal producing white spots, and various kinds of noise produced by the preliminary model of signal processor. This should be considerably improved in a near future. Even so, these pictures should demonstrate the high resolution capability of the technique and its future possibilities. Right now, the theoretical limit of resolution is practically attained. Finally, these images were formed on the convertor in about 10 seconds, but the type of liquid crystal used and the actual operating temperature were not favorable: we expect to reach about 1 second.

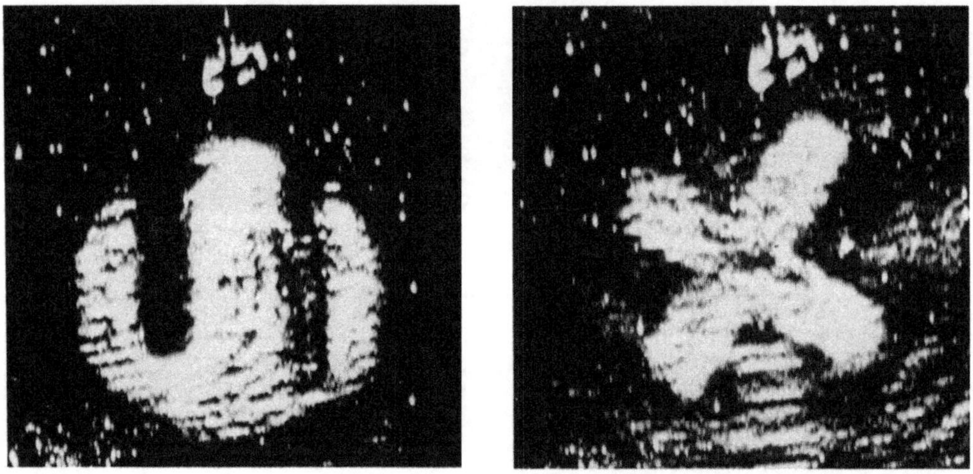

Fig. 10 Acoustical images of test objects after amplitude detection processing.

5. CONCLUSION

We have shown that an acoustically transparent nematic liquid crystal cell in the homotropic structure using a new acousto-optical effect makes possible an original acoustical holography technique. This liquid crystal convertor (LCC) needs only incoherent light for instant conversion of an acoustical hologram into a visible hologram which can be simply processed to give intensity images, with very high resolution and contrast. We have presented preliminary results which indicate the practicability of the present system, particularly in the field of non-destructive testing. The most immediate application is in aerospace construction for observation of various defects in layered panels, composite materials, aircraft wings, etc.

At the moment, we are developing an improved prototype of an ultrasonic camera using a LCC, for acoustic intensity imaging.

It will also incorporate features allowing phase imaging and interferometric acoustical holography.

As compared to a liquid surface levitation system, a LCC system needs no expensive laser, has the possibility of operating in any position and may have reasonable weight and dimensions permitting various types of field operation.

REFERENCES

[1] JEAN-LUC DION, "Un nouvel effet des ultrasons sur l'orientation d'un cristal liquide", C.R. Acad. Sc. Paris, t. 284, série B (1977), 219-222.

[2] J.L. DION, "Anisotropie ultrasonore et réorientation moléculaire dans un nématique: nouvelle confirmation", C.R. Acad. Sc. Paris, t. 286, série B (1978), 383-385.

[3] J.L. DION and A.D. JACOB, "A new hypothesis on ultrasonic interaction with a nematic liquid crystal", Appl. Phys. Lett., 31 (1977), 490-493.

[4] JEAN-LUC DION, "The orienting action of ultrasound on liquid crystals related to the theorem of minimum entropy production" J. Appl. Phys., 50 (1979), 2965-2966.

[5] J.L. DION, R. SIMARD, A.D. JACOB and A. LEBLANC, "The acousto-optical effect in liquid crystals due to anisotropic attenuation: new developments and applications", IEEE Ultrasonics Symposium Proc., sept. 1979, IEEE cat. No. CH1482-9/79, p. 56-60.

[6] O.A. KAPUSTINA and V.N. LUPANOV, "Experimental investigation of a liquid crystal acousto-optical image convertor", Soviet Phys. Acoust. vol. 23, No. 3 (1977), 218-221.

[7] S. LETCHER, J. LEBRUN and S. CANDAU, "Acousto-optic effect in nematic liquid crystals", J. Acoust. Soc. Am., vol. 63, No. 1 (1978), 55-59.

[8] H.G. BRÜCHMÜLLER, "Die Flüssigkristallzelle als akusto-optischer Wandler", Acustica, v. 40 (1978), 155-166.

[9] A.E. LORD Jr. and M.M. LABES, "Anisotropic ultrasonic properties of a nematic liquid crystal", Phys. Rev. Lett., v. 25 (1970).

[10] F. JÄHNIG, "Dispersion and absorption of sound in nematic", Z. Physik, v. 258 (1973), 199-208.

[11] P. GREGUSS, "A new liquid crystal acoustical-to-optical display", Acustica, vol. 29 (1973), 52-58.

[12] M.D. FOX, W.F. RANSON, J.R. GRIFFIN and R.H. PETTEY, "Theory of Acoustic holographic interferometry", in Acoustical Holography, vol. 5, (Plenum Press, New York, 1974).

[13] K. SUZUKI and B.P. HILDEBRAND, "Holographic interferometry with acoustic waves", in Acoustical Holography, vol. 6, (Plenum Press, New York, 1975).

[14] J. PASTEUR and Y. SEIZERIAT, "Holographie acoustique; application au traitement optique de l'information acoustique", Optica Acta, 24 (1977), 859-875.

[15] B.B. BRENDEN and G.L. FITZGERALD, "Acoustical Holography and Imaging" in Acoustical Imaging and Holography, 1 (1978), 1-48.

COMPUTERIZED RECONSTRUCTION OF ULTRASONIC FIELDS BY MEANS OF ELECTROSTATIC TRANSDUCER ARRAY

Bernard Hosten and José Roux

Laboratoire de Mécanique Physique, E.R.A. CNRS N° 769
Université de Bordeaux I
351 coursde la Libération 33405 TALENCE CEDEX (France)

0) Introduction

Thanks to the development of many kinds of arrays, (mechanical sweeping [1, 2], piezoelectric [3, 4] or electrostatic [5] transducers arrays), the ultrasonic fields can be sampled with a more and more large apertures and a smaller and smaller sample interval. So the angular spectrum algorithm is becoming perfectly adapted to the field's numerical reconstruction.

In the first section, it will be given a review of the theory's foundation from the Green's theorem [6, 7] and the validity's hypothesis will be specified. We show how the use of a square wave guide reduces the effects of finite aperture and artificial periodisation of the field by spectrum sampling.

In the second section, we describe an acquisition process in real time from an electrostatic array of 256 x 256 transducers.

Finally, we present some images of acquired fields and reconstructed fields in the transducer plane or object plane.

I. Theory

a) Angular spectrum algorithm

Angular spectrum algorithm can be directly derived from Green's theorem parallel planes.

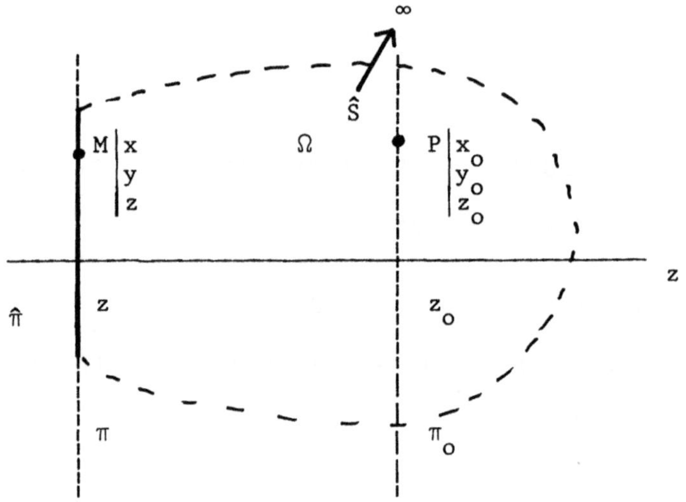

Figure 1.

Let us denote $\phi(M)$ the velocity potential, regular solution of Helmholtz's homogeneous equation $(\Delta + k^2)\,\phi = 0$, $k = \omega/c$, in a regular domaine Ω, without sources. π and π_o are two planes with a common perpendicular $\vec{z}$. $\Sigma = \hat{\pi} \cup \hat{S}$ is the frontier of Ω, where $\hat{\pi}$ in the part of plane for z constant, on which a regular surface $\hat{S} \to \infty$ leans.

From the Green's theorem :

$$\phi(P) = \iint_\Sigma \left[\phi(Q)\cdot\frac{\partial}{\partial n_Q} G(P-Q) - G(P-Q)\cdot\frac{\partial}{\partial n_Q}\phi(Q) \right] d\Sigma(Q)$$

with $G(P-Q) = -\dfrac{\exp i\,k|\vec{PQ}|}{4\pi\,|\vec{PQ}|}$: Green's function of the unlimited space.

Hypothesis : 1) It does not exist source in the half-space right side of π.

2) The Sommerfeld's conditions of regularity to infinity (when $S \to \infty$) are verified.

Thus :

$$\phi(P) = \iint_\pi \left[\phi(M)\,\frac{\partial}{\partial n_M} G(P-M) - G(P-M)\,.\,\frac{\partial}{\partial n_M}\phi(M) \right] dS(M)$$

with : $\dfrac{\partial}{\partial n_M} = -\dfrac{\partial}{\partial z}$, $dS(M) = dx\,dy$ and

$$g(z_o - z) = -1/4\,\pi \left[x_o^2 + y_o^2 + (z_o - z)^2 \right]^{-\frac{1}{2}} \exp ik \left[x_o^2 + y_o^2 + (z_o - z)^2 \right]^{\frac{1}{2}}$$

We obtain : $\phi/\pi_o = g(z_o - z) * \frac{\partial\phi}{\partial z}\Big|_\pi - \phi/\pi * \frac{\partial g}{\partial z}(z_o - z)$

* is the convolution product on x and y.

Let us denote : $\mathcal{F}f(x,y,z) = F(u,v,z)$, the Fourier transform of f in x and y. The operators $\mathcal{F}$ and $\partial/\partial z$ can be commuted.

In the Fourier space, the previous equation becomes :

$$\mathcal{F}\phi/\pi_o = \mathcal{F}g.\frac{\partial}{\partial z}\mathcal{F}\phi/\pi - \mathcal{F}\phi/\pi.\frac{\partial}{\partial z}\mathcal{F}g$$

By using the development in plane waves [7] of :

$$\frac{1}{r}\exp ikr = \frac{ik}{2\pi}\iint_{-\infty}^{+\infty}\frac{1}{p}\exp ik\left[mx + ny + p\,|z_o - z|\right]dm\,dn$$

where : $r = \left[x^2 + y^2 + (z_o - z)^2\right]^{\frac{1}{2}}$ and $p = \left[1 - (m^2 + n^2)\right]^{\frac{1}{2}}$

with $u = m/\lambda$, $v = n/\lambda$, $q = kp$

we obtain : $g(z_o - z) = -\frac{i}{2}\iint_{-\infty}^{+\infty}\frac{1}{q}\exp iq|z_o - z| . \exp 2i\,(ux_o + vy_o)\,du\,dv$

This relation defines by its inverse the Fourier transform of g :

$$\mathcal{F}g = \frac{-i}{2q}\exp iq\,|z_o - z|$$

$$\mathcal{F}\frac{\partial g}{\partial z} = \frac{\partial}{\partial z}\mathcal{F}g = \begin{cases} -\frac{1}{2}\exp iq|z_o - z| & \text{if } z_o > z \\ \frac{1}{2}\exp iq|z_o - z| & \text{if } z_o < z \end{cases}$$

For a backward propagation : $z_o > z$

$$\mathcal{F}\phi/\pi_o = \frac{1}{2}\left[\mathcal{F}\phi/\pi - \frac{i}{q}\frac{\partial}{\partial z}\mathcal{F}\phi/\pi\right]\exp iq|z_o - z|$$

If we take the derivative according to z_o, we obtain :

$$\frac{\partial}{\partial z_o}\mathcal{F}\phi/\pi_o = iq\,\mathcal{F}\phi/\pi_o \;\forall\; z_o$$

This property is also true for z, because the choice of z_o is arbitrary :

$$F\phi/\pi = F\phi/\pi_o \quad \exp - iqd \qquad \text{with } d = |z_o - z|$$

angular spectrum algorithm is resumed by :

$$\phi/\pi = F^{-1}_{uv} \; \{ H^{-1}(u,v,d) . F\phi/\pi_o \}$$

with $H(u,v,d) = \exp iqd$

b) Finite aperture and sample effects

The sample of data ϕ/π_o makes its spectrum $\psi(u,v,z_o) = F\phi/\pi_o$ periodical. But if the sample interval of ϕ/π_o is less than $\lambda/2$, the Shannon's theorem is respected and the result is not affected by the limitation in Fourier space.

In the other hand, the limitation in the π_o plane with a L x L square window leads to an important lost of information. This limitation corresponds to the spectrum sampling with an interval 1/L.

$F\phi/\pi_o$ is replaced by the distribution :

$$\hat{\psi}(u,v,z_o) = \psi(u,v,z_o) . \sum_n \sum_m \delta(Lu-n) . \delta(Lv-m)$$

The result of angular spectrum algorithm is given by :

$$F^{-1} \hat{\psi}(u,v,z) = \hat{\phi}/\pi = \phi/\pi * \sum_n \sum_m \delta(\frac{x}{L} - n) . \delta(\frac{y}{L} - m)$$

This periodisation of result has a null effect when ϕ/π and ϕ/π_o are bounded support (L x L) or periodical (L x L) fonctions.

That can be experimentally realized by propagation in a wave guide with a (L x L) square section.

The regularity conditions at infinity are replaced by the homogeneous Neumann's conditions $\frac{\partial\phi}{\partial n} = 0$ on the guide's walls. The innerfield in each cross-section is:

$$\hat{\phi}/\pi = \phi/\pi * \sum_n \sum_m \delta(\frac{x}{L} - n) . \delta(\frac{y}{L} - m)$$

where ϕ/π is the free space solution.

The Green's fonction of the problem is :

$$G = -\frac{1}{4\pi r} \exp ikr * \sum_n \sum_m \delta(x-nL) \cdot \delta(y-mL)$$

The angular spectrum algorithm is thus becoming :

$$\hat{\phi}/\pi = FFT^{-1}\{ \hat{H}^{-1}(u,v,d).FFT\ \hat{\phi}/\pi_o \}$$

where FFT (FastFourier transform) replaces *F*.

Thus the guide's use avoids the lost of information caused by a finite aperture. The only error which subsists is the limitation of the samples'number. A guide the cross-section of which is a L x L square and a N x N samples are equivalent to a NL x NL aperture in the free field.

II. Data acquisition

This algorithm, first improved with success on ultrasonic aerian fields, [9] treats now some data from an electrostatic array.

This array [5] is simply built by setting a dielectric paper sheet between two printed circuits which have 256 parallel metalized strips of 0.7 mm width, 1 mm spaced, and are placed orthogonally (fig. 2). The front one is built on a plastic sheet which is moved by the acoustic radiation, and is applied against the rigid back plate by means of a partial vacuum.

Polarizations of one row among the 256 ones enables the 256 corresponding transducers. Unfortunately, each of ones is charged by the 255 others condensators of its column. To reduce this lost of signal, we have placed, on each column, a current-voltage converter realized with an operationnel amplifier to obtain a very low input impedance.

The signal, issued from one converter,is led to a pair of synchronous 90° shifted detections, which are acting during the 125 µs burst of the ultrasonic signal at 1 Mc/s.

The 512 tracks come into an analog multiplexer from 2 x 256 to 2, driven by a 2 Mc/s clock. Two A/N converters supply the numerical values with four bits, in less than 200 ns. Four bits is sufficient at the present time because the dynamical range is not better than 30 db.

The 64 K data bytes are stored in a dynamical memory. The circulation of its address allows to visualise, with a cycle duration of 1/25s, its content on a TV monitor which shows the real or imaginary part of ultrasonic field's image in the acquisition plane.

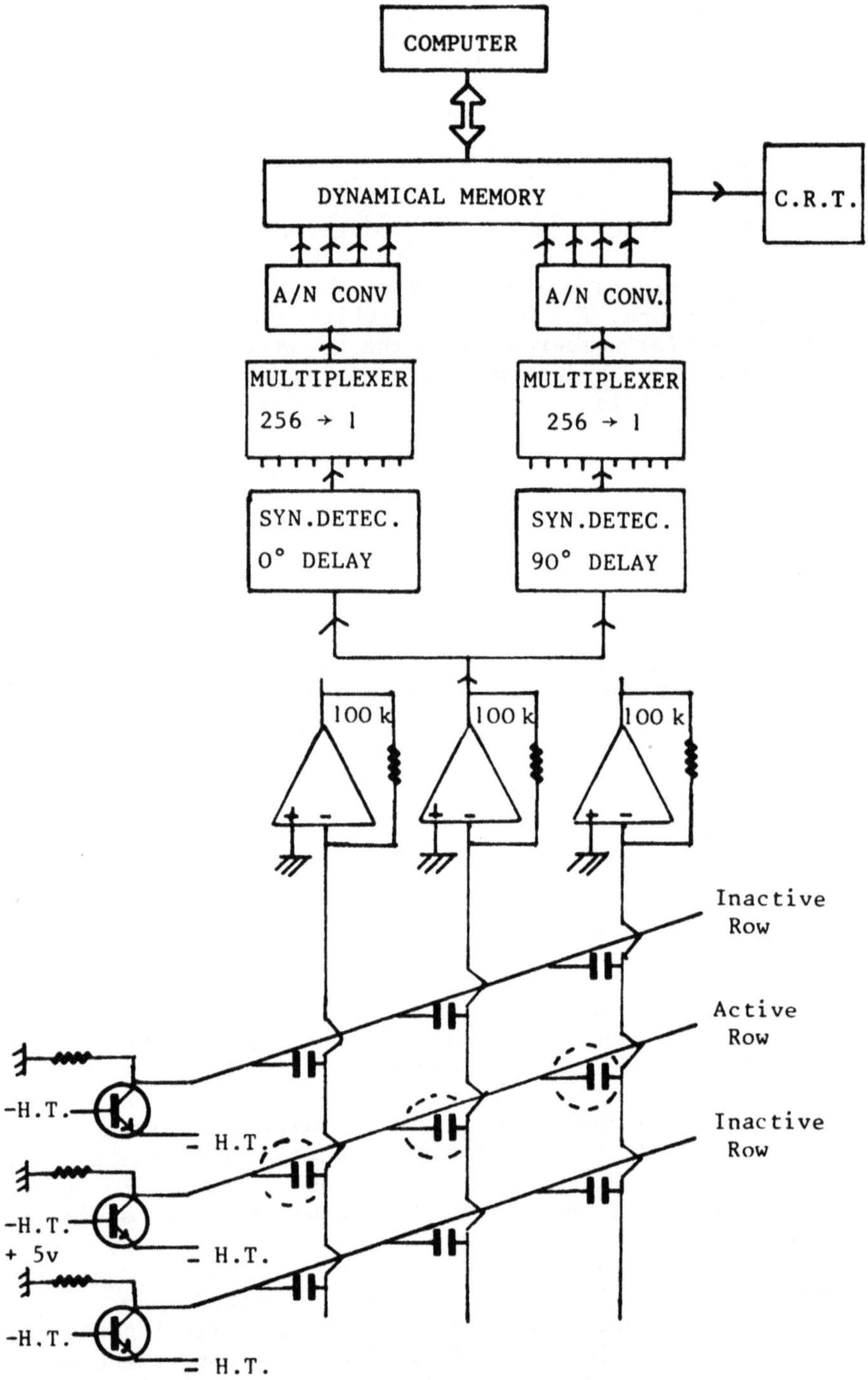
COMPUTER
DYNAMICAL MEMORY
C.R.T.
A/N CONV
A/N CONV.
MULTIPLEXER
256 → 1
MULTIPLEXER
256 → 1
SYN.DETEC.
0° DELAY
SYN.DETEC.
90° DELAY
100 k
100 k
100 k
Inactive Row
Active Row
Inactive Row
-H.T.
H.T.
-H.T.
+ 5v
H.T.
-H.T.
H.T.

Figure 2.

Simultaneously, the input of a row is achieved in three steps of 125 μs each : emission, polarization-synchronous detection, multiplexing. Thus the writing cycle duration is about 1/11 s.

A desk-computer allows data's recording, some easy numerical treatment and the data's exchange with an another big computer with which we reconstruct the field in an another plane.

III. Results

The results are the images of acquired or reconstructed fields with free space propagation. The system is just finished and we have not yet checked the wave guide propagation. For the same reason, the emettor is a simple ceramic, with a diameter of 50 mm, which does not fit to the array's large aperture. The emission frequency is about 1 Mc/s.

On the ceramic's field (real part fig. 3, imaginary part fig. 4), we have applied the angular spectrum algorithm. Fig. 5 shows the field's amplitude in the plane source which is located at 730 mm from the array. The ceramic's outline is perfectly visible when we compute the phase : fig. 6.

We have interposed a R letter made of common plastic, the largest lateral dimension of which is 20 mm, 3 mm thickness. The field which is obtained on the array is shown, fig. 7 (imaginary part). The backward propagation in the R plane 230 mm far from the array gives the fig. 8 (imaginary part).

IV. Conclusion

These first results show that this system owns a good lateral resolution (Fig. 8). The most important noise's causes are the large differences of sensibility between the transducers and the remanent polarization of the dielectric. This two defaults made some vertical strips (Fig. 3, 4).

Digital enhancement of images will now be performed by appropriate computation sequences to make suitable corrections sensibility's variations of the transducers and their residual crossmodulation.

Shortly, we will use a square wave guide to multiply the aperture and superpose multi-frequency acquisitions to increase the range revolution.

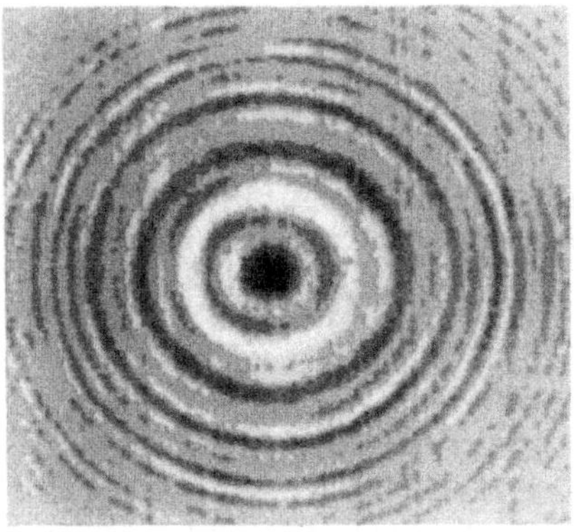

Fig. 3: Acquisition Plane
Real Part

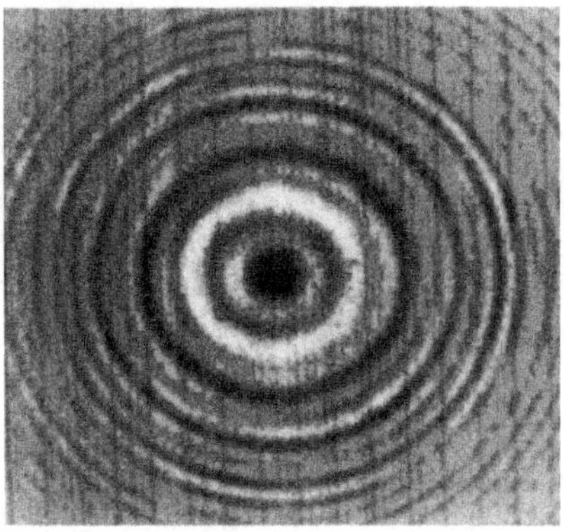

Fig. 4: Acquisition Plane
Imaginary Part

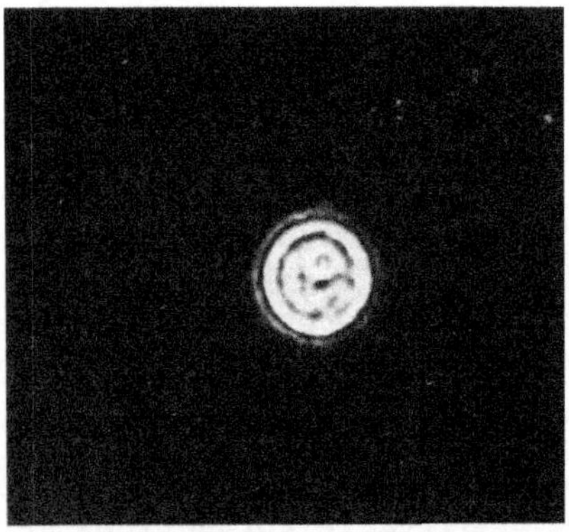

Fig. 5: Source Plane
Amplitude

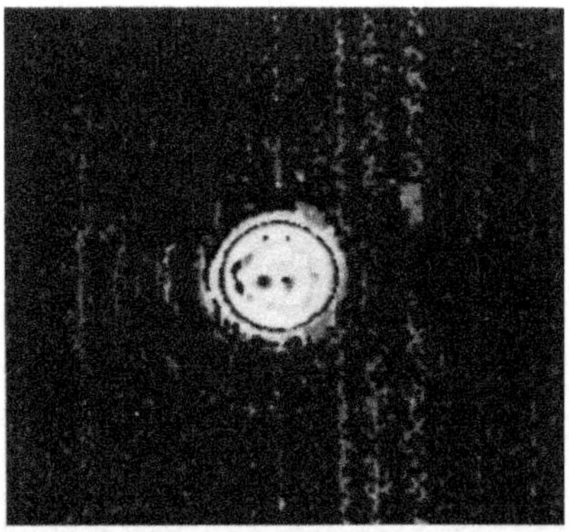

Fig. 6: Source Plane
Phase

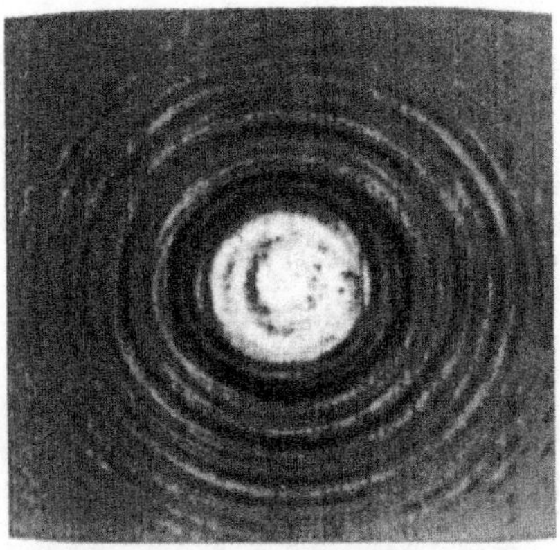

Fig. 7: Acquisition Plane
Imaginary Part

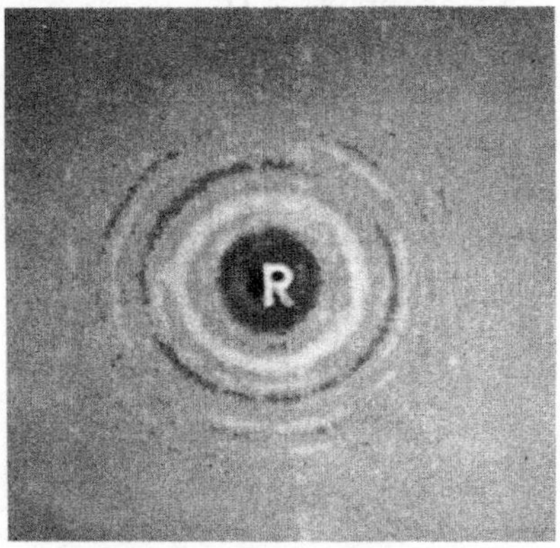

Fig. 8: Object Plane
Phase

REFERENCES

1. Y. AOKI - Acoustical Holography, vol. 5 (Plenum Press, 1974) p. 551.

2. B. HOSTEN et J. ROUX - Acustica, 40, n° 4, 1978.

3. M.D. EATON, R.D. MELEN and J.D. MEINDL - Acustical Holography, vol. 8 (Plenum Press, 1980) p. 55

4. C. DAS, G.A. WHITE, B.K. SINHA, C. LANZL, H.F. TIERSTEN and J.F. Mc DONALD - Acoustical Holography, vol. 8 (Plenum Press, 1980) p. 119

5. P. ALAIS - Acoustical Holography, vol. 5 (Plenum Press, 1974) p. 671.

6. J.W. GOODMAN - Introduction to Fourier Opties, New York, Mc Graw-Hill, 1968.

7. B. HOSTEN et J. ROUX - Comptes-Rendus, t. 285, Série B, 1977, p. 317.

8. J.R. SHERWELL and E. WOLF - J.A.S.A., 58, n° 12, 1968, p. 1596.

9. B. HOSTEN - Revue du CETHEDEC, 16e année, 3ème trimestre 1979, numéro 60.

NEW PROCESSING TECHNIQUES IN ULTRASOUND IMAGING SYSTEMS

G.F.Manes, C.Susini, P.Tortoli, and C.Atzeni

IROE, Consiglio Nazionale delle Ricerche, Firenze
Istituto di Elettronica, Università di Firenze
via Panciatichi, 64, 50127 Firenze, Italy

ABSTRACT

Novel techniques are described that allow echo signals from ultrasound imaging systems to be processed at a reduced bandwidth. The basic principle consists in processing the complex envelope of the signal, independently of the carrier frequency. However, rather than at baseband, requiring phase-quadrature channels, processing is performed at i.f. by a single channel.

Even though various techniques are possible, surface-acoustic wave (SAW) transversal structures have been employed for the first time in a number of real-time processors.

Programmable delay-lines are described, exhibiting more than 50 dB dynamic range. Based on independent processing of echo carrier and envelope, a phased-array configuration requiring one single delay-line for all the array channels is introduced.

An Hilbert transformer allowing the exact echo envelope to be obtained is demonstrated, and the design of a real-time programmable inverse filter is described.

INTRODUCTION

Electronic sector scanning in ultrasound phased-array echographic equipments is today a well-established tool for real-time medical imaging, especially for dynamic visualisation of very rapidly moving structures, such as the heart [1,2]. A noticeable effort has been directed towards the investigation of variable delay lines, that can perform the key function of equalising the time-delays of echo pulses received at the array elements from different spatial resolution cells. The currently employed systems make use of one delay-line for each array channel, capable of providing the required delays in a discrete quasi-continuous manner. The delay increment is dimensioned with reference to a suitable fraction of the cycle corresponding to the operation frequency, so as to establish the phase coherence needed to ensure an adequate constructive interference of the echo pulses to be summed. Each delay line must typically provide hundreds of delay steps to cover the overall delay to be compensated, which depends on the array aperture and on the extent of the scanned sector.

The complexity of the delay system is due to the fact that its design depends on the value of the operation frequency, even though the pulse energy is typically concentrated in a narrow band around this frequency. The alternative techniques developed in this paper are based on the concept of providing separate and independent processing of the carrier oscillation and of the envelope of the echo pulses. Once the carrier has been processed, the bandwidth to be handled depends no longer on the carrier frequency, but merely on the bandwidth of the pulse envelope. This feature yields to the design of an extremely simplified delay system as well as of other important processing filters. Reduced-bandwidth processing can be performed using a number of different configurations and devices. We describe here the use of surface-acoustic-wave (SAW) devices, especially designed for this purpose.

THE NEW PROCESSING TECHNIQUES

As well known, in a phased-array system the signals received at the array elements from a reflecting target exhibit a mutual delay depending on the target spatial position and on the element separation. The reconstruction of the image requires the equalisation of this delay along the array channels for each resolution cell, so as to produce constructive summation of all element contributions.

In CW systems delay equalisation is accomplished merely by properly changing the phase of each signal, as the echo is simultanrously present at all the N array elements, the maximum delay being negligible in front of the signal length. It is therefore sufficient to control N variable phase-shifters, one for each array channel, in order to establish the phase coherence needed for a constructive summation. Phase shifts are quantised to multiples of an increment

$$\Delta\phi = \frac{2\pi}{M} \qquad (1)$$

where M is high enough that the approximation of the quantised shifts to the exact values ensures an adequate constructive interference of summed signals. Values of M as low as 10 satisfy this requirement.

Conversely, in pulsed ultrasound phased-arrays the spatial length of the signal in the propagation medium is small in comparison with the array aperture. As a consequence, the mutual delay of pulses received at different array elements is no longer negligible with respect to the pulse length. Constructive summation thus requires that, in addition to phase equalisation, simultaneity of pulses along the array is restored. Both phase- and group-delay are usually changed using delay-lines with variable discrete delay: analogue lumped-constant delay-lines are most commonly employed, although different solutions have been investigated [3]. The maximum delay to be compensated, usually of the order of some microseconds, must be covered in multiples of the small delay increment

$$\Delta\tau = \frac{1}{Mf_o} \quad \text{(with } f_o \text{ the carrier frequency)} \qquad (2)$$

corresponding to the phase-shift increment (1), which is typically of the order of tens of nanoseconds. Delay lines must therefore provide a large number of delay steps, which increases with increasing carrier frequency.

As an example, an array of 25 mm aperture for diagnostic cardiological application requires a maximum delay of about 10 usec to scan a sector of 80°. At 2.5 Mhz operation frequency, a delay increment of the order of 40 nsec is required, thus yielding 250 delay steps; at 5 MHz frequency, an equivalent delay quantisation yields to a doubled delay step number. As an array contains typically 16 - 32 elements, a tremendous complexity of control to rapidly switch among the different delay configurations ensues.

The alternative technique introduced by the authors [4,5,6] is based on the separation of signal processing into the two independent operations of carrier phase equalisation and group delay equalisation. The needed phase shift is communicated first to the echo signals, like in CW arrays, while the simultaneity of the pulses is successively restored by changing their group delay.

Let the echo pulse be represented by

$$s(t) = a(t) \cos (2\pi f_0 t + \theta) \tag{3}$$

where f_0 is the carrier frequency, $a(t)$ the pulse envelope and θ a phase angle. The pulse is previously phase-shifted by the amount

$$\phi (\tau) = 2\pi f_0 \tau \qquad (\text{ modulo } 2\pi) \tag{4}$$

where τ is the desired delay, yielding

$$s'(t) = a(t) \cos\{ 2\pi f_0 (t - \tau) + \theta \} \tag{5}$$

Then, a delay line is used to introduce a group delay τ , shifting the pulse envelope $a(t)$ without altering the preprocessed carrier, yielding the required delayed pulse

$$s''(t) = a(t-\tau) \cos\{ 2\pi f_0 (t - \tau) + \theta \} = \\ = s(t - \tau) \tag{6}$$

According to the phase quantisation step (1), carrier phasing requires programmable selection among a set of M incremental values covering a 2π shift, which can be impressed to the signal using a number of electronic techniques. Once phase coherence of signals has been established, the increment Δt of the delay line required for time shifting the signal envelope is no longer dependent on the carrier frequency value f_0, but on the extent of the signal bandwidth around f_0. As the fractional bandwidth B/f_0 of ultrasound pulses is generally low, due to the high Q of piezoelectric transducers, a noticeable reduction of the bandwidth to be processed ensues. As a consequence, the delay increment can be noticeably larger than that required by the previously described approach.

The extent of the required group-delay quantisation was evaluated on the basis of computer simulation of the receiver directivity pattern resulting for different values of the ratio α between the quantisation step Δt and the pulse width T, i.e.

$$\alpha = \frac{\Delta T}{T} \tag{7}$$

The pulse was assumed of gaussian envelope, and T conventionally corresponded to the gaussian width between its e^{-1} points.
A 16 element array was assumed, with element separation equal to 1.5 wavelength for a propagation velocity in the medium of 1,500 m/sec and a 2 MHz operation frequency.

Computation of the directivity pattern was performed for a number of focalisation points at various distances d from the array and various angles Θ over a scanning sector of $\pm 40^{o}$ with respect to the normal to the array. For each case,two complementary graphic representations have been plotted. The first one is a 3-dimensional pattern G(t,r), representing the envelope of the signal resulting from summation of the pulses received at the 16 channels, after delay quantisation, as a function of time t and r, the distance from the focal point along lines parallel to the array. The second one is a plane section of the above 3-d plot, using the density of plotted lines as a measure of the pattern amplitude [7].

The inspection of the results shows, as expected, a degradation of the array directivity properties when increasing the quantisation step, evidenced by a reduction of the pattern peak-to-sidelobe ratio and a widening of the mainlobe. The peak attenuation, moreover, increases with the scanning angle. With respect to the ideal value, the peak attenuation is contained within 10 % for $\alpha = 0.3$, 20 % for $\alpha = 0.5$, 40 % for $\alpha = 1$.

A reasonable compromise can be assumed the choice $\alpha = 0.4$, for which the dispersion of peak amplitude is lower than 20%.Two examples of the receiver directivity pattern G(t,r) obtained for this value of α are shown in fig. 1 b and 2 b for focus at d = 70 mm, $\Theta = 0$ and d = 70 mm, $\Theta = 30^{o}$ respectively. These must be compared with the plots in fig. 1 a and 2 a, showing, for the two cases,the patterns resulting from an exact delay compensation.The comparison shows that, apart from the expected peak reduction and a small mainlobe widening, no appreciable change of the directivity properties results.

Assuming 1 μsec as a realistic pulse width, the above choice of the quantisation step means that a 400 nsec delay increment is sufficient to provide an adequate group-delay compensation. With reference to the example reported above, the required 10 μsec overall delay can be now covered in 25 steps, rather than 250, so drastically reducing the delay line requirements.

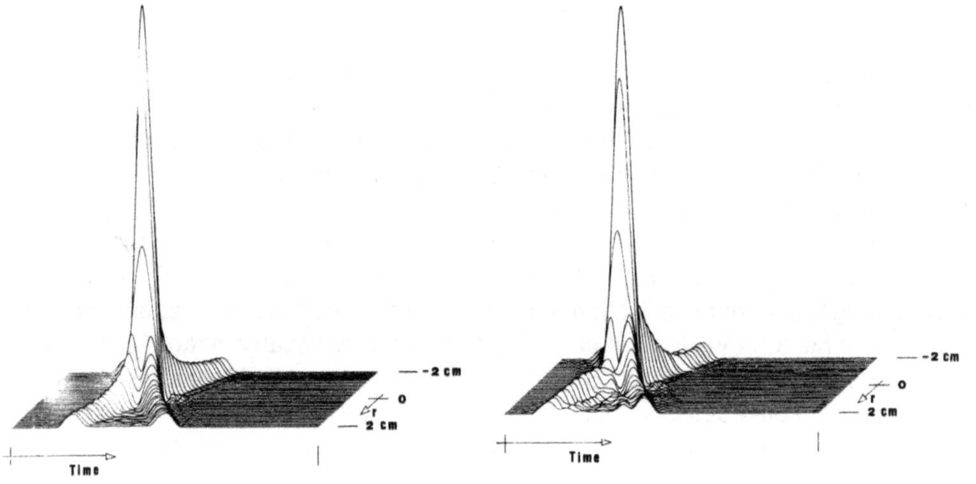

a b

Fig. 1 - Array directivity patterns with focus at d = 70 mm, Θ= 0, resulting from a) exact group-delay compensation b) group-delay quantisation with increment $\Delta t = 0.4$ T.

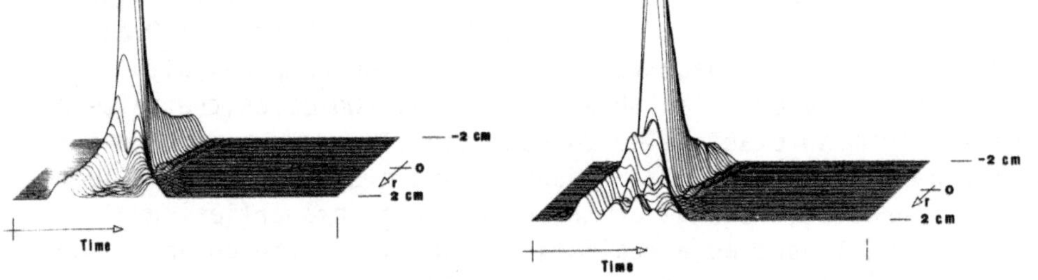

a b

Fig. 2 - Array directivity patterns with focus at d = 70 mm,Θ = 0, resulting from a) exact group-delay compensation b) group-delay quantisation with increment $\Delta t = 0.4$ T.

DESIGN OF THE DELAY SYSTEM

Even though various techniques are possible, carrier phasing was simply accomplished by mixing the signal with a coherent local oscillator, providing M digitally selected phase shifts in increments of $2\pi/M$. Mixing produces two sidebands, each retaining the desired information. Further processing can be accomplished at baseband, provided that the in-phase and quadrature components of the signal are independently handled. According to the classical architecture in fig.3, this requires mixing of signal (3) with a coherent reference oscillation of the same frequency f_o and with the same oscillation shifted by 90°, followed by low-pass filtering.

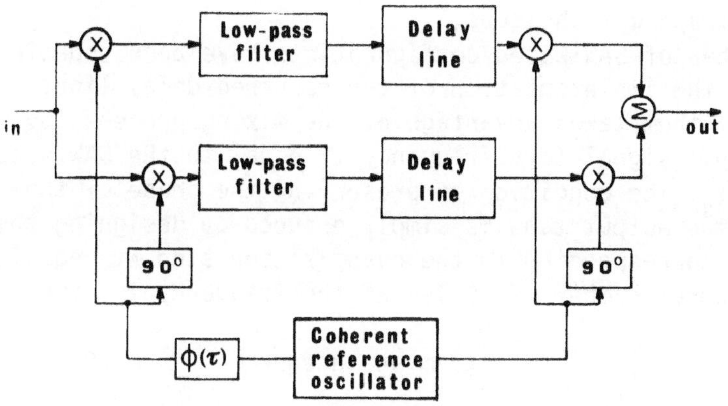

Fig. 3 - Architecture for signal baseband processing.

Each of the resulting quadrature components

$$a(t)\cos\left(\theta + k\frac{2\pi}{M}\right) \qquad (8)$$

$$a(t)\sin\left(\theta + k\frac{2\pi}{M}\right) \qquad k = 0,1,\ldots,M-1$$

centered at zero frequency, extends over a bandwith B/2. Group delay must be impressed to both components along the two channels, and this operation could be performed, for example, by digital techniques, taking advantage of the reduced bandwidth.

The need of two processing channels can be avoided, however, if processing is carried out on the bandpass component resulting from the mixing process. This single-channel configuration requires the selection of the information-carrying sideband around a suitable intermediate frequency f_i. A bandpass delay-line is now needed, that does not alter the phase of the preprocessed input signal.

An attractive solution for this purpose is the use of tapped surface-acoustic-wave (SAW) delay-lines [8]. SAW devices have a bandpass response centered at a 'synchronous' frequency f_s typically above ten MHz. They cannot therefore directly coupled to signals from ultrasound arrays. On the other hand, they possess a number of desirable properties, such as a compact size (10 μsec delay requires approximately 35 mm-length piezoelectric substrate), large bandwidth, high dynamic range, reproducibility, low cost. Tapping is easily implemented by metal interdigital electrodes, directly deposited on the plane surface of a piezoelectric crystal, such as $LiNbO_3$, using photolithographic techniques.

A number of SAW-based configurations have been especially developed for the implementation of the required delay line.

One of them takes advantage of the mixing process, by upconverting the input signal to a frequency f_i equal to the SAW synchronous frequency f_s. The condition of preserving the phase of the input signal at the output taps is simply ensured by designing the intertap delay, corresponding to the quantisation step Δt, equal to a proper integral number of cycles at the frequency f_s, i.e.

$$\Delta t = \frac{m}{f_s} \qquad \text{with m an integer} \qquad (9)$$

In a prototype system based on this technique, a SAW delay line, fabricated on a highly polished $LiNbO_3$ substrate, was designed with 400 nsec intertap. As the SAW synchronous frequency was 17.5 MHz, the input pulse at 1.75 MHz was mixed with a 15.75 MHz reference oscillation. Discrete phase shifts multiple of $2\pi/14$ were selected by applying the reference oscillation to a digital delay line. Digital control of a data selector provided the reference oscillation having the required phase shift to be applied at the LO port of a double-balanced mixer. After mixing, the upper sideband was bandpass filtered and injected into the SAW delay line.

Special techniques, described elsewhere[4], were used in the design of the SAW device in order to minimise multiple-path reflections between the taps, cross-talk and feedthrough coupling signals and other undesirable effects that could reduce the dynamic range. The output signal retains the impressed delay information, but is translated in frequency with respect to the original pulse. The original format can be recovered, if desired, by mixing the output with the same 15.75 MHz coherent oscillator. In fig. 4 a the variable delay capability of the system is demonstrated. Details of the achieved dynamic range, of the order of 50 dB, are shown in fig. 4 b.

An alternative technique makes use of a novel SAW-based configuration, which allows baseband signals to be directly compatible with a SAW structure [9,10]. The interface is simply constituted by a dual-gate MOSFET, through which the incoming low-frequency signal is impressed as amplitude-modulation to an oscillation equal to the SAW synchronous frequency f_s. The original signal format is recovered at each output tap by a peak-detector circuit (fig. 5). In this manner the oscillation at f_s serves merely to enter the SAW delay line, but does not affect in any manner the transfer characteristic of the device. The device accept baseband signals up to a maximum frequency depending on its bandwidth, and outputs at each tap a delayed signal replica in the same format. The most important novelty of this technique is that the SAW processor results compatible with a large class of baseband-operating analogue and analogue/digital arithmetic units. This feature is at the basis of the real-time processors described in the next section. When this SAW device is used as delay line, the intertap delay Δt must meet the condition, analogous to (9),

$$\Delta t = \frac{m}{f_i} \qquad \text{with m an integer} \qquad (10)$$

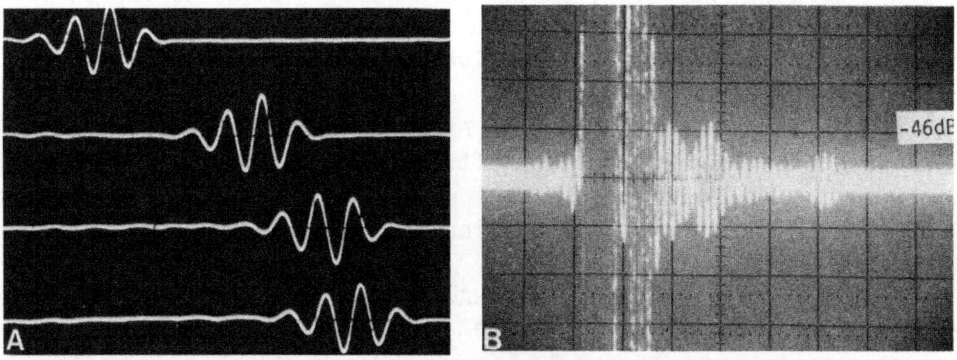

Fig. 4 - Variable delay experiments.
a - Delayed pulses (horiz. 1 μsec/div)
b - Dynamic range details (vert. 20 mV/div;horiz.2μsec/div)

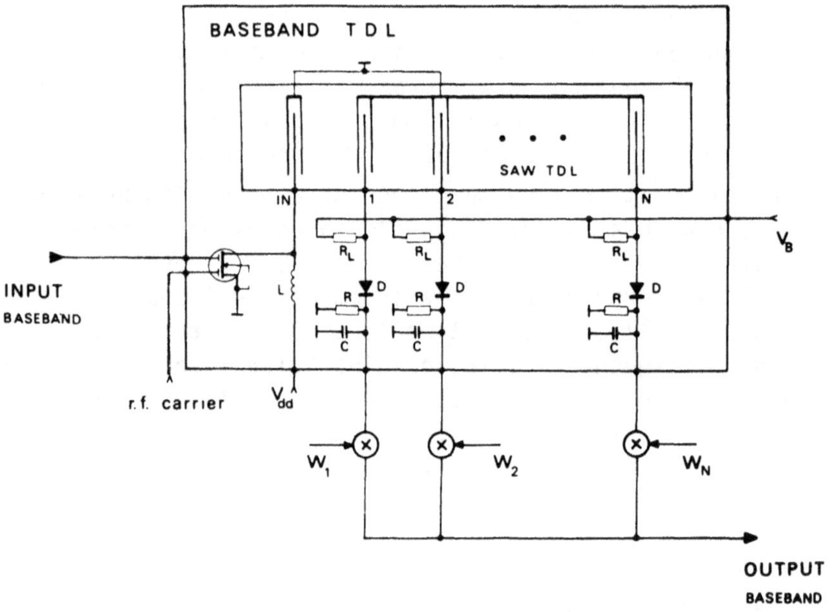

Fig. 5 - Schematic diagram of the SAW baseband processor. When used as delay line, external weights w_n are unitary and tap outputs are separated.

where now f_s is the frequency of the input signal. To take advantage of low-frequency operation, it is convenient f_s to be as low as possible, consistently with the condition that the down-converted signal band is contained within the positive frequency region. The limit value is therefore $f_s = B/2$. In the case that the desired value of f_s and the delay quantisation step Δt are not consistent with condition (10), the undesired phase shift due to an intertap delay other than (10) can be taken into account when programming the phase shifts to be impressed to the signal carrier in the mixing process, before entering the delay line.

Another important feature of SAW tapped delay-lines is that they can be used in a parallel-in serial-out (PISO) configuration. This feature, associated with the low-frequency operation and with the required low tap number, gives rise to the unique possibility of employing one single delay-line to process the signals from all the array channels, yielding a tremendous simplification with respect to previous systems. CMOS digitally controlled multiplexers can be used to address the signal from each array element to the appropriate tap of the SAW device, operated in a PISO configuration, ensuring high tap insulation and low cross-talk.

An experimental simulation of this system is shown in fig. 6. The four pulses in fig. 6 a (2 μsec/div) simulate echoes at 1.75 MHz from four array channels, which are adressed through a multiplexer to the appropriate taps of a baseband-operating 570 nsec-intertap SAW delay line. After delay compensation, they sum at the output tap, giving rise to the pulse shown in fig. 6 b (horiz. scale 2 μsec/div; vert. scale 100 mV/div). In fig. 6 c this pulse is shown in a 20 times expanded vertical scale (5 mV/div) to evidence the 50 dB dynamic range achieved.

The scheme of the complete simplified delay system is shown in fig. 7, where the signal from each array channel is preprocessed by mixing with a digitally phase-controlled reference oscillator and successively applied to the proper tap of a PISO operated SAW delay line through a multiplexer address matrix.

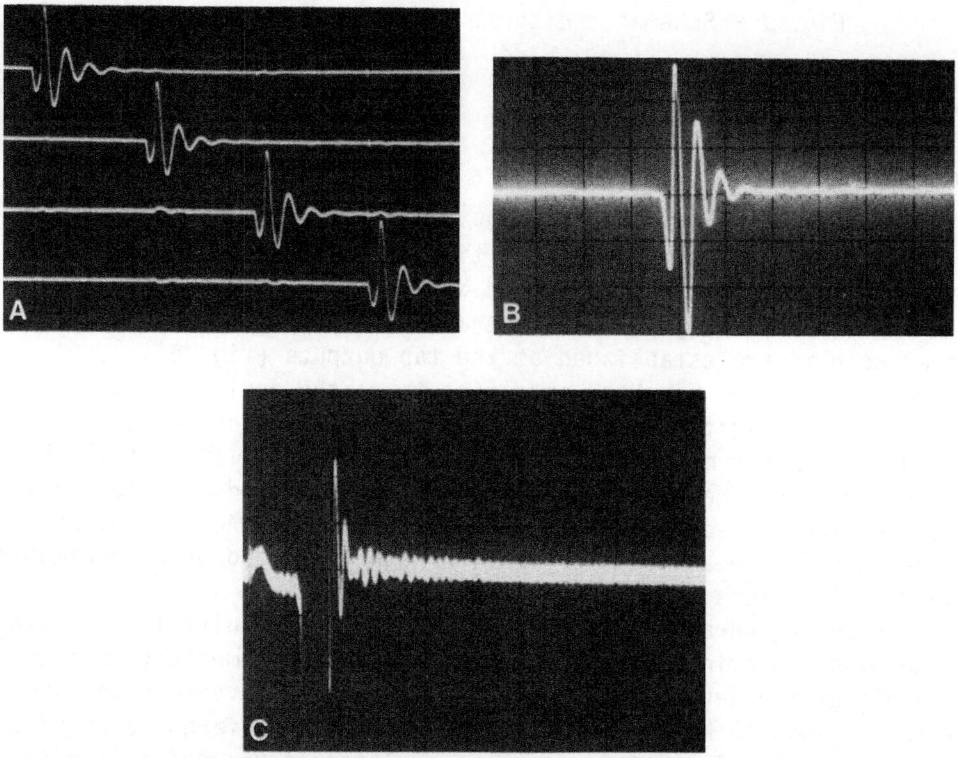

Fig. 6 - Experiment of echo delay equalisation using a SAW baseband delay-line. See text for explanation.

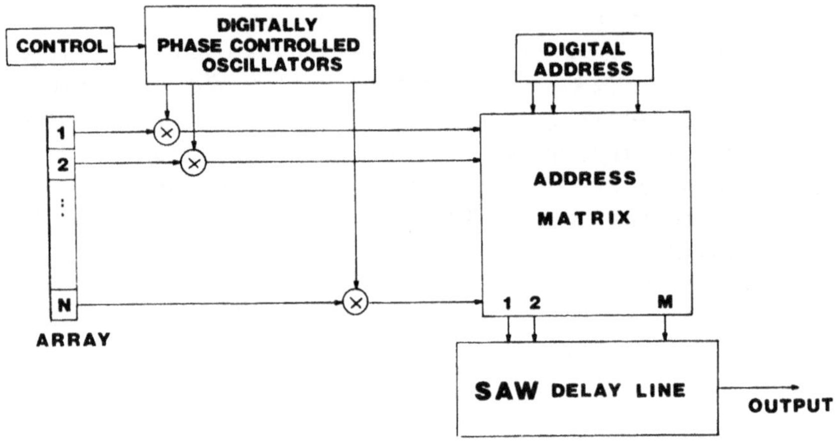

Fig. 7 - Schematic diagram of the delay system

REAL TIME PROCESSORS

The SAW baseband-compatible device described above offers a novel means of performing in real time a number of processing operations on echo signals. SAW tapped delay lines in fact represent a powerful means for implementing transversal filters, provided that proper weights are established at the tap outputs (fig. 5).
As mentioned above, baseband operation makes the device compatible with analogue/ digital arithmetic units. In particular, digital/ analogue converter multipliers can be used to digitally control the desired weights. The high data rate and high B.T product capability of SAW devices, associated with the flexibility of DAC multipliers, offers an unique means of implementing both fixed and/or programmable transversal filters.

The four-quadrant operation required for DAC multipliers could be achieved, in principle, by subtracting the complementary outputs, available at the device taps, through operational inverters and summators [11]. This technique, however,gives rise to different electrical paths for the complementary signals, resulting in unacceptable offset levels increasing with frequency in excess of a few hundreds KHz. An alternative configuration, which takes advantage of the SAW tapped

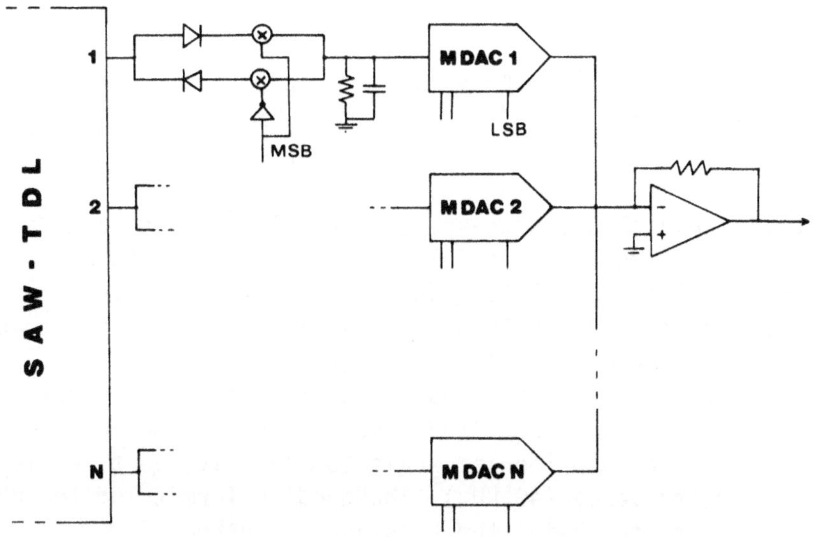

Fig. 8 - Configuration of digitally programmable weights.

delay-line capability of providing opposite polarity outputs, is illustrated in fig. 8. Two opposite polarity detectors are connected to individual tap outputs, thus making available two opposite polarity signals, which can be selected by analogue switches under control of the digital weighting coefficient MSB. After polarity selection, a two-quadrant configuration performs the absolute value multiplication. One single operational amplifier was employed for summing eight different DAC multiplier outputs, resulting in a feedthrough offset level of approximately -40 dB at 1.5 MHz and -34 dB at 3 MHz.

The Hilbert transformer

Hilbert transform [12], providing a 90^0 phase shift over the bandwidth of a signal, represent the ideal signal processing operation required to extract the signal envelope. For signals of very low fractional bandwidth, alternative processing techniques, such as full-wave linear rectification followed by low-pass filtering, produce a practically exact envelope. This is no longer true, however, when the fractional bandwidth is increased, as is the case of ultrasound pulses used in many medical diagnostics and nondestructive testing applications.

A prototype of an Hilbert transformer has been carried out using the described transversal filter structure. Weights to obtain the

Hilbert transform were established according to the sampled impulse response

$$h(n) = \frac{2}{\pi} \frac{\sin^2(\pi n/2)}{n} \qquad n \neq 0$$
$$= 0 \qquad n = 0 \qquad (11)$$

The SAW device had 15 570 nsec-spaced taps, corresponding to a sampling rate of 1.75 MHz. The useful bandwidth was therefore approximately 800 KHz. The central tap was used to obtain the in-phase replica of the input signal delayed by the same amount as the transform signal resulting from the Hilbert filter.

The transformer impulse response is shown in fig. 9 a (horiz. scale 2 µsec/div). The spectral amplitude over the useful 800 KHz bandwidth is displayed in fig. 9 b (horiz. scale 100 KHz/div; vert. scale 10 dB/div; center frequency 440 KHz). The good uniformity of the 90^{0} phase shift was tested using the Lissajous's method.
Extraction of the signal envelope requires squaring and summing of the delayed signal from the central tap and of its Hilbert transform, and taking the square root of the result. Particular care was required to mantain the phase quadrature relationship through the squaring circuits.

Fig. 10 shows from top to bottom a low fractional-bandwidth input pulse, its delayed replica from the central tap and its Hilbert transform. In fig. 10 b the pulse envelope obtained by peak detection (top) is compared with that obtained by Hilbert transform (bottom).

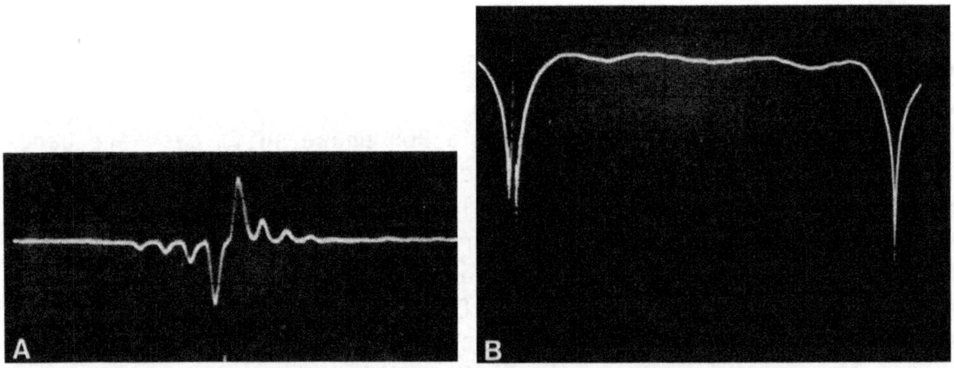

Fig. 9 - Impulse response (a) and spectral amplitude (b) of the SAW Hilbert transformer.

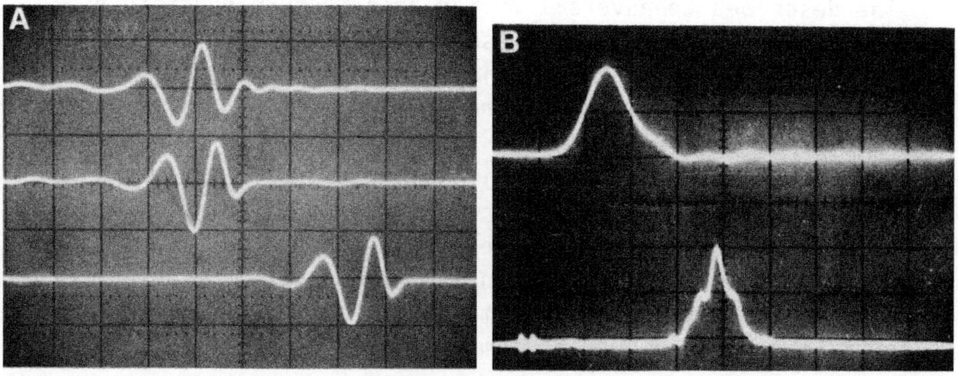

Fig. 10 - Experiments of Hilbert processing.
a - Top: input signal; center: delayed in-phase replica; bottom: Hilbert transform. (2 μsec/div)
b - Envelope detection by Hilbert processing (bottom) compared with peak-detection (top).

Fig. 11 shows two closely spaced pulses (top), whose envelope is extracted by peak detection (center) and by Hilbert processing (bottom). A clear improvement in resolution can be remarked.

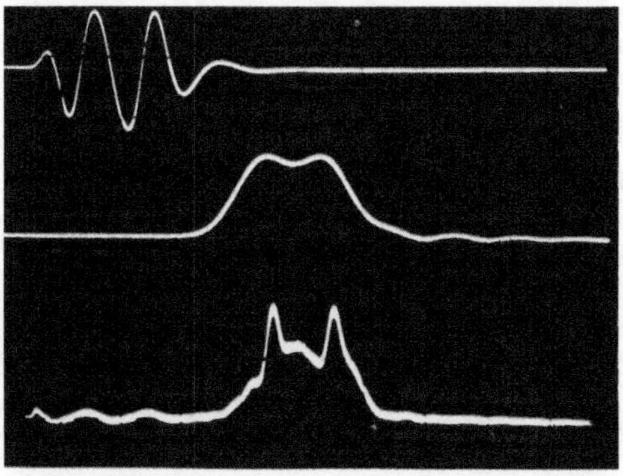

A OVERLAPPING ECHOES

B PEAK DETECTION

C HILBERT PROCESSING

? USEC / DIV

Fig. 11 - Envelope detection of closely spaced pulses.

Programmable inverse-filtering

The described transversal filter architecture constitutes a programmable processing unit, whose transfer function depends on the established weighting coefficients. A microprocessor can be used to rapidly alter the weights according to a desired filtering function.
An attractive capability is the implementation of a programmable inverse filter[13], for application to nondestructive testing and medical imaging. Work is in progress to implement a microprocessor controlled inverse filter based on the novel system. Other important applications are enviseaged in time-domain adaptive filters based on an iterative technique such as l.m.s. algorithm[14].

Acknowledgements

This work has been supported by italian National Research Council, Special Program on Biomedical Engineering.

References

1. J.C.Somer, Electronic sector scanning for ultrasonic diagnosis, Ultrasonics, 6: 153 (1968).
2. O.T.Von Ramm, and F.L.Thurstone, Cardiac imaging using a phased array ultrasound system, Circulation, LIII: 258 (1976).
3. R.E.McKeighen, and M.P.Buchin, New techniques for dinamically variable electronic delays for real time ultrasonic imaging, in "Proc. IEEE Ultrasonics Symposium", Cat 77CH 1264-1 SU : 250 (1977)
4. G.Manes, S.Gasperini, and C.Atzeni, A novel variable delay system for ultrasound beam steering/focusing, in "Septième colloque sur le traitement du signal et ses applications", Nice,103/1 (1979).
5. G.Manes, C.Atzeni, C.Susini, and J.C.Somer, A new delay technique with application to ultrasound phased-array imaging system, Ultrasonics, 17: 225 (1979).
6. G.Manes, C.Atzeni, and S.Gasperini, A single-channel reduced-bandwidth SAW-based processor for ultrasound imaging, in "Proc. IEEE ltrasonics Symposium", Cat 79CH 1482-9 SU: 179 (1979).

7. C.Susini, Focalisation properties of a phased-array ultrasound imaging system based on a new type of delay compensation, Atti della Fondazione G.Ronchi, 3: 335 (1980).
8. "Surface Wave Filters", H.Matthews, ed., J.Wiley & Sons, New York (1977).
9. G.Manes, and C.Atzeni, Baseband compatible SAW processors, Electron. Lett., 15: 661 (1979).
10. G.Manes, et al., Baseband compatible SAW processors: preliminary experiments, in "Proc. IEEE Ultrasonics Symposium", Cat 79CH 1482-9 SU : 748 (1979).
11. "Application guide to CMOS multiplying D/A converters", D.H.Sheingold, ed., Analog Devices, Inc. (1978).
12. A.V.Oppenheim, and R.W.Shafer, in "Digital signal processing," Prentice Hall, Englewood Cliffs, New Jersey (1975).
13. D.Behar, et al., Use of a programmable filter for inverse filtering, Electron. Lett., 16 : 88 (1980).
14. J.M.McCool, and B.Widrow, Principles and applications of adaptive filters, IEE Conf. Publ. 144 : 84 (1976).

MEASUREMENTS AND ANALYSIS OF SPECKLE IN ULTRASOUND B-SCANS

S.W. Smith[1/2], J.M. Sandrik[1], R.F. Wagner[1], and O.T. van Ramm[3]

[1]Div. of Electronic Pdts., Bureau of Radiological Health Food and Drug Admin., Rockville, MD 20857
[2]Dept. of Radiology, Duke University, Durham, NC 27706
[3]Depts. of Biomedical Engineering and Medicine Duke University, Durhan, NC 27706

I. Introduction

The issue of image texture in ultrasound B-scans of soft tissue organs has become an important topic. If B-scan texture is a good descriptor of tissue parenchyma, it can be used as the basis for clinical diagnosis. Several attempts have been made to relate B-scan image textures to disease states of soft tissue.[1,2,3] However, an alternative source of image texture is coherent acoustic interference, i.e. speckle, which is primarily a function of the imaging system rather than the tissue medium. Burckhardt[4] first presented a simple analysis of acoustic speckle assuming many scatterers within the system resolution cell. These targets scatter wavelets with random phases. Abbott and Thurstone[5] reviewed the difference between traditional laser speckle and broadband pulse echo acoustic speckle in images of diffuse structures. Bamber and Dickinson have developed a computerized tissue model[6] and have made statistical measurements of texture from simulated B-scans based on that model.[7] Flax et al. have also recently shown results from numerical modeling and measurements of B-scan image texture.[8]

The objective of the present research was to measure the first and second order statistical properties of ultrasound B-mode images which exhibited texture but no discernible structures. The images were obtained from several phantoms which mimic soft tissue with two commercial compound B-scanners used for normal clinical applications. Table 1 lists the transducer combinations and the scatterer size and concentration of tissue phantoms used in obtaining the speckle images for analysis. The goal was to

TABLE I

Tissue Mimicking Phantoms / Scanner/ Transducer	RMI Phant Agar Gel	ATS Phantoms Oil Based Gel						
	7 μm Part .lg/ cm^3	150 μm 1 part/mm^3	150 μm 2 part/mm^3	180 μm 1 part/mm^3	212 μm 1 part/mm^3	250 μm 2 part/mm^3	400 μm .004g/cm^3	400 μm .006g/cm^3
Searle Phosonic								
1.6 MHz, 19mm, LIF	X							
2.25 MHz, 19mm, LIF	X							
3.5 MHz, 19mm, LIF	X							
Unirad GZD								
1.6 MHz, 19mm, MF				X				
2.25 MHz, 19mm, MF				X				
3.5 MHz, 13mm, SF		X	X	X	X	X	X	X

utilize scatterers from a small fraction of a wavelength (7μm) up to the approximate size of a wavelength (400μm) at more than one concentration for both oil and water based gelatins. By using such a wide variety of phantoms we intended to include the same scattering mechanisms present in tissue. Although the origin of diffuse echoes from tissue is not well understood, organs such as the liver and heart exhibit uniform texture which seems to arise from small scatterers.[9,10] To the extent that phantoms simulate the scattering mechanisms of tissue, the statistical measurements and models which are derived for phantoms can also be applied to tissue. In evaluating these images, the first order statistics, equivalent to the brightness variations at a single point in space, were characterized by the "signal to noise ratio" (SNR) over long scan lengths within the image. Second order statistics, the coarseness of the spatial structure of the texture, were described by the autocorrelation function (ACF) and its Fourier transform, the noise power spectrum (NPS). These parameters were measured as a function of range, transducer frequency and the size and concentration of scatterers within the phantoms. The measurement data were then compared to the predictions of a simple analytical model of a pulse-echo imaging system applied to a homogeneous medium containing a large number of scatterers. Such analysis and measurement techniques could ultimately be applied to clinical images in the hope of resolving the question of the origin of ultrasound B-scan texture.

II. Measurement Techniques

Figure 1 shows a linear scan, ultrasound image made with a Searle Phosonic B-scan system at 2.25MHz using a 19 mm focused transducer. The system included a digital scan converter and displayed 40 dB of echo dynamic range in 32 shades of gray. Images were also obtained with a Model GZD Unirad scanner which utilized an analogue scan converter and displayed 30 dB of dynamic range in 11 shades of gray. The image in Figure 1 is typical of those obtained for the texture analysis with a variety of combinations of transducers and tissue mimicking phantoms. The water based gelatin phantom was fabricated by Radiation Measurements Inc. and has been described previously.[11] The oil-based gelatin phantoms, fabricated by ATS laboratories, were composed of styrene-butadiene resin and mineral oil similar to that recently used in the development of an anthropomorphic ultrasound phantom[12].

The images were recorded on x-ray film using a multi-format camera or on Polaroid type 665 transparency film for later analysis. Figure 2 shows a simple block diagram of the image analysis system which has been used extensively at the Bureau of Radiological Health for radiographic screen-film analysis. A point by point measurement of film optical density was made along selected lines in the ultrasound images in directions corresponding to both

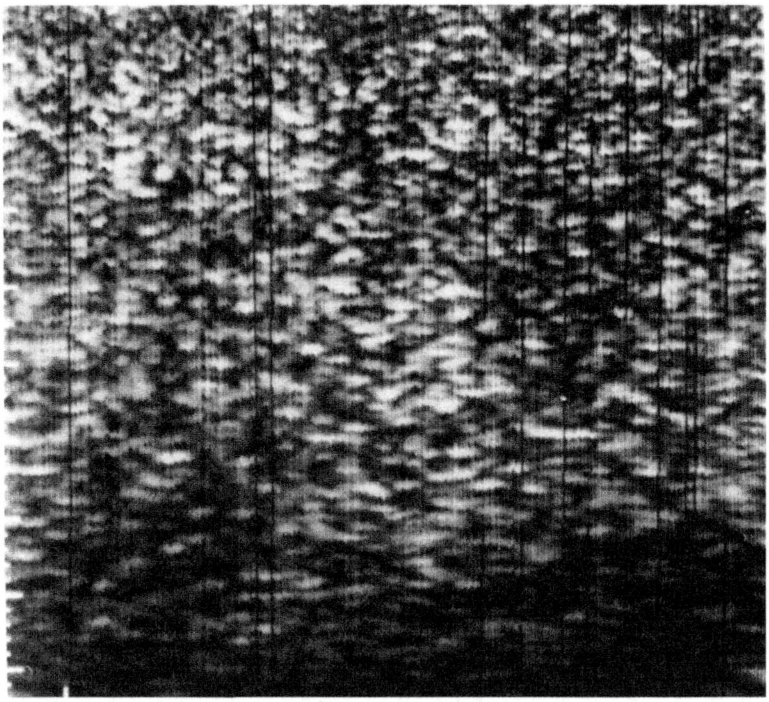

Figure 1. Linear Scan of Ultrasound Phantom Containing 7 µm Scatterers; 19 mm, 2.25 MHz Transducer

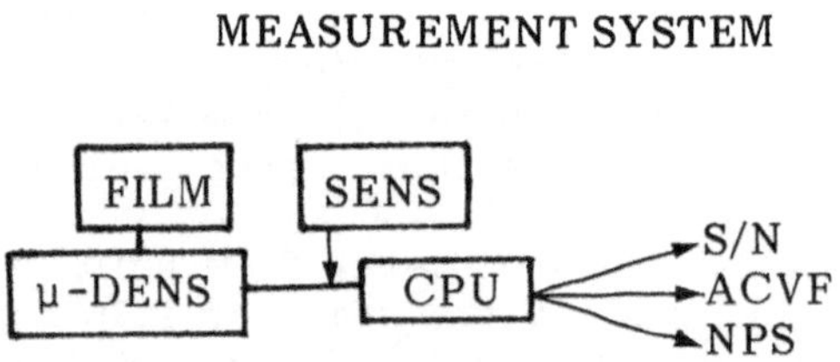

Figure 2. Block Diagram of Image Analysis System

the axial and lateral image dimensions. A Perkin-Elmer Model PDS microdensitometer was used for these measurements with careful attention to the correct choice of optical slit dimensions. The optical density data were stored in computer memory.

Throughout the image formation the ultrasound echoes had been subject to several non-linear processes including logarithmic amplification, display on the television monitor as affected by the gamma of the CRT, and the conversion of monitor brightness to film density. Hence, it was necessary to convert the optical density values to the original echo signal amplitudes by applying sensitometric transfer curves. The transfer function between film optical density and ultrasound input signal was determined from the manufacturers specifications and through the use of an r.f. signal source and a calibrated attenuator to measure the echo input signals equivalent to the gray shades of the bar test pattern displayed with every ultrasound image.

Calculations of SNR, ACF, and NPS were then made from echo data for sampled image scans in both the lateral and axial directions.

III. First Order Statistics

The SNR as a function of slit length was calculated where

$$SNR = \frac{\bar{A}}{\left[\overline{(A - \bar{A})^2}\right]^{\frac{1}{2}}} \qquad (1)$$

A = Echo amplitude

and $\bar{A}$ = Mean amplitude

In calculating SNR, one must be careful in the choice of the width of the image scan.[13] Too long a microdensitometer slit will result in an integration effect which decreases the standard deviation thus increasing the SNR. For these measurements the slit length was reduced until the value of SNR was independent of slit length. Measurements were made at six locations in images obtained with the Searle scanner at 1.6, 2.25 and 3.5 MHz using the 7 μm water based gel phantom.

Table II lists average values of SNR for 6 image lines in the axial and lateral dimension. The standard error of the mean for each set of measurements is also tabulated. The data indicate that the SNR in the images varies over a range from 1.5 to 2.1.

TABLE II

	1.6 MHz	2.25 MHz	3.5 MHz
Axial Direction	2.06 ± .14	1.48 ± .09	1.89 ± .12
Lateral Direction	1.78 ± .06	1.78 ± .10	2.13 ± .05

There seemed to be no correlation between the SNR measurement results and the range in the image or the transducer frequency. Previous models of coherent interference due to point scatterers in ultrasound B-scans predict a Rayleigh probability distribution of echo amplitudes resulting in a SNR of 1.91.[4] These measurements are consistent with the prediction of a Rayleigh distribution for 7 μm scatterers.

The deviation of individual measurements of SNR from the predicted value of 1.91 can be attributed to non-uniform B-mode scanning motion as well as variation in mean echo amplitude due to structural inhomogeneities in the phantom. The phantom inhomogeneities are evident in the images as regions of greater brightness.

IV. Second Order Statistics

In order to examine the coarseness of the granularity of image texture or speckle pattern, one must consider the second order statistics. A suitable measure of the average distribution of speckle sizes is the autocorrelation function (ACF) or its Fourier transform the noise power spectrum or Wiener spectrum.[14] An analysis of laser speckle by Goodman[14] has shown a direct relation between the PSF of the imaging system and the average speckle size as determined by the ACF. The granularity of pulse-echo ultrasound images in the range direction is distinctly different from that of the lateral direction. This observation is consistent since the point spread function (PSF) of the imaging system differs markedly between the range and lateral directions. The PSF in the range direction is determined by the pulse length of the transducer impulse response; while the PSF in the lateral direction is determined by the transducer diffraction pattern as predicted by physical optics. From a descriptive viewpoint, for the case of a pulse-echo ultrasound system one can assume a collection of many scatterers randomly positioned in the resolution cell in the field of the transducer.

The random walk summation of the phasors over the transducer face resulting from the propagating echoes returning from each of the scatterers gives rise to the echo amplitude at the transducer output. At ranges near the focal point and beyond, the variation in path lengths from the scatterers to the transducer is sufficiently small so that the collection of scatterers resembles a single target, and the transducer response in the lateral dimension from the scatterers is close to that of the point spread function.

At near ranges, for small lateral displacements of the transducer, the differences in path lengths from the scatterers to the transducer face undergo large changes and produce large variations in the echo resultant sum of phasors. Hence, at short ranges the granularity of peaks and nulls in the speckle image is finer. Thus, despite the relatively large size of the transducer point spread function at near ranges the speckle pattern is fine grained and the speckle size will continue to decrease as the range is decreased until a lower limit is reached at the spot size of the CRT display. Fine grain speckle patterns at near ranges have been noted for both clinical images[15] and images of a phantom containing 7 μm scatterers.[8]

Examining the speckle size from a more rigorous basis, one can use analytical techniques similar to those of Goodman. Assuming a one dimensional focused transducer, assuming no refraction effects of the intervening medium and no broad band effects, one can write an expression for the image auto-correlation function in the lateral direction at ranges near the focal point and beyond as

$$\mathrm{ACF}\ (\Delta x) = \frac{\kappa}{\lambda z}\int_{-\infty}^{\infty} |P\ (\xi)|^2 \left[\exp\left(2\pi i\ \frac{\xi}{\lambda z}\ \Delta x\right)\right] d\xi \qquad (2)$$

where z is the range, λ, the transducer center wavelength, ξ the position on the face of the transducer and κ, a constant of proportionality,. Δx is the displacement in the image. P is the amplitude transmission of the lens pupil, i.e., the pupil function for the transducer of dimension D. For the case of pulse-echo ultrasound the scatterers are both insonified and viewed through the same lens. We may define a new effective pupil function.

$$P = P_1 * P_1 \qquad (3)$$

where * denotes a convolution, and where the lens pupil function P_1 is given by

$$P_1 = \mathrm{rect}\ (\xi/D) = \begin{cases} 1, & |\xi| \leq D/2 \\ 0, & \text{otherwise} \end{cases} \qquad (4)$$

That is, we replace the system of insonification and detection by a system with equivalent resolution determined by P as just defined.

Writing P explicitly and making the substitution $\frac{\xi}{\lambda z} = f_x$ in the expression for ACF we obtain

$$ACF(\Delta x) = \kappa \int_{-\infty}^{\infty} \left[rect\left(\frac{f_x \lambda z}{D}\right) * rect\left(\frac{f_x \lambda z}{D}\right) \right]^2 \exp(2\pi i f_x \Delta x) d f_x \quad (5)$$

Since the noise power spectrum and ACF are Fourier transform pairs we may read off the NPS, $W(f_x)$ directly as the integrand without the Fourier kernel

$$W(f_x) = \left[rect(f_x/f_o) * rect(f_x/f_o) \right]^2$$

$$= (1-|f_x|/f_o)^2 , \qquad |f_x| \leq f_o \quad (6)$$

$$= 0, \qquad \text{otherwise}$$

where

$$f_o = D/\lambda z$$

is the highest frequency which appears.

The Fourier transformation $\mathbf{F}$ required to obtain the ACF can be carried out using the convolution theorem. First, note that

$$\mathbf{F}[rect(f_x/f_o)] = \frac{\sin(\pi \Delta x\ f_o)}{\pi \Delta x\ f_o} \equiv sinc(\Delta x\ f_o) \quad (7)$$

and therefore

$$\mathbf{F}[rect(f_x/f_o) * rect(f_x/f_o)] = sinc^2(\Delta x\ f_o) \quad (8)$$

where Δx is the variable conjugate to f_x. Second, using the convolution theorem in reverse

$$ACF(\Delta x) = \kappa \mathbf{F}[rect(f_x/f_o) * rect(f_x/f_o)]^2 \quad (9)$$

$$= \kappa\ sinc^2(\Delta x\ f_o) \circledast sinc^2(\Delta x\ f_o)$$

Since the point spread function (PSF) for insonification and detection is $sinc^2$ (Δx fo) we have simply

$$ACF\ (\Delta x) = \kappa\ PSF * PSF \quad (10)$$

This is the classical result for the output autocorrelation function when a white noise pattern (here the scatterers which are fine grained compared to the spread function) is transferred through a linear system with a symmetric spread function, PSF[16]. That is, the PSF of the system "smooths" out the input granularity pattern.

We have carried out the convolution graphically and note that the approximation

$$ACF\ (\Delta x) \approx \mathrm{sinc}^4\ (\Delta x\ f_{o/2}) \tag{11}$$

is very good over the range $1 \leq ACF < 0.1$

At the latter point the actual ACF has an inflection point at a displacement,

$$\delta x = 2/f_o = 2\ \frac{\lambda z}{D} \tag{12}$$

We consider this point as the average size of the speckle or the correlation length for this one dimensional analysis.*

V. Speckle Size Measurements

ACF's were calculated from independent micro-densitometer scans of the B-mode image in the lateral and axial directions. The sample ACF or the ACF estimator of the echo data of an image line was defined.[17]

$$ACF(\Delta x) = \frac{\frac{1}{N} \sum_{t=1}^{N-t} x_t . x_{t+k}}{\sigma^2} \tag{13}$$

where X_t are the echo amplitudes along an image scan
k = the lag distances
N = number of samples along the image scan
σ^2 = the variance of the echo amplitudes

Images of the oil based gelatin phantoms (Table I) were obtained with the Unirad GZD scanner using 1.6, 2.25 and 3.5 MHz focused transducers.

For measurements of 2nd order statistics a long narrow micro-densitometer slit was used. The narrow diminsion provides the necessary spatial resolution along the length of the scan; while the long dimension averages over fluctuations in the direction perpendicular to the scan. In other words it was necessary to decouple the axial versus lateral direction.

---

*The above analysis can be extended to two-dimensions in a straightforward way by using cylinder functions instead of the rectangle function. The resulting power spectrum is then proportional to the square of the optical transfer function of a diffraction limited lens, which is well approximated by a parabola, like Eq. (6), with a slightly expanded scale.

Figure 3 shows a typical autocorrelation function in the axial direction for an image obtained at 1.6 MHz using a phantom containing 180 μm particles. The high frequency modulation of the ACF was due to the raster lines of the television monitor. It served as a calibration check, but was filtered out in measurements of axial direction speckle size.

As noted above the size of the speckle pattern in the axial direction was expected to correspond roughly to the pulse length of the transducer impulse response. Table III compares the -6 dB pulse width of the transducer with the average -6 dB width of the axial direction ACF. The standard deviation is also listed for the ACF measurements along six axial lines within each image.

The table shows reasonable agreement between pulse width and ACF width. There were no trends of variations or the ACF with range for the image axial direction.

TABLE III

Frequency (MHz)	-6 dB Pulse width (mm)	-6 dB ACF width (mm)
1.6	1.20	1.12 ± .22
2.25	1.33	.99 ± .19
3.5	1.48	1.10 ± .25

Since the r.f. pulses contained varying number of cycles it was impossible to detect direct correlation of axial speckle size versus transducer frequency.

Figure 4 shows a typical ACF in the lateral direction for an image obtained at 2.25 MHz using a phantom containing 180 μm particles. Focal points for each of the transducers were located between 3 and 5 cm. Similar plots were obtained at 1.6 and 3.5 MHz at six different ranges out to 8 cm into the image. Analogous measurements were also obtained for all the phantoms described in Table I. Each plot of ACF showed the distinct features of a narrow main lobe, first inflection point, and a low frequency ripple out to several centimeters.

To determine the accuracy of the model described above these features of the ACF data were compared to the predictions. It was felt that the long range correlations demonstrated in the low frequency ripple were due to structures, i.e. regional inhomogeneities, in the phantom. The width of the main lobe and locations of first inflection varied with range and transducer frequency. Figure 5 compares the location of the first inflection with the predicted average speckle size, $2\frac{z\lambda}{D}$, from Eqn. (12)

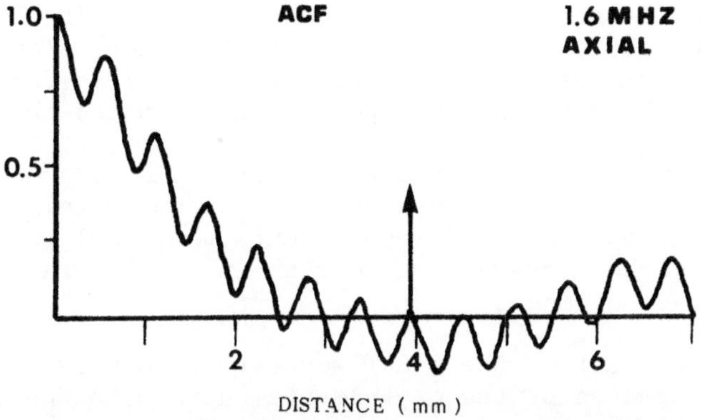

Figure 3. Autocorrelation Function in Axial Direction from Image of Ultrasound Phantom Containing 180 μm Scatterers

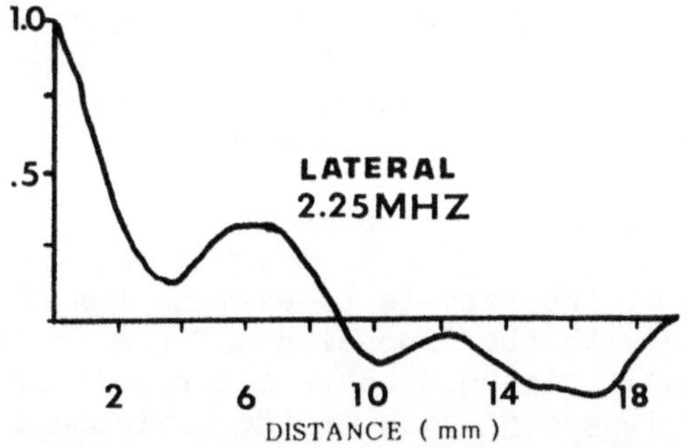

Figure 4. Autocorrelation Function in Lateral Direction from Image of Ultrasound Phantom Containing 180 μm Scatterers

as a function of range at three transducer frequencies.

Figure 5(a) demonstrates considerable scatter of the experimental points about the line. A linear fit through these points, however, is close to the predicted line. Figures 5(b) and 5 (c) show quite good agreement for 2.25 and 3.5 MHz. The location of the first inflection of the ACF appears to reach a lower limit value of approximately 2 mm in the very near field for these frequencies. This was not unexpected as was noted above.

Similar results were obtained at the same frequencies for the Searle scanner in the case of the phantom containing 7 μm particles. In both sets of measurements the model performed well in predicting the average speckle size in the lateral dimension of the pulse-echo images.

A further test of the model was carried out by comparing the noise power spectra of the experimental data with the predicted quadratic relationship of Eq. (6). The NPS were calculated by applying the FFT algorithim to the ACF data. Figure 6 shows a typical spectrum plotted on a log-log scale. All spectra exhibited three dominant features. (1) A large amplitude low frequency region was present at approximately .03 cyc/mm which corresponded to the low frequency ripple noted in the AFC as described above. (2) A region of negative slope existed out to frequencies of approximately 1 cyc/mm. The form $(1 - f_x/f_o)^2$ was generally measured indicating a parabolic function in good agreement with the previous analysis. (3) At spatial frequencies greater than 2 cyc/mm low level white noise was measured. This noise resulted from the electronic noise of the imaging system and the film granularity of the recorded ultrasound images. It is interesting to note that the amplitude of the white noise was up to three orders of magnitude less than the amplitude of the spatial frequencies corresponding to the image speckle.

VI. Speckle Size vs. Scatterer Size

A final area of interest is to examine the variation of image speckle size with the dimensions of the scatterers in the ultrasound phantoms. Figure 7 shows the result of preliminary ACF measurements from images of each of the ultrasound phantoms made at 3.5 MHz. Again the ACF first inflection in the lateral direction was used as a measure of average speckle size as described in Figure 4. Data was taken at ranges from 6.5 mm to 8 mm for both low concentrations of scatterers (open circles) as well as high concentrations (solid dots). Although the scatter of the data points is significant, there seems to be no correlation between the size or concentration of scatterers within the ultrasound phantoms and the average speckle size as described by the ACF.

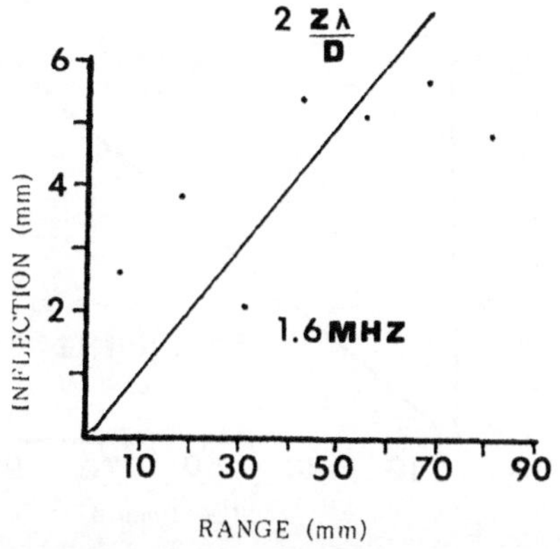

Figure 5a. ACF First Inflection in Range Compared to Predicted Speckle Size at 1.6 MHz

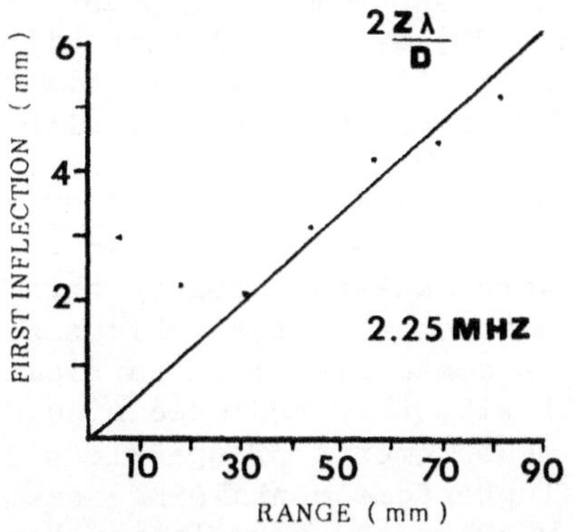

Figure 5b. ACF First Inflection vs. Range Compared to Predicted Speckle Size at 2.25 MHz

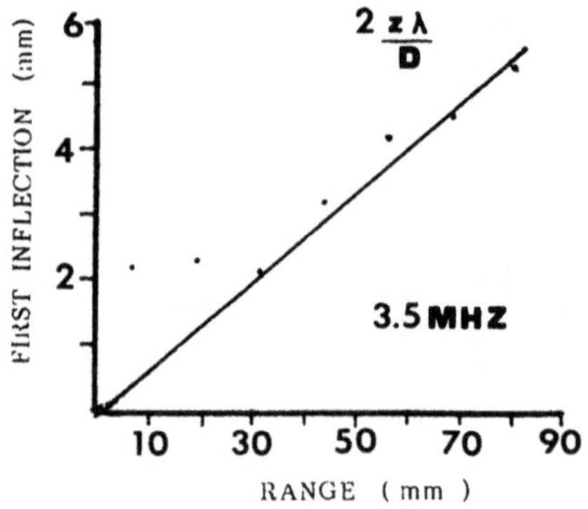

Figure 5c. ACF First Inflection vs. Range Compared to Predicted Speckle Size at 3.5 MHz

VII. Summary

In this paper an examination of the 1st and 2nd order statistics has been made for texture images obtained with a variety of ultrasound tissue mimicking phantoms. Measurements of 1st order statistics (SNR) were consistent with previous predictions that Rayleigh statistics are applicable to the formation of ultrasound images of homogeneous material with large numbers of small scatterers (7 μm). Utilizing analysis similar to that for laser speckle, an analytical model was presented to predict the average speckle (2nd order statistics) size in ultrasound images. Several assumptions were made to simplify this very complex problem. However, measurements of speckle size in images produced from phantoms containing 7 μm particles and 180 μm particles agreed well with the predicted size dependence of $\frac{2z\lambda}{D}$. Finally preliminary measurements of speckle size in images obtained from several phantoms containing scatterers from 7 μm to 400 μm in diameter showed no correlation between speckle size and scatterer size. The results presented here are still preliminary and much work still remains before final conclusions can be made about the origin and diagnostic value of image texture of clinical ultrasound scans. However, it does seem evident that the measurement techniques and analyses of this research will eventually be extended to soft tissue images.

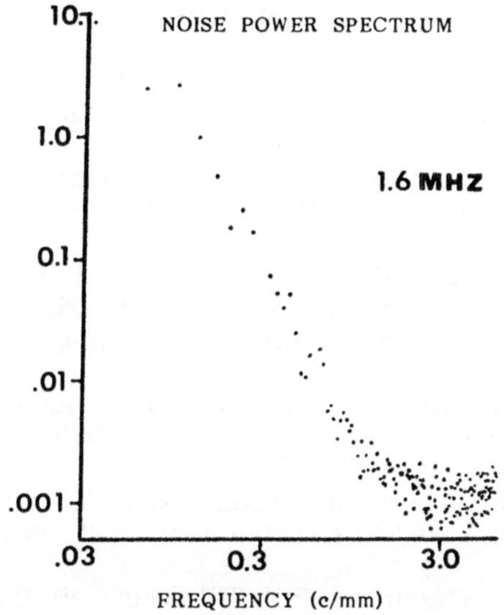

Figure 6. Noise Power Spectrum of Lateral Direction from Image of Ultrasound Phantom Containing 180 μm Scatterers

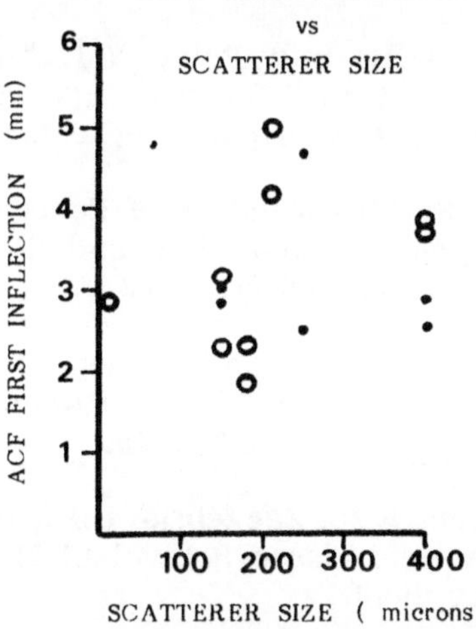

Figure 7. ACF First Inflection vs. Scatterer Size at 3.5 MHz
o = Low Concentration of Scatterers
• = High Concentration of Scatters

The mention of commercial products, their source, or their use in connection with material reported herein is not to be construed as either an actual or implied endorsement of such products by the Department of Health and Human Services, Public Health Service.

References

1. Ultrasonic Tissue Characterization, M. Linger, ed., NBS Pub. 453, USDOC, Washington, DC 20234, October, 1976.

2. Ultrasonic Tissue Characterization II, M. Linger, ed., NBS, Pub. 526, USDOC, Washington, DC 20234, April, 1979.

3. Kossoff, G., Garret, W.J., Carpenter, D.A., Jellins, J. and Dadd, M.J., "Principles and Classification of Soft Tissues by Gray Scale Echography," Ultrasound in Medicine and Biology, 2, 1976, pp. 89-105.

4. Burkhardt, C.B., "Speckle in Ultrasound B-Mode Scans," IEEE Trans. on Sonics and Ultrasonics, SU-25, January, 1978, pp. 1-6.

5. Abbott, J.G. and Thurstone, F.L., "Acoustic Speckle: Theory and Experimental Analysis," Ultrasonic Imaging, 1, 1979, pp. 303-324.

6. Bamber, J.C. and Dickinson, R.J. "Ultrasonic B-Scanning: A Computer Simulation," Phys. Med. Biol., 25, 1980, pp. 463-479.

7. Dickinson, R.J., "A Computer Model for Speckle In Ultrasound Images: Theory and Application," Acoustic Imaging, Vol. 10, Plenum Press, New York, 1981.

8. Flax, S.W., Glover, G.H. and Pelc, N.J. "A Stochastic Model of Ultrasound Image Texture," Proceedings American Institute of Ultrasound in Medicine, New Orleans, September 15-19, 1980, p. 81.

9. Fields, S. and Dunn, F., "Correlation of Echographic Visualizability of Tissue with Biological Composition and Physiological State," J. Acoust. Soc. Amer., 54, pp. 809-812, 1973.

10. O'Donnell, M., Mimbs, J.W. and Miller, J.G., "The Relationship Between Collagen and Ultrasonic Attenuation in Myocardial Tissue," J. Acoust. Soc. Amer., 65, pp. 512-517, 1979.

11. Burlew, M.M., Madsen, E.L., Zagzebski, J.A., Banjavic, R.A. and Sum, S.W., "A New Ultrasound Tissue-Equivalent Material," Radiology, 134, pp. 517-520, February, 1980.

12. Scherginger, A.L., Carson, P.L., Carter, W., Clayman, W., Johnson, M.L., Rashbaum, C. and Smith, S.W., "A Tissue Equivalent Upper Abdominal Phantom For Training and Equipment Demonstration," Proceedings American Institute of Ultrasound in Medicine, Montreal, August, 1979, p. 118.

13. Arrenault, H.H. and April, G., "Properties of Speckle Integrated With a Finite Aperture and Logarithmically Transformed," J. Optical Soc. Amer., 66, pp. 1130-1163, November, 1976.

14. Goodman, J.W., "Statistical Properties of Laser Speckle Patterns," Laser Speckle and Related Phenomena, J.C. Dainty, ed., Springer-Verlag, Heidelberg, 1975, pp. 9-75.

15. Jaffe, C.C. and Harris, D.J., "Sonographic Tissue Texture: Influence of Transducer Focusing Pattern," Amer. Jnl. of Roentgenology, pp. 343-347, August, 1980.

16. Dainty, J.C. and Shaw, R., Image Science, Academic Press, London, 1974, p. 222.

17. Wagner, R.F. and Sandrik, J.M., "An Introduction to Digital Noise Analysis," The Physics of Medical Imaging: Recording System Measurements and Techniques, A.G. Haus, ed., Am. Inst. Phys., New York, 1979, p. 528.

NOTE ADDED IN PROOF: The function deduced for the speckle auto-correlation function in this paper applies directly to the echo amplitude in the RF signal. The autocorrelation function in a non-trivial way, but may be approximated by the autocorrelation function of intensity. This function has about the same scaling and spatial extent as the function studied here. The results of this further study are the subject of a manuscript in preparation.

CALIBRATION OF IMAGING SYSTEMS BY MEANS OF SPHERICAL TARGETS

Michel Auphan, Roger H. Coursant, and Claude Méquio

Laboratoires d'Electronique et de Physique Appliquée

3, avenue Descartes - 94450 Limeil-Brévannes (France)

ABSTRACT

The problem of calculating the impulsive signal scattered on a target of spherical symmetry, when directivity functions of both transmitting and receiving transducers are known, is mathematically expressed.

In cases where the rigid spherical target is not very small, computations of the received signal become difficult because we deal with quotients of ill-conditioned polynomials. But as these polynomials have integral coefficients and particular properties, a solution has been found on condition of a frequency limitation of the signal.

An experimental set up has been built in order to drop the ball. The echo on the falling ball is then recorded by means of a storage oscilloscope. Various ball diameters have been used and results are compared with the computerized curves. The discrepancy is within the relatively large experimental errors. They are mainly due to the non piston-like behaviour of transducers and to the assumption of a rigid target (Rayleigh surface waves on the ball are neglected).

INTRODUCTION

The study of acoustic imaging systems requires accurate calibration methods for transducers. Unfortunately no satisfactory measurement devices are presently proposed. The small available hydrophones have complicated responses and radiation patterns ; they are often too large to provide a field measurement in one point.

The power measurement (via a radiative force or a calorimetric test) gives only the total transmitted power, whereas the meaningful information is the pressure induced in a given point by the transducer.

As to the reciprocity method, it is generally carried out with a plane reflector [1]. It provides a kind of information taking into account the power radiated into a half space and it is impossible to deduce from this information an estimate of the local behaviour of the transducer, unless unrealistic assumptions such as piston-like modelling are made.

There is indeed a need for a calibration method in relation with just the pressure induced somewhere by the transducer. That is the reason why we have thought of a small feasible target giving a calculable echo, hence a spherical rigid target. This kind of target has been much treated in the literature but generally from a continuous wave point of view, an exhaustive bibliography is included in [2] . This bibliography could be updated by adding [3,4,5]. As for the impulse response we have treated it in a previous paper [8] where we have calculated the induced acoustic pressure all around a rigid sphere striken by a plane (δ-shaped) impulse wave and we shall discuss here how this first result can be used to calculate the echo when directivity functions of both transmitting and receiving transducer are known. Of course, one could expand the transmitted wave in plane waves, then calculate the total echo by summing up the pressure terms on the receiving transducer surface. An integration on this surface would then lead to the force acting on this transducer, hence to the electrically delivered signal. This would not be elegant. Furthermore, we can know the directivity function of a transducer in a certain volume without knowing the distribution of pressure on its surface for instance thanks to some measurements. In view of these considerations, we propose here a more general method of echo calculation relying on the reciprocity theorem developed in connection with the network theory.

POSITION OF THE PROBLEM

The method described here is not at all restricted to piston-like transducer taken as a starting point here but is relevant of any kind of transducers. For the sake of simplicity, we have written down the equations for the case of a piston in order to refer to the force and velocity on this piston to avoid the introduction of electric inputs or outputs of the transducers and to remain independent of the transducer modelling. When the transducer acts as a piston, the total transfer function is the convolution product of the transfer function of the transducer itself up to the window and the geometric window transfer function. If the transducer does not act as a piston this separation is no more possible and the function ϕ should include all the transfer from an electrical excitation. This transfer can easily be derived provided a minor change of equations is made.

Let us suppose that we deal with one emitting and one receiving transducer (which obviously could be the same) and let $\phi(x^1,x^2,x^3,t)$ and $\phi'(x^1,x^2,x^3,t)$ be their respective radiative impulse responses. These responses are defined as the velocity potential induced at point of coordinates x^1,x^2,x^3 and at time t when the transducer face motion corresponds to a δ-function for its velocity according to a piston mode behaviour. The problem is then to calculate the force induced on the face of the receiving transducer when the emitting transducer face has undergone a velocity δ-impulse motion. Such a force might be defined as the mechanical response of the scattering system. This mechanical response should have to be convolved by the acousto-electric responses of both emitting and receiving transducers in order to give the total electric impulse response of the measuring system.

SOLUTION AS A SERIES IN THE FREQUENCY DOMAIN

The solution is easier to express in the frequency domain. We shall then use the Fourier transforms $\hat{\phi}(x^1,x^2,x^3,\nu)$ and $\hat{\phi}'(x^1,x^2,x^3,\nu)$ of the radiative impulse responses ϕ and ϕ', ν being the frequency, $\hat{\phi}$ and $\hat{\phi}'$ are generally referred to as the radiative transfer functions of the transducers.

Then the mechanical transfer function of the system which is the Fourier transform $\hat{F}(\nu)$ of the mechanical response F(t) of the scattering system can be rigorously expressed in terms of ϕ and ϕ' and their space derivatives of various orders at the point (x_o^1,x_o^2,x_o^3) which is the center of the spherical target. The mathematical developments leading to this relation are to be published elsewhere [6]; we give here only the result :

$$\hat{F}(\nu) = -\ 2\pi C\mu \sum_{n=o}^{\infty} (2n+1)\ C_n(k) \sum_{m=o}^{n} a_n^m\ K^{-2m}\ O_m(\hat{\phi},\hat{\phi}') \qquad (1)$$

where :

- C is the sound velocity in the medium ;
- μ is the density of the medium ;
- $K = \dfrac{2\pi\nu}{C}$
- a_n^m are the coefficient of the n-th Legendre polynomial $P_n(Z)$ such that :

$$P_n(Z) = \sum_{m=o}^{n} a_n^m\ Z^m \qquad (2)$$

- $O_m(\hat{\Phi},\hat{\Phi}')$ is a differential operator defined by :

$$O_m(\hat{\Phi},\hat{\Phi}') = \sum_{\ell_1,\ell_2,\ldots\ell_m=1} \frac{\partial^m\hat{\Phi}(x_0^1,x_0^2,x_0^3,\nu)}{\partial x^{\ell_1}\partial x^{\ell_2}\ldots\partial x^{\ell_m}} \frac{\partial^m\hat{\Phi}'(x_0^1,x_0^2,x_0^3,\nu)}{\partial x^{\ell_1}\partial x^{\ell_2}\ldots\partial x^{\ell_m}} \quad (3)$$

- $C_n(K)$ is a set of functions of the frequency characterizing the spherical target and independent of the transducer parameters.

This relation (1) holds for any kind of target of spherical symmetry. For a rigid spherical target of radius R we have :

$$C_n(K) = - \frac{j'_n(KR)}{h'_n(KR)} \quad (4)$$

j_n and h_n being respectively the spherical Bessel and Hankel function according to an usual terminology[7]. Henceforth we shall consider only rigid targets according to (4).

SOLUTION AS A SERIES IN THE TIME DOMAIN

Theoretically if we are able to calculate the inverse Fourier transform of (1), i.e. the inverse Fourier transform of functions $C_n(\frac{2\pi\nu}{C})$ we should obtain a mathematical expression of the mechanical response F(t) in the time domain. This expression would obviously be a double series as (1) but unfortunately it does not converge. Nevertheless, it is actually a time response that we measure with an oscilloscope and for our calibration purpose we need the calculation of the shape of return pulse clearly in terms of time. The difficulty stems from the fact that the bandwidth of every physical system is bounded. With the assumption of an upper frequency limit ν_M of the system passing band then the inverse Fourier transform of (1) becomes convergent.

It can even be shown that the functions $C_n(\frac{2\pi\nu}{C})$ have a negligible contribution to the echo for sufficiently large values of the order n, i.e. :

$$n > 2\pi\nu_M \frac{R}{C} \quad (5)$$

if the system does not transmit signals of frequencies larger than ν_M.

Hence practically the series can be replaced by a bounded sum and problems of convergence vanish. But the new problem is the difficulty of calculation of the Fourier transform of functions C_n.

In the frequency domain these functions do not tend towards zero when the frequency goes to infinity. Therefore they are not relevant of a direct numerical Fast Fourier Transform, unless we apply some arbitrary frequency filtering process, which is always hardly compatible with causality.

Thus we have to find another way of calculating the inverse Fourier transform of $C_n(K)$. It is helpful to notice that in view of (4) C_n can be written :

$$C_n(K) = \frac{1}{2}\left[(-1)^{n+1}\, e^{2iKR}\, \frac{\omega_n\left(\frac{-1}{iKR}\right)}{\omega_n\left(\frac{1}{iKR}\right)} - 1\right] \qquad (6)$$

where $\omega_n(x)$ are polynomials of integral coefficient related to Bessel polynomials[11]. Then the ratio :

$$\frac{\omega_n\left(\frac{-1}{iKR}\right)}{\omega_n\left(\frac{1}{iKR}\right)}$$

can be split into simple elements corresponding to each of its poles. The inverse Fourier transform can then be written in terms of a sum of complex time exponential functions as explained in [8]. But this method can hardly be applied for $n > 14$ because we deal with exponential of very large imaginary arguments resulting in significant errors during numerical calculations.

That is why we have developed another method of calculation of the Fourier transform of $C_n(K)$. This method relies on another series than (1) obtained by various manipulations taking into account some particular properties of polynomials ω_n. It would be out of the scope of this paper to describe this method which is to be published elsewhere[6].

The final formula makes use of functions $G_m(x)$ which can be numerically calculated. This formula is :

$$F\left(t - \frac{2R}{C}\right) = -\pi\mu C \sum_m G'_m\left(\frac{ct}{R}\right) * L_m(\psi,\psi') \qquad (7)$$

G'_m being the derivative of G_m and $L_m(\psi,\psi')$ is the inverse Fourier transform of $K^{-2m}O_m(\hat{\Phi},\hat{\Phi}')$, i.e. :

$$L_o(\psi,\psi') = \psi(x_0^1, x_0^2, x_0^3, t) * \psi'(x_0^1, x_0^2, x_0^3, t)$$

$$\frac{\partial^2}{\partial t^2} L_1(\psi,\psi') = C^2 \sum_{\ell=1}^{3} \frac{\partial\psi(x_0^1,x_0^2,x_0^3,t)}{\partial x^\ell} * \frac{\partial\psi'(x_0^1,x_0^2,x_0^3,t)}{\partial x^\ell} \tag{8}$$

$$\frac{\partial^4}{\partial t^4} L_2(\psi,\psi') = C^4 \sum_{j,\ell=1}^{3} \frac{\partial^2\psi(x_0^1,x_0^2,x_0^3,t)}{\partial x^j\,\partial x^\ell} * \frac{\partial^2\psi'(x_0^1,x_0^2,x_0^3,t)}{\partial x^j\,\partial x^\ell}$$

and so on... But practically the three first operators suffice. Then only three functions G_m have to be calculated. An interesting case is that of two circular transducers aiming at the spherical target, their axis which intersect on the centre of the target making an angle of 90°. In this case it can be shown that operator L_1 vanishes.

By means of our program we have calculated with the accuracy of the computer (error smaller than the maximum of G time 10^{-9} for the absolute value) the functions G_0, G_1 and G_2 by summing a series equivalent to (1) up to n = 36 (G_0 and G_2 are shown on figure 1).

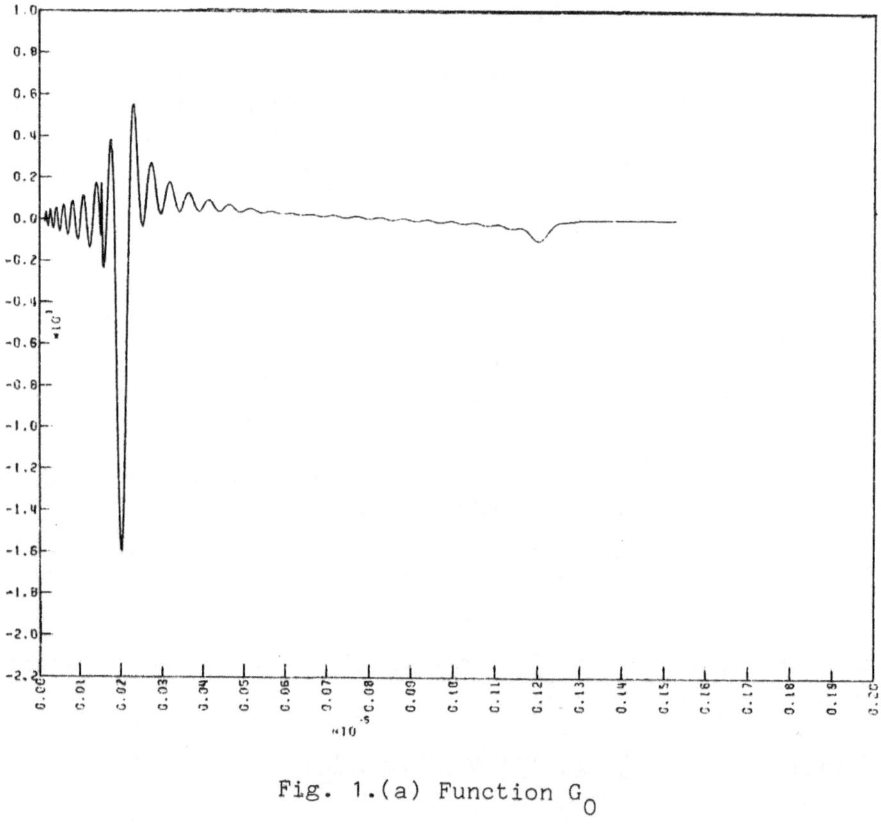

Fig. 1.(a) Function G_0

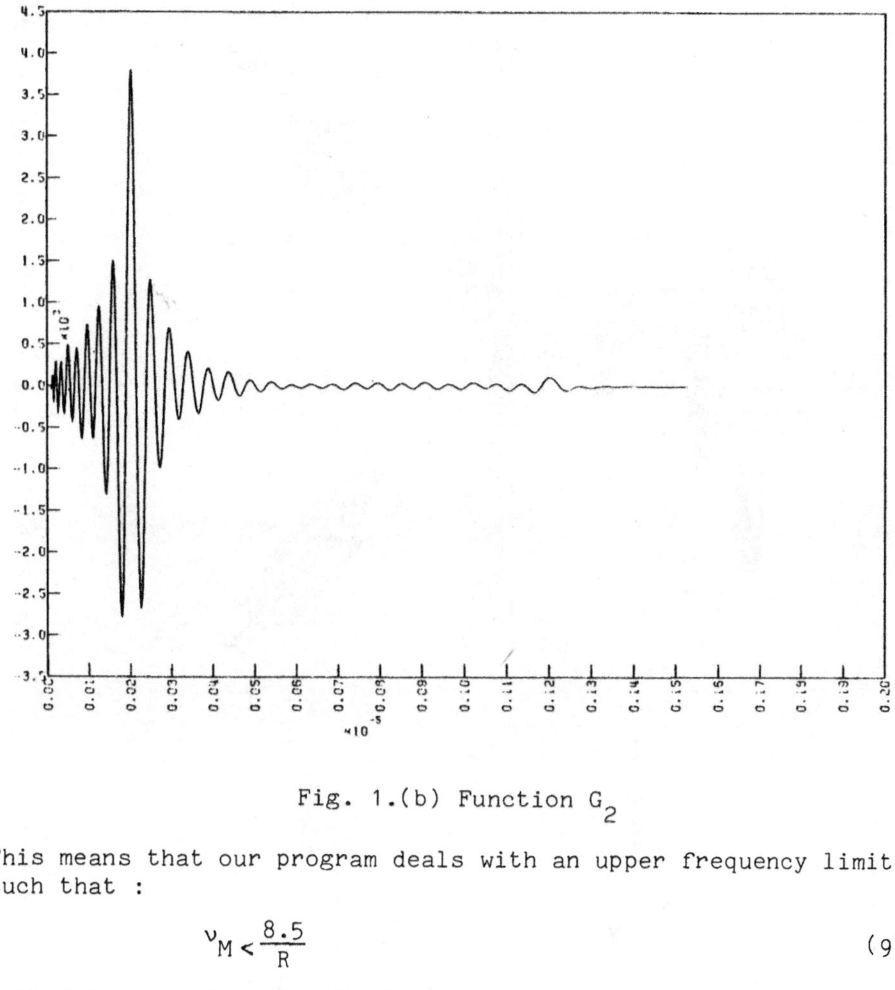

Fig. 1.(b) Function G_2

This means that our program deals with an upper frequency limit ν_M such that :

$$\nu_M < \frac{8.5}{R} \tag{9}$$

with R in mm and ν_M in MHz for propagation in water. Then we are able to calculate the echo on a rigid target having a diameter equal to 11 wavelengths or less.

COMPARISON WITH EXPERIMENT

The only comparison between calculations and experiments yet published are relative to continuous wave[12,4,5]. It is then of main interest to carry out this comparison in the case of an impulse mode.

Our first attempt to store the shape of reflected acoustic pulses made use of nylon or tungsten wire to support a steel ball. But it turned out that the echo on the wire had not a negligible amplitude. Therefore we have built a dropping set up (fig. 2a, 2b, 2c) in order to collect an echo from a ball falling by gravity in water.

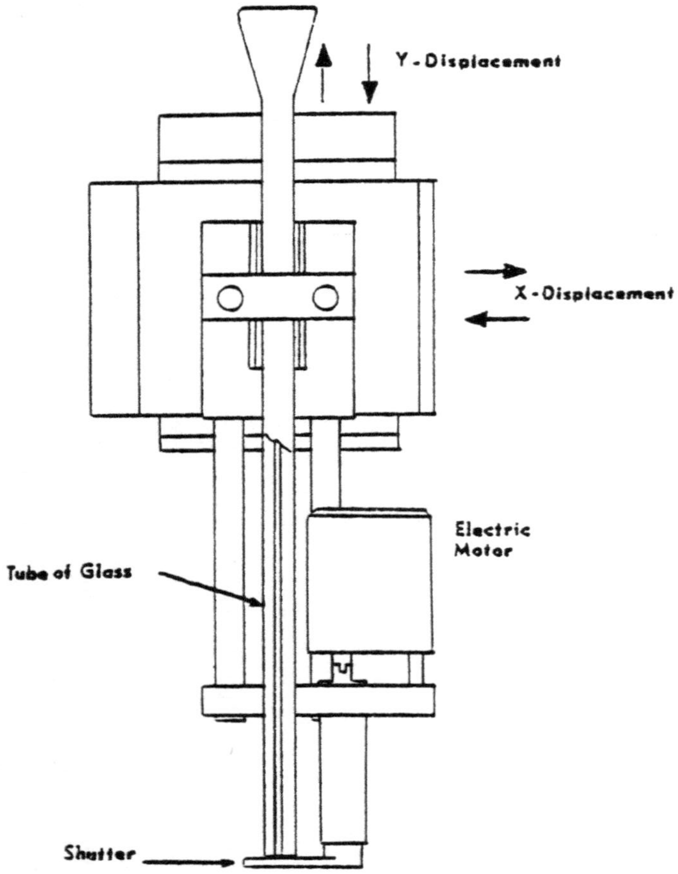

Fig. 2.(a) Front view of the dropping set up

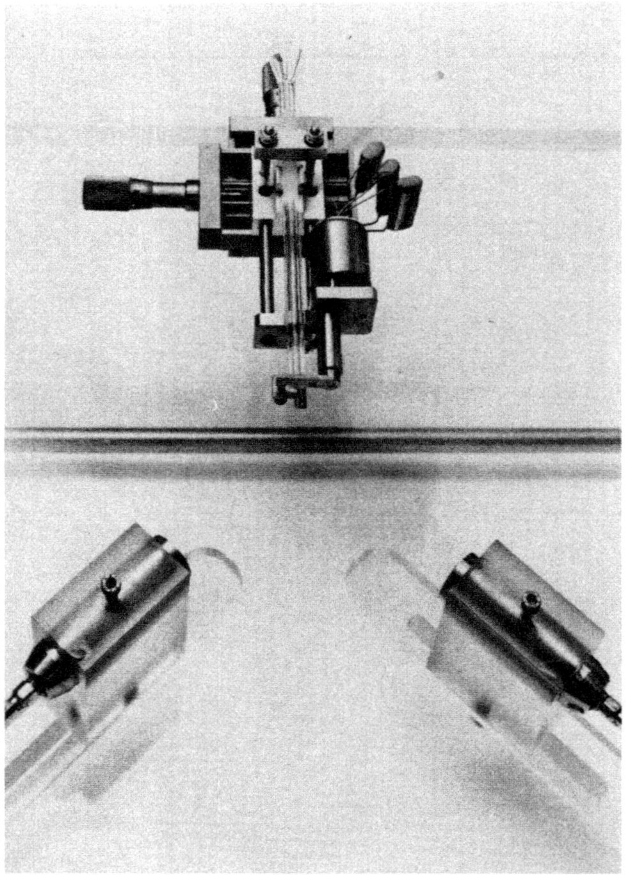

Fig. 2.(b) Photograph of the dropping set up with the two transducers

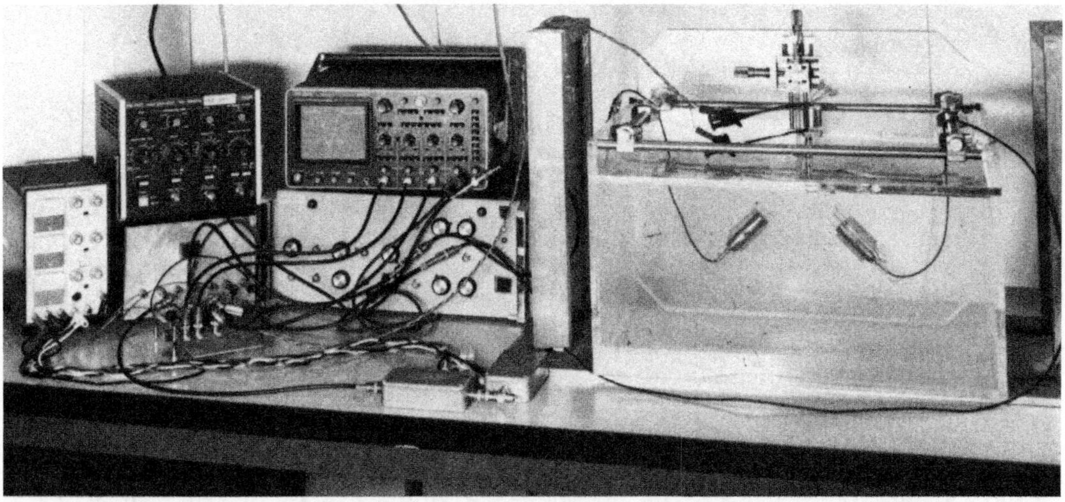

Fig. 2.(c) Photograph of the complete experimental system

The echo pulse was then recorded by means of a storage oscilloscope triggered when the ball appeared to be at the right place.

Practically the ball is initially kept inside the bottom of a tube closed by a shutter. This shutter can be removed with the help of a small electric motor. The location of the falling down ball is tracked by the echoes issued from it since the emitting transducer is excited by short pulses of a repetition rate of 1 kHz.

As the ball falls down the interval between the transmission shot and the time of echo collection decreases. This time of collection is detected by a comparator used to trigger up the storage oscilloscope when the ball, supposed to fall along a straight vertical line, is just at the crossing point of the transducer axis about 1 centimeter below its dropping place.

In order to be sure that the bandwidth of the system does not exceed the frequency taken into account for computation according to relation (9) a filter has been placed between the transducer and the storage oscilloscope.

For the computation of the stored electric signal the mechanical response given by (7) has to be convolved by both acousto-electric responses (from electric to mechanic terminals and vice versa) of the transducers by the pulse response of the filter and by the shape of the excitation pulse.

The study of these acousto-electric responses have yet been set forth in [9], pages 130-133, and is relative to pure thickness mode transducers without loss and with a small electromechanical coupling coefficient. The patterns of radiation ϕ and ϕ' have been calculated with the assumption of a piston behaviour according to[10] Two identical circular transducers with axis perpendicularly to each other were used.

All the convolution products have been carried out directly in the time domain. Though more costly in computer time it gives more accurate results than Fast Fourier Transforms.

As to the excitation pulse it has been stored separately through the same filter. Hence by taking this digitized shape of pulses for calculations we had no more to care for the pulse response of the filter. It is indeed well known that if we had excited the emitting transducer with a filtered shape of pulse it would be equivalent to filtering the collected signal in view of the linearity of phenomena involved.

RESULTS

We have selected three photographs taken from the storage oscilloscope (figures 3a, 4a, 5a). They correspond to tungsten carbide balls of diameters 0.79, 1, 2 mm. The transducers have a ceramic diameter of 10 mm.

Figures 3a to 5a are juxtaposed to figures 3b to 5b corresponding the computation in the same cases with the same scale (20 or 50 millivolts/division). It can be seen that the agreement between experimental and simulated results are better for the small balls of 0.79 and 1 mm than for the ball of 2 mm.

DISCUSSIONS

There are at least four reasons to explain the discrepancy of the results with simulated data.

1. Uncertainty of the location of the ball. We are not sure that the ball path is a straight vertical line since the opening of the shutter induces in the water motions which can deviate the ball. This can be proved by the dispersion of the amplitude of signals stored with the same ball. We have actually selected the storage of largest amplitude with the idea that they should correspond to the right location.

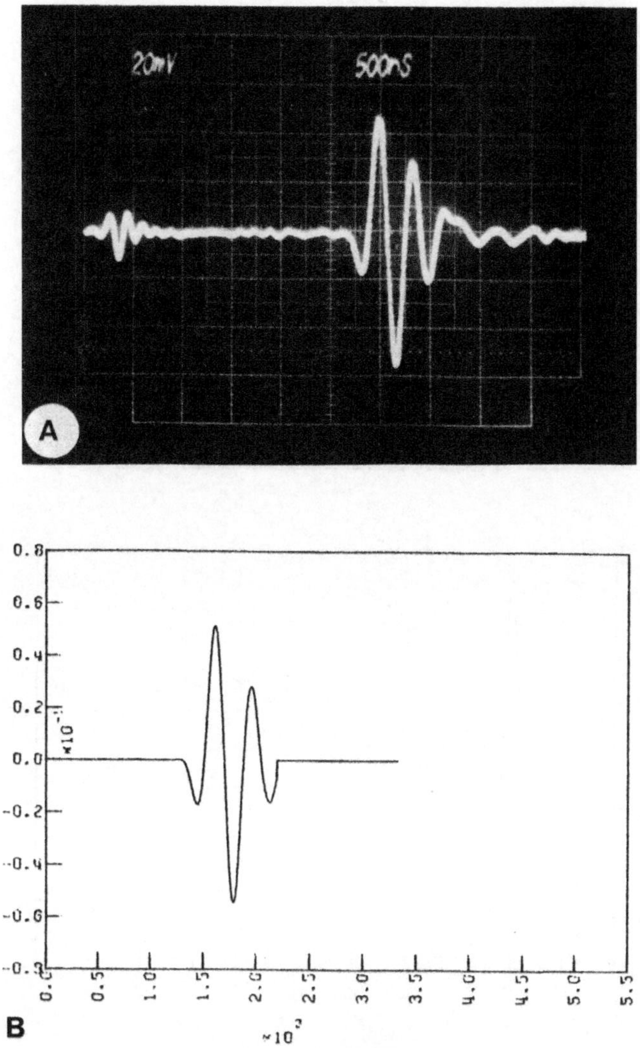

Fig. 3. Echo on tungsten carbide ball diameter 0.8 mm

(a) measured
(b) simulated

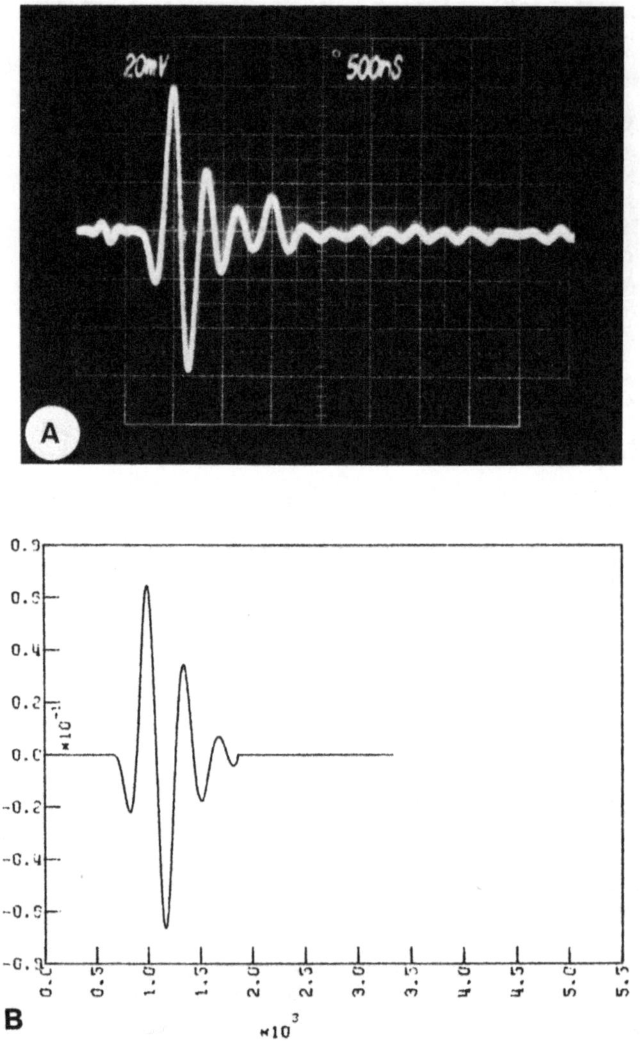

Fig. 4. Echo on tungsten carbide ball diameter 1 mm

(a) measured

(b) simulated

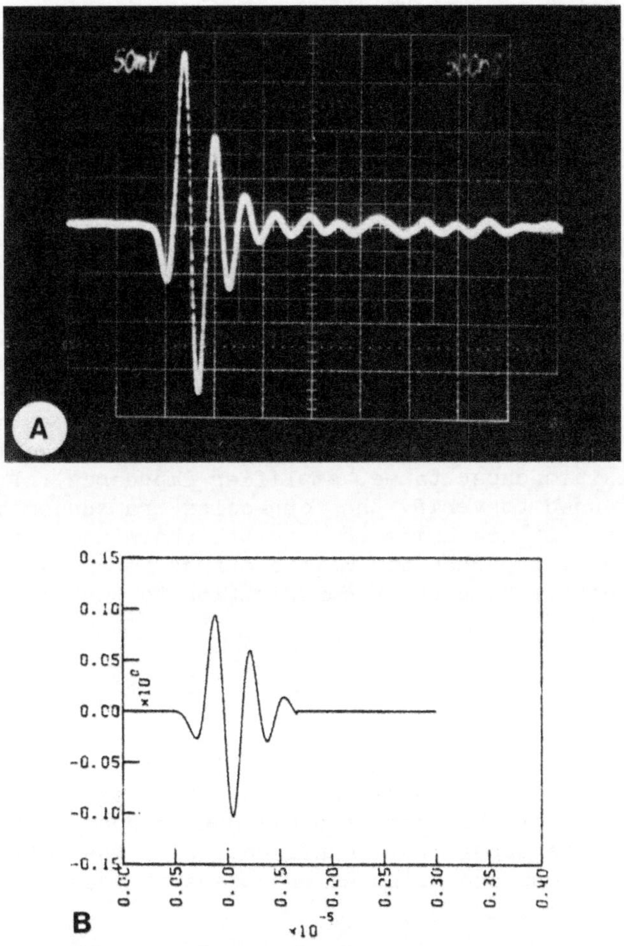

Fig. 5. Echo on tungsten carbide ball diameter 2 mm

(a) measured
(b) simulated

2. Assumption of the piston mode which has been proved to be erroneous by 30 % for the axial pressure as evaluated by measurements of the pattern of radiation. This assumption is responsible for a large part of the absolute discrepancy. The relative discrepancy for the balls of .79 or 1 mm is less relevant of this hypothesis since for small ball the contribution of term L_2 (taking into account the second derivative of the field ψ) is small compared to that of $L_0 = \psi * \psi'$. But for the ball of 2 mm both contributions of L_0 and L_2 are opposite and of the same order and this may explain the larger discrepancy of this case.

3. Assumption of target rigidity. We know that the tungsten carbide has an acoustic impedance of 92.10^6 kg $m^{-2}s^{-1}$. This large value compared to 1.5 10^6 for water should make negligible the amount of wave inside the ball. Nevertheless even a small amount of energy inside the ball and appearing as generalized Rayleigh waves[2] will remain a long time inside this ball since the difference of acoustic impedance between tungsten carbide and water prevents also its way out. Then the elasticity of the ball is likely to be responsible of a longer trail of the pulse.

4. Electric load of the receiver. Our modelization does not take completely into account the electric load of the receiver (coaxial cable, parasitic capacitance, amplifier impedance). Though we have endeavoured to verify the four-poles transducer transfer functions used by measuring an acoustic chain made of two transducers facing each other at small distance we are sure that non negligible errors have their contribution to the total discrepancy.

CONCLUSIONS

We are still in a stage where more efforts have been devoted to theoretical or computational tasks than to experimental measurements. Despite of the large experimental errors implied by our methods the results appear as a confirmation of the validity of our theoretical equations and computer processes at least for rigid target.

In order to offer a good calibration tool, the use of an echo on spherical target requires three conditions :

1. Incorporation, in the theoretical model, of Rayleigh waves.
2. Realization of an improved method of supporting the ball (maybe a magnetic levitation).
3. Realization of transducers acting accurately in piston mode or in well known mode.

When these three conditions are satisfied, the association of a small ball and a known transducer is equivalent to a small calibration hydrophone but its pattern of radiation is completely known and stable which is not the case of a small calibration transducer.

Then techniques using spherical targets are likely to undergo a large development in the ultrasonic field.

REFERENCES

1. R.H. Hill and N.L. Adams, Reinterpretation of the reciprocity theorem for the calibration of acoustic emission transducers operating on a solid, Acoustica 1979, vol. 43, pp. 305-312.
2. W.P. Mason and R.N. Thurston, Physical Acoustics, vol. XII, Academic Press 1976.
3. G. Deprez and R. Hazebroock, Mesure d'une réponse impulsionnelle en acoustique. Application à la diffraction d'une onde sonore par une sphère, Acoustica 1980, vol. 45, pp. 96-102.
4. F.B. Stumpf and A.M. Junit, Effect of a spherical scatterer on the radiation reactance of a transducer at an air-water surface, JASA 1980, vol. 67(2), pp. 715-716.
5. P.L. Edwards and J. Jarzynski, Use of a microsphere probe for pressure field measurements in the megahertz frequency range, JASA 1980, vol. 68(1), pp. 356-359.
6. M. Auphan, R.H. Coursant and C. Mequio, Paper to be published in the Philips Journal of Research.
7. P.M. Morse and K.U. Ingard, Theoretical Acoustics, Mc Graw Hill book Company, 1968.
8. M. Auphan and J. Matthys, Reflection of a plane impulsive acoustic pressure wave by a rigid sphere, Journal of Sound and Vibration 1979, vol. 66(2), pp. 227-237.
9. R.H. Coursant, Les transducteurs ultrasonores, Acta Electronica 1979, vol. 22(2), pp. 129-141.
10. M. Auphan and H. Dormont, Pulse acoustic radiation of plan damped transducers, Ultrasonics 1977, vol.15(4), pp. 159-168.
11. E. Grosswald, Bessel polynomials, Springer-Verlag 1978.
12. R. Hickling, Analysis of echoes from a solid elastic sphere in water, JASA 1962, vol. 34, n° 10, pp. 1582-1592.

ANALYSIS OF VIBRATING SURFACES USING ACOUSTIC HOLOGRAPHY

Masahide Yoneyama*
Nippon Columbia Co., Ltd.
5-1, Minato-Cho, Kawasaki-Ku
Kawasaki 210, JAPAN

Carl Schueler and Glen Wade
Dept. of Electrical and Computer Engineering
University of California
Santa Barbara, CA 93106 USA

ABSTRACT

Laser holographic interferometry has proved to be a useful tool for analysis of oscillating surfaces. Unfortunately, optical interferometry is limited to vibration amplitudes of microns or less because of the short wavelength of laser light, and to small objects because of coherence requirements and the limited power in a dispersed laser beam. We examine an acoustic holographic method to overcome the limitations of laser holography for vibrational analysis. Since powerful, coherent ultrasound can be dispersed over a large area, large objects may be analyzed. The long acoustic wavelength allows millimeter vibrational amplitudes, a thousand times larger than laser holography can handle. Acoustic holographic detection is linear, whereas optical interferometry is a non-linear process, in practice. Therefore, acoustic holography should allow the object's vibrational amplitude to be obtained for every point on the object surface, whereas, in practice, laser interferometry yields the vibrational amplitude only at discrete points (the null points) on the object surface. Using theoretical analysis and computer simulations, we consider vibrational analysis of objects with vibrational amplitudes in the millimeter range. We present the theory for surface vibration with variation in two dimensions, and with both plane-wave insonification as well as the more general case of spherical wave insonification.

---

*Dr. Yoneyama collaborated on this work with Dr.'s Schueler and Wade while he was a visiting research scholar at UCSB during the 1979-80 academic year.

INTRODUCTION: LIMITATIONS OF LASER INTERFEROMETRY

Acoustic vibrational analysis is inspired by the well-known method of laser holographic vibrational analysis, first reported by Powell and Stetson [1]. In their work, a laser hologram was exposed for a time period much greater than the period of vibration of the oscillating surface. The result was a hologram that produced an image resembling the image obtained for the stationary surface, except for bright and dark contour fringes superimposed. Powell and Stetson showed that the contour fringe intensity is related to the vibrational amplitude through the square of the zero-order Bessel function.

The zero-order Bessel function, J_0, is shown in figure 1. The maximum value of the Bessel function occurs when the argument is zero. As the argument increases in value, the Bessel function passes through zero to a minimum value, and then oscillates about zero with decreasing amplitude. As the amplitude of vibration increases, the intensity of the image of the vibrating surface oscillates between maxima and minima, with decreasing overall brightness. This result has been of practical use to measure vibrational patterns of oscillating surfaces [2], but it has at least the following restrictive limitations:

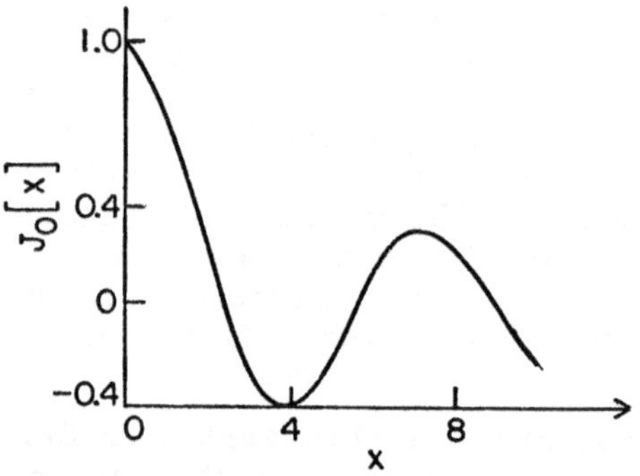

Figure 1 Zero-order Bessel function of the first kind, plotted for argument x, ranging from 0 to 10.0.

1) In both laser and acoustic interferometry, the argument of the J_0 function is inversely proportional to the radiation wavelength. (See equation 6). Therefore, the shorter the radiation wavelength, the more densely spaced the contour fringes become. This limits the maximum vibrational amplitude on the object to about ten wavelengths of the radiation [3]. He-Ne laser light has a wavelength of about 0.6 microns, so that approximately 6 microns is the maximum vibrational amplitude that can be measured with laser interferometry.

2) In practice, photographic film is a non-linear recording medium, which makes it impossible to accurately measure vibrational amplitude at every point of the image. At the points on the image where the intensity is zero, the zero-order Bessel function also has the value zero, and only there is the vibrational amplitude accurately determined.

By increasing the wavelength and spatial spread of the coherent radiation, and linearizing the detection system, the limitations of laser interferometry would be overcome. Acoustic transducers have linear response, and acoustic radiation can conveniently have wavelengths of the order of mm., compared to the 0.6 micron wavelength of He-Ne laser light.

Several researchers have investigated acoustic interferometry [4-8]. However, the investigations were limited to applications for non-vibrating objects, and include several types of interferometry. The one most closely related to time-exposure interferometry of a vibrating object is double-exposure interferometry. It has been shown, but only for laser interferometry [9], that a double-exposed hologram of an object at two extreme positions of displacement yields approximately equivalent results to a time-exposed hologram of a vibrating object with the same maximum displacement amplitude. Even if this result is valid for acoustic interferometry, there are many situations for which it is not possible to obtain instantaneous holograms at two extremes of vibration of an object, and double-exposure interferometry would not be applicable. Lohmann [10] has suggested that acoustic holography might be useful for vibrational analysis of large objects, such as ship hulls. The method proposed here allows analysis of a vibrating object, and appears to be the first realization of his suggestion.

The development of acoustic vibrational analysis, although motivated in concept by Powell and Stetson's work on optical vibrational analysis, cannot be assumed at the outset to be equivalent to the optical case. This is primarily because the speed of sound is so much less than the speed of light. The development here takes into account the speed of sound, as well as the type of acoustic detection and image reconstruction system that may be used to obtain interpretable vibrational interferograms in the acoustic case.

PRINCIPLE OF ACOUSTIC VIBRATIONAL ANALYSIS

Figure 2 shows an acoustic receiver in the $z = 0$ plane. Plane insonifying waves propagate in the plus z direction from a point source at $z = -\infty$. Let the phase of the plane-wave insonification be zero at $z = 0$ and at $t = 0$. A point object is placed at $z = z_0$ in front of the receiver as shown. The object has sinusoidal motion given by $z = z_0 - a \sin\nu t$, where a is the amplitude of the oscillation, and ν is the angular frequency of the oscillation. Let ω be the angular frequency, and c the velocity, of the acoustic waves. Denote the time at which a

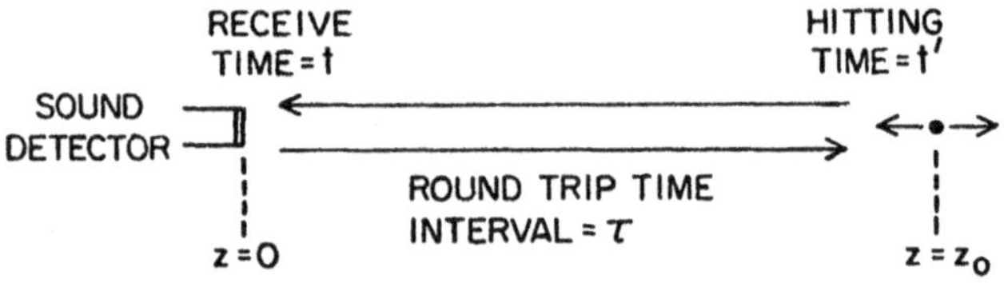

Figure 2 A point object at equilibrium position z_0, vibrates sinusoidally along the z axis. Plane waves insonify the object at time t'. The detector receives reflected sound at time t. The round trip is τ.

wavefront strikes the point object as $t = t'$, the "hitting" time in figure 2. The round-trip time interval required for the wavefront to leave the detector at $z = 0$, strike the point object, and return to the detector is τ. Finally, let t denote the time at which the wavefront returns to the detector, the "receive" time in figure 2.

With this notation, the round trip time interval can be written:

$$\tau = 2(z_0 - a \sin \nu t')/c. \tag{1}$$

The hitting time t' becomes:

$$t' = t - \tau/2. \tag{2}$$

Substitution of t' in the expression for τ yields:

$$\tau = \frac{2z_0}{c} - (\frac{2a}{c}) \sin \nu[t - \frac{(z_0 - a \sin \nu t')}{c}]$$

$$= \frac{2z_0}{c} - \frac{2a}{c} \{[\sin \nu(t - z_0/c) \cos\{(\nu a \sin \nu t')/c\}]$$

$$- [\cos \nu(t - z_0/c) \sin\{(\nu a \sin \nu t')/c\}]\}.$$

Maximum values of ν and a are:

$\nu = 1000 \times 2\pi$ rad/sec;

$a = 2.0$ mm;

$c = 150{,}000$ cm./sec in water.

Therefore, $[\nu a/c]max = 0.0084$ radian. This allows the approximations:

$$\sin[(\nu a \sin \nu t')/c) \cong (\nu a \sin \nu t')/c \tag{3a}$$

$$\cos[(\nu a \sin \nu t')/c] \cong 1 \tag{3b}$$

τ becomes:

$$\tau \cong 2z_0/c - (2a/c)\sin \nu(t - z_0/c)$$

$$+ (2\nu a^2/c^2)(\sin \nu t')\cos\nu(t - z_0/c).$$

With the maximum values of ν, a, and c, $(2\nu a^2/c^2)$ has about one-hundredth the value of the second term in τ above, and can be ignored. Finally, the practical expression for the round-trip time becomes:

$$\tau \cong 2z_0/c - (2a/c) \sin \nu(t - z_0/c). \tag{4}$$

With expression (4) for τ, the phase at the receiver becomes:

$$\phi(t) = \omega(t - \tau)$$
$$= \omega t - \omega[2z_0 - 2a \sin \nu(t - z_0/c)]/c. \tag{5}$$

The received signal is given by $\exp[j\phi(t)]$. When that signal is passed through the processing illustrated in figure 3, the following result is obtained:

$$[\int_0^T e^{j[\phi(t)-\omega t]} dt]/T \propto J_0[4\pi\ a/\lambda], \tag{6}$$

where λ is the wavelength of the sound, and J_0 is the zero-order Bessel function.

Eqn. (6) is the fundamental expression for acoustic holographic interferometry of a vibrating object, and was developed for the case of a point object and a single detector. The detection process requires a reference signal to measure the phase $\phi(t)$. If there were a two-dimensional array of detectors, rather than only one detector, the information measured by the array, and processed as indicated in figure 3, would constitute a record of the time-averaged complex-amplitude of the wavefield at the aperture defined by the array. This record could be thought of as a time-averaged complex-amplitude hologram of the wavefield at the aperture. (6) is the time-averaged complex-amplitude at one detector, and hence represents a time-averaged, spatial sample at one point of an aperture situated in a plane normal to the z axis of figure 2, and located at z = 0. From (6) one sees that the time-averaged complex amplitude of the sound received from the vibrating point object is a Bessel function of zero order whose argument is proportional to the vibrational amplitude of the point object. This result is similar to that of Powell and Stetson for laser holographic interferometry, even though in that analysis propagation time is not a significant factor, and here it must be taken into account.

Time averaging in (6) can be done in the computer by summing many instantaneous received signals from the existing equipment in the water. Figure 4 is a graph of 100 simulated time-average complex-amplitude hologram samples for the case considered. A single detector repetitively samples the sound received from a vibrating point object as illustrated in figure

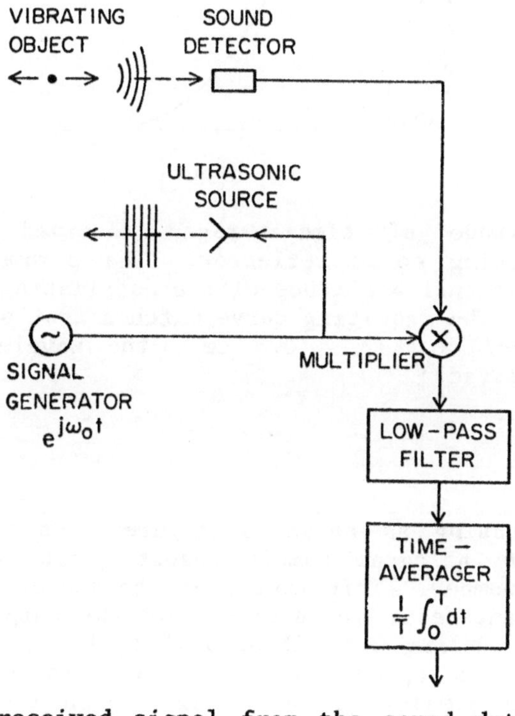

Figure 3 The received signal from the sound detector is processed as shown. The output of the time averager has the form of a zero-order Bessel function, whose argument is proportional to the amplitude of vibration of the object.

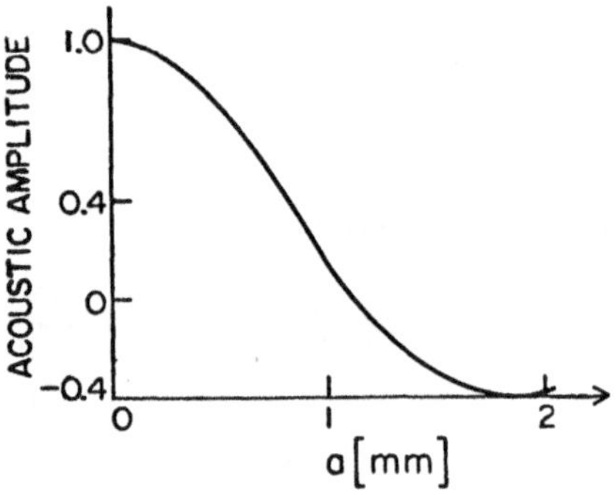

Figure 4 Amplitude of time-averaged received data from a vibrating point reflector. The parameter a is the vibrational amplitude of the oscillating point reflector. The resulting curve matches that of $J_o[x]$, for $x = (4\pi/\lambda)a$. (λ = 0.6 cm., the wavelength of sound simulated.)

2. After processing as shown in figure 3, a time-averaged, complex-amplitude hologram sample results that consists of a single complex number. This number is the sum of 100 uniformly sequenced, instantaneous measurements of the complex amplitude at the detector during one vibrational cycle of the point object. This sum corresponds to the time integral of (6). It is possible in the simulation to obtain a good result by time-averaging over a single cycle because the instantaneous measurements are made at uniformly sequenced times. This may not be possible in practice, so that it may be necessary to average over many cycles in a practical experiment.

Each of the point hologram samples is for a different vibrational amplitude a of the point object. One goal of the simulation was to see if the complex number representing each time-average hologram sample for the vibrating point object was equal to the magnitude of the zero order Bessel function of the vibrational amplitude as indicated in (6). The vibrational amplitude a was varied from 0 to 2.0 mm. in 0.02 mm steps, at an oscillation frequency of 1000 Hz. The simulation was done using equations (4) and (5). The result matches the J_0 function as specified in (6).

EXTENSION TO ANALYSIS OF A TWO-DIMENSIONAL SURFACE VIBRATION

The previous analysis was for a point source vibrating along the z axis a distance z_0 from a single detector. In the present analysis, there are a large set of points on the xy plane, with an unknown pattern of vibration, as illustrated in figure 5. The equilibrium position of each point on the surface lies exactly on the xy plane a distance z_0 from a set of detectors lying in the x_1y_1 plane. Let the unknown vibrational pattern be represented by the function a(x,y) over the xy plane. The vibration of any point of the vibrating surface is assumed to be sinusoidal, and in the z direction. (As in Powell and Stetson's work, the negligible lateral motion that would exist for a vertically vibrating surface is ignored [11]). Therefore, the vibration at each point is given by a(x,y)sinνt, where ν is the constant vibrational frequency of the surface in radians/ sec. For simplicity, the surface is assumed to be insonified

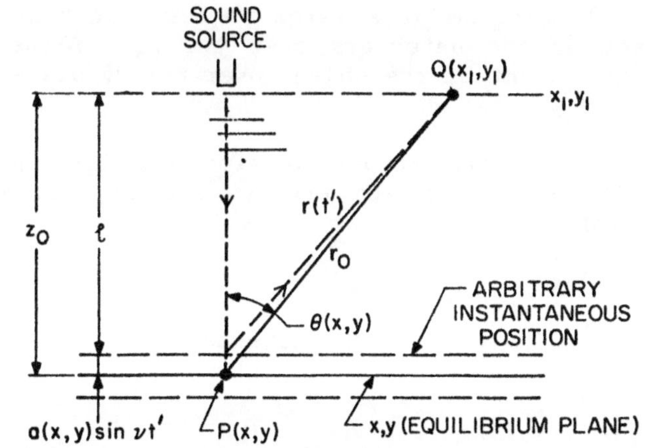

Figure 5 Vibrational analysis with plane-wave insonification. The dotted line above the equilibrium plane represents the position of P(x,y) at time t'. The lower dotted line acts as a reminder that P(x,y) will have a corresponding location below the equilibrium plane at another time during the vibration cycle.

from above by a normally incident plane wave, whose phase is zero at the x_1y_1 plane. The distance to the equilibrium position of an arbitrary point of the surface P(x,y) from an arbitrary detector is denoted r_0. The distance to the instantaneous position of P(x,y) at time t' is denote r(t'). The vertical distance from the instantaneous position of P(x,y) at t' to the detector plane x_1y_1 is denoted $\ell(t')$.

Based on the definitions, and on figure 5, the following relations can be written:

$$\ell = z_0 - a(x,y)\sin\nu t' \tag{7}$$

$$r_0 = (z_0^2 + [x - x_1]^2 + [y - y_1^2]^{1/2} \tag{8}$$

Since $z_0 >> a(x,y)$:

$$r \cong r_0 - a(x,y)(\cos\theta)\sin\nu t' \tag{9}$$

θ depends on both x,y and x_1y_1. However, one would usually want to make a small hologram of a large object. In that case, the spatial extent of the detectors over the x_1y_1 plane is small compared to the extent of the object over the xy plane, and θ is independent of x_1y_1.

From figure 5, the roundtrip time for the radiation to travel from the x_1y_1 plane to P(x,y), and back to a detector $Q(x_1,y_1)$ becomes:

$$\tau = (\ell + r)/c. \tag{10}$$

The hitting time, in terms of the detected time t, is:

$$t' = t - r/c. \tag{11}$$

Equations similar to (4) and (5) are written with relations (7) - (9):

$$\tau = [z_0 + r_0 - a(x,y)(1 + \cos\theta)\sin\nu t']/c \tag{12}$$

$$t' = t - (r_0 - a(x,y)(\cos\theta)\sin\nu t')/c. \tag{13}$$

The approximations (3) and substitution of (13) into (12) yields:

$$\tau \cong [z_0 + r_0 - a(x,y)m(x,y)\sin\nu(t - r_0/c)]/c \tag{14}$$

where $m(x,y) \equiv 1 + \cos\theta$.

The phase angle of the reflected sound at the detector at point $Q(x_1,y_1)$ in figure 5 becomes:

$$\phi(t) = \omega(t - \tau)$$

$$\cong \omega t - \frac{\omega}{c}[z_0 + r_0 - a(x,y)m(x,y)\sin\nu(t - r_0/c)]. \tag{15}$$

This is the phase angle of a wave measured at Q in response to a reflected wave from a very small area element on the surface at P in figure 5. Since the points P and Q are arbitrary, (15) specifies the phase angle at Q from any point of the surface. The detector at Q receives information from the entire surface, and the received signal at Q can be written as a superposition of the signals from all points on the surface [12]:

$$h'(x_1,y_1) = \frac{1}{j\lambda}\iint_{-\infty}^{\infty}\frac{e^{j\phi(t)}}{r}\,dx\,dy. \tag{16}$$

In this expression, it is possible to allow $r \cong r_0$ in the denominator without significant error. Substituting (15) into (16) gives:

$$h'(x_1,y_1) = \frac{e^{j\omega(t-\frac{z_0}{c})}}{j\lambda}\iint_{-\infty}^{\infty}\frac{e^{-j\frac{\omega}{c}[r-am\sin\nu(t-\frac{r_0}{c})]}}{r_0}\,dx\,dy$$

where $a(x,y)$ and $m(x,y)$ have been replaced by a and m, for brevity.

The received signal is processed as illustrated in figure 3, with a low-pass filter to eliminate the ω carrier. The low-pass filter yields the real and imaginary parts of the signal:

$$h(x_1,y_1) = \frac{e^{-jkz_0}}{j\lambda}\iint_{-\infty}^{\infty}\frac{e^{-jk[r_0-am\sin\nu(t-\frac{r_0}{c})]}}{r_0}\,dx\,dy \tag{17}$$

where $k = 2\pi/\lambda$.

where $k = 2\pi/\lambda$. This result can be obtained using the back-propagation reconstruction technique.[13]

A computer simulation based on (15) and (18) for the case of a one-dimensional surface illustrates the procedure. The simulation is for the case of a one-dimensional set of point reflectors that are insonified by a normally incident plane-wave from above. The point reflectors are aligned along the x axis,

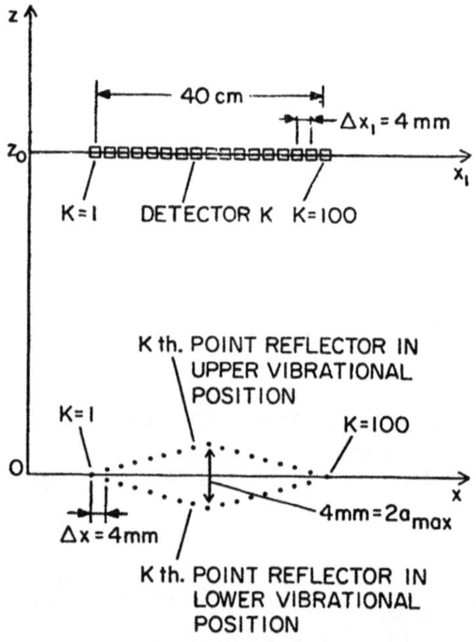

Figure 6 The geometry of the simulation to test the plane-wave theory. The point reflectors are precisely located under corresponding detectors in the receiving array.

$h(x_1,y_1)$ is time-averaged to obtain the acoustic interferogram. If the vibrational frequency corresponds to a period $2\pi/\nu$, then (17) should be time averaged for a period $T >> 2\pi/\nu$. The time-averaged signal can be expressed as:

$$\bar{h}(x_1,y_1) = \frac{1}{T}\int_0^T h(x_1,y_1)\, dt$$

$$\cong \frac{e^{-jkz_0}}{j\lambda} \iint_{-\infty}^{\infty} \frac{e^{-jkr_0}}{z_0}\left[\frac{1}{T}\int_0^T e^{jkam\sin\nu(t-\frac{r_0}{c})} dt\right]dx\, dy$$

$$= A_0 \iint_{-\infty}^{\infty} f(x,y)g(x-x_1,y-y_1)\, dx\, dy \qquad (18)$$

where

$$A_0 = e^{-jkz_0}$$

$$f(x,y) = \frac{1}{T}\int_0^T e^{jkam\sin\nu(t-\frac{r_0}{c})} dt$$

$$g(x-x_1,\ y-y_1) = \frac{1}{j\lambda}\frac{e^{-jkr_0}}{z_0}$$

and r_0 is given by (8).

The approximation $r_0 \cong z_0$ in the denominator of (18) is commonly made, and $g(x-x_1,y-y_1)$ is recognized as a Green's function of wave propagation from the xy plane to the x_1y_1 plane [12]. Therefore, the function $f(x,y)$ is the time-averaged object wavefield distribution in the x,y plane that results in the received distribution $\bar{h}(x_1,y_1)$ in (18). Furthermore, equation (18) is in the same form as for the received wave distributions for a stationary object. The only difference is the form of $f(x,y)$. Therefore, $f(x,y)$ can be obtained with image reconstruction techniques that are normally used to reconstruct acoustic holograms of stationary objects [13]. In the appendix, it is shown that:

$$f(x,y) \cong J_0[k\ m(x,y)a(x,y)] \qquad (19)$$

as shown in figure 6, and the detectors are aligned along the x_1 axis, a distance z_0 = 100 cm. from the reflectors. The vibration amplitude is zero at both ends of the point reflector set, and increases linearly to a maximum value of 2.0 mm. at the center reflector. Note that in this case, the object and hologram are the same size. Therefore, the angle θ in figure 5 will depend on both x and x_1. This causes only slight error in the result, because $\cos\theta_{max} > 0.9$.

The initial result of the simulation is shown in figure 7. This result is to be compared to the result of back-propagating an image of the non-vibrating reflectors (figure 8). Note that there is an initial dip in the image at both edges. Ideally, the image of the stationary reflectors would be completely flat. However, back-propagation reconstruction with a finite aperture is space-variant [14]. Therefore, the image data that produced figure 7 is corrected by performing a complex division by the image data that produced figure 8. This results in the dashed curve shown in figure 9. The solid curve in figure 9 is the result of a direct calculation of the J_0 function of the vibrational amplitude for each point reflector in the simulation. The agreement is remarkably good, considering the reconstruction space-variance, and the θ variation in the simulation.

Figure 10 illustrates one way that unknown vibrational amplitude at each point of a curve such as that in figure 9 could be obtained. Figure 10 is a graph of the vibrational amplitude as a function of the zero-order Bessel function value. Because the zero-order Bessel function has been shown to be proportional to the acoustic amplitude of the image, this graph should yield accurate conversion of the image acoustic amplitude to object vibrational amplitude.

SPHERICAL INSONIFICATION

The extension of the two-dimensional analysis to the more general case of spherical wave insonification from a small insonifying transducer is presented in this section. It should be noted that even in this case, the analysis presented in the previous section will usually be adequate because spherical waves become almost planar if z_0 is large enough.

Figure 11 is a modification of figure 5 for the general case of a point source of sound rather than plane wave insonification. The position of the point insonifier is held fixed in the x_1y_1 plane at location (x_1', y_1'). The vector r_1 denotes the distance from the point insonifier to the instantaneous position of an arbitrary point P(x,y) on the vibrating surface at time

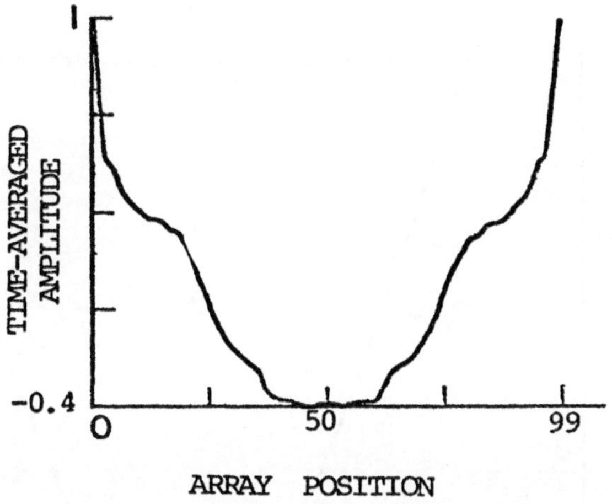

Figure 7 Initial result of the computer simulation, without correction for system space-variance.

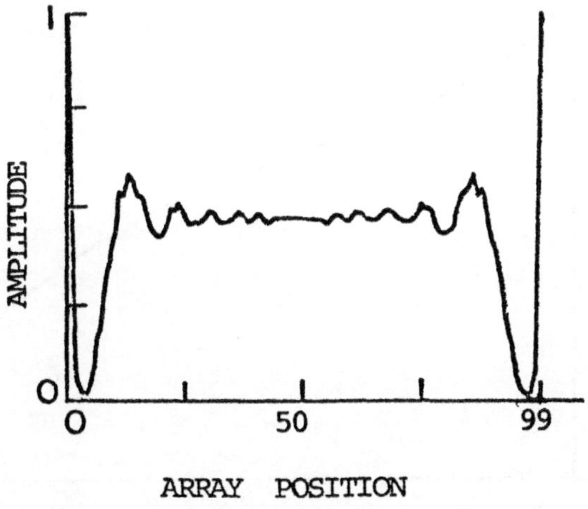

Figure 8 Simulated image of non-vibrated sources of figure 6.

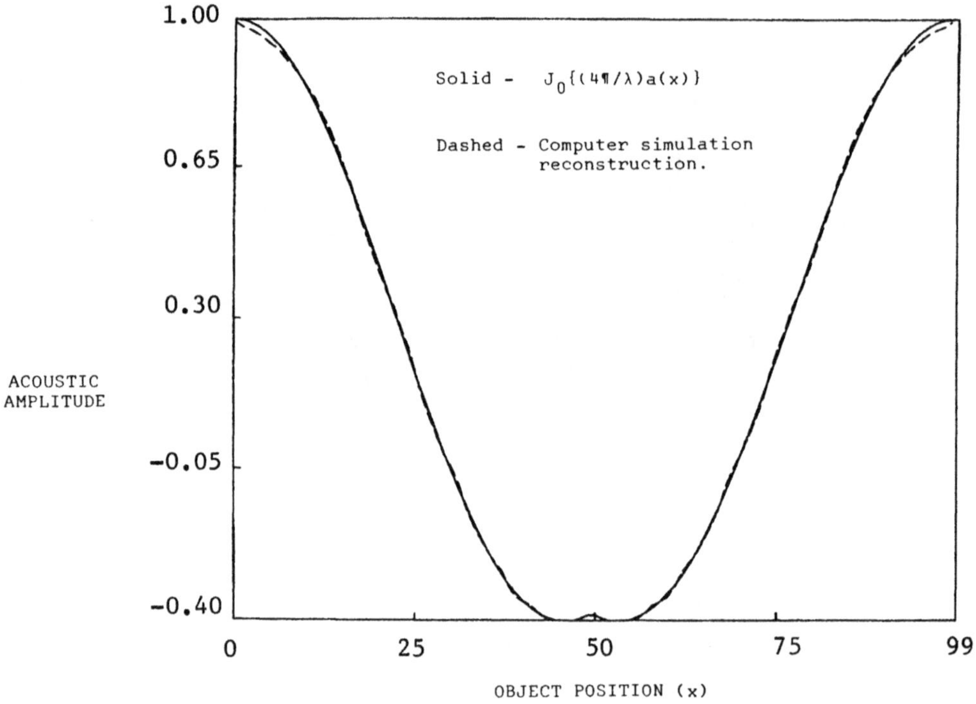

Figure 9 Corrected computer simulation compared to the J_o function.

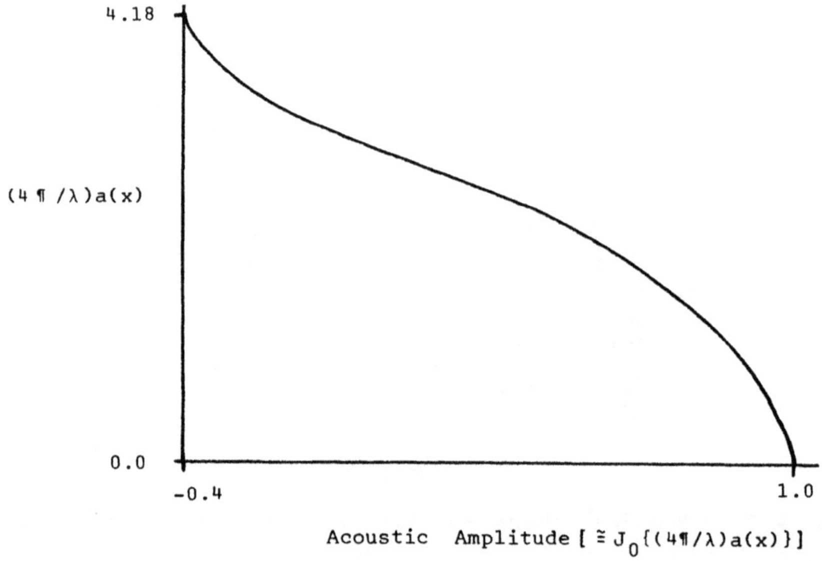

Figure 10 Conversion of acoustic amplitude to vibrational amplitude, a(x).

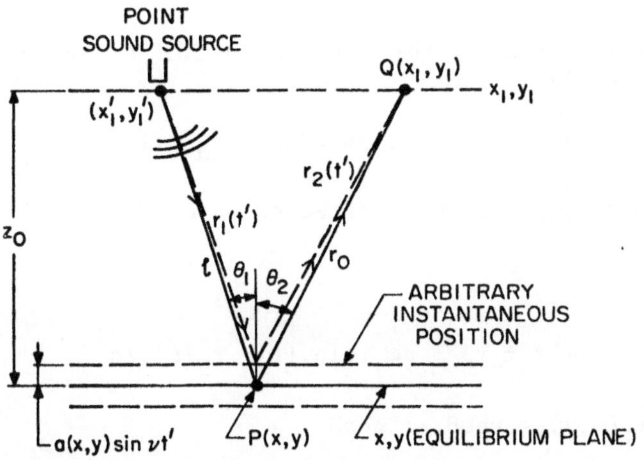

Figure 11 Vibrational analysis with spherical-wave insonification.

t'. ℓ denotes the distance from the point insonifier to the equilibrium position of P(x,y) in the xy plane. r_0 again denotes the distance from the equilibrium position P(x,y) to an arbitrary detector at $Q(x_1,y_1)$. Finally, r_2 denotes the distance from the instantaneous position of P(x,y) at time t' to $Q(x_1,y_1)$.

Therefore, the following set of relations can be written:

$$\ell = (z_0^2 + [x - x_1']^2 + [y - y_1']^2)^{1/2} \qquad (20a)$$

$$r_0 = (z_0^2 + [x - x_1]^2 + [y - y_1]^2)^{1/2} \qquad (20b)$$

$$r_1 = \ell - a(x,y)(\cos\theta_1)\sin\nu t' \qquad (20c)$$

$$r_2 = r_0 - a(x,y)(\cos\theta_2)\sin\nu t'. \qquad (20d)$$

θ_1 and θ_2 are assumed to be independent of x_1,y_1 as θ was in the previous section. The equations relating t, τ, and t' become:

$$\tau = [\ell + r_0 - a(x,y)(\cos\theta_1 + \cos\theta_2)\sin\nu t']/c$$

$$t' = t - [\ell - a(x,y)(\cos\theta_1)\sin\nu t']/c.$$

Substitution of t' into the expression for τ, and use of the approximations (3), yields:

$$\tau \cong [\ell + r_0 - a(x,y)m_1(x,y)\sin\nu(t - r_0/c)]/c. \tag{22}$$

The phase angle of the reflected wave at point Q of figure 11 becomes:

$$\begin{aligned}\phi(t) &= \omega(t - \tau)\\ &= \omega t - \omega[\ell + r_0 - am_1 \sin\nu(t - r_0/c)]/c\end{aligned} \tag{23}$$

where $m_1 = m_1(x,y) = \cos\theta_1 + \cos\theta_2$.

The received signal at Q in figure 11 becomes a superposition of signals from the entire surface:

$$h'(x_1,y_1) = (1/j\lambda) \iint_{-\infty}^{\infty} \frac{e^{j\phi(t)}}{(r_1+r_2)}\, dx\, dy. \tag{24}$$

If $z_0 >> a(x,y)$:

$$r_1 + r_2 \cong \ell + r_0.$$

Therefore, $h'(x_1,y_1)$ becomes:

$$h'(x_1,y_1) \cong \frac{e^{j\omega t}}{j\lambda} \iint_{-\infty}^{\infty} \frac{e^{-jk[\ell+r_0-am_1\sin\nu(t-\frac{r_0}{c})]}}{4\pi(\ell+r_0)}\, dx\, dy.$$

This is processed and time-averaged as illustrated in figure 3 to obtain:

$$\bar{h}(x_1,y_1) = \iint_{-\infty}^{\infty} \partial(x,y)g(x-x_1,y-y_1)\, dx\, dy \tag{25}$$

where

$$\partial(x,y) \equiv \left(\frac{z_0}{\ell+r_0}\right)e^{-jk\ell} \frac{1}{T}\int_0^T e^{jkam_1 \sin\nu(t-\frac{r_0}{c})} dt \quad (26)$$

$$g(x-x_1,y-y_1) \equiv \frac{1}{j\lambda} \frac{e^{jk(z^2+[x-x\ ]^2+[y-y\ ]^2)^{1/2}}}{z_0} \quad . \quad (27)$$

The back-propagation reconstruction technique can be used to obtain $\partial(x,y)$, since $g(x-x_1,y-y_1)$ is the Green's function, and (25) has the same form as (18). Let $\partial(x,y) = \alpha(x,y)f(x,y)$, where:

$$\alpha(x,y) \equiv \frac{z_0}{\ell+r_0}\ e^{-jk\ell} \quad (28)$$

$$f(x,y) \equiv \frac{1}{T}\int_0^T e^{jkam_1 \sin\nu(t-\frac{r_0}{c})} dt. \quad (29)$$

Since $\alpha(x,y)$ is directly computable from the known geometry of the system, $f(x,y)$ can be obtained by the back-propagation reconstruction process [15]. The vibrational amplitude information $a(x,y)$ can be extracted once $f(x,y)$ is reconstructed. Since (29) is of the same form as $f(x,y)$ in the previous section, $f(x,y) \cong J_0[k\ m_1\ a(x,y)]$.

DISCUSSION

Acoustic vibrational analysis depends on the phase-measuring ability of a holographic recording system. A pulse- echo technique [16] cannot perform vibrational analysis, because phase is not measured. Acoustic vibrational analysis also depends on linear acoustic detection [or at least a known nonlinear relationship] to yield an accurate estimate of surface vibrational amplitude at every point on the surface, in the absence of speckle degradation. Since acoustic waves have very long wavelength compared to laser light, this technique will allow measurement of vibrational amplitudes many orders of magnitude greater than laser interferometry. The wide spread of the coherent acoustic radiation should allow analysis of very large vibrating surfaces.

Also, if the surface of interest is large compared to a resolution cell of the imaging system, speckle should not be too troublesome. Therefore, the technique that has been developed here is limited in application to such large objects. The speckle limitation may not restrict the technique in practice because the cases for which the object size is not large compared to an image resolution cell may be amenable to laser interferometry.

Finally, note that the J_0 function oscillates about zero as its argument increases, as in figure 1. This behavior causes the well-known contour patterns on laser interferograms of vibrating objects. It would also cause similar patterns in the reconstructed images from acoustical interferograms of vibrating surfaces. In the case of figure 9, each half of the symmetric reconstruction begins to become multi-valued at the center. Were the maximum vibrational amplitude at the center greater than 2.0 mm., then the center of the reconstruction would repeat J_0 function values that correspond to smaller vibrational amplitude toward the edge of the reconstruction. In this case, it is possible that the same reconstruction could also result for a more complicated vibrational pattern in which the vibrational amplitude increases from the edge part way to the center, and then decreases to the center to repeat some of the J_0 function values.

In such a case, it is still possible to ascertain the correct vibrational pattern if the vibrational amplitude is adjustable. The vibrational pattern is independent of the vibrational amplitude of the driving oscillator that causes the surface of interest to oscillate. The J_0 function argument is proportional to vibrational amplitude. Therefore, one can reduce the argument of the J_0 function to an arbitrarily small value at all points on the vibrating surface by reducing the driving oscillator's vibrational amplitude. If the maximum amplitude of vibration on the surface is less than the amplitude for which the J_0 function becomes multi-valued, then the locations of the vibrational minima and maxima on the surface of interest may be unambiguously determined by a time-exposed interferogram. The driving oscillator amplitude may then be increased to give a contour pattern in another interferogram, so that the detailed vibrational amplitude pattern between the minima and maxima may be determined.

The use of such a procedure has allowed complicated vibrational patterns to be analyzed successfully with laser holography at Nippon Columbia Co., in Japan, and should be applicable in the case of acoustic vibrational analysis, as well. In addition, note that since the argument of the J_0 function is inversely proportional to the wavelength of the imaging radiation, increasing the wavelength of the radiation is equivalent to reducing the vibrational amplitude of the driving oscillator. Laser holographers cannot take advantage of this observation, since they are usually limited to the use of short wavelength laser light. Acoustic vibrational analysis has the advantage of offering control over the wavelength of the insonifying radiation. For this reason, the use of acoustic vibrational analysis is not restricted to the case of surfaces driven by controllable driving oscillators, but may be applied when the driving oscillator is not under human control. Examples might include underwater structures undergoing oscillation due to water current, or wind-driven oscillations of a structure in air.

APPENDIX: Derivation of Eq. (19)

Two independent derivations are given.

1) $$f(x,y) = \frac{1}{T}\int_0^T e^{jkamsin\nu(t-\frac{r_0}{c})}dt$$

$$= \frac{1}{\nu T}\int_{-\nu r_0/c}^{\nu(T-r_0/c)} e^{jkamsin\kappa}d\kappa$$

$$= \{ \int_0^{2\pi f_\nu(T-r_0/c)} e^{jkamsin\kappa}d\kappa + \int_0^{2\pi f_\nu r_0/c} e^{jkamsin\kappa}d\kappa\}/(2\pi f_\nu T)$$

where $f_\nu = \nu/2\pi$, and $\kappa = \nu(t - r_0/c)$. If $T >> 1/f_\nu$, we can write:

$$f(x,y) \cong [f_\nu(T - r_0/c)$$

$$\cdot \int_0^{2\pi} e^{jkamsin\kappa}dk + f_\nu\frac{r_0\varepsilon}{c} \int_0^{2\pi} e^{-jkamsin\kappa}d\kappa/2\pi f_\nu T]$$

where $\varepsilon \leqq 1$, and has whatever value is necessary to make the above expression valid. Since the two terms in f(x,y) above have different signs in the exponents, only the real parts can be combined:

$$\mathrm{Re}[f(x,y)] = [f_\nu(T-r_0/c) + f_\nu r_0\varepsilon/c]\int_0^{2\pi}\cos(k\ am\sin\kappa)d\kappa/(2\pi f_\nu T)$$

$$= [T - r_0/c + r_0\varepsilon/c]\ J_0[k\ a(x,y)m(x,y)]/T$$

$$\cong J_0[k\ a(x,y)m(x,y)].$$

Therefore, the real part of f(x,y) yields a zero-order Bessel function of the vibration amplitude a(x,y), with the variation m(x,y) as a slight error.

$$\mathrm{Re}[f(x,y)] \cong J_0[k\ m(x,y)a(x,y)] \quad (19)$$

where $k = 2\pi/\lambda$.

2) Expand the integrand in f(x,y) as follows [17]:

$$e^{jz\sin\theta} = J_0(z) + 2jJ_1(z)\sin\theta$$

$$+ 2J_2(z)\cos 2\theta + 3jJ_3(z)\sin 3\theta + \ldots$$

where $z \equiv km(x,y)a(x,y)$, and $\theta \equiv \nu(t-r_0/c)$. Therefore, f(x,y) becomes:

$$f(x,y) = \int_0^T e^{jz\sin\theta}\ dt/T$$

$$= J_0(z) + 2jJ_1(z)\int_0^T \sin\theta\ dt/T$$

$$+ 2J_2(z)\int_0^T \cos 2\theta\ dt/T + \ldots$$

If $T \gg 2\pi/\nu$, then the second term of the above equation can be approximated as follows:

$$2jJ_1(z)\int_0^T \sin\theta\ dt/T \cong 2jJ_1(z)[\nu/2\pi]\int_0^{2\pi/\nu} \sin\theta\ dt = 0.$$

Similarly, the third and succeeding terms can be neglected, and we obtain:

$$f(x,y) \cong J_0[k\ a(x,y)m(x,y)] \quad (19)$$

where $k = 2\pi/\lambda$.

ACKNOWLEDGEMENT

The authors wish to acknowledge the Research Committee of the University of California, Santa Barbara, for a seed grant to support parts of the work reported. Tracy Hamilton prepared the manuscript.

REFERENCES

[1] R. L. Powell and K. A. Stetson, "Interferometric Vibration Analysis by Wavefront Reconstruction," J. Optical Society of America, Dec. 1965, vol. 55, no. 12, pp. 1593-1598.

[2] Proceedings of the Topical Meeting on Hologram Interferometry and Speckle Metrology, Optical Society of America, June, 1980.

[3] A. F. Metherell, "Linearized Subfringe Interference Holography," Acoustical Holography, vol. 5, P. Green, ed., Plenum Press, New York, 1974, pp. 41-58.

[4] W. E. Kock, "New Forms of Ultrasonic and Radar Imaging," Ultrasonic Imaging and Holography, G. W. Stroke, W. E. Kock, Y. Kikuchi, and J. Tsujiuchi, ed's., Plenum Press, New York, 1974, pp. 324-328.

[5] M. D. Fox, W. F. Ranson, J. K. Griffin, and R. H. Pettey, "Acoustical Holographic Interferometry," Acoustical Holography, vol. 5, P. Green, ed., Plenum Press, New York, 1974, pp. 103-120.

[6] W. S. Gan, "Fringe Localization in Acoustical Holographic Interferometry," Acoustical Holography, vol. 6, Plenum Press, 1975, pp. 621-636.

[7] K. Suzuki and B. P. Hildebrand, "Holographic Interferometry with Acoustic Waves," Acoustical Holography, vol. 6, N. Booth, ed., Plenum Press, New York, pp. 577-596.

[8] H. D. Collins, "Acoustical Interferometry Using Electronically Simulated Variable Reference and Multiple Path Techniques," Acoustical Holography, vol. 6, N. Booth, ed., Plenum Press, New York, pp. 597-620.

[9] R. J. Collier, C. B. Burckhardt, and L. H. Lin, Optical Holography, Academic Press, New York, 1971, pp. 437.

[10] A. W. Lohmann, "Some Aspects of Optical Holography That Might be of Interest for Acoustical Imaging," Ultrasonic Imaging and Holography, cf. ref. 6, pp. 372-373.

[11] cf. ref. 3, pp. 1597, above eq. (10).

[12] J. W. Goodman, Introduction to Fourier Optics, McGraw-Hill, 1968, pp. 45-46, and 58.

[13] J. L. Sutton, "Underwater Acoustic Imaging," Proc. of the IEEE, vol. 64, no. 4, April 1979, pp. 554-566.

[14] C. Schueler, Development and Applications of Computer-assisted Acoustic Holography, Ph.D. Thesis, UCSB, October 1980.

[15] Incidentally, if a(x,y) = 0, this analysis reduces to the re-construction of a stationary image of an object with spherical-wave insonification.

[16] K. R. Erikson, B. J. O'Loughlin, J. J. Flynn, E. J. Pisa, J. E. Wreede, R. E. Greer, B. Stauffer, and A. F. Metherell, "Through-Transmission Acoustical Holography for Medical Imag ing - A Status Report," Acoustical Holography, vol. 6, N. Booth, ed., Plenum Press, New York, 1975, pp. 15-55.

[17] G. Arfken, Mathematical Methods for Physicists, Academic Press, New York, 1968, pp. 375.

ROLE PLAYED BY CELLULAR COHERSION AND ORGANIZATION IN THE INTERACTION OF ULTRASOUND WITH SOFT TISSUES

G. Berger, J. Perrin, D. Bourgoin, A. Salesses*,
A. Agneray, Y. Darlas.

Laboratoire de Biophysique, UER Cochin Port-Royal
ERA CNRS n°498
24, rue du Fbg St Jacques, 75674 Paris
* Laboratoire d'histologie

I) INTRODUCTION

The absorption of soft tissues has so far been explained by relaxation phenomena on the macromolecular scale. That these phenomena are involved cannot be denied since this is proved by many experiments carried out on pure solutions of macromolecules. But these mechanisms do not justify the significant differences in attenuation measured in different tissues or the different pathological states of the same tissue.

Our hypothesis is that the architectural organization of the tissue in cells must be taken into account, these cells being held together by surface forces and junction mechanisms of the "desmosome" type. These hypotheses have been checked "in vitro", in transmission by tissue perturbation modifying cell cohesion, the exoskeleton and the surface forces. As these data are to operate as feature elements "in vivo" at a second stage it was necessary to establish the acquisition of reflected signals and to test the methods for measuring the attenuation.

II) ATTENUATION MEASURED IN TRANSMISSION AND CORRELATED WITH TISSUE PERTURBATION

21-Apparatus - Signal treatment method

Two probes (Fig 1) : a transmitter and a receiver, brought in line with a great accuracy, are placed in an ultrasonic tank filled with an isotonic solution maintained at PH=7.4 and thermostatically controlled at 25°C. The sample to be studied is placed on the borderline between the zones of Fresnel and Fraunhofer; in this zone incident energy falls by 6 dB on a diameter of 2.5 mm and by 20 dB on a diameter of 4.3 mm. The transmitter is stimulated by a "Metrotek" module emitting impulses whose band width is 11 MHz at 3 dB. The resulting transmission spectrum is focused on 5 MHz with a frequency range of 6 MHz at 6 dB. The stationary nature of the transmitted signal makes possible its stroboscopic sampling. On a Fourier Transformer 5 µs are converted into 1024 points and the power spectrum is obtained in 512 points. All the data acquired can be fed into a Hewlett-Packard 9835 computer to be processed later.

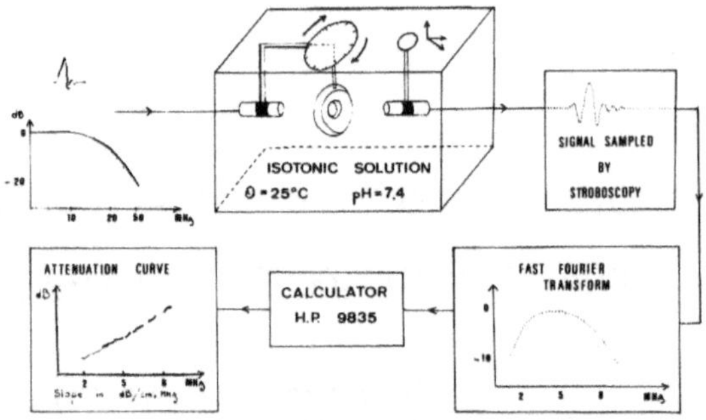

Fig. 1. Equipment for "In Vitro" Experiments

The attenuation is obtained by a classical spectroscopic method. The difference in dB between the logarithm of the reference specturm (without any structure on the ultrasound beam) and the logarithm of the spectrum obtained with the sample gives the attenuation versus frequency curve. In the frequency band where the standard deviation of the curve is small the shape of the curves matches a linear regression of attenuation as a function of frequency; the regression index gives the loss of intensity coefficient in dB/cm.MHz.

22- Methodology of the biological experiments

The main perturbing solutions used in our experiments were enzymes, fixers, chelating agents. Their action was studied mainly on rabbit-liver with some experiments on the liver of Wistar rats. The method of introduction chosen was a natural one - that of the vascular system of the organ. Before being killed by an injection of Nesdonal, the animal receives an intravenous injection of "heparine" to prevent post-mortem coagulation. After the abdomen has been slit open a solution is injected by means of a catheter in the portal vein. In this way the solution is propagated through the vascular system soaking all or at any rate a considerable quantity of the cells, and emerging through the supra-hepatic veins (that have been severed). The reference sample is obtained by merely washing the liver with an isotonic solution maintained at PH=7.4 and then severing one hepatic lobe. The rest of the liver continues to be injected with the perturbing solution. Each sample, both reference and perturbed, is taken from the corresponding hepatic lobe in a perfectly reproducible cylindrical shape and then submitted to insonification after five minutes' degassing. All the operations are timed so that the evolution in time can be followed. After the experiment, the sample is fixed with formalin 10% for histological examination.

These ultrasound measurements are made in conjunction with an attempt to find rheological correlations. On the macroscopic level the aim is to evaluate the moduli of longitudinal elasticity (Young's modulus) and the transversal elasticity moduli. Microscopically the aim of one micromanipulation is to evaluate deformation energy of cultures of tissues and the forces of intercellular cohesion when the deformation results in the separation of cells.

23- Results obtained according to the various perturbations

a- Cellular dissociation

a-1- Trypsine

The perturbing enzyme injected by the method already described is a solution of trypsine 0.5% maintained at PH=7.4. In all the experimental cases (fig. 2) the attenuation coefficient falls significantly between the control sample and the perturbed sample. Histology enables the degree of efficiency of the perturbation to be confirmed (fig 3) The elements permitting a judgment to be made are the optically white spaces observed between the cells that do not exist in the control sample and the more globulous appearance of the cells. A parallel experiment enabled the dissociation which was sometimes doubtful on slides to be made more objective, since a slight mechanical contraint which was quite inadequate for normal tissue was able to separate them distinctly. It should also be noted that the small percentage of necrosis, inevitable in any sample, is not significantly different in the control tissue and the perturbed tissue. Dissociation is therefore the only parameter to which the fall in the attenuation coefficient is due. Two interpretations seem possible. There may be either absorption by resonance on the cellular or pluricellular level or viscous relaxation phenomena due to the movement of the cells in a visco-elastic envi-

t = 0 mn

EXPERIMENTS	FREQUENCY LIMITS (MHz)	ATTENUATION COEFFICIENT OF α_R REFERENCE SAMPLE (dB/cm.MHz)	α_P PERTURBED SAMPLE (dB/cm.MHz)	$\frac{\alpha_P - \alpha_R}{\alpha_R}$
n·1	3.5 – 6	1.71	1.13	0.34
n·2	4 – 7	1.12	0.50	0.55
n·3	3 – 7	1.31	0.72	0.45
n·4	3 – 8	1.10	0.77	0.30

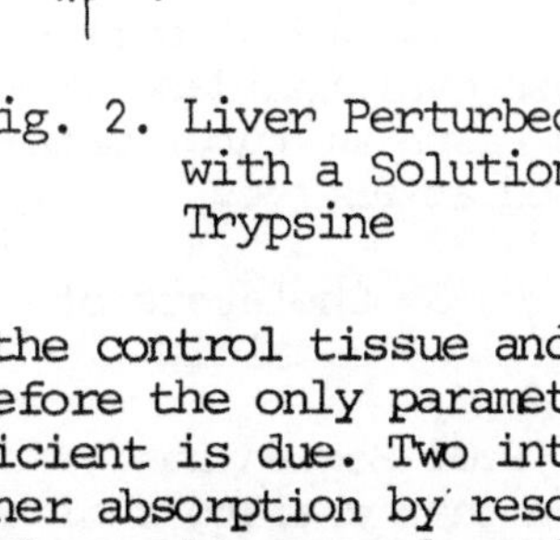

Fig. 2. Liver Perturbed with a Solution of Trypsine

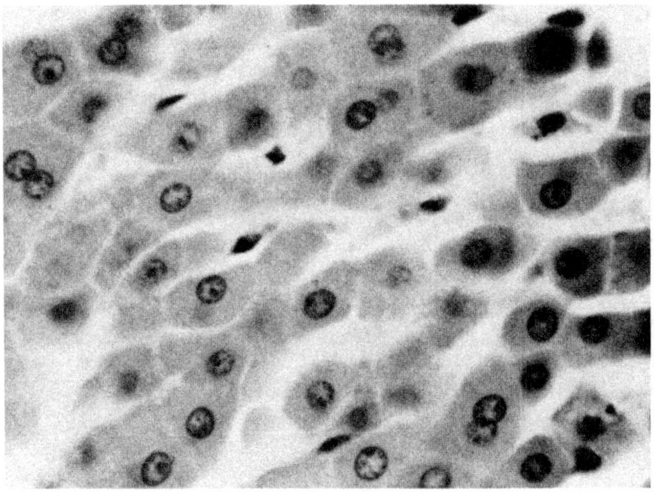

Sample injected with a solution of enzyme (collagenase-hyaluronidase) coloured with Hematoxyline of Weigert-Eosine.

Fig. 3 Exapmle of cells dissocation

ronment and these may occur in addition to the relaxation phenomena on the macromolecular level. With time (Fig.2), the fall in the attenuation coefficient is more pronounced which is explained by the continuation of the enzyme action which increases dissociation. Its rise is more difficult to interpret but may be connected with the increased necrosis and cell sedimentation.

a-2- Collagenase-hyaluronidase

In the same spirit as in the previous experiment but with even more satisfactory results (Fig.4 Experiment A) we used a mixture of collagenase-hyaluronidase (Howard-Pesch mixture) maintained at PH=7.4.

a-3- Chelators of calcium

P.O. Seglen has shown that chelators of Ca^{++} encourage the action of enzymes. We therefore

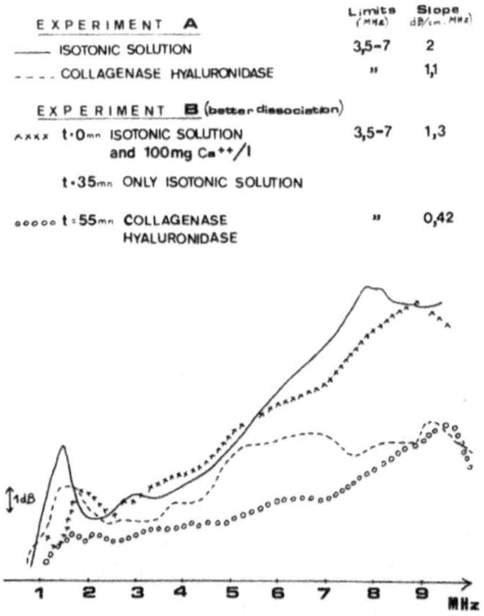

Fig. 4 Livers Injected With Experiment A and B

carried out a perfusion of normal saline solution containing 100 mg/l of Ca^{++}, followed 35 mn later by a perfusion of normal saline solution, then 50 mn later by a perfusion of collegenase-hyaluronidase. The results obtained for each stage are presented Fig.4, Experiment B.

The aim of this process was to increase dissociation, and we obtained a fall in the attenuation coefficient that was considerably greater than with the enzyme alone. This confirms the part played by intercellular liaison forces in the attenuation.

b- Action of fixatives.

An hour after the perfusion of the formol into the vascular network, the attenuation falls by an average of 50%. This stage corresponds to the fixing of the intercellular exoskeleton and suggests that only absorption by relaxation is apprehended. After a few days, when all the tissue is rigidified, the attenuation rises to its normal level.

c-Osmolar perturbations

The injection of a hyperosmolar or hypoosmolar solution should in theory lead to a decrease or an increase of the cell mass, and therefore vary the frequency of the cell resonators. The curves obtained are very perturbed and in particular have a much less linear appearance according to the frequency. In this case, it seems preferable to characterize them by the differential according to the frequency.

d- Experimental cancerization

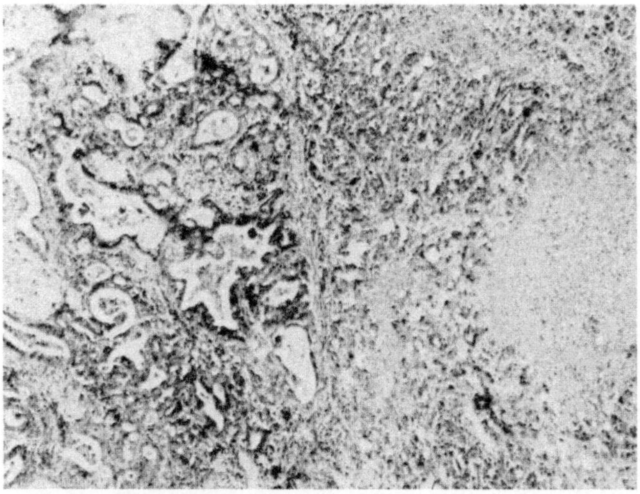

Fig.5 : Cancer presenting different histological sites : important necrosed places and tubulated formations. Coloration : Hematoxyline of Weigert-Eosine.

As rabbit-liver does not lend itself to cancerization, we studied maturation cancers induced in Wistar rat livers by butter additive. The cancers are characterized histologically (Fig.5) as being cellular hepatocholangioma with compact parts and a tubulated structure plunged in a collageneous stroma. Several seats of necrosis and considerable sclerosis are observed. This hepatome, compared with a normal rat liver, shows an increase in the global attenuation (from 0.5 to 0.9 dB/cm.MHz on average (Fig.6). This increase does not indicate anything about the absorption mechanisms, for the histological existence of macrovesicles like the experiment show an increase in the scattered energy.

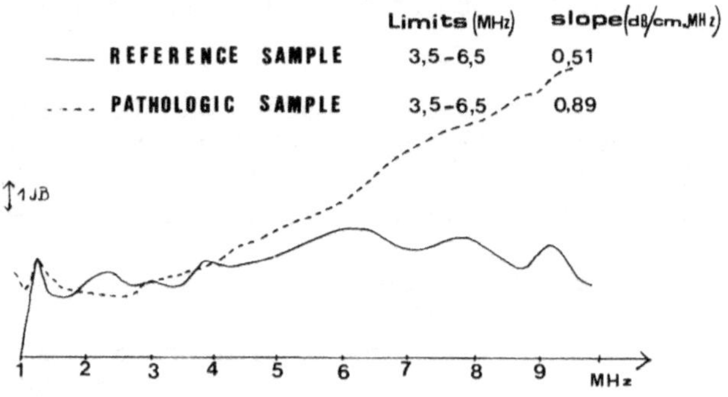

Fig. 6 Experimental Cancer (Liver of Wistar Rat)

e- Striated muscle tissue

The anisotropism of the psoas muscle enabled two types of sample to be taken according to the direction of the fibres. In the first case, the striated muscular cells were presented parallel to the direction of the incident beam, in the second perpendicular to it. If relaxation phenomena alone were involved, the attenuation curves should be identical; however the curves do not have the same linearity of homogeneous tissues and, when we fit in low frequency bands, have very different coefficients (Fig. 7). This "non-linearity" of the tissue can be explained by the appearance of shearing waves taking the place of the longitudinal ones with a higher rate when the cells are placed perpendicular to the beam, rather then parallel.

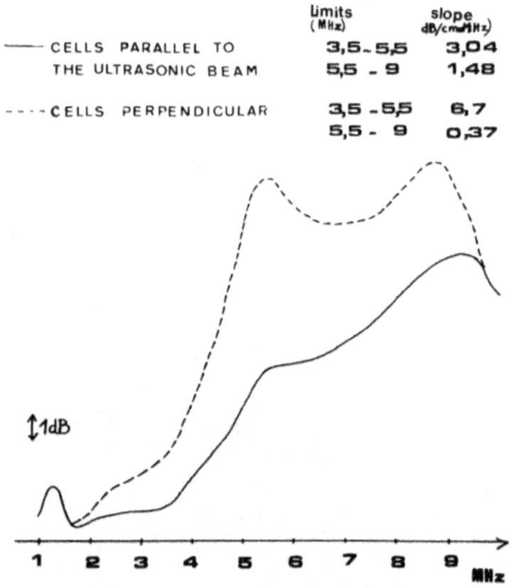

Fig. 7 Anisotropic Tissue (Rabbit Muscle)

III) REFLECTION : ATTENUATION WITH A VIEW TO EXPERIMENT "IN VIVO"

The operations of amplification, detection of signals that are carried out on a classical echoBscan lose the real frequency components of the signal, parameters whose importance has been proved by the experiments "in vitro'. Since the signal is not stationary owing to circulation and breathing the first stage was to digitize the signal with a high sampling rate and to adapt this product to classical echographic equipment. The second part concerns tests for the measuring of attenuation on reflected signals.

31-Signal sampling

The conversion into points is carried out by a digital analogical converter TRW 10073 with a sampling frequency tuned to 24 MHz; I2 MHz is in fact sufficient for a medical diagnosis except where the eye is concerned. In order to prevent a contraction of the spectrum, a filter was inserted that causes an attenuation of 3 dB at 11 MHz and 70 dB at 13 MHz. The signal is converted into 256 steps coded into 8 bits, giving an 8 mm depth of exploration. Details of this are given in Fig.8.

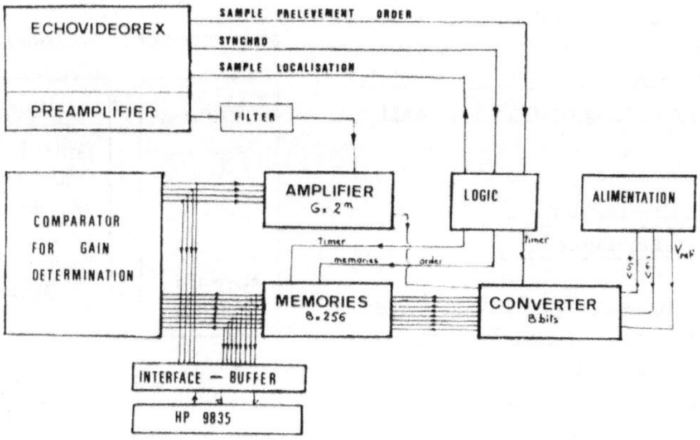

Fig. 8 Details of the Sampler

In order to make the best use of the dynamic of the converter, each sample is subjected to an amplification throughout the 8 mm determined by an automatic control system. This amplification is controlled by 4 bits value n, the method of optimal determination of which is given Fig 9, and corresponds to value 2^n. Its theorical dynamics, 90 dB, is in practice limited to 78 dB by the noise of the pre-amplifier.

32- Complete system

The converter is inserted into a classical echographic system (Fig.10). On the echo B a preliminary scanning localizes a zone of interest , on which a very bright marker is placed. During the next scanning a depth of 8 mm is converted in 256 steps. A computer, a Fourier transformer and a plotter enable the data to be processed. This exploration can then comprise several echoes isolable by a Gaussian window focussed on amplitude peaks equal in width to the lenght of 3 impulses.

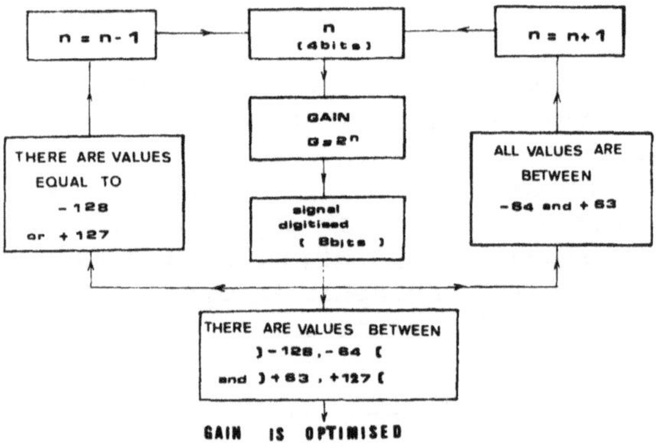

Fig. 9 Automatic Gain Control

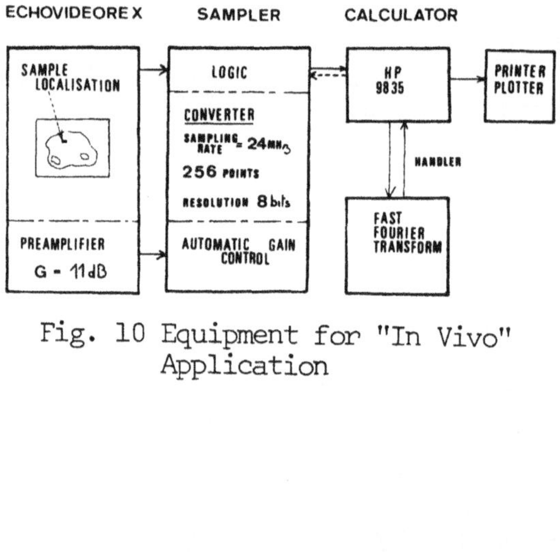

Fig. 10 Equipment for "In Vivo" Application

33-Correlation reflection-transmission

We study, simultaneously, the signals transmitted and the signals reflected by the same sample. Here we present the example of a sample of 1 cm of normal rabbit-liver (Fig.11) whose attenuation curve and slope were obtained by the method described in II. The transmitter probe serves simultaneously as receiver to form an echographic image which shows the echoes of the entry and exit sides, and the intermediate echoes probably due to collagenic structures. As an exploration is carried out on 8 mm, the whole of the reflected signal is obtained when 3 samples are digitised. The echoes of the entry and exit sides (E_1 and E_2) are obtained with an automatically optimised amplification on **18 dB** while the intermediate echoes appear clearly with 30 dB. The result obtained by the deconvolution of the spectra between E_1 and E_2 agrees perfectly with the slope measured in transmission and is explained by the perfect plane shape of the entry and exit sides. But the values obtained by the deconvolution between the other pairs of echoes, such as (E_1,E_4) presented in Fig.11 are very different from the previous results.

This experiment clearly shows the important variations obtained according to the shape of the reflector and indicates a statistical method to measure attenuation in the reflection mode. Statistical methods can easily be applied on a great distance where the number of reflectors is important. On a small distance, there is still to find a way for measuring attenuation as one of the tissue characteristic parameter which is needed.

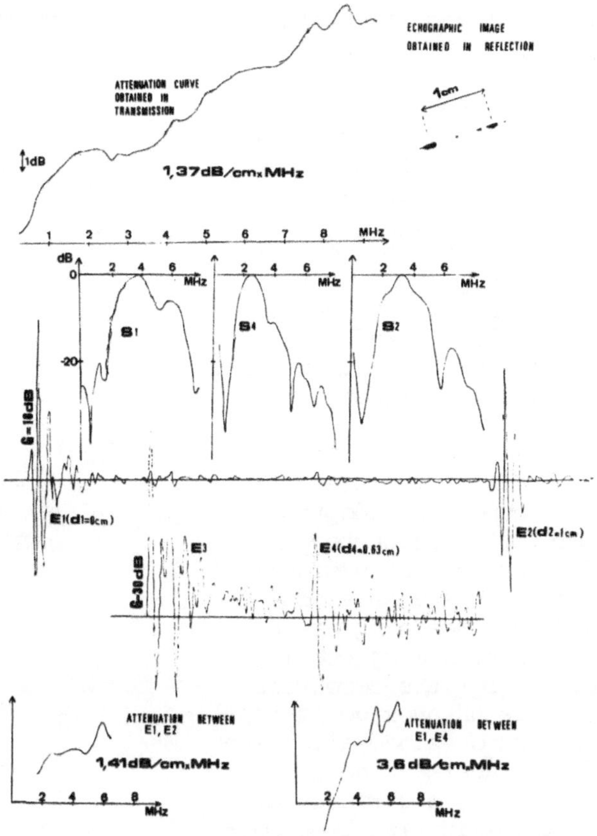

Fig. 11 Results Compared Between Reflection and Transmission

34- Tests on glove fingers, attenuation measurement

We used plastic gloves filled with a macro-molecular solution (lacto-globulin) of varying concentration and therefore of varying attenuation. A macro-molecular solution is interesting in that it gives linear attenuation curves according to the frequency as many authors have shown by experiment. The attenuation was measured by using the entry and exit sides of a glove finger. These echoes are obtained in the following manner : a first image is used to pick out the structures on the screen and to place the index either slightly above the gloves' entry side or slightly below the exit side. Each converted sample is transfered to the Hewlett Packard computer.

The first technique used consisted in taking the logarithmized spectrum of the entry signal after its multiplication by the Gaussian interval and in subtracting the spectrum of the exit signal. The attenuation curves thus obtained were rarely linear, and the attenuation coefficient obtained when allowing a linear regression showed considerable distorsion. It can be explained by the importance of the shape of the reflector and its incidence with the ultrasonic beam, which cannot be controlled accurately enough.

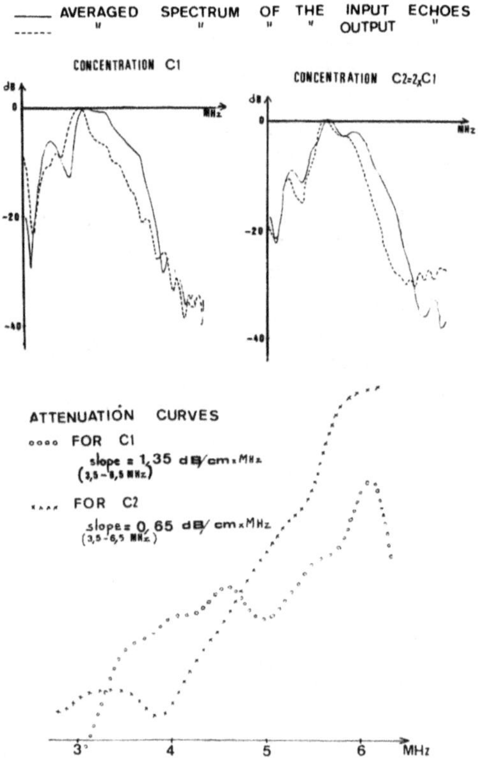

Figure 12

Finally, a statistical method was adopted (Fig.12). Structure and index for the entry signal remain in the same position and many scanning are carried out. The spectrum retained for the deconvolution is an average of the spectra obtained after each data transfer. The same operation is carried out on the exit signal. According this process the attenuation curves are effectively linear and the attenuation coefficient is nearly double when doubling the macro-molecular concentration.

IV) CONCLUSION

The authors have confirmed by "in vitro" experiments the importance of cellular organisation as an ultrasonic absorption parameter. Through the action of enzymes, which dissociate cells without affecting them, the slope of attenuation drops in proportion corresponding to the degree of dissociation; the same result is obtained after having fixed the exoskeleton. Let us stress that the mechanism of interest may be either multimodale resonance operating at the cellular or pluricellular level or viscous relaxation also operating at the cellular level. In the case of anisotropic tissues, the cells orientation according to the direction of the ultrasonic beam may modify vibration modes and therefore attenuation curves. Experiments for measuring attenuation in the reflection mode indicates a statistical method due to influence on frequency components of the reflector shape. An apparatus, easily usable "in vivo" has been set up and the authors are now making experiments based on animal organs and classical human echoscan.

REFERENCES

Berger G.,Perrin J.,1979,Attenuation principles and measurements for Tissue characterization, in : "First E.E.C. Workshop on Tissue Characterization".

Foldy L.L., Multiple scattering of Waves : General theory of isotropic scattering by randomly distributed scatterers, 1945, Phys. Rev. 67.

Fraser J.,Kino G.S.,and Birnholz J., 1977, Cepstral signal processing for tissue signature analysis, in "Second International Symposium on Ultrasonic Tissue Characterization" National Bureau of Standards. Gaithersburg,Md.

Hill C.R.,Chivers R.C.,Huggins R.W.,and Nicholas D., Scattering of US by human tissue, in:"Ultrasound: its applications in medecine and biology", pp 441-482, F.J. Fry, ed.

Hill C.R., 1976, Frequency and Angular dependance of ultrasonic scattering from tissue, in: "Ultrasonic Tissue Characterization" pp 197-206, NBS Publication 453.

Kak A.O., and Dines K.A., 1978, Signal Processing of broad band pulsed ultrasound : Measurement of attenuation of soft biological tissues, IEEE Transact. on Biomedical Engineering, Vol. BME 25, n°4, pp 321-344.

Kuk R., Schwartz M., and von Micsky L., Parametric estimation of the acoustic attenuation coefficient slope for soft tissues, IEEE Ultrasonics Symposium, Proc. 1976 - CH 1120, 5 SU, pp 44-47.

Kuk R., and Schwartz M., Estimating the acoustic attenuation coefficient slope for liver from reflected ultrasound signals, IEEE Transact. on Sonics and Ultrasonics, 1979, Vol. SU 26, n°5, pp 353-361.

Kuk R., Clinical application of an ultrasound attenuation coefficient estimation technique for liver pathology characterization, IEEE Transact. on Biomedical Engineering, 1980, Vol. BME 27, n°6.

Lele P.P.,Senapati N., and Murphy A., Ultrasonic frequency domain analysis and studies on acoustical scattering for diagnosis of tissue pathology, 1975, in: "28th. Conf. on Eng. in Medecine and Biology",

Pauly , and Schwan, Mechanism of absorption of ultrasound in liver tissue , 1971, J. Acoust. Soc. America 50,pp 692-699.

Perrin J., Berger G., Salesses A., and Agneray M., Role played by coherent and incoherent diffusion in ultrasonic attenuation of biological tissues, 1979, in : "Ultrasonics International 79", IPC Science and Technology Press Ltd., pp 518-523.
Salesses A., Perrin J., Berger G., and Agneray A., in "Communication 5th European Anatomical Congress, Prague", 1979.
Seglen P.O., Preparation of rat liver cells, 1972, Experimental Cell Research 74, pp 450-454.
Seglen P.O., Preparation of rat liver cells 2. Effect of ions and chelators on tissue dispersion, 1973, Experimental Cell Research 76, pp 25-30.

IMPROVEMENTS OF IMPEDIOGRAMS ACCURACY

Alain Herment[x], Pierre Péronneau[x], Michel Vaysse

ERA CNRS N° 785
Hôpital Broussais 75014 Paris (France)
x INSERM

1. INTRODUCTION

Echography is limited by its principle to provide cartographic data concerning the geometry of the media observed. Impediography[1, 2, 3, 4, 5, 6], by the way of more complete treatment of the echographic signal, provides impediograms on which the acoustic impedance characterizing the media is displayed according to observation depth.

The echographic signal from which the acoustic impedance of the medium will be extracted depends at least upon two types of parameters :

i) the geometric and acoustic properties of the insonified medium,

ii) the characteristics of the ultrasonic equipment used to collect the signal.

As a consequence a correct determination of acoustic impedance implies to take in account these parameters.

Untill now impediography mainly relies on the elimination of the equipment imperfections ; the filtering effect of the probe is noticely removed by deconvolution to obtain the medium impulse response from which the impediogram is calculated. In order to increase the application field of impediography we propose a method of deconvolution based on a process identification technique. The process not only corrects the equipment imperfections but also allows one to take in account some of the geometric and acoustic characteristics of the examined medium. A better knowledge of the medium will then be obtained and more accurate impediograms will be calculated.

2. DESCRIPTION OF THE METHOD

The principle of the method is schematized on figure 1. The signal generated by the ultrasonic equipment is modified by the medium impulse response during its propagation through the acoustic medium and the received echographic signal e(t) is digitized for later numerical treatments.

On the other hand, the acoustic device impulse response is used, after digitization to numerically modelize a synthetized signal S(t) taking in account a number of equipment and medium characteristics.

This modelized signal is then compared to the sampled echographic signal by a convenient test, adjusted through an identification algorithm and characterized by a set of parameters. Finally, the medium impulse response is reconstructed from these parameters.

In order to completely define the identification algorithm we must precisely determine :

i) the generation of S(t) in the model,
ii) the test of comparison and
iii) the medium impulse response reconstruction.

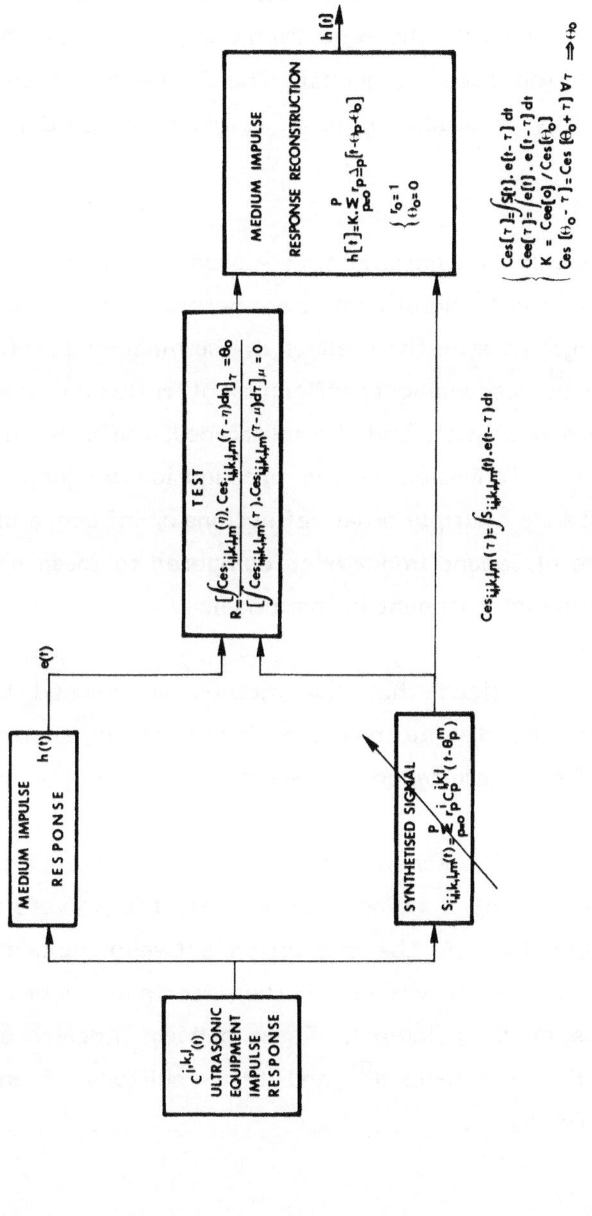

Fig. 1 : Principle of the method.

2.1. The model

In order to modelize the medium impulse response we must define which are the most significant equipment and medium characteristics contributing to generate the echographic signal and how they interfere in the genesis of the received signal. The influence of each of these parameters must then be studied in order to get a precise definition of the model.

2.1.1. The significant parameters : A study of the biological medium indicates that the most significant parameters for the obtention of confident impediograms are : the position of the medium interfaces and the amplitude of the corresponding coefficients of reflexion, the angulation between the ultrasound beam and the interfaces, the curvatures of these interfaces and the attenuation in the propagation medium. Parameters describing for instance multiple order reflections or influence of impedance discontinuities are of second importance compared to these previous ones and will not be taken into account in the model.

It has to be noticed that the method is reduced to a simple deconvolution when only the interfaces positions and the amplitudes of the coefficients of reflexion are taken into account.

2.1.2. Genesis of the echographic signal : The received signal can be expressed as the sum of a number of echoes, the waveform of them $C^{j,k,l}(t)$ is determined by the angulation between incident beam and interfaces (index j), the curvature of the interfaces (index k) and the attenuation in the medium (index l). These echoes interfere according to the position of the interfaces θ^m and the amplitude of the different coefficients of reflexion r^i.

For a medium composed of N+1 layers, the received signal may be written in the form :

$$e(t) = \sum_{n=0}^{N} r_n^{\,i} \cdot C_n^{j,k,l} (t-\theta_n^m) \qquad (1)$$

$C_n^{j,k,l}(t)$ can be considered as the convolution product between the ultrasonic device impulse response C(t), which is the signal obtained on reception when the transmitted signal is reflected on a flat acoustic mirror perpendicular to the ultrasonic beam, and a function f(j,k,l,t) which represents the echo shape deformations generated by the medium. It has to be noticed that function f only takes into account the part of attenuation that modify the echo waveform and subsequently only the frequency-dependant attenuation.

As a matter of fact the function f is extremely complex. For sake of simplicity, the influence of each parameter will be studied independantly from the others.

2.1.3. Sensitivity of the echographic signal to the parameters : The modification of echo shape induced by the three parameters j, k and l must be studied

i) In order to verify if they are of different nature : if two different parameters induce the same sort of echo waveform modification it will not be possible to determine for this echo the respective contribution of each parameter ;

ii) In order to construct the model : for this purpose, the sensitivity of the echo shape to these parameters must be precisely studied to define their law of variation in the algorithm.

2.1.3.1. Influence of the incident angle : In order to follow the variations of the echo shape as a function of the angle of incidence, we have recorded with transducers (1) and (2) and for angles varying from zero to five degrees the signals obtained on a flat brass reflector. The corresponding echoes are presented on figures 2 to 7 for probe 1.

i) the comparison of the different signals to the reference echo (figure 2) indicates that even for an angulation as weak as one degree the received signal is significantly modified, the first period of the signal is lenghtened. That modification of the echo shape increases along with the angle leading to important deformations of the beginning of the echo for angles of 2, 3 and 4 degrees. For a 5 degrees incidence the signal is so distorded that it seems to be composed of two successive echoes.

ii) the amplitude of the echo also decreases when the angulation increases. That decrease is automatically taken in account in our algorithm. These changes in amplitude are not put in evidence on figures 2 to 7 because they have been recalibrated by a variation of the ultrasonic device amplification.

iii) finally, for each angle and for both of the probes, the echo deformation with respect to the distance between the probe and the interface has been studied. No variation on echo shape and amplitude are noticeable for distances smaller than 7.5 cm. In these angulation and distance ranges, function f is quite independant of the sample-probe distance.

. Probe 1 Krautkramer H5K Ø 10 mm center frequency 5 MHz

. Probe 2 Aerotech α Ø 12,7 mm center frequency 2,25 MHz

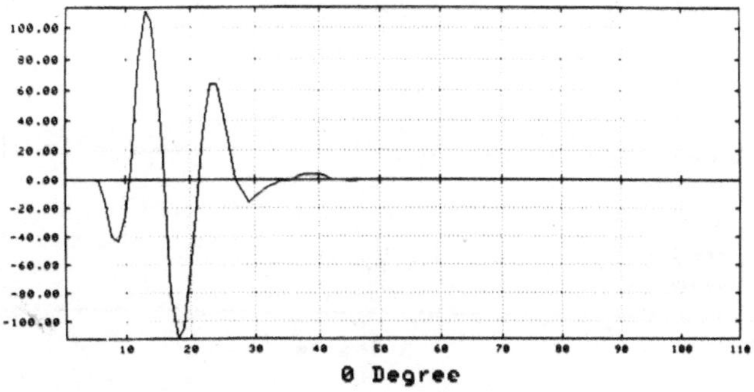

Fig. 2 : Influence of incidence angle. Received echo : the plate is parallel to the transducer front surface.

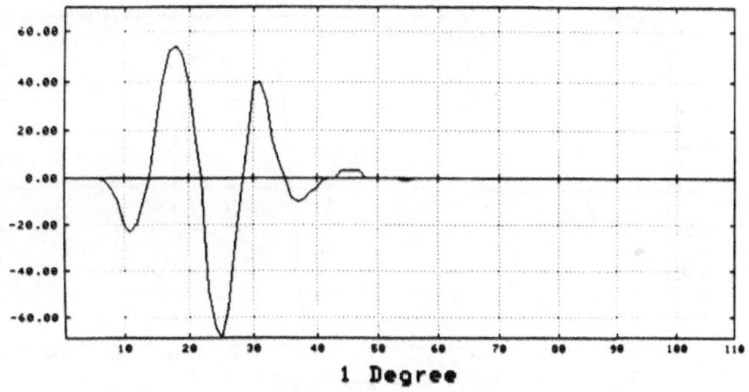

Fig. 3 : Influence of incidence angle. Received echo : the plate is tilted of 1 degree with respect to the transducer front surface.

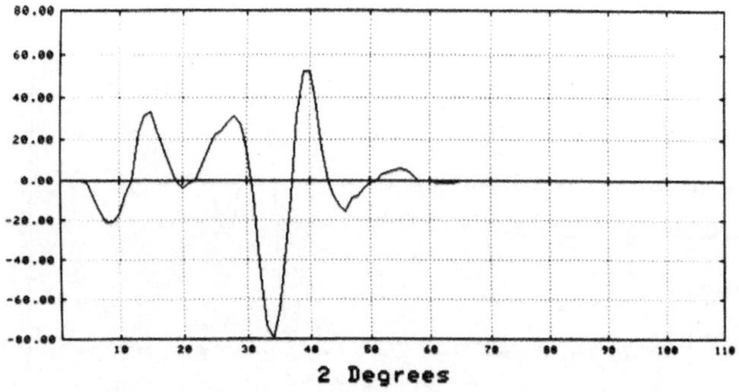

Fig. 4 : Influence of incidence angle. Received echo : the plate is tilted of 2 degrees with respect to the transducer front surface.

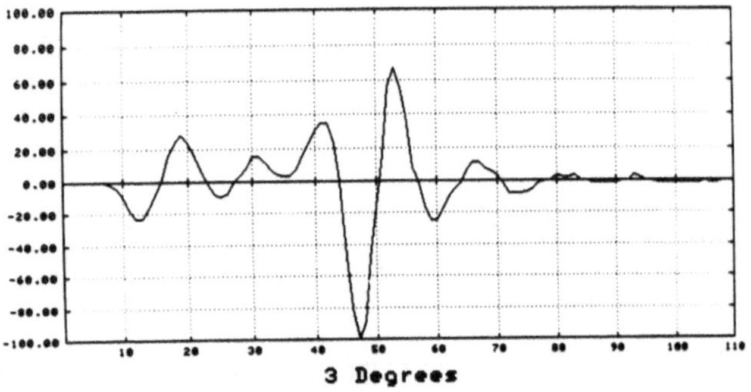

Fig. 5 : Influence of incidence angle. Received echo : the plate is tilted of 3 degrees with respect to the transducer front surface.

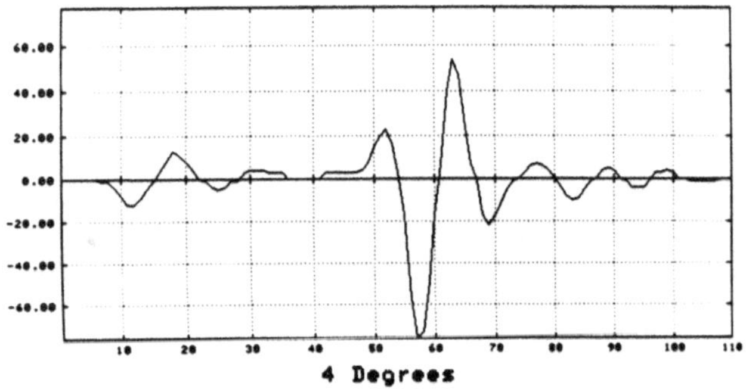

Fig. 6 : Influence of incidence angle. Received echo : the plate is tilted of 4 degrees with respect to the transducer front surface.

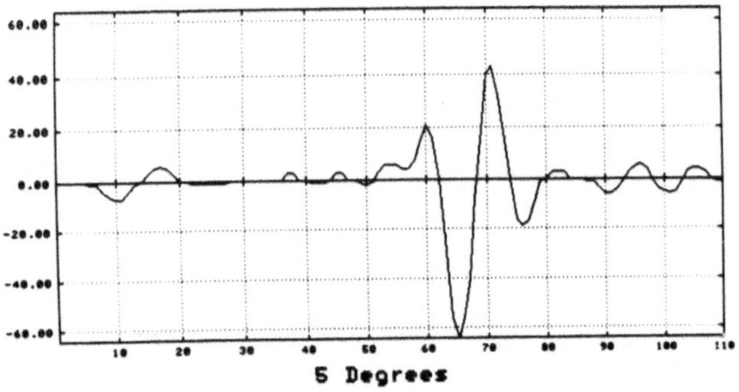

Fig 7 : Influence of incidence angle. Received echo : the plate is tilted of 5 degrees with respect to the transducer front surface.

2.1.3.2. Influence of interface curvature : In order to simulate curved interfaces, cylindric targets with radii ranging from 2.5 mm to 100 mm have been used. The received echoes have been recorded for the two probes with an incident beam normal and centered on the cylinder axis. Figures 8 and 9 present the received echoes using probe (2) for cylinders of infinite radius (plane surface) and of 2.5 mm radius. As previously done the amplitude variations of the echoes have been corrected on these figures. A slight modification of the echo shape is only observed for such different curvatures.

i) However the curvature parameter has been introduced in the model in order to determine the actual sensitivity of the numerical test : the sensitivity of the test is limited by the equipment and measurement noises, and some specific measurements and computations have been done on the same series of cylinders to determine the practical limit of the process for our experimental device. The results show that radius modifications less than 5 mm cannot be detected.

ii) Considering the poor sensitivity of the test to the curvature, a rough correction of the echo amplitude will only be possible.

iii) Finally, for our experimental arrangment and for both of the probes, the results do not depend on the transducer-cylinder distance when it ranges between 3 cm and 10 cm.

2.1.3.3. Influence of dispersive attenuation : A series of lucite plates the thickness of which ranging from 0 to 50 mm have been used to generate dispersive attenuation. Signals received with probes (1) and (2) have been recorded. Figures 10, 11 and 12 respectively correspond to echoes obtained without attenuation and with attenuations induced by 20 mm and 40 mm of lucite.

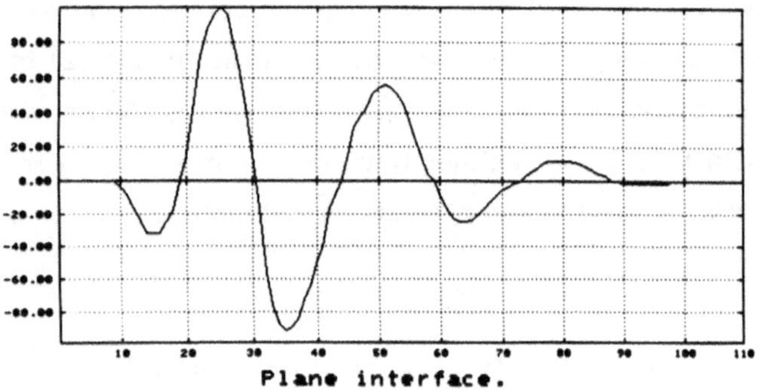

Fig 8 : Influence of interface curvature.
The target is a flat reflector.

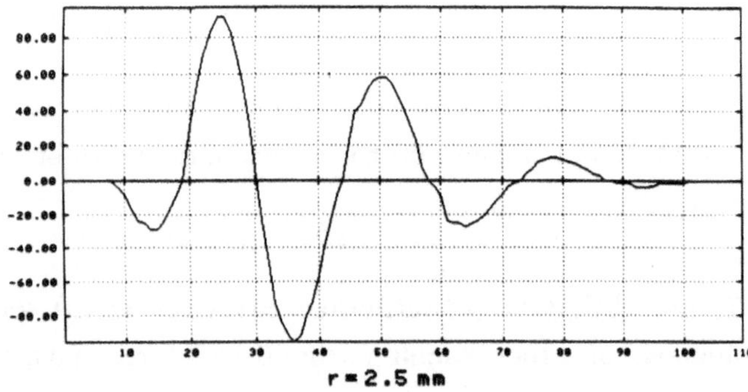

Fig. 9 : Influence of interface curvature.
The target is a cylinder with a radius of 2.5 mm.

The echo displayed figure 10 has been measured on a water/lucite interface whereas the others have been generated by a lucite/water transition, this explaining the phase opposition between them.

i) Observation of these figures indicates that the echo waveform lenghtens with the increase of attenuation while its central frequency slightly decreases. That is the consequence of the weakening of the highest frequencies of its spectrum by dispersive attenuation.

ii) The changes in amplitude due to attenuation are again automatically taken into account by the algorithm.

iii) A series of experiments have shown that the results are independant of the probe position with respect to the dispersive layer as it may be expected.

2.1.4. Law of variation of the parameters : Considering the previous results it appears that the type of echo shape modifications induced by angulation is very different from the one due to attenuation. The respective contribution of these parameters to the genesis of an echo should be easily differentiated. Unfortunately it is difficult to precise the type of waveform modification due to curvature because of its smallness.

Taking into account the different sensitivities of echo shape with respect to these medium parameters, we used : steps of 0.5 degree for angle variations,steps in attenuation corresponding to the one induced by 10 mm of lucite and steps of 15 mm for the bend radius. In addition, the parameters r_n^i and θ_n^m determining the deconvolution itself are varying by steps of 0.01 for r_n^i and of 20 ns for θ_n^m, the latter corresponding to a distance of 15 μm in tissues.

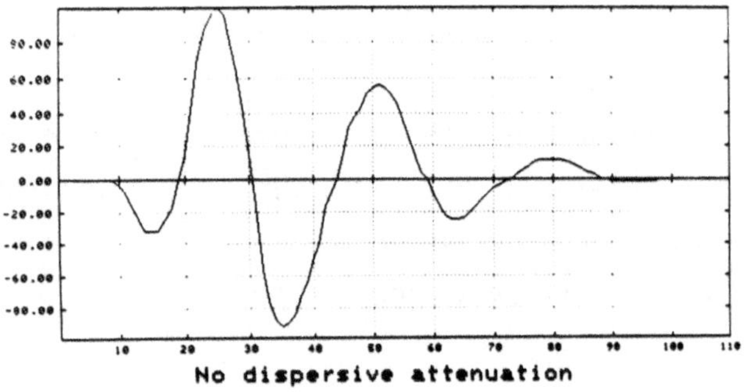

Fig. 10 : Influence of dispersive attenuation.

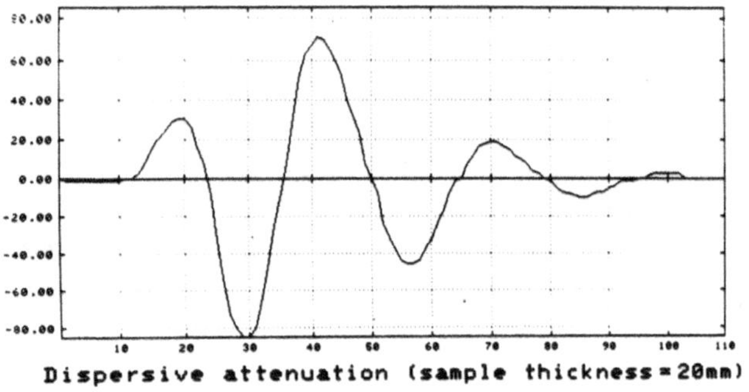

Fig. 11 : Influence of dispersive attenuation.

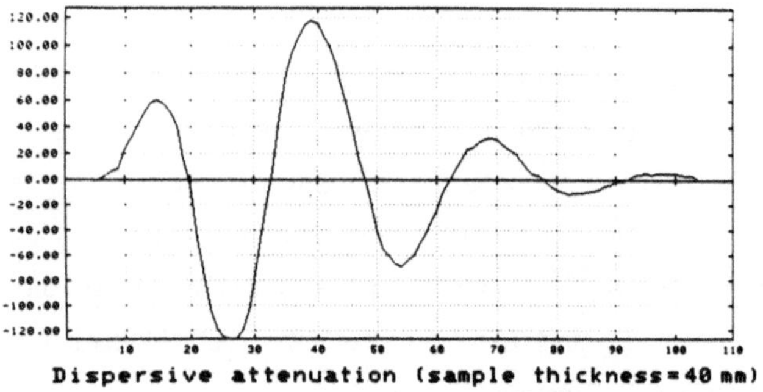

Fig. 12 : Influence of dispersive attenuation.

The same variations of parameters has been used for both transducers (1) and (2).

Values of r^i and θ^m are generated by the computer, the different values $C^{j,k,l}(t)$ have been recorded from a series of calibrated targets and fed into the computer memory.

2.1.5. The model : As illustrated on figure 1, a signal $S_{i,j,k,l,m}(t)$, image of the received signal, is synthetized under the following form in the model :

$$S_{i,j,k,l,m}(t) = \sum_{p=o}^{P} r_p^i \cdot C_p^{j,k,l}(t - \theta_p^m) \qquad \text{with} \begin{cases} r_o^i = 1 \\ \theta_o^m = 0 \quad m \end{cases} \qquad (2)$$

for different values of r_p^i, $C_p^{j,k,l}$ and θ_p^m determined by the parameters i, j, k, l and m.

This signal is a digitized form of the expression (1) of e(t) where p corresponds to the order of the digital sample and P represents the number of samples of the digitized signal.

The constraints on the values r_o^i and θ_o^m correspond to a calibration and a time shift of the signal in order to decrease further computation time.

2.2. The test of comparison

The cross-correlation function between $S_{i,j,k,l,m}(t)$ and e(t) :

$$C_{es_{i,j,k,l,m}}(\tau) = \int S_{i,j,k,l,m}(t) \cdot e(t-\tau)\, dt$$

is computed. Let us call $S_{io,jo,ko,lo,mo}(t)$ the value of $S_{i,j,k,l,m}(t)$ such as $C_{es_{io,jo,ko,lo,mo}}(\tau)$ the corresponding correlation function, possesses a vertical axis of symetry. It can then be demonstrated from the general properties of the correlation that $S_{io,jo,ko,lo,mo}(t)$ is in this case proportional in the ratio of K to e(t) with a time translation of θ_o.

The values θ_o and K are easily extracted from $C_{es_{io,jo,ko,lo,mo}}(\tau)$ and from $C_{ee}(\tau)$ the auto correlation function of the received signal.

$$\left.\begin{array}{l}\theta_o = C_{es}(\tau): \text{ symetry axis position,} \\ K = C_{ee}(0)/C_{es_{io,jo,ko,lo,mo}}(\theta_o)\end{array}\right\} \quad (3)$$

The symetry of the correlation function is tested by means of a coefficient R defined as :

$$R = \frac{\left[\int C_{es_{i,j,k,l,m}}(\eta) \cdot C_{es_{i,j,k,l,m}}(\tau-\eta)d\eta\right]_{\tau=\theta_o}}{\left[\int C_{es_{i,j,k,l,m}}(\tau) \cdot C_{es_{i,j,k,l,m}}(\tau-\mu)d\tau\right]_{\mu=0}}$$

R represents the ratio between the autoconvolution function of $C_{es_{i,j,k,l,m}}(\tau)$ for $\tau = \theta_o$ to the auto-correlation function of $C_{es_{i,j,k,l,m}}(\tau)$ for $\mu = 0$. It can be shown that the denominator of the ratio represents the energy of the signal and that its numerator is generally inferior to the denominator except when $C_{es_{i,j,k,l,m}}(\tau)$ is a symetric function, in this case $R = 1$.

In actual conditions, because of the sampling noise, the value of $C_{es_{i,j,k,l,m}}(\tau)$ that maximizes the ratio will be considered as fitting the test condition.

2.3. Impulse response reconstruction

The medium impulse response can be reconstructed from (2) and (3) :

$$h(t) = K \cdot h_s(t - \theta_o)$$

with $h_s(t) = \sum_{p=o}^{P} r_p^{io} \cdot \Delta_p^{jo,ko,lo}(t - \theta_p^{mo})$ $\quad \begin{cases} r_p = 1 \text{ for } p = 0 \\ \theta_p = 0 \text{ for } p = 0 \end{cases}$

$\Delta_p^{jo,ko,lo} = (\delta_{jo,p}, \delta_{ko,p}, \delta_{lo,p})$ is a set of three numbers characterizing respectively the beam/interface angulation, the interface curvature and the dispersive attenuation of the medium in front of the considered interface.

From that impulse response it is possible to deduce the corresponding impediogram which includes more information concerning the examined medium than classical impediograms.

3. RESULTS

Let us now test the performances of the proposed method from the point of view of :

- . depth discrimination,
- . noise immunity, and
- . sensitivity to echo shape deformation due to geometric and acoustic properties of the medium.

As for the model construction each of these parameters will be considered independantly from the others.

3.1. Depth discrimation

To illustrate the type of results that may be obtained with this process, we have conducted an experiment on a 50 μm mylar sheet and the curve a) on figure 13 shows the corresponding received signal. The application of the method gives for the second interface : $r_1 = 0.89$ and $\theta_1 = 50$ ns. Curve b) on the same figure, shows the reconstructed impulse response which is in good agreement with the theoretical one. In addition, it has to be remarked that plane and perpendicular interfaces and no dispersive attenuation have been detected by the algorithm in accordance with the experimental arrangment.

The theoretical depth discrimination of this method is limited by the accuracy of the sampling device and by the bandwidth of the reference waveform $C^{j,k,l}(t)$. Depth discrimination as good as 15 microns may be obtained with our arrangment. Unfortunately, we could not verify it on so thin samples. Nevertheless, taking into account the respective sound velocities in mylar and in biological tissues, the results of the previous experiment indicate that a depth discrimination in living medium as good as 30 μm could be obtained.

3.2. Noise immunity

We will here compare the results obtained with the proposed method, for signals with and without noise. Firstly we process the signal received from a 1 mm lucite plate submerged in water. We have then repeated our measurement with analog addition of white noise in the bandwidth of the equipment to these same echoes and processed the composite signals, for signal to noise ratio varying from 20 dB to 2.5 dB. Results are presented on figure 14.

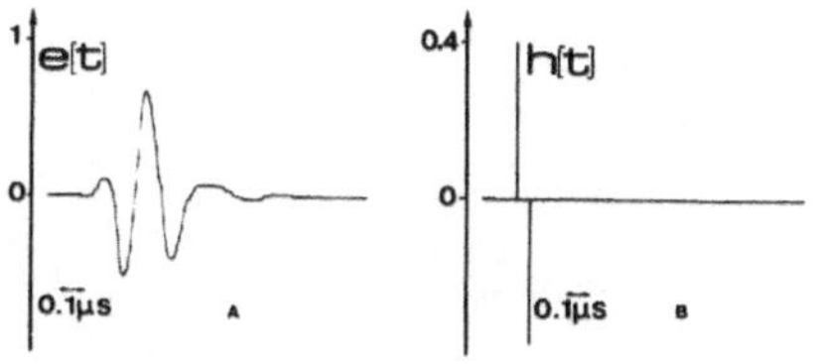

Fig. 13 : Depth resolution :
A : received signal
B : computed medium impulse response

We may observe that no noticeable error is introduced on the impulse responses compared to the one obtained without noise as far as the S/N ratio is greater or equal to 10 dB. For lower values of this ratio, an error on the interfaces position together with a modification of the value of the coefficients of reflexion can be observed. Others series of measurements for various positions of interfaces and different values of their coefficients of reflexion also indicate a limit for the S/N ratio around 10 dB. This value seems to be a good estimation for the performances of the process.

3.3. Sensitivity to echo shape deformation

For this study, we have treated the signal received from a lucite sample composed of two layers. The first plate has a thickness of 1 mm and the second of 10 mm. They are separated by 1.6 mm of water on the beam axis. The first plate is tilted of 3 degrees and the second one remains

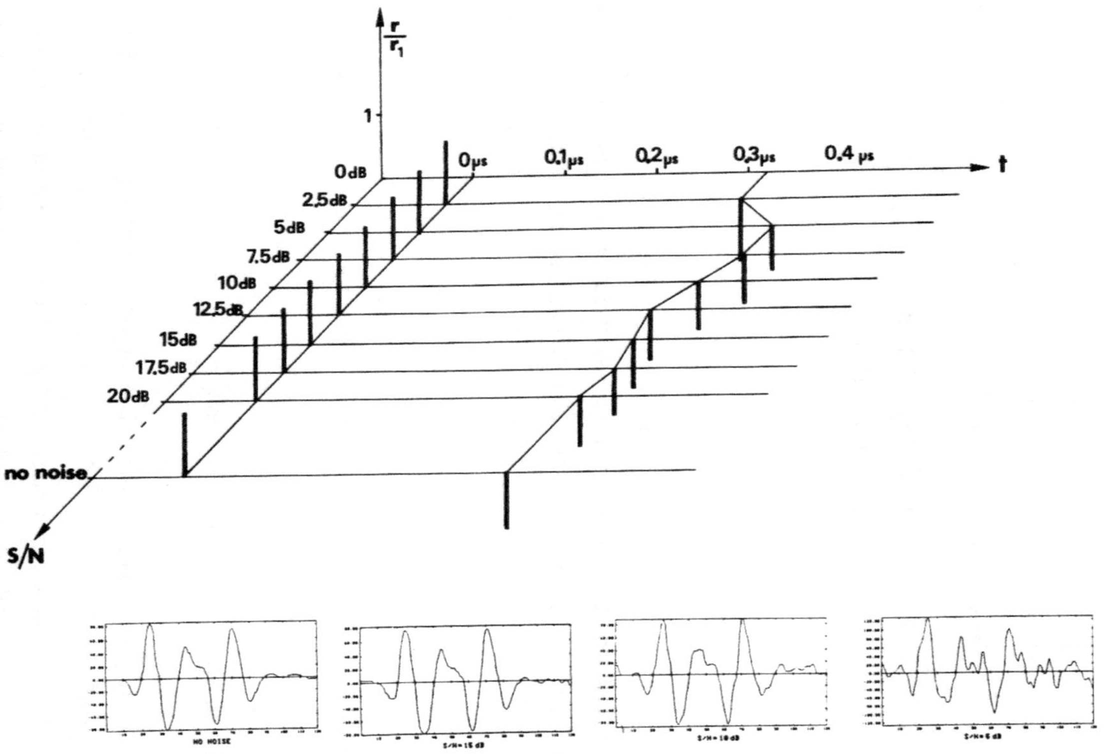

Fig. 14 : Noise immunity. Influence of noise on the medium impulse response. The amplitude of the coefficients of reflexion and the position of the interfaces are significantly modified for S/N ratio lower than 10 dB.

perpendicular to the ultrasound beam. The results are reported on figure 15 which displays the position of the interfaces and the amplitude of the corresponding coefficients of reflexion together with angulation and dispersive attenuation.

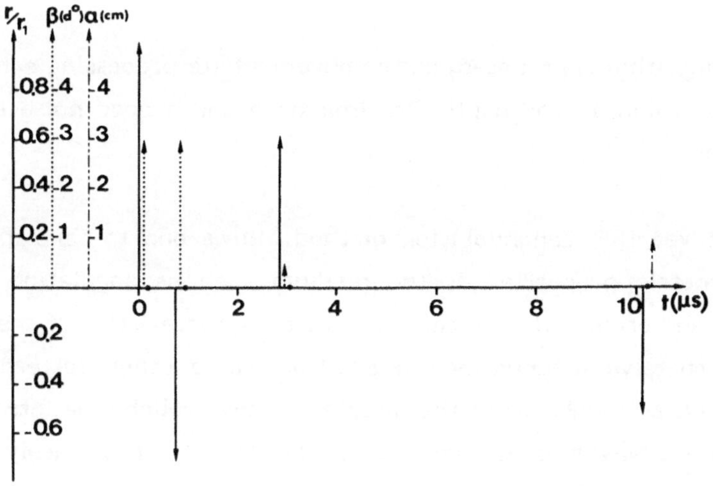

Fig. 15 : Computed impulse response for a complex sample :

abcsissa : time in μs

ordinate : r/r_1 normalized coefficient of reflexion

β incidence angle in degree,

α attenuation, expressed in terms of dispersive attenuation induced by a 1 cm lucite plate.

These results well correlate with the sample theoretical impulse response. The two sides of the first plate are correctly detected with an orientation of 3 degrees, the front surface of the second layer is found tilted of 0.5 degree while the rear surface is found to have no angulation. These results are probably due to a slight angulation of the plate due to an imperfect adjustment of the sample. No noticeable dispersive attenuation is pointed out for the first three interfaces. For the last interface a dispersive attenuation corresponding to 10 mm of lucite has been detected by the method.

4. CONCLUSION

An algorithm is proposed, quite convenient for processing echographic signals, for which a good depth discrimination and a good noise immunity are required.

Moreover this deconvolution method allows one to take in account some geometric properties of the medium such as angulation between interfaces and probe axis and some acoustic characteristics of the medium such as dispersive attenuation. In addition, some other patterns of the medium such as roughness of the interfaces may probably be introduced in the algorithm. Nevertheless the process does not seem precisely quantify the curvature of the interfaces.

Anyway the method appears to be able to widden the possibilities of impediography which is untill now very dependant of medium geometry.

ACKNOWLEDGMENTS

Work supported by the I.N.S.E.R.M. (ATP n° 83.79.115/3).

REFERENCES

1) I. Beretsky, Detection and characterization in a human arterial wall by raylographic technique, and in vitro study, in "Ultrasound in Medicine", D. White and R. Brown Ed., Plenum Press Publ., New-York, Vol. 3B : p. 1597 (1977).

2) A. Herment, P. Péronneau, M. Vaysse, A new method of obtaining an acoustic impedance profile for characterization of tissue structures, Ultrasound Med. Biol. 5 : 321 (1979).

3) J.P. Jones, Current problems in ultrasonic impediography, in "Ultrasonic Tissue Characterization", M. Linzer Ed., US Department of Commerce, National Bureau of Standards Publ., Washington, p. 253 (1976).

4) J.P. Jones, Ultrasonic Impediography and its applications to tissue characterization, in "Recent Advances in Ultrasound in Biomedicine", D. White Ed., Research Studies Press, Forest Grove, Vol. 1 : p.131 (1977).

5) S. Leeman, The impediography equations, paper presented at the 8th Int. Symp. Acoustical Imaging 29 may-2 june 1978, Key Biscayne (Florida), in "Acoustical Imaging and Holography", A. Metherell Ed., Plenum Press Publ., Vo. 8.

6) J.P. Lefebvre, New trends in wide band echography : ultrasonic impedography, Intern. Conf. on Signals and Images in Med. and Biol., Paris, April 24-28 1978, in "Biosigma 78", Vol. 2, p. 140 (1978).

THEORETICAL BASIS OF AN ULTRASONIC INVESTIGATION METHOD OF STRATIFIED ABSORBING MEDIA

Jean-Pierre Lefebvre

Laboratoire de Mécanique et d'Acoustique - CNRS

B.P. 71 - 13277 Marseille cedex 9 - France

INTRODUCTION

This paper is devoted to the extension of Jones' ultrasonic impediography to absorbing media. We show that this extension leads to a new technique which allows to reach two parameters : the acoustic impedance profile (generalization of Jones' impediography), and the dissipative parameter profile.

So, firstly, what is Jones' ultrasonic impediography ? One knows that it is the first natural quantitative step of ultrasonic echography : the use of very broadband signals theoretically allows to reach the impulse response in reflection of the medium. Then, this impulse response has to be connected with the intrinsic parameters of the medium, so that the measure of the impulse response leads to a measure of these last quantities. Jones found that the main parameter, which generates the backscattered signals, is this acoustic impedance profile, more precisely the variations of the intrinsic parameter [1]. Jones found a very simple formula connecting the impulse response in reflection of the medium H(t) with its acoustic impedance profile $\mathbf{Z}(\xi)$ (i.e. the acoustic impedance $\mathbf{Z}$ as a function of the travel-time ξ).

Then a simple formula inversion allows to reach this impedance profile from the impulse response measure.

The scheme of the metrology is therefore :

(wideband back-scattered measures) → (Impulse response in reflection) → (acoustic impedance profile)

The Jones' impediography limits are :

- The recovery of the impulse response from wideband measures, is a technical problem (wideband transducer technology) plus a numerical problem (deconvolution). We are not concerned with this problem here.

- The tissue characterization from the impulse response is a physical problem (tissue modelization) plus a theoretical and numerical problem (inverse problem). We are concerned by this because Jones' tissue model is too idealistic, as the model is :
. lossless (biological tissues are dissipative media : they are the source of acoustic absorption-dispersion phenomena, especially since impediography needs wide band signals),
. stratified, with layers arranged perpendicular to the beam direction ; the problem is thus one dimensional,
. weakly varying, so that a mono-scattering approximation may be used. However, such an approximation is too rough, it cannot take in account the multiple reflections that often happen.

In a previous study, we were concerned with this last problem. We solved the inverse problem for strongly varying media by mean of use of a Gelfand-Levitan procedure, which is a procedure for inverse quantum scattering problems [2].

In an other study, we were concerned with three dimensional structures [3], in a different maner than Leeman [4].

Now, we are interested with dissipation. It raises two problems : the modelization problem and the inverse problem.

ACOUSTIC MODELLING OF DISSIPATIVE MEDIA

We choose the most simple model : that of viscosity. We assume an idealized viscous layered fluid arranged perpendicular to the acoustic beam (one dimensional case).

The propagation equation is obtained by the classical linearization method, from the equations of the mechanics of continua, in the viscous fluid case [3]. It is:

$$-\frac{1}{c^2}\frac{\partial^2 p^A}{\partial t^2} + \rho\frac{\partial}{\partial x}\left[\frac{1}{\rho}\frac{\partial}{\partial x}\left(p^A + \tau\frac{\partial p^A}{\partial t}\right)\right] = 0 \qquad (1)$$

where $p^A(x,t)$ is the acoustic pressure field (t tissue coordinate, x space coordinate), c(x) the isentropic speed of sound (space varying), ρ(x) the density (space varying), τ(x) the dissipative parameter (space varying) : with τ connected with the first and second viscosities (μ,λ) by : $\tau = \frac{\lambda+2\mu}{\rho c^2}$

REMARK 1 : τ has a time dimension.

REMARK 2 : For a homogeneous medium and a mono-frequency plane-wave excitation (with a pulsation ω), this formulation leads to an absorption coefficient $\alpha = \omega^2\tau/2c$, i.e. a "classical" square frequency varying absorption (no relaxation effect).

The chosen modelization is still too idealistic, but is a first step in the dissipation characterization. It is more physical than the usual formulation with complex frequency varying celerity and density, which is only realistic for homogeneous media, and it is an exact formulation for a viscous heterogeneous fluid.

THE TWO PARAMETERS CHARACTERIZATION

The present equation needs three parameters : the celerity c, the density ρ, and the dissipative parameter τ.

We decrease this number, by the introduction of the well-known travel-time ξ, in substitution of the space-coordinate x :

$$\xi = \begin{cases} \dfrac{x}{c_0} & ; x \leq 0 \quad \text{(where } c(x) = c_0\text{: homogeneous space)} \\ \displaystyle\int_0^x \frac{dx'}{c(x')} & ; x > 0 \quad \text{(heterogeneous space)} \end{cases}$$

This coordinate-change leeds to :

$$(2) \qquad -\frac{\partial^2 p^A}{\partial t^2} + Z \frac{\partial}{\partial \xi}\left[\frac{1}{Z} \frac{\partial}{\partial \xi} \left(p^A + \tau \frac{\partial p^A}{\partial t}\right)\right]$$

with $Z = \rho c$ the classical impedance.

REMARK 3 : For the lossless case ($\tau = 0$), we recover the Pekeris equation, which is the basis of our generalization of the Jones' technique to strongly varying media [2].

We may write also :

$$(3) \qquad -\frac{\partial^2 p^A}{\partial t^2} + \frac{\partial^2}{\partial \xi^2}\left(p^A + \tau \frac{\partial p^A}{\partial t}\right) - q \frac{\partial}{\partial \xi}\left(p^A + \tau \frac{\partial p^A}{\partial t}\right) = 0$$

with $q = \dfrac{d}{d\xi} \operatorname{Log} Z$

Or :

$$(4) \qquad -\frac{\partial^2 p^A}{\partial t^2} + \frac{\partial^2 p^A}{\partial \xi^2} - A_{(\xi,t)}[p^A] = 0$$

with an operator
$$A_{(\xi,t)}[\bullet] = q \frac{\partial}{\partial \xi}[\bullet] + \left[q \frac{d\tau}{d\xi} - \frac{d^2\tau}{d\xi^2}\right] \frac{\partial}{\partial t}[\bullet] + \left[q\tau - 2\frac{d\tau}{d\xi}\right] \frac{\partial^2}{\partial \xi \partial t}[\bullet] - \tau \frac{\partial^3}{\partial \xi^2 \partial t}[\bullet]$$

We now have only two parameters : (Z, τ) or (q, τ).

THE INVERSE PROBLEM : GENERAL CASE

As in a previous paper [2], a direct attempt to solve the inverse problem is given - i.e. the identification of the characteristic parameters by mean of scattering measures - by transforming the acoustic problem into a quantum scattering one. Indeed, inverse problems of quantum scattering are the most studied of the litterature on inverse problems. One could hope to find the most advanced inverse procedures in that context.

Let : $\Psi = \eta\left(p^A + \tau \dfrac{\partial p^A}{\partial t}\right) \; ; \; \eta = Z^{-1/2}$

and (time-Fourier transform) :

$$\Psi(\xi, t) \rightleftharpoons \psi(\xi, \omega) = \int_R \Psi(\xi, t)\, e^{-i\omega t}\, dt$$

The time-independant Schroedinger equation is obtained :

$$\frac{\partial^2 \psi}{\partial \xi^2} + \omega^2 \psi - \mathcal{V}(\xi,\omega)\,\psi = 0 \qquad (5)$$

with a scattering potential : $\mathcal{V}(\xi,\omega) = V(\xi) + i\,\dfrac{\omega^3 \tau(\xi)}{1+i\omega\tau(\xi)}$

where $V(\xi) = \dfrac{1}{\eta}\dfrac{d^2\eta}{d\xi^2}$ as in the lossless case [2].

We have a scattering potential with cubic frequency-variation for imaginary part.

This is an inverse problem not yet solved. The only avaiable solutions of the inverse problem are these with ω^0 [5], ω^1 [6], [7] and ω^2 [8] variations.

It is still an inverse problem to solve in the general case, which ogurs much theoretical difficulties (problems of functionnal and numerical analysis).

We try to solve it, like Jones in the lossless case, in the ideal case of weakly varying media.

THE INVERSE PROBLEM : WEAKLY-VARYING MEDIA

We need the classical scheme : approximate solution of the direct problem, followed by a formula inversion.

However, with regard to the lossless case, we have here two functionnal parameters to identify : the acoustic impedance $Z(\xi)$ (or the parameter $q(\xi) = \frac{d}{d\xi} \mathrm{Log}\, Z$) and the dissipative parameter $\tau(\xi)$

Two measures are thus necessary to solve the inverse problem. We choose the two impulse response in reflection of the medium (with the hypothesis of limited extension of the inhomogeneous medium) : the impulse response in reflection to the left $H_{12}(t)$ and the impulse response in reflection to the right $H_{21}(t)$.

These two impulse response are obtained by probing the specimen by its two sides (two echographs) or by turning it over (only one echograph).

REMARK 4 : One may note that the two impulse response $H_{12}(t)$, $H_{21}(t)$ are the inverse Fourier transforms of the two reflection coefficient $S_{12}(\omega)$, $S_{21}(\omega)$, of the scattering matrix $S(\omega)$.
(While $S_{11}(\omega)$ and $S_{22}(\omega)$ are the two transmission coefficient, with the reciprocity relation $S_{11}(\omega) = S_{22}(\omega)$).

$$S(\omega) = \begin{bmatrix} S_{11}(\omega) & S_{12}(\omega) \\ S_{21}(\omega) & S_{22}(\omega) \end{bmatrix}$$

In the lossless case, one shows that S (ω) is unitary, so the S matrix is then uniquely determined by the relfection coefficient to the left S_{12} (ω). It is not the case here.

Calculation of the approximate impulse response in reflection to the left

The impulse response $H_{12}(t)$ is obtained when the medium is probed with an acoustic impulse $\delta(t - \xi)$ incoming (at $t=0$, for $\xi=0$) from the left homogeneous space ($\xi < 0$), and propagating to the right : a part is transmitted to the right and other part is reflected, giving rise to a term $H_{12}(t + \xi)$, defining the impulse response in reflection to the left $H_{12}(t)$.

So, the scattering solution for this problem, in the left homogeneous space ($\xi < 0$) is : $\Psi_1(\xi,t) = \delta(t-\xi) + H_{12}(t+\xi)$

When the medium is weakly inhomogeneous, a Born approximation may be used (the monodiffusion) to calculate Ψ_1 and H_{12}.

One shows that the first order approximate of the direct problem is, with the formulation (4) :

(6) $$H_{12}^{(1)}(t) = \frac{1}{4}\left[q - \frac{1}{2} q \frac{d\tau}{d\xi} + \frac{1}{2}\frac{dq}{d\xi}\tau + \frac{1}{4}\frac{d^2\tau}{d\xi^2}\right]_{\xi = t/2}$$

REMARK 5 : In the lossless case ($\tau = 0$), we recover the Jones formulation

$$H_{12}^{(1)}(t) = \frac{1}{4}\left[q(\xi)\right]_{\xi=t/2} = \frac{1}{4}\left[\frac{d}{d\xi}\operatorname{Log} Z\right]_{\xi=t/2}$$

REMARK 6 : When there is no impedance variation ($q = \frac{d}{d\xi}\operatorname{Log} Z = 0$), but only a dissipative-parameter variation, then : $H_{12}^{(1)}(t) = \frac{1}{16}\left[\frac{d^2\tau}{d\xi^2}\right]_{\xi=t/2}$

So, a dissipative variation (viscosity variation) can give rise to a backscattered signal, like an impedance variation. This point was also noted by Leeman [9], in a different manner.

Calculation of the approximate impulse response in reflection to the right

Now we probe the medium with an acoustic impulse $\delta(t+\xi-\xi_L)$ incoming (at $t=0$, for $\xi = \xi_L$) from the right homogeneous space ($\xi > \xi_L$) and propagating to the left : a part is transmitted to the left, while the other part is reflected, giving rise to a term $H_{21}(t - \xi + \xi_L)$, defining the impulse response in reflection to the right $H_{21}(t)$.

The scattering solution in the right homogeneous space ($\xi > \xi_L$) is then : $\Psi_2(\xi,t) = \delta(t+\xi-\xi_L) + H_{21}(t-\xi+\xi_L)$

With the same assumption of weakly varying medium, the Born approximation for the direct problem leads, to first order :

$$H_{21}^{(1)}(t) = -\frac{1}{4}\left[q + \frac{1}{2} q \frac{d\tau}{d\xi} - \frac{1}{2}\frac{dq}{d\xi}\tau - \frac{1}{4}\frac{d^2\tau}{d\xi^2}\right]_{\xi = \xi_L - t/2} \tag{7}$$

The two approximate formulations (6 - 7) of the impulse responses in reflection of the medium will allow to reach the two parameters Z (or q) and τ.

Approximate solution of the inverse problem

By simple linear combinations of the two impulse responses, we have :

$$q^{(1)}(\xi) = 2\left[H_{12}^{(1)}(2\xi) - H_{21}^{(1)}(2\xi_L - 2\xi)\right] \tag{8}$$

and

$$\frac{d^2\tau^{(1)}}{d\xi^2} - 2q^{(1)}(\xi)\frac{d\tau^{(1)}}{d\xi} + 2\frac{dq^{(1)}}{d\xi}\tau^{(1)}(\xi) = 8\left[H_{12}^{(1)}(2\xi) + H_{21}^{(1)}(2\xi_L - 2\xi)\right] \tag{9}$$

The relation (8) leads to $q^{(1)}(\xi) = \left[\frac{d}{d\xi}\text{Log}\,Z\right]^{(1)}$ and so to the first order impedance profile $Z^{(1)}(\xi)$.

This is a generalization of Jones' impediography to absorbing media.

After solving (8) and so determining q (ξ), the solution of the ordinary differential equation with varying coefficients (9) leads to $\tau^{(1)}(\xi)$, first approximation of the dissipative parameter profile $\tau(\xi)$.

We reach a new parameter, the dissipative parameter profile (in our case, the viscosity profile).

One can hope by exploiting this technique, to obtain new information about biological tissues.

Moreover, it is obvious that this technique is a simple double impediography. So, the technical difficulties will be the same as those encontered in impediography : wide-band transducers, deconvolution procedure. No new technical problem will arise.

CONCLUSION

We made a tentative of extension of the impediography technique to absorbing media. As a first step, we treated the simplest model of dissipation : the viscosity case. This is still an unrealistic modelization of biological tissue, but it is better than no dissipation at all (absorption varying as the square of the frequency).

In the general case - strongly varying media - no solution is as yet avaiable.

For the simpler case of weakly varying media, an approximate solution has been found, which leads :

- firstly, to a generalization of the impediography formulas,
- secondly, to a new technique, leading to the dissipative parameter profile.

This generalization of Jones' impediography is a simple combination of conventional impediography : it only needs the probing of the medium by its two sides.

[1] J.P. Jones, Ultrasonic impediography and its application to tissue characterization, in "Recent advances in ultrasound in biomedicine" (D.N. White Ed.), Research studies Press, LOndon, (1977).

[2] J.P. Lefebvre, Théorie d'une méthode quantitative d'investigation des milieux stratifiés : l'impédographie acoustique, Acustica, vol. 41, n° 1, (1978).

[3] J.P. Lefebvre, Quelques méthodes quantitatives d'investigation par ultrasons - Problèmes de modélisation et d'inversion de la rétrodiffusion, Thèse d'Etat, Université de Provence (Aix-Marseille 1),(1981)

[4] S. Leeman, The impediography equations, in : Acoustical imaging, vol. 8, A. Metherell Ed. (Proceedings of the 8th Int. Symposium on acoustical imaging Key Biscayne, Florida, May 1978).

[5] K.Chadan, P.C. Sabatier, INverse problems in quantum scattering theory, Springer Verlag, New York, (1977).

[6] M. Jaulent, Inverse scattering problems in absorbing media, Journ. Math. Physics, vol. 17, n° 7, (1976).

[7] A.C. Schmidt, The one dimensional inverse scattering problems Ph. D. Thesis, Harvard University, Cambridge, (1974).

[8] M. Razavy, Détermination of the wave velocity in an inhomogeneous medium from the reflection coefficient, Journ. Acoust. Soc. Amer., vol. 58, n° 5, (1975).

[9] S. Leeman, P. Vaughan, The inverse backscattering problem - a different approach, in : Cavitation and inhomogeneities in underwater acoustics, W. Lauterborn Ed. / Springer series in Europhysics n° 4, Springer Verlag, Berlin Heidelberg, (1980).

AN ANALYSIS OF THE PARAMETERS AFFECTING TEXTURE IN A B-MODE ULTRASONOGRAM

Joie Pierce Jones and Carolyn Kimme-Smith

Department of Radiological Sciences
University of California Irvine
Irvine, California 92717 USA

ABSTRACT

The complex echo-intensity patterns observed in gray-scale B-mode ultrasonograms are known as texture. Texture is functionally related to the type and state of tissue interrogated, to the statistics of the scattering process, to the characteristics of the interrogating ultrasound pulse, to the scanning technique utilized, and to the type of display instrumentation. We have recently begun a program to examine the nature of texture, as it applies to medical ultrasonics, and are in the process of investigating the various tissue, acoustical, and instrumentation parameters which affect and determine texture. This paper presents the results of our investigation to date. As a part of our study we have simulated on a computer (Perkin-Elmer 8/32) the chain of electronic processing from individual A-line ultrasound waveforms to the B-mode display. By changing in software the various processing parameters we have investigated their effect on texture. Several statistical measures of texture have been developed for a quantitative comparison and analysis of our results. The implications of this study to the design of B-mode ultrasound devices are noted and reviewed.

INTRODUCTION

Texture in a B-mode ultrasonogram results from the mapping, into an appropriate display format, of the intensities of ultrasound pulses which are reflected and scattered from tissue. Each pixel of the conventional CRT rectangular grid is assigned a gray level value which corresponds in some way to the echo intensity values for that region. Strong echoes from specular reflecting tissue boundaries serve as major anatomical markers while much weaker echoes which are scattered from structures within tissue provide texture patterns which in many cases are uniquely characteristic of the interrogated pathology. If detailed correlations can be established between texture in a B-mode ultrasonogram and tissue pathology then the implications for and impact on diagnostic medicine would be significant. The establishment of such correlations will require the processing and display of texture in such a fashion that the resulting texture is determined by tissue structures rather than other non-biological factors which often contribute to, and may prodominate over, the texture pattern.

Here we seek to evaluate the influence of parameters other than tissue properties on texture. In particular we explore the impact of changes in B-mode ultrasound instrumentation parameters on texture. To undertake our study we use a computer to simulate the functions performed by the ultrasound hardware. Using our ultrasonic data acquisition system(1) we record the rf waveform associated with each individual A-line; the computer then processes this A-line data set into a simulated B-mode ultrasonogram. Thus, we simulate in software what the ultrasound instrumentation does in hardware. This computer simulation of an ultrasound scanner allows us to easily investigate the effect of changes in instrumentation parameters on texture without making expensive and time-consuming changes in hardware. Moreover, by conducting a series of simulations on a given A-line data base (thereby producing a series of simulated B-mode ultrasonograms representing various changes in instrumentation parameters) we are assured that our comparisons are valid and reasonable.

EXPERIMENTAL METHODS

Figure 1 is a block diagram of the detection and display process of a typical B-mode ultrasound system. The various blocks in the figure beginning with rectification represent those elements of hardware we wish to simulate in software. This is accomplished as indicated by our computerized data acquisition system which records a digital representation of the rf ultrasound waveform after it has been detected and subjected to time gain compensation (TGC) but before it has been envelope detected and the phase information lost.

Figure 2 is a block diagram of the data acquisition and analysis process we follow. This summarizes how our computerized data acquisition system allows us to simulate in software the functions of our ultrasound instrumentation.

Figure 3 is a block diagram of the major components of our computerized data acquisition system. All the data used in this paper was taken on a large natural sponge which was scanned under computer control by the Klinger positioning system. A standard 2.25MHz, 15mm unfocused transducer was used. A total of 600 separate

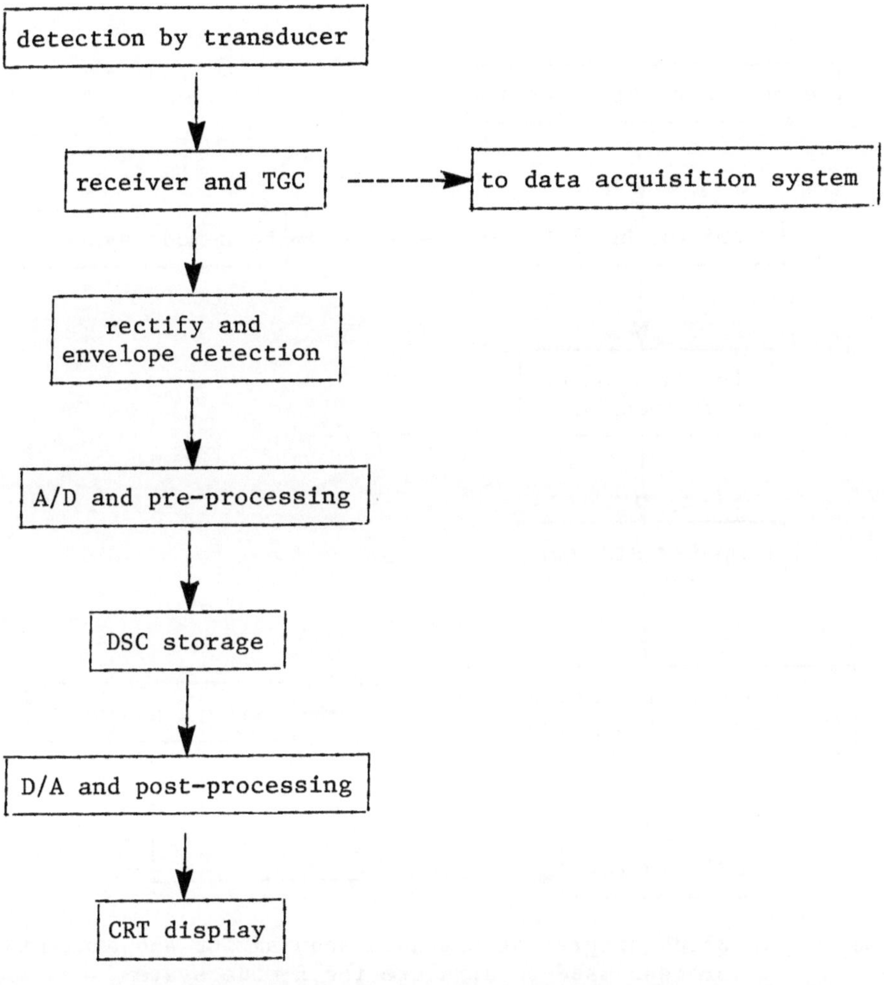

Figure 1. Block diagram of the detection and display process of a typical B-mode ultrasound system we wish to simulate.

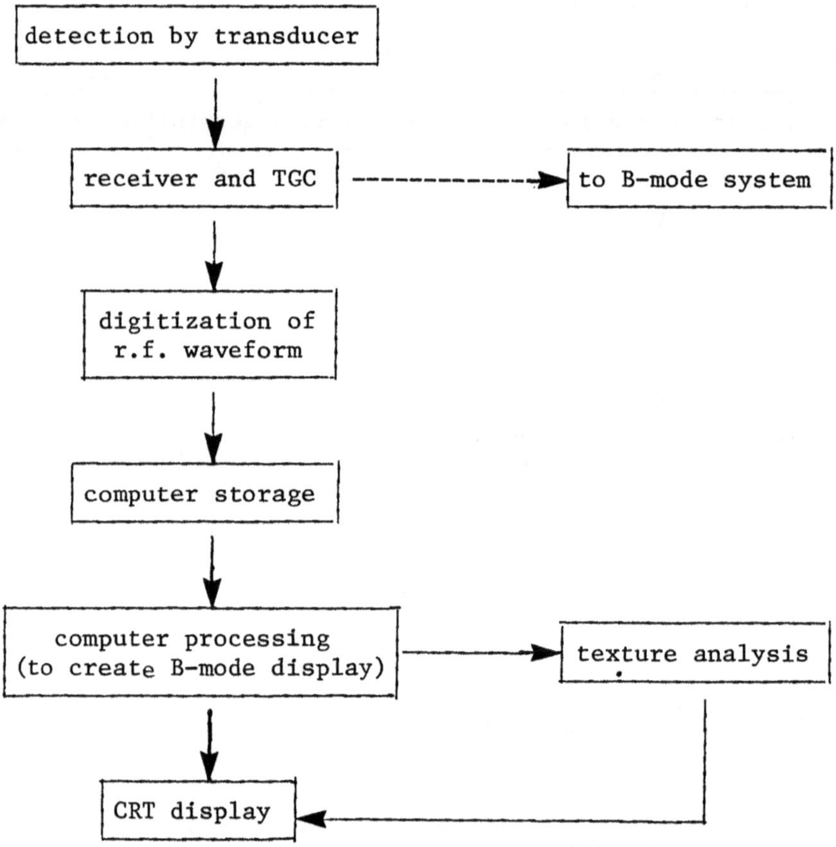

Figure 2. Block diagram of the data acquisition and analysis process used to simulate the B-mode system.

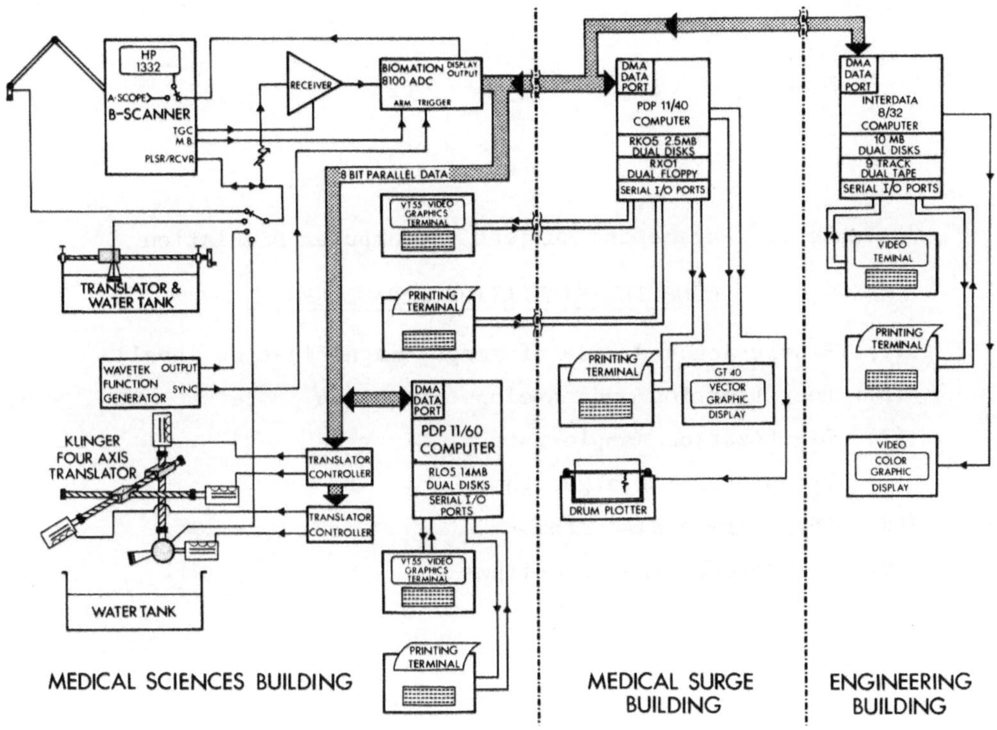

Figure 3. Block diagram of the computerized ultrasound data acquisition system implemented at the University of California Irvine.

A-lines were taken in 0.1mm intervals. The rf waveform associated with each A-line was sampled at 20MHz with the Biomation 8100 A/D converter with 6 to 7 bit resolution. Each A-line was 2048 points long (the maximum size of the Biomation buffer memory). This data set of 600 A-lines represents over a million bytes of information and can be constructed, via computer simulation, into a 6cm x 6cm mapping of echo intensities.

Table 1. Parameters subject to computer simulation

COMPUTER SIMULATION PARAMETERS

(1) Resolution and size of image (magnification level)

(2) Rectification and envelope detection process

(3) Digitization sample rate

(4) Pre-processing algorithms

(5) DSC storage algorithms

(6) Post-processing algorithms

In this study the digital data recorded by the Biomation 8100 is automatically transferred to the PDP 11/40 for storage on magnetic disc. A completed data set is then transferred to the Perkin-Elmer 8/32 computer for processing and the simulation of the B-mode instrumentation process. The various hardware parameters subject to computer simulation are detailed in Table 1.

EXAMPLES OF COMPUTER SIMULATION OF INSTRUMENTATION PARAMETERS

Figure 4 is a B-mode ultrasonogram of the large "elephant-ears" natural sponge used in all of the computer simulation studies reported in this paper. Six hundred individual A-lines, about 7cm in depth and with a 0.1mm lateral spacing, were recorded in this same B-scan plane, as described in the previous section. Prior to digitization each A-line was amplified by a linear receiver and subjected to an identical linear TGC. This single data set (of 600 A-lines) was used for all the simulation studies described here.

Figures 5 through 14 illustrate the various processing options we have applied to our data base, simulating various changes in the B-mode instrumentation. Here each example is the result of varying a single parameter, as will be described below. All photographs of the B-mode simulations were taken within a half-hour of one another to insure that CRT brightness drift would not adversely effect the results. The occasional streaks seen in all the photographs are a result of digitization failure when collecting the original data.

One of the first decisions a user must make, when beginning a B-mode scan, is the anatomical area the scan will cover on the display. If transducers of a higher frequency or a more shallow focus are used, the anatomical area displayed will be reduced, but the image will contain more detail. If the standard 3cm/division option is selected, then 16 samples from 8 A-lines will contribute to the gray-level value assigned to a single pixel. This assumes that the A/D converter which digitizes the envelope detected ultrasound signal for input to the digital scan converter runs at 15MHz, as is commonly the case. If the user selects the 1cm/division option, then 5 samples from 3 A-lines contribute to a single pixel. Figures 5 and 6 illustrate the differences between the resulting two textures. The texture in the display taken with the 1cm/division option looks less grainy and in fact has a lower mean gray level and gray level standard deviation than the texture in the display taken at 3cm/division. (Note: in all of the computer simulated displays, white is the gray level of highest intensity.) Even if a larger transducer with a lower center frequency is used (which gives reduced spatial resolution), we still seem to see more detail at the lower cm/division setting because fewer of the samples are used for each pixel. In this case each sample has a proportionally greater impact on the pixel value. Thus, a gradually increasing amplitude in a given direction, when represented by the 1cm/division map, might have a gray level ramp of pixels before the maximum pixel value is achieved; the 3cm/division option, on the other hand, is likely to represent the increasing amplitude as a single pixel at the maximum amplitude.

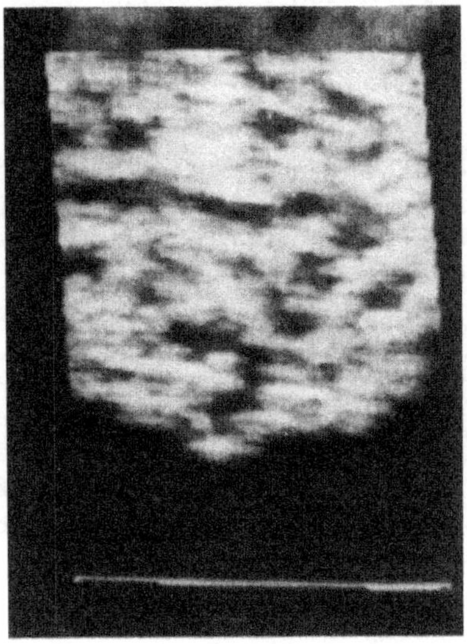

Figure 4. B-mode ultrasonogram of section of natural sponge.

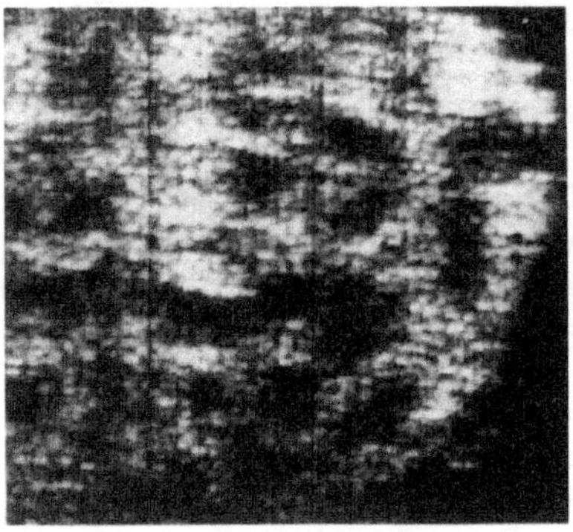

Figure 6. Simulated B-mode ultrasonogram with 1cm/division option.

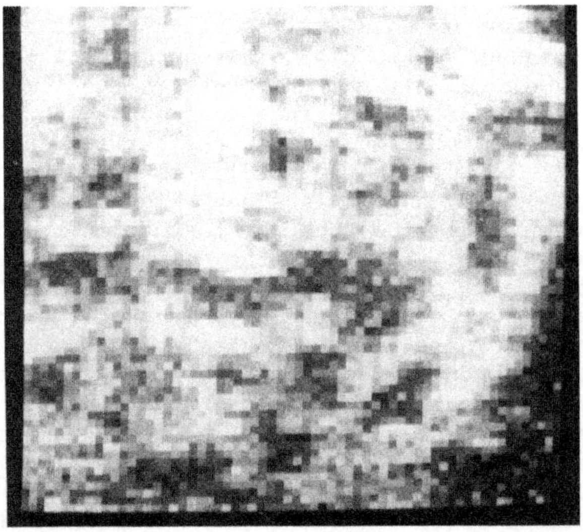

Figure 5. Simulated B-mode ultrasonogram with 3cm/division option (ADC sampling rate for input to DSC = 15MHz; "standard" gray-scale mapping; peak-weighting DSC algorithm).

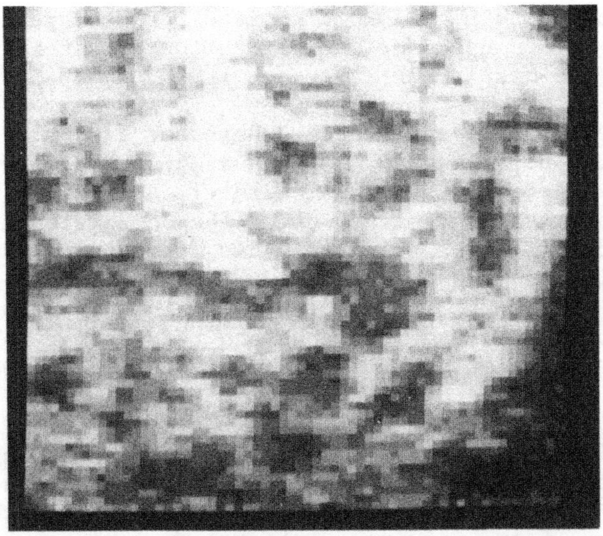

Figure 7. Simulated B-mode ultrasonogram, ADC sampling rate for input to DSC = 10MHz.

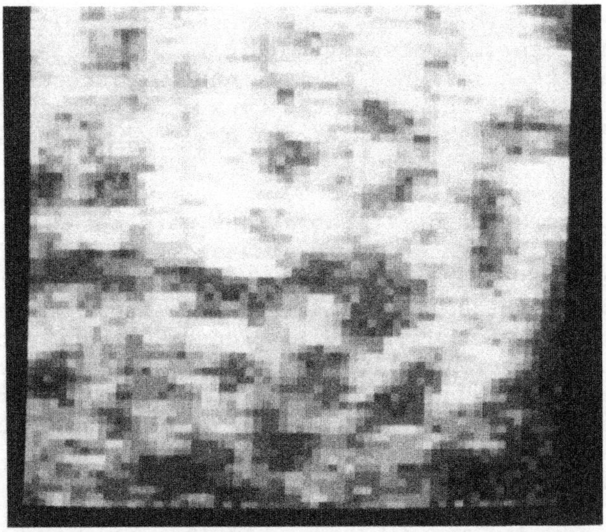

Figure 8. Simulated B-mode ultrasonogram, ADC sampling rate for input to DSC = 20MHz.

If the B-mode instrumentation uses a digital scan converter (DSC), as most current units do, then the rate at which the envelope detected ultrasound signal is digitized for storage in the DSC as well as the method used for data selection and storage in the DSC will affect the display of texture.

By interpolating our standard data set (recorded at 20MHz) using Newton's three point method, we have simulated B-mode units with 10, 15, and 20MHz A/D converters. The results are shown in Figures 7, 5, and 8 and are all displayed using the 3cm/division option. The image recorded at 10MHz (Figure 7) has missed some of the isolated peaks seen in the images recorded at 15MHz (Figure 5) and at 20MHz (Figure 8). The image recorded at 20MHz is surprisingly less grainy, with more subtle shading, than the one recorded at 15MHz; however, it does not represent a 33% improvement even though we had to store 33% more points.

In addition to the ADC sampling rate, the mapping of intensity amplitudes to gray levels in the digitizer can have a strong impact on texture. Many units have a preprocessing option which controls the selection of the mapping function. These curves are often devised empirically by adjusting voltage levels until the texture in a particular test object appears appropriately grainy or even. These visual adjustments seem to produce mappings which are approximately logrithmic, so that low intensity echoes are assigned to most of the gray-level values, while strong echoes are assigned to only a few values. Figure 5 illustrates one particular gray-scale mapping which is standard with at least one conventional ultrasound device. Figure 9 is an identical display to Figure 5 except even more gray-levels have now been assigned to low intensity echoes. The resulting image appears lighter than Figure 5 because echoes in Figure 5 which were assigned to dark gray values are now assigned to lighter gray values.

All of the preceeding illustrations were based on a 4-bit gray scale, that is, 15 shades of gray plus one gray level for labeling. If instead we have a system with a 6-bit dynamic range (63 gray levels plus one for labeling) and if we map these gray levels with a curve equivalent to that used for Figure 9, then we obtain the display shown in Figure 10. We should remember that most CRT's cannot display more than 16 shades of gray, but that the increased gray shades may be needed for numerical processing. We could also use the increased gray-levels for gray-level windowing, as is now done in computerized tomography.

The DSC storage algorithm controls the selection of one gray level from several A-lines. At 15MHz, 3cm/division, and with a scanning rate of 1000 A-lines/second, we have 128 intensity values associated with one pixel. If we envelope detect each A-line, we

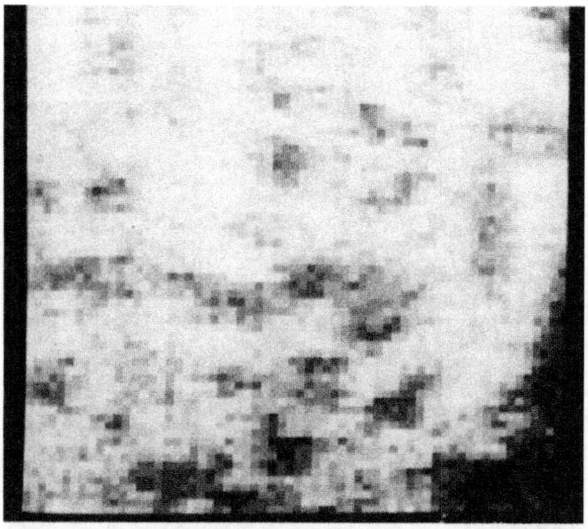

Figure 9. Simulated B-mode ultrasonogram identical to Figure 5 except more gray-levels have been assigned to low intensity echoes.

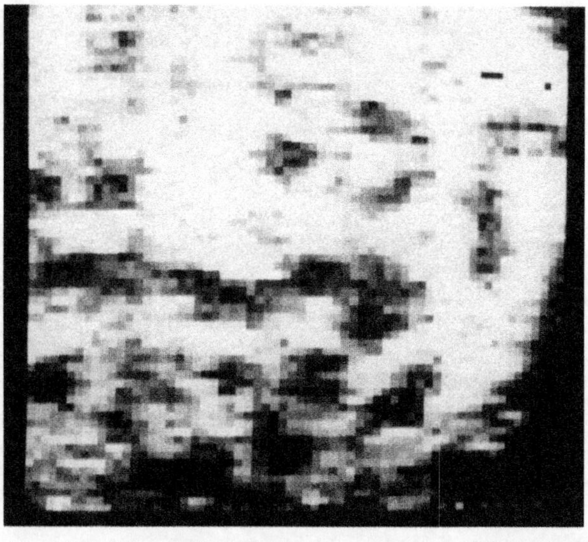

Figure 10. Simulated B-mode ultrasonogram. Identical to Figure 9 except based on a 6-bit rather than a 4-bit gray-level.

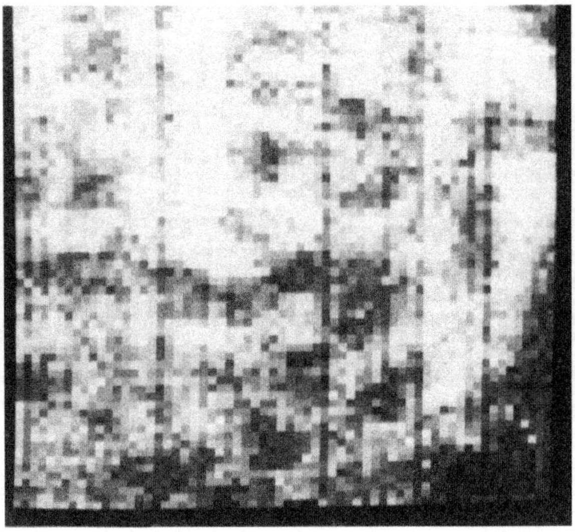

Figure 11. Simulated B-mode ultrasonogram using last-value DSC storage algorithm.

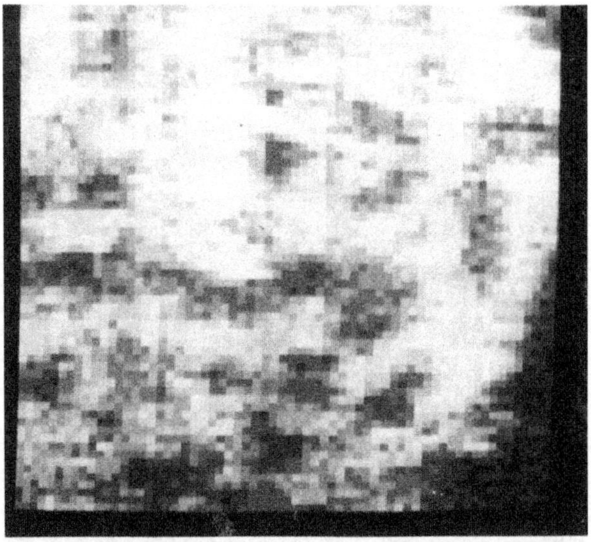

Figure 12. Simulated B-mode ultrasonogram using averaged-intensity DSC storage algorithm.

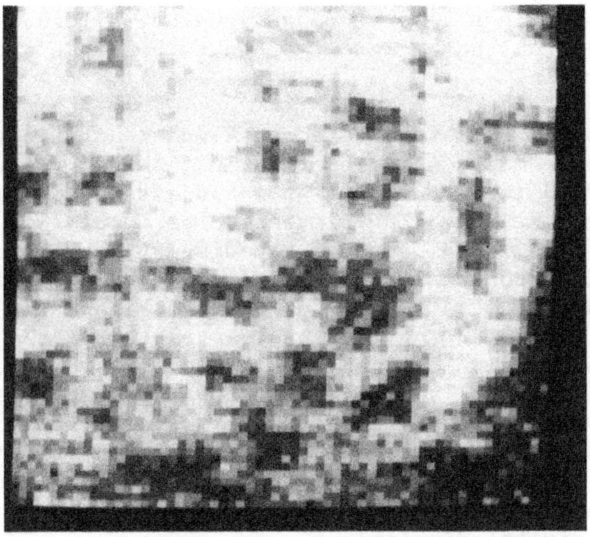

Figure 13. Simulated B-mode ultrasonogram. Here more gray-levels have been assigned to low intensity echoes by a post-processing gray-scale map.

still have 8 values which can contribute to the gray level assigned to a given pixel. A so-called survey option on the B-mode unit selects the last A-line value as the assigned gray-level value. This gives a very grainy image (see Figure 11) compared to Figure 5 where the maximum A-line value was used. The selection of the maximum A-line value as the assigned gray-level value is commonly known as a peak-weighting algorithm. If instead of these two methods we envelope detect the A-lines and then average the points within the boundaries of each pixel to select a gray-level value, the resulting algorithm produces a smoothed image (as shown in Figure 12). Note the increased detail seen in Figure 12, particularly in the central region. Since this region is predominately a light (or high intensity) area, preserving the maximum A-line value tends to reduce contrast for lower intensity echoes. The averaged-intensity DSC storage algorithm, by storing a lower intensity value, allows us to see smaller changes from the background then is possible with the peak-weighting algorithm.

The last option usually offered on B-mode units is a post processing gray-scale map. After storage in the DSC, the digital image is converted to an analog signal and displayed on a gray-scale CRT. The assignment of voltage levels to particular gray-levels depends on the dynamic response of the display, the room lighting, the brightness and contrast settings of the display, and the viewer's preferences. We illustrate this option with Figure 13; here the gray-scale map has moved the mean gray-level toward the dark end of the scale and has assigned more gray-levels to low intensity signals. While this is conceptually similar to the gray-scale assignment illustrated by Figure 9, the results are not at all comparable. Some of these differences are a result of the application of a particular storage algorithm; others are a result of differences in gray-level display on the CRT (here we are simulating an analog CRT with a Genesco digital display).

CONCLUSIONS

These few simple examples only serve to illustrate some of the many variables which contribute to texture in B-mode ultrasonograms. Work now in progress seeks to develop a more quantitative measure of texture and to compare such texture descriptors with our visual evaluation of the same images. We also wish to use our computer simulation system as a design aid and test various B-mode processing algorithms which are not presently implemented in hardware.

REFERENCES

1. J.P. Jones and R. Kovack, "A Computerized Data Analysis System for Ultrasonic Tissue Characterization", in Acoustical Imaging, Vol. 9, K. Wang (ed), Plenum (1980).

THE CHARACTERISTIC ECHOSTRUCTURES OF THE DIFFERENT COMPONENTS OF MAMMARY TISSUE

J.L. Lamarque, M.J. Rodiere, J. Attal*, J.M. Bruel, J.P. Rouanet, A. Djoukhadar, E. Boubals, J. Mariotti, J. Roustan

Service de Radiodiagnostic Hopital ST ELOI

34 059 MONTPELLIER CEDEX FRANCE

* Universite des sciences et techniques du Languedoc
3 4 060 MONTPELLIER CEDEX FRANCE

A mammographic picture is a plain picture in gray scale made up from the various geometric projection of the volume of mammary elements which has a heterogenous structure. It has to be considered according to the three-dimensional anatomic/histologic structure to obtain the best information. This plain image hardly allows the radiologist to imagine the elementary shape of structures behind pathological opacities and transparencies.

We have investigated the effects of histological structures and how image information is superimposed or hidden and finally which kind of opacities and x-ray densities might correspond with each histological structure. This work was performed using correlations between plain film mammography and radiographs of macroscopic and microscopic sections.

This study allowed us to consider several essential points:

- The epithelium of the galactophores is not shown because of its low radiopacity.

- Mammographic densities are due above all to fibrous connective tissue of various roentgenologic density.

As ductal epithelium was not perceptible by x-rays, and since pathology of this key structure is the main problem in breast disease, we had tried to determine if this structure as well as other histological structures could be perceptible by ultrasonography. Ultrasonography has undergone a great development in all fields of medicine and especially in radiology.

This technique has been applied to clinical senology, it has many advantages such as a low cost, repeatability, and a non-invasive method of exploration. The diagnosis in ultrasonography is based on two factors: the morphological study of the explored organ, and the variations of echostructures that are more or less characteristic of some tissue modifications (edema, necrosis, fatty infiltration, liquid filled lesions...), but ultrasonography has many limitations because of the lack of an effective study of ultrasound's variables that could be estimated and analyzed according to the variations of structure of the different tissues. The evidence of a characteristic echostructure of a tissue type (either normal or pathologic) had to be demonstrated in order to give ultrasonographers reliable and specific signs of a tissue and its pathology.

Material and Method:

We had studied histological section images obtained by acoustic microscopy and performed correlations between these two types of imaging in an attempt to find at a microscopic level the characteristic echostructures of the different component of mammary tissues: samples of mammary tissue removed during surgery were prepeared by essentially the same procedure used to prepare samples for light microscopy; excised tissue is first stabilized by fixation in a formalin solution, subsequently a piece of the tissue is embedded in paraffin. The paraffin infused throughout the material providing the stiffness required to cut thin sections with standard microtome techniques. The sections cut must have a 5 microcons thickness for acoustic microscopy, a very thin layer of gelatin is used to fasten the microtome section onto the 2 microns mylar support membrane then the paraffin is removed by rinsing the sample in Xylene and a series of ethanol-water solution.

We have to remember that all the samples examined are unstained so the acoustic response is typical of what can be expected from the mammary tissues as altered only by the procedures of fixation.

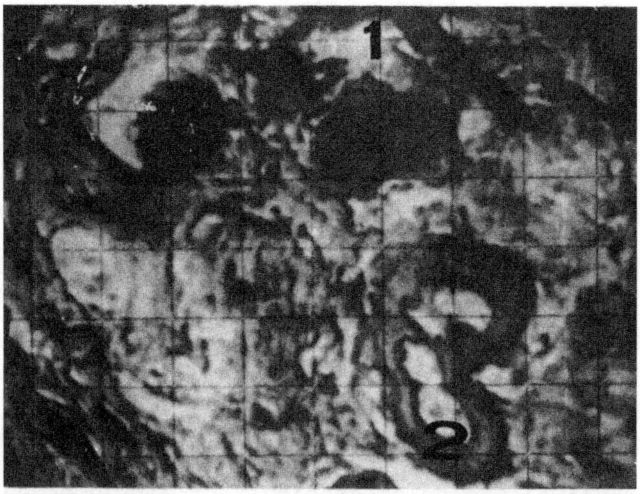

A

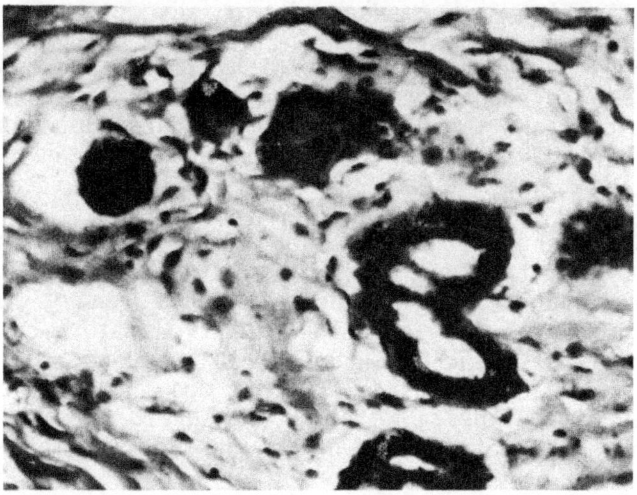

B

A: Acoustic micrograph (magnification = 280) showing the characteristic echostructures of collagen fibers.
1 - Collagen fibers
2 - Ductal epithelium

B: Light micrograph showing the same structures (X = 250).

Figure No. 1

All the images were obtained with an acoustic frequence of 600 MHz, (resolution = 1 micron), and samples were explored by transmission acoustic micrography.

The information that is provided from the acoustic microscope is distinct from the light microscope because acoustic microscopy responds to the elastic properties of the cells that are examined; and differences between the elastic properties of normal or abnormal cells can be visualized with the acoustic microscope.

After acoustic microscopy, samples were stained and studied in light microscopy and correlations were performed.

Results:

The picture obtained by acoustic microscopy (transmission method) are in a gray scale with white and black areas, white areas correspond to areas of acoustic transparency with little or no reflection, little or no absorption, and little or no diffusion. Black areas may correspond to areas of high acoustic impedence (great reflection) or to areas of high absorption or diffusion.

It appears from our correlations with light microscopy that collagen fibers are dense and black which is explained by the fact that collagen tissue has a high acoustic impedence and collagen areas are characterized by areas of great reflection (Fig. 1).

Another important fact is that in light microscopy in samples stained by haematoxyline eosine, it is not possible to distinguish young collagen fibers that are present in mammary lobules surrounding the galactophores from mature collagen that work as a fixed tissue of support, but in acoustic microscopy there is an obvious acoustic contrast and young collagen is less dense than fixed fibers; we think that it is due to the fact that young fibers are active and undergo hormonal modifications (Fig. 2).

Ductal epithelium is also perceptible in mammary ducts, and it is possible to distinguish between the two types of ductal epithelium (secretory epithelium in the lobules and excretory epithelium in large collecting ducts). It is important to note that the dense line that surrounds ductal epithelium is what we call the basal membrane, its visualization is explained by the fact that it consists of hyaline collagen of high acoustic impedence, these fibers separate ductal epithelium from myoepithelial cells throughout the breast.

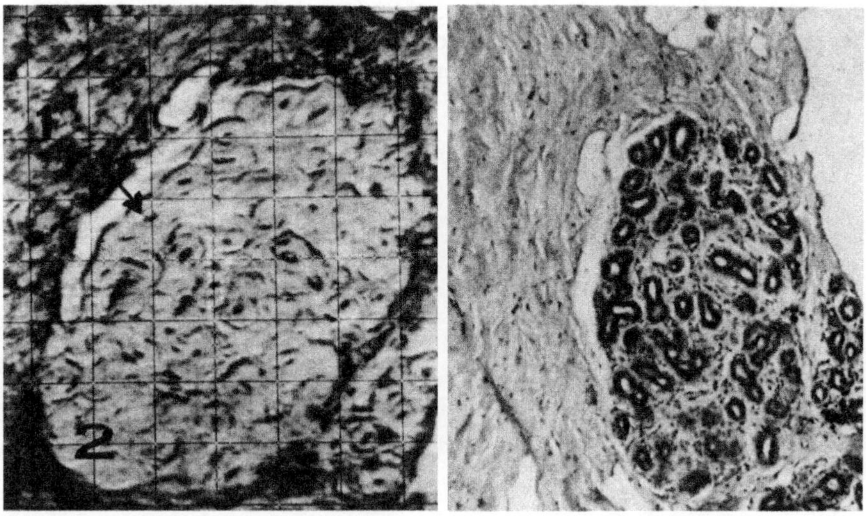

A B

A: Acoustic micrograph of a lobule (X = 130) showing the acoustic contrast between young and fixed collagen.

1 - Mature collagen fibers
2 - Young collagen
3 - Basal membrane (arrow)

B: Light micrograph of the same area (X = 200).

Figure No. 2

The preparation of samples dissolves fat, thus making it impossible to get information presently about fatty tissue. We do, however, see fat cells easily, but without containing any fatty material.

Conclusion:

It is obvious that the intrinsic acoustic contrast of mammary tissues is more than sufficient to allow a clear visualization of the structures, so that correlation between acoustic and light micrographs can be easily performed, in an attempt to visualize characteristic echostructures. Acoustic microscopy shows characteristic pictures of connective tissue that were not perceptible in light microscopy.

Important clinical applications are hoped for in the near future.

ULTRASONIC BLOOD CHARACTERIZATION

M. Boynard and M. Hanss

Laboratoire de Biophysique - Faculté de Médecine

45, rue des Sts-Pères 75006 PARIS et 74, rue Marcel Cachin 93000 BOBIGNY

ABSTRACT: The scattering coefficient of blood, χ, is usually very low in the MHz range. However, it is significantly increased for blood samples with high sedimentation rate (ESR) values. A detailed account is given on the derivation of χ from A. mode echographic signals. Then, it is related to the organization of the erythrocytes (RBC) in the blood, through a simplified statistical model. The mean number of RBC per aggregate can thus be determined.

Experimental results on the value of the back-scattering coefficient are given : correlation with the theoretical model for the hematocrit dependence, studies on high ESR samples, studies on artificially aggregated samples.

1) INTRODUCTION

The well-known shear-dependence of the whole blood viscosity is explained in the low-shear region by the red blood cell (RBC) aggregates (1,2) . The study of this aggregation is interesting particularly in relevance with some pathological states (3). Many methods are used for in vitro characterization of RBC aggregates (low-shear viscosity, microscopic examination, light back-scattering ...). None of these can be applied simply for in vivo determination, morever they cannot give a response for the no-shear region. An ultrasound-based method would, in principle, not be subject to these inconveniences.

In this work, we will describe how to obtain information on the aggregation state of blood from back-scattered intensity measurements and give some experimental results.

In all ultrasound scattering experiments, the measured quantity is a voltage given by a probe. From this voltage, an average back-scattered intensity can be deduced. This average intensity is given by all the individual pressure waves back-scattered by the RBC (or their aggregates) located in a given interrogated volume. These pressure waves can be determined from the incident wave pressure amplitude.

These different steps will be now developped following a treatment given by SIGELMANN (4) and modified according to our model. A more complete analysis will be given elsewhere (5,6).

2) RELATION BETWEEN THE BACK-SCATTERED INTENSITY AND THE SCATTERING COEFFICIENT χ

A simplified diagram of the probe and the blood sample are shown on figure 1.

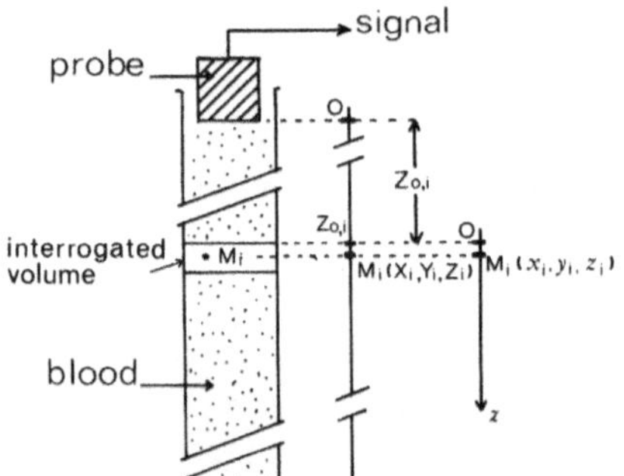

Figure 1

The probe is connected to an A-mode echograph (figure 2).

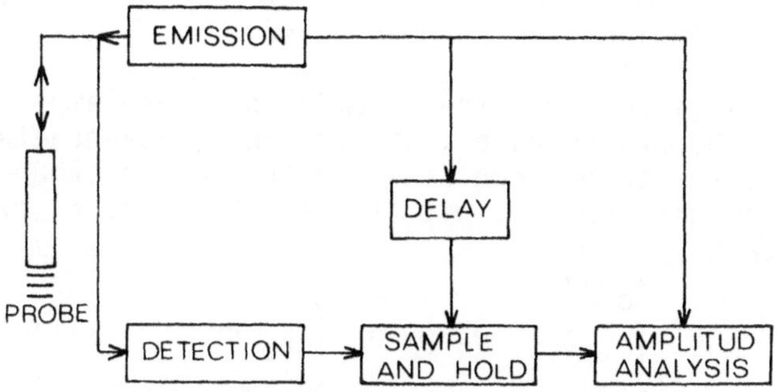

Figure 2 : Block diagram

The probe acoustical signal can be described, as a first approximation, by a pressure sinusoidal wave burst with amplitude p_o, frequency $\omega/2\pi$, duration τ, repetition rate 1/T, modulated by the time function F(t') (figure 3).

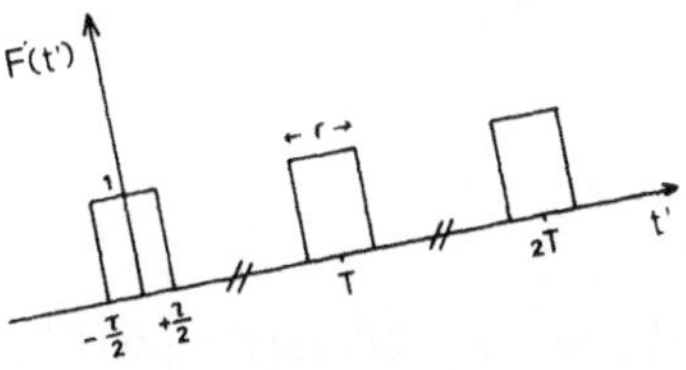

Figure 3

The incident pressure wave p_i at a point M(X,Y,Z) in the blood sample is given by :

$$p_i = p_o \exp(-\alpha Z).\cos \omega t'.F(t') \qquad (1)$$

(α : attenuation coefficient ; $t' = t - Z/c$; c : sound velocity ; t = 0 at the beginning of each wave burst).

At a given time, the incoming pressure amplitude, P_d, is the sum of the pressures p_d, at the probe, back-scattered by the elementary scatterers contained in the volume around point M, situated at the distance Z_o, with the thickness $\Delta z = \tau c/2$ and sectional area S :

$$P_d = \int_{Z_o}^{Z_o+\Delta z} dz \int_S dS\, p_d = \int_{S\Delta z} p_d\, dv \qquad (2)$$

Assuming spherical back scattered waves and the condition $\Delta Z << Z_o$, p_d is given by :

$$p_d = \frac{p_o}{Z_o} \exp(-2\alpha Z_o) A(M,t').\cos\left[\omega(t' - \frac{Z}{c}) + B(M,t')\right] F(t' - \frac{Z}{c}) \qquad (3)$$

In this equation A(M,t') and B(M,t') are two functions depending on the scattering properties of point M and on its random position in the interrogated volume.

Let $M_1(X_1=x_1, {}_1=y_1, Z_1=Z_{o,1}+z_1)$ and $M_2(X_2=x_2, Y_2=y_2, Z_2=Z_{o,2}+z_2)$ be two points located in two slabs with thickness Δz at the distance $Z_{o,1}$ and $Z_{o,2}$ from the probe. The scattered waves by these slabs have the pressure amplitudes $P_{d,1}$ and $P_{d,2}$ given by eq.(2 :

$$P_{d,i} = \int_{S\Delta z} p_{d,i}\, dv_i \qquad (4)$$

dv is an elementary volume around each point M of slab i (i=1 or 2). The pressure amplitude correlation function is given by :

$$C_{(1,2)} = < \int_{S\Delta z} p_{d,1}\, dv_1 \int_{S\Delta z} p_{d,2}\, dv_2 > \qquad (5)$$

Introducing the p_d values given by eq.(3 for the case $F(t' - Z/c) = 1$, and with the approximations :

$Z_{o,1} \simeq Z_{o,2}$ (which will be called Z_o)

$Z_1 - Z_2 << Z_o$

eq. 5 can be developped and the following result obtained :

$$C_{(1,2)} = \frac{p_o^2}{Z_o^2} \exp(-4\alpha Z_o) \int dv_1 \int dv_2 < A_1 \cos(- \frac{2z_1}{c} \omega + B_1) \ldots$$

$$\ldots A_2 \cos(- \frac{2z_2}{c} \omega + B_2) > \qquad (6)$$

If the two slabs 1 and 2 are well separated, the correlation function remains equal to zero. If the two slabs are superimposed ($Z_{o,1} = Z_{o,2}$), the correlation function is different from zero and can be calculated after the following transformation :

$$< A_1 \cos(- \frac{2z_1}{c} \omega + B_1) A_2 \cos(- \frac{2z_2}{c} \omega + B_2) >$$

$$= \chi \delta(x_1 - x_2) \delta(y_1 - y_2) \delta(z_1 - z_2)$$

where $\delta(r_i - r_j)$ is the Dirac distribution and χ is called the scattering coefficient. This calculation gives the mean square scattered pressure amplitude $< P_d^2 >$, from which the mean back-scattered intensity, $< I_d >$, can be obtained by the relation (7) :

$$< I_d > = \frac{< P_d^2 >}{2\rho c} \qquad (7)$$

The final result is :

$$< I_d > = \frac{p_o^2 \exp(-4\alpha Z_o)}{4Z_o^2 \rho} \tau S \chi \qquad (8)$$

The scattering coefficient χ can be transformed (6) into the usual differential back-scattering cross-section $d\sigma/d\Omega$:

$$\frac{d\sigma}{d\Omega} = \chi S \Delta z \qquad (9)$$

3) RELATION BETWEEN χ AND THE BLOOD AGGREGATION

A first approximation model of blood structure has been previously defined (8): transient spherical aggregates containing each m erythrocytes mechanically correlated for a time much longer than the ultrasound pulse duration (figure 4).

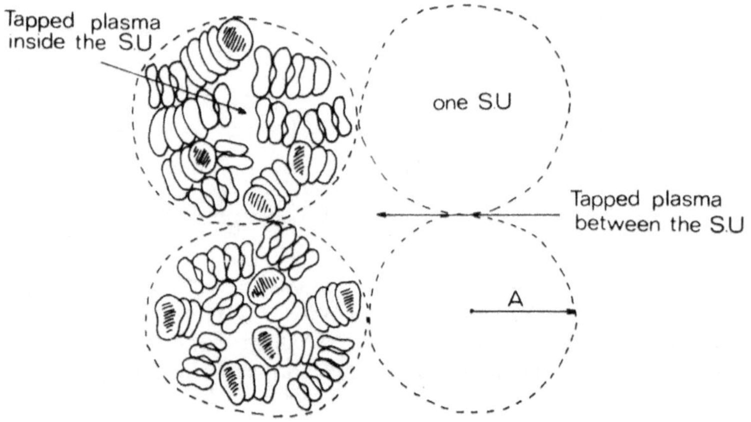

Figure 4 : Blood structure model, showing the transient spherical aggregates which are the scattering units (S.U.) with radius A.

By using a simplified statistical model, the following result has been obtained :

$$\chi = \frac{1}{S\Delta z}\frac{d\sigma}{d\Omega} = \left(\frac{3}{4\pi}\right)^2 \frac{\omega^4}{c_o^4} K\ m\ v_h\ p_h^2\ H(1 - p_h\ p_{su}\ H) \quad (10)$$

where v_h is the RBC volume, p_h the RBC packing factor inside the aggregate, p_{su} the aggregate packing factor inside the blood. K is a factor depending only on the acoustic impedance of the RBC and of the plasma.

By determining χ, we can then calculate m by using known or reasonable values of the different parameters occuring in eq. 10.

4) EXPERIMENTAL

A non-focused broad band 8,5 MHz probe is used (Panametrics 18419) in conjunction with a laboratory-made A-mode echograph. The electric signals generated by the probe when activated by the back-scattered pressure waves are time-gated for depth and sample volume selection. The corresponding electric pulses are selected by a multichannel amplitude analyzer. Thus the average voltage pulse amplitude $< V >$ can easily be measured.

The blood sample is contained in "Altuglass" tubing

(inner diameter : 9 mm ; length : 90 mm). In order to eliminate the sedimentation,the blood is stirred before each measurement. A polished metal cylinder can be fitted into the plastic tubing (top surface situated at distance Z_o from the probe),to calibrate the incident acoustic pressure, as is explained elsewhere (6).

The scattering coefficient is obtained from the following relation :

$$\chi = 2\,\frac{(<V_{d,m}^2> + <V_{r,m}^2> - 2<V_{d,m}><V_{r,m}>)}{V_R^2}\;\frac{Z_o^2}{S\tau c} \qquad (11)$$

where : $V_{d,m}$ and $V_{r,m}$ are the measured voltage pulse amplitudes generated respectively by the scattering suspension (blood) and by the suspending medium (water or plasma).

V_R is the voltage pulse amplitude generated by the reflexion from the metal cylinder top surface, the tubing cavity between this surface and the probe being filled with the blood.

Different blood samples have been studied :

- fresh high sedimentation rate (ESR) blood
- fresh and outdated stored blood
- artificially aggregated blood by the use Dextran 500 at various concentration (2 to $3{,}5.10^{-5}$M).

The hematocrit study has been performed by changing the RBC concentration through the usual centrifugation technique.

5) RESULTS AND DISCUSSION

One of our first results is the fact that in our experimental conditions, normal (non aggregated) blood has a too low scattering coefficient to be measured with confidence. Therefore only results using aggregated blood will be presented.

Typical results are presented on figures 5 and 6. They show the hematocrit dependence of χ for Dextran-treated and for fresh high-ESR blood samples. On these curves, each point represents the average of 7 to 10 measurements. Each measurement is obtained after 15.000 to 20.000 pulses.

Theoretical curves have been calculated by assuming a parabolic function and least-square fitted to the experimental points. As shown on the figures, the agreement between the experimental points and the theoretical curves is good. Therefore, as a first order approximation, the theoretical result given by

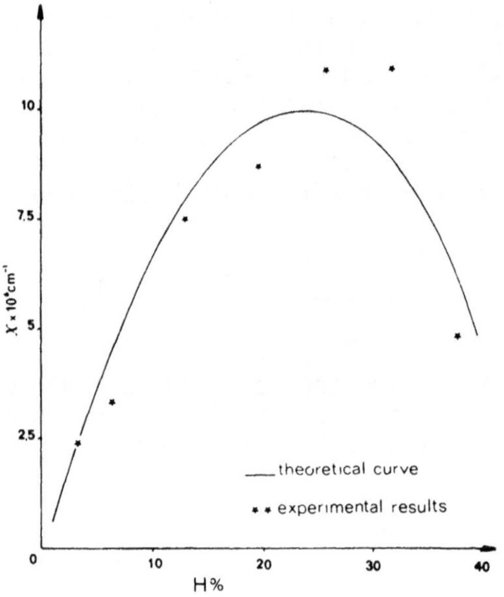

Figure 5 : Back scattering coefficient of artificially aggregated blood as a function of hematocrit

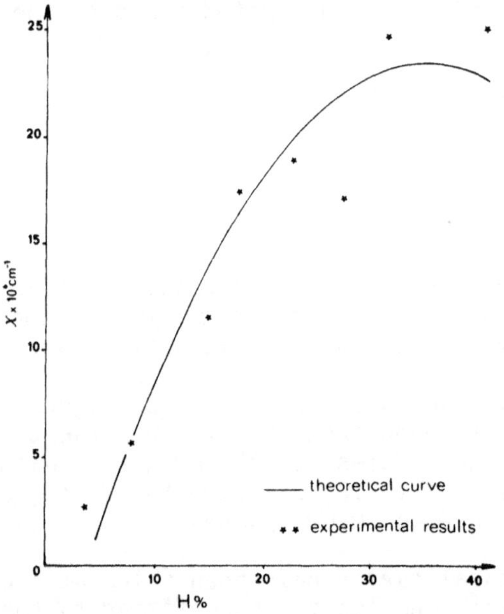

Figure 6 : Back scattering coefficient of high E S R blood as a function of hematocrit

eq. (10) can be used to calculate the blood structure parameters.

From the maximum value of $\chi = f(H)$, one can determine the $p_h\ p_{su}$ by using the following known values :

$F = 8{,}5$ MHz
$\alpha = 2$ dB.cm^{-1}
$Z_o = 1{,}9$ cm
$S^o = 0{,}567$ cm^2
$c = 1550$ m.s^{-1} (plasma)
$c_o = 1620$ m.s^{-1} (RBC)
$\rho^o = 1{,}078$ (plasma)
$\rho_o = 1{,}223$ (RBC)
$v_h^o = 0{,}87.10^{-10}$cm^3

From the initial slope of the $\chi = f(H)$ curve, the product mp_h can be determined. We will assume $p_h \simeq p_{su}$ to obtain an order of magnitude of m. These determinations are presented on table 1.

Table 1

	H for χ_{max} (%)	$\sqrt{p_h\ p_{su}}$	Slope (cm^{-1})	m
Artificially aggregated blood	23,5	1,46	8.10^{-5}	15
High ESR blood	34,5	1,20	15.10^{-5}	30

6) CONCLUSION

The scattering coefficient of blood is low (in the 1-10MHz frequency range) and has been measured only for aggregated samples. The hematocrit dependence can be explained by a theoretical model from which the average number of RBC per aggregate, m, and the average packing factor, p, can be determined. The value of m is in good accordance with figures estimated from rheological and light-scattering methods (9-10). The values of p are also reasonable (a cubic packing of spheres would give a packing factor of 1,65).

These results show that ultrasound back-scattering of blood can give some informations on the aggregation state of blood.

REFERENCES

(1) H. SCHMID - SCHONBEIN and R.E.WELLS Jr., Rheological propecties of human erythrocytes and their influence the "anomalous" viscosity of blood, Ergebn.physiol., vol.23, 146,1971.

(2) S. CHIEN, in "the red blood cell", D.N.SURGENOR ed.,Academic Press, N.Y., 1975, pp 1075-1095.

(3) M.H. KNISELY, in "Handbook of physiology", circulation,sect.2 vol.3, W.F. HAMILTON ed., 1969, chp 63.

(4) R.A. SIGELMANN and J.M. REID, Analysis and measurement of ultrasound back scattering from an ensemble of scatterers excited by sine wave bursts, J. Acoust.Soc.Am.,vol 53, 1351, 1973.

(5) M. BOYNARD, Etude des interactions entre hématies par rétrodiffusion ultrasonores, thesis, Université Paris VI, 1979.

(6) M. BOYNARD, M. HANSS, Mesure de la section efficace de diffusion ultrasonore des hématies, in preparation.

(7) P.M. MORSE and L.U. INGARD, in "Theoretical acoustics", Mac Graw Hill, N.Y., 1968.

(8) M. HANSS, M. BOYNARD, Ultrasound back scattering from blood: hematocrit and erythrocyte aggregation dependence, ultrasonic tissue characterization II, M. LINZER ed.,N.B.S. spec. publ. N° 525, 1979, pp 165-169.

(9) J.C. HEALY, Etude expérimentale des associations reversibles entre les globules rouges, thesis, Université Paris VI, 1973.

(10) P. MILLS, D. QUEMADA and J.DUFAUX, Etude de la cinétique d'agrégation erythrocytaire dans un écoulement de Couette, rev. phys. Appl., vol.15, 1357, 1980.

ALGORITHM FOR ON LINE DECONVOLUTION OF ECHOGRAPHIC SIGNALS

Alain Herment[x], Guy Demoment, Michel Vaysse

ERA CNRS N° 785 Hôpital Broussais 75014 Paris (France)
and LSS-CNRS ESE Gif sur Yvette (France)
x Att. de Recherches à l'INSERM

1. INTRODUCTION

The information contained in the signal received from an insonified medium is only partly extracted in conventional echography.

Deconvolution may contribute to restore this information by providing the medium impulse response which characterizes the medium itself, eliminating for this purpose the filtering action of the ultrasonic equipment which mainly depends on the smearing effect of the transducer.

Significant literature has been published on deconvolution [1, 2, 3, 4, 5, 6] and some performant methods have been proposed but they involve long and time consuming calculations. A deconvolution algorithm is suggested which gives accurate results and only involves a few number of calculations. The simplicity of this method allows an easy implementation with a microprocessor for quasi-real time deconvolution.

2. DESCRIPTION OF THE METHOD

The ultrasonic device and the acoustic medium are schematized on figure 1 together with the principle of the algorithm.

The signal e(t) applied on the transducer by the electric emitter is firstly convolved by $h_e(t)$, the transducer impulse response for emission. This signal is then modified during its propagation through the insonified medium : it is convolved by $e_c(t)$, the medium impulse response. Finally, the signal is convolved by $h_r(t)$, the transducer impulse response for reception. The signal delivered by the ultrasonic equipment can therefore be expressed as :

$$S_c(t) = e(t) \bullet h_e(t) \bullet e_c(t) \bullet h_r(t)$$

or in a more simple way :

$$S_c(t) = e_c(t) \bullet h(t) \qquad (1)$$

where $\bullet$ denotes the convolution operator and $h(t) = e(t) \bullet h_e(t) \bullet h_r(t)$ the ultrasonic device impulse response.

Equation (1) can be discretized as :

$$S_c(n) = \sum_{i=o}^{N} h(i) \,.\, e_c(n-i)$$

where $S_c(n)$ is the sampled received signal,

h(i) is the sampled impulse response of the ultrasonic device, composed of N+1 discrete values,

$e_c(n)$ is the sampled medium impulse response.

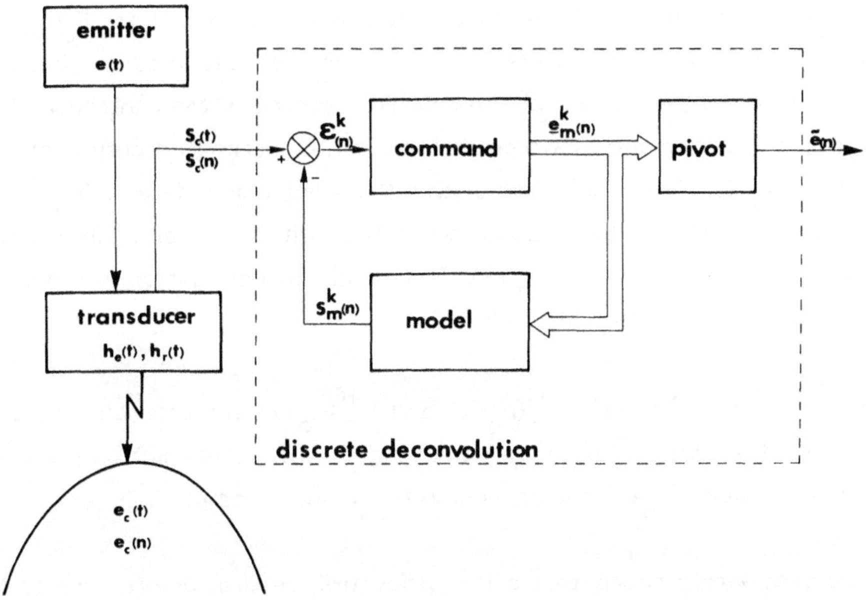

Fig. 1 : Principle of the ultrasonic device and of the algorithm.

This received signal will be treated by the deconvolution procedure to get $\tilde{e}(n)$, estimation of the unknown medium impulse response $e_c(n)$.

For this purpose, the different sampled values of the signal $S_c(n)$ will be successively introduced in the deconvolution algorithm. For each of these values a sample of $\tilde{e}(n)$ will be delivered by the process.

Let us now consider the very instant when the value $\tilde{e}(n_o)$ has just been delivered by the algorithm. The next value $S_c(n_o+1)$ is then fed into the process and an error signal $\epsilon^o(n_o)$ is calculated from that sample and from $S_m^K(n_o)$, the previously estimated output of the model for $S_c(n_o)$. A vector $\underline{e}_m^o(n_o)$ that represents an estimation of the medium impulse response is generated by the command. This vector is then introduced in the ultrasonic device model which delivers a new estimated output of the model, $S_m^o(n_o+1)$. From this value and the unchanged data $S_c(n_o+1)$, a new error $\epsilon^1(n_o+1)$ may be calculated together with a vector $\underline{e}_m^1(n_o+1)$ and a model estimated output $S_m^1(n_o+1)$. That cycle of calculation is repeated up to the chöosen iteration indice k = K.

The values $\epsilon_o^K(n_o+1)$, $\underline{e}_m^K(n_o+1)$ and $S_m^K(n_o+1)$ are then obtained. A coordinate of vector $\underline{e}_m^K(n_o+1)$ is extracted by the deconvolution pivot, it provides the value $\tilde{e}(n_o+1)$ corresponding to the input sample $S_c(n_o+1)$.

To completely characterize the procedure, several points have to be defined :

(i) the action of the model on $\underline{e}_m^k(n)$,
(ii) the action of the command on $\epsilon^K(n)$ and
(iii) the action of the pivot on vector $\underline{e}_m^k(n)$.

2.1. The model

The model output will be of course the convolution product at instant n and for the iteration k between the estimated input vector $\underline{e}_m^k(n)$ and the ultrasonic device impulse response h(i) :

$$S_m^k(n) = \sum_{i=o}^{N} h(i) \cdot e_m^k(n-i)$$

This relationship can be written using a matrix notation :

$$s_m^k(n) = \underline{h}^t \cdot \underline{e}_m^k(n) \qquad (2)$$

where $\underline{h}^t = (h(o),......, h(n),.......h(N))$ is the transposed vector of $\underline{h}$, ultrasonic device impulse response, and $\underline{e}_m^k(n)$ the vector defined by $(\underline{e}_m^k)^t = (e_m^k(n), e_m^k(n-1),.......e_m^k(n-N))$

2.2. The command

The command is designed to minimize an input distance defined by a quadratic norm in the space of input.

$$Q(n) = \| \underline{e}_m(n) - \underline{e}_c(n) \|_M^2 = (\underline{\Delta e}(n))^t \cdot M \cdot \underline{\Delta e}(n) \qquad (3)$$

where M is a symetric positive definite matrix and $(\underline{\Delta e}(n))^t = (\underline{e}_m(n)-\underline{e}_c(n))^t$ the transposed vector of $\underline{\Delta e}(n)$ which can be developed as :

$$(\underline{\Delta e}(n))^t = (e_m(n)-e_c(n), e_m(n-1)-e_c(n-1),.......e_m(n-N)-e_c(n-N)) \qquad (4)$$

The deconvolution must lead to a decrease of the norm Q(n) between two consecutive iterations. In other words, the quantity :

$$\Delta Q^{k+1}(n) = Q^{k+1}(n) - Q^k(n)$$

must be negative. Taking in account equation (3), it is possible to write :

$$\Delta Q^{k+1}(n) = (\underline{\Delta e}^{k+1}(n))^t \,.\, M \,.\, (\underline{\Delta e}^{k+1}(n)) - (\underline{\Delta e}^{k}(n))^t \,.\, M \,.\, (\underline{\Delta e}^{k}(n))$$

Let us now introduce the vector $\underline{\delta e}^{k+1}(n)$ defined as follows :

$$\underline{\delta e}^{k+1}(n) = \underline{\Delta e}^{k+1}(n) - \underline{\Delta e}^{k}(n) \qquad (5)$$

A new expression is then obtained for $\Delta Q^{k+1}(n)$:

$$\Delta Q^{k+1}(n) = (\underline{\Delta e}^{k}(n) + \underline{\delta e}^{k+1}(n))^t . M . (\underline{\Delta e}^{k}(n) + \underline{\delta e}^{k+1}(n)) - (\underline{\Delta e}^{k}(n))^t . M . (\underline{\Delta e}^{k}(n))$$

$$= 2\,(\underline{\Delta e}^{k}(n))^t . M . (\underline{\delta e}^{k+1}(n)) + (\underline{\delta e}^{k+1}(n))^t . M . (\underline{\delta e}^{k+1}(n)) \qquad (6)$$

The differences $\underline{\delta e}^{k+1}(n)$ are unknown values : they effectively depend on the unknown actual medium impulse responses $\underline{e}_c(n)$ as shown by relationships (4) and (5). In order to eliminate these unknown differences, a law of variation must be choosen for $\underline{\delta e}^{k+1}(n)$. This law has been defined as a linear function of the ultrasonic system impulse response $\underline{h}$:

$$\underline{\delta e}^{k+1}(n) = \mu . M^{-1} . \underline{h} \qquad (7)$$

where μ is a scalar for sake of simplicity of computation.

Equation (6) can then be re-written :

$$\Delta Q^{k+1}(n) = 2(\underline{\Delta e}^{k}(n))^t . M . (\mu . M^{-1} . \underline{h}) + (\mu . M^{-1} . \underline{h})^t . M . (\mu . M^{-1} . \underline{h})$$

$$= 2\,\mu\,(\underline{\Delta e}^{k}(n))^t . \underline{h} + \mu^2 . \underline{h}^t . M^{-1} . \underline{h} \qquad (8)$$

Let us now introduce in that equation $\epsilon^k(n)$, the error for iteration indice k :

$$\epsilon^k(n) = S_m^k(n) - S_c(n)$$

$$= (\underline{e}_m(n))^t . \underline{h} - (\underline{e}_c(n))^t . \underline{h}$$

$$= (\underline{e}_m^k(n) - \underline{e}_c(n))^t . \underline{h}$$

$$= (\underline{\Delta e}^k(n))^t . \underline{h}$$

Equation (8) may then be expressed under the following form :

$$\Delta Q^{k+1}(n) = 2\mu . \epsilon^k(n) + \mu^2 . \underline{h}^t . M^{-1} . \underline{h}$$

Let us recall that the convergence of the algorithm will be ensured as far as $\Delta Q^{k+1}(n)$ will be negative. $\Delta Q^{k+1}(n)$ is a quadratic function of μ, this condition is therefore verified for any value of $\mu = \alpha\tilde{\mu}$ with $0<\alpha<2$ where :

$$\tilde{\mu} = - \frac{\epsilon^k(n)}{\underline{h}^t . M^{-1} . \underline{h}} \qquad (9)$$

Moreover $\Delta Q^{k+1}(n)$ takes a maximum negative value for $\mu = \tilde{\mu}$.

It is now possible to describe the action of the command by writing the recurrence law of input vector $\underline{e}_m^{k+1}(n)$ as a function of the error $\epsilon^k(n)$.

It can be written, taking into account relationships (4) and (5) :

$$\underline{\delta e}^{k+1}(n) = (\underline{e}_m^{k+1}(n) - \underline{e}_c(n)) - (\underline{e}_m^{k}(n) - \underline{e}_c(n)) = \underline{e}_m^{k+1}(n) - \underline{e}_m^{k}(n)$$

so that $\underline{e}_m^{k+1}(n) = \underline{e}_m^{k}(n) + \underline{\delta e}^{k+1}(n)$

Equations (7) and (9) lead to :

$$\underline{e}_m^{k+1}(n) = \underline{e}_m^{k}(n) - \frac{\alpha . M^{-1} . \underline{h}}{\underline{h}^t . M^{-1} . \underline{h}} . \epsilon^{k}(n)$$

which can be written as :

$$\underline{e}_m^{k+1}(n) = \underline{e}_m^{k}(n) - \underline{\lambda} . \epsilon^{k}(n) \qquad (10)$$

where $\underline{\lambda}$ which is a constant deterministic vector for a given ultrasonic device, leads to a very simple implementation of the method (2N multiplications per iteration and no matrix calculation).

2.3. The pivot

The last step is to define the action of the pivot on vector $\underline{e}_m^{K}(n)$. The pivot will extract a given coordinate of the vector $\underline{e}_m^{K}(n)$ and deliver it as the value $\tilde{e}(n)$. This operation can be mathematically defined as the multiplication of $\underline{e}_m^{K}(n)$ by a vector $\underline{P}$:

$$\tilde{e}(n) = \underline{P} . \underline{e}_m^{K}(n) \qquad (11)$$

where $\underline{P} = (P_o, P_1, \ldots\ldots, P_\pi, \ldots\ldots, P_p)$ with $\begin{cases} P_\pi = 1 \\ P_i = 0 \text{ for any } i \neq \pi \end{cases}$

The general method is now fully described by relationships (2), (10) and (11) but these equations depend on some parameters : M^{-1}, $\underline{P}$, K, $\underline{e}_m^o(n)$ and α which must now be determined.

3. DETERMINATION OF M^{-1}, $\underline{P}$, $e_m^o(n)$, K and α

A convenient choice of these different parameters is necessary to ensure a good fitting of the algorithm to the deconvolution of echographic signals.

3.1 Choice of M^{-1}

For sake of simplicity and computation time, M^{-1} is choosen as a diagonal matrix ; moreover M^{-1} must be time-invariant for a fast deconvolution procedure. A good convergence of the algorithm implies that the matrix elements m_{ii} decrease when index i increases. The swiftness of that decrease characterizes the importance which is given by the algorithm to recent samples $S_c(n)$ relatively to the others. A fast decreasing law allows the algorithm to accept fast variations of the received signal and consequently affords a good depth resolution but contributes to make it more sensitive to noise.

In our experimental conditions a typical compromise is obtained for the values $m_{ii} = 1/(1+i)^2$ when the number of samples of the ultrasonic device impulse response N+1 is about 32.

3.2 Choice of P

The algorithm appears as an information extraction procedure which needs some learning delay. This delay corresponds to the index π of the non zero coordinate of vector $\underline{P}$.

Experience indicates that the best compromise between sensitivity to noise and learning delay is obtained when π is the index associated to the first relative maximum of $\underline{h}$, the ultrasonic equipment impulse response.

3.3 Initialization of vector $\underline{e}_m^o(n)$

The performances of the deconvolution method strongly depend on the initialization of the first coordinate of the vector $\underline{e}_m^o(n)$. Before the first iteration and with no information on the evolution of the received signal between instants n-1 and n, the simplest solution consists in giving to that first coordinate the preceeding estimated value. As a consequence : if

$(e_m^K(n-1))^t = (a, b, c, ..., n, 0)$,

$(e_m^o(n))^t = (a, a, b, c, ..., n)$.

3.4 Number of iterations K

Taking into account the precedent choice of M^{-1}, $\underline{P}$ and $\underline{e}_m^o(n)$ and giving to α the value 1 ($\mu = \tilde{\mu}$), experiment shows that the number of iterations may be reduced to one with no significant alteration of the process performances in usual conditions. This choice contributes of course to significantly decrease the time of calculation. However we must notice that for other couples of values α and K the noise immunity of the method and unfortunately the calculation time can be increased.

4. EXAMPLE OF APPLICATION OF THE METHOD

To illustrate the deconvolution performances in real experimental conditions, we have treated by the algorithm the echographic signal received from a lucite sample composed of two parallel plates separated by 0.3 mm of water. The thickness of the lucite layers was respectively 0.5 mm and 1 mm. Figure 2 presents the received signal which have been sampled at a frequency of 50 MHz with a dynamic of 8 bits.

The impulse response delivered by the algorithm is drawn on figure 3. The four transitions between water and lucite are pointed out by the four impulses which characterize the coefficients of reflexion of the medium interfaces. This impulse response fits quite well with the theoretical one ; the decrease of the impulses amplitude with depth is due to transmission losses at each interface.

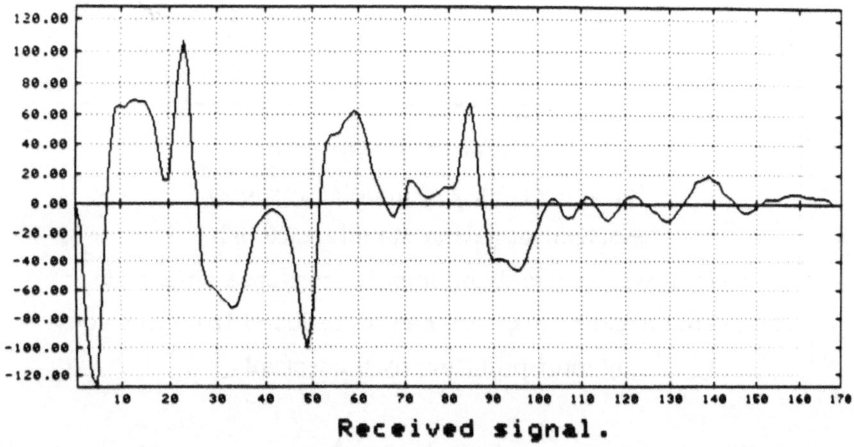

Fig. 2 : Echographic signal received from two parallel lucite plates (0.5 mm and 1 mm thick) separated by 0.3 mm of water. Horizontal axis is graduated in number of samples (1 sample = 20 ns) Vertical axis corresponds to a signal digitization on 8 bits.

5. LIMITS OF THE METHOD

Three points are of importance for application of the method to biological tissues examination :

i) depth resolution, in order to separate echoes generated by very thin layers,

ii) robustness with respect to noise, in order to get significant results for deep examinations,

iii) robustness with respect to some alterations of the echo waveform, generated for instance by dispersive attenuation of the medium.

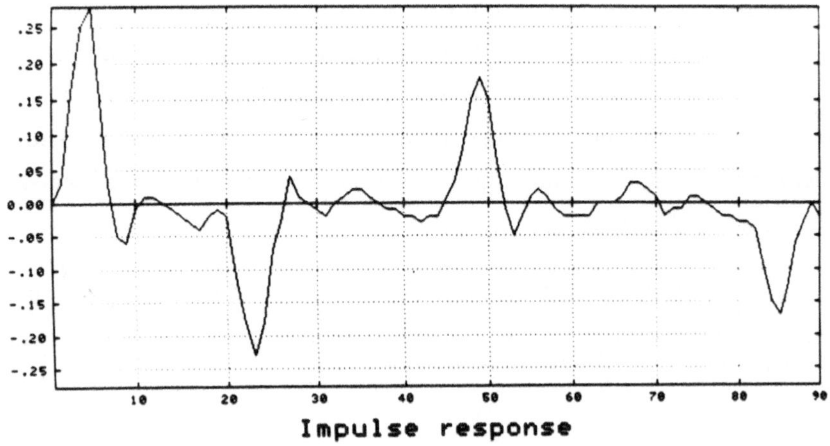

Fig. 3 : Normalized impulse response
Horizontal axis (1 sample = 20 ns)
Vertical axis : normalized impulse response amplitude
The actual impulse response can be deduced from this one
by multiplication by a constant.

As we have previously explained, the choice of the set of algorithm parameters will influence the performances of the deconvolution and a particular performance can be improved but at the detriment of the others. To test the possibilities of the method on the three previous points, an unique set of parameters have been choosen and the three following tests will be done in identical conditions.

5.1. Depth resolution

In order to test the resolution of the process, we simulated a series of biological layers (with perfect geometry) of different thicknesses by synthetizing the corresponding echographic signals from an echo obtained on a water/lucite interface.

Figures 4 to 7 present the impulse responses corresponding to layers with a thickness of respectively 0.3 mm, 90 μm, 60 μm and 30 μm (bandwidth of the equipment : 3.5 MHz). These responses fit well with the

expected results for a thickness greater or equal to 60 μm. One can notice that for thinner samples (e.g. 30 μm) the impulse response is degraded :
i) by an underestimation of the values of the coefficients of reflexion and
ii) by an over estimation of the delay between the two echoes. Nevertheless even for these very thin layers the algorithm will never diverge.

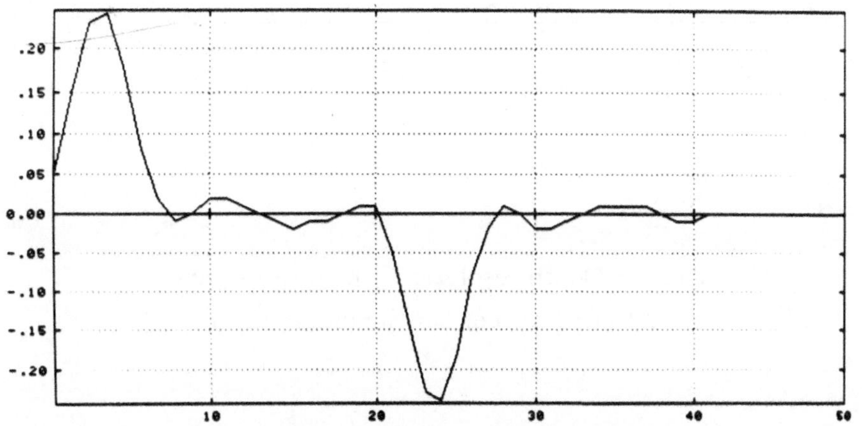

Fig. : 4 : Depth resolution : impulse response obtained on a 0.3 mm simulated layer.

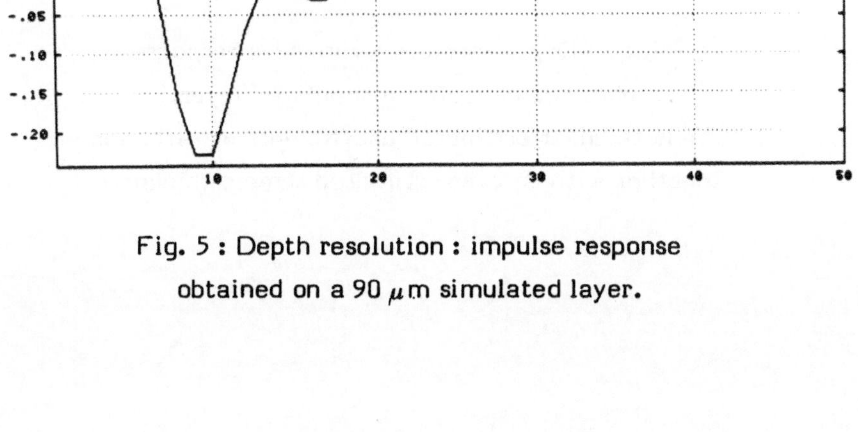

Fig. 5 : Depth resolution : impulse response obtained on a 90 μm simulated layer.

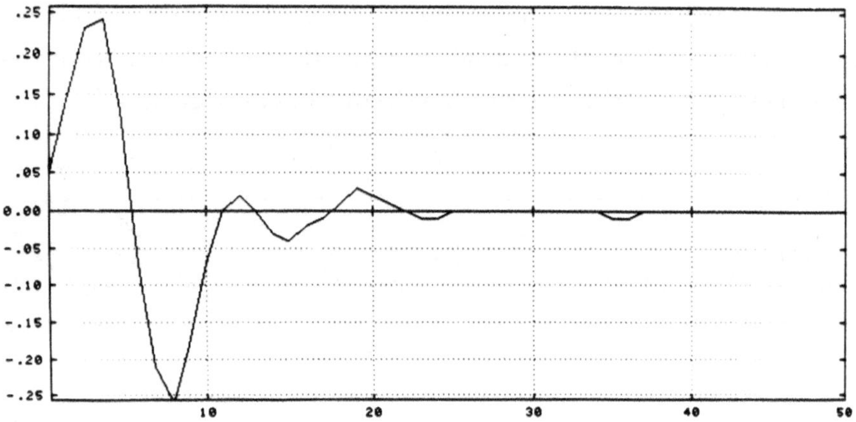

Fig. 6 : Depth resolution : impulse response obtained on a 60 μm simulated layer.

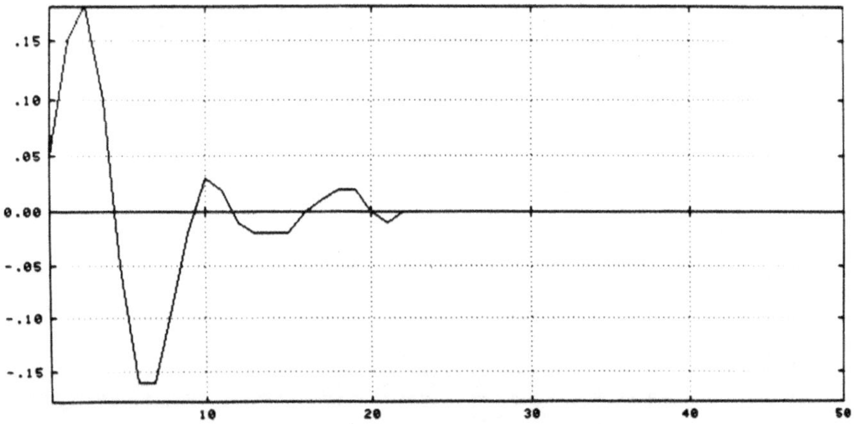

Fig. 7 : Depth resolution : impulse response obtained on a 30 μm simulated layer. We note under estimated coefficients of reflexion together with an over estimated sample thickness.

5.2. Robustness with respect to noise

In order to test the sensitivity to noise, the deconvolution process has been applied to signals synthetized with a variable amount of white noise in the bandwidth of the system. The results are drawn on figure 8 (no noise) and on figures 9 to 11 for respective signal to noise ratios of 30 dB, 20 dB and 10 dB.

From these results it appears that for S/N ratios higher or equal to 10 dB, no major incidence of the noise is pointed out on the shape and the amplitude of the impulses. However, it must be noticed that the base line of the restored response is drifting with depth as the S/N ratio decreases.

From that point of view it seems realistic to consider that, for the choosen set of algorithm parameters, the proposed system gives correct results down to a S/N ratio of about 25 dB.

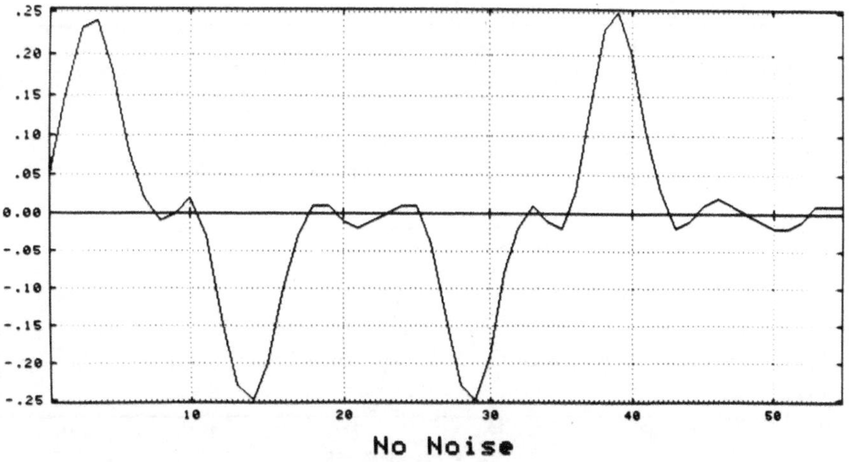

Fig. 8 : Sensitivity to noise : impulse response (no noise).

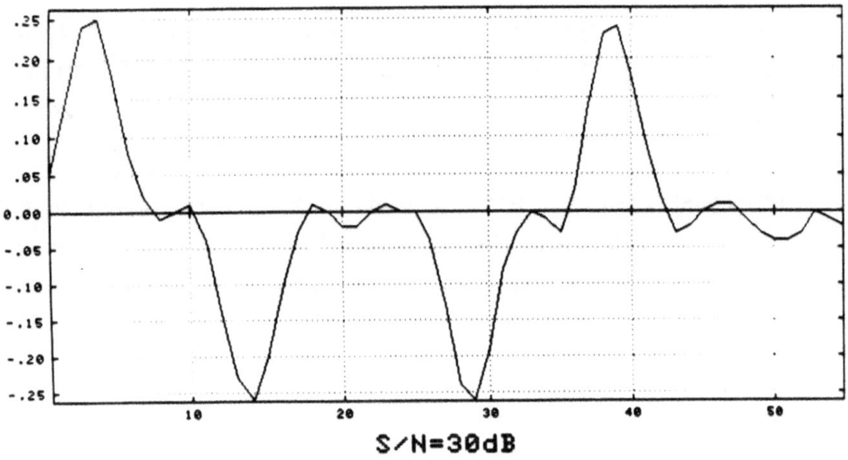

Fig. 9 : Sensitivity to noise : impulse response (S/N = 30 dB).

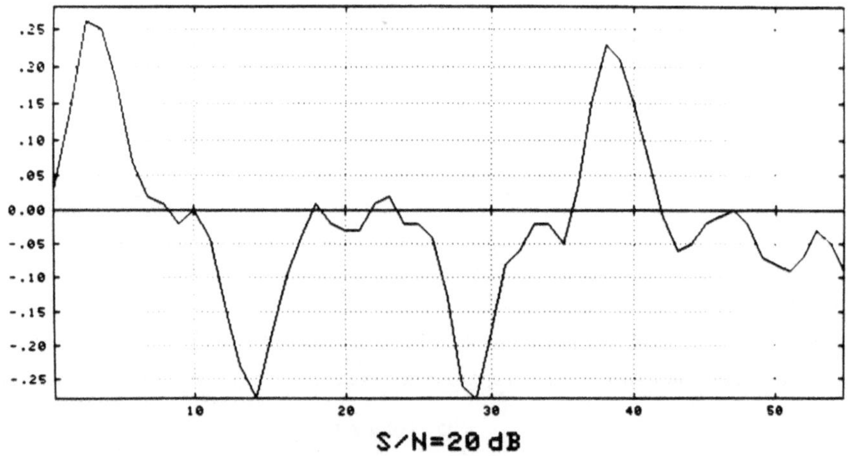

Fig. 10 : Sensitivity to noise : impulse response (S/N = 20 dB).

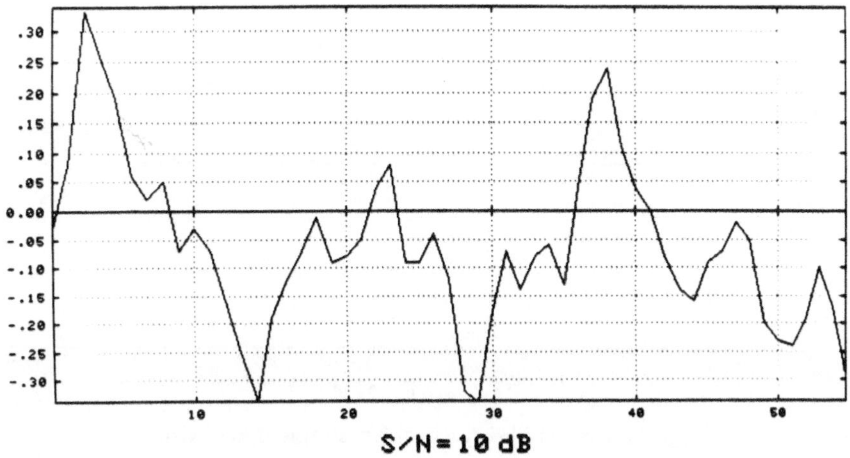

Fig. 11 : Sensitivity to noise : impulse response (S/N = 10 dB).

5.3. Robustness with respect to echo shape modification

In order to simulate the effects of dispersive attenuation on the echoes we sampled the received signal at higher frequencies (50 MHz to 65 MHz) than the ultrasonic device impulse response (50 MHz) in order to mimic the dispersive attenuation which lengthens the echo waveform.

For comparison, figure 12 presents the deconvolution results for a single echo obtained on a water/lucite interface, sampled at the usual frequency of 50 MHz. Figures 13 to 15 respectively correspond to the impulse responses obtained on the same interface but for signals sampled at respective frequencies of 55 MHz, 60 MHz and 65 MHz. We observe that relative differences as important as 30 % between the sampling frequency of the system impulse response and the sampling frequency of the received signal do not alter overmuch the deconvolution.

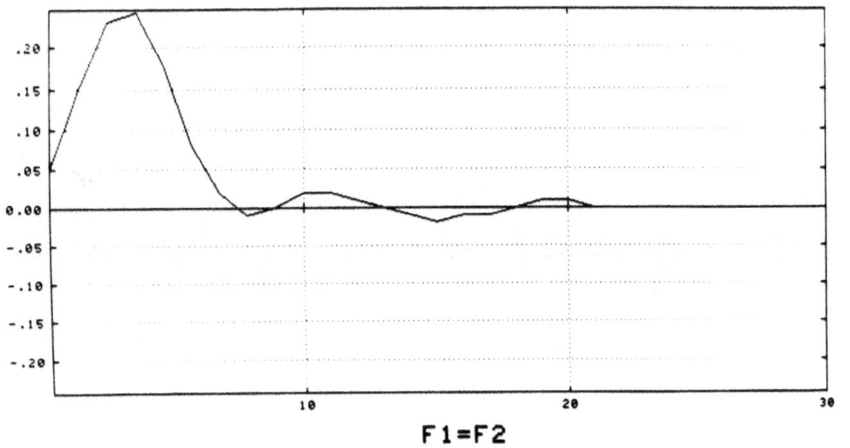

Fig. 12 : Sensitivity to echo shape distortion :
Impulse response ($\Delta f/f = 0$).

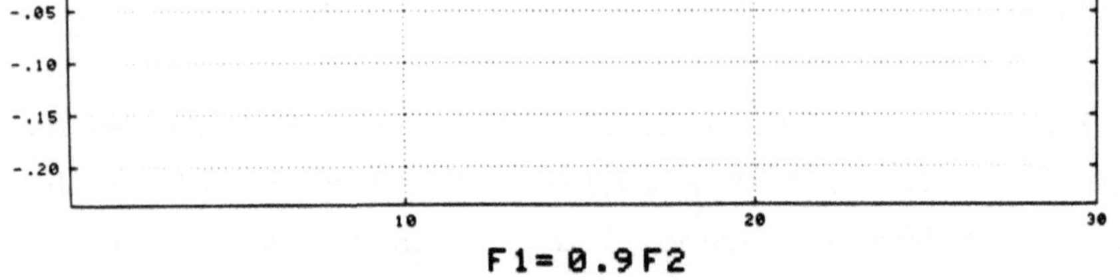

Fig. 13 : Sensitivity to echo shape distortion :
Impulse response ($\Delta f/f = 10$ %).

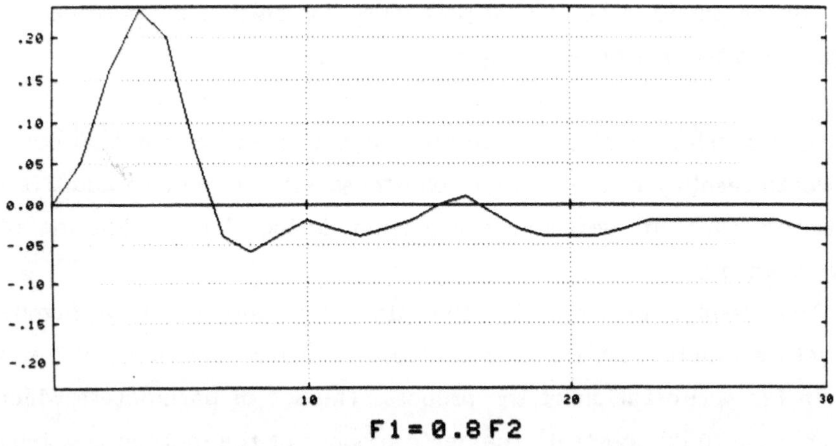

Fig. 14 : Sensitivity to echo shape distorsion :
Impulse response ($\Delta f/f$ = 20 %).

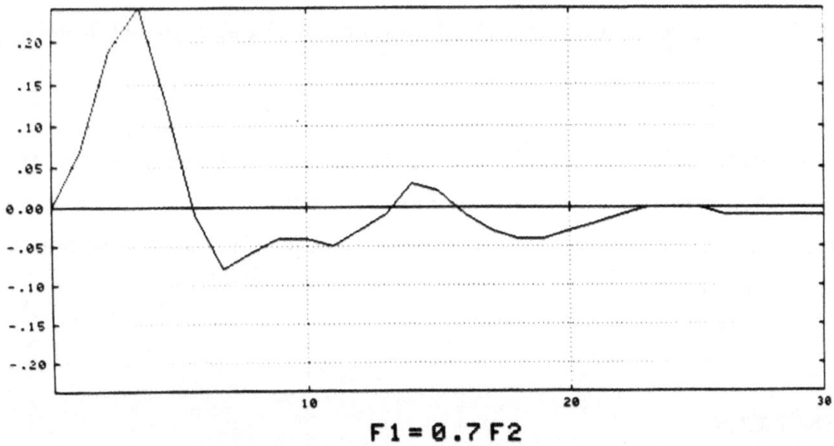

Fig. 15 : Sensitivity to echo shape distorsion :
Impulse response ($\Delta f/f$ = 30 %).

6. CONCLUSION

The propounded numerical procedure for on-line discrete deconvolution presents two major advantages :

(i) A possible application to biological tissues examination due to its good depth resolution and correct robustness with respect to additive noise and echo waveform deformations arising during the propagation of the acoustic wave.

The results provided by the algorithm depend on a number of parameters which must fit to the ultrasonic equipment and the insonified medium for optimization of the process. The set of parameters which has been choosen to demonstrate the performances of the method is convenient for shallow examinations in tissues. For deeper measurements, another set of parameters will allow an improvement of the noise immunity, but will of course induce a decrease of the depth resolution.

(ii) An easy implementation of the method by means of a microprocessor due to its simplicity and the few number of calculations involved.

However before a generalized application of the process, for instance to fine ultrasonic imaging, some work is still to be done to make the method still more suitable to the usual echographic systems, and to the biological tissues.

REFERENCES

1. E. S. Furgason, R. E. Twyman, V. L. Newhouse, "Deconvolution processing for flaw signatures", Advanced Research Projects Agency, project n° 33615-75-C-5252.

2. B. R. Hunt, The inverse problem of radiography, Mathematical Biosciences 8 : 161 (1970).

3. A. J. Niemi, On discrete deconvolution, Med. and Biol. Engng 14 (5) : 582 (1976).

4. A. Papoulis, C. Chamzas, Improvement of range resolution by spectral extrapolation, Ultrasonic Imaging 1 : 121 (1979).

5. D. L. Phillips, A technique for the numerical solution of certain integral equations of the first kind, J. Assoc. Comput. Mach. 9 : 84 (1962).

6. J. Richalet, A. Rault, R. Pouliquen, "Identification des processus par la méthode du modèle", Gordon and Breach Ed., Paris (1971).

TOMOGRAPHIC RECONSTRUCTION OF B-SCAN IMAGES

Dietmar Hiller, and Helmut Ermert

Institut für Hochfrequenztechnik
Universität Erlangen-Nürnberg
D-8520 Erlangen, West-Germany

ABSTRACT

This paper explains an imaging system which uses a linear transducer array and special data processing to reconstruct echo data in a tomographic manner. A conventional parallel-scan is performed by pulse-exciting the array elements one at a time or in small groups. The received signals are recified, low-pass filtered, and then stored digitally. The array is then moved to the next position on a circle around the object. This procedure is repeated until the object is totally surrounded. The recorded data which represents a set of B-scan images from different aspect angles is processed in either of the following two ways:

1) Each B-scan image is integrated over the aperture-axis for each point of the time-axis. These "pseudo-projections" are reconstructed as it is done in conventional computerized tomography.

2) Each B-scan image is convolved with a two dimensional filter function and then all filtered images are superimposed to give the final image.

The main advantage of the first technique is its ability to use the readily available and fast processing algorithms of computerized tomography (e.g. the convolution or ART algorithms for parallel beam geometry). The second technique however is less sensitive to measurement errors due to specular reflections or absorption. The results of experiments carried out with both of the described techniques are presented, and possible improvements are discussed.

INTRODUCTION

With the great success which X-Ray Computerized Tomography has had, it was apparent to apply this method to those parameters in the ultrasonic region which also can be measured in transmission such as the absorption coefficient or the velocity. However a large disadvantage of this technique is, that for error-free reconstruction every object point must be intersected by rays from every aspect angle. In many applications this is difficult or impossible, for example due to the presence of bone material in biological specimens. Therefore the most successful and highly developed ultrasonic imaging methods utilize pulse-echo techniques. It was however pointed out by several workers (/1/,/2/,/3/) that pulse-echo data can also be interpreted as a line integral over the reflectivity of a medium under certain circumstances. Therefore with a special measurement geometry the reconstruction techniques of Computerized Tomography (CT) can be applied. Initial experiments using this method have also been published /1/. The algorithms of transmission CT must be modified corresponding to the special geometry of reflection, but the principle of reconstruction remains essentially unaltered. The reconstructed parameter, however, is now the distribution of reflectivity; therefore this method has been termed "reflectivity tomography".

There are some special problems associated with reflectivity tomography. The theory of reconstruction is based on the assumption of a perfectly scattering medium, i.e. a collection of point scatterers having no mutual interaction. In the case where the object contains reflecting surfaces the result becomes erroneous. Another problem with currently published methods is the fact that the integration over the unknown parameter is performed over curved paths (especially circles) rather than over straight lines as done in CT. This fact must be taken into account in the reconstruction process. The additional amount of computations, which are necessary to backproject the measured data along curved paths, results in reconstruction times which are about an order of magnitude higher than with comparable algorithms for conventional CT. In addition to this, the integration along curved paths requires the viewing of the object over a full 360 degrees by the transducer in order to achieve uniform object sampling. Thus one of the main advantages of pulse-echo measurement over transmission CT is lost.

This paper will describe two methods for reconstruction of pulse-echo data using a tomographic algorithm. The first method is based upon a plane wave approximation instead of a spherical wave to interrogate the object. This eliminates the problems of integration along curved lines as well as the necessity of viewing the object over 360 degrees. The second method is an improvement of the plane wave approximation to incorporate transducer directivity characteristics. This second method reduces imaging errors caused by non-perfectly scattering media, such as objects containing reflecting surfaces or exhibiting absorption.

BACKGROUND OF METHODS

Reconstruction of Pseudoprojections

Figure 1 shows the geometry of reflectivity tomography as described by Norton /3/ (Only the two dimensional case of this problem is considered). An omnidirectional transducer which emits an acoustical pulse is located at point T. This wave propagates outward and intercepts an object, consisting of a distribution of point scatterers which can be described by a reflectivity $a(\rho,\varphi)$. Point scatterers which lie on a circle of radius r centered at T are reached by the wavefront at the same time. Each point scatterer reflects a spherical wave, which propagates back toward the transducer. These waves again arrive at the transducer simultaneously and are superimposed. The measured echo signal is therefore proportional to the integral over the reflectivity of all point scatterers lying on a circle with radius r. This is the previously mentioned case where the measured data results from an integration of the unknown parameter along a curved path. In order to allow integration along straight paths (parallel or diverging) which is the basis of conventional transmission CT, it would be necessary to use a plane wave instead of a spherical wave.

In practice this plane wave generation poses a problem which is non-trivial. Operation in the nearfield of a transducer, as sometimes proposed /1/, is impractical due mainly to the strong spatial variation of pulse amplitude in this region.

Another possibility is to operate in the transducer's far field so that spherical waves can be approximated with neglegible error as plane waves in the object area. To achieve this the distance between

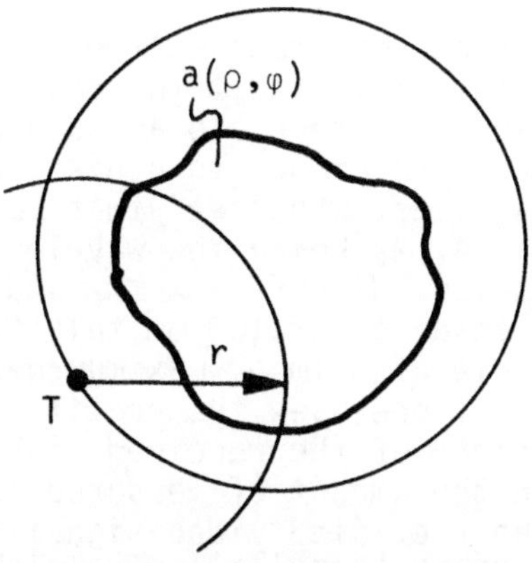

Fig. 1 geometry of conventional reflectivity tomography

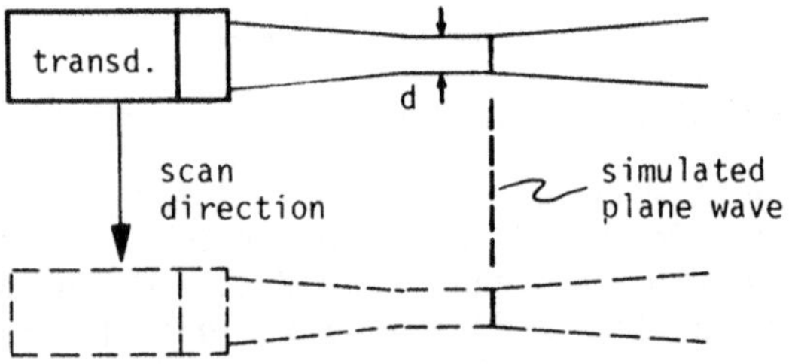

Fig. 2 Simulation of a plane wave

transducer and object has to be much greater than the maximum dimension of the object itself. For reasonably sized biological objects such a setup becomes inconveniently large. Furthermore, the $1/r^2$ drop of the received signals may cause noise problems.

One can also operate in the focal region of a weakly focussed transducer. In this region the beam width varies only slowly with axial distance, and the wavefronts can also be approximated as plane waves. The problem in this case is, however, that the transducer has to be much larger than the object, in order to obtain a beam of sufficient extension. This leads also to a rather large setup.

As an alternative possibility, a plane wave can be synthesized as shown in Fig. 2. Again a transducer with weak focusing is used. The acoustic beamwidth in the focal plane is denoted by d. If the transducer is scanned over an aperture with a stepsize Δu and all received signals from the different positions are superimposed, then each transducer position contributes a part to a simulated plane wave whose size would correspond to the size of the whole aperture. If the directivity of the transducer is sufficiently good, so that it has a sufficiently small beam angle, then the approximation of a plane wave is also good in the farfield of the transducer.

The stepsize Δu has in practice to be smaller than the beamwidth d, to achieve a uniform amplitude of the simulated plane wave over the whole aperture. This means that the beampatterns of two adjacent positions overlap each other. To avoid unwanted phase cancellations, the round-trip-distance to a distinct point in the object plane from all positions, who "see" that object point, may not differ by more than $\lambda_0/4$, λ_0 being the wavelength in water at center frequency of the pulse. With f_0 = 2 MHz and a distance from an object point to the transducer of 100 mm this for instance leads to a maximum focal beamangle of 8 mm or maximum beamangle in the farfield $\pm$ 2.5°. However, if one uses the rectified and low pass filtered video-signal instead of the received high frequency signal, than a great reduction in the amount of measured data is realized. Since integration over the rectified video-signal is an incoherent process and therefore no phase cancellation results, the beam angle

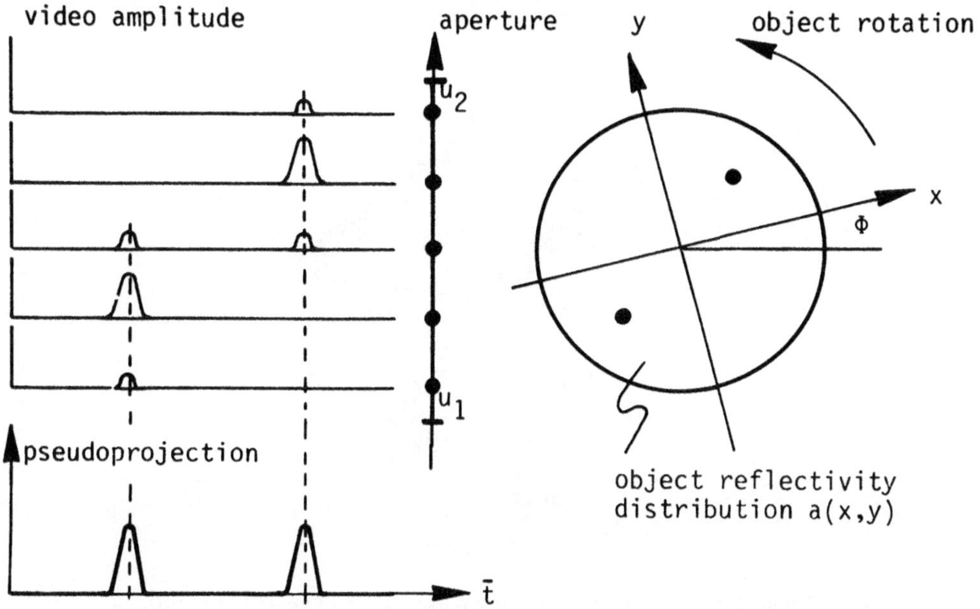

Fig. 3 Geometry of pseudoprojection measurement

can then be made larger. For example, if a difference in round-trip-distance of half the video-pulse length is permitted, then in the case of a 1 μs pulse the beamangle may have a maximum value of ± 5°. With that pulse length the axial resolution is approximately 1.5 mm, which should be sufficient for a large number of practical applications.

Mathematically the superposition of signals from different aperture positions can very easily be expressed as:

$$p(\bar{t},\Phi) = \int_{\bar{u}_1}^{\bar{u}_2} s(\bar{t},u,\Phi)du \qquad (1)$$

where: $s(\bar{t}, u, \Phi)$ is the measured (video) signal at aperture position u and at time $t = 2\ \bar{t}/c$ (The spatial coordinate $\bar{t}$ denotes the spatial equivalent to the echo arrival time).

This function s represents nothing more than a conventional B-scan image under the aspect angle Φ (Fig. 3). The resulting (spatial) function p will be referred to a a "pseudo-projection" because of its close similarity to a measured projection in transmission-CT. The difference is, that in transmission-CT, integration over the

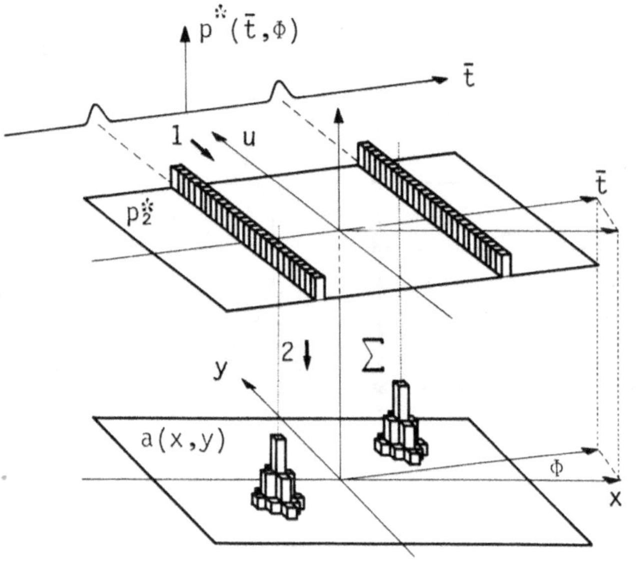

Fig. 4 Geometry for backprojections of filtered pseudo-projections. The (filtered) projection from aspect angle Φ is spread into a two-dimensional function (step 1). This function is then rotated by the angle Φ and summed to the image under reconstruction (step 2).

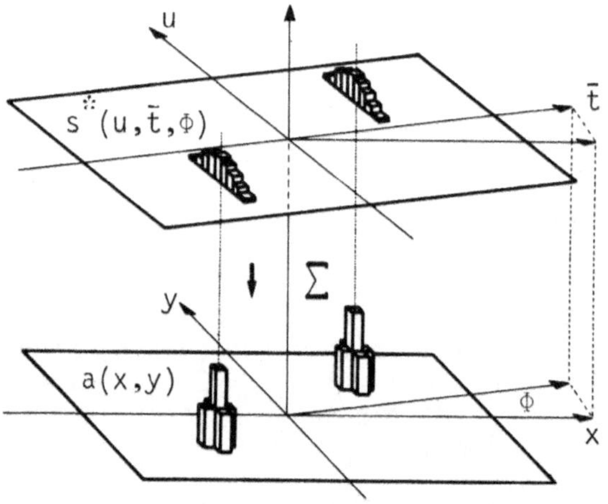

Fig. 5 Geometry for summation of filtered B-scans. The (filtered) B-scan from aspect angle Φ is rotated by the angle Φ and summed to the image under reconstruction.

unknown parameter is performed implicitly during the measurement; whereas a pseudo-projection results from a numerical integration of the received signals perpendicular to the sound propagation direction.

If the same procedure of aperture-scanning and signal integration is repeated for different angles Φ, by rotating either the object or the measurement setup in equally spaced angular steps after each complete scan, then the reflectivity can be reconstructed from the pseudo-projections by using one of the well known methods of CT such as the convolution or ART algorithms. This kind of tomographic reconstruction of B-scan images shall be referred to as "reconstrution of pseudo-projections" or simply as "method 1". Similar to transmission-CT and different from other proposed reflectivity tomography systems only a rotation of 180 degrees is necessary in either the object or the transducers. A further rotation only yields the time-inverted signals of the position $\Phi - \pi$ and therefore gives no additional information.

Reconstruction by the convolution method, which is sometimes also called filtered backprojection, should be briefly described. (Extensive treatment of this and other methods can be found in the literature, see for instance /5/). As the name implies, this technique consists of two steps, namely a convolution of the projections with a filter function $f(\bar{t})$:

$$p^*(\bar{t}, \Phi) = p(\bar{t}, \Phi) * f(\bar{t}) \qquad (2)$$

and backprojection of these filtered projections to the image plane:

$$a(x, y) = \int_0^{\pi} p^*(x \cos \Phi + y \sin \Phi, \Phi)\, d\Phi \qquad (3)$$

As a filter function, one of the popular kernels like the Ramachandran/Lakshminarayanan-kernel or the Shepp/Logan-kernel may be used. It is, however, also possible to introduce special properties of the imaging system, for instance the axial resolution of the B-scan-image, into the kernel.

Summation of filtered B-scans

It is an interesting question, whether or not it is possible to make use of the lateral resolution that is present in the B-scan images . Figure 4 is a graphical interpretation of equation (3) and shows, how the backprojection works during the reconstruction process. For each aspect angle the filtered pseudoprojection p^* (which is shown in the upper part in Figure 4) is spread in the direction of the u-coordinate and then this two-dimensional function p_2^* is added to the image a.

If one neglects the pre-filtering for the moment, then $p_2(\bar{t}, u, \Phi)$ (the unfiltered version of p_2^*) with $\partial p_2/\partial u = 0$ could be interpreted as a B-scan image with absolutely no lateral resolution. Therefore it seems reasonable to use a function, which more directly relates to the original B-scan image than the pseudoprojection, instead of summing the p_2^*'s. As an analogy to method 1 we again use a two-step reconstruction process. The first step is to filter the B-scan image with a filter function

$$s^*(\bar{t}, u, \Phi) = s(\bar{t}, u, \Phi) * f(\bar{t}) \qquad (4)$$

and then to rotate and sum s^* to the image a

$$a(x, y) = \int_0^{\pi} s^*(x \cos \Phi + y \sin \Phi, \; y \cos \Phi - x \sin \Phi, \Phi)\, d\Phi \qquad (5)$$

This is demonstrated in Fig. 5. As the convolution kernel f(t) we use here a slightly modified Shepp/Logan kernel. Again, some features of the imaging system can be incorporated in thát kernel to further improve the reconstruction performance. This method will be referred to as "summation of filtered B-scans" or simply as "method 2". A similar technique of summing B-scans from different aspect angles, but without filtering, has been described by Greenleaf /7/.

If the point-spread function (PSF) of the imaging system is calculated using each of the two methods, then it can be recognized, that they match well in the near vicinity of the actual position of the point-object. But the PSF of method 2 converges much faster to 0 than with method 1 as distance increases. In practice this means no improved resolution. Nevertheless, a much better signal-to-noise ratio in the reconstructed image, and a higher immunity to measurement errors is achieved. The reason can be seen by looking at Fig. 4 und 5: Due to the "spreading" of the pseudoprojections during the reconstruction process, the whole image area is influenced by each object point in method 1; whereas, in method 2 only a circle of a diameter corresponding to the lateral resolution is influenced. The same applies as well to measurment errors, which will cause a streak over the whole image in method 1 and only a relatively small dash in method 2.

The image reconstruction according to method 2 can be considered as if a small tomogram would be reconstructed for the vicinity of each image point with the diameter of the tomogram being of similar size as the lateral resolution in the B-scan image. Method 2 might also be interpreted as a sophisticated compound-scan.

Comparison between reconstruction methods

Both methods 1 and 2 have the disadvantage compared with other proposed reflectivity tomography techniques, that at least one mechanical movement is necessary, namely the stepped rotation of the object. This motion is relatively time-consuming. The parallel scan of the transducer is suitably carried out by an electronic scan of a linear transducer array, which works much faster. The disadvantage of a mechanical movement is however compensated for by a much faster reconstruction algorithm. Parallel-beam geometry algorithms are very simple and fast to reconstruct. Estimates show, that CT algorithms generally require approximately the same number of projections as there are sample points per projection, provided the spatial sampling interval of the projections is of about the same dimension as the pixel size /4/. This is necessary to avoid angular undersampling, which causes streak artifacts in the reconstruction. Those artifacts first appear in the peripheral region of an image and spread out to the center when the number of projections decreases. Therefore it follows that, for a 128 x 128 pixel image, one should take about 120 projections spaced by 1.5^0 for a reconstruction with method 1. This means a fairly large amount of data to be measured and stored. Because the data is summed to form the pseudoprojections, the amount of data is thus reduced by a factor of N (N being the number of transducer positions). Only this amount of data has to be handled during reconstruction. Actually, this data reduction could already be performed during the measurement procedure by using some relativly simple hardware, which properly sums the data as it is collected. This could be done in real time.

If one again wishes to reconstruct a 128 x 128 pixel image (this time, using method 2), then the number of projections depends on the lateral resolution of the B-scan. This is because of the fact, that method 2 can be interpreted as the reconstruction of a small tomogram for each image point. If the B-scan image has for instance a lateral resolution of 15 pixels, which is about a tenth of the whole image, then the number of aspect angles can also be reduced by a factor of 10 compared to the case of method 1, but at least 10 transducer positions are necessary to give a complete B-scan. Thus the amount of data actually used in the reconstruction program is roughly the same with both methods. The following reconstruction times were measured for both the methods (all numbers are per aspect angle, with a 128 x 128 pixel image, 32 transducer positions and 128/samples/projection):

backprojection of filtered pseudoprojection (convolution method)	1 s
summation of filtered B-scan-images	6 s

SETUP

The block diagram of the measurement setup, with which the experiments were carried out is shown in Fig. 6. The mechanical part consists of a turntable, to which a linear transducer array is mounted. A second identical array can be mounted facing the first one, to simultaneously perform transmission measurements. The arrays are submerged in a watertank. The turntable has a 200 mm diameter hole at its center to allow testobjects to be brought into the water. This also permits in vivo-measurements. The array has a length of 160 mm and consists of 54 elements, with only the 32 center elements being involved in the measurement at the moment. The diameter of the object-area hence measures only 93 mm. This was sufficient for all preliminary experiments. The center frequency of the array elements is 2.3 MHz, the length of the transmitted pulses is approximately 2 μs. These arrays are commercially available ones, which are used in real-time scanners.

The selection of array elements is done by use of reed-relais. A single element or a small group of elements connected in parallel can be active at one time. The received signal is rectified, low-pass filtered and digitized with a Biomation transient recorder. Data is then stored on a floppy disk. The measurement is fully controlled by a desktop computer. After the measurement is completed, the data set is sent to a computer center via a telephone link. The reconstruction is done offline. The reconstructed image is transferred back to the desktop computer and again stored on a floppy disk. The image can then be visualized on a TV-monitor.

The measurement procedure and data transfer is the same for both methods 1 and 2. Only at runtime of the job is it decided, which method will be taken. This gives the advantage of a high flexibility and an easy comparison of performance. Also transmission data such as time-of-flight or absorption measurements can be taken with the same setup and be transferred and reconstructed in the same way. Thus comparison can also be made between transmission mode reconstructions and reflection mode reconstructions.

EXPERIMENTAL RESULTS

Computer simulated and experimentally measured data were reconstructed using both methods 1 and 2. In all of the following experiments, 45 aspect angles and 32 transducer positions were used. The aspect angles covered the full 360°, although for perfectly scattering objects this would not be necessary as pointed out before. The number of 45 pseudoprojections might be too small for a noise-free reconstruction in method 1, but it is the maximum amount of B-scans, that fit on a single floppy disk.

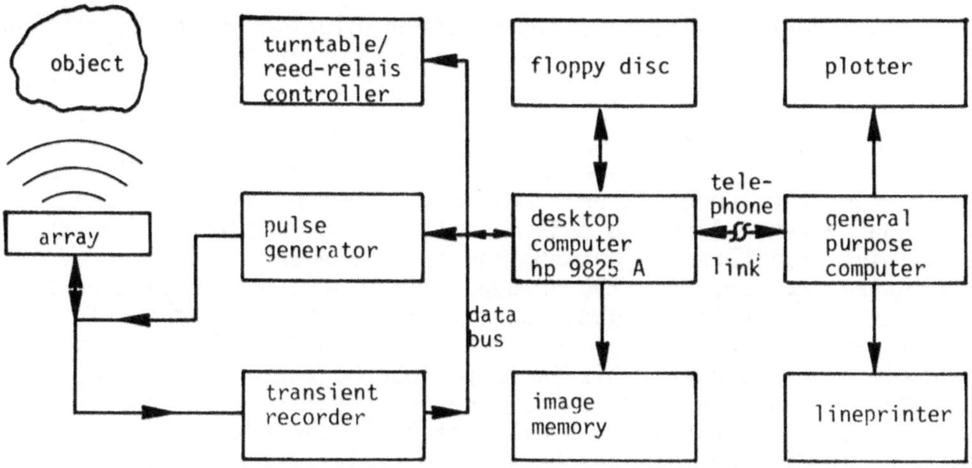

Fig. 6 Block diagram of the measurement system

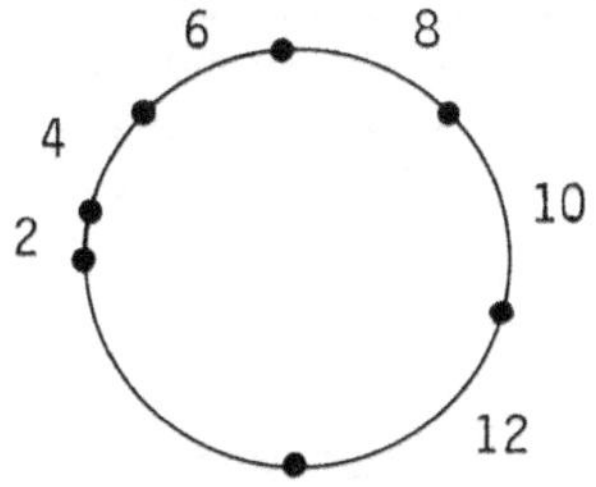

li.:Schaumstoffobjekt
re.:PVC-Formstück

(a) (b)

Fig. 7 Testobjects used for the experiments
a) Sketch of the arrangement of steel-needles. All dimensions are in mm.
b) Photograph of the sponge-phantom and the PVC-piece

To experimentally investigate the resolution, test objects consisting of 0.6 mm diameter steel-needles were imaged. One of those objects is sketched in Fig. 7a. These needles represent fairly good point-scatterers because of their diameter of approximately λ. Fig.8a shows a single B-scan of that object. (The array was on the left side of the object). This figure shows clearly the good axial and the characteristical poor lateral resolution common to conventional B-scan imaging systems. Also the data contains noise, resulting from the sensitivity-differences between the array elements (which were not corrected for) and from the trigger-jitter of the transient recorder. A reconstruction of the same object using method 1 is shown in Fig.8b. As it can be seen, all needles down to the smallest distance of 2 mm are clearly separated. This is very close to the axial resolution of the B-scan image. Fig. 8c shows a reconstruction of the same data set as before, however this time with method 2. The resolution is the same as in Fig. 8b, but the background artifacts, noticable in Fig.8b, disappeared in Fig. 8c, except for the near vicinity of the object points.

In earlier experiments with transmission CT /6/, sponge phantoms were sucessfully used as models for "soft tissue". Their acoustical properties such as velocity and absorption agree well with that of biological specimens, provided care has been taken to remove all airbubbles from the sponge. To simulate inhomogeneities, holes can be punched through these phantoms. Fig. 7b (left) shows one of those objects containing holes with diameters of 2,4,6,8 and 10 mm. Fig. 9a is the reconstruction with method 1, Fig. 9b was reconstructed using the same data but method 2. As can be seen from those images, sponge is also well suited for modeling an ideal scattering medium. The single scattering points of which the sponge consists are reconstructed well, but because the average distance between these scatterers is relatively large, the punched holes can only be recognized down to 6 mm diameter. Noticable again is the significant improvement of image quality compared to a single B-scan (Fig. 9a), where single scatterers are not at all evident, and the holes are just barely recognizable.

To investigate the performance of both methods when applied to data from objects containing reflecting surfaces, test objects cut from PVC were imaged. One of those irregularly shaped pieces is shown on the right hand side of Fig. 7b. As can be noticed from the single B-scan (Fig. 10a), this object is indeed strongly reflecting, so that only a few points of the circumference appear in each B-scan. After reconstructing this object with method 1 or 2 the total circumference is visible. However, method 1 (Fig. 10b) yields a much worse result than method 2 (Fig. 10c). This is mainly because of the fact, that the data set is highly inconsistent with the reconstruction concept due to the specular reflections. The fact that method 2 is relatively insensitive to such measurement errors now becomes important.

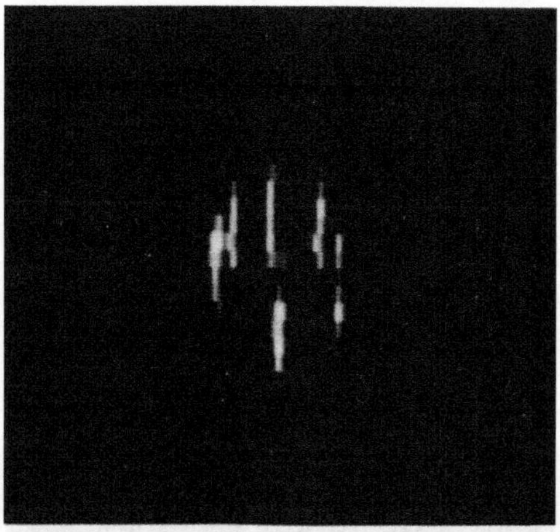

(a)

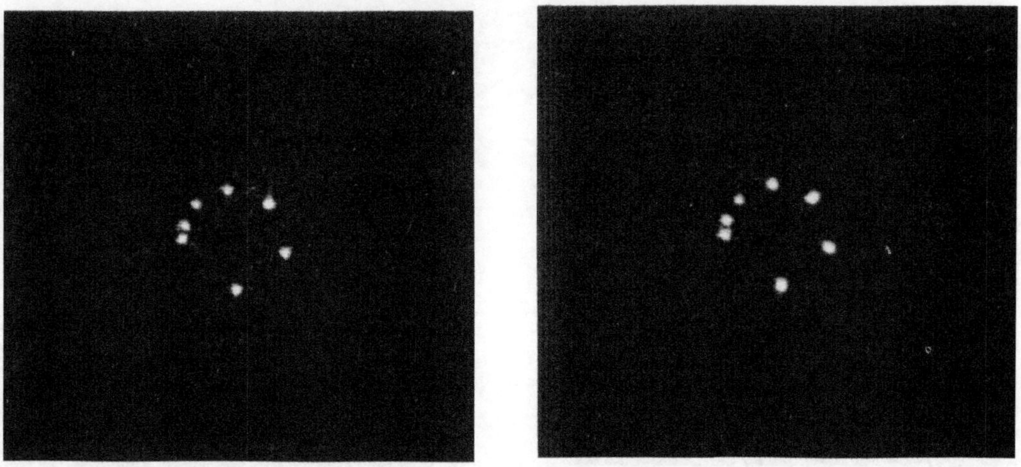

(b) (c)

Fig. 8 Reconstruction of the steel-needles

(a) One single B-scan
(b) Reconstruction of filtered pseudoprojections
(c) Summation of filtered B-scans

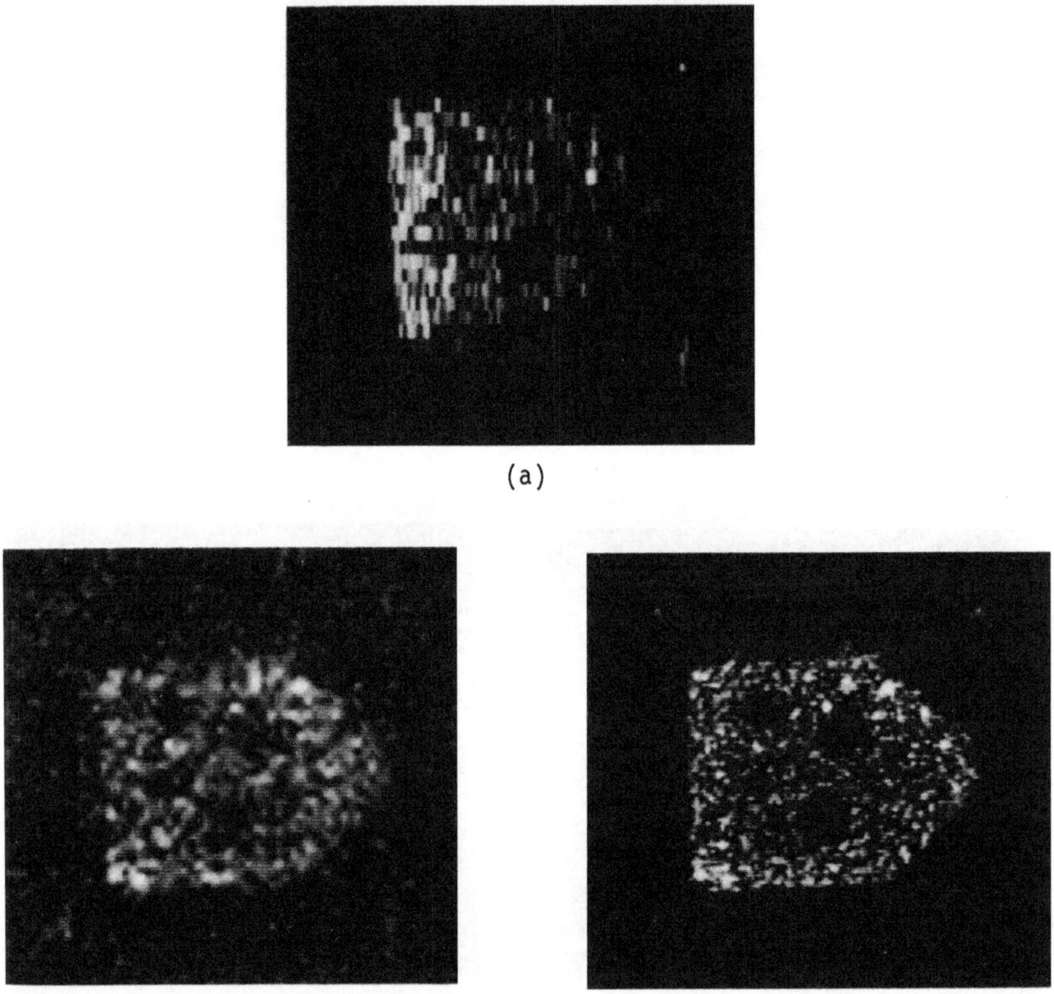

(a)

(b) (c)

Fig. 9 Reconstruction of the sponge phantom

(a) One single B-scan
(b) Reconstruction of filtered pseudoprojections
(c) Summation of filtered B-scans

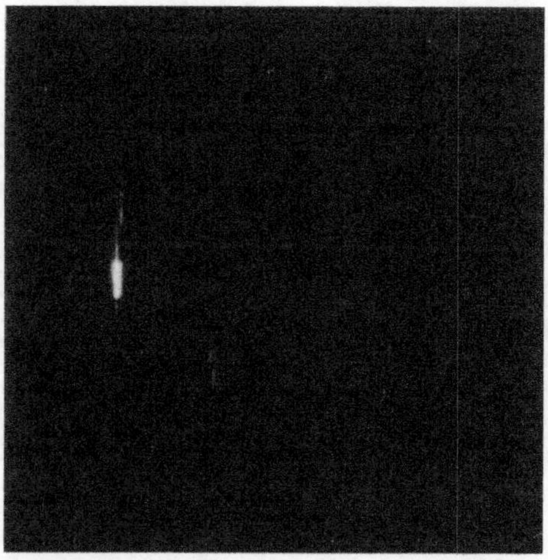

(a)

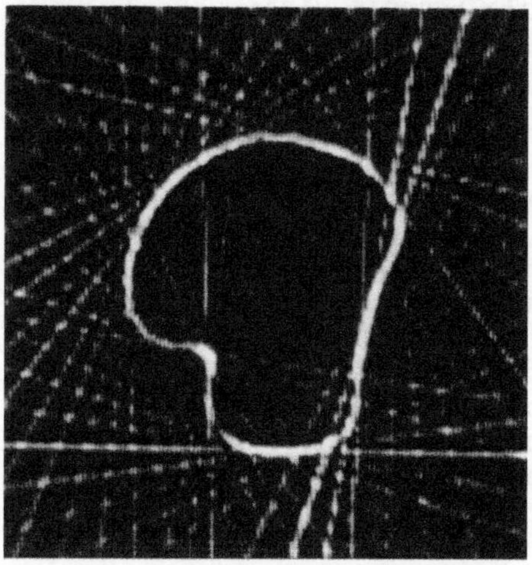

(b)

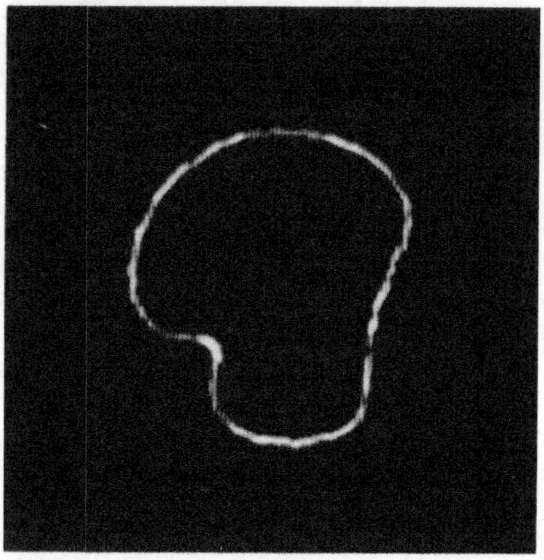

(c)

Fig. 10 Reconstruction of the PVC-piece

(a) One single B-scan
(b) Reconstruction of filtered pseudoprojections
(c) Summation of filtered B-scans

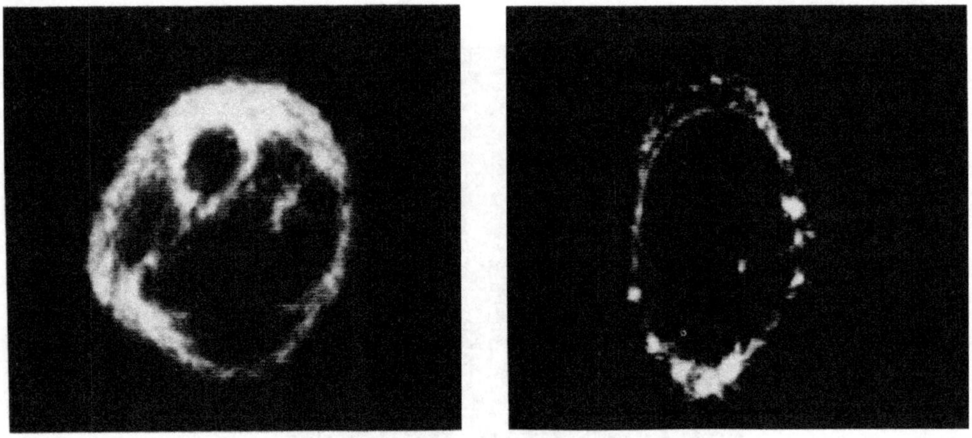

(a) (b)

Fig. 11 Reconstruction of biological specimens by summation of filtered B-scans.
a) left human forearm (in vivo)
b) freshly excised human testicle (longitudinal cross-section)

Some experiments were also carried out on biological specimens. So Fig. 11a shows a cross-section of a human forearm (in vivo), taken approximately 7 cm above the wrist and reconstructed with method 2. Both of the bones (ulna, lower left side, and radius, upper side) can be clearly distinguished, also some internal structures inside the soft tissue can be noticed, which probably correspond to tendons and blood vessels. This measurement was performed without depth-dependent gain control. So no compensation for the tissue absorption was made, as it is usually done in commercial equipment. Therefore the bone surfaces deep in the tissue are not as clearly seen as the outer ones. Another problem with the image is the blurring due to the necessity of keeping the arm in exactly the same position during the whole time of data aquisition and storage, which takes about 2 minutes with the present setup.

Fig. 11b, as another example of the application of the above described techniques to soft tissue, shows a longitudinal cross-section through a freshly excised human testicle reconstructed with method 2. The pathological investigation revealed a cancer, that occupied nearly the lower third part of the testicle. Clearly visible is the outside circumference along with the crescent-shaped epididymis at the upper part. The irregular, diffuse structures at the lower part probably correspond to a section of the tumor. It is strange however, that echos from the rest of the tumor are totally missing. The single point in the middle results from a needle, with which the specimen was fixed to the setup.

CONCLUSIONS

Both the methods that have been described in this paper show several advantages over other proposed or realized reflectivity tomography techniques. Initial results seem slightly in favor of the second method, because of its lower sensitivity to deviations from the ideal case of perfectly scattering objects. Even in the worst case of specular reflecting surfaces, the image quality and the resolution will be at least as good as that of a conventional parallel or compound-scan system.

Experiments with biological specimens demonstrate the potential usefulness for medical diagnosis. A combination of information may be possible in cases, where both transmission and reflectivity CT techniques are applicable, such as breast and testicle examination. Such a combination might allow a correction of transmission data with the knowledge of reflectivity properties, as well as the correction of echo data with knowledge of attenuation and the spatial variations of velocity.

ACKNOWLEDGEMENT

These investigations were supported by Research Grant No. DFG-ER 94-1 from the Deutsche Forschungsgemeinschaft, Bonn-Bad Godesberg.

REFERENCES

1. G. Wade et al: Acoustic echo Computer Tomography. in: Acoustic Imaging, Vol. 8 (ed.: A. F. Metherell), p. 565-576, Plenum Press, New York, 1978

2. S. A. Johnson, J. F. Greenleaf, B. Rajagopalan and M. Tanaka: Algebraic and analytic inversion of acoustic data from partially or fully enclosing apertures. in: Acoustic Imaging, Vol. 8 (ed.: A. F. Metherell), p. 577-598, Plenum Press, New York, 1978

3. S. J. Norton and M. Linzer: Ultrasonic reflectivity tomography: Reconstruction with circular transducer arrays. Ultrasonic Imaging 1, p. 154-184, 1979

4. R. A. Brooks, G. H. Weiss and A. J. Talbert: A new approach to interpolation in Computed Tomography. J. of Computer Assisted Tomography 2, p. 577-585, 1978

5. R. Gordon and S. W. Rowland: Three-dimensional reconstruction from projections: A review of algorithms. Int. Rev. of Cytol. 38, p. 111-151, 1974

6. D. Hiller and H. Ermert: The application of transducer arrays in ultrasound Computerized Tomography. in: Ultrasonics Int. 1979 Conf. Proceedings, p. 540-544, IPC Science and Technology Press, Guildford

7. J. F. Greenleaf, S. A. Johnson, W. F. Samayoa and C. R. Hansen: Refractive index by reconstruction: Use to improve compound B-scan resolution. in: Acoustical Holography, Vol. 7 (ed.: L. W. Kessler), p. 263-273, Plenum Press, New York, 1977

TOMOGRAPHY FROM MULTIVIEW ULTRASONIC DIFFRACTION DATA:

COMPARISON WITH IMAGE RECONSTRUCTION FROM PROJECTIONS

M.F. Adams and A.P. Anderson

University of Sheffield
Dept. of Electronic and Electrical Engineering
Mappin Street
Sheffield, U.K.

INTRODUCTION

The tomographic mode of imaging is a powerful aid to non-invasive diagnostics. By computerised data assembly, image slices of a 3-D object showing internal structure can be presented to the observer since the probing radiation penetrates through the object. This imaging procedure is unusual since it would appear not to have any close analogue in Nature.

The requirement for improved diagnostic interpretations of 3-D structures and their functions has led to the development of projection techniques for computed assembly of multiview data from electron microscopes[1] X-ray scanners and more recently, γ-ray cameras[2]. Clearly the medical applications of tomography are dominant.

To fulfill similar and complementary roles, ultrasonic diagnostic imaging requires a tomographic capability. The B-scan procedure provides this form of imaging, but the resolution is poor if compared with synthetic aperture/holographic imaging systems. In the X-ray regime, for example, tomographic images are obtained using a variety of back projection algorithms whose basis is the Radon transform. These methods are not strictly appropriate for ultrasonic diffraction data although their direct application has been attempted[3]. However this approximate solution not only loses accuracy but has severe penalties if image reconstruction from a small number of views is attempted.

PRINCIPLE OF THE METHOD

For simplicity of presentation we shall describe the principle in terms of generating a two-dimensional (2D) image field from one-dimensional (1D) views (or scans) although the theory can easily be extended to the general case of 3D objects. Consider the data acquisition scheme shown schematically in Fig.1. where the scan is along a line, s, at a distance z_o from the central region of the object. If the transmitter and receiver are both located at s then a single point scatterer at co-ordinates (x,z) with reflectivity g(x,z) will give rise to a signal $g(x,z)\exp[jk2r]$ at the receiver.

Considering a single line of the object at a given z and neglecting any modification of wavefronts scattered from this line while passing through g(x,z), the total received signal due to this line is given by

$$G(s) = \int_x g(x,z)\exp[jk2r]\,dx \tag{1}$$

If the scan is in the Fresnel region of the object[4], i.e. the scan dimensions satisfy $L^4 < 8\lambda z_o^3$, then the distance r can be approximated to

$$r \simeq z_o + z + \frac{(s-x)^2}{2(z_o+z)} \tag{2}$$

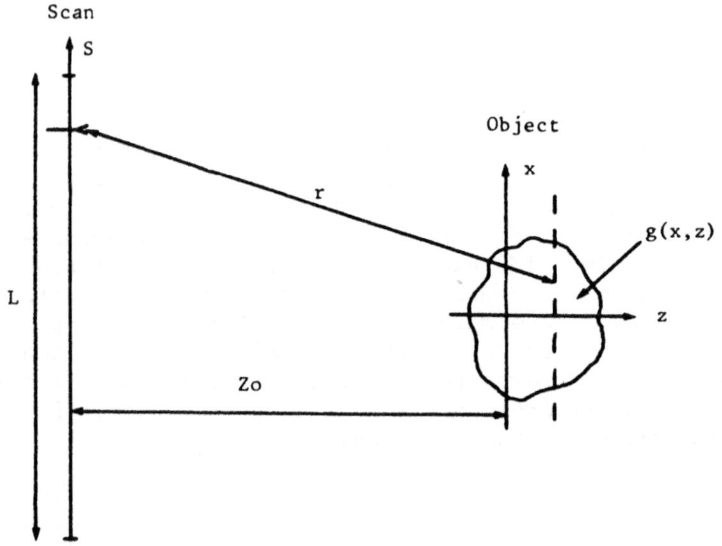

Fig.1 : Co-ordinate System

Substituting in (1) yields

$$G(s) = \mathrm{rect}(\frac{S}{L})\exp\left[jk(2z + 2z_o + \frac{S^2}{z+z_o})\right].\int_x g(x,z).$$

$$.\exp\left[\frac{jk}{z+z_o}(x^2-2sx)\right]dx. \quad (3)$$

If we let

$$g'(x,z) = g(x,z)\exp\left[\frac{jkx^2}{z+z_o} + jk2(z+z_o)\right] \quad (4)$$

then $G_z(s)$ represents the Fourier transform relationship,

$$G_z(s) = \int_x g'(x,z)\exp\left[\frac{-jk2sx}{z+z_o}\right]dx \quad (5)$$

$$= F_1\left[g'(x,z)\right],$$

where F_1 denotes the 1D Fourier transform.

Then equation (3) can be rewritten

$$G(s) = \mathrm{rect}(\frac{S}{L}).\exp\left[\frac{jks^2}{z+z_o}\right].G_z(s) \quad (6)$$

Equations (4), (5) and (6) represent the scan signal due to one line of the object in terms of a Fourier transform relationship. The total scan signal from all such lines is then given by the integral over z,

$$G_T(s) = \int_z \mathrm{rect}(\frac{S}{L})\exp\left[\frac{jks^2}{z+z_o}\right]G_z(s)\ dz. \quad (7)$$

If the amplitude and phase of the signal $G_T(s)$ is recorded over an aperture L then an image can be reconstructed at any depth by focusing the data and performing an Inverse Fourier transform. The focusing at a depth z_r is achieved by premultiplication with a quadratic phase factor $\exp\left[-jks^2/(z_o+z_r)\right]$ before performing the Inverse Fourier transform. Therefore the image $I(x,z_r)$ at depth z_r is given by,

$$I(x,z_r)=F_1^{-1}\{\int_z \mathrm{rect}(\frac{S}{L})\exp\left[jks^2(\frac{1}{z_o+z} - \frac{1}{z_o+z_r})\right]G_z(s)dz\} \quad (8)$$

where F_1^{-1} denotes the 1D Inverse Fourier transform.

From equation (5),

$$g'(x,z) = F_1^{-1}\left[G_z(s)\right]$$

so that (8) can be rewritten, using the convolution theorem and changing the order of integration,

$$I(x,z_r) = \int_z \{F_1^{-1}\left[\text{rect}(\tfrac{S}{L})\exp\left[jks^2(\frac{1}{z_o+z} - \frac{1}{z_o+z_r})\right]\right] \underset{x}{*} g'(x,z)\}dz \tag{9}$$

where $\underset{x}{*}$ represents a 1D convolution in the x direction. If the scan is in the far field of the object so that $z_o \gg z_r$ then

$$\frac{1}{z_o+z} - \frac{1}{z_o+z_r} \simeq \frac{z_r-z}{z_o^2}$$

and define

$$p(x,z_r-z) = F_1^{-1}\left[\text{rect}(\tfrac{S}{L})\exp\left[jks^2(\frac{z_r-z}{z_o^2})\right]\right] \tag{10}$$

then substituting in (9) yields,

$$I(x,z_r) = \int_z p(x,z_r-z) \underset{x}{*} g'(x,z)dz \; . \tag{11}$$

Expanding the convolution in x

$$I(x,z_r) = \iint_{zx} p(x-x_o,z_r-z)g'(x_o,z)dx_o\, dz \; . \tag{12}$$

Thus the image $I(x,z_r)$ formed from one linear 'view' of the object can be written as the 2D convolution,

$$I(x,z_r) = p(x,z_r) \underset{2D}{*} g'(x,z_r) \; . \tag{13}$$

The form of the convolving function $p(x,z_r)$ is best examined by a computer simulation of the response of the system to a point scatterer. The result is shown in Fig.2 in which 32 samples of phase and amplitude data were assumed to be recorded over a 170λ aperture, 250λ distant from the point. The image field was reconstructed at each depth z_r simply by premultiplying the data

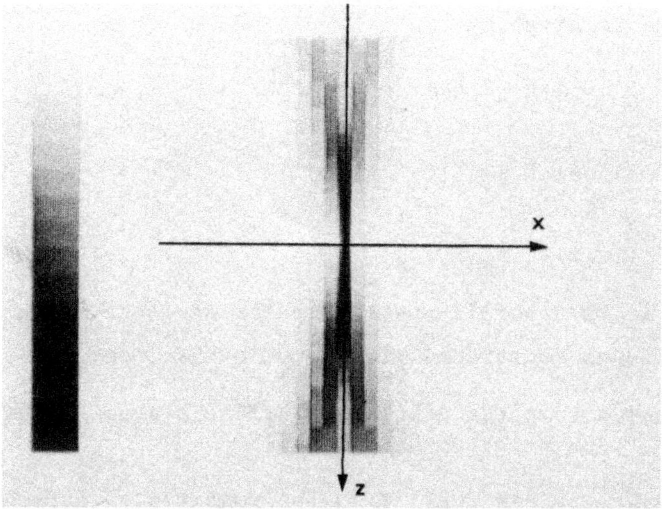

Fig.2 : Impulse response for one view

with the quadratic phase focusing factor* and performing the Inverse Fourier transform using an FFT algorithm. The above scan dimensions yield a theoretical lateral resolution of 0.7λ at the point, and the overall dimension in depth of Fig.2 is 22λ.

Equation (13) states that the system response to a general object is given by the convolution of the function of Fig.2 with the object. Thus a 'pseudo' 2D image is obtained from one 1D scan although the resolution in depth is considerably worse than that obtainable in the lateral dimension. In order to improve the depth resolution of the system it is beneficial to combine several views of the object.

Each image from a single view, at angle Θ, is given by

$$I_\Theta(x,z) = g(x,z) * p_\Theta\,(x,z), \tag{14}$$

where $p_\Theta(x,z)$ is the convolving function, $p(x,z_r)$ rotated to correspond to the new direction Θ.

---

* The foregoing theory and example are not intended to imply a restriction to Fresnel region scanning. A more precise transformation between the scan data and any image line (z) is readily applicable if the Fresnel field approximation in (2) is not valid[5].

If N different views are combined additively then the resultant image is given by

$$I_N(x,z) = \sum_{\Theta} \{g(x,z) * p_{\Theta}(x,z)\}$$

$$= g(x,z) * \{\sum_{\Theta} p_{\Theta}(x,z)\} \quad (15)$$

$$= g(x,z) * p_N(x,z), \quad (16)$$

where $p_N(x,z)$ is the overall system impulse response given by the combination of each individual view impulse response.

Fig.3 shows a computer simulation of the system impulse response for 6 views equispaced through 180°.

From (16) we can see that it may be possible to further improve the resultant image, I_N, by deconvolution with the known system impulse response p_N.

The complete tomographic technique is demonstrated by computer simulation for a simple object consisting of point scatterers using the scan dimensions shown in Fig.4. Six views at intervals of 30° rotation of the target were used.

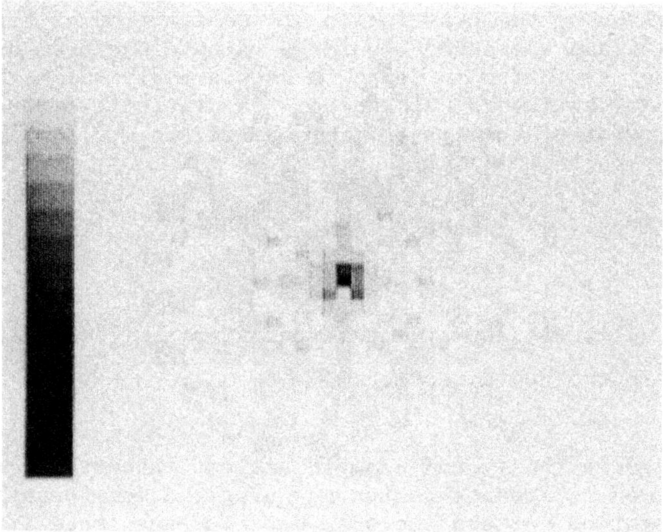

Fig.3 : Impulse response for six views

Each view was used to produce a 'pseudo' 2D image as described above and then the 6 views were combined additively. The resultant image is shown in Fig.5a. This image was then deconvolved with the known system impulse response by applying a weighting function, based on p_N, to the 2D transform of I_N. The resultant deconvolved image is shown in Fig.5b.

COMPARISON WITH PROJECTION APPROACH

It is instructive to compare the foregoing technique based on synthetic aperture imaging with other techniques in which only projection data is available. In these cases it is only possible to generate from one view a 2D image field which is given by the 'back projection'[1] of the data as shown in Fig.6. The convolving function for one image is a 'line response' in the direction of view with no depth information. In particular for the view at angle Θ in Fig.6b the image can be written

$$I_\Theta(x,z) = g(x,z) \underset{2D}{*} \delta(x^1), \tag{17}$$

where $x^1 = x\cos\Theta - z\sin\Theta$ and $\delta(x^1)$ represents the line response at view angle Θ.

In order to obtain more information about the object it is necessary to use many views combined in a manner analagous to that shown in Equation (15) for the focused technique, i.e. the 'back-projected' images from each view are combined additively. Since the convolving function for each view is a 'line response' the overall system response for N views will be a 'star-shaped' function as shown in Fig.6c for N = 4 views.

For a large number of views, the projection impulse response p'_N tends to $1/r$ where r is distance from the point i.e.

$$\lim_{N\to\infty} p'_N\ (x,z) = \frac{1}{r} = \frac{1}{\sqrt{x^2+z^2}}\ . \tag{18}$$

In that case the image is given by

$$\lim_{N\to\infty} I_N(x,z) = g(x,z) * \frac{1}{r} \tag{19}$$

so that, in principle, the object is recovered perfectly from $I_N(x,z)$ by deconvolution.

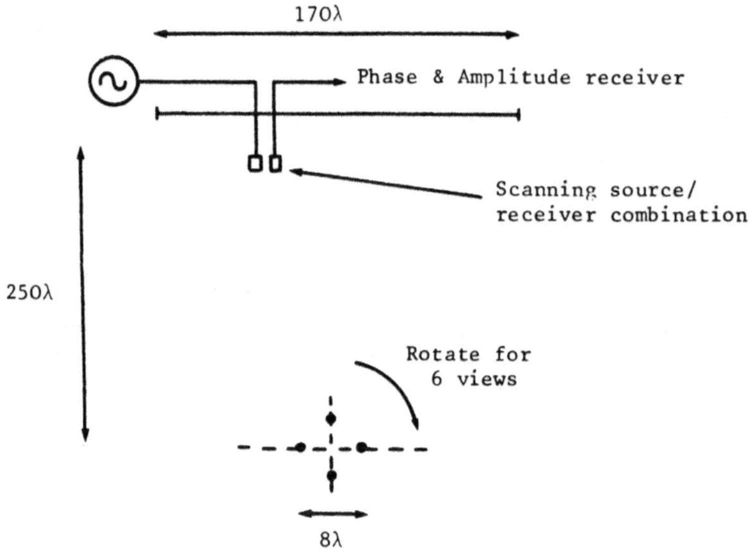

Fig.4 : Scan dimensions for computer simulation

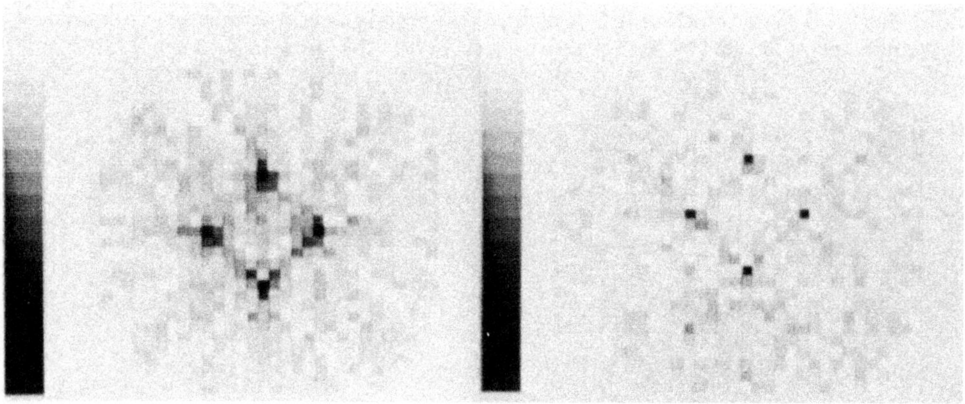

a) Addition of 6 views b) Deconvolved image

Fig.5 : Simulation of focused technique

The 2D Fourier transform of equation (19) yields

$$\tilde{I}_N\ (u,\omega) = \tilde{G}(u,\omega) . \frac{1}{\rho} \qquad (20)$$

where $\rho = \sqrt{u^2+\omega^2}$,

$$\tilde{G}\ (u,\omega) = F_2\left[g(x,z)\right],$$

$$\frac{1}{\rho} = F_2\left[\frac{1}{r}\right],$$

and F_2 denotes the 2D Fourier transform operation.

Thus by applying a weighting of ρ to $\tilde{I}_N$ and performing an inverse Fourier transform the object g(x,z) is obtained.

This is the basis of the convolution-back-projection method[7] widely used in medical CAT scanners[6].

As a comparison with the focused technique an attempt was made to reconstruct the object of Fig.4 using only its projection onto 6 views equispaced through 180°. Fig.7a demonstrates simple additive reconstruction of the object from the 6 back projected views. Fig.7b is an attempt to deconvolve this image with the 1/r function as described above. We see that because each view possesses depth information, the overall response of the focused system has improved much more rapidly than that of the equivalent back projection method for this small number of views.

CORRESPONDENCE TO DIFFRACTION TOMOGRAPHY IN FREQUENCY SPACE

For the case of projection data the reconstruction techniques can be analysed in Frequency Space by the use of the Projection Theorem which states : the 1D Fourier transform of the projection of a 2D object onto a line yields a central cross section of the 2D Fourier transform of the object[6].

This follows from the 2D transformation of equation (17) which yields

$$\tilde{I}_\Theta(u,\omega) = G(u,\omega) . \delta(\omega^1), \qquad (21)$$

where $\omega^1 = u \sin \Theta + \omega \cos \Theta$ and $\delta(\omega^1)$ is the 2D transform of $\delta(x^1)$ i.e. a line orthogonal to the direction of view.

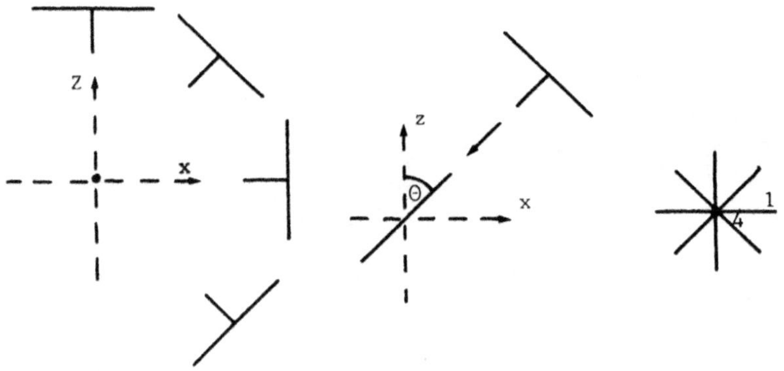

a) Record projections of object.
b) Back projection image from one view.
c) Sum all such Back projections.

Fig.6 : Back Projection Technique

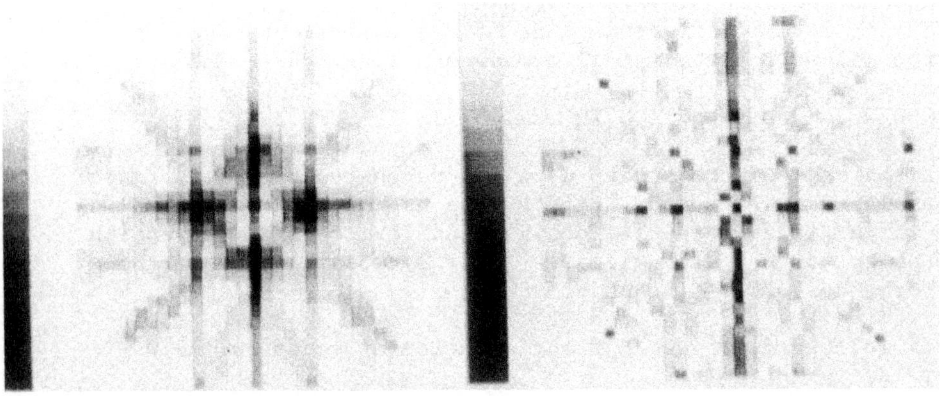

a) Addition of 6 views b) Filtered image

Fig.7 : Simulation of back projection technique

Similarly the 2D Fourier transform of the focused system impulse response given by equation (16) will yield the region of the 2D FT of the object which is sampled by one view. 2D Transforming equation (10) yields

$$\tilde{P}(s,\omega) = \int \mathrm{rect}(\frac{S}{L})\exp\left[\frac{jks^2 z_r}{z_o^2}\right]\exp\left[-j2\pi z_r \omega\right]dz_r \quad (22)$$

$$= \mathrm{rect}(\frac{S}{L})\ \delta(\omega - \frac{s^2}{\lambda z_o^2}) \ . \quad (23)$$

From equation (5) we see that the scan coordinate (in distance) ,s, is converted to spatial frequency u, by the relationship

$$u = \frac{2s}{\lambda z_o} \quad (24)$$

so that equation (23) becomes

$$\tilde{P}(u,\omega) = \mathrm{rect}(\frac{u}{L_u}).\delta(\omega - \frac{\lambda}{4}u^2). \quad (25)$$

Thus the 2D Fourier transform of the impulse response yields an arc in Fourier Space given by equation (25) and shown in Fig.8. The arc has been 'naturally' interpolated onto the square grid by the transformation process.

The image from one view given by equation (13) yields the 2D transformation

$$\tilde{I}(u,\omega) = \tilde{P}(u,\omega).G^1(u,\omega). \quad (26)$$

Thus the 2D FT of the object, $G^1(u,\omega)$,is known along an arc in Fourier space as described by equation (25). The function $G^1(u,\omega)$is given by the transform of equation (4) which, for the scanning source/receiver reflection imaging system is (ignoring the relatively small quadratic phase term as we are in the far field of the object),

$$G^1(u,\omega) = F_2\{g(x,z)\exp(jk2z)\} \quad (27)$$

$$= G(u,\omega - \frac{2}{\lambda}), \quad (28)$$

where $G(u,\omega) = F_2\left[g(x,z)\right].$ (29)

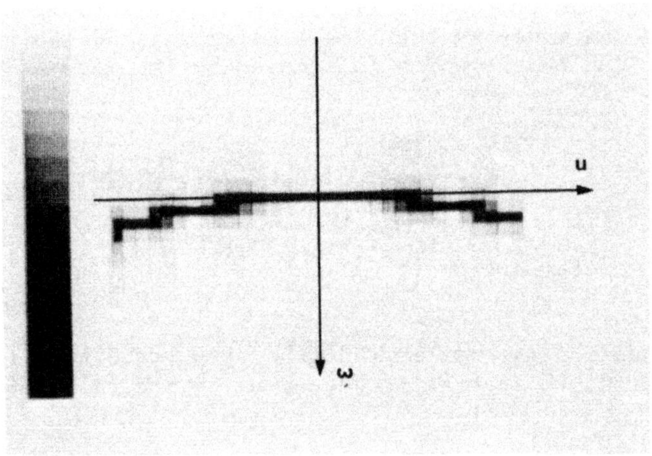

Fig.8 : 2D Fourier transform of impulse response for one view

In order to remove the 2/λ shift in the transform in equation (28) it is necessary to consider a transmission imaging system with a fixed point source located at $z=z_o$. It can be shown that this removes the linear phase term, $\exp[jk2z]$, from equation (27) and also that the use of a fixed source rather than scanning source/receiver effectively doubles the wavelength so that equation (26) becomes

$$\tilde{I}(u,\omega) = \tilde{P}_T(u,\omega).G(u,\omega) \tag{30}$$

$$\text{where } \tilde{P}_T(u,\omega) = \text{rect}\left(\frac{u}{L_u}\right)\delta\left(\omega - \frac{\lambda}{2}u^2\right) \tag{31}$$

We can now see a comparison between the sampling function $\tilde{P}_T(u,\omega)$ in equation (31) with the result derived by Mueller[8] et al who were considering plane wave illumination of a weakly scattering target.

Their sampling function from one view can be written

$$\tilde{P}_M(u,\omega) = \delta\left[\omega - \frac{1}{\lambda}\left(1-\sqrt{1-\lambda^2u^2}\right)\right], \tag{32}$$

which is the equation of a circle centred on $\emptyset$, $1/\lambda$ in Fourier space.

The functions $\tilde{P}_T$ and $\tilde{P}_M$ are shown in Fig.9 for comparison purposes.

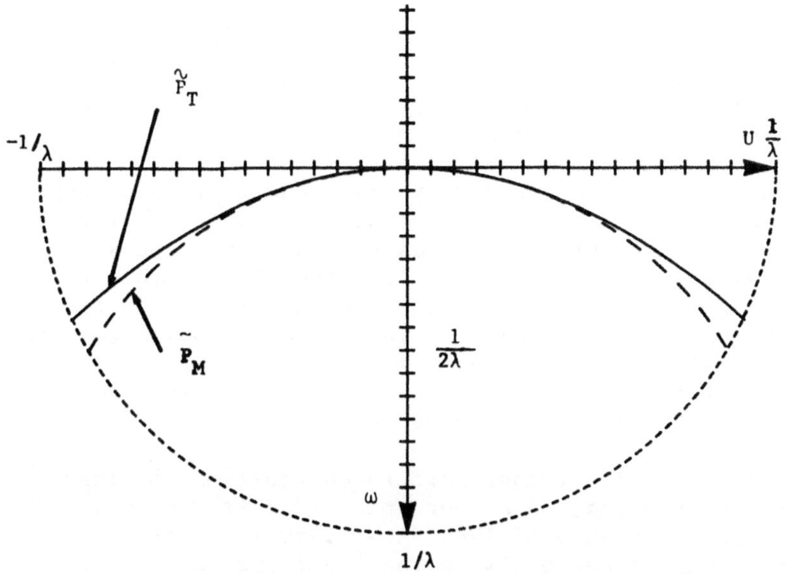

Fig.9 : Comparison of focused technique ($\tilde{P}_T$) with diffraction tomography ($\tilde{P}_M$) in spatial frequency space

The result (31) derived from the imaging approach seems to be equivalent to that derived by Mueller, but applicable to spherical wave insonification and scanning in the Fresnel region of the object. If the precise transformation referred to in (5) is employed, the results converge. Our method differs by assembling the image directly in image space as a focusing and deconvolution operation,and it does not require the approximation of interpolating the measured data from the arc $\tilde{P}_M$ onto frequency space.

EXPERIMENTAL RESULTS

Some results from the initial study of this subject have been reported for the microwave regime[9]. Subsequently, an ultrasonic scanner employing a precision mechanical positioner has been constructed which can perform raster scans of a transmit/receive transducer pair in a water tank. The transducers are lensed so that they function as 'point sources'. The scanning operation, data acquisition, and recording, are controlled by a TMS 9900 microprocessor system.

To investigate the tomographic reconstruction technique experimentally, line scans of a simple test object were obtained from an arrangement with the dimensions shown in Fig.10. The test object was similar to the computer model and consisted of four metal rods positioned orthogonally to the scan line and able to be rotated about their central axis.

Because our system for direct phase and amplitude recording of the ultrasonic field is not yet operational, the data for this preliminary validation exercise was recorded and processed to yield phase and amplitude values by the 'in-line' holographic technique[10]. The recording system for this technique is very simple requiring a constant signal derived from the transmitter to be mixed with the received signal as indicated on Fig.10. However the resultant phase and amplitude values are inaccurate due to the unwanted residual outputs from the holographic process.

Fig.11a shows the tomographic reconstruction obtained from six views, or linear scans, each containing 32 data samples, and an insonification frequency of 1MHz (λ = 1.5mm in water). Fig.11b shows the result of deconvolving the ideal system response (Fig.3). Both these results exhibit considerable 'noise' due mainly to the limitations of the in-line holographic technique for multiview image assembly. The effect of shadowing of the two rods on the left side is noticeable and due to the views being taken around a 180° sector on the right side. Nevertheless, this first result is encouraging and the resolution approaches that expected from the synthetic aperture. Of course, data could also be acquired by a transducer array with appropriate element spacing and recording circuitry, but the present experimental arrangement is flexible and has the advantages of using only two large aperture transducers readily available from manufacturers.

CONCLUSIONS

A synthetic aperture imaging approach to ultrasonic tomography from multiview diffraction data has been presented. Image quality improves more rapidly with a small number of views compared to back projection data processing. The approach corresponds closely to an alternative method of diffraction tomography using data assembly in frequency space. However there does appear to be some difference due to interpolation procedures inherent in each approach although this point has not yet been investigated.

Preliminary experimental results using data recorded from backscattered ultrasonic fields are encouraging.

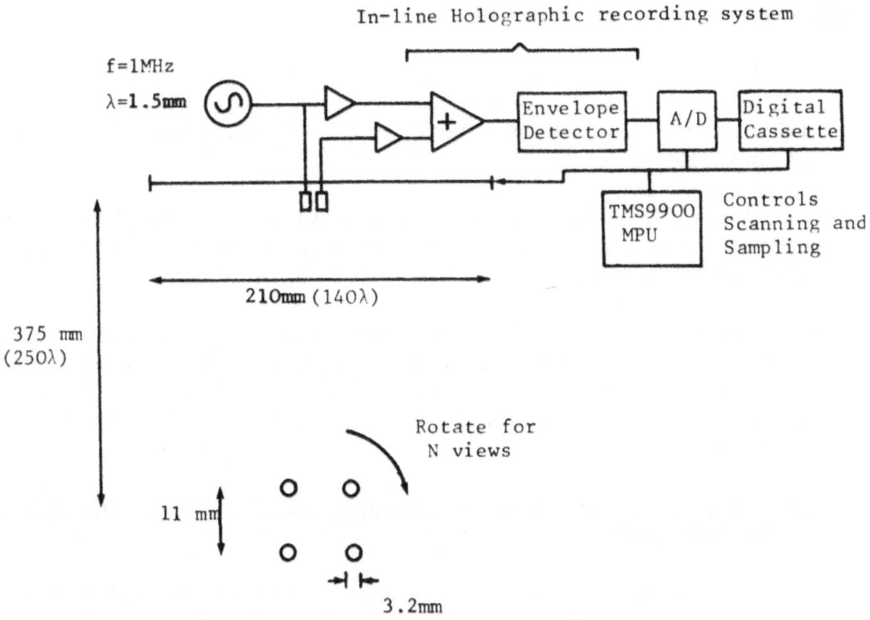

Fig.10 : Experimental system

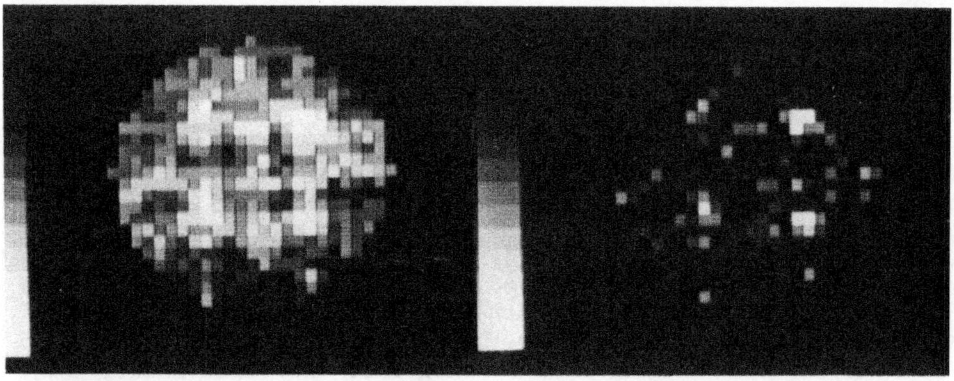

a) Addition of 6 views b) Deconvolved image

Fig.11 : Experimental results

REFERENCES

1. R.A. Crowther, D.J. DeRosier, and A. Klug, The reconstruction of a three dimensional structure from projections and its application to electron microscopy, Proc.R.Soc. London Ser.A 317, pp 319-40,1970

2. P.H. Jarritt, P.J.Ell, M.J. Myers, N.J.G. Brown, and J.M. Deacon, A new transverse-section brain imager for single-gamma emitters, J. Nucl.Med., 20, pp 319-327,1979

3. J.G. Greenleaf, S.A. Johnson, W.F. Samayoa, and F.A. Duck, Algebraic reconstruction of spatial distributions of acoustic velocities in tissue from their time-of-flight profiles, in "Acoustical Holography Vol.6" N. Booth Ed, Plenum Press, New York, 1975

4. A.P. Anderson, Microwave Holography, Proc. IEE, 124 (11R), pp 946-962, 1977

5. J. Shewell, and E. Wolf, Inverse diffraction and a new reciprocity theorem, J. Opt. Soc.Am., 58, 1596, 1968

6. H.J. Scudder, Introduction to computer-aided tomography, Proc. IEEE, 66, pp 623-37, 1978

7. G.N. Ramachandran, and A.V. Lakshminarayanan, Three dimensional reconstruction from radiographs and electron micrographs; II Application of convolutions instead of Fourier transforms, Proc.Nat.Acad.Sci, 68, No.9, pp 2236-40, 1971

8. R.K. Mueller, M. Kaveh, and G. Wade, Reconstructive tomography and applications to ultrasonics, Proc. IEEE,67, No.4, pp 567-587, 1979

9. M.F. Adams, and A.P. Anderson, Three dimensional image construction technique and its application to coherent microwave diagnostics, Proc.IEE pt H, 127, No.3, pp 138-142, 1980

10. K.H.S. Marie, A.P. Anderson, J.C. Bennett, Digital processing technique for suppressing interfering outputs in the image of an in-line hologram, Electron.Lett., 15, No.8 pp 241-243, 1979. Microwave images with reduced background effects from digitally processed in-line holograms, Electron.Lett., 16, No.13, pp 493-494, 1980

NONLINEAR IMAGE RECONSTRUCTION FROM ULTRASONIC TIME-OF-FLIGHT PROJECTIONS

Hermann Schomberg

Philips GmbH, Forschungslaboratorium Hamburg

D-2000 Hamburg 54, F.R.G.

INTRODUCTION

This paper is concerned with reconstructive ultrasound tomography. More specifically, it deals with that variation of reconstructive ultrasound tomography which is based on time-of-flight projections and produces images which represent the acoustic refractive index. The method has been studied since a while; see Mueller et al. (1979) for a review. The main application envisaged is breast imaging, but still the images obtained with this method do not show the details required for a medical application. We shall expound that this is mainly due to shortcomings in the mathematical model employed. Then we derive a more appropriate model and a pertinent reconstruction algorithm. The efficiency of both is demonstrated using real data obtained with a phantom. We start with a short description of the underlying experiment.

THE BASIC EXPERIMENT

The projections may be measured e.g. with an apparatus as sketched in fig. 1. This apparatus consists of a tank filled with water, two opposite ultrasound transducers, and auxiliary electronic and mechanical equipment not shown in the figure. The transducers act as emitter and receiver, resp. The tank is assigned a (ξ_1, ξ_2, ξ_3)-coordinate system as indicated. The object, e.g. a female breast, is assigned an (x_1, x_2, x_3)-

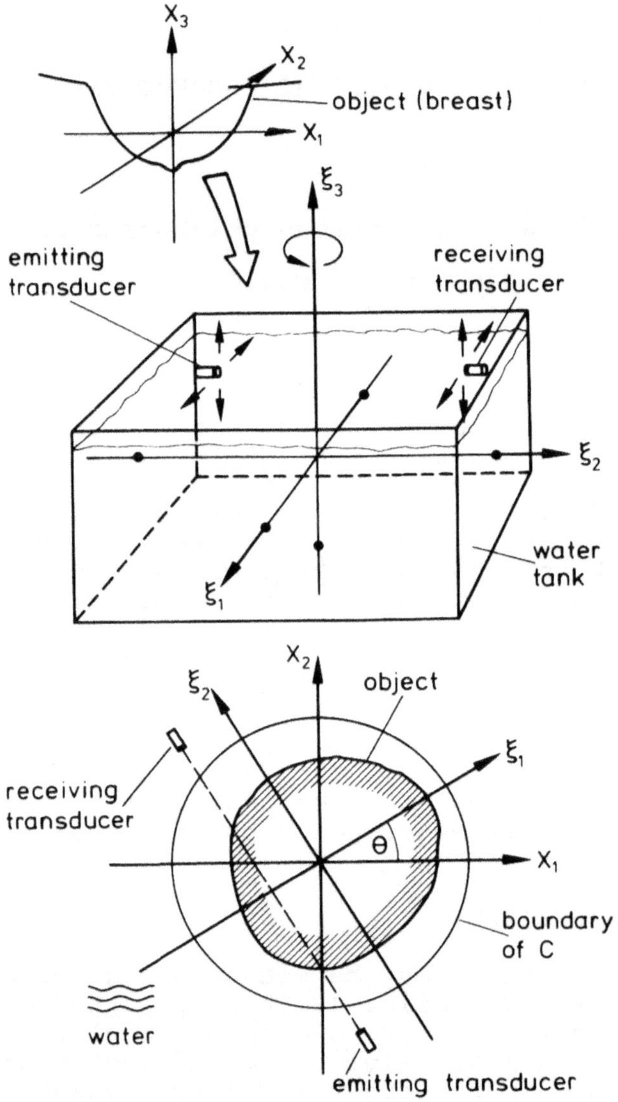

Fig. 1. Illustrating the basic experiment. The lower panel shows a horizontal cross-section.

coordinate system and immersed into the tank such that the origins and the vertical axes of the two coordinate systems coincide. The object has to fit into an imaginary cylinder of the form

$$C = \{(x_1,x_2,x_3) \mid x_1^2 + x_2^2 \leq r,\ \underline{x}_3 \leq x_3 \leq \bar{x}_3\}$$

which does not contain the transducers. While the object remains fixed, the tank may be rotated about its vertical axis. Also, the transducers may be moved horizontally and vertically as indicated by the arrows in fig. 1, but their ξ_1- and ξ_3- coordinates remain the same throughout. Thus, we can characterize the position of the transducers with respect to the object by the coordinate triple $\zeta := (\theta,\xi_1,\xi_3)$, where θ denotes the angle between the positive ξ_1- and x_1-axes.

To make a measurement, an ultrasound pulse is launched at the emitter. The pulse travels through water and object and, after some delay, arrives at the opposite receiver. There may be several arrivals due to multiple paths. The delay between emission and first arrival is measured and represents the 'time-of-flight' of that pulse. Such measurements are taken for a great number of positions $\zeta^{klm} := (\theta^k, \xi_1^l, \xi_3^m)$, where

$$0 = \theta^1 < \theta^2 < \ldots < \theta^K < 2\pi \text{ (or } \pi)$$

$$-r \leq \xi_1^1 < \xi_1^2 < \ldots < \xi_1^L \leq r,$$

$$\underline{x}_3 \leq \xi_3^1 < \xi_3^2 < \ldots < \xi_3^M \leq \bar{x}_3.$$

For each position ζ^{klm}, we thus obtain an associated time-of-flight, T_{klm}.

The data acquisition may considerably be speeded up by using e.g. a 'fan beam' emitter or an array of emitters, respectively combined with an opposite array of receivers. In such a case the following considerations apply as well.

THE CONVENTIONAL MODEL

The measured projection data do not yet represent the wanted image of the object. Rather, this image consists of the map of the acoustic refractive index

$$\bar{n} = c_o/\bar{c} \quad .$$

Here, $\bar{c}$ is the speed of sound, assumed to be a

function of space, and c_o is the constant speed of sound in water. Thus $\bar{n} = 1$ outside the object. In a breast, the refractive index varies between 0.95 and 1.05 or so (Glover and Sharp, 1977). We also call $\bar{n}$ the 'true image' of the object.

Of course, the time-of-flight of a pulse is strongly influenced by the speed of sound along its path. In the conventional approach to reconstructive ultrasound tomography one assumes that the apertures of the transducers are ideal points and that the detected portions of the pulses travel along the straight lines connecting these points at the respective positions. Under these assumptions one is led to describe that influence by the equations

$$\int_{\bar{\gamma}(\zeta^{klm})} \bar{n}\, ds = c_o T_{klm} + \bar{\epsilon}^{mod}_{klm} + \epsilon^{exp}_{klm} , \qquad (1)$$

$$k = 1,\ldots,K , \quad l = 1,\ldots,L , \quad m = 1,\ldots,M .$$

Here, the left hand side means the line integral of the true image $\bar{n}$ along the straight line $\bar{\gamma}(\zeta^{klm})$ connecting the centres of the emitting and receiving apertures at position ζ^{klm}. The ϵ's are small error terms which are often omitted in the literature but in fact must be added to make the equations (1) true. More precisely, the 'model error' $\bar{\epsilon}^{mod}_{klm}$ is to account for the error which comes from the fact that the assumptions leading to (1) are not exactly satisfied, and the 'experimental error' ϵ^{exp}_{klm} is to absorb the inaccuracies in measuring the time-of-flight of a pulse. In general, these error terms are unknown, and the model error is not random.

Associated with the equations (1) is the integral operator $\bar{M}$ defined by

$$(\bar{M}n)(\zeta) := \int_{\bar{\gamma}(\zeta)} n\, ds \quad . \qquad (2)$$

This operator represents the conventional model of reconstructive ultrasound tomography. $\bar{M}$ is closely related to the Radon transform: Because the lines $\bar{\gamma}(\theta,\xi_1,\xi_3)$ are completely contained in the horizontal plane through $(0,0,\xi_3)$, the function

$$f := n-1 \tag{3}$$

satisfies

$$R(f(\cdot,\cdot,\xi_3)) = (\bar{M}f)(\cdot,\cdot,\xi_3) \ , \tag{4}$$

where R denotes the Radon transform. Since $f(\cdot,\cdot,\xi_3)$ is uniquely determined by $R(f(\cdot,\cdot,\xi_3))$, it follows that n is uniquely determined by $\bar{M}n$. The experiment provides finitely many 'contaminated' samples of $\bar{M}n$ in form of the numbers c_oT_{klm}.

INVERTING THE CONVENTIONAL MODEL

As with reconstructive X-ray tomography it should therefore be possible to reconstruct, for each m, the slice $\bar{n}(\cdot,\cdot,\xi_3^m)$ of the true image from the measured data $T_{\cdot\cdot m}$, using (1) - (4). Mathematically, this endeavour amounts to estimate an unknown function $\bar{f}$ of two variables and zero outside some bounded domain, given finitely many numbers g_{kl} such that

$$|(R\bar{f})(\theta^k,\xi_1^l) - g_{kl}| \leq \epsilon_{kl} \ , \ \begin{matrix} k = 1,\ldots,K, \\ l = 1,\ldots,L. \end{matrix} \tag{5}$$

Here the ϵ_{kl} are supposed upper bounds for the errors in the data.

Although $\bar{f}$ is uniquely determined by $R\bar{f}$, the inequalities (5) do not specify $\bar{f}$ completely, simply because they do not even specify $R\bar{f}$. But one might hope that all the f's compatible with (5) are nevertheless close to $\bar{f}$. Unfortunately however, there exist rather arbitrary functions in this class (Smith et al., 1977). Hence we need additional information to rule out the unacceptable ones. Usually one postulates that the wanted function be 'smooth' or 'regular' in some sense. Methods to constrain the solutions of (5) in this way are referred to as regularization.

Of course we also need a 'reconstruction algorithm' to compute a regularized solution of (5). A number of well proven methods are available (Gordon and Herman, 1974). Seemingly they all belong to either of two classes. Algorithms of the first class find their results by solving properly discretized versions of the equation Rf = g, where g has been estimated from the g_{kl}. Such a discrete version consists of a linear system of equations which is advantageously solved by the

Kaczmarz method (Tanabe, 1971), also known as ART. Algorithms of the second class exploit the equation $f = R^+g$, where R^+ is the inverse of the Radon transform. This operator is explicitly known and hence the expression R^+g may be discretized and evaluated on a computer. The second class contains the computationally most efficient methods and in particular the filtered back-projection method. All reconstruction algorithms require some kind of discretization which gives rise to an additional discretization error. Also, some kind of regularization has been built into every successful reconstruction algorithm. As experience and to some extent also theory show, properly implemented algorithms of both classes give good approximations to $\bar{f}$, provided:

1. the sampling pattern of $R\bar{f}$ is sufficiently dense and even;
2. the errors caused by model, experiment and discretization are sufficiently small;
3. $\bar{f}$ itself is sufficiently 'regular'.

The better these conditions are satisfied, the better the computed solutions will estimate $\bar{f}$ and hence the more useful they will be in medical applications.

Algorithms of both classes have been applied to data g_{kl} derived from time-of-flight projections of phantoms and breasts in the manner described above (e.g. Carson et al., 1977; Glover and Sharp, 1977; Greenleaf and Johnson, 1978; Jakowatz and Kak, 1976). The results, however, do not show the fine details one is accustomed to e.g. in reconstructive X-ray tomography. Hence at least one of the stated conditions was not satisfied. Since in the reported cases sampling pattern and discretization were all right and the true images 'regular', the dominating reason must have been an intolerably large error caused by model and/or experiment.

To confirm this conclusion, we constructed an object with a known true image $\bar{n}$. Then we measured time-of-flight data T_{klm} of this object and evaluated $\bar{\tau} := M\bar{n}$. The true image is cylindrically symmetric and shown in fig. 2. In this case $\bar{\tau}$ is independent of θ and ξ_3. The function $\bar{\tau}(0,\cdot,0)$ is displayed and compared with the measured data $c_o T_{1\cdot 1}$ in fig. 3.

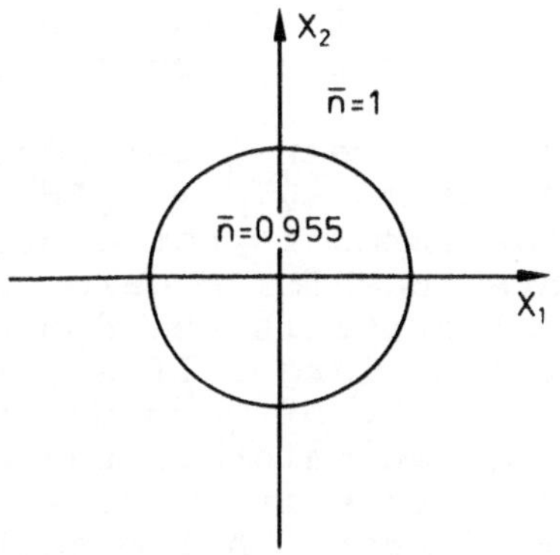

Fig. 2. The sample true image. This $\bar{n}$ depends on $(x_1^2 + x_2^2)^{1/2}$ only. The diameter of the circle is 35 mm.

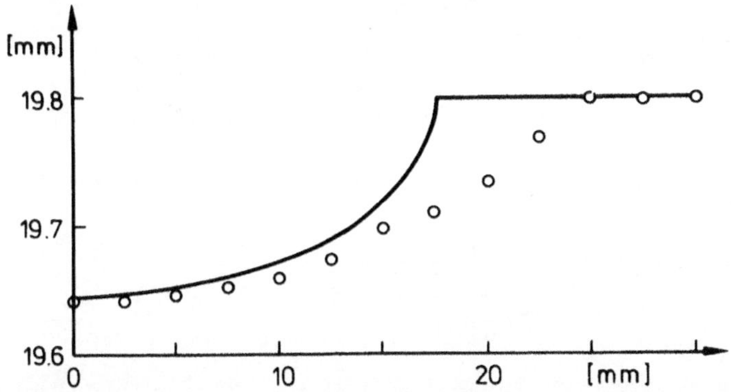

Fig. 3. Comparing experiment and conventional model. The full line illustrates the right half of the projection $(\bar{M}\bar{n})(0,\cdot,0)$ for the $\bar{n}$ shown in fig. 2. The circles mark the time-of-flight data actually measured with a phantom having that $\bar{n}$ as true image.

Indeed, experiment and conventional model (2) do not agree very well. A reconstruction of $\bar{n}$ based on the measured data and (2) would result in a shallow 'bowl' with a considerably larger diameter than the true 'pot'.

The discrepancy between experiment and model (2) is expressed by the terms $\bar{\epsilon}^{mod}_{klm} + \epsilon^{exp}_{klm}$ in (1). The individual size of these terms is not a priori clear, but there are many hints that for a carefully designed experiment the model error is the dominating one, whenever the total error is large. If so, we cannot really improve the image quality of reconstructive ultrasound tomography as long as we maintain the conventional model. Fortunately there exist models with a significantly reduced model error. In the following we shall exhibit one.

THE IMPROVED MODEL

We still consider emitting and receiving apertures as ideal points, but now assume that the detected portions of the pulses have travelled along the rays, in the sense of geometric acoustics, connecting these points. If there are several such rays, we take a fastest. With $\tilde{\gamma}(\zeta^{klm};\bar{n})$ denoting the selected ray at the position ζ^{klm}, we now are led to

$$\int_{\tilde{\gamma}(\zeta^{klm};\bar{n})} \bar{n}\, ds = c_o T_{klm} + \tilde{\epsilon}^{mod}_{klm} + \epsilon^{exp}_{klm} \,, \qquad \begin{array}{l} k = 1,\ldots,K \,, \\ l = 1,\ldots,L \,, \\ m = 1,\ldots,M \,. \end{array} \tag{6}$$

Associated with (6) are the integral operators $\tilde{\tilde{M}}$ and $\tilde{M}$, defined by

$$(\tilde{\tilde{M}}(n,f))(\zeta) := \int_{\tilde{\gamma}(\zeta;n)} f\, ds \tag{7}$$

and

$$\tilde{M}n := \tilde{\tilde{M}}(n,n) \quad .$$

$\tilde{M}$ is the model announced. The use of geometric acoustics has been suggested before, e.g. by Glover and Sharp (1977), Johnson et al. (1977), Jakovatz and Kak (1976), McKinnon and Bates (1980). The hope seems justified that $\tilde{M}$ leads to a smaller model error than $\bar{M}$ does, so that $\tilde{M}$ is in fact an improved model. To demonstrate how large the improvement can be, we also computed $\tilde{\tau} := \tilde{M}\bar{n}$ for the $\bar{n}$ of fig. 2. The function $\tilde{\tau}(0,\cdot,0)$ is displayed, together with the old $\bar{\tau}(0,\cdot,0)$ and the measured data $c_0T_{1,1}$, in fig. 4. The agreement between $\tilde{\tau}$ and the measured data has become much better. The re-

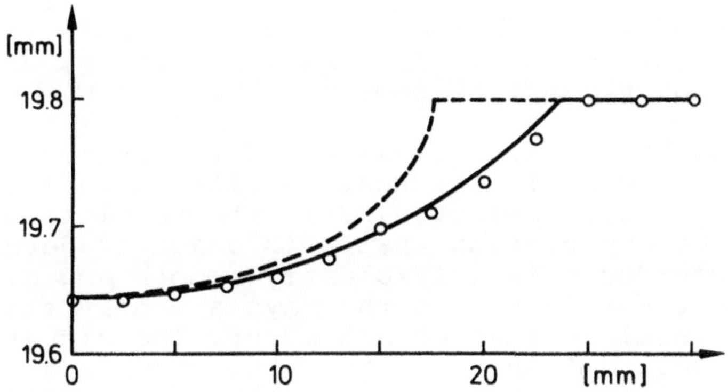

Fig. 4. Comparing experiment and improved model. The full line illustrates the right half of the projection $(\tilde{M}\bar{n})(0,\cdot,0)$ for the $\bar{n}$ shown in fig.2. The circles mark the same measured data as in fig. 3. The dashed line here corresponds to the full line in fig. 3.

maining discrepancy can to some extent be explained by the finite width of the transducers which in this case was 2.5mm: rays starting and ending at the edges of the transducers can be a little bit faster than the central rays. Indeed, if in fig. 4 we shift the measured data to the left by half the transducer width, the agreement with the predicted values becomes still better.

The operator $\tilde{M}$ is nonlinear, because the integration paths in (7) depend on n. However, $\tilde{M}$ is still linear in the second argument. If $n_o \equiv 1$, then $\tilde{M}(n_o,\cdot)$ is identical with the conventional model $\bar{M}$. Thus we can regard the conventional model $\bar{M}$ as the 'linearized' version $\tilde{M}(n_o,.)$ of the improved model $\tilde{M}$; the model error in (1) may be split up according to

$$\bar{\epsilon}^{mod}_{klm} = \tilde{\epsilon}^{mod}_{klm} + \epsilon^{lin}_{klm} ,$$

where $\tilde{\epsilon}^{mod}_{klm}$ is the model error in (6) and ϵ^{lin}_{klm} represents the linearization error. It can be shown that this error tends faster to zero than $n - n_o$ (Lavrentiev et al., 1970). Hence the conventional model may well be acceptable whenever $n - n_o$ is small enough. If n is independent of x_3, then the rays $\tilde{\gamma}(\zeta;n)$ remain in the horizontal plane containing their endpoints, and $\tilde{M}$ acts slicewise on n, just as does $\bar{M}$.

Slightly modified versions of the improved model have been studied in the Russian literature (e.g. Muhometov, 1977; Romanov, 1974). The results obtained justify the expectation that n is uniquely determined by $\tilde{M}n$, provided n is sufficiently smooth and close to a constant. For such an n the rays are only slightly bent and multiple rays do not occur. The situation changes, when n is not so. However, chances are that the true images representing breasts are sufficiently smooth and close to a constant.

INVERTING THE IMPROVED MODEL

So far we have been cutting down the model error, but we do not yet know how to actually obtain a better reconstruction of the true image. This problem has been treated at some length by Schomberg (1980). Here we only give the ideas.

The inverse operator of $\tilde{M}$ is not explicitly known, and there is no chance to alter this. We therefore decided to discretize the equations (6) and then to solve, in some way or another, the resulting system of equations.

There is no unique way of discretizing (6), but it should be done with care to achieve a small discretization error. In matrix-vector notation, the discrete version of (6) will take the form

$$A(\bar{u})\bar{u} + b(\bar{u}) = s + \epsilon \quad . \tag{8}$$

In this equation, $\bar{u}$, $b(\bar{u})$, s, and ϵ are vectors, and $A(\bar{u})$ is a matrix. The vector $\bar{u}$ represents a discrete version of the true image $\bar{n}$, s contains the measured data, and ϵ is the total error caused by model, experiment and discretization. (8) summarizes K·L·M individual equations. Each of these may be written in the form

$$a_i(\bar{u})^T\bar{u} + b(\bar{u}) = s_i + \epsilon_i \quad , \tag{9}$$

where $a_i(\bar{u})^T$ denotes the i-th row of $A(\bar{u})$. The left hand side in (9) is a discrete version of one of the line integrals in (6). $A(u)$ and $b(u)$ depend on u but can be computed from u. The evaluation of $a_i(u)^T$ and $b_i(u)$ requires the numerical computation of a ray connecting two given points.

In (8), $\bar{u}$ and ϵ are unknown, but ϵ is hopefully small. So we try to find an approximation to $\bar{u}$ from the approximating equation

$$A(u)u + b(u) = s \quad .$$

If we happen to know a good approximation u^o to $\bar{u}$, then we might content ourselves with an approximate solution of the linearized equation

$$A(u^o)u + b(u^o) = s \quad . \tag{10}$$

This system can be solved iteratively using the Kaczmarz method with u^o as initial guess. Let u^1 denote an approximate solution of (10). If u^1 is already good enough, we can stop. Otherwise we may evaluate $A(u^1)$ and $b(u^1)$ and find a possibly better u^2 as approximate solution of

$$A(u^1)u + b(u^1) = s \quad , \tag{11}$$

say again with the Kaczmarz method and with u^1 as initial guess. We could continue in this way. Such a method has been proposed by Johnson et al. (1977). Alternatively, we could find a (different) u^2 as approximate solution of the linear equation

$$A(u^o)u+b(u^o) = s+A(u^o)u^1+b(u^o)-A(u^1)u^1-b(u^1) . \tag{12}$$

Here the left hand side is identical with that of (10), but the right hand side has changed. This procedure may be iterated as well. Such a strategy has been proposed by McKinnon and Bates (1980).

If u^o is the discrete version of $n_o \equiv 1$, then the left hand side in (10) or (12) is a discrete version of the conventional model. In this case we can find approximate solutions of (10) and (12) by the faster filtered back-projection method.

When properly implemented, both strategies might work. In the r-th step, both require the evaluation of $A(u^{r-1})$ and $b(u^{r-1})$, and hence in particular the computation of K·L·M rays, which is the dominating load.

We suggest a scheme which does not avoid the computation of the rays but possibly makes a better use of them. Our method is a straightforward, nonlinear modification of the Kaczmarz method. It works like this:

Choose u^o;
for r = 1,2,... do:
  if u^{r-1} is all right, then stop. Otherwise:
  $v^o := u^{r-1}$;
  for i = 1,..., K·L·M do:

$$v^i = v^{i-1} - \frac{a_i(v^{i-1})^T v^{i-1} + b_i(v^{i-1}) - s_i}{a_i(v^{i-1})^T a_i(v^{i-1})} a_i(v^{i-1}) \; ; \qquad (13)$$

  end;
  $u^r := v^{K\cdot L\cdot M}$;
end.

Regularization is accomplished by an additional smoothing procedure.

If $a_i(v^{i-1})$ and $b_i(v^{i-1})$ in (13) are replaced by $a_i(u^{r-1})$ and $b_i(u^{r-1})$ throughout, then the inner loop of this method reduces to one cycle of the linear Kaczmarz method for the equation $A(u^{r-1})u+b(u^{r-1}) = s$, and we obtain a method of the type (11). However, in (13) always the latest information is used in updating the iterates. Hence this scheme can be expected to be faster, provided it converges at all.

Under mild conditions, the sequence of the u^r generated by the method (13) does converge. This is ultimately due to Fermat's principle obeyed by the rays. See Meyn (1980) and Schomberg (1980) for details.

To investigate the efficiency of the method (13) we applied it to simulated and real time-of-flight data. So far the true images chosen have been independent of x_3. The problem then reduces to two dimensions. The real data were obtained with phantoms consisting of condomes and finger cots filled with saline. The true image of one of our phantoms is sketched in fig. 5a). Time-of-flight data were measured at K = 75 equidistant angles and L = 64 equidistant lateral positions at each angle. Fig. 5b) shows a reconstruction of the true image based on the measured data and the conventional model. This reconstruction is overlaid with artifacts contaminating the finer details of the true image. Also, the diameter of the central spot is larger than it should be. Fig. 5c) shows a reconstruction based on the same data and the improved model. The artifacts of this image are so small that the low contrast spots can be identified, and the diameter of the central spot is now correct.

Our experience with the improved model and the modified Kaczmarz method for its numerical inversion indicates that good reconstructions are obtainable under essentially the same conditions that were stated above for the conventional model, as long as the true image is sufficiently close to a constant. However, the computational effort for inverting the improved model is definitely higher than that for the conventional model. This is the price for the reduction of the model error.

CONCLUDING REMARKS

Finally we try to estimate the feasibility of the time-of-flight version of reconstructive ultrasound tomography for breast imaging.

The experience gained with the conventional model and the arguments of the first half of this paper show that one cannot expect ever to obtain sufficiently good images based on the conventional model. Highly accurate time-of-flight measurements are necessary, of course, but not sufficient.

On the other hand, the improved model discussed here offers the chance to obtain much better images. It will certainly be necessary to use transducers with

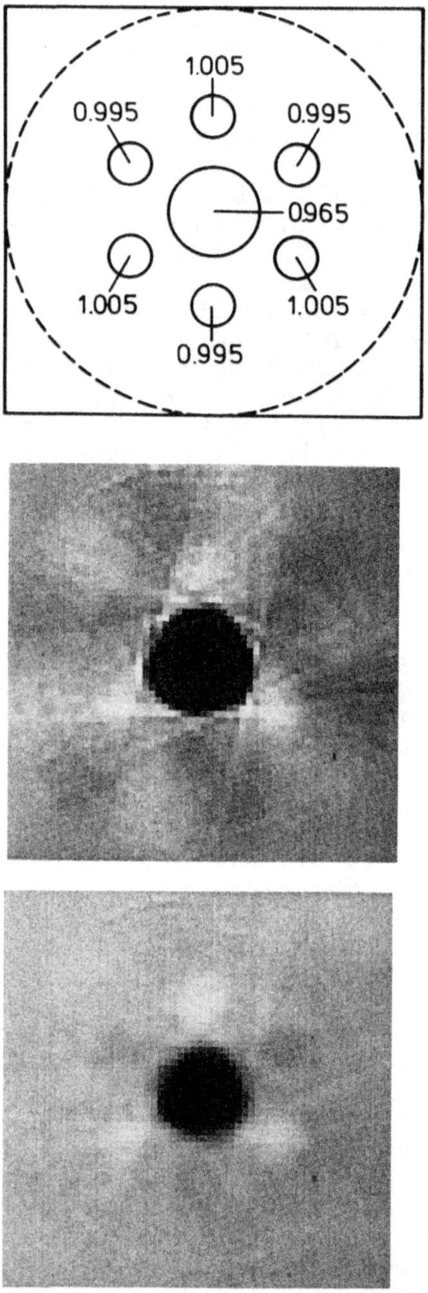

a)

b)

c)

Fig. 5. Halftone pictures of true and reconstructed images: (a) true image, (b) reconstruction based on conventional model, (c) reconstruction based on improved model.

small apertures, an accurate detection scheme for the first arrival of a pulse, and the three-dimensional version of the model. But then sufficiently accurate images may well be obtainable. The computation times can be cut down by using a proper high-speed special purpose computer.

ACKNOWLEDGEMENT

The author thanks his colleagues K. H. Meyn, H. J. Schneider and M. Tasto for discussions of the mathematical and technical aspects, and K. Benthien and P. Klinger for writing the programs and conducting the numerical experiments.

REFERENCES

Carson, P. L., Oughton, T. V., Hendee, W. R., Ahuja, A. S., 1977, Imaging Soft Tissue through Bone with Ultrasound Transmission Tomography by Reconstruction, Med. Phys. 4, 302-309.

Glover, G. H., Sharp, J. C., 1977, Reconstruction of Ultrasound Propagation Speed Distributions in Soft Tissue: Time-of-Flight Tomography, IEEE Trans. Sonics & Ultrason. SU-24, 229-234.

Gordon, R., Herman, G. T., 1974, Three-Dimensional Reconstruction from Projections: A Review of Algorithms, Int. Rev. Cyt. 38, 111-151.

Greenleaf, J. F., Johnson, S. A., 1978, Measurement of Spatial Distribution of Refractive Index in Tissues by Ultrasonic Computer Assisted Tomography, Ultrasound Med. & Biol.3, 327-339.

Jakowatz, C. V., Kak, A. C., 1976, Computerized Tomographic Imaging Using X-Rays and Ultrasound, School of Electrical Engineering, Purdue Univ., West Lafayette, Indiana, Technical Report TR-EE 76-26.

Johnson, S. A., Greenleaf, J. F., Tanaka, M., Flandro, G., 1977, Reconstructing Three-Dimensional Temperature and Fluid Velocity Vector Fields from Acoustic Transmission Measurements, ISA Transactions 16, 3-15.

Lavrientiev, Romanov, Vasiliev, 1970, 'Multidimensional Inverse Problems for Differential Equations', Springer, New York.

McKinnon, G. C., Bates, R. H. T., 1980, A Limitation on Ultrasound Transmission Tomography, Ultrasonic Imaging 2, 48-54.

Meyn, K. H., 1980, A Generalization of a Theorem of Ostrowski and its Application to a Nonlinear Extension of the Method of Kaczmarz, to be published.

Mueller, R. K., Kaveh, M., Wade, G., 1979, Reconstructive Tomography and Applications to Ultrasonics, Proc. IEEE 67, 567-587.

Muhometov, R. G., 1977, The Problem of Recovery of a Two-Dimensional Riemannian Metric and Integral Geometry, Soviet. Math. Dokl. 18, 27-31.

Romanov, V. G., 1974, On the Uniqueness of the Definition of an Isotropic Riemannian Metric inside a Domain in Terms of the Distances between Points of the Boundary, Soviet Math. Dokl. 15, 1341-1344.

Schomberg, H., 1980, Nonlinear Image Reconstruction from Projections of Ultrasonic Travel Times and Electric Current Densities, in "Mathematical Aspects of Computerized Tomography', Herman, G. T. and Natterer, F. eds., Springer, New York, to appear.

Smith, K. T., Solmon, D. C., Wagner, S. L., 1977, Practical and Mathematical Aspects of the Problem of Reconstructing Objects from Radiographs, Bull. Amer. Math. Soc. 83, 1227-1270.

Tanabe, K., 1971, Projection Method for Solving a Singular System of Linear Equations and its Applications, Numer. Math. 17, 203-214.

COMPUTERIZED ULTRASONIC TOMOGRAPHY BY ELECTRONIC SCANNING AND STEERING OF A RING ARRAY

M. Clément[+], P. Alais[++], J.C. Roucayrol[+] and J. Perrin[+]

[+] Lab. Biophysique, ERA 498, Univ. Paris-5.
[++] Lab. Mécan. Phys., ERA 537, Univ. Curie, St Cyr l'Ecole, France

ABSTRACT

In order to realize a rapid and accurate in-vivo scanner for breast imaging, a novel technique using a 450 elements -360° ring array has been developed. Every mechanical motions during recording have been thus entirely eliminated, all the tomographic projections being obtained by electronically scanning the array. The electronics implemented feature a computer-controlled steering of the ultrasonic beam as previously reported and an electronic ring multiplexer having now reached completion. Although data are recorded sequentially with one projection at a time, tomographic records can be achieved within two seconds by scanning first the tomographic angle θ (constant K mode) rather than the projection indice K (constant θ mode). Shorter recording times could be achieved with the realized multiplexer which can accept several recording channels for fan beam tomography. A 3.5 MHz ring array is reaching completion and has been built in five separate 72°-90 elements arrays which are then assembled to realize the whole ring array. One of this 90 elements sector has already been fully realized and its manufacturing steps are described. Tomographic computer reconstructions of a rubber phantom insonified with a 12 elements test array are reported.

INTRODUCTION

Ultrasonic echographic imaging has long been used in the diagnosis of breast tumors. Indeed, the first reported applications have been published by Wild and Neal (1951). Numerous improvements in technology have been made since these early works but there still remains difficulties to identify and accurately measure tumors and cysts respectively smaller than 1 cm and 3 mm (Griffiths, 1978; Dale et al, 1980).

An alternative approach to echographic imaging can be found in computerized transaxial tomography. If there is with ultrasound the adverse effect of refraction and diffraction, tomography can nevertheless be applied to relatively homogeneous medium such as breast tissues, where the variations of ultrasound velocities and impedances are kept to a minimum. Quantitative ultrasonic imaging of breast tissues has clearly been demonstrated by the study of ultrasound attenuations (Chivers and Hill, 1975; Wells, 1975; O'Brien, 1977) and ultrasound velocities in breast tissues (Kossof et al, 1973; Greenleaf et al, 1974, 1975, 1980; Glover, 1977). These studies have shown that the ultrasound attenuation is small in fatty tissues, medium in parechymatous tissues and large in collagenous and finally calcified tissues.Similarly ultrasonic velocities have been found small in fatty tissues, intermediate in parenchymatous tissues and high in collagenous and calcified tissues (Kossof et al, 1973 ; O'Brien, 1977; Glover, 1977). Furthermore, if most breast tissues have shown a relatively strong correlation between velocity and attenuation, infiltrating medullary carcinomas have on the contrary shown high velocities (Greenleaf et al, 1980; Glover, 1977) but low attenuation (O'Brien, 1977). Benign lesions such as fibroadenoma and fluidic cyst showed small velocities while scirrhous carcinoma exhibited the highest attenuation and velocity (O'Brien, 1977). Clearly, all these investigations strongly indicate that breast tissues could be characterized with a two-dimensional quantitative reconstruction of their distribution of acoustic refractive index and attenuation.

In the first reported works on ultrasonic transaxial tomography (Greenleaf et al, 1975, 1980; Glover 1977) data recording was carried out with mechanical scanners. Besides the lost of time this implies, mechanical scanning introduces vibration and position errors on the ultrasonic transducers which cannot entirely be eliminated. In order to avoid all these mechanical movements during scanning, tomographic data recording using a 360° ring array has been proposed (Clément and Alais, 1978 ; Clément et al, 1979). With such an array, all the required projections can be obtained by an electronic multiplexing of the transducer elements of the array. As well as eliminating vibration errors, this offers also the advantage of fast scan for real time recording and imaging. It is also particularly well suited for a breast imaging scanner, as it can be easily fitted around the breast patient.

This paper describes the various techniques which have been developed for the realization of an in-vivo breast imaging scanner using a ring array. The multiplexing method used to scan such an array for transaxial tomographic records is described. The progress achieved in the construction of a 450 elements ring array are reported.

METHOD

In conventional recording of a transaxial tomogram, one has to perform a rectilinear scan of two transducers on each side of the tissue being investigated. Propagation times or attenuations through the tissue are measured at different K positions along the rectilinear scanning loci AB and A'B' at angle Θ (figure 1). Then the scanning rig is rotated to another angle to record a new set of different K projections.

The equations governing the propagation time T (k,Θ) and the attenuation Io/I (k,Θ) between the transmitter at T and the receiver at R are respectively :

$$T(k,\theta) = \sum_{i,j\in[k,\theta]}^{N} \frac{1}{V_{ij}} l_{ij} + \frac{1}{V_w}\left[D - \sum_{i,j\in[k,\theta]}^{N} l_{ij}\right] \qquad (1)$$

$$\mathrm{Log}\,\frac{I_o}{I(k,\theta)} = \sum_{i,j\in[k,\theta]}^{N} \alpha_{ij} l_{ij} + \alpha_w\left[D - \sum_{i,j\in[k,\theta]}^{N} l_{ij}\right] \qquad (2)$$

where D is the distance between the transducers, N the total number of pixels in the object grid, Vw and Vij respectively the ultrasound velocities in water and within the object pixel i, j, α_w and α_{ij} respectively the attenuation coefficients in water and within the object pixel i, j. Equations (1) and (2) can be rearranged as follows :

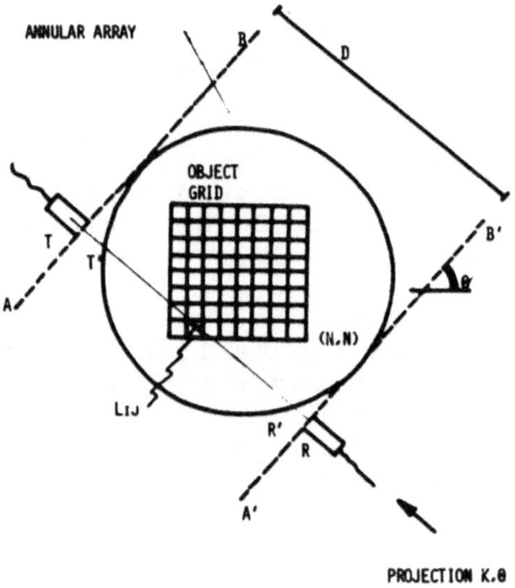

Fig. 1. Geometry of tomographic record with a ring array

$$\Delta T(k,\theta) = \left[T(k,\theta) - \frac{D}{V_w}\right] = \sum_{i,i\in[k,\theta]}^{N} \left[\frac{1}{V_{ij}} - \frac{1}{V_w}\right] l_{ij} \tag{3}$$

$$P(k,\theta) = \left[\text{Log}\,\frac{Io}{I(k,\theta)} - \alpha_w\, D\right] = \sum_{i,j\in[k,\theta]}^{N} \left[\alpha_{ij} - \alpha_w\right] l_{ij} \tag{4}$$

The two systems of equations (3) and (4) can be solved using for example matrix inversion techniques or iterative methods such as one of the well known ART algorithms. The differential time of flight $\Delta T(k,\theta)$ will lead to a reconstruction of the refractive indices in each pixel relative to water $(1/V_{ij}-1/V_w) \# \Delta V_{ij}/V_w^2$, while P (k,θ) will lead to a reconstruction of the difference of attenuation coefficients in each pixel and water $(\alpha_{ij}-\alpha_w) = \Delta\alpha_{ij}$ .If times of flight or attenuations are now recorded along a ring array with the transmitter and the receiver respectively at T' and R' (figure 1), the differential time of propagation $\Delta T^R(k,\theta)$ and the attenuation measurement $P^R(k,\theta)$ can be espressed by the following equations :

$$\Delta T^R(k,\theta) = \left[T^R(k,\theta) - \frac{2R}{V_w} \sin\left(\frac{\pi}{N_R}|I-J|\right) \right] = \sum_{i,j\in[k,\theta]}^{N} \left[\frac{1}{V_{ij}} - \frac{1}{V_w}\right] l_{ij} \quad (5)$$

$$P^R(k,\theta) = \left[\mathrm{Log} \frac{Io}{I^R(k,\theta)} - 2\alpha_w R \sin\left(\frac{\pi}{N_R}|I-J|\right) \right] = \sum_{i,j\in[k,\theta]}^{N} [\alpha_{ij} - \alpha_w] l_{ij} \quad (6)$$

$k = 1,2,\ldots K$

$\theta = 1,2,\ldots \Theta$

Where R is the ring array radius, I and J the transmitter and receiver array indices ranging from 1 to N_R , the total number of array elements.

Comparison of equations (3) and (5) as well as (4) and (6) shows that again $\Delta Vij/Vw^2$ and $\Delta \alpha_{ij}$ can be reconstructed from ring array measurements of respectively time of flight and attenuation . There needs to be however to correct the times of propagation T^R (k,θ) and the ratio $\mathrm{Log}[Io/I^R(k,\theta)]$ measured along the ring array, by respectively $\frac{2R}{Vw} \sin\left(\frac{\pi}{N_R}|I-J|\right)$ and $2\alpha wR \sin\left(\frac{\pi}{N_R}|I-J|\right)$. These terms compensate in fact for the variation in times of propagation and attenuations in water along T'R' (k,θ). They can be substracted from the measured data at the reconstruction step or more conveniently in real time during recording using a look up table.

RING ARRAY ELECTRONIC SCANNING

There are several ways one can electronically scan a ring array for transaxial tomographic records.

1 Transmitter - N Receivers.

One way of scanning a ring array is to select a single transmitter element firing ultrasound within a fan beam and record N projections using N receivers simultaneously. (figure 2a). Each of these recorded data has to be stored and multiplexed before it is read by the computer. This mode of scanning is probably the fastest as N data are recorded using a single transmitted pulse. Typically, for a 5 KHz pulse repetition rate and N = 32, 10 000 data can be recorded in just over 62 ms which would be quite suitable for real time imaging. If this approach is very attractive from the point of view of scanning time, it is not so from the point of view of the limited ultrasonic power which can be transmitted out of a single array element. An alternative approach is to use several transmitting elements so as to improve somewhat on transmitted power and signal to noise ratio.

M Transmitters - 1 Receiver

In order to have a greater ultrasonic transmitted power, M array elements can de used as transmitters, a single element being used as the receiver. If transmitted power is roughly M times larger than in the previous approach, the whole data recording is N times slower and the insonifying beam is narrower. This may create diffi-

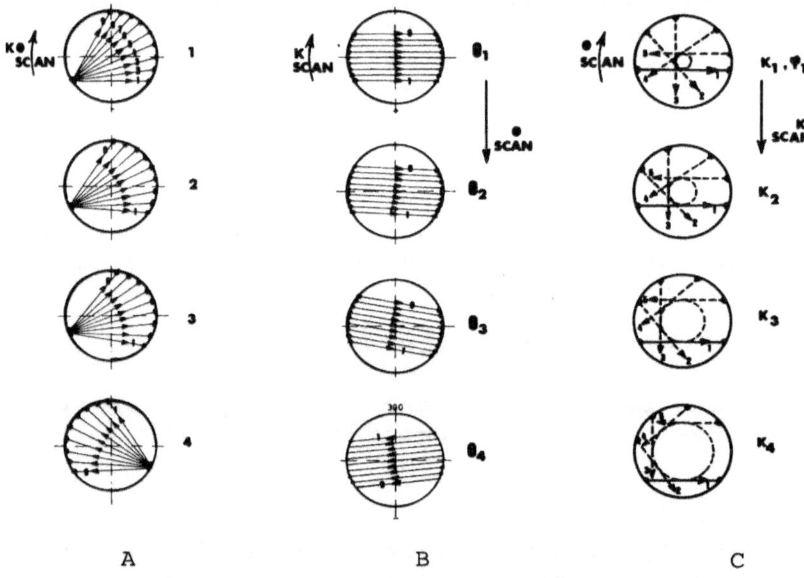

Figure 2. Some ring array scanning modes.

culties when using a ring array where the offset angle φ between a tomographic projection and the array surface varies between 0° and say + 40°. This difficulty can be mostly overcame using an electronic steering of the narrow beam in order to maintain a high level of transmitted power for large offset angles (Clément et al,1979). Reduced scanning times can be achieved by using N narrow beams, i.E. N x M transmitting and N receiving array elements. Howeyer N cannot be taken as large as in the fan beam mode, as the N narrow beams must not be ambiguously detected by several receivers. For a 5 KHz repetition rate an N = 3, recording time of 10 000 data could be achieved in the order of 700 ms.

There are two main fashions of scanning narrow beams using a ring array. One is to increment first the tomographic projection indice K for a given tomographic angle $\Theta 1$ and then increment Θ. This way of scanning which is illustrated in figure 2 B is the analog of raster scanning used in the first generation of mechanical X rays tomographic scanners. With that type of scanning, contiguously scanned projections have identical θ angles ("constant - θ" scanning mode) while the projection indice K and the offset angle φ differ. Furthermore, such scannings require separate multiplexers for transmitting and receiving elements as these are electronically scanned in opposite directions along the ring array.

A more attractive way of scanning a narrow beam along a ring array is described in figure 2 c. This time, it is the tomographic angle which is scanned first from 0° To 180° at constant $K = K_1$("constant-K" mode), K being then incremented before recording another set of projections. The transmitting and receiving array elements are now scanned in the same direction along the ring array, so that a single multiplexer can be used. Moreover, contiguously scanned projections (say for k = ki) have identical offset angles φ_i as well as identical transmitter - receiver distances D_i . That property is most useful in systems where the firing offset angles and the range gating of received signals are computer controlled. Indeed, a considerable input-output computer time is saved this way by latching these two variables only every block of data of identical K_i (and φ_i and D_i) rather than for every single projection as it would be the case in the "constant- θ " mode of scanning of figure 2B.

EXPERIMENTAL SYSTEM

The computer controlled electronics which has been constructed for steering and multiplexing a ring array is described in figure 3. It includes a mini computer Hewlett Packard 9835 A with a 64 K byte memory interfaced onto the electronic system via a single 16 bit interface. Before each record of blocks of 100 data having the same projection indice K (refered as "constant-K" scanning mode in the previous chapter), 4 latches L 1 - L4 are succeedingly loaded with status words using fast direct memory access output technique (table 1). The first latch L 1 stores on 9 bits the firing angle magnitude and sign. Latch L 2 stores on 14 bits the range gating delay ΔT as well as the amplifier gain G. Finally, latches L3 and L4 store a 22 bits word which is used by a 450 outputs ring multiplexer to select the firing group of transmitters as well as the receiver corresponding to an initial projection. A real time control of status words 3 and 4 gives a great versatility in scanning as the number of transmitters and receivers can be programmed. Fan beam scannings,as well as single or multi-narrow beam scannings can thus be easily selected simply by software instrutions.

Table 1. Status word.

WORD NB	FUNCTION	Nb of BITS
1	Steering angle	9
2	Range delay Gain	8 6
3	Nb of TX Nb of RX RX INDEXING	3 3 8
4	TX INDEXING	8

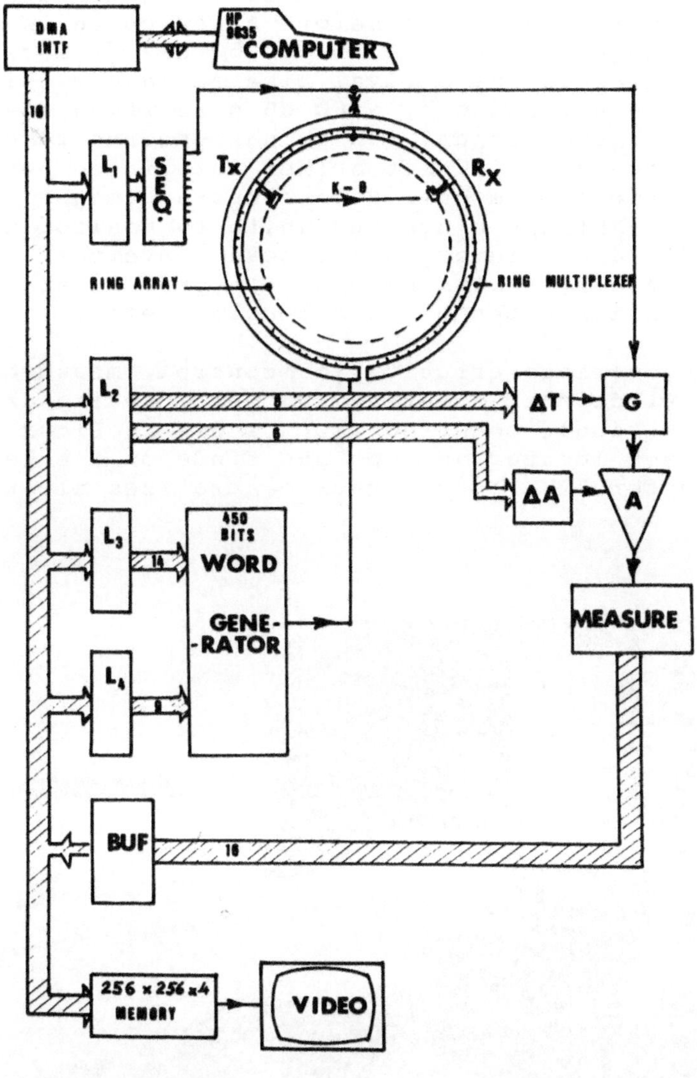

Figure 3. System architecture.

Once the 4 status latches and the ring multiplexer have been initialized, a read D.M.A. transfer instruction is executed to start the fast record of a block of constant K data. A 5 kHz clock is used to fire every 200 µs an ultrasonic pulse. Then the 450 channels ring multiplexer selects another projections before a new pulse is transmitted. A triggering circuit and a 250 MHz digital counter are used to measure the arrival time of each received ultrasonic pulse amplified by a 60 dB gain. Times-of-flight up to 655,35 microseconds with 10 nanoseconds resolution can be latched into a 16 bit buffer interfaced between the counter and the system bus. Reconstructed images are displayed onto a Philips black and white tv monitor through a 256 x 256 x 4 bit memory and a scan converter. This video memory is actually 4 bits deep for 16 grey levels imaging but can be easily extended to 256 color levels.

All the various circuits for control, measurement , interface, electronic steering and multiplexing are located within a single mounting rack shown in Figure 4. In the first rack located on top one finds on the left the 250 MHz counter. The second rack centralizes all the

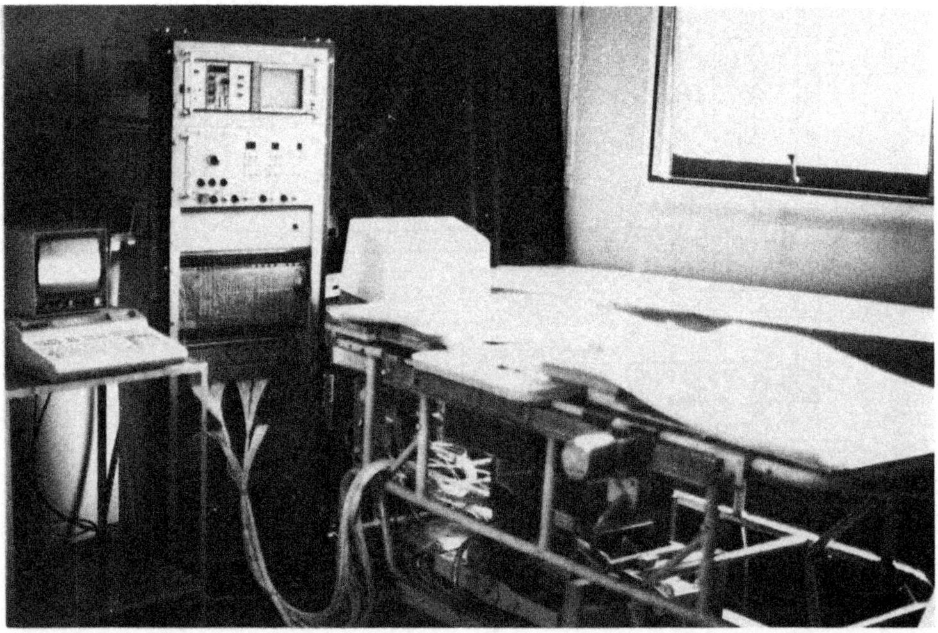

Fig. 4 Experimental apparatus for breast imaging.

controls of the in-vivo breast examination bed, including 3 programmable stepping motor controls for x,y,z automatic array positionning. It includes also water temperature control and pump control for emptying the immersion tank. A third rack contains the video memory and will be able to accept a 4 bit extension for a 256 color level display . All the various computer control and interface boards are located within the fourth rack, which also includes the 4 status latches, the electronic steering circuits, the rf amplifier and the 16 bits digital counter latch. The 450 channels ring multiplexer is located onto the last two racks, each countaining 26 all c-mos multiplexer boards packed with high density spacing (10.5 mm).

RING ARRAY CONSTRUCTION

In a previous article (Clément et al, 1979), the various fabrication stages of a cylindrical ring array was described. Using a similar manufacturing procedure, the construction of a second prototype has been initiated and is now reaching completion. This new ring array features now a toroidal shape which will give an additional slice focusing superposed onto the lateral focusing already provided by the previous cylindrical ring array.

In principles, the construction of a toroïdal ring array do not greatly differ from a cylindrical one, apart from the casting moulds which must have different shapes. There are however two main difficulties which are introduced. First, the manufacturing requirements onto the piezo-electric ceramics are more stringent, as it is not easy to obtain numerous spherical sectors with an accurate radius of curvature. Secondly, one must use special purpose tools for sawing the inter-electrode grooves of the array, which are no longer flat as for cylindrical arrays. These difficulties are not at all insurmontable, they simply will tend to increase somewhat the manufacturing cost of the array.

The toroïdal ring array currently under construction is made of five entirely self contained 72° toroïdal sectors of 90 transducer elements. Once realized, these sectors will be assembled to make the complete toroïdal array. One of these sectors has already been entirely completed. It was built using a loaded araldyte toroidal sector which the ceramics were fasten with glue before the electrodes were sawn. Each electrode was then soldered to a mini-coaxial cable (figure 5A). The sector was then completed by casting its toroidal front face with araldyte (figure 5B) which was de-gased under a primary vacuum.

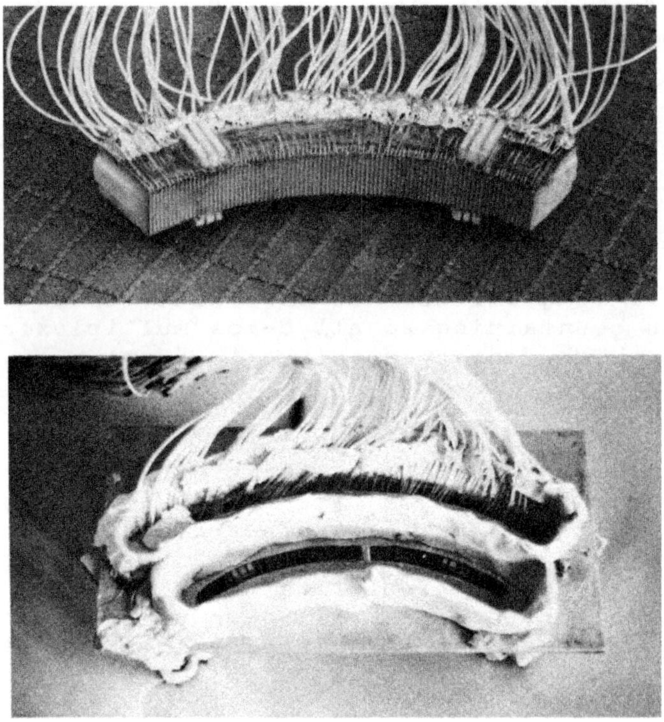

A

B

Figure 5. Manufacturing stages of a toroïdal ring sector

Figure 6. 72° - 90 éléments toroïdal ring sector for tomographic breast imaging.

Figure 6 shows the toroidal 72° sectorial array in its final stage. The other 4 sectors will be realized following the same procedure, the front face thickness being carefully controlled from one sector to another. The nominal frequency of this sector has been selected to 3.5 MHz, each electrode being 18 mm high, 0,53 mm wide and with an angular spacing along the circumference of 0.8 degree. It has a total number of 90 transducing elements so that the complete ring array will have 450 line transducers.

EXPERIMENTAL RESULTS.

In order to access experimentally a ring array in terms of the image quality obtained from a fan beam (single transmitter) or a narrow beam (several transmitters) tomographic record, an electronic steering of the ultrasonic beam has been implemented with computer controlled time delays. The kind of steering which has been obtained is illustrated in figure 7 showing the received signals at 0° and 30°, respectively without steering (figure 7a) or with a 30° steering (figure 7b).

As the number of sequenced transmitters increases, the electronic steering becomes more effective, the beam

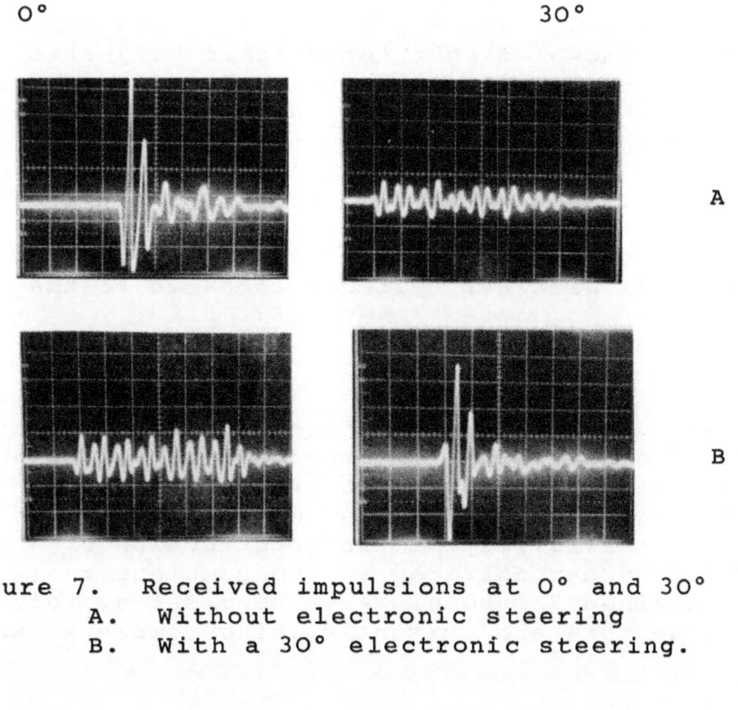

Figure 7. Received impulsions at 0° and 30°
A. Without electronic steering
B. With a 30° electronic steering.

is less divergent in the Fresnel region, and the transmitted power increases. However, at constant element spacing, the width of the beam in the central region of the ring array decreases in a first time with the number of transmitters, but increases in a second time if this number is increased further. This suggests that there may be an optimum number of line transmitters which can be used to record a given tomographic projection, in terms of transmitted power and signal-to-noise ratio in one hand and image resolution in the other one.

We have not been able yet to carry out experiments with the complete ring array from which we could experimentally find out the influence of the number of steered transducers onto the reconstructed image quality. However, preliminary results of images reconstructed numerically using an A.R.T. algorithm of a rubber phantom seem to indicate that refraction effects are reduced for certain ultrasonic beam shapes and dimensions.

The phantom used for these tomographic reconstructions was realized with two concentric cylinders made out of rubber with ultrasonic velocity closed to that of water (figure 8A). As the object was with circular symmetry, one serie only of parallel projections at constant tomographic angle was recorded by mechanically scanning a test array of 12 line transmitters and a diametrically opposed line receiver (F = 3.5 MHz, H = 18 mm, w = 0,53 mm). A complete set of time-of-flight tomographic projections was then obtained by numerically introducing for all other angles a noise of $\pm$ 5 ns. Comparison of the ART reconstructions obtained respectively from a single transmitter (figure 8B) and from several line elements grouped into a 7.5 mm wide transmitter (figure 8C), shows clearly the influence of the transmitted beam. Reconstructions obtained with wider transmitter groups exhibited poorer definitions of the inner 6 mm diateter cylinder compared to that of figure 8C.

CONCLUSIONS

A method for recording without mechanical movements ultrasonic computerized tomograms of the breast has been proposed. The developed techniques use a 450 elements toroïdal ring array and a programmable electronic multiplexer. The feasibility of the construction of such an array has been demonstrated by the complete realization of a self contained 72°-90 elements toroïdal sector. Other 4 similar sectors are currently being completed and will

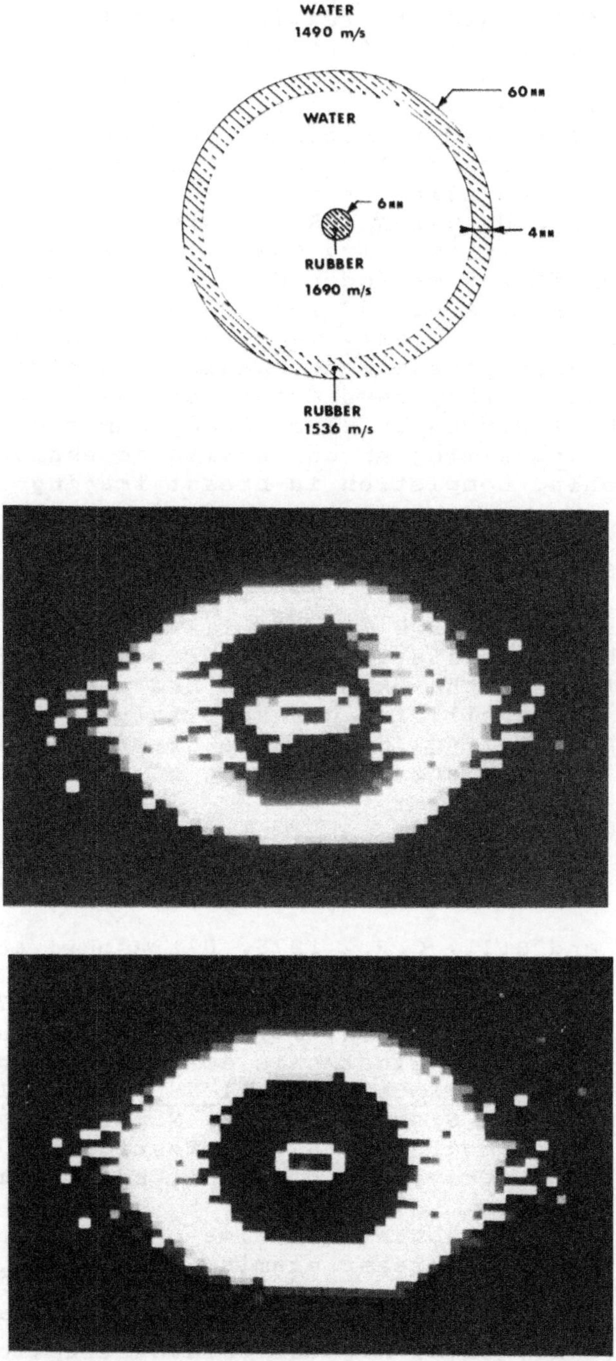

A

B

C

Figure 8. Rubber phantom and video displays of its 3.5 MHz tomographic reconstructions.

be assembled side by side to realize the full 360° ring array. The system described features a 450 channel programmable multiplexer allowing a computer control of the scanning methods suitable for fan beam, single or multi-narrow beam transaxial tomographic records. It features also an electronic steering of the transmitted beam. The computer interface implemented allows the record of 10 000 projections within 2 seconds using DMA transfers of data blocks of same k indices.This is quite suitable for in-vivo breast imaging although it can be reduced at least two times by the implementation of a second recording time-of-flight measuring channel. Preliminary results on tomographically reconstructed images of a rubber phantom insonified at 3,5 MHz with a test array have suggested that the image quality might be improved by electronic beam steering. Our future work will be aiming at an in-vivo accessment of the ring array reaching completion in breast imaging applications.

ACKNOWLEDGMENTS

One of the author (M.C.) would like to thank most sincerely the technical assistance of M. Fruchard for mechanical construction of the examination bed and casting moulds, and M. Cassanelli for his help in printed circuit lay out and manufacturing. He is also greatly indebted to Mrs A. Clement for component soldering of the multiplexer cards.

REFERENCES

Chivers, R.C., and Hill, C.R., 1975, Ultrasonic attenuation in human tissue, Ultr. Med. & Biol,2 : 25-29.

Clément, M. and Alais, P., 1978, French Patent n°78 18 424.

Clément, M., Alais, P., Perrin, J., 1979, Ultrasonic computed tomography by electronic scanning of an annular array, Ultrasonics International 79 Proc., 511-517.

Dale, G., Gairard, B., Gros, D., 1980, Numerical analysis of breast echotomograms, in : "Investigative Ultrasonology", C.R. Chivers and Alvisi, ed., Pitman medical Press, London.

Glover, G. H., 1977, Computerized time-of-flight ultrasonic tomography for breast examination, Ultra. Med & Biol, 3 : 117-127.

Greenleaf, J. F., Johnson, S.A., Lee, S.L., Herman, G.T., and Wood, E.H., 1974, Algebraic reconstruction of spatial distributions of acoustic absorption within tissue from their 2-D acoustic projections, in : "Acoustical Holography," vol 5, P.S. Green, ed.,

Plenum Press, New York.

Greenleaf, J.F., Johnson, S.A., Samayoa, W.F., and Duck, F.A., 1975, Algebraic reconstruction of spatial distributions of acoustic velocities in tissue from their time-of-flight profiles, "Acoustical Holography", Vol 6, N. Booth, ed., Plenum Press, New York.

Greenleaf, J.F., Kenue S.K., Rajagopalan, B., Bahn, R.C., and Johnson, S.A., 1980, Breast imaging by ultrasonic computer assisted tomography, "Acoustical Imaging", Vol 8, A. Metherell, ed., Plenum Press, New York.

Griffiths, K., 1978, Ultrasound examination of the breast, Med. Ultr., 2 : 13-19.

Kossof, G., Fry, EM., and Jellins, J., 1973, Average velocity of ultrasound in the human female breast, J.A.S.A., 53 : 1730-1736.

O'Brien, W.D., 1977, The role of collagen in determining ultrasonic propagation properties in tissue, "Acoustical Holography", Vol 7, L.W. Kessler, ed., Plenum Press, New-York.

Wells, P.N.T., 1975, Absorption and dispersion of ultrasound in biological tissue, Ultr. Med. & Biol, 1 : 369-376.

Wild, J.J., and Neal, D., 1951, The use of high frequency ultrasonic waves for detecting changes of texture in living tissues, Lancet 1 : 655-657.

EXPERIMENTAL RESULTS OF COMPUTERIZED ULTRASOUND ECHO TOMOGRAPHY

G. Maderlechner, E. Hundt, E. Kronmüller,
and E. Trautenberg

Siemens AG, Forschungslaboratorien
Otto-Hahn-Ring 6
D-8000 Müchen 83, Germany

ABSTRACT

In a series of experiments on test objects and excised organs we investigated the influence of several physical effects on image quality in the procedure of computerized ultrasound echo tomography. The main results are the following: 1. High resolution of less than one wavelength is demonstrated with simple test objects. This resolution is independent of direction and constant over the whole image plane. 2. Image distortions by interference of waves, well known from B-scans, are strongly reduced. 3. The sensitivity and dynamic range are high and are shown at objects with high contrast. 4. Images of excised organs have essentially better quality than conventional B-scans.

The procedure of computerized tomography is modified for ultrasonic features. Several scanning modes to form the projections are presented. Filtering by convolution is applied in one or two dimensions. Lateral information is included resulting in a superposition of B-scan like images. The image quality is essentially improved.

INTRODUCTION

From all types of ultrasonic imaging the B-scan method is most popular. In medical applications it is of high diagnostic value. But still there are disturbances, which reduce the image quality. We refer to the following three distortions: The resolution in lateral direction is much less than in axial direction. Interference effects lead to disturbances like speckle or granulation in homogeneous scattering regions. Outlines of organs or curved surfaces often appear incompletely.

The aim of our work is to improve the quality of ultrasonic images by digital methods [1]. We apply the procedure of computerized tomography and show by experiments how these disturbances are removed.

We decided to use reflexion mode. This allows interpretation of the reconstructed images similar to well known B-scans. Furthermore reflexion mode has several advantages over transmission mode. This is demonstrated by experiments especially concerning resolution, dynamic range and image distortions.

The method of X-ray computerized tomography is adapted to ultrasound [2] . This is straightforward as long as the ultrasound beam is wide compared to the object [3,4]. But usually the beamwidth is much smaller than the object. In this case, lateral averaging is applied to produce the projections of a tomographic image [5]. If the lateral averaging is omitted, B-scan like images are recorded and these are processed to a tomographic image.

All three methods were developed in our laboratory and applied on several objects. We compare the methods and describe their influence on the image quality by means of our experiments.

PRINCIPLE OF TOMOGRAPHY AND SCANNING

Computer tomography is a reconstruction of a 2-dimensional image from a set of projections. For ultrasound four different procedures are explained in Figure 1:

1. The simple case of "infinite" beamwidth is shown in Figure 1a. The object is entirely covered by the beam of the transducer, which is used as transmitter and receiver. All reflection points on a circular arc around the transducer contribute to the echo peak. Therefore each peak of the echo sequence represents a projection along an arc. Thus the whole echo sequence is a 1-dimensional projection of the 2-dimensional cross section. A set of projection is recorded for various rotation angles φ , stepped in small increments $\Delta\varphi$, covering a full circle. The image is reconstructed from this set of projections by backprojection as well known from X-ray tomography. The reconstruction along circular arcs instead of straight lines does not cause severe problems. For convolution we use a modified Shepp-Logan kernel [6]. We can do this convolution alternatively before superposition at the one dimensional echo sequences or after superposition at the blurred two dimensional picture.

2. Usually the beamwidth is much smaller than the object. This case of "finite" beamwidth is shown in Figure 1b. We scan the

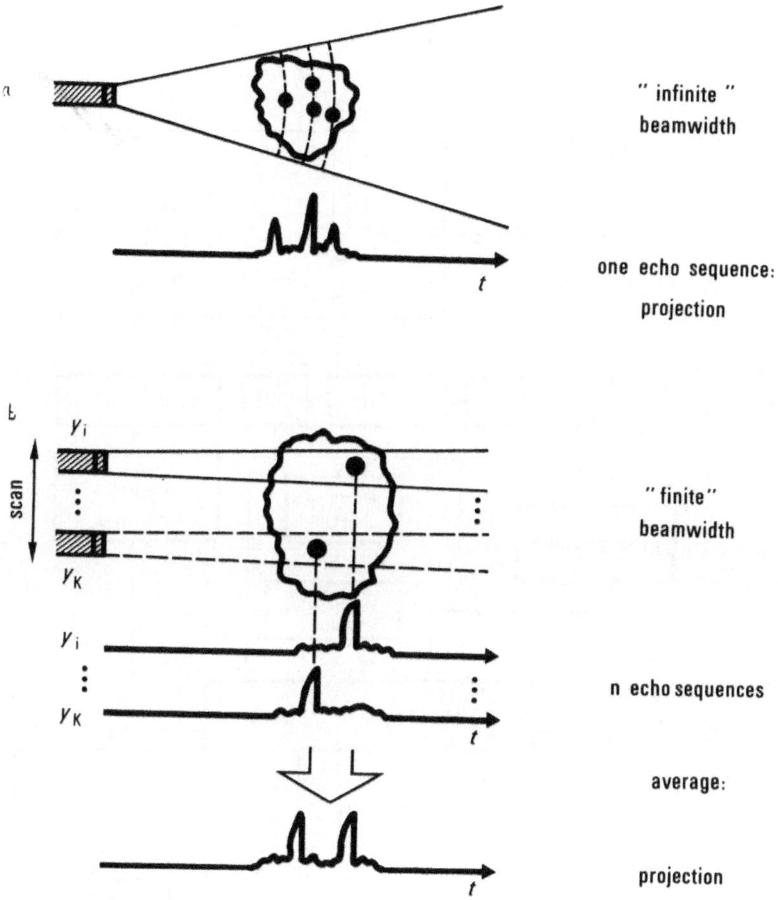

Fig. 1. Scanning modes of echo tomography: a) At "infinite" beamwidth the object is covered by beam. One echo sequence forms a projection. b) The transducer has finite beamwidth which is much smaller than the object. Lateral scanning and averaging of n echo sequences form the projection. These projections are measured for each rotation angle φ around the object.

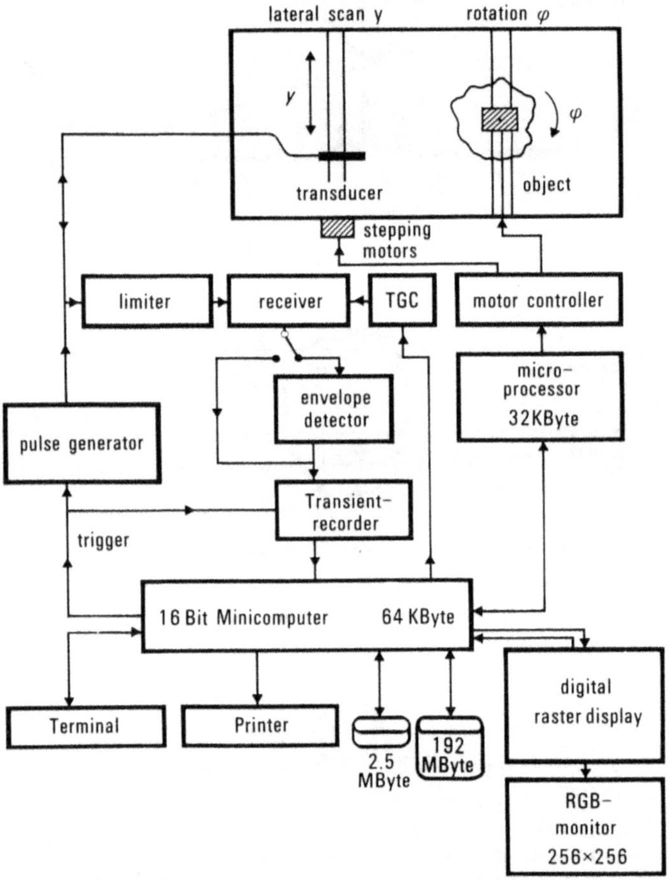

Fig. 2. Experimental arrangement of computerized echo tomography.

object like for a B-scan image. The echo sequence from the lateral scanning positions y_i are averaged to form our echo sequence. Again this sequence represents a 1-dimensional projection corresponding to the case of "infinite" beamwidth. But here the projection is along straight lines instead of circular arcs. This procedure is repeated for several rotation angles φ. The reconstruction corresponds to the way of "infinite" beamwidth.

3. An extension of the simple procedure of computerized tomography is introduced by inclusion of lateral information. In the case of "finite" beamwidth we can evaluate the echo sequences from each lateral scanning position. This leads to a B-scan image and preserves the lateral information. We use these B-scan images from several rotation angles φ instead of the 1-dimensional projections. This procedure is similar to the compound scan. We apply reconstruction and processing algorithms of computerized tomography.

4. A set of full B-scans from every rotation angle results in a large amount of data, which needs high storage capacity and long processing time. These data contain a lot of redundant information. To reduce the amount of data, we choose a way between the two extremes of lateral averaging (mode 2) and B-scan superposition (mode 3). We average over a few lateral positions and process the sequences as in mode 3.

EXPERIMENTAL ARRANGEMENT

The ultrasound transducer and the object are immersed in a water bath (Fig. 2) to allow good coupling and free motion. The object is freely rotable by a stepping motor. The transducer is mechanically scanned in lateral direction. We use a heavily damped leadmetaniobate transducer of 10 mm diameter. The spectrum is very broad and centered at 2 MHz. This corresponds to a wavelength of 0.75 mm in water. The two stepping motors are controlled by a microprocessor. This allows simultaneous processing in the minicomputer during motion of the motors.

Control of measurement and calculation for reconstruction of the tomographic image are performed on-line with the minicomputer. The time for measuring one projection and calculating backprojection on a 100 x 100 pixel image is about 0.8 seconds.

The analog part of our system is a standard equipment for B-scans (Fig. 2). The transducer is excited by a very short needle shaped pulse of 600 V. The echo signals are received by the same transducer. The time gain control allows a coarse compensation of attenuation in the object. In standard case the echo signals are rectified before A/D conversion. The transient recorder has a resolution of 50 ns and 8 bit. During data recording the current

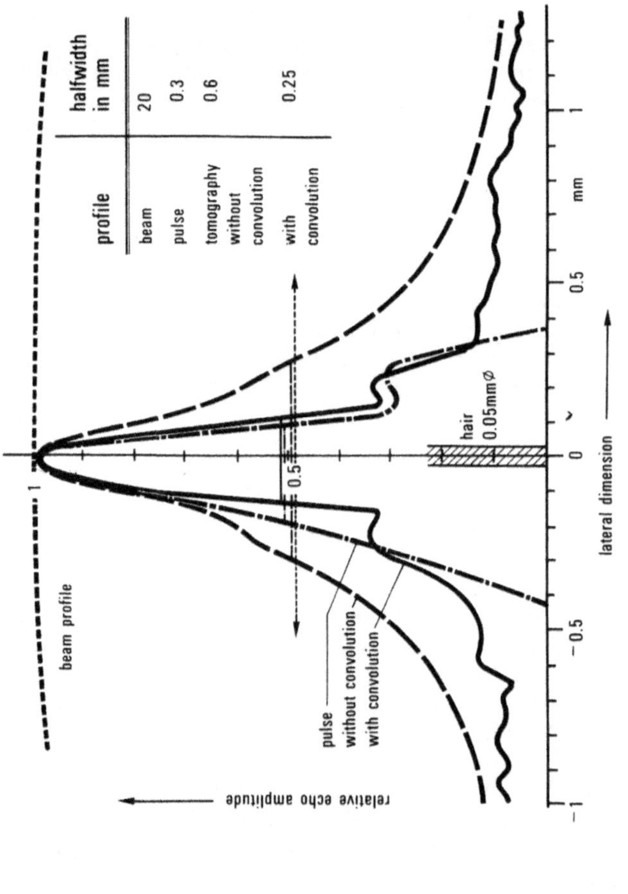

Fig. 3. Resolution of ultrasound echo tomography with pulse of 2 MHz center frequency represented by various profiles.

measurement and the image reconstructed so far are displayed on a RAMTEK digital display, which has colour and black-and-white monitors.

RESOLUTION OF ECHO TOMOGRAPHY

We demonstrate the resolution power of our procedure with a hair, which has a diamecer of 50 µm (Fig. 3). It has been carefully adjusted orthogonal to the image plane. Thus it represents an ideal point-like reflector.

We indicate the beam profile of the transducer measured with this hair. Its halfwidth is 20 mm and corresponds to lateral resolution of a B-scan at this distance. This represents the case of infinite beamwidth.

The echo pulse is shown after rectification. The halfwidth is 0.3 mm. With this pulse the image of the hair has been reconstructed by simple backprojection without convolution. The cross section through the image is shown as a profile. The halfwidth is 0.6 mm or about a factor 2 larger than the pulse width. This is in accordance with the 1/r smearing effect of simple superposition.

The smearing is removed if the echo signals are filtered with our modified Shepp-Logan kernel. The resulting profile has a halfwidth of 0.3 mm, about the value of the pulse itself.

With 0.3 mm we have reached the physical limit of half a wavelength at 2 MHz in water.

Furthermore the resolution does not change with the distance of the hair to the transducer. Of course it has rotational symmetry. This means that resolution in lateral and axial direction are equal. By transmission tomography it would not be possible to image the hair. In this case resolution is mainly limited by the diameter of the beam.

SENSITIVITY AND DYNAMICS

Compared to usual B-scan technique sensitivity and dynamics are drastically increased by echo tomography. This is shown for the case of scanning mode 4.

We used a sponge with holes and steel needles (Fig. 4). The sponge is a homogeneous scattering medium similar to tissue scattering. Its impedance is near to water. The sponge has a rectangular cross section of 110 x 80 mm. The holes have diameters from 5 to 25 mm. The needles generate high contrast to the weakly scattering sponge. The echoes from the needles are about a factor of 100 stronger than the sponge echoes.

Backprojection without convolution shows large smearing around the needles. The holes can be recognized, but the sponge does not appear homogeneous.

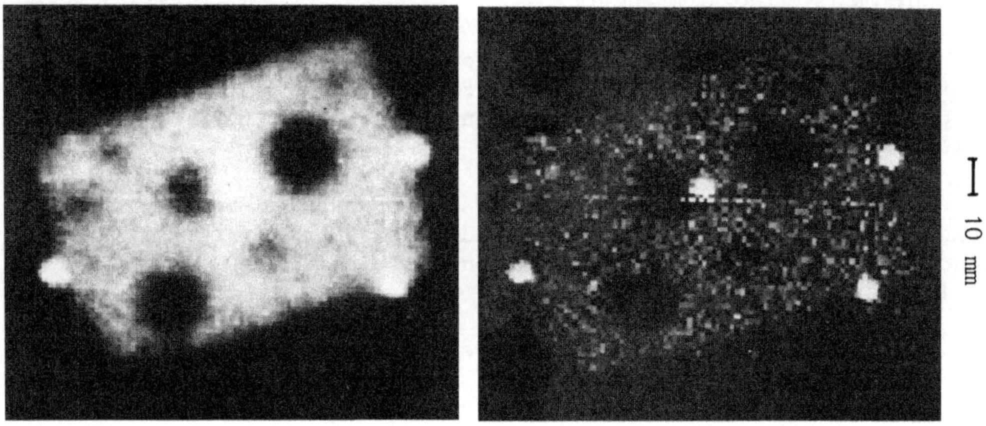

(a) (b)

Fig. 4. Dynamic range of ultrasound echo tomography. Sponge with holes and 4 steel needles. 60 measurements every 6 degrees with pixel size of 1.5 mm. (a) without convolution (b) after convolution with 2-dimensional modified Shepp-Logan kernel.

After convolution the needles appear 100 times stronger than the sponge area and reproduce the original high contrast. The needles appear sharply and the scattering area of the sponge appears homogeneously. This experiment demonstrates high sensitivity and contrast of echo tomography. This is because in biological tissue the reflection coefficient varies over several orders of magnitude. In transmission mode the absorption coefficient or the velocity of sound have only small variations. This leads to lower contrast and sensitivity.

In Fig. 4 the sponge is shown homogeneously. Speckles and granularity are drastically reduced compared to a single B-scan. The remaining granulation is mainly caused by the large pixel size of 1.5 mm and numerical rounding errors.

This image was made with "finite" beamwidth according to mode 4 with 10 lateral positions. The number of measurements was 60 every 6 degrees.

DISTURBANCES BY INTERFERENCE

Disturbances as well known from usual B-scans are strongly reduced by computerized tomography. This is caused by an averaging effect of the superposition of several projections. We investigated the influence of interference.

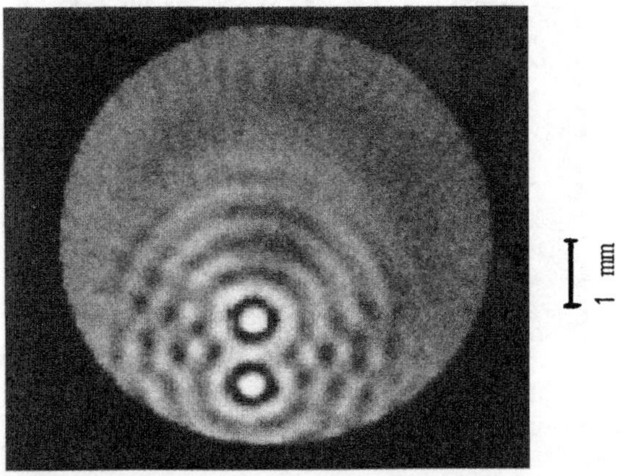

Fig. 5. Distortions by interference. Echo tomography of 2 steel needles of 0.3 mm diameter. The echo pulse has a length of 3 mm or 6 wavelengths. 90 measurements every 4 degrees.

Tomographic images made with the short pulse used in the resolution test did not show interference disturbances. Therefore we used a four times larger pulse, shown in Fig. 5. We reconstructed the image of 2 needles of 0.3 mm diameter and 1.1 mm separation. In this case the echo signals were not rectified. The wavecrests, shown in white, and the wavetroughs, shown in dark, indicate the pulse train even after superposition. We recognize that interference has only local effects. It produces dark and white interruptions in the wave pattern. This typical interference pattern is restricted to the neighbourhood of the needles. This interference pattern is very weak compared to the main signal of the object. But generally the phase cancellation effects are averaged away by the large number of measurements from different directions. The necessary number of directions depends on the object. In this case more than 90 measurements did not improve the image.

EXCISED ORGANS

The image quality for the various scanning modes is demonstrated with excised organs.

Fig. 6 is the echo tomogram of an excised pig's heart. It proves the high image quality compared to usual B-scans. The outlines of the surfaces are complete and clear. The heart muscle is a mainly isotropic scattering region which appears homogeneously.

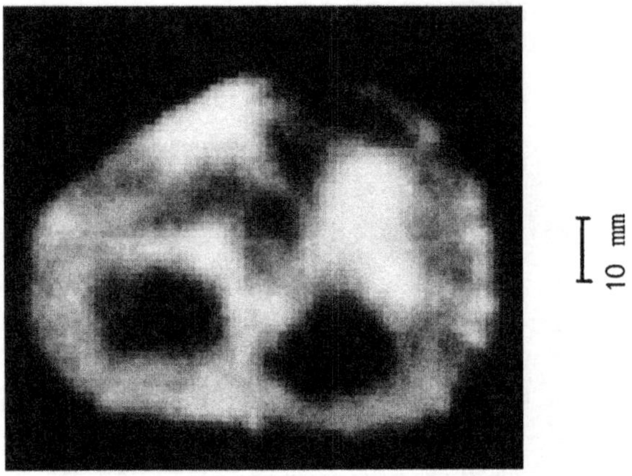

Fig. 6. Ultrasound echo tomography of a pig's heart. 60 measurements every 6 degrees with scanning mode 3 and pixel size 1.5 mm.

The strong scattering centres are caused by coagulated blood. The inner structure is given by vessels and ventricels which are free of echoes. The reflection coefficient of these three regions varies very much in strength and isotropy. They are separated very well and demonstrate the high dynamic range.

The image was reconstructed from 60 measurements every 6 degrees in scanning mode 3 with finite beamwidth.

With a chicken's heart we achieved the same results. In this case the organ is very small and entirely covered by the beam. It represents scanning mode 1 with infinite beamwidth. Therefore with conventional B-scans no outlines or structure would be seen.

SUMMARY

We conclude by summarizing the advantages of echo tomography over B-scan and transmission tomography. Resolution is proved to be better than one wavelength or less than 0.5 mm. In contrast to B-scans this resolution remains constant over all directions and is independent from the distance to the transducer. Compared to transmission tomography finer structures can be resolved. The sensitivity and dynamic range are improved. Interference effects leading to speckle like granulation of homogeneous regions are reduced. Outlines of organs and curved surfaces are displayed completely and with high contrast. The contrast of backprojected images is improved by a modified Shepp-Logan kernel, which is applied in one or two dimensions.

ACKNOWLEDGEMENTS

The authors gratefully acknowledge the discussions with Dr. Marschall and the assistance of Ms. Arnold in coding and running the computer programs.

REFERENCES

1. E. Hundt and E. Trautenberg, "Digital Processing of Ultrasonic Data Deconvolution", IEEE Trans. Sonics and Ultrasonics, Vol. SU-27, 249-252 (1980).
2. S. A. Johnson, J. F. Greenleaf, M. Tunaka, R. Rajagopolan and R. C. Bahn, "Reflection and Transmission Techniques for High Resolution Quantitative Synthesis of Ultrasound Parameter Images", IEEE Ultrasonic Symposium Proceedings (IEEE Cat. No. 77 CH1264-ISU) (1977).
3. S. J. Norton and M. Linzer, "Ultrasonic Reflectivity Tomography: Reconstruction with Circular Transducer Arrays", Ultrasonic Imaging , Vol. 1, 154-184 (1979).
4. G. Wade, S. Elliot, I. Khogeer, G. Flesher, J. Eisler, D. Mensa, N. S. Ramesh and G. Heidbreder, "Acoustic Echo Computer Tomography", Acoustical Imaging, Vol. 8, A. F. Metherell, editor, Plenum Press, New York, 565-576 (1980).
5. E. Hundt, G. Maderlechner, E. Kronmüller and E. Trautenberg, "Resolution and Image Quality by Ultrasonic Echo Tomography: Experimental Approach", Fifth International Symposium on Ultrasonic Imaging and Tissue Characterization in Gaithersburg, National Bureau of Standards, Washington, D. C., June 1-3 (1980).
6. L. A. Shepp and B. F. Logan, "The Fourier reconstruction of a Head Section", IEEE Trans. Nucl. Sci., Vol. NS-21, 21-43 (1974).

BOUNDED PULSE PROPAGATION

S. Leeman, Lynda Hutchins and *J. P. Jones

Department of Medical Physics, Royal Postgraduate Medical School, London, W12 OHS, U.K. and *Department of Radiological Sciences, University of California Irvine Irvine, California 92717

INTRODUCTION

A full understanding of the propagation of bounded pulses in dispersive media is fundamental to most developments in ultrasonic imaging and materials characterisation. Our interests lie in the medical applications of ultrasound, and, since it is by no means easy to make direct measurements of the fields in tissues, it may be argued that theoretical predictions are at least as useful as time-consuming and delicate measurements conducted within synthetic tissue-like materials.

In order to achieve a realistic theoretical prediction of pulse propagation in tissues, the following approximations may be incorporated: Most diagnostic pulse-echo imaging applications of medical ultrasound are carried out with pulses centered on 2 - 7 MHz, and only the frequency band ∿ 1 - 10 MHz need concern us. At present, there is no compelling evidence to suggest that shear waves may be supported over any significant distances in tissues, so that only longitudinal waves need be considered in studies of pulse propagation. A linear theory is probably adequate, even though peak intensity levels in diagnostic pulses may be rather high. In any event, it does not seem unreasonable to first explore to the full the advantages of linear theory, and to venture into non-linearity only when forced to do so by a demonstrated inability to fit the experimental facts. Single element pulse-echo systems invariably have circular transducers, which suggests the simplifying inclusion of cylindrical symmetry into the treatment.

A number of papers exist on the prediction of transient fields radiating from circular pistons (the list is quite long, but, recently, attention has focused on the methods developed by Tupholme, 1969, and Stephanishen, 1971), and direct observation (Weight and Hayman, 1978) has confirmed that the methods are adequate to describe both the "near" and "far" fields of piston-like transducers radiating into loss-less, uniform media (water-tank measurements). These results, however, are only of limited value when considering pulse propagation within tissues, since an important feature of the latter is their severe attenuation, which is clearly dispersive. Moreover, the restriction to piston-like radiators looks less appropriate as transducer design and construction become more advanced. Both these effects are incorporated in the treatment presented below, but we will assume that the radiation is into a uniform, lossy medium. Given that the level of inhomogeneity within tissues (as evidenced by the weakness of the scattering) is often quite small, we may regard our results as being relevant to at least a reasonable subset of soft tissues.

THE TISSUE MODEL

An essential step in any treatment is the specification of the wave equation governing the propagation of longitudinal waves in tissues. One such proposal (Leeman, 1980) seems remarkably successful in describing the dispersive absorption of tissues over a reasonable frequency range; however, the extension of this model to the full three-dimensional case is not entirely trivial. We therefore adopt a simplified version, which adequately describes the dispersive attenuation of many soft tissues:

$$\nabla^2 p - \frac{1}{C^2}\frac{\partial^2 p}{\partial t^2} - 2A\frac{\partial p}{\partial t} = 0$$

where $p(\underline{r},t)$ is the pressure at point $\underline{r}$, at time t, and A and C are parameters characterising the medium. Since the medium we are considering is a uniform, isotropic one, A and C are constants, and, since there is no scattering, their values should be fixed by the ultrasound absorption and velocity. However, it is not inappropriate to consider the model as describing attenuation in tissues, provided we are concerned only with pulse transmission, and provided that the attenuation is uniform.

The wave equation adopted here has been employed for describing the attenuation of (scalar) electromagnetic fields, and has been extensively researched. However, the method of solution we propose does not seem to have been used in this context before. The extent to which this model can fit measured attenuation values (Gross, Frizzell and Dunn, 1979) is shown in Fig. 1. The predicted signal velocity (which is the appropriate quantity to consider), will then be in error, and can be matched to experimental data only if additional terms, involving mixed time and space derivatives, are introduced into the model (Leeman, 1980). The additional mathematical complexity these terms contribute argues strongly against retaining them in this presentation, since the results of the simplified model show all the main features of pulse propagation in dispersive media, and are approximately valid for human tissues provided an appropriate rescaling is carried out to take account of the velocity error.

SOLUTION OF THE WAVE EQUATION

We assume axial symmetry, and work in cylindrical polar coordinates,

$$p(\underline{r},t) \equiv p(\rho,z;t)$$

For a pulse incident, at $t = 0$, on the lossy medium, which is considered to fill the half space $z > 0$, the following boundary conditions are self-evident.

$$p(\rho,0;t) = B(\rho)\, f(t) \qquad \text{for } t \geqslant 0$$

$$= 0 \qquad \text{for } t < 0$$

Also,

$$p(\rho,z;0) = \frac{\partial p}{\partial t}(\rho,z;0) = 0 \qquad \text{for } z > 0$$

$B(\rho)$ is the (incident) beam profile, assumed normalised to $B(0) = 1$, and $f(t)$ is the (incident) pulse shape.

The equation is readily solved by the method of Laplace transforms. Introduce

$$\Pi(\rho,z;s) \equiv \mathcal{L}\{p(\rho,z;t)\}$$

where $\mathcal{L}$ indicates the Laplace transform operation, and the wave equation simplifies to

$$\nabla^2\Pi - \left(\frac{s^2}{C^2} + 2As\right)\Pi = 0$$

Assume separability,

$$\Pi(\rho,z;s) = G(\rho)\, F(z,s)$$

and we find, in the usual way, that, in general,

$$\Pi(\rho,z;s) = \int_0^\infty d\lambda.\lambda.b(\lambda)\ J_o(\lambda\rho)\ F_o(s)\ e^{-\sqrt{\frac{s^2}{C^2}+2As+\lambda^2}\,z}$$

where b(λ) is some weight function. The boundary conditions may once more be invoked to establish that b(λ) is nothing but the Hankel transform of the incident beam profile,

$$b(\lambda) = \int_0^\infty d\rho.\rho.B(\rho)\ J_o(\rho\lambda)$$

and also that F_o is the Laplace transform of the incident pulse shape,

$$F_o(s) = \mathcal{L}\{f(t)\}$$

In order to regain p(ρ,z,t), the inverse Laplace transform of Π has to be carried out - this is easily performed by having recourse to published tables (Abramowitz and Stegun, 1965). Finally, we obtain

$$p(\rho,z;t) = P(\rho,z;t) * f(t)$$

where * denotes convolution, and where

$$P(\rho,z;t) = B(\rho)\ e^{-\alpha z}\ \delta(t - z/C)$$

$$+\ Cz\ e^{-\alpha\tau}\ H(t - z/C)\ \frac{1}{\xi}\frac{\partial}{\partial\xi}\int_0^\infty d\mu.\mu B(\mu)\int_0^\infty d\lambda.\lambda J_o(\mu\lambda)J_o(\rho\lambda)J_o(\xi\sqrt{\lambda^2-\alpha^2})$$

with $\tau = Ct$

$\alpha = AC$

$\xi = \sqrt{\tau^2 - z^2}$

and where J_o denotes the Bessel function of order 0, and H the unit step function.
P(ρ,z;t) is clearly the propagating pulse for the case that f(t) = δ(t), and contains all the information necessary to predict the structure of the field in the medium. In order to proceed further, more explicit assumptions about the form of B(ρ) have to be made.

We will examine a particularly simple case.

FINITE BEAM-WIDTH, NO ABSORPTION

It is instructive to examine the loss-less case, $\alpha = 0$, in order to validate our result against that of the Green's function method used by Tupholme, 1969, and Stephanishen, 1971. We do not, however, assume that $B(\rho)$ is constant, as is done in other treatments, but merely that $B(\rho) = 0$ for $\rho > R$. To avoid confusion, we denote this function by $B_R(\rho)$. The loss-less impulse response reduces to

$$P_o(\rho,Z;t) = B_R(\rho)\,\delta(t - Z/C)$$

$$+ CZ\,H(t - Z/C)\,\frac{1}{\xi}\frac{\partial}{\partial\xi}\int_0^R d\mu.\mu.B_R(\mu)\int_0^\infty d\lambda.\lambda\,J_o(\mu\lambda)J_o(\rho\lambda)J_o(\xi\lambda)$$

The λ-integration is standard (see Watson, 1929), and there results

$$P_o(\rho,Z;t) = B_R(\rho)\,\delta(t - Z/C)$$

$$+ CZ\,H(t - Z/C)\,\frac{1}{2\pi\xi}\frac{\partial}{\partial\xi}\int_0^R d\mu.B_R(\mu).\frac{1}{\Delta}$$

where Δ is the area of the triangle with sides μ, ρ and ξ. Where these three lengths do not form the sides of a triangle, the integral vanishes. The geometric relationships are shown in Fig. 2, and the relationship of our method to that of Tupholme-Stephanishen now becomes clear. Transform the μ-integration to one over the angle Θ (defined in the figure, and constrained to $0 \leqslant \theta \leqslant \pi$) to obtain:

$$P_o(\rho,Z;t) = B_R(\rho)\,\delta(t - Z/C)$$

$$+ \frac{2}{\pi}\,CZ\,H(t - Z/C)\,\frac{1}{\xi}\frac{\partial}{\partial\xi}\int_0^{\Theta_{max}} B_R(\Theta)\,d\Theta$$

where $B_R(\Theta)$ is the value the function $B_R(\mu)$ takes along "ξ-circle" in Fig. 2. Note that if the "ξ-circle" does not intersect the edge of the transducer, then $\Theta_{max} = \pi$. If it does, then

$$\Theta_{max} = \cos^{-1}\left\{\frac{\rho^2 + \xi^2 - R^2}{2\xi\rho}\right\}$$

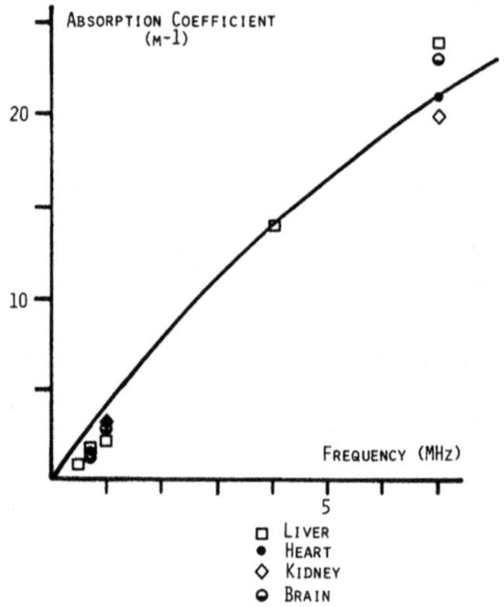

Fig. 1. Model fit to absorption data of Gross, Frizzell and Dunn (1979).

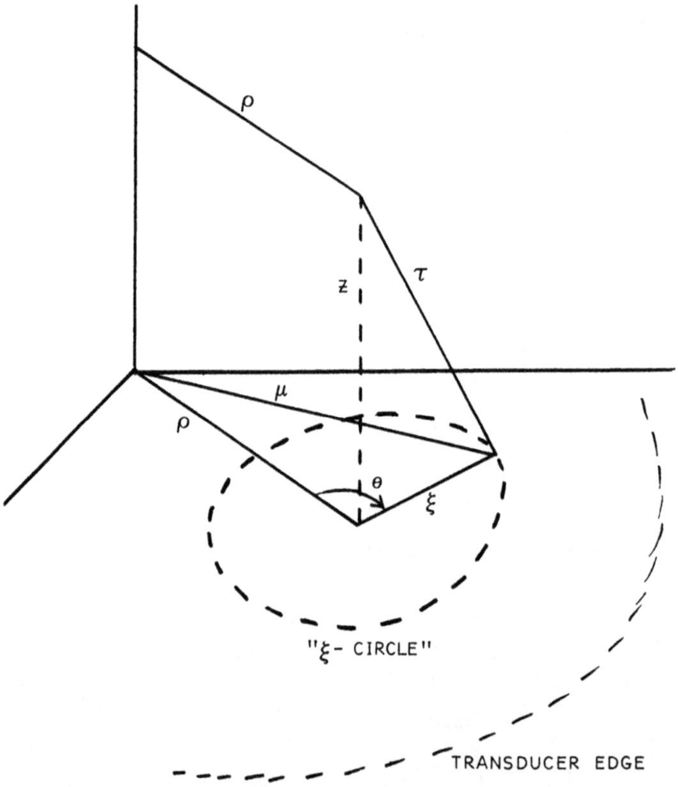

Fig. 2. Geometric relationships between variables appearing in calculation of "edge-wave".

Consider now a transducer of radius R, in the plane $z = 0$, and excited to emit a δ-function pressure pulse at $t = 0$. At a given field point (ρ, z) nothing is measured until time $t = z/C$, at which the "direct-wave" $B_R(\rho)\ \delta(t - z/C)$, will be seen to arrive. (For simplicity, we consider $\rho < R$, but the arguments are easily modified for $\rho > R$). This is a perfect replica of the initial emission pulse. As t increases beyond z/C, contributions from an ever-increasing ξ-circle are picked up. These follow immediately in the wake of the direct wave for $\rho \leqslant R$, but are non-zero even beyond $\rho = R$, provided that sufficient time has elapsed. If $B_R(\rho)$ is a slowly-varying function, these contributions may be expected to be rather weak. After some time, however, the ξ-circle will intersect the edge of the transducer disc, so that $\Theta_{max} < \pi$. In the circumstance that $B_R(\rho)$ drops discontinuously to zero at $\rho = R$, a rather stronger contribution to the field may be expected when the ξ-circle intersects the transducer edge.

For the extreme case of constant excitation,

$$B_R(\rho) = 1 \text{ for } \rho \leqslant R$$

$$= 0 \text{ for } \rho > R$$

which appears to be the only case treated in the literature, there is no contribution unless the ξ-circle intersects the transducer edge (since $\partial/\partial\xi\{\text{constant}\} = 0$). A finite time thus elapses after the passage of the direct wave before another pressure wave is felt. This has given rise to the notion that the pressure impulse response consists of two components: the "direct-wave" and the "edge-wave". We have seen, however, that this convenient assignment is less obvious in the case of non-uniform excitation, but we will nevertheless retain this convenient nomenclature; and will refer to the first term in the expression for P_ρ as the direct-wave, and to the second term as the "edge"-wave. The absence of a finite delay between the arrival of the direct- and edge-waves in the case of non-uniform excitation, seems to be confirmed by the observations of Weight and Hayman, 1978: their Fig. 5 shows this effect, and we speculate that its origin lies in a non-uniform excitation of their transducer.

FINITE TRANSDUCER, LOSSY MEDIUM

In this case, the direct wave is given by $B_R(\rho)e^{-\alpha z}\delta(t - z/C)$ and is again a replica of the initial excitation, but is damped with the (constant) loss coefficient, α. Thus, even in the presence of dispersive attenuation, a recognisable component of the propagating pulse has the same shape as the emission pulse. Following immediately on the direct wave is the dispersively modified edge wave:

it is, in fact, possible to establish a geometrical construction to indicate how the "edge-wave" is built up (similar to the "ξ-circle" method for the loss-less case), but this is rather involved, and is not presented here. It can also be shown that, at $t = Z/C$ (i.e. at the arrival time of the direct wave) the edge wave does not vanish for $\rho < R$, even in the case of constant excitation; in this respect, therefore, there is a significant difference to the loss-less case.

It is interesting also to examine the axial pulse shape. As in the one-dimensional case (Leeman, 1980) this consists of a (non-dispersively) damped replica of the incident axial pulse, travelling at a constant (i.e. non-dispersive) velocity, accompanied by a distance-dependent "rumble", which tends to dominate after some time. The velocity at which the pulse-front propagates is, in this model, independent of the carrier frequency of the initial pulse, $f(t)$, and is appropriately called the signal velocity. As might be expected, the on-axis rumble turns out to be dependent on the form of the incident beam profile, $B(\rho)$.

It is clear that, despite the dispersive nature of the medium considered here, a component of the propagating pulse (arising from the direct wave) exhibits all the features associated with non-dispersive attenuation. At small Z-values, when the edge-wave component to the axial pulse can be relatively small, careless measurements may tend to underestimate the dispersive character of the medium.

The time of first arrival of the signal, provided that this is carried out for $\rho < R$, is, in this model, a non-dispersive constant, the so-called signal velocity. This may explain why velocity measurements carried out by time of flight transmission procedures fail to detect velocity dispersion in tissues. Indeed, a possible test for velocity dispersion in tissues is suggested by this phenomenon: since the characteristic impedance (density x velocity) may be measured, in principle, by both a reflection or pulse transmission method, we may expect the results of the former to have a stronger dependence on the carrier frequency of the interrogating pulse than the latter - provided that tissues are dispersive in accordance with our simple model!

REFERENCES

Abramowitz, M. and Stegun, I. A. (eds), 1965, Handbook of Mathematical Functions, Dover Publications, New York.

Leeman, S., 1980, Ultrasound Pulse Propagation in Dispersive Media, Phys. Med. Biol., 25(3):481

Stephanishen, P. R., 1971, Transient Radiation from Pistons in an Infinite Planar Baffle, J. Acoust. Soc. Amer., 49:1629.

Tupholme, G. E., 1969, Generation of Acoustic Pulses by Baffled Plane Pistons, Mathematika, 16:209.

Watson, G. N., 1929, A Treatise on the Theory of Bessel Functions, Cambridge University Press, Cambridge.

Weight, J. P. and Hayman, A. J., 1978, Observations of the Propagation of Very Short Ultrasonic Pulses and their Reflection by Small Targets, J. Acoust. Soc. Amer., 63(2):396.

THEORETICAL STUDY OF PULSED ECHOGRAPHIC FOCUSING PROCEDURES

Mathias Fink

Laboratoire de Mécanique Physique
Université P. et M. Curie
78210 , Saint-Cyr , France

ABSTRACT

This paper is devoted to a theoretical evaluation of the pulsed echographic response of spherical focusing apertures used in B-scan imagery. Loss-less as well as absorbing medium are investigated. The effects of transducer bandwidth on lateral resolution and side lobe level is studied. The differences observed between the pressure pattern and the complete echographic response pattern is emphasized for well damped and for narrow-band transducers. The influence of a frequency dependant attenuation on these patterns is shown, and an explanation of the defocusing process observed in biological medium [1] is given.

INTRODUCTION

The quality of B-scan images depends mainly upon the resolution of the imaging system. The lateral resolution is proportional to the width of the main lobe and the useful dynamic range is directly proportional to the side lobe level. Besides, the axial resolution is dependant on the bandwidth of the transducer.

In the study of the focusing processes adapted to the B mode echography, where the irradiating signals are relatively brief, the usual monochromatic approach is not sufficient. The various unavoidable transient periods that exist in transmission as well as in reception are forgotten.

Even for relatively long excitations, the acoustic pressure felt by a small target will have a really monochromatic behavior only after the lighting of this target by the entire transmission aperture. During the reception, there exists also a similar transient period during which the wave front reflected by the target aimed at have not yet the time to cover the entire reception aperture.

By neglecting these transient phenomena, the monochromatic approximation can under-estimate the importance of the response of targets located outside of the axis of the transducer. In particular, if one considers that the electronic reception processing identifies itself with a peak detection of the signal received by the reception aperture, it can happen that the signal observed from off axis targets during the transient period will have a peak value higher than the one observed later during the "monochromatic" period, because the destructive interferences, which are effective during the monochromatic state are less important in the transient state. Moreover, for very well damped transducers, these transient periods are only observed and the directivity pattern may then be quite different from that obtained with narrow band transducer.

PULSED ECHOGRAPHIC RESPONSE

In this paper, we used an extension of the classical impulse response approach [2, 3, 4, 5] to non uniform and non planar apertures (delay correction)for loss-less and for absorbing medium. The impulse response in transmission $h(\vec{r}, t)$ corresponds to the transient velocity potential pattern associated with a transmitter whose surface is subjected to a velocity impulse.

In a loss-less medium, the solution of the diffraction problem using a Green's function development permits to write this impulse response at a field point as a function of time

$$h(\vec{r}, t) = \int P(\vec{m}_o, t) \frac{\delta(t - R/c)}{2\pi R} ds_o \qquad (1)$$

$$R = |\vec{r} - \vec{r}_o|$$

where $\delta(t - R/c) / 2\pi R$ is the green function representing the spherical wave generated by a point source located in $\vec{r}_o$. $P(\vec{m}_o, t)$ is the "generalized" pupil function that characterizes the geometrical behavior of the transmission pupil through the distribution

$$P(\vec{m}_o, t) = A(\vec{m}_o)\, \delta(t - \tau(\vec{m}_o)) \qquad (2)$$

It is worth noting that such a coupling between t and $\vec{m}_o$ holds only in the case where all the points of the transducer move in the same vibration mode. $A(\vec{m}_o)$ is the amplitude modulation of the aperture and $\tau(\vec{m}_o)$ is the delay correction which takes into account the geometry of the transducer and disappears for a planar aperture.

For absorbing medium, the evaluation of the impulse response must be modified to take into account the new green function of the problem.
In biological tissue, for frequency in the range of one to ten megahertz, the attenuation is approximately proportional to the frequency

$$\alpha(f) = \beta |f| \qquad (3)$$

Thus, by neglecting the velocity dispersion linked to this attenuation (the relative velocity dispersion is of the order of 10^{-4} [6]), we may replace the usual monochromatic spherical wave by an attenuated spherical wave of the form

$$\frac{e^{i(kr - 2\pi ft)}}{2\pi R} \; e^{-\beta|f|R} \qquad (4)$$

Thus, in pulsed mode, the green function which is the time Fourier transform of this attenuated wave may be written

$$\frac{\delta(t - R/c)}{2\pi R} \; \frac{\beta' R}{(\beta' R)^2 + t^2} \qquad \beta' = \beta/2\pi \qquad (5)$$

where the time spreading of the impulse wave caused by the attenuation has the shape of a Lorentzian curve.
With these conditions, the diffractive impulse response in absorbing medium is given by

$$h_A(\vec{r},t) = \int P(\vec{m}_o,t) \, \frac{\delta(t - R/c)}{2\pi R} \; \frac{\beta' R}{(\beta' R)^2 + t^2} \, dso \qquad (6)$$

which is a relatively complicated expression.
It must be noticed that we have neglected causality in the determina- of the impulse response. This is justified by the fact that, in our problem, the impulse response may be obtained to within any temporal translation.

By doing the assumption that all the points of the receiving (or radiating) aperture are approximately equidistant from the observation point $\vec{r}$ (i.e. the duration of the impulse response for a loss-less medium is small compared to $\overline{R}/c$), we may write $h_A(\vec{r}, t)$ as the time convolution product

$$h_A(\vec{r},t) = h(\vec{r},t) \oplus \frac{\beta'\overline{R}}{(\beta'\overline{R}) + t^2} \tag{7}$$

where $\overline{R}$ is the mean distance between the aperture and the observation point $\vec{r}$. The reducing of expression (6) to a time convolution is justified by the fact that the time spreading of all the waves illuminating the point $\vec{r}$ is approximately the same, and so the attenuation effect on the impulse response is time invariant and may be expressed as the convolution product.

If one has evaluated the impulse response $h(\vec{r}, t)$ (or $h_A(\vec{r},t)$) the velocity potential created by a transducer subjected to an arbitrary velocity excitation $V_n(t)$ is then obtained by the convolution product

$$\phi(r, t) = V_n(t) \oplus h(r, t) \tag{8}$$

Thus a simple time derivative will give the acoustical pressure $p(\vec{r},t)$. If one takes into account the relation

$$V_n(t) = E(t) \oplus i_E(t) \tag{9}$$

where E(t) designates the electrically input signal used for exciting the transducer whose electro acoustic transfer function is $i_E(t)$, the incident pressure is expressed as

$$p(\vec{r},t) = -\rho_\circ \frac{dE(t)}{dt} \oplus i_E(t) \oplus h(\vec{r},t) \tag{10}$$

This theory of the pulsed response can, of course, be extended to the study of the reception of the acoustical signals reflected by a punctual target. We shall do the assumption that such a target acts as small spherical source having a radial velocity directly proportional to the incident pressure.
Thus this "scattering" center will generated a reflected pressure wave. Its intersection with the transducer aperture will give an electrical signal directly proportional to the instantaneous surface

of this intersection zone. The determination of this surface is obtained by means of the same spatial integration than in formula (1) A similar impulse response $h(\vec{r}, t)$ to an impulse velocity source may be defined at reception, which is completely equivalent to the one observed in transmission, if the same aperture is used.
If ones takes into account the acousto electric transfer function $i_R(t)$ locally relating the incident pressure with the output electrical signal, actually delivered, identifies itself with

$$E_{echo}(\vec{r},t) = \frac{d^2E(t)}{dt^2} \otimes i_E(t) \otimes i_R(t) \otimes h(\vec{r},t) \otimes h(\vec{r},t) \quad (11)$$

This result brings to evidence the concept of pulsed echographic response $H(\vec{r}, t)$ characterizing the radioactive behavior of the pupil in transmit-receive mode by means of the autoconvolution product of the simple impulse response

$$H(\vec{r}, t) = h(\vec{r}, t) \otimes h(\vec{r}, t) \quad (12)$$

The selection of $i_E(t)$ and $i_R(t)$ depends on many parameters : the type of construction of the transducers (front face, back face, thickness...), the electrical characteristics of the various interfaces (pulse generator, impedance matching transformer, reception preamp).

THE SPHERICAL TRANSDUCER

The most usual pupil function to consider is obviously the one that is associated with a spherical delay correction adapted to the focal distance F that one seeks to obtain. Such a pupil function simulates the geometry of a single spherical transducer. The pupil function associated with this focusing aperture (of diameter D = 2A) is written under the form

$$P(\vec{m}_\circ,t) = \delta(t - (F + \sqrt{F^2 + m_\circ^2})/c) \cdot \text{circ}_A(\vec{m}_\circ) \quad (13)$$

where the function $\text{circ}_A(\vec{m}_\circ)$ is 1 when $\vec{m}_\circ < A$ and 0 otherwise. We shall now study the theoretical impulse response of this transducer.

The numerical results which will be shown are coming from calculation that have been performed without any approximation. However, we shall use the Fresnel approximation in the analytical calculation of the impulse response

$$R = |\vec{r} - \vec{r}_\circ| \simeq z + (\vec{m} - \vec{m}_\circ)^2 / 2z \quad (14)$$

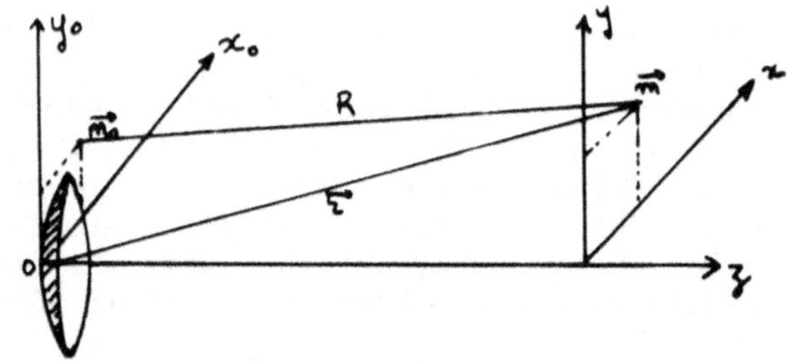

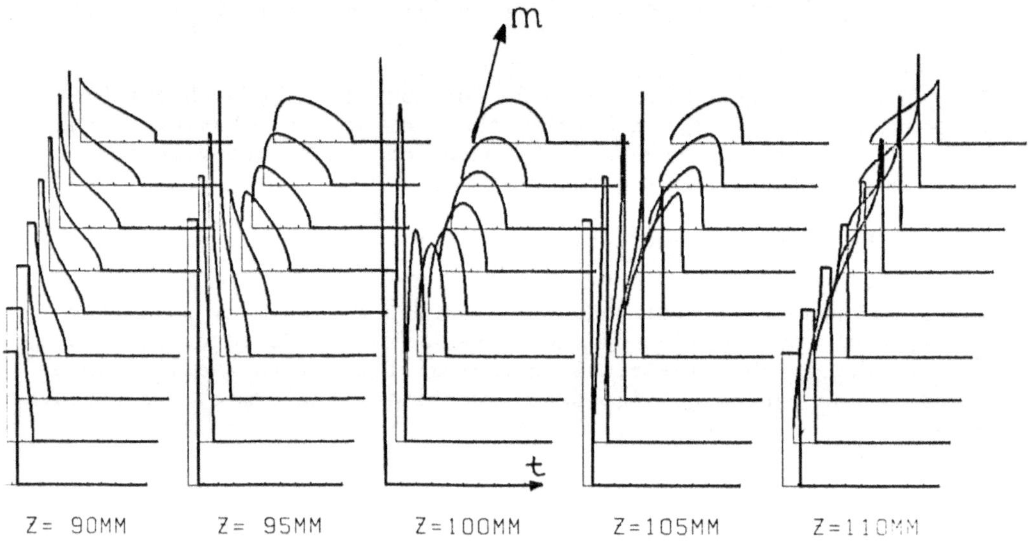

Fig. 1 Impulse velocity potential $h(z,m,t)$ for a spherical transducer (F=100mm, A=20mm)

This approximation, which will permits us to efficiently simplify the calculation, is not, a priori, at all justified to the extent that one is interested in the case of the response to an ideal punctual source of the form δ (t) with a spectrum infinitely spread (the error made is always large as compared with the shortest wave lengths). However, the fact that the function $h(\vec{r},t)$ is to be used for the calculation of the response to real sources which are far from themselves behaving as $\delta(t)$ and whose spectrum is limited on the upper side by a frequency f max (which in our application does not go higher than 6 or 7 MHz), allows us to justify the value of this approximation, if the errors committed on the determination of R are smaller than a fraction of the wave lengtn λ_{min}. With these conditions, $h(\vec{r},t)$ may be obtained except for a temporal translation z/c under the form

$$h(\vec{r},t) \propto \frac{1}{z} \int P(\vec{m}_o,t)\ \delta(t - (\vec{m}-\vec{m}_o)^2/2cz)ds_o \qquad (15)$$

The pupil function may be written by approximating the spherical delay correction with a quadratic correction of the form $\tau(m_o)=-m_o^2/2cF$ The pulsed response varies then according to

$$h(\vec{r},t) \propto \frac{1}{z} \int \delta(t+\vec{m}_o{}^2/2cF - (\vec{m}-\vec{m}_o)^2/2cz) \, . \, \mathrm{circ}_A(\vec{m}_o)ds_o \quad (16)$$

I. FOCAL RESPONSE (z = F)

By simplifying the formula (16), the focal response, except for a temporal translation $m^2/2cF$ is then written under the form

$$h(z=F,m,t) \propto \frac{1}{F} \int \delta(t+\vec{m}\vec{m}_o/cF).\mathrm{circ}_A(\vec{m}_o)ds_o \quad (17)$$

The delay law of the form $\delta(t+\vec{m}\vec{m}_o/cF)$ linear according to m, which is the result of the combination of two quadratic delay laws of the same curvature $\delta(t+m_o^2/2cF)$ and $\delta(t-(\vec{m}-\vec{m}_o)^2/2cF)$ can, in fact be conceived, by analogy with the behavior of transmission lenses as being associated with an oblique plane wave, of slope m/F, emerging from a point source located at infinity in the direction determined by $O\vec{m}$. The integral value then represents the instantaneous surface of the intersection zone enclosed between this oblique wave front and the disc plane of radius A located in the plane $z = 0$. This intersection is a section of a straight line parallel, for example, to the y_o axis which moves along the x_o axis with a constant speed cF/m, and whose instantaneous surface varies as $2\sqrt{A^2 - m_o^2}$ which entails for h a dependance of the form

$$h(F,m,t) \propto \frac{A^2}{\pi F} \, . \, \sqrt{1-(t/\Delta')^2} \; \frac{\mathrm{rect}_{2\Delta'}(t)}{\Delta'} \quad (18)$$

where $\Delta' = Am/cF$. This a semi circular time function whose duration is proportional to m, and whose amplitude decreases as $1/m$. This is a function whose integral is constant, and that when m goes towards 0, identifies itself with $A^2/2F\ \delta(t)$ (See Fig. 1).

II. IMPULSE RESPONSE ($z \neq F$)

One shall study now the evaluation of this response outside of the focal plane. By defining an apparent curvature radius ε by $1/\varepsilon = 1/z - 1/F$, one can then relate the behavior of such a spherical aperture to the behavior of a disc plane aperture. By simplifying the formula (16), the response, except for a temporal translation, $m^2/2cz$ is then written under the form

$$h(z\neq F,m,t) \propto \frac{1}{z} \int \delta\,(t - m_o^2/2c\varepsilon + \vec{m}\vec{m}_o/cz) \, .\mathrm{circ}_A(\vec{m}_o)ds_o \quad (19)$$

One can associate with this delay law, a spherical wave of curvature ε and of oblique axis of slope m/z, whose intersection with the disc of radius A defines a zone whose instantaneous surface is proportional to h.
One then notices that by supposing $\gamma = \varepsilon/z$, the changing of variables $\vec{M}_o = \vec{m}_o/\gamma$ reduces the expression of the formula to :

$$h(z,m,t) \propto \gamma(\gamma/z)\int \delta(t - M_o^2/2c(z/\gamma) - \vec{M}\vec{M}_o/c(z/\gamma))\,\mathrm{circ}_{A/\gamma}(M_o)\,dS_o$$

Where $dS_o = ds_o/\gamma^2$. One then reduces the evaluation of the impulse response of a spherical transducer to a target located at the distance z, except the factor γ, to the determination of the response of a plane disc of aperture $A_\gamma = |A/\gamma|$ to a target located in $z_\gamma = |z/\gamma|$
By using the results obtained by different authors, one then obtains two determinations of $h(z,m,t)$ according to the value of m with respect to A_γ and to the value of z with respect to F

$$z < F \quad h(z,m,t) = \left|\begin{array}{lll} \gamma c & t<t_1 & — \\ \gamma \frac{c}{\pi} \arccos\left(\frac{2cz_\gamma t + m^2 - A\gamma^2}{2m\sqrt{2cz_\gamma t}}\right) & t_1<t<t_2 & t_1<t<t_2 \\ 0 & t>t_2 & t<t_1 \;;\; t>t_2 \end{array}\right. \qquad \begin{array}{l} m \leqslant A_\gamma \quad m>A_\gamma \end{array}$$

$$z > F \quad h(z,m,t) = \left|\begin{array}{lll} \gamma c & t_2>t>t_2-t_1 & — \\ \gamma \frac{c}{\pi} \arccos\left(\frac{2cz_\gamma (t_2-t) + m^2 - A\gamma^2}{2m\sqrt{2cz_\gamma (t_2-t)}}\right) & t<t_2-t_1 & t<t_2-t_1 \\ 0 & t>t_2 & t>t_2 \end{array}\right.$$

where $t_1 = (m-A\gamma)^2/2cz_\gamma$ and $t_2 = (m+A\gamma)^2/2cz_\gamma$
One can then by replacing γ with $F/|F-z|$ transform the equation $m = A/\gamma$, which defines the dividing surface between the two areas, in the form $m = (A/F).|F-z|$ which is the equation of a conical surface. Inside this cone, the peak value of the pulsed response remains constant for a given depth z and is equal to $cF/|F-z|$, whereas outside this cone, this peak value decreases in a monotonous manner with m.

The limitation of the impulse response to on axis field points may be written according to (17), with the changing of variable $\tau = m_o/2c\varepsilon$

$$h(z,0,t) \propto \frac{A^2}{2z}\,\frac{\mathrm{rect}_\Delta(t)}{\Delta} \qquad (22)$$

This is a rectangular function of duration $\Delta = A^2/2c\varepsilon$, whose amplitude varies as $\gamma = \varepsilon/z$.
This is a function whose integral is constant, and that when z goes towards F, identifies itself with $A^2/2F.\delta(t)$ which is the impulse response at the center of curvature.

These different results are going to enable us to find the amplitude $\phi\omega(\vec{\varepsilon})$ of the monochromatic response of pulsation ω
For on axis point the Fourier transform of (21) depends on z according to

$$\phi\omega(z,0) = \frac{F}{|F - z|\omega} \quad \sin \frac{A^2\omega \; |F - z|}{4c \; F \; z} \qquad (23)$$

For the focal response, the Fourier transform of the semicircular function (18) varies as a Bessel function of the first order.

$$\phi\omega(F,m) = \frac{J_1 \; (\frac{\omega A m}{cF})}{\omega m} \qquad (24)$$

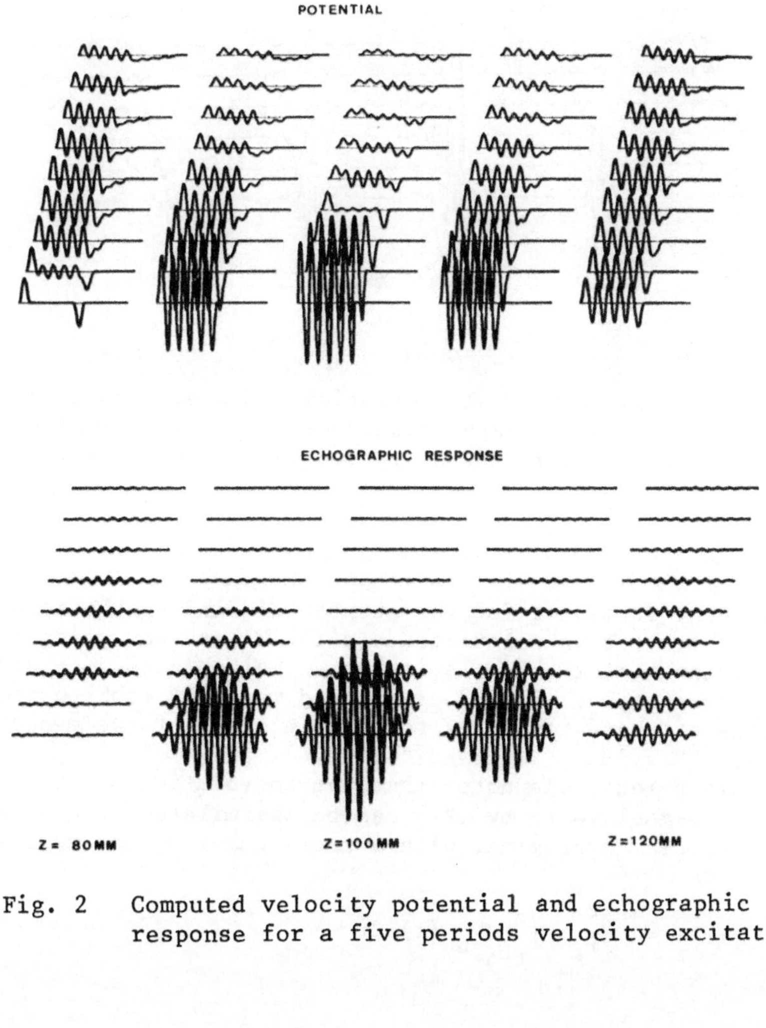

Fig. 2 Computed velocity potential and echographic response for a five periods velocity excitation.

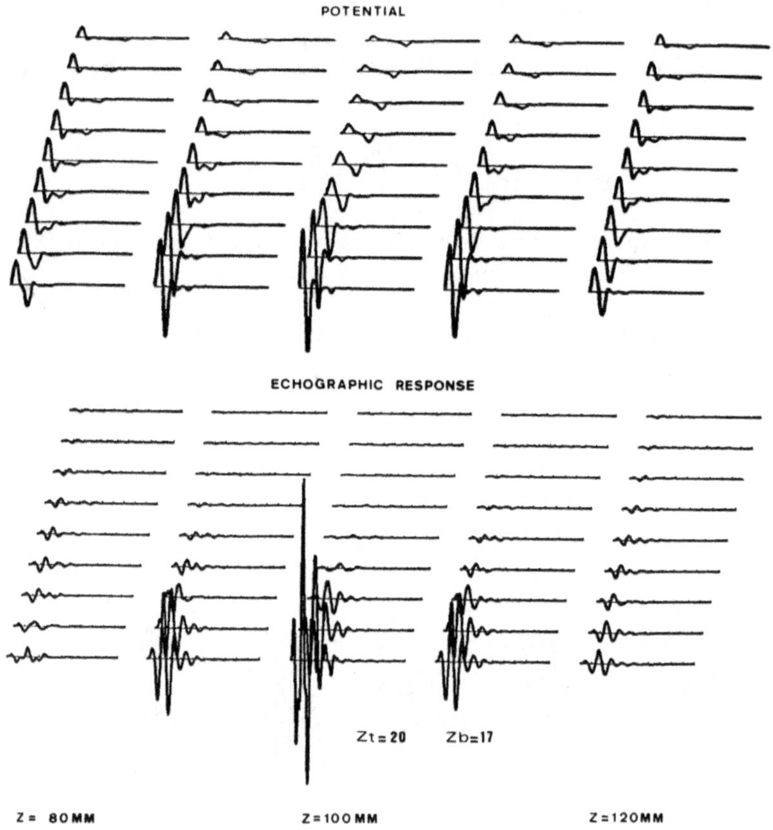

Fig 3 Computed velocity potential and echographic response for a well damped transducer. (Backing impedance 17.10^6 Rayleigh and transducer impedance 20.10^6 R.)

However, it is not for the purpose of finding the monochromatic response that we have calculated $h(r,t)$. We are now going to study the determination of the velocity potential obtained for different velocity excitations (Fig. 2 et 3), and we shall seek to compare the peak value of these responses to the amplitude of the monochromatic response.

For on axis points, one notes that the convolution of the velocity excitation signal $V_n(t)$ by (22) can be assimilated to a simple operation of temporal averaging, with a rectangular window moving along the function

$$\phi(t). = \frac{A^2}{2f_\Delta} \int_{t-\Delta}^{t} V_n(\tau)\, dr$$

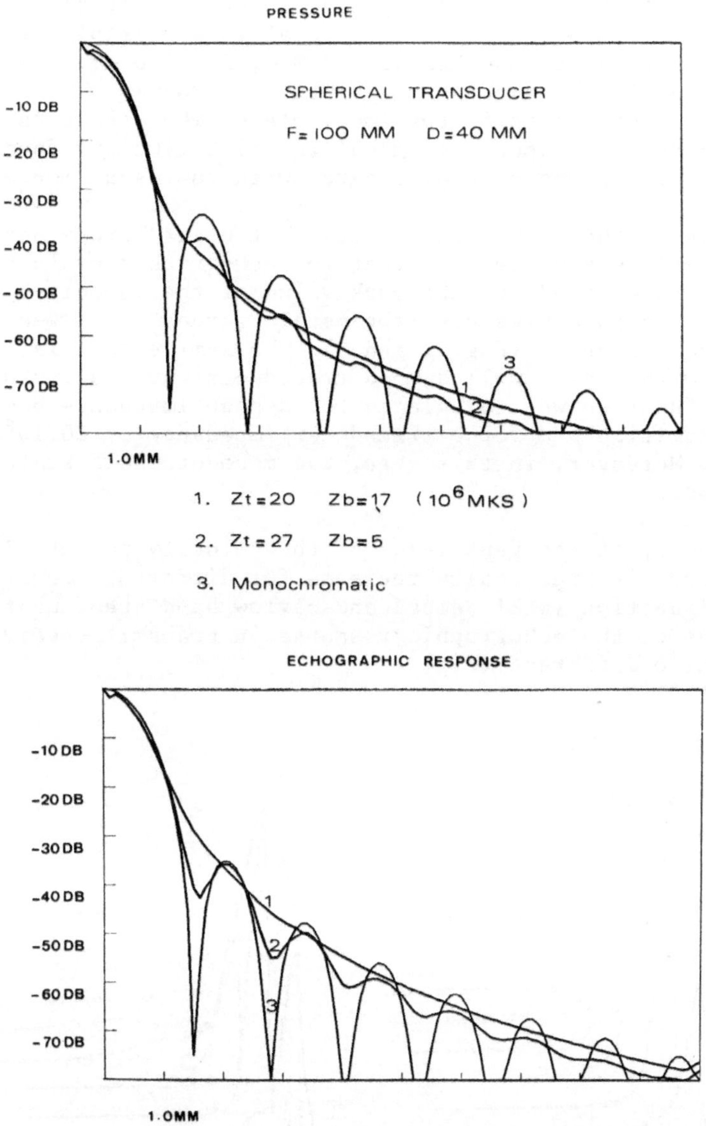

Fig. 4 Peak values of the pressure and of the echographic response in the focal plane for 1 : a well damped transducer
2 : a narrow band transducer
3 : the monochromatic case.

For example, with a sinusoidal velocity containing five oscillations of period T, it is observed that after a transcient period, the averaging of the signal may give a zero potential, if the duration Δ of the impulse response is equal to a multiple of T .
For our reference transducer (f = 3 MHz, F = 100 mm, A = 20 mm), the first "zero" is observed at the depth z = 80 mm.
The same effect occurs in the focal plane, when the duration 2Δ' of the semicircular response is equal to 1.22 time the fundamental period T. Thus, these results agree with the usual monochromatic study.

However, the transient effects that occur before and after the monochromatic state are important for echographic evaluation.
If one is interested in the peak value of the velocity potential, or of the acoustic pressure, the usual "zeroes" observed in monochromatic mode disappear (Fig. 2 and 3). The same effect is, of course, observed with a very well damped transducer whose acoustoelectric transfer function was calculated for a high impedance backing of 17.10^6 Raileigh and for a transducer impedance of 20.10^6 Raileigh (Fig. 3). Moreover, in this case, the monochromatic state has no time to occur.

However, if the peak value of the velocity potential and of the pressure is practically the same for different acoustoelectric transfer function (well damped and narrow band transducer), the peak value of the echographic response in transmit-receive mode may be quite different (Fig. 4).

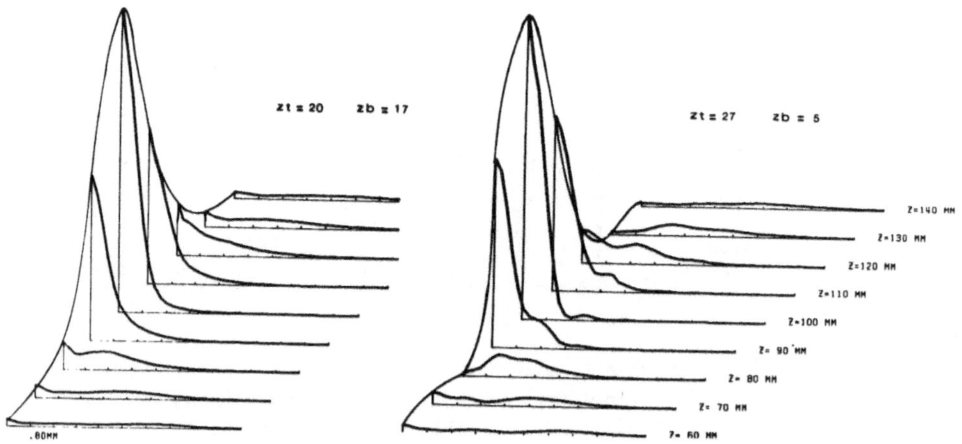

Fig. 5 Echographic response at different depths.
Left figure : well damped transducer
Right figure : narrow band transducer

This is due to the fact that the echographic response is dependant on the autoconvolution product of the velocity potential through the formula 11 , and that the peak value of the autoconvolution product of symmetrical functions is equal to the energy of these functions.

A study of this energy around the "zeroes" of the velocity potential shows, that if this energy presents a strong minimum around these zeroes for long excitations, it is not the case for short signals (there is no time for the destructive interference of the monochromatic state). Thus, the echographic response of a well damped transducer has a smooth pattern, compared to the one of a narrow band transducer.

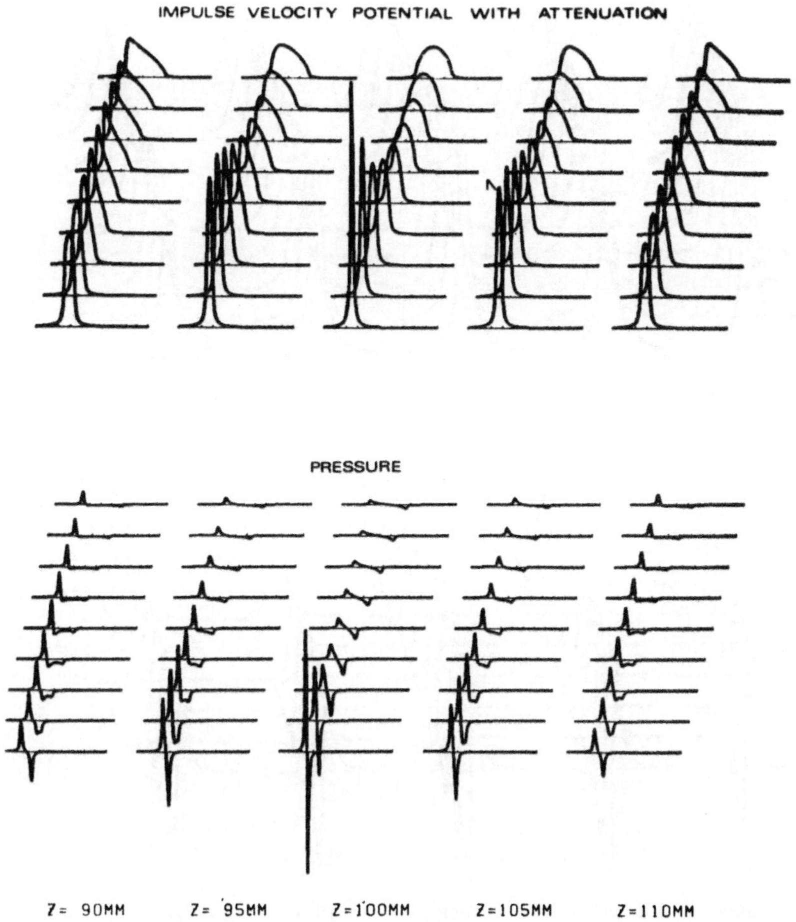

Fig. 6 Impulse potential and impulse pressure for an attenuation of .1 db by cm and by MHz.

Effect of the attenuation

We have evaluated the modification of the impulse response due to the acoustic losses observed in biological medium.
On the figures 6 and 7, we have represented respectively the impulse velocity potential and the impulse pressure response for two different attenuation coefficient. These results are obtained by computing the convolution product of the Lorentzian spreading function with the usual loss-less impulse response.

In figure 6, the attenuation coefficient is equal to .1 db by cm and by MHz. This is a small value compared with the one observed in tissue like medium (≃1 db by cm and by MHz). However, the interest

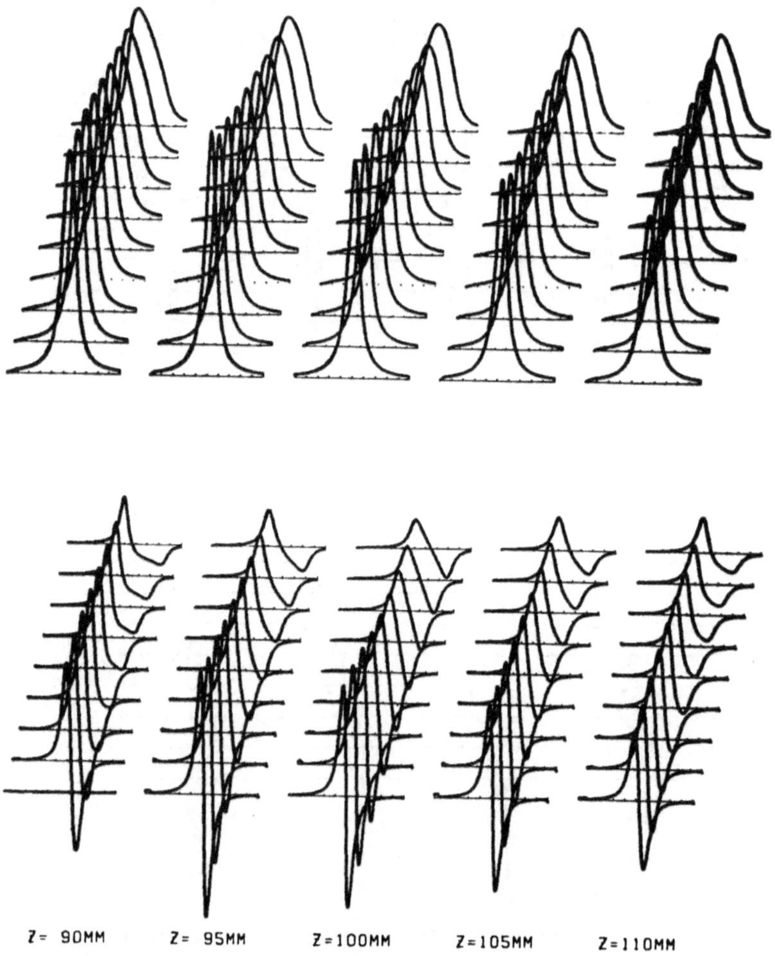

Fig. 7 Impulse potential and impulse pressure for an attenuation of .5 db by cm and by MHz.

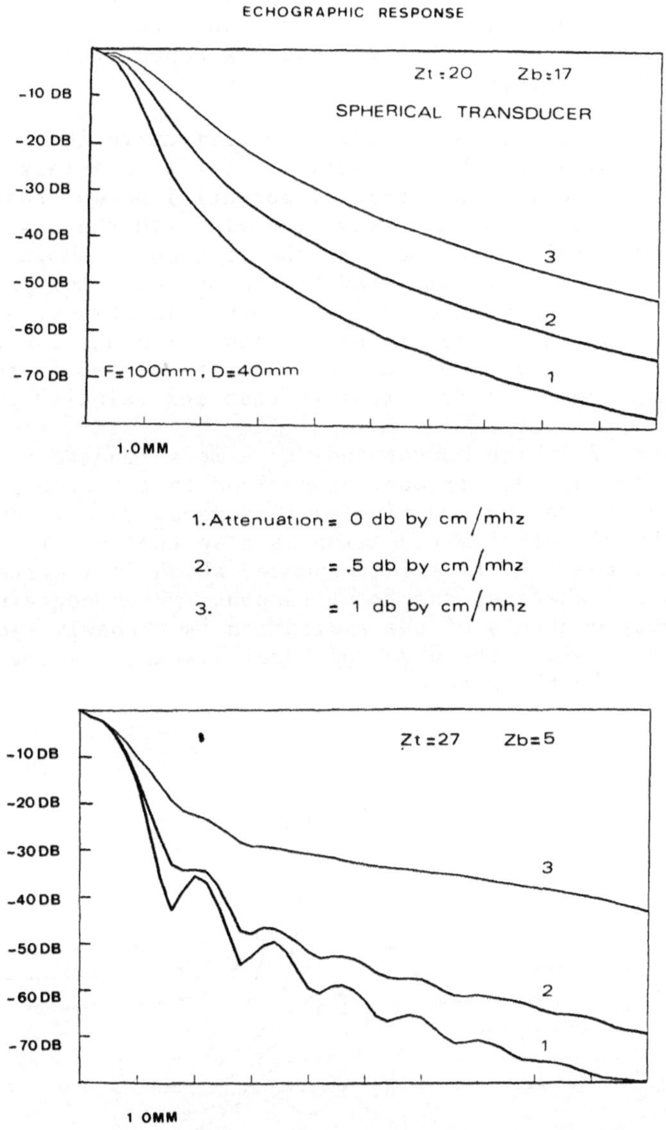

Fig. 8 Echographic response in the focal plane for different attenuations
Top figure : well damped transducer
Bottom figure : narrow band transducer.

of this representation comes from the fact, that even with the small filtering attenuation effect, the singularities of the impulse response (velocity potential and pressure) disappear. Moreover, on the figure 6, it may be observed that the transverse pattern of the impulse pressure presents a better focusing aspect than for the one of the impulse velocity potential.

This effect is linked to the time derivative involved in the pressure determination. More precisely, for a loss-less medium, the observation of the impulse velocity potential shows that the duration of these responses increases linearly with the distance m from the focal point (18), and thus, in the frequency domain, the effect of diffraction may be assimilated to the one of a low pass filter whose cutoff frequency becomes lower when m increases. As the time derivative involved in the pressure calculation has a high pass filtering effect, the combined results of these two filters involves a better focal response for pressure than for velocity potential.

In figure 7, which corresponds to a more important attenuation .5 db by cm and by MHz, it must be noticed that a strong defocusing phenomena occurs on the two impulse responses. This is due to the fact that the effect of attenuation is also the one of a low pass filter. Thus, the far off axis response, which is low frequency, is not perturbed ; whereas, the focal response which contains all the high frequency contents of the excitation is strongly reduced by these filter. Besides the observed focal distance is then reduced as it is shown in the figure.

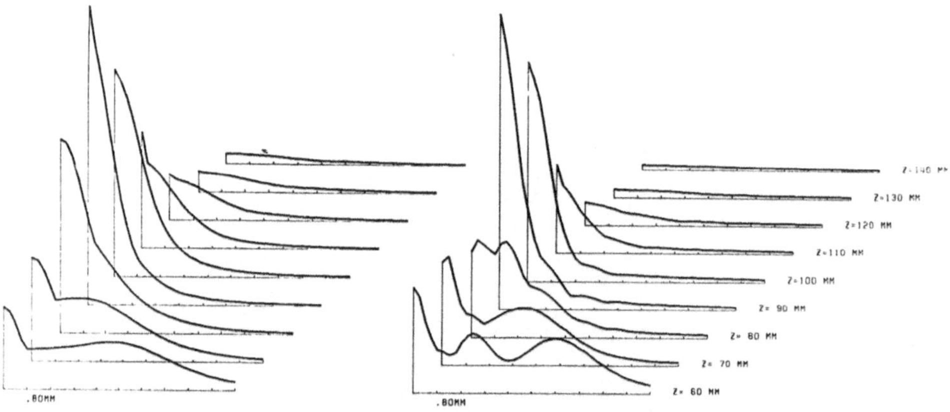

Fig. 9 Echographic response at different depth for an attenuation of 1 db by cm and by MHz
left figure : well damped transducer
Right figure : narrow band transducer

In figure 8, we have compared the peak value of the echographic response calculated in the focal plane for different values of the attenuation coefficient.
It must be checked that in the case of a well damped transducer (top figure), the curves 2 and 3 corresponding respectively to an attenuation of .5 and 1 db by cm and by MHz show a strong increasing of the 6 db beamwidth and of the side lobe level.
The bottom figure, which corresponds to a narrow band transducer shows a different aspect. The 6 db beamwidth changes a little, whereas the side lobe level is strongly increased.

The figure 9 shows that for these two kinds of transducers, there is also a very strong side lobe level coming from the nearest field points, that is involved by the low frequency contents of this zone.

To conclude this study, we may notice that even in the case of narrow band transducers strong effects are observed in absorbing medium. In all the cases, the side lobe level seems to increase hardly ; which is not the case for the 6 db beamwidth. The extension of these investigations to different focusing procedures is now under study.

REFERENCES

1 F.S. FOSTER, and J.W. HUNT
Transmission of ultrasound beams through human tissue - Focusing and attenuation studies.
Ultrasound in Med. and Biol., Vol. 5, 257-269, 1979.

2 P.R. STEPANISHEN
Wide bandwidth Acoustic near and far field transient from baffled pistons.
IEEE, Ultrasonics Symposium Proceedings, 113-118, 1977.

3 D.E. ROBINSON, S. LEE, L. BESS
Near field transient radiation patterns for circular pistons.
IEEE Transactions on acoustics, speech and signal processing ASPP 22, N° 6, 395-403, 1974

4 D.R. DIETZ, S.J. NORTON and M. LINZER
Wideband annular array response.
IEEE Ultrasonics symposium proceedings, 206-211, 1978.

5 M. FINK and M.T.LARMANDE
Wideband Fresnel focusing array response
IEEE Ultrasonics Symposium Proceedings, 204-209, 1979.

6 M.O. DONNEL, E.T. JAYNES, J.G. MILLER
Mechanisms : Relationship between ultrasonic attenuation and dispersion.
Tissue characterization meeting, Gaithesburg, 1978.

MATCHED GAUSSIAN APODIZATION OF PULSED ACOUSTIC PHASED ARRAYS

Andre J. Duerinckx

Computer Sciences Department
IBM Thomas J. Watson Research Center
P.O.Box 218, Yorktown Heights, NY 10598

ABSTRACT:

Gaussian apodization weighting functions matched to the time delay distribution for focusing of phased arrays are introduced. The matching concept is based upon the theory of eigenmodes of freely propagating acoustic beams with a finite cross-section. The effect of this matched gaussian apodization on the beam cross-section is analysed for several acoustic phased array configurations generating acoustic wavefronts in a weakly attenuating and homogeneous medium. A strength of focusing parameter is introduced. The relation between apodization parameters and the strength of focusing is analyzed and illustrated. For strong focusing the parameters of the gaussian matched apodization become independent of the focal range. A simulator was developed which allows the study of acoustic pressure waves in the near-field and far-field generated by a linear phased array excited with short pulses. The simulator provides 2-D and 3-D representations of acoustic wave fronts generated by a pulsed phased array, illustrating the effects of apodization for pulsed acoustic beams in the time-space domain.

Key words: Phased array, pulsed operation, apodization, matched gaussian apodization, strength of focus, modelling acoustic wavefronts by computer.

1. Introduction

Electronically steered and/or focused acoustic transducer arrays, referred to as phased arrays, are emerging as essential parts of modern medical ultrasonic imaging systems [1 – 8], mainly because they provide a method of rapidly sequentially addressing an area or volume of interest without the mechanical scanning of a transducer. Each transducer element of the array has a controlled delay element added to it. For a linear array, a linearly varying time delay causes steering of the acoustic beam, and a quadratically varying time delay causes focusing. The acoustic beam generated by an array consists of a main lobe and grating lobes [1], with side lobes of varying intensity. Amplitude apodization of phased arrays is a well known technique to control the shape and level of side lobes for cw operation [9]. Apodization of pulsed phased arrays is a less documented technique [2,10].

In semi-cw operation, using long sinusoidal bursts, the lateral resolution of a phased array is governed by classical diffraction and the depth (or range) resolution by the pulse response. In the focal zone, which is part of the "unfocused array" near-field, the beam cross-section or beampattern can be approximated as the Fourier transform of the array geometry, taking into account the amplitude (apodization) and phase (time delays) modulation. In the "unfocused array" far-field the acoustic beam cross-section becomes complicated, except for those arrays that generate a simple combination of normal modes of freely propagating acoustic beams with a finite cross-section.

In pulsed operation the array elements are pulse excited and wide-band diffraction theory shows the interaction between temporal and spatial response. For a pulsed array with or without apodization it becomes very difficult to make accurate predictions about the acoustic beam cross-section based upon cw behaviour for the same geometry. Consequently, there seems to be a need for additional analysis of array near-field patterns [10,11] for pulsed arrays with narrow long PZT elements [18 – 21] as used in medical applications.

In this paper we study a very simple but important type of amplitude apodization, namely "matched gaussian apodization". Gaussian apodization is an effective way to eliminate side lobes in the focal zone of a cw phased array. There is only one parameter to be adjusted for linear arrays: the gaussian width w_a of the gaussian amplitude profile at the transmitter array. A natural question to ask is as follows: is there an optimal w_a value? In order to answer this question, one has to first define one of the following parameters as the key to an optimization criterion: the amount of side-lobe suppression versus beam width increase at the focal point, the beamwidth in the far-field and the departure from gaussian shape in the far-field, to list a few.

Criteria based on an individual or a combination of these parameters are rather arbitrary. However, this should not prevent us from choosing one of these parameters, if we keep in mind that our definition of "optimum beam" is not absolute. We will try to match the gaussian beam width w_a to the distribution of time delays used to focus the array at a focal range F, in such a way that the acoustic beam cross-section stays gaussian over the total range of interest (near-field and far-field), and we define such a beam as "optimal" for our purposes. In order to explain this apodization matching, we will review the theory of eigenmodes of freely propagating optical (and acoustical) beams with a finite cross-section. As opposed to plane waves which are the eigenmodes of freely propagating beams with an infinite cross-section, the eigenmodes of freely propagating beams with a finite cross-section are gaussian beams. We will present a derivation of a "matched gaussian apodization" scheme that allows us to launch focused gaussian beams from a cw phased array. We will study how well this scheme works for pulsed phased arrays. The purpose of this paper is to investigate the effects of this one particular type of apodization in the time-space domain for pulsed phased arrays, by following an acoustic pulse while it propagates through an homogeneous medium with negligeable acoustic attenuation. Other researchers have looked at the effect of apodization in the focal zone of a cw phased array [2,9].

In section 2 we present a model to study transient effects in the acoustic wavefronts generated by pulsed phased arrays in an area covering the near-field of the corresponding unfocused arrays. The near-field of individual array elements is not modeled. The model is a compromise between a wide-band diffraction approach and practical considerations such as the trade-off between accuracy and complexity of a computer simulation.

In section 3 we introduce the concept of "strength of focusing" which will be a key parameter in the calculation of the parameters for matched gaussian apodization of phased arrays.

In section 4 we introduce the concept of matched gaussian apodization. In order to do this we first review the cw wave theory for freely propagating acoustic beams with a finite cross-section, and redefine the concept of "gaussian beam" in acoustics. We also review the approximations for deriving beamprofiles in the focal zone of a cw focused array. We will contrast cw models (pulse-echo mode) of beam profiles to computer simulations of wavefronts (transmission mode) generated by pulsed phased arrays generating polychromatic acoustic beams.

In section 5 we study the instantaneous pressure wave generated by a transmitter array and use advanced computer graphics techniques to display the pressure wave. Transient effects, such as the change in lateral and depth resolution, the structure (geometrical location and amplitude) of side lobes and grating lobes, are analyzed and displayed in both the "unfocused array" Fresnel (near-field) zone and the transition between "unfocused array " Fresnel and Fraunhofer zone. These transient effects are shown for both uniform and matched gaussian apodization, which allows us to comment on the global effects of apodization in the time-space domain.

2. Modelling Pulsed Acoustic Phased Arrays.

We will consider phased arrays sending out acoustic energy into a homogeneous medium, such as a water-tank, and operated in a way similar to the way arrays are being used for medical B-scan imaging. We will consider pulsed operation with a center frequency f = 2.25 MHz and a bandwidth at 50 percent response of 35 percent typical for echocardiography [7]. We assume a homogeneous medium with frequency independent acoustic attenuation. The frequency dependence of the acoustic attenuation $\alpha(f)$ could easily be taken into account, at a cost in computing time. An inhomogeneous index of refraction ditribution would require the addition of a two-boundary condition ray tracing algorithm, and could also be taken into account.

2.1 Transmitter arrays

The geometry of the two central elements of an acoustic array is shown in figure 1. Figure 1 explains the definition of the position vectors $\bar{r}_i$, $i = -N/2, ..., +N/2$ (for even N), centered at the middle of each transducer. Figure 1 also indicates our choice of polar coordinate system in the azimuthal scan plane (the z-y plane). The acoustic pressure wave propagates in the z direction. The focal point range (F) and steering angle (θ), are combined as a vector $\bar{r}_f = (F,\theta)$ in polar coordinates. For a transmitter array we will use the notation $\bar{r}_{tr,f}$ instead of $\bar{r}_f$ whenever confusion is possible with the focusing parameters of a receiver array that is part of the same imaging system.

Using the Huygens principle [12], and referring to figure 1 for the geometry and coordinate definitions, we calculate the instantaneous pressure $P(\bar{r},t;\bar{r}_{tr,f})$ generated by a pulsed transmitter phased array as the summation of the contribution from all transducer elements for a point $\bar{r}$ at a fixed range R from the center of the array within the azimuthal scan plane:

$$P(\bar{r},t;\bar{r}_{tr,f}) =$$

$$\sum_{i=\frac{-N}{2}}^{\frac{+N}{2}} \left\{ S_i(\bar{r}-\bar{r}_i)\; p_{tr,i}[t - \frac{|\bar{r}-\bar{r}_i|}{v_0} - \Delta t_i(\bar{r}_{tr,f})]\; e^{-\alpha_o(|\bar{r}-\bar{r}_i|)} \right\} \quad (1)$$

where t is the time, the function $S_i(\bar{r}-\bar{r}_i)$ is the radiation pattern generated by an individual array element, and a frequency independent acoustic attenuation $\alpha(f) = \alpha_0$ and acoustic velocity $v(f) = v_0$ are assumed. The time delays Δt_i are a function of the desired steering angle and focal range. The function $p_{tr,i}(t)$ is the acoustic pressure wave generated by an individual array element after shock excitation. We will assume that $p_{tr,i}(t) = p_{tr}(t)$, i.e., all transducers have the same temporal output. Figure 2 shows a typical temporal response $p_{tr}(i)$ and the corresponding frequency spectrum. Figure 3 shows the array geometry for $N = 16$ transducer elements.

For small steering angles and frequency independent attenuation Eq. (1) is a good wide-band diffraction model to describe a pulsed transmitter phased array. The use of a short temporal pressure wave $p_{tr,i}(t)$, only a few cycles long, makes it definitely a wide-band model. However, Eq. (1) is a combination of a wide-band and narrow-band model, very different from the approach taken in [11], as will be explained in section 2.2 .

The frequency dependence of the radiation beam pattern $S_i(\bar{r}-\bar{r}_i)$ and acoustic attenuation $\alpha(f)$ causes temporal stretching of the pressure wavefront due to high frequency cut-off. Also, the beam divergence increases due to beam softening, i.e., a loss of high frequencies, resulting from increased attenuation with frequency in biological tissues [13,14]. We can neglect these effects for transmitter arrays excited with very long sinusoidal pulses where most of the energy is concentrated in a very narrow frequency band (narrow-band operation). For short pulse excitation, or wide-band operation, neglecting these frequency dependencies, i.e. using a narrow-band model, will limit the validity of our simulation. The pressure wavefront $p_{tr,i}(t)$ is a function of the angle θ_i (indicated in figure 1) for large angles [15,16]. The acoustic beam experiences a decrease in intensity due to spherical wave propagation, a minor effect in the far-field for small steering angles. But for small steering angles and frequency independent attenuation none of the above mentioned effects is significant enough to change the conclusions to be drawn from our simulations.

The purpose of our acoustic wavefront simulator based upon this model is not to study wide angle steering and strong attenuation with resulting beam softening, but rather to investigate general properties of pulsed phased arrays. Therefore we decided to limit the complexity of the model. We believe that our hybrid (wide- and narrow-band) model is probably a good compromise between the wide-band nature of the acoustic pressure waves generated by a pulsed array, a wide-band diffraction approach and the complexity of an efficient and fast computer algorithm.

The description of a receiver phased array is different from that used to describe a pulsed transmitter array, but the resulting equations are almost identical in form, if one uses the array to measure scattered signals (as in echocardiography). We will use the notation $p_{rec,i}(t)$ for the electrical response to an acoustic pressure impulse of the detection system consisting of element i of a receiver array and the additional electronics.

2.2 Far field of narrow transducers

As stated earlier we want to describe the near-field of the total array. The near-field of the unfocused array generally extends to a larger range than that of any individual array element. Consequently we can use a model for the far-field of individual transducer elements to cover most of the near-field of the unfocused array. When we focus the array the focal zone lies within the near-field of the unfocused array. The radiation pattern of any transducer element is given by the following far-field radiation pattern:

$$S_i(\bar{r}-\bar{r}_i) = A_i \ G_i(|\bar{r}-\bar{r}_i|) \ \gamma_i(\theta_i) \quad for\ i = -\frac{N}{2},\ldots,\frac{N}{2} \tag{2}$$

where N is the number of transducers in the array (N assumed even here), θ_i is the angle between the perpendicular to the transducer element and the line connecting the transducer element and the point $\bar{r} = (x = x_0,y,z)$ as shown in figure 1, A_i are the apodization parameters, $G_i(|\bar{r}-\bar{r}_i|)$ is a geometrical spread-factor (beam expansion), and $\gamma_i(\theta_i)$ is the beam profile for individual transducers measured at the center frequency. The frequency dependence of $S_i(\bar{r}-\bar{r}_i)$ has been left out because we use a narrow-band model to describe the effect of the beam pattern of individual array elements.

There exists an extensive body of literature analyzing the field characteristics of pulsed and cw piezoelectric radiators. However, the majority of the work has considered transducer apertures several wavelengths in size. In phased array applications using long, narrow piezoelectric elements, the width to thickness ratio w/t is on the order of one. The mechanical [17 – 19,21] and electrical [20] coupling between such elements and the resulting narrowing of the radiation beamprofile have been documented. This difference in beampattern between wide and narrow transducers is the reason why we decided not to take the approach in [11], which uses models for transducer beam patterns as developed by Stephanishen and Freedman [15,16] that are only valid for very wide transducer elements.

In our simulations we consider transducer elements with a width w on the order of a wavelength. This means that we can consider most objects, when placed at a distance of several inches from the array, to be in the far-field of the individual transducers. We do not know of any published analytical expressions for the beampattern of narrow transducers, and will therefore approximate the beamprofile $\gamma_i(\theta_i)$ of a transducer element with width w by the following analytical expression, which is a valid far-field model when $w >> \lambda$ and $w = w_{eq}$, but is used here as an heuristic model for the case where $w \cong \lambda$:

$$\gamma_i(\theta_i) = \mathrm{sinc}(\frac{w_{eq}}{\lambda} \sin \theta_i) \quad , \tag{3}$$

where w_{eq} is called the equivalent width, λ is the wavelength, and the sinc-function is defined as $\mathrm{sinc}(x) = \sin \pi x / \pi x$. Our choice of equivalent width is such that our analytical beam pattern matches experimentally measured patterns [18]. Typically w_{eq} is larger than the real element width w (see figure 1). The use of Eq. (3) will of course limit the validity of our results to small steering angles, because we neglect large angle effects as descibed in [15,16].

2.3 Steering

Linearly varying delays Δt_i are used to steer the acoustic beam over an apgle θ. We will limit our steering angle to small values (from $+ 20$ to $- 20°$). This limitation will be sufficient to stay within the validity bounds of our model. The side lobe structure (in spatial and temporal domain) is strongly dependent upon the steering angle θ as is illustrated in figure 6 of reference [22*b*]. We do not need to go to very large angles to study some of the effects of non-zero angle steering.

2.4 Focusing

The delays Δt_i are a function of the desired focal point position and beam steering angle. Quadratically varying delays are used to focus the acoustic

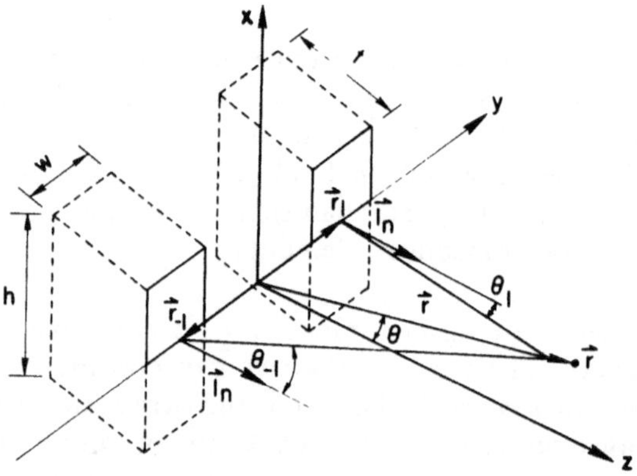

Fig. 1 Geometry of a linear array and choice of origin for polar coordinate system.

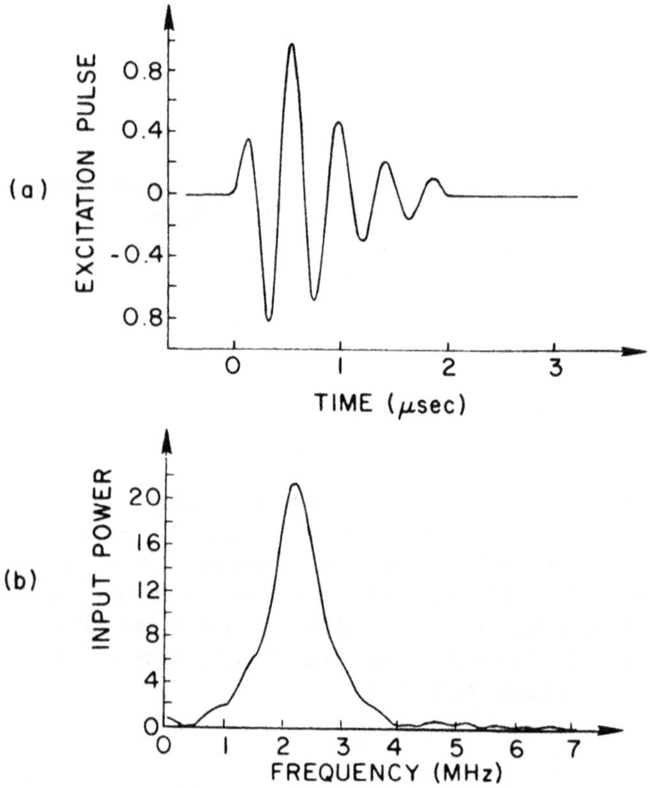

Fig. 2 Typical temporal impulse response (a) and corresponding frequency spectrum (b) of a single transducer as used in our simulations. The impulse response is an experimentally measured electrical output, using a single transducer.

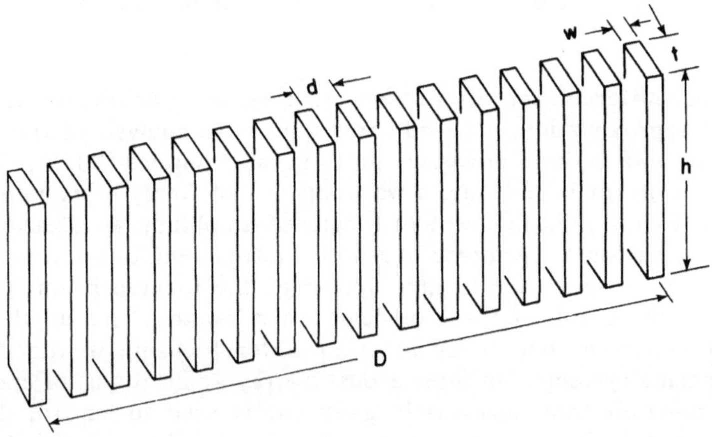

Fig. 3. Geometry of a linear array with 16 transducer elements, as used in our simulations.

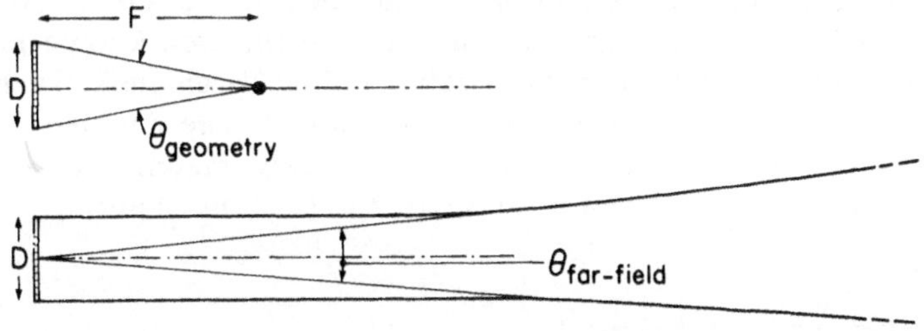

$$K = \frac{\theta_{geometry}}{\theta_{far\text{-}field}} = \frac{D^2}{2F\lambda}$$

Fig. 4 Strength of focusing parameter of a phased array.

beam. The calculation of the contribution to Δt_i for focusing is straightforward [1 – 3] using a spherical wave model. The conventional approach, spherical waves converging towards a point, assumes a beam cross-section at the array with constant amplitude (*or* $A_i = 1$ for all i) and time delays (or phase delays) that vary quadratically with the position of the transducer in the array.

A slightly different approach (see section 4), using diffraction theory and the Fresnel approximation, is based upon a cw wave analysis of freely propagating beams with a finite cross-section as in laser physics [23,24]. The most important reason for introducing a wave analysis of freely propagating beams is that it will clarify the concept of "matched amplitude apodization" to be introduced in section 4. Another reason for introducing this approach is that it allows us to estimate the length over which the focused beam stays collimated, called the depth of focus or beam waist length. This depth of focus can be an important parameter for transmitter focusing in the design of acoustic imaging systems. In most acoustic array applications an apodization weighting function (not necessarely gaussian) is used to control the shape (i.e., the side lobe levels and extent) of the acoustic beam in the lateral direction. The 2-D shape of the beam cross-section can be decomposed as a superposition of 2-D Hermite-Gaussian polynomials, which allows us to use the gaussian beam theory [25].

In the case of uniform apodization one can derive the parameters of the gaussian beam that most closely matches the beam with uniform apodization. One can do this by requiring that both beams, i.e., the beam with uniform apodization and the gaussian beam, have the same 3 dB beam width in the lateral direction (i.e., in a cross-section) at the focal range. Once the parameters of the gaussian beam are known, one can use the well-known properties of gaussian beams [24] to approximately predict parameters, such as the depth of focus, of a beam with uniform apodization. The detailed derivation of the parameters of a gaussian beam that closely matches a beam with uniform apodization will be given in subsection 4.2 of this paper.

3. Strength of focusing

An important parameter that determines the level of complexity of the pressure wavefront generated by a pulsed phased array is the degree of focusing [22]. This strength of focusing plays also an important. role in the study of matched gaussian apodization for pulsed arrays, as will be shown in this paper. We therefore define, as a measure of the degree of focusing, the following parameter [6,12]:

$$K = \frac{D^2}{2\ F\ \lambda}, \tag{4}$$

where F is the focal distance. The definition can be generalized for non-zero steering angles. Our measure of focal strength K has a very simple geometrical interpretation. We can redefine K as

$$K = \frac{\theta_{geometry}}{\theta_{\text{far-field}}}, \tag{5}$$

where the geometric optics aperture is defined as

$$\theta_{geometry} = \frac{D}{F}, \tag{6}$$

the angle spanned by the array when viewed from the focal point, and the diffraction limited focal zone angle is

$$\theta_{\text{far-field}} = 2\frac{\lambda}{D}, \tag{7}$$

for our choice of a uniform unity apodization weighting function. We also have that

$$\theta_{geometry} = \frac{1}{f}, \tag{8}$$

where f is the f-number of the focusing system. This physical interpretation of the definition of the K-number is clarified in figure 4. A typical K-value for f_0 = 2.25 MHz, D = 25.4 mm (1 in.) and F = 152.4 mm (6 in.) is K = 3.218 . Small K-values, i.e., K-values smaller than 4, correspond to weak focusing, and large K-values indicate strong focusing. The separation line of K = 4 is rather arbitrary and was chosen after analyzing many different wavefronts (see ref. [22]). Strong focusing is characterized by a more complicated acoustic wavefront shape, with the useful acoustic energy spread over a much larger area in the plane of propagation, whereas for weak focusing, there is a greater spatial concentration of the acoustic energy. This difference in wavefront complexity is illustrated in the upper parts of figures 12, 14 and 15. In figure 12 we have K = 3.218 and weak focusing. In figures 14 and 15 we have strong focusing with respectively K = 12.9 and K = 9.7 . The change with K of the spatial extent, amplitude and complexity of the side lobes surrounding the main lobe is clearly visible. A more detailed explanation of these features will be given in section 5.1 .

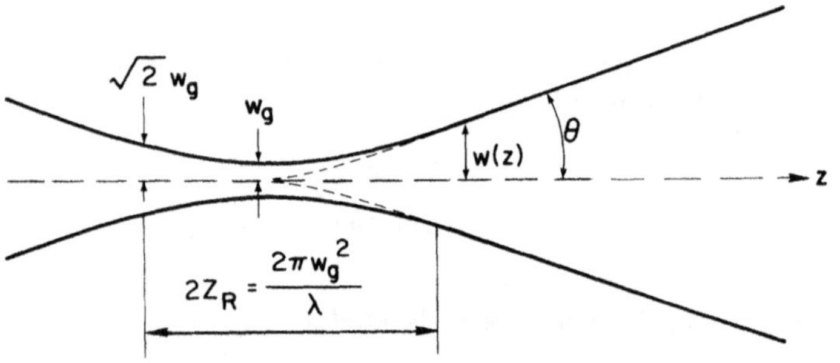

Fig. 5 Contour of gaussian beam and graphical representation of the Rayleigh range Z_R and gaussian beam width w(z).

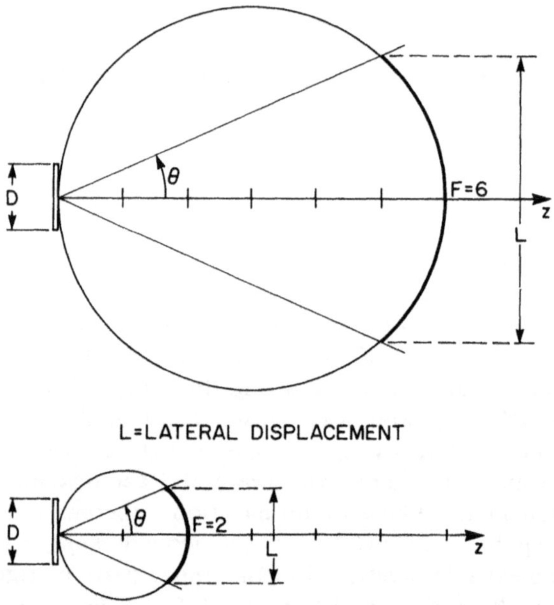

Fig. 6 Focal zone of an acoustic cw phased array, with width D and : (top) focal range $F = 6D$, (bottom) $F = 2D$.

The parameter K allows us to define the Fresnel and Fraunhofer zones [1,12] for a linear unfocused array. The Fresnel (near-field) zone for the unfocused array starts at a distance z such that

$$\frac{z^3}{D^2\ F} >> \frac{\pi}{32}\ \mathrm{K} \tag{9}$$

The Fraunhofer (far-field) zone starts at a distance z such that

$$\frac{z}{F} >> \frac{\pi}{2}\ \mathrm{K}. \tag{10}$$

A more general definition for the Fraunhofer zone at non-zero steering angles can be found in [26]. When we focus the array, the focal zone where $z \cong F$ lies in the "unfocused array" Fresnel zone and not in the Fraunhofer zone. It is important to emphasize that the wavefronts generated using our model (as shown and discussed in section 5, figures 12, 14 and 15) are acoustic wavefronts in the near-field of the unfocused array and in the transition zone between near-field and far-field. However, for the individual transducers the wavefronts are in the far-field.

4. Matched Gaussian Amplitude Apodization

Amplitude apodization of phased arrays is a well known and proven technique for controlling the beam cross-section, i.e., the shape and levels of side lobes, for cw operated arrays [9]. Digital filtering techniques are often used for this purpose [27]. The usefulness of apodization for pulsed phased arrays is not as clear [2,10] as it is for cw phased arrays. Compared to cw systems, apodization reduces resolution (a larger 3 dB beamwidth) and provides significantly less sidelobe suppression for wide-band systems [2]. In this paper we will concentrate on one particular scheme, that we call "matched gaussian apodization". In order to explain this apodization matching, we will review the theory of eigenmodes of freely propagating optical (and acoustical) beams with a finite cross-section. We will study how well this apodization scheme works for pulsed phased arrays. The purpose of this paper is to investigate the effects of this one particular type of apodization in the time-space domain for pulsed phased arrays, by following an acoustic pulse while it propagates through an homogeneous medium, as opposed to only looking at the effect of apodization in the focal zone of a cw phased array.

4.1 CW analysis of acoustic beams.

The theory of freely propagating light beams or "beam waves" with a finite cross-section is well known [23,24] and has been applied to predict the behaviour of optical beams in laser resonators, optical waveguides and other optical elements [23]. A similar theory can be readily applied to acoustic beams generated by an infinitely baffled plane transducer or transducer array of finite dimensions [12,21]. The far-field pattern prediction given using a Huygens integral approach [1,12] is identical to the prediction of a "gaussian beam" analysis [23,24] for a coherently excited transducer (i.e., no time delays in the case of an array). For the purposes of this analysis we will neglect grating lobes and concentrate on the main lobe and side lobes. Therefore we can replace the array with a long continuous transducer. We will very closely follow the analysis given in [23] for 3-D optical beams. In this paper we only study apodization of 1-D linear arrays, and therefore a model for 2-D acoustic beam propagation would be sufficient. The derivation and equations for 2-D beams and 3-D beams are identical. We nevertheless decided to derive most of the equations for 3-D beam propagation.

If one assumes an homogeneous loss-free acoustic medium with constant density and constant velocity, one can use the following simplified Helmholtz equation for the acoustic pressure p :

$$\nabla^2 p + k^2 p = 0, \tag{11}$$

where $k = 2\pi/\lambda$ is the propagation constant in the medium. Let us assume, for an acoustic wave travelling in the z direction (see figure 1), that

$$p(x,y,z) = P(x,y,z) \ \exp(-jkz) \tag{12}$$

where P(x,y,z) is a slowly varying complex function which represents the differences between an acoustic beam and a plane wave, namely: a nonuniform intensity distribution, expansion of the beam with distance of propagation, curvature of the phase front, and other differences discussed below. By inserting Eq. (12) into Eq. (11) one obtains

$$\frac{\partial^2 P}{\partial x^2} + \frac{\partial^2 P}{\partial y^2} - 2jk\frac{\partial P}{\partial z} = 0, \tag{13a}$$

where it has been assumed that P varies so slowly with z that its second derivative $\partial^2 P/\partial z^2$ can be neglected. To study 1-D linear phased arrays one can separate P(x,y,z) as $P(x,y,z) = P_x(x,z)P_y(y,z)$, and, by making identical assumptions, obtain the equations for 2-D beam propagation in the y-z plane as

$$\frac{\partial^2 P_y}{\partial y^2} - 2jk\frac{\partial P_y}{\partial z} = 0. \tag{13b}$$

The differential Eq. (13) for P has a form similar to the time dependent Schrodinger equation. It is easy to see that

$$P(x,y,z) = \frac{w_g}{w(z)} \exp\left[j\phi(z) - \frac{jkr^2}{2R(z)} - \frac{r^2}{w(z)^2}\right] \tag{14}$$

is a solution of Eq. (13), where, for 3-D wave propagation,

$$r^2 = x^2 + y^2. \tag{15}$$

For 2-D wave propagation in the y-z plane, the equations are identical, except for the replacement of P(x,y,z) by $P_y(y,z)$ and $r^2 = y^2$. The meaning of the constant parameter w_g (g stands for "gaussian") will become clearer soon. The parameter $w(z)$ describes the gaussian variation in beam intensity with the distance r from the central axis (the z-axis for zero angle steering). The parameter $R(z)$ describes the curvature of the phase front which is spherical near the axis. The parameter $\phi(z)$ represents a phase shift which is associated with the propagation of the acoustic beam. The factor $w_g/w(z)$ gives the expected intensity decrease on the axis due to the expansion of the beam. The two real beam parameters $R(z)$ and $w(z)$ can be combined into one complex beam parameter q(z) as

$$\frac{1}{q(z)} = \frac{1}{R(z)} - j\frac{\lambda}{\pi w(z)^2}. \tag{16}$$

Introducing this complex beam parameter allows us to rewrite the solution [*Eq.*(14)] as

$$P(x,y,z) = \frac{w_g}{w(z)} \exp\left[j\phi(z) - jk\frac{r^2}{2q(z)}\right] \tag{17a}$$

for 3-D beam propagation, or as

$$P_y(y,z) = \frac{w_y}{w(z)} \exp\left[j\phi(z) - jk\frac{y^2}{2q(z)}\right] \tag{17b}$$

for 2-D beam propagation in the y-z plane.

We now define the distance Z_R as

$$Z_R = \frac{\pi w_g^2}{\lambda}. \tag{18}$$

The Rayleigh range, Z_R, is the 3 dB depth of focus for a gaussian beam with both gaussian amplitude and quadratic phase profile, as is indicated in figure 5. For a detailed derivation of this property we refer to [23,24]. Up until now we have not defined the origin of the z-axis. In our analysis of phased arrays the z = 0 origin does correspond to the position of the array. We will now define a new z-axis origin (indicated as z') such that $z' = 0$ corresponds to the waist of the gaussian beam, i.e., that part of the gaussian beam where the cross-section is the narrowest. Substituting Eq. (14) into Eq. (13) will allow us to calculate $R(z)$, $w(z)$ and $\phi(z)$ as follows:

$$w(z') = w_g \sqrt{1 + \left(\frac{z'}{Z_R}\right)^2}, \tag{19}$$

$$R(z') = z' \left[1 + \left(\frac{Z_R}{z'}\right)^2\right], \tag{20}$$

and

$$\phi(z') = \arctan\left(\frac{z'}{Z_R}\right) \tag{21}$$

where the origin $z' = 0$ corresponds to the beam waist (located at the focal point). A coherent acoustic beam with a gaussian intensity profile as obtained above is not the only solution of Eq. (13), but is perhaps the most important one. This beam is often called the "fundamental mode" as compared to the higher order modes to be discussed later. Figure 5 illustates the contours of the gaussian beam in Eq. (14) and gives a graphical representation of $w(z')$ in Eq. (19) . We now see how w_g can be interpreted as the beam waist. The gaussian beam contracts to a minimum diameter $2w_g$ at the beam waist where the phase front is plane. The 3 dB beam waist is $(w_g)_{3dB} = 1.20 w_g$ for the gaussian beam.

Thus far we only discussed one solution of Eq. (11). There are other solutions with similar properties, called the higher order modes. They have, in cartesian coordinates, the following general form :

$$P(x,y,z) = K_{nm} \; H_m(\sqrt{2}\; \frac{x}{w(z)}) \; H_n(\sqrt{2}\; \frac{y}{w(z)})$$

$$\exp\left(-j\phi(z) + k\, \frac{r^2}{2\, q(z)}\right), \tag{22}$$

where K_{nm} is a constant, the functions $H_m(x)$ and $H_n(y)$ are Hermite polynomials, the parameter q(z) is the same as for the fundamental mode, but

$$\phi(z') = (m + n + 1)\; arctan(\frac{z'}{Z_R}). \tag{23}$$

In the remainder of this section we will only consider the fundamental mode.

4.2 Matched gaussian apodization.

For a cw operated array of N elements, positioned as shown in figures 1 and 3 along the y-axis, with width D and sending out acoustic energy along the z-axis, one uses the following quadratic time delays to focus the array at a range $z = F$:

$$T(y_i) = \frac{y_i^2}{2cR_a}\; \cos^2\theta \tag{24}$$

where y_i is the position of the individual array elements along the y-axis, c is the speed of sound, θ is the steering angle, and R_a is the radius of curvature of the phase front. For a zero steering angle the corresponding phase delays are

$$\Delta\phi(y_i) = -j\omega T(y_i) = -jk\frac{y_i^2}{2R_a} \tag{25}$$

The conventional approach (spherical waves converging towards a point, i.e., $R_a = z$) assumes a beam cross-section at the array with constant amplitude and quadratically varying time delays. This causes the generation of side lobes.

We now introduce "matched gaussian apodization", a type of apodization where the amplitude weighting function

$$A_i = A(x = 0, y = y_i), \tag{26}$$

and, for 1-D phased arrays,

$$A(x,y) = A(x)\, exp[-\frac{y^2}{w_a^2}] \tag{27}$$

matches the phase delays $\Delta\phi(y_i)$, such that only the fundamental gaussian mode is generated. In order to do that one has to choose the parameters R_a and w_a as

$$w_a = w(z' = -F) = w_g \sqrt{1 + (\frac{F}{Z_R})^2}, \tag{28}$$

and

$$R_a = |R(z' = -F)| = F\ [1 + (\frac{Z_R}{F})^2], \tag{29}$$

where w_g is the beam waist of the fundamental gaussian mode, and Z_R as defined in Eq. (18). The problem here is that we do not know w_g.

At the focal point $z' = 0$ or $z = F$, for uniform apodization and R_a = F, the diffraction limited 3 dB beam width (narrow-band or monochromatic) w_o, is [1 – 3]

$$w_o = 0.88\ F\frac{\lambda}{D}. \tag{30}$$

The 3 dB beam width at the waist of the fundamental gaussian beam is $(w_g)_{3dB} = 1.20\ w_g$, as stated earlier. From a physics point of view it seems reasonable to require that

$$(w_g)_{3dB} > w_o, \tag{31}$$

in order not to generate a gaussian beam that is focused beyond the diffraction limits. If we focus the fundamental gaussian mode up to the diffraction limit, we have that

$$w_g = 0.733\ F\frac{\lambda}{D}, \tag{32}$$

or

$$w_g = 0.367 \frac{D}{K}, \tag{33}$$

where K was defined as a measure of the degree of focusing. Using Eqs. (4) and (33) we can rewrite the expressions for w(z) and R(z) as follows :

$$R(z') = z' \; [1 + 0.0445 \frac{1}{K_o^2} \; (\frac{F}{z'})^2], \tag{34}$$

and

$$w(z') = 0.092 \frac{D}{K_o} \sqrt{1 + 22.46 \; K_o^2 \; (\frac{z'}{F})^2}, \tag{35}$$

where

$$K_o = \frac{K}{4}. \tag{36}$$

For strong focusing ($K_o > 1$) we have that :

$$R_a = | R(z' = -F) | \cong F, \tag{37}$$

and

$$w_a = w(z' = -F) \cong 4.739 \; w_g K_o, \tag{38}$$

or

$$w_a = w(z' = -F) \cong 0.434 \; D. \tag{39}$$

From Eqs. (26), (27) and (39) we thus obtain that, for strong focusing,

$$A(x,y) = A(x) \exp [-1.327 \; (\frac{y}{D/2})^2], \tag{40}$$

and therefore, for an even number of tranducer elements N , we have

$$A_{N/2} = A_{-N/2} = A(x = 0, |y| = \frac{D}{2}) = 0.265. \tag{41}$$

The gaussian apodization function A_i becomes independent of the geometry for strong focusing. This has an important consequence for dynamic focusing, because it means that one can stay optimally apodized without changing the A_i's as long as the focal point is not too far away, and K_o remains large.

This shows the importance of knowing the strength of focusing in order to calculate the parameters w_a and R_a used to generate focused beams with matched gaussian apodization. By using matched gaussian apodization we are guaranteed that there will be virtually no side lobes for cw arrays and that the beam remains gaussian over a large propagation distance. It is the purpose of this paper to investigate the practical uses of matched gaussian apodization for pulsed phased arrays and see if similar conclusions apply.

Typical parameter values for f_0 = 2.25 MHz, D = 25.4 mm (1 in.), z = 101.6 mm (4 in.) and in water are: w_0 = 2.315 mm, w_g = 1.929 mm, K_o = 1.207, w_a = 11.2 mm, R_a = 104.7 mm, and Z_R = 17.781 mm. This order of magnitude should be kept in mind when one talks about "focusing the acoustic beam in a point". More correctly one should think of a focal area of dimensions $w_0 \; x \; Z_R$, which is usually not negligible in size.

4.3 CW focused arrays (transmission-reflection mode)

We will first analyze the beam cross-section (or beamprofile) around the focal point for a cw excited phased array with gaussian apodization functions. We will then compare these cw beam profiles to the pulsed beamprofiles obtained by measuring the shape of the beam cross-section of a pulsed phased array with a peak amplitude and an energy detector. (see subsection 4.4).

We use the same derivation as Shott [2] for the phasor $F_m(R,\theta)$ which describes the monochromatic response of a coherent-summation imaging system (a phased array) to a point-source reflector located at (R,θ), where the time dependent field pattern $F(t,R,\theta)$ is given as

$$F(t,R,\theta) = Re[\; F_m(R,\theta) \; \exp\,(j2\pi\nu_o t)\;], \tag{42}$$

and ν_o is the illumination frequency. We introduce a change of coordinates:

$$\overline{R} = \frac{R}{\lambda} \tag{43}$$

and

$$u = \sin\theta, \tag{44}$$

and we now write $F_m(R,\theta)$ as $F_m(\overline{R},u)$. The expression for $F_m(\overline{R},u)$ becomes very simple along the focal line [2] defined by the equation

$$\frac{1-u^2}{2\overline{R}} = \frac{1-u_f^2}{2\overline{R}_f}, \tag{45}$$

where R_f and $u_f = \sin\theta_f$ define the focal point. The shape of this focal line is indicated in figure 6 for a 16 element array with width D = 25.4 mm (1 in.) and focal range (a) F = 152.4 (6 in.) and (b) F = 50.8 mm (2 in.). Along such a focal line, the phasor $F_m(\overline{R},u)$ is equal to the scaled Fourier transform of the product of the apodization function with the array aperture. Figure 7 shows the amplitude of the phasor along the focal line as a function of angle $\theta = \sin^{-1}u$ for the array and focal line shown in figure 6(a) (D = 25.4 mm, F = 152.4 mm), for (a) uniform apodization and (b) matched gaussian apodization. One can notice a widening of the main lobe and a net reduction of the sidelobe level. Both plots shown in figure 7 have been normalized.

4.4 Pulsed focused arrays (transmission mode)

We have simulated the output of a peak amplitude and an energy detector that are moved laterally across the focused beam of a 16 element pulsed phased array, similar to the array shown in figure 3. The profile thus obtained is a measure of the acoustic pressure profile in a beam cross-section, along a line parallel to the array. Figures 8 to 11 show these simulated detected signals for different array and focal point geometries, at two different propagation distances: (a) $z = F$ and (b) $z = 2F$. The profile shown in figure 8(a) can be compared to the profile of the phasor $F_m(R,\theta)$ of the cw beam along the focal line in figure 7(b). The $\theta = 0°$ angular position in figure 7 corresponds to the zero (0) lateral position in figure 8. However, the reader should be aware of the differences between the two simulations.

We analyzed three cases. Figure 8 shows the peak amplitude profile at two different propagation (axial) distances for an array with width $D = 25.4$ mm, $F = 152.4$ mm and $K_o = 0.804$ (weak focusing). Figure 9 shows the total energy profile for the array of figure 8. Figure 10 shows the peak amplitude profile at two different propagation (axial) distances for an array with width $D = 50.8$ mm, $F = 152.4$ mm and $K_o = 3.218$ (strong focusing). Figure 11 shows the peak amplitude profile at two different propagation (axial) distances for an array with width $D = 25.4$ mm, $F = 50.8$ mm and $K_o = 2.413$ (strong focusing).

One immediately notices the substantial reduction of side lobes when the pulse is detected with conventional detectors (peak amplitude and energy). In

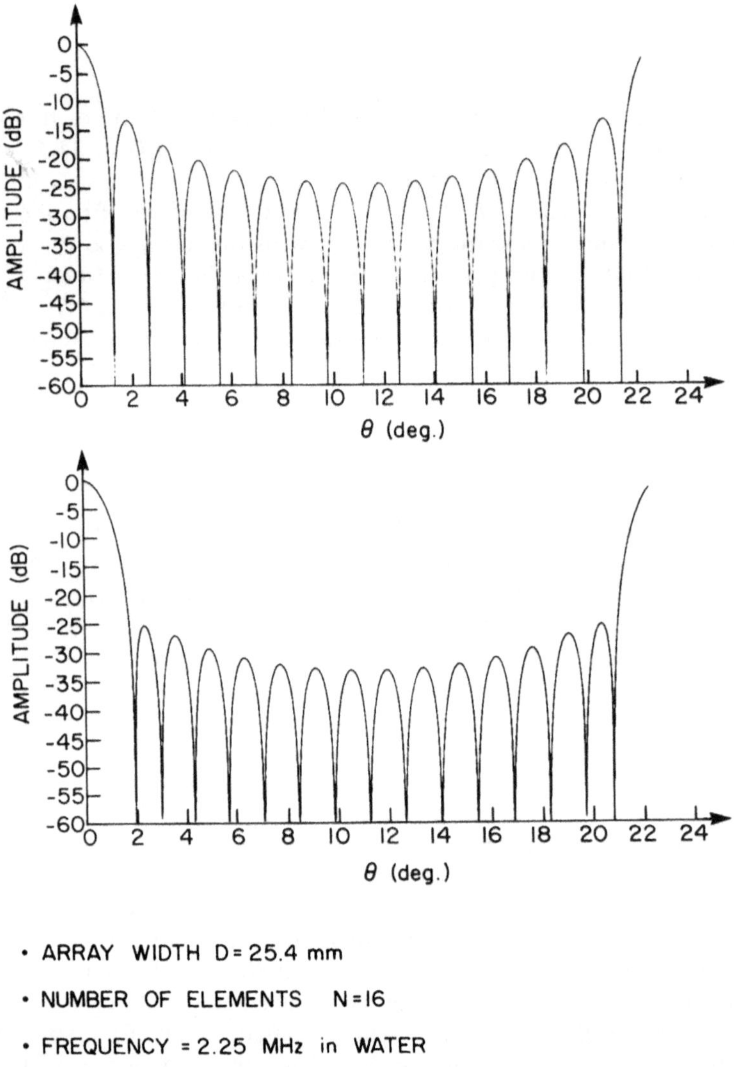

Fig. 7 Predicted amplitude profile in the focal zone for an acoustic cw phased array with width $D = 25.4$ mm and $N = 16$ elements, and (a) uniform apodization or (b) matched gaussian apodization for $F = 152.4$ mm (width $w = 7.99$ mm).

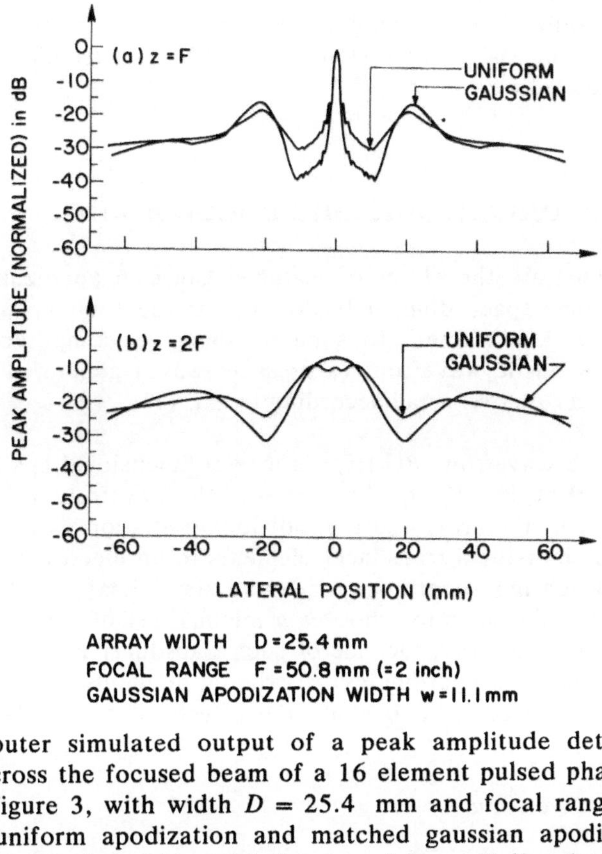

ARRAY WIDTH D = 25.4 mm
FOCAL RANGE F = 50.8 mm (= 2 inch)
GAUSSIAN APODIZATION WIDTH w = 11.1 mm

Fig. 8 Computer simulated output of a peak amplitude detector moved laterally across the focused beam of a 16 element pulsed phased array, as shown in figure 3, with width $D = 25.4$ mm and focal range F = 152.4 mm for uniform apodization and matched gaussian apodization. The beamprofiles are shown for axial distances (a) z = F (focal range) and (b) z = 2F (transition to far-field).

order to generate these plots we used an approximation [$Eq.$(37)] for R_a, but the correct expression [$Eq.$(28)] for w_a. As is well known from other studies of amplitude apodization of pulsed phased arrays [2], the side lobe level suppression is worse for a pulsed system than for a cw system. Our plots (Figures 7 and 8) show a sidelobe level suppression for peak amplitude detection of typically 10 to 15 dB for the pulsed arrays, whereas for the cw array the sidelobe level suppression is closer to 30 dB. We will study, in section 5, what happens in the time-space domain, and analyze where exactly in the wavefront the sidelobe suppression occurs. The advantage of this type of gaussian apodization, matched to the time delays, is that the side lobe suppression is still visible at distances $z = 2F$. This is not guaranteed for unmatched apodization.

5. Acoustic wavefronts and apodization

We will illustrate the effect of matched gaussian apodization for pulsed arrays in the time-space domain by looking at the total acoustic wavefront generated. The key element to simulate this is a computer program that calculates the acoustic wavefront (a pressure wave) generated in water by a pulsed transmitter phased array according to Eq. (1).

The acoustic wavefront $P(\bar{r},t;\bar{r}_f)$ is a two-dimensional spatial function of $\bar{r} = (x = 0, y, z)$ at time $t = t_0$. The shape of the wavefront will depend upon: a) the geometry of the array and the individual radiation patterns; b) the time response of the individual transducer elements to an electrical impulse; c) the apodization weighting function and d) the time delays. The simulator is written such that the operator chooses a minimal set of parameters for each run: the width of the array, the size of each transducer element and the focal point position (in polar coordinates). It should be mentioned that our simulator has also been used [22] to study the effects of time delay quantization [29,30].

5.1 Simulated wavefronts

The simulator typically produces snapshots of the acoustic wavefront while it propagates in time over a distance of 304.8 mm (12 in.), which, in water, takes about 204 μs. In order to illustrate some of the features of the simulator, and at the same time study transient focused acoustic wavefronts in the near-field for uniform apodization (see also [22]), we chose a linear array of width $D = 25.4$ mm (1 in.) with 16 transducer elements. The limitation to 16 array elements was selected after considering the increase in cost and complexity of a real system for a higher number of elements (see also [33]). The width $D = 25.4$ mm is a compromise between lateral resolution at the

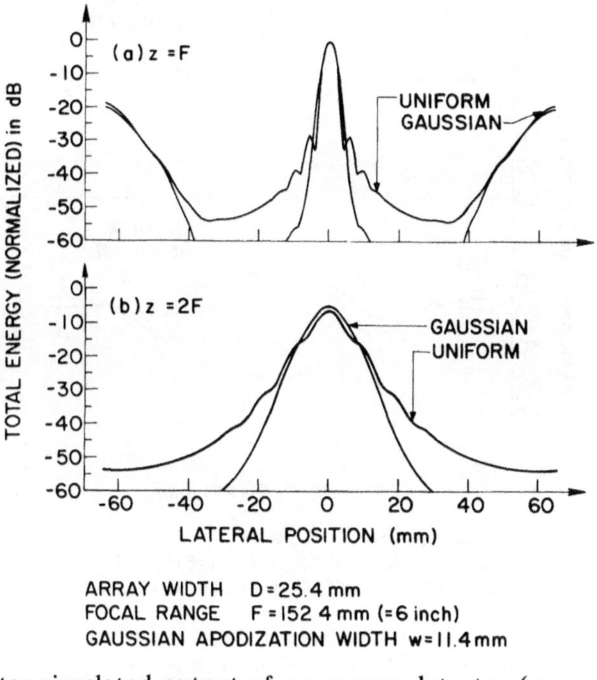

Fig. 9 Computer simulated output of an energy detector (same parameters as in figure 8).

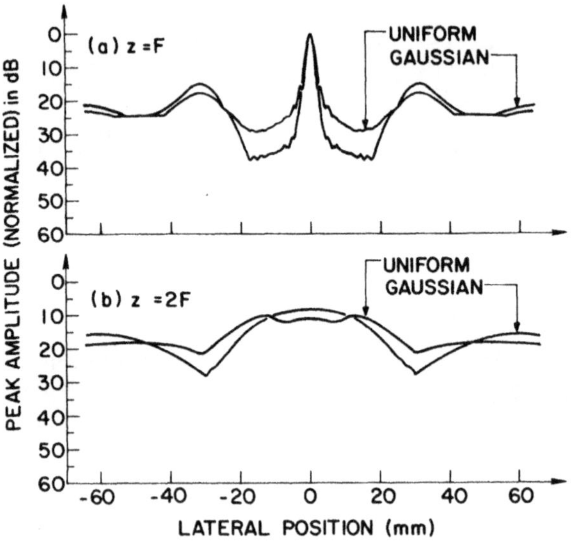

ARRAY WIDTH D = 50.8 mm
FOCAL RANGE F = 152.4 mm (=6 inch)
GAUSSIAN APODIZATION WIDTH w = 22.1 mm

Fig. 10 Computer simulated output of a peak amplitude detector moved laterally across the focused beam of a 16 element phased array, as shown in figure 3, with width $D = 50.8$ mm and focal range F = 152.4 mm for uniform apodization and matched gaussian apodization. The beamprofiles are shown for axial distances (a) z = F and (b) z = 2F.

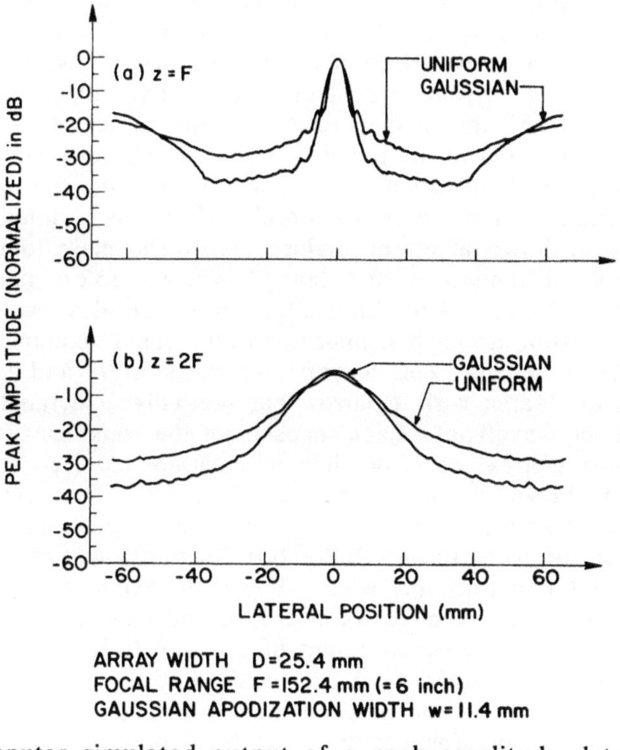

Fig. 11 Computer simulated output of a peak amplitude detector moved laterally across the focused beam of a 16 element phased array, as shown in figure 3, with width $D = 25.4$ mm and focal range F = 50.8 mm for uniform apodization and matched gaussian apodization. The beamprofiles are shown for axial distances (a) z = F and (b) z = 2F.

focal range and grating lobe levels. Figure 2 shows the normalized temporal impulse response $p(t)$ and the corresponding frequency spectrum. Figure 3 shows the array geometry. The 16 identical PZT transducers are shown. Each transducer has the following geometrical dimensions: width $w = 0.635$ mm, thickness t (half a wavelength in PZT) , and height $h = 12.7$ mm. The total width of the array of 16 transducers is $D = 25.4$ mm. The distance between the centers of two adjacent transducers is given by $d = 1.693$ mm.

Figure 12 illustrates typical output for the array shown in figures 2 and 3, for a zero steering angle and focal range of 152.4 mm (6 in.). What we see is the shape of the spatial distribution of the acoustic pressure wavefront at three different times during its propagation for two different apodization functions (uniform and matched gaussian). Only the positive wavefront is shown to avoid confusion in the pictures. Negative pressure values have been set to zero. We can then normalize each wavefront, take the logarithm, and display a logarithmic version between 0 and -30 dB, as is done in figure 13. The wavefront is shown after the major part of the main lobe energy has propagated over a distance z = 50.8 mm (2 in.), z = 152.4 mm (6 in.) and z = 304.8 mm (12 in.). The three different temporal views of the same propagating wavefront are each accompanied by a small diagram on the upper right side of the wavefront. This diagram shows the array and a 152.4 x 254 mm^2 (6 x 10 $in.^2$) area with a narrow bar (actually a rectangle) indicating the position of the wavefront. Each snapshot of the wavefront is limited to a rectangular viewing area, corresponding in position and size to the narrow rectangular bar shown in the upper right diagrams. The rectangular area surrounding the pressure wave is 5.54 mm wide in the direction of pulse propagation referred to as the axial direction. In the transversal direction the rectangle is 139.7 mm (5.5 in.) wide. The scale extension along the axial direction is required to visualize the shape of the wavefront without losing information about the transversal extent of the wavefront after it propagates beyond the focal area. Indeed, for many focal point ranges the pulse becomes extremely wide in the direction perpendicular to the direction of propagation after passing through the focal zone. The focal point area is indicated on figure 12 with a circle of 5.1 mm (0.2 in.) radius. The circle can be seen as an ellipse in the picture with the wavefront corresponding to a propagation distance $z = F =$ 152.4 mm (= 6 in.). It should also be mentioned that the sampling density is nonuniformly spread over the image, in order to decrease the CPU time required by the graphics routines.

The upper part of figure 14 illustrates, as does figure 12, typical output of the simulator for an array similar to the one shown in figures 2 and 3 with diameter $D = 50.8$ mm, focal range $F = z = 152.4$ mm (6 in.) and zero steering angle. The lower part of figure 13 shows the simulated wavefront for matched gaussian apodization. The wavefront is shown at three different times during the pulse propagation as in figure 12.

Figure 15 shows simulated wavefronts for the same array as in figure 12, but with focal range now at F = 50.8 mm (2 in.). However, we only show the wavefront at two different propagation times.

5.2 Analysis in the time-space domain

We will now analyze figures 12, 14 and 15 in terms of the effect of apodization upon the main lobe, side lobe and grating lobe structures. Due to our choice of display, with only a 5.54 mm viewing distance in the axial direction (the z-axis), we are not able to show grating lobes in their full extent. We only see the front part of the grating lobes, because they follow the main lobe with a time delay too large to fit into our viewing window. The side lobe structure, however, is readily visible. During the major part of the acoustic pulse propagation, the main lobe and first order side lobes are hard to seperate because they partially overlap. The higher order side lobes are visible as the leading wavefronts in front of and laterally next to the main lobe wavefront. But in the upper part of figure 12, for the propagation distance z = 152.4 mm (6 in.) we can distinguish the main lobe and first order side lobes clearly. The side lobe structure that we see does remind us of the sinc-function behaviour predicted by a narrow-band diffraction theory for the focal zone. For other propagation distances we can no longer make such clear distinction between main and side lobes. A more detailed discussion of the shape of the grating lobes and side lobes when the wavefront has propagated over a distance equal to the focal range can be found in [11,22].

Figures 12, 14 and 15 are snapshots taken at one particular time of an acoustic wavefront generated by a phased array. Therefore they give us the acoustic pressure along a line parallel to the array (the y-axis in figure 1) in different beam cross-sections or x-y planes (i.e., at different positions along the z-axis) at the time of the snapshot. The shape of the acoustic pressure profile along a line perpendicular to the array (the x-axis in figure 1) in an x-y plane (or beam cross-section) is irrelevant here, because we are interested in studying arrays and not single transducers. The acoustic pressure profile along the z-axis or axis of wave propagation is also shown. However, apodization will mainly affect the pressure profile in the y-axis direction because we are studying a 1-D linear array parallel to the y-axis.

When we compare the acoustic pressure profile in different x-y planes for uniform and matched gaussian apodization in all three figures (12, 14 and 15) we notice a few general features. Along the y-axis direction, the sidelobes immediately adjacent to the main lobe are substantially reduced for matched gaussian apodization, and cause the general impression of sidelobe reduction in the profiles shown in figures 8 to 11. In figure 12, for z = 152.4 mm, the acoustic pressure in the lateral (or y-axis) direction drops very nearly as a gaussian function. The acoustic pressure profile in an x-y plane (or a beam

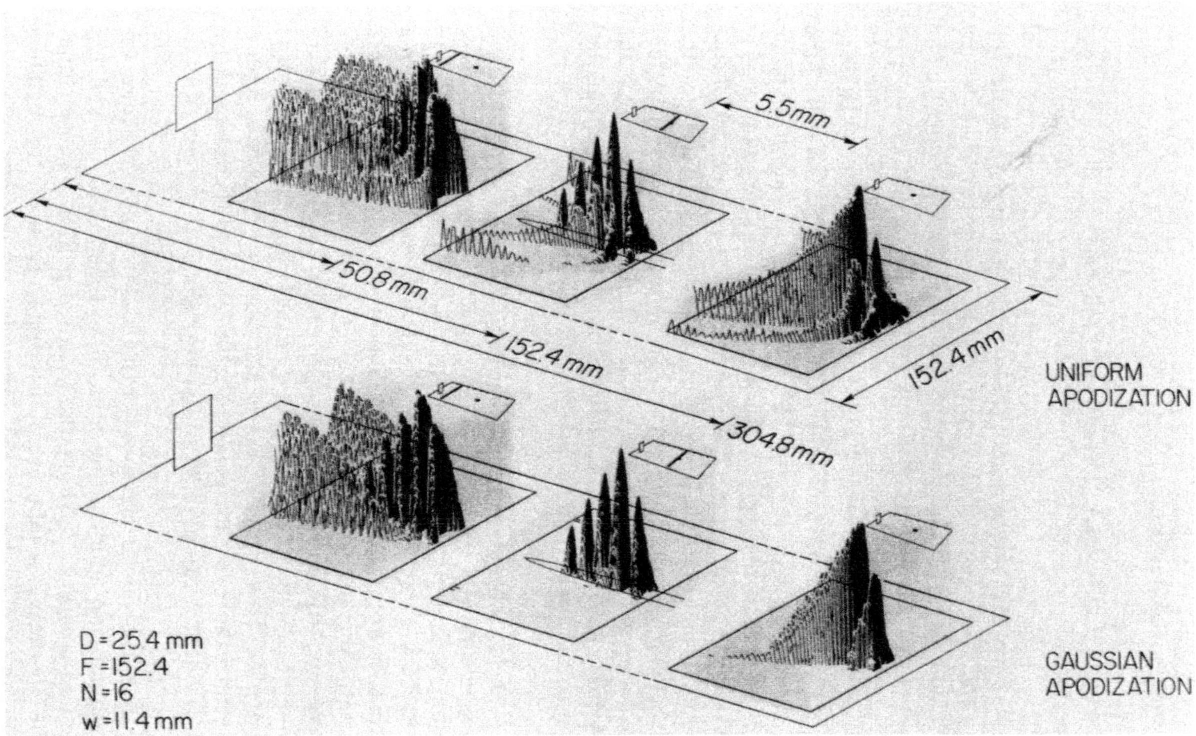

Fig. 12 Computer simulated snapshot of an acoustic pressure wavefront generated by the array shown in figure 3, with width $D = 25.4$ mm, focused at a range of 152.4 mm (6 in.), as it propagates in time: upper wave fronts are for uniform apodization; lower wave fronts are for matched gaussian apodization. Timing: after 34.3 μs (left), after 103 μs (middle), and after 203 μs (right).

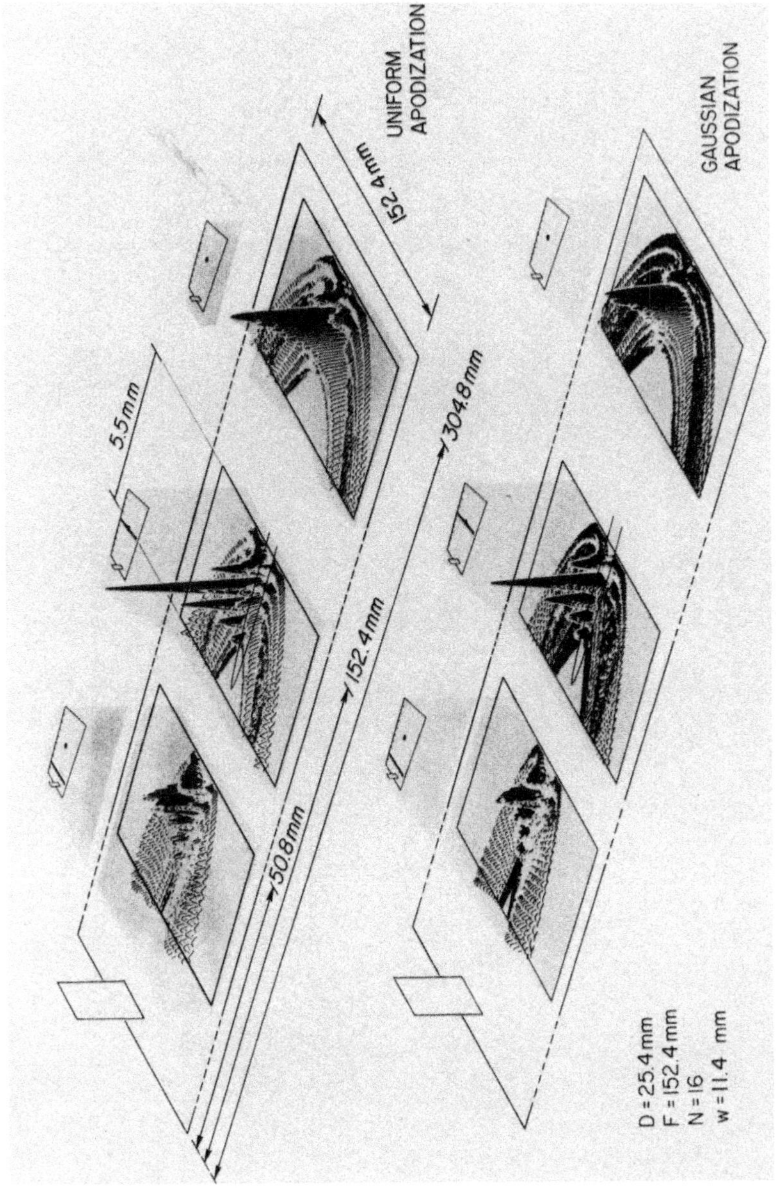

Fig. 13 Same computer simulation as in figure 12, but using a logarithmic display (0 to -30 dB).

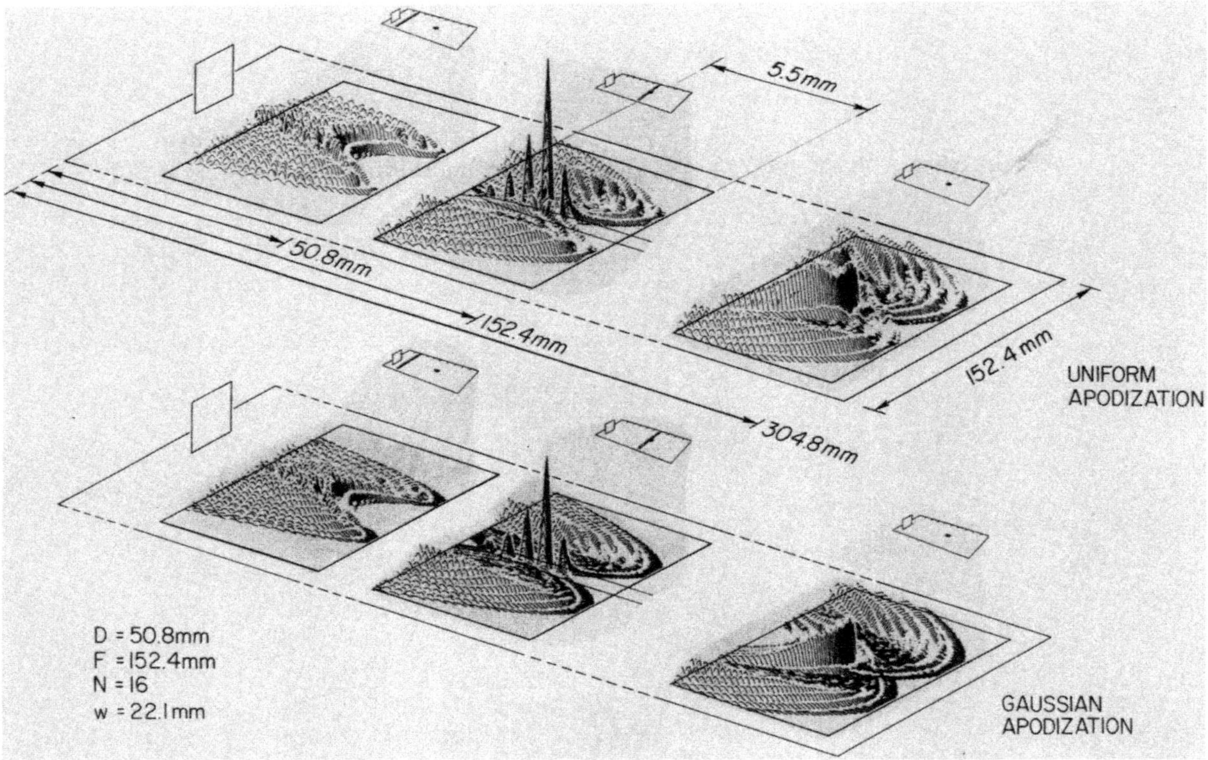

Fig. 14 Computer simulated snapshot of an acoustic pressure wavefront generated by the array shown in figure 3, with width $D = 50.8$ mm, focused at a range of 152.4 mm (6 in.), as it propagates in time: upper wave fronts are for uniform apodization; lower wave fronts are for matched gaussian apodization. Timing: after 34.3 μs (left), after 103 μs (middle), and after 203 μs (right).

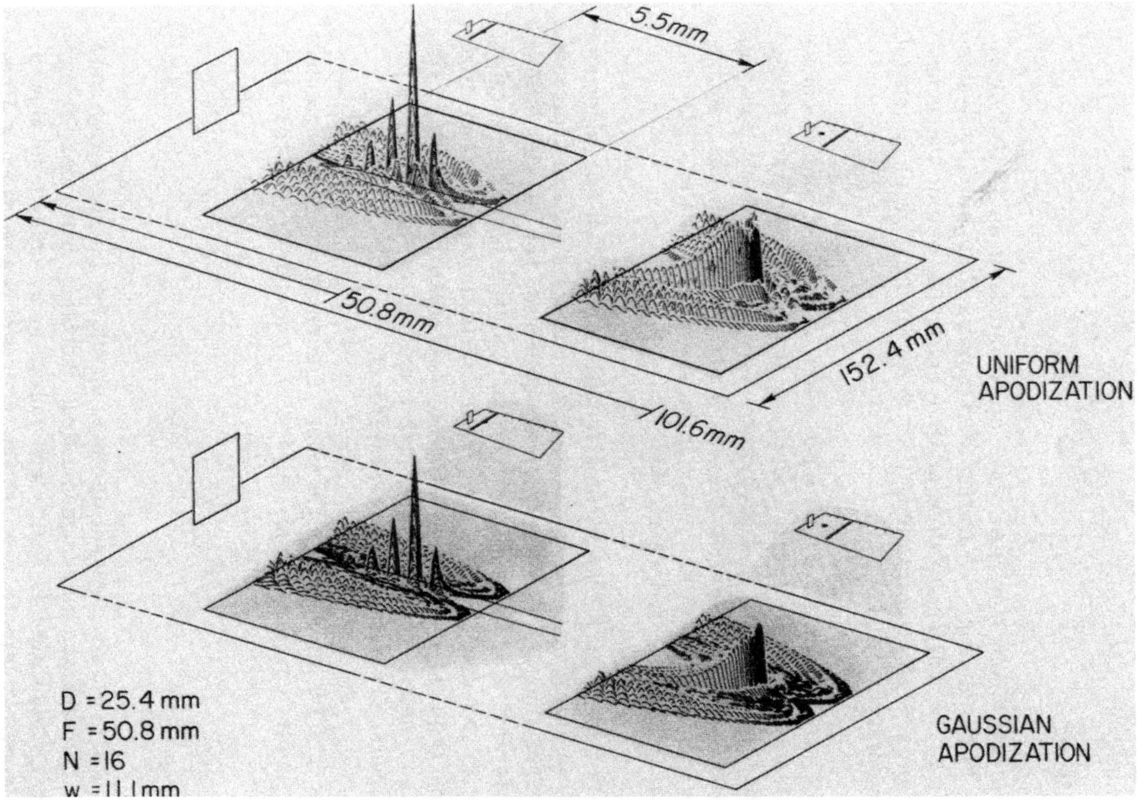

Fig. 15 Computer simulated snapshot of an acoustic pressure wavefront generated by the array shown in figure 3, with width $D = 25.4$ mm, focused at a range of 50.8 mm (2 in.), as it propagates in time: upper wave fronts are for uniform apodization; lower wave fronts are for matched gaussian apodization. Timing: after 34.3 μs (left), and after 68.6 μs (right).

cross-section) through the largest wave of the wavefront for z = 2 F goes through a local minimum at the center of the beam as is shown in both figures 13 and 14. This local minimum, called a "dip" in the wavefront, is discussed in [22], and disappears when gaussian apodization is used.

5.3 Implementation

Our acoustic wavefront simulator was implemented on an IBM VM370 computer using PL/I language. We used the Graphics Compatibility System (GCS), generalized for 3-D applications at IBM. Graphics output was generated on a Tektronix 4025 type terminal. For more information on a general software implementation see [22*b*].

Our computer code consists of two basic parts. A first subroutine calculates the instantaneous acoustic pressure $P(\bar{r},t;\bar{r}_f)$ at any location $\bar{r}$ and time t, given data on the array position and geometry. A second subroutine does the computer graphics work (3-D projections,etc...) and the hidden-line calculations. It it the second routine that requires the major part of the CPU time used to generate a picture of one wavefront.

The simulation of acoustic wavefronts requires an hidden-line algorthm for the graphics and display part, and about 9600 sample points to represent the wavefront. The flow chart of the computer code used for our simulations of acoustic wavefronts is given in [22*b*]. This computer code mainly consists of a DO LOOP to calculate the acoustic pressure distribution at a given time t. The CPU time required to do the calculation of the acoustic pressure in the DO LOOP of the computer code is very small compared to the time required for the computer graphics and display of wavefronts.

The amount of CPU time required is about 2 minutes per 10,000 sample points for our implementation on an IBM 370 computer, using the GCS package and an in-house hidden line algorithm [31].

Conclusions

We introduced matched gaussian apodization of cw phased arrays and investigated its dependence upon geometrical factors. The idea behind matched gaussian apodization is that a cw phased array should be able to generate eigenmodes of freely propagating acoustic beams with a finite cross-section. For acoustic imaging our main interest is in pulsed phased arrays, and therefore an analysis in the time-space domain of wavefronts generated by a pulsed array is required. We first derived an expression for the apodization function for cw phased arrays, and used it for pulsed arrays, which is an approximation very similar to the one made when steering and focusing parameters derived for cw arrays are used with focused arrays. In order to facilitate the derivation, the strength of focusing parameter K was defined and its relevance to matched gaussian apodization was explained. We did not discuss what happens to a gaussian beam when it propagates through strongly attenuating and/or inhomogeneous media such as biological tissues.

In order to study the effects of apodization on wavefronts in the time-space domain, we developed a model to study transient beam-forming in the near-field of pulsed phased arrays. The model is a good approximation to a wide-band diffraction model for small steering angles and propagation of acoustic energy in a medium with frequency independent acoustic attenuation. We analyzed the effect of matched gaussian apodization upon acoustic wavefront shape. The comments on the changes in acoustic beam cross-section, i.e. the substantial reduction of side lobes immediately surrounding the main lobe, are fairly general and apply to many types of apodization. However, the fact that the acoustic pressure profile in a beam cross-section is a gaussian function with no "dip" or local minimum at the center of the beam, should only be expected with matched gaussian apodization. A more detailed comparison of matched gaussian apodization with other types of apodization, such as Hamming, Kaiser, Dolph-Chebychev, Parzen, Blackman, to name a few, will be the subject of another paper.

Acknowledgements

I would like to thank C.N. Liu and A. Stern for their many suggestions and discussions that helped improve the acoustic wavefront simulator. A. Stern, now at the University of California in Berkeley, in the mathematics department, wrote part of the original software. C.N. Liu was extremely helpful in improving the presentation and wording of the ideas expressed in this paper. A.E. Siegman at Stanford University introduced me to the concept of gaussian laser beams. E. J. Pisa at Rohe Scientific Corporation, and C. W. Barnes at the University of California at Irvine provided useful comments and additional references.

Art Appel, Cliff Nass and Art Stein were very helpful in teaching me how to use the graphics facilities at IBM. C. Nass provided a modified 10,000 point halo algorithm. Louis Kristianson and Jeanne Berlin were very helpful in making clear figures from computer output.

LIST OF REFERENCES.

[1] Macovski A., Ultrasonic imaging using acoustic arrays, *Proc. IEEE 67* , 484-495 (1979).

[2] Shott J.D., Charge-coupled Devices for Use in Electronically Focused Ultrasonic Imaging Systems, Ph.D. Dissertation, EE Dept, Stanford University, *ICL Technical Report* No. 4957-1, May, 1978.

[3] Eaton J., High Performance Preprocessor Electronics for Ultrasound Imaging, Ph.D. Dissertation, EE Dept, Stanford University, *ICL Technical Report* No. G557-3, May, 1979.

[4] von Ramm O.T and Thurstone F.L., Cardiac imaging using a phased array ultrasound system. I : System design, *Circulation 53* , 258-... (1976).

[5] Somer J.C., Electronic sector scanning for ultrasonic diagnosis, *Ultrasonics 6* , 153-159 (1968).

[6] Iinuma K., Kidokoro T., Ogura I., Takamizawa K., Seo Y., Hashiguchi M. and Uchiumi I., High resolution electronic-linear scanning ultrasonic diagnostic equipment, *Ultrasound Med. Biol. 5* , 51-60 (1979).

[7] von Ramm O.T. and Smith S.W., A multiple frequency array for improved diagnostic imaging, *IEEE Trans. Sonics Ultrasonics SU-25* , 340-345 (1978).

[8] Somer J.C., Oosterbaan W.A. and Freund H.J., Ultrasonic tomographic imaging of the brain with electronic sector scanning system: Electroscan, in *1973 IEEE Ultrason. Symp. Proc.* pp. 43-48 (IEEE Cat. No. 73CHO 807-8 SU).

[9] Steinberg B.D., Principles of Aperture and Array System Design, (Wiley, New York, NY, 1976).

[10] Norton S.J., Theory of Acoustic Imaging, Ph.D. Dissertation, EE Dept, Stanford University, *SEL Technical Report* No. 4956-2 ,Dec., 1976.

[11] Tancrell R.H., Callerame J., and Wilson D.T., Near-Field, transient acoustic beam-forming with arrays, in *1978 IEEE Ultrasonics Symp. Proc.*, pp. 339-343 (IEEE Cat. No. 78CH 1344-ISU).

[12] Goodman J.W., Introduction to Fourier Optics, (McGraw-Hill, New York, NY, 1968).

[13] Banjavic R.A., Zagzebski J.A., Madsen E.L. and Goodsitt M.M., Distortion of Ultrasonic beams in Tissue and Tissue-Equivalent Media, *Acoustical Imaging and Holography 1* , 165-177, 1979.

[14] Foster F.S. and Hunt J.W., Transmission of ultrasound beams through human tissue- focussing and attenuation studies, *Ultrasound Med. Biol. 5*, 257-268 (1979).

[15] Freedman A., Sound field of plane and gently curved pulsed radiators, *J. Acoust. Soc. Amer. 48*, 221-227 (1970).

[16] Freedman A., Far-field of pulsed rectangular acoustic radiator," *J. Acoust. Soc. Amer. 49* , 738-748 (1971).

[17] Sato J., Fukukita H., Kawabuchi M. and Fukumoto A., Farfield Angular radiation pattern generated from arrayed piezoelectric transducers, *J. Acoust. Soc. Amer. 67* , 333-335 (1980).

[18] Smith S.W., von Ramm O.T., Haran M.E., and Thurstone F.L., Angular response of piezoelectric elements in phased array ultrasound scanners, *IEEE Trans Sonics Ultrasonics SU-26* , 185-190 (1979).

[19] Kino G.S. and DeSilets C.S., Design of slotted transducer arrays with matched backings, *Ultrasonic Imaging 1*, 189-209 (1979).

[20] Bruneel C., Delannoy B., Torguet R., Bridoux E. and Lasota H., Electrical coupling effects in an ultrasonic transducer arrays, *Ultrasonics 17*, 255-260 (1979).

[21] Delannoy B., Lasota H., Bruneel C., Torguet R. and Bridoux E., The infinite planar baffles problem in acoustic radiation and its experimental verification, *J. Appl. Phys. 50* , 5189-5195 (1979).

[22*a*] Duerinckx A.J., Computer simulator for acoustic phased arrays, Presented at the 5th International Symposium on Ultrasonic Imaging and Tissue Characterization, National Bureau of Standards, Gaithersburg, Maryland, June 1-6, 1980. *Ultrasonic Imaging 2* , 190 (1980) (abstract only).

[22*b*] Duerinckx A.J., Modelling wavefronts from acoustic phased arrays by computer, *IEEE Trans Biomed. Eng. BME-28* (Feb. 1981).

[23] Kogelnik H. and Li T., Laser beams and resonators, *Applied Optics 5* , 1550-1567 (1966).

[24] Siegman A.E., Introduction to Lasers and Masers, pp 304-342, (McGraw-Hill, New York, NY, 1971).

[25*a*] Siegman A.E. and Sziklas E.A., Mode calculations in unstable resonators with flowing saturable gain. 1:Hermite-Gaussian Expansion, Appendix B, *Applied Optics 13* , 2775-2792 (1974).

[25*b*] Cavanagh E. and Cook B.D., Gaussian-Laguerre description of ultrasonic fields- Numerical example: circular piston, *J. Acoust. Soc. Am.* 67:1136-1140 (1980).

[26] Dameron D.H., Diagnostic improvements in medical ultrasonic imaging systems, Appendix A: The generalized Fraunhofer approximation, Ph.D. Dissertation, EE Dept, Stanford Univ, *ICL Technical Report* No. 4956-4, March, 1978.

[27] Oppenheim A.V. and Schafer R.W., Digital Signal Processing, (Prentice-Hall, New York, NY, 1975).

[28] Bracewell R., The Fourier Transform and Its Applications, Chapter 18, (Mc Graw-Hill, New York, NY, 1965).

[29] Bates K.N., Effects of phase and amplitude quantization on the focused aperture, in *Acoustical Holography* Vol. 9 (1979) (In press).

[30] Eaton M.D., Bardsley B., Melen R.D. and Meindl J.D., Effects of coarse phase quantization in ultrasound scanners, in *1978 IEEE Ultrasonics Symposium Proc.* pp. 784-788. (IEEE Cat. No. 78CH 1344-ISU).

[31] Appel A., Rohlf F.J. and Stein A.J., The haloed line effect for hidden line elimination, *IBM Research Report* , RC 7458, Dec. 1978 (order copies by writing to the IBM Thomas Watson Research Center); Reprinted in *Computer Graphics 13*, 151-157 (1979) as part of SIGGRAPH'79 Proceedings (1979 ACM 0-89791-004).

[32] Vogel J., Bom N., Ridder J. and Lancee C., Transducer design considerations in dynamic focusing, *Ultrasonics Med. Biol. 5* , 187-193 (1979).

[33] Assenza D. and Pappalardo M., Echographic imaging with dynamically focused insonification, *Ultrasonics 18* , 38-42 (1980).

APPLICATION OF TIME-SPACE IMPULSE RESPONSES TO CALCULATIONS OF ACOUSTIC FIELDS IN IMAGING SYSTEMS

Henryk Lasota, and Roman Salamon

Institute of Telecommunication
Technical University of Gdańsk
80-952 Gdańsk, Poland

INTRODUCTION

Analytical problems arise in acoustic imaging systems when calculating pressure field created by complicated radiating apertures, scattering objects etc. as well as modifications of the field by various system elements as lenses, diaphragms etc. Difficulties are still greater when wideband signals are being used.

The paper proposes a general approach to these problems by describing an acoustic system by a set of time-space impulse responses. The idea of the method is to decompose the system into basic elements such as radiating or modulating surfaces and space layers. Each of these elements has an impulse response of its own,the final result being obtained by applying convolution operations to input signals, radiating apertures and impulse responses just mentioned.

Impulse responses being introduced allow to describe clearly and exactly functioning of system elements and their influence on time-space pressure distribution for arbitrary exciting signals.

The method has been applied to the analysis of radiation of linear apertures excited by a Dirac impulse. The problem has an analytic solution in the whole field - neither Fraunhofer nor Fresnel approximations are necessary in this case.

As an example of capabilities of the method the pressure field has been calculated of a linear aperture excited by a rectangular pulse and of a linear lens excited by a Dirac impulse. The harmonic wave field of the linear aperture has been calculated for verification of the method.

LINEAR SYSTEM APPROACH TO THE BROADBAND DIFFRACTION PROBLEM

In optics linear systems approach known as Fourier optics is in common use when solving problems of pressure field of spatially complicated quasi-plane harmonic sources. The radiating system is treated as a space linear system.[2]

The method is based on the Helmholtz integral formula for the case of planar radiating surface put down in the rectangular coordinates.

The method presented in this paper uses as basis the Kirchhoff integral formula being more general. It leads to the solution of a form of cascade combination of time-space linear systems what allows in consequence to calculate the time-space pressure distribution for arbitrary excitation.

The Kirchhoff integral formula relates the time-space pressure distribution in the field $p(\vec{r},t)$ with the distribution of the pressure $p(\vec{r}_o,t)$ and its normal gradient $\partial p/\partial n(\vec{r}_o,t)$ on a closed surface $S(\vec{r}_o)$ that includes the radiating surface S_o:

$$p(\vec{r},t)=-\int_{S(\vec{r}_o)}\left[p(\vec{r}_o,t)*\frac{\partial g}{\partial n}(\vec{r}-\vec{r}_o,t)-\frac{\partial p}{\partial n}(\vec{r}_o,t)*g(\vec{r}-\vec{r}_o,t)\right]dS(\vec{r}_o) \qquad (1)$$

where:

$$g(\vec{r}-\vec{r}_0,t) = \frac{\delta(t - \frac{|\vec{r}-\vec{r}_0|}{c})}{4\pi\,|\vec{r}-\vec{r}_0|}$$

is the free space Green function.[3]

Let us consider a system consisting of an input plane having pressure distribution $p_0(x,y,t)$, a space layer of thickness z and an output plane with resulting pressure distribution $p_z(x,y,t)$. After some simple transformations of the Kirchhoff formula, the relation between the two distributions can be written as follows:

$$p_z(x,y,t) = -\frac{1}{4\pi c}\,\frac{d}{dt}\iint_{-\infty}^{\infty} p_0(\xi,y,t) * \frac{\delta[t\,\frac{r_z(x-\xi,y-\eta)}{c}]}{r_z(x,\xi,y-\eta)}\,\alpha(x-\xi,y-\eta)\,d\xi\,d\eta \qquad (2)$$

where $r_z(x,y) = |\vec{r}| = z\sqrt{1+(\frac{x}{z})^2+(\frac{y}{z})^2}$

and $\alpha(x,y)$ is an obliquity factor of value depending on boundary conditions in the input plane.[4]

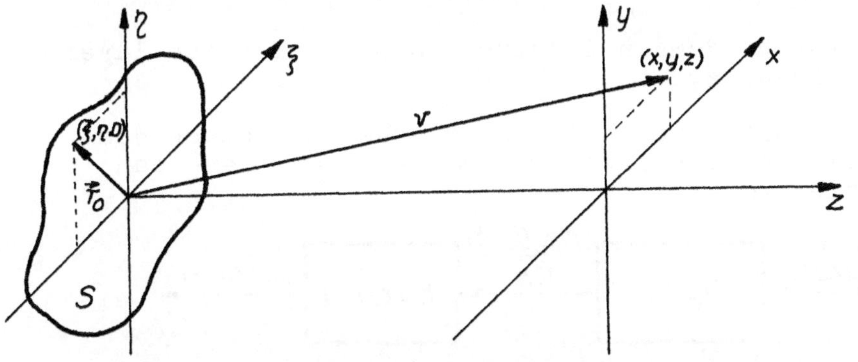

Figure 1. Coordinates of radiation problem.

For the rigid source in the rigid baffle the factor is $\alpha_R(x,y)=2$, equation (2) being a broadband version of the Rayleigh formula. The soft source in the pressure-release baffle has the factor: $\alpha_S(x,y)= 2z/r_z(x,y)$, equation (2) corresponding to the Rayleigh-Sommerfeld formula. When the source of a surface matched to the medium radiates in a free space, the factor has the value: $\alpha_K= 1+z/r_z(x,y)$. This corresponds to the case of the so called Kirchhoff simplified formula.

Equation (2) is in fact a three-argument time-space convolution and can be written, therefore, as:

$$p_z(x,y,t)=h_p(t) \underset{t}{*} h_z(x,y,t) \underset{x\,y\,t}{***} p_o(x,y,t) \qquad (3)$$

The input and output plane pressure distributions $p_o(x,y,t)$ and $p_z(x,y,t)$ can be treated as three dimensional input and output signals, respectively, functions $h_p(t)$ and $h_z(x,y,t)$ being impulse responses of the radiation system /Fig. 2/:

$$h_p(t) = -\frac{1}{4\pi c}\,\delta'(t)$$

- time domain impulse response of radiating surface,

$$h_z(x,y,t) = \frac{\alpha(x,y)}{r_z(x,y)}\,\delta\left[t-\frac{r_z(x,y)}{c}\right]$$

- time-space domain impulse response of space layer of thickness z.

In most interesting cases the time-space distribution of the pressure in the input plane can be expressed as:

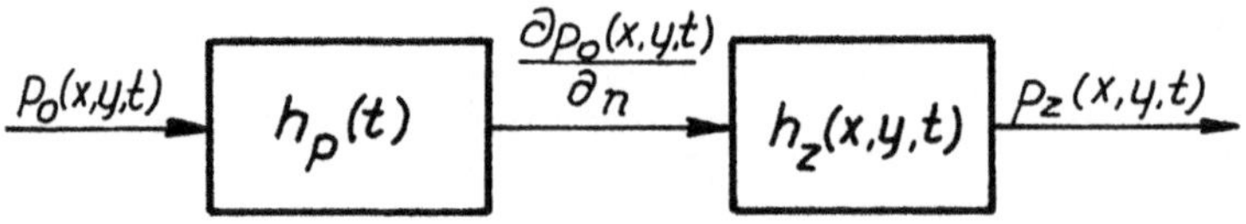

Figure 2. Radiation as a linear system

$$p_0(x,y,t) = p_0(t) \underset{t}{*} h_A(x,y,t) \tag{4}$$

with:

$$h_A(x,y,t) = A(x,y)\ \delta[t - d(x,y)]$$

- weighting function having the sense of impulse response of the aperture.

Function A(x,y) determines amplitude weighting /apodization/ of the transducer, scattering properties of object being observed or transmission properties of transparent object. Function d(x,y) represents space distribution of delays of radiated, reflected or transmitted signals that corresponds to phase shift distribution in narrowband case.

Finally, the relation between the exciting signal $p_0(t)$ and the resultant output distribution $p_z(x,y,t)$ will be as follows:

$$p_z(x,y,t) = h_p(t) \underset{t}{*} h_A(x,y,t) \underset{x\,y\,t}{***} h_z(x,y,t) \underset{t}{*} p_0(t) \tag{5}$$

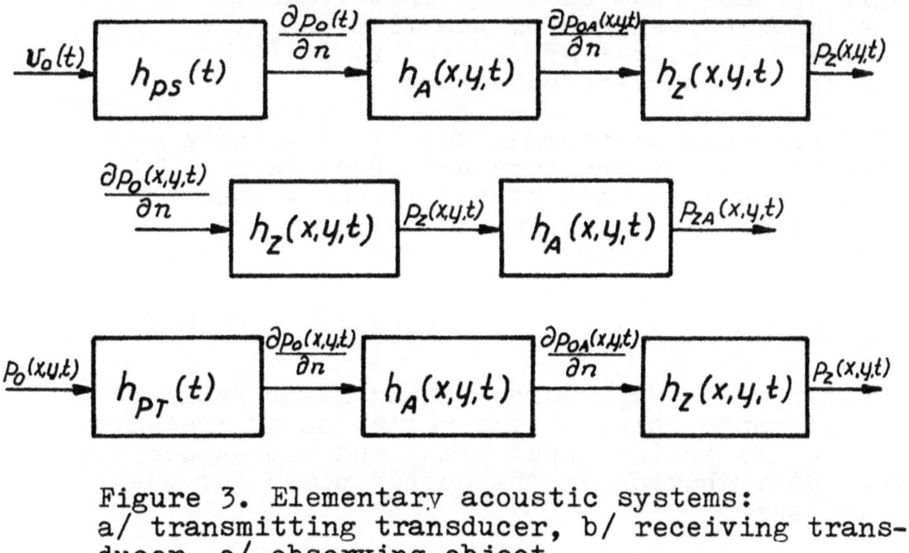

Figure 3. Elementary acoustic systems: a/ transmitting transducer, b/ receiving transducer, c/ observing object

The structure of the equation given above shows that the acoustic system can be considered as a cascade combination of three linear systems described by impulse responses previously defined /Fig. 3/. This fact reflects the possibility of decomposition of the wide-band diffraction problem into elementary effects: 1/derivation taking place on the radiating surface, 2/modulation introduced by space distribution of amplitudes and delays in the aperture and 3/effect of spreading in space.

Figure 3 shows three elementary acoustic systems composing all the echographic systems: transmitting transducer, receiving transducer and object being observed /either scattering or transparent/. As the surface velocity $v_o(t)$ is generally treated as input quantity in acoustic radiation, the surface impulse response of transmitting transducer is: $h_{PS}(t) = -\rho/4\pi \cdot \delta'(t)$. The receiving piezoelectric transducer is sensitive to acoustic pressure itself, so the surface impulse response equals unity /$h_{PR}(t) = 1$/, pressure gradient in the input plane being the input signal. In the case of observing object, being secondary source, the pressure is the input quantity what leads to the surface impulse response of the form: $h_{PT}(t) = \mp 1/4\pi c \cdot \delta'(t)$, with sign (-) for rigid targets and (+) for pressure-release targets.

Two elements are common for all the three systems: modulating aperture and space layer, both having impulse responses of simple and evident interpretation. It is useful to define a function that is the three argument convolution of these responses:

$$k_z(x,y,t) = h_A(x,y,t) \underset{x}{*} \underset{y}{*} \underset{t}{*} h_z(x,y,t) \qquad (6)$$

This combined aperture - space layer /ASL/ impulse response relates time variation of pressure gradient $g_o(t)$ in the input plane and time-space pressure distribution in the output plane for the given aperture:

$$p_z(x,y,t) = g_o(t) \underset{t}{*} k_z(x,y,t) \quad (7)$$

The ASL impulse response concept is essential in broadband diffraction problems analysis. It allows to calculate the time shape of the pressure in the whole space when the pressure gradient of the input plane wave is known. In the case of radiation or scattering, when velocity or pressure is the input quantity, the additional derivation /convolution with hp (t) / gives the final result.

APPLICATION OF THE METHOD TO LINEAR APERTURES

The method just described has been applied to analysis of radiation of a linear aperture /slit/ excited by a Dirac impulse.

Let us consider an aperture characterised by the general weighting function /Fig. 4/:

$$h_A(x,y,t) = \Pi\left(\frac{x}{2a}\right) A(x)\, \delta(y)\, \delta[t - d(x)] \quad (8)$$

where: $\Pi\left(\frac{x}{2a}\right) = \begin{cases} 1 & \text{for} \quad |x| \leq a \\ 0 & \quad\quad\; |x| > a \end{cases}$

is the gating function. In this case the combined ASL impulse response has the following form:

$$k_z(x,t) = \int_{-\infty}^{\infty} \frac{\Pi\left(\frac{\xi}{2a}\right) A(\xi)\, \alpha(x-\xi)}{r_z(x-\xi)}\, \delta\left[t - \frac{r_z(x-\xi)}{c} - d(\xi)\right] d\xi \quad (9)$$

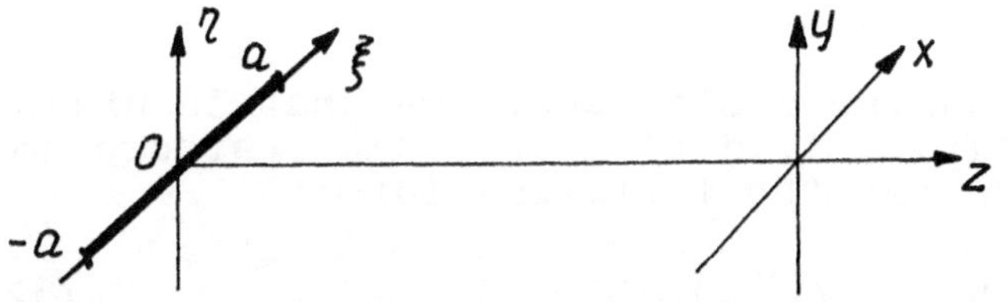

Figure 4. Unidimensional aperture /slit/

The above equation can be put down in a general form as follows:

$$k(t) = \int_{-\infty}^{\infty} f(\xi)\ \delta[t - \tau(\xi)]\, d\xi \qquad (10)$$

that expresses convolution with a Dirac function of composed argument. As it can be shown, the convolution, being a Stjeltjes integral, has the following analytic solution:

$$k(t) = \frac{f[\xi_1(t)]}{|\tau \cdot [\xi_1(t)]|} \cdot 1\!\!1[-\xi_1(t) + c_1] +$$

$$+ \frac{f[\xi_2(t)]}{|\tau'[\xi_2(t)]|} \cdot \Pi \left[\frac{\xi_2(t) - \frac{c_2 - c_1}{c_2}}{c_2 - c_1} \right] + \cdots$$

$$\cdots + \frac{f[\xi_n(t)]}{|\tau'[\xi_n(t)]|} \cdot 1\!\!1[\xi_n(t) - c_{n-1}]$$

where $1\!\!1(\cdot)$ is a Heavyside step function, $(-\infty, c_1), (c_1, c_2), \cdots c_{n-1}, \infty)$ are intervals of monotony of the function $\tau(\xi)$ and $\xi_1(\tau), \xi_2(\tau), \ldots$ $\ldots \xi_n(\tau)$ are the inverse functions $/\tau^{-1}(\xi)/$ in these intervals.

Although the solution could seem to be complicated, its evidence and simplicity will be shown in some examples.

UNIFORMLY EXCITED SLIT

Let us analyse a slit aperture that is uniformly excited $/A(x) = 1,\ d(x) = 0/$. This aperture impulse response has the following form:

$$h_A(x,t) = \Pi\left(\frac{x}{2a}\right) \delta(t) \qquad (12)$$

and the ASL impulse response is as follows:

$$k_z(x,t) = n(x,t)\cdot\frac{\alpha(t)}{\sqrt{t^2-t_z^2}} \qquad (13)$$

where: $t_z = z/c$ - time of propagation between input and output planes, $\alpha_R(t) = 2$, $\alpha_S(t) = 2t_z/t$, $\alpha_K(t) = 1+t_z/t$ - obliquity factors, $n(x,t)$ - modified gating function taking three values: 2,1 and 0 that depends on time and position with respect to paraxial zone $/|x| \leqslant a/$:

when $|x| \leqslant a$:

$$n(x,t) = \begin{cases} 2 & \text{for } t_z \leqslant t \leqslant t_1 \\ 1 & t_1 < t \leqslant t_2 \\ 0 & \text{elsewhere} \end{cases}$$

when $|x| > a$:

$$n(x,t) = \begin{cases} 1 & \text{for } t_1 \leqslant t \leqslant t_2 \\ 0 & \text{elsewhere} \end{cases}$$

where $t_{1,2} = t_z\sqrt{1+\left(\frac{a \mp |x|}{z}\right)^2}$ - time limits.

It is worth noting that the form of the ASL impulse response is the same for all the points along x-axis /z=const/ without regard to the aperture width 2a. Only gating coefficient values /time limits of the wave passing by the field point/ depend on the distance from the z-axis and aperture dimensions.

There is a very simple explanation of the above mentioned effect. The ASL impulse response of an infinitely long aperture would depend only on the distance z, contributions coming to the field point from both sides simultaneously /n=2/. Limiting the aperture implies limits of time intervals in which contributions appear. Figure 5 explains the arising of ASL impulse responses in two field points A and B being within and outside paraxial zone.

Figures 6 and 7 show the ASL impulse responses $h_z(x,y,t)$ in cases of pistonlike sources in rigid and pressure-release baffles, respectively.

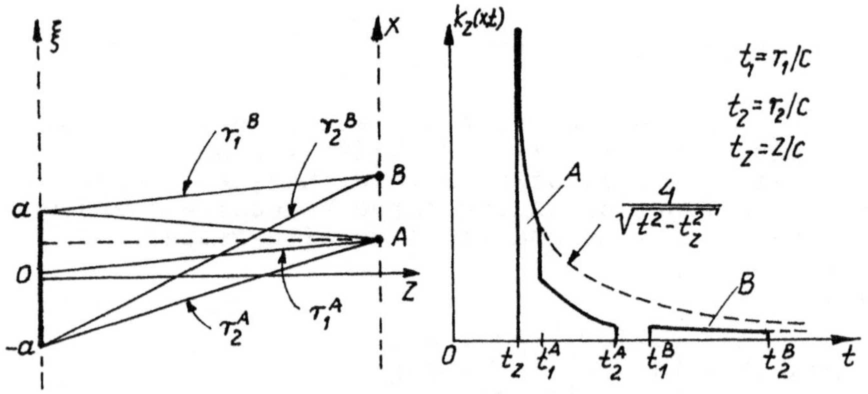

Figure 5. Arising of ASL impulse response

In the paraxial zone all the curves start at time instant t_z with infinite value and end relatively quickly, especially at long distances. The farther from z-axis the more the shape of the curves approximates the rectangular pulse the duration of which tends to $2t_a = 2a/c$ for tangential direction. In pressure-release baffle case /Figure 7/ the field diminishes quicker in this direction, moreover, there is a slight difference in the near field with respect to the rigid baffle case that vanishes at longer distances.

It is noteworthy that the solution we have obtained needs neither paraxial approximation commonly used in Fourier optics nor geometric approximation of Fresnel / $\sqrt{1 + (x/z)^2} \approx 1$ for amplitudes and $\sqrt{1 + (x/z)^2} \approx 1 + x^2/2z^2$ for phase-shifts or delays/ as well as Fraunhofer far field approximation / $(x - \xi)^2 \approx x^2 - 2x\xi$ / being all in common use in harmonic field analysis.[2,3]

The qualitive differences between the far field approximation and exact solution can be seen from Figure 8 presenting the ASL impulse response calculated with this approximation. The differences

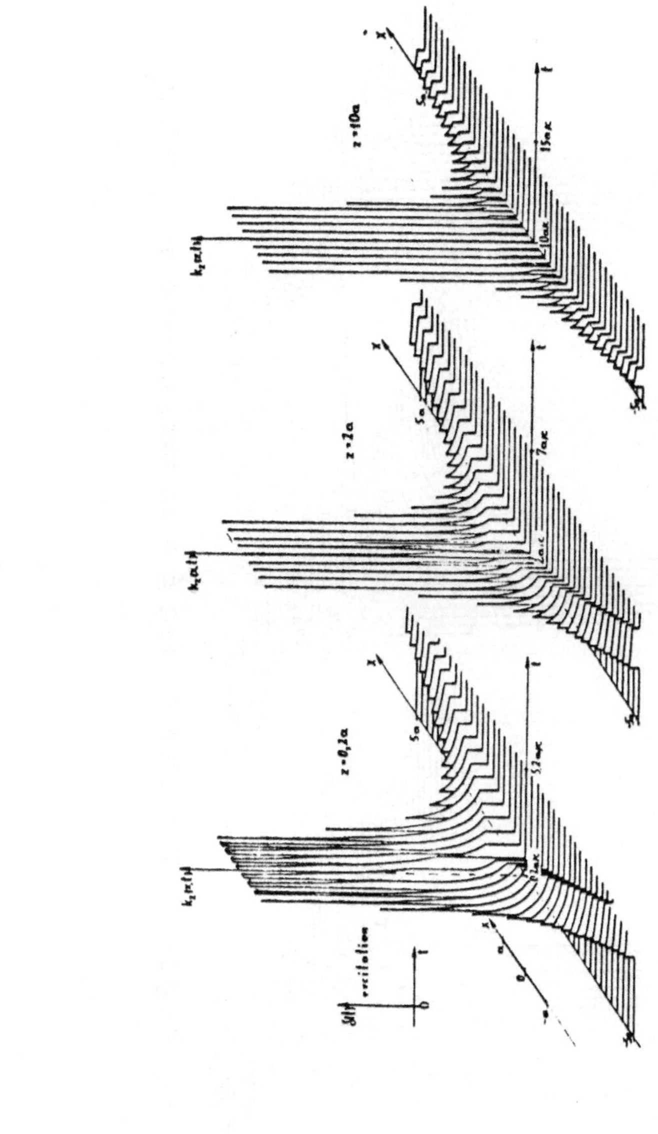

Figure 6. ASL impulse response of unweighted slit in rigid baffle for z = 0,2a, 2a, 10a

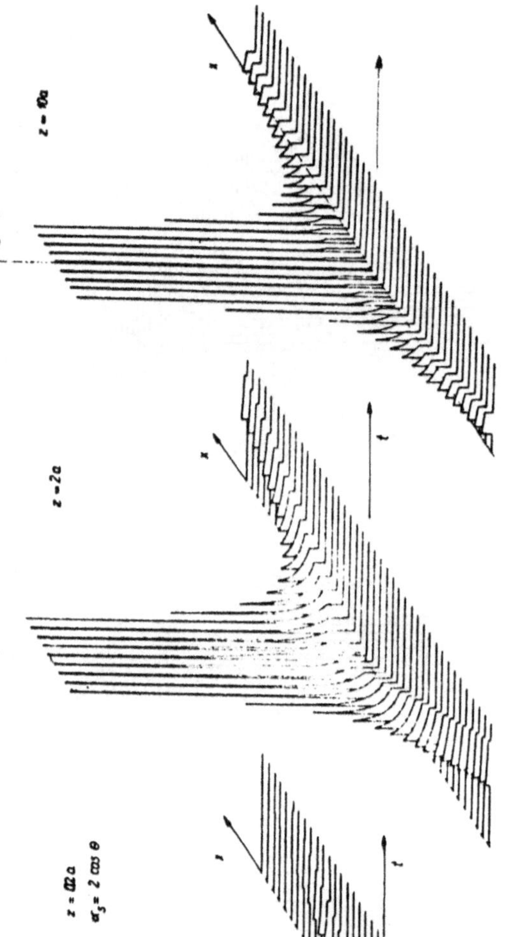

Figure 7. The same as Fig.6 - in pressure-release baffle

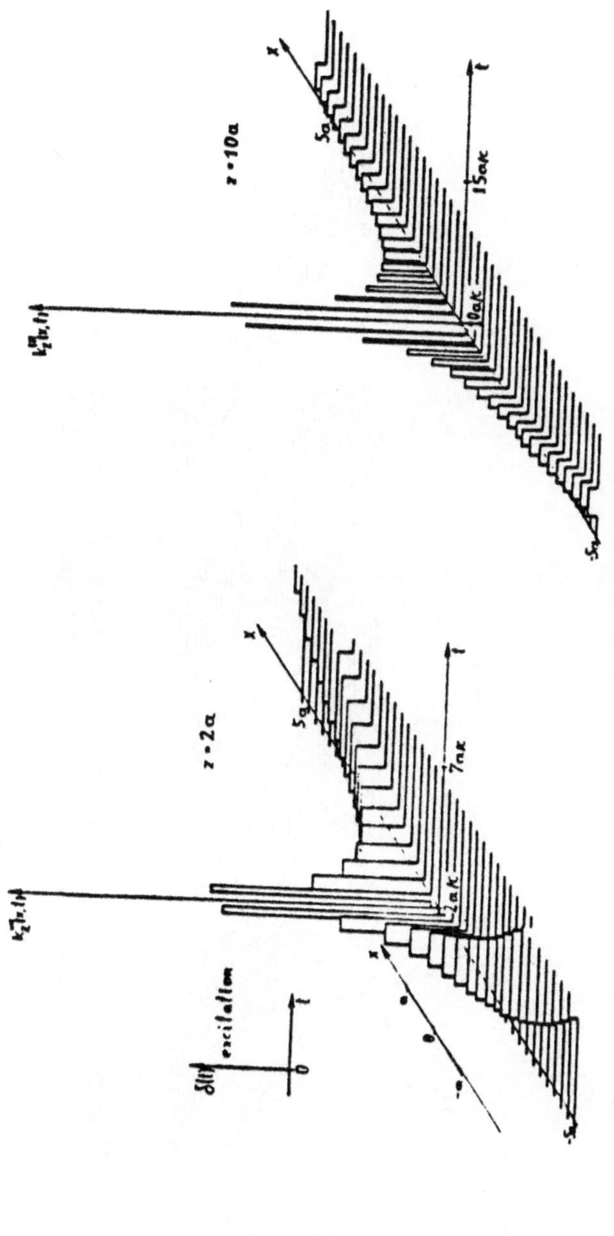

Figure 8. The same as Fig.6 calculated with Fraunhofer approximation

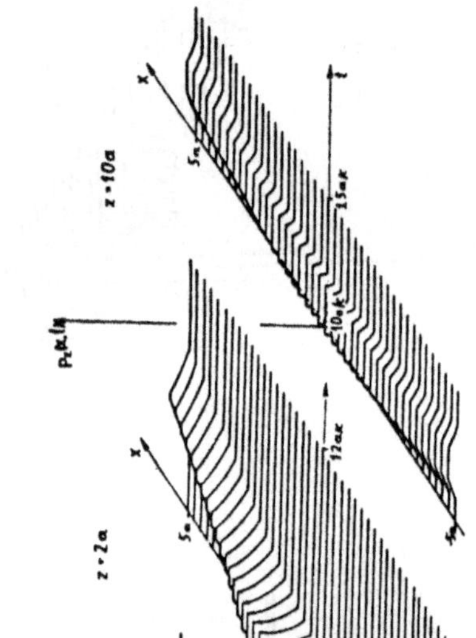

Figure 9. Time-space pressure distribution of unweighted slit excited by rectangular pulse / $T = 2\ t_a$ /

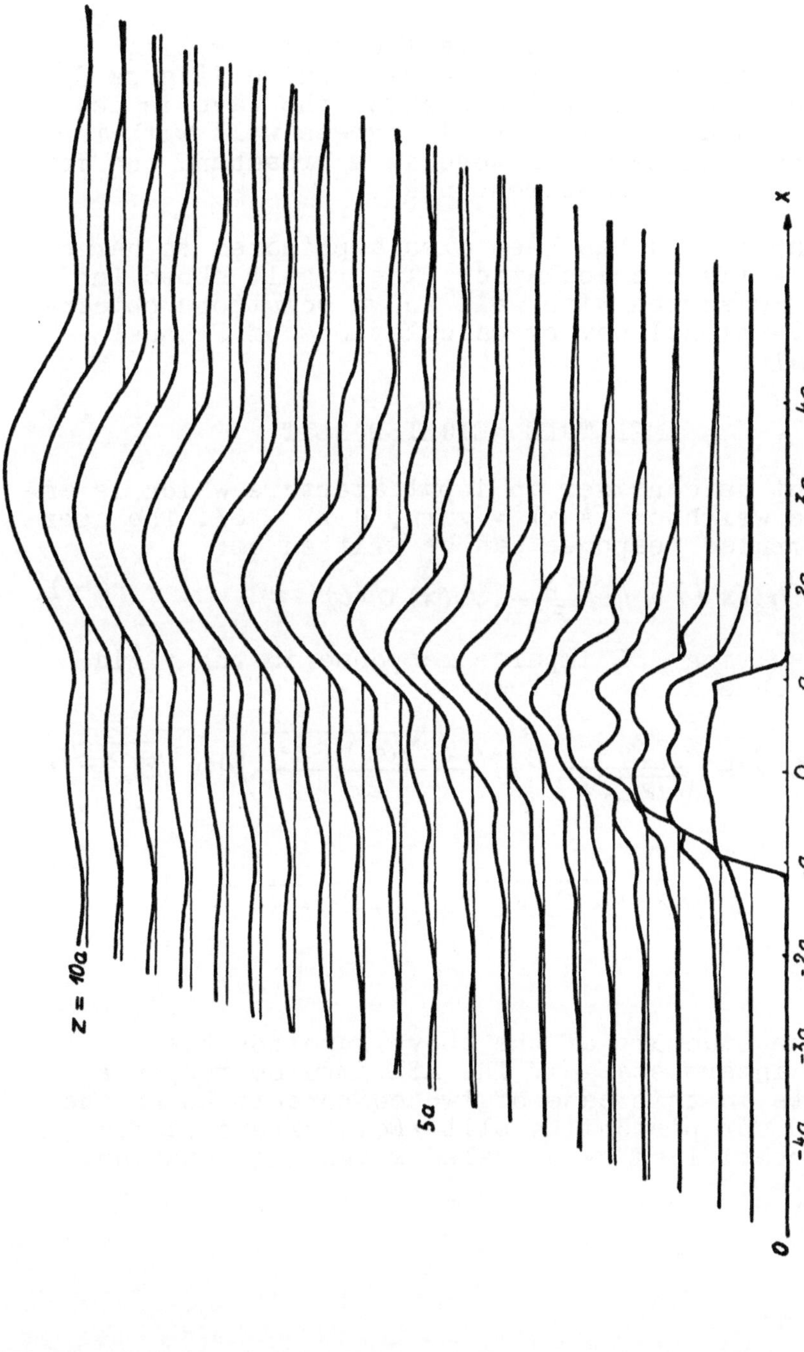

Figure 10. Field of harmonically excited slit /2a = 5λ/

are evident, especially in paraxial zone.

As an example of capabilities of the method the ASL impulse response has been used to calculation of acoustic pressure field of the slit excited uniformly by rectangular pulse /Figure 9/. As it can be seen in the figure, the wave is least deformed near the z-axis. It agrees well with the fact that the paraxial zone is a broadband one for having short impulse responses.

The method has been also applicated to harmonic wave field calculation. The result shown in Figure 10 agrees with well known solutions obtained in conventional way by calculations of Fresnel integrals.

AMPLITUDE WEIGHTED SLIT

Let us consider an input aperture which is amplitude weighted /A (x) = var , d (x) = 0/. The aperture impulse response can be written as:

$$h_A(x,t) = \Pi\left(\frac{x}{2a}\right) A(x)\, \delta(t) \qquad (14)$$

The resulting ASL impulse response in the rigid baffle case is as follows:

$$k_z(x,t) = \frac{2}{\sqrt{t^2 - t_z^2}} \left\{ \Pi\left(\frac{x - \sqrt{(ct)^2 - z^2}}{2a}\right) A\left(x - \sqrt{(ct)^2 - z^2}\right) + \right.$$

$$\left. + \Pi\left(\frac{x + \sqrt{(ct)^2 - z^2}}{2a}\right) A\left(x + \sqrt{(ct)^2 - z^2}\right) \right\} \qquad (15)$$

The stucture of the above function has a simple interpretation. The ASL impulse response consists in this case of two components being the same as for pistonlike slit /Eq. (13)/ except for being multiplied by a scaled weighting function.

The scaling functions depend on x coordinate of the field point, time limits staying the same as in equation (13).

UNIDIMENSIONAL LENS

The lens is a delay-weighted object . Unidimensional lens of the focal length f is represented by the aperture impulse response of the form /$A(x) = 1$, $d(x) = -x^2/2fc$ /:

$$h_A(x,t) = \Pi\left(\frac{x}{2a}\right)\delta\left(t + x^2/2fc\right) \tag{16}$$

We have calculated the ASL impulse response of the lens using Fresnel approximation. Although not necessary, it simplifies the final form of the solution that is as follows:

$$k_{zf}(x,t) = \frac{2n_f(x,t)}{\sqrt{2t_z(t-t_z)(1-\frac{z}{f})+t_x^2\frac{z}{f}}} \tag{17}$$

where $n_f(x,t)$ is the gating coefficient with time limits modified due to input delays and $t_x = |x|/c$.

As it can be noticed in Figure 11, the above solution has interesting features. While ahead of the focal distance /$z < f$/ the ASL impulse response of the lens resembles in general sense that of the unweighted slit, its shape behind the focal distance /$z > f$/ undergoes a sort of time inversion.

In the focal distance itself /$z = f$/ the mentioned response has the form:

$$k_{zf}(x,t) = \frac{2n_f(x,t)}{t_x} \tag{18}$$

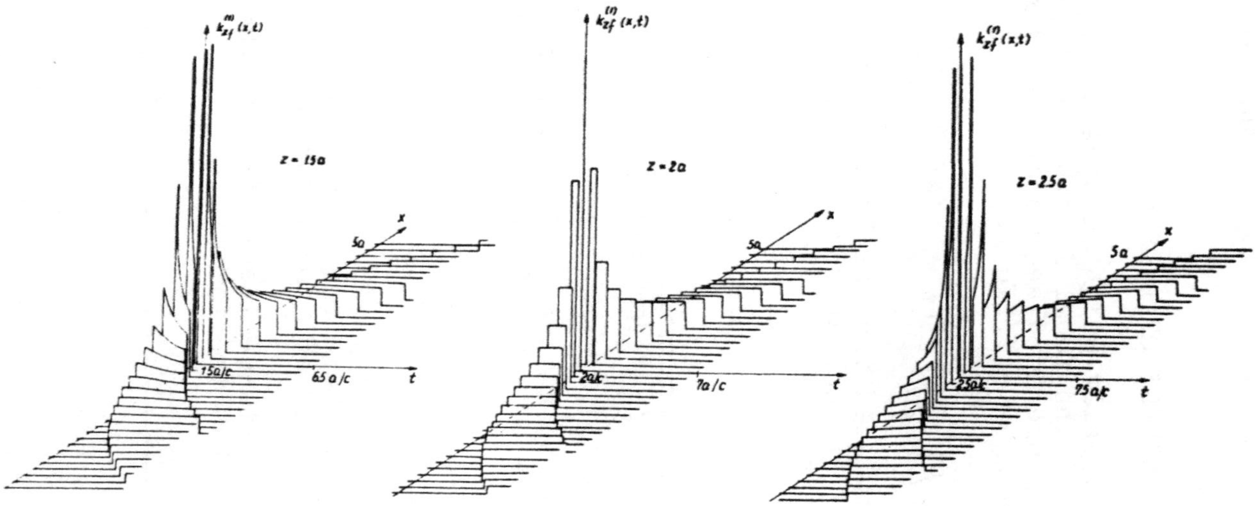

Figure 11. ASL impulse response of unidimensional lens in vicinity of focal distance / z < f, z = f, z > f /

that is the same as for the Fraunhofer approximation ASL response of unweighted slit. It agrees well with the known feature of lenses of transforming far-field distribution to the focal plane.

CONCLUSION

The method described in the paper represents a generalization of two methods being used in analysis of acoustic fields. The first is known as Fourier optics[2] and the other is the impulse response method developed by Stepanishen [6]. The Fourier optics method makes it possible to analyse effectively complex acoustic systems that use narrowband signals,while the second one concerns analysis of simple acoustic systems for broadband signals.

As it can be easily seen,the method being proposed unifies features of both allowing to study complex acoustic systems that use broadband signals.

The elimination of the harmonic function from the description of acoustic radiation problem accentuates the basic physical and geometric effects constituting the diffraction phenomenon. Therefore results are easy to interpret.

At the same time the simple analytic form of the solutions is advantageous for the effective solving of problems of complex acoustic systems.

The unity of the analytic description of the whole field that does not depend on zones /Fraunhofer, Fresnel nor shade zone/ is really noteworthy. In this way no approximations are needed that have been usually necessary for obtaining clear results.

Moreover, when using our method, taking into account the obliquity factors does not complicate the problem, paraxial approximation commonly used in Fourier optics is not actual anymore.

In conclusion we can state that the time-space impulse response method can be applied successfully while analysing complex acoustic systems in the case of broadband signals. This method can be also used effectively for narrow-band signals after having acquired certain experience in interpretation of impulse responses we have introduced.

Though we have presented the method in application to unidimensional systems, it can be used in fact in bidimensional systems analysis without any fundamental modification.

REFERENCES

1. H. Lasota, "Time-space impulse response method with application to analysis of acoustical imaging systems", Ph.D. Dissertation, Technical University of Gdańsk, 1978.

2. J. W. Goodman, "Introduction to Fourier Optics", Mc Graw-Hill, New York, 1968.

3. E. Skudrzyk, "Fundamentals of Acoustics", Springer, Wien, 1970.

4. B. Delannoy, H. Lasota et al., "The infinite planar baffles problem in acoustic radiation and its experimental verification", J. Appl. Phys. 50, 8, 5189, 1979.

5. R. Berkowitz, "Modern Radar", J. Wiley and Sons, New York, 1965.

6. P. R. Stepanishen, "Transient radiation from pistons in an infinite planar baffle", J. J. Acoust. Soc. Am. 49, 5 /2/, 1628, 1971.

ACOUSTIC IMAGING BY WAVE FIELD EXTRAPOLATION

PART I: THEORETICAL CONSIDERATIONS

A.J. Berkhout, J. Ridder, L.F. v.d. Wal

Delft University of Technology
Department of Applied Physics
Group of Acoustics
P.O. Box 5046, 2600 GA Delft, The Netherlands

ABSTRACT

In this paper it is shown that the focussing problem can be approached from the theory of inverse filtering. First a forward model is derived with the aid of the acoustic wave equation:

$$Q(z_o) = \sum_{i=o}^{N} W(z_i, z_{i+1})\, R(z_{i+1}) W(z_{i+1}, z_i);$$

$$P(z_o) = S(z_o) Q(z_o) D(z_o). \tag{1}$$

Each row of complex-valued data matrix $P(z_o)$ contains the response of one source (or one source array) at a given position in acquisition plane z_o. Transducer matrices $S(z_o)$ and $D(z_o)$ define the acquisition geometry and the transducer configurations. Propagation matrices $W(z_i, z_{i+1})$ and $W(z_{i+1}, z_i)$ are defined by the wave equation and quantify the propagation effects in layer (z_i, z_{i+1}) for downward and upward travelling waves respectively. Scattering matrix R determines the acoustic reflectivity properties of the medium. In the *forward* problem ('modeling') $P(z_o)$ is computed for a given set of $R(z_i)$. In the *inverse* problem ('focussing') $R(z_i)$ is computed for a given $P(z_o)$. The inverse operator, being derived from physical model (1), simulates a suite of new unfocussed images for fictitious recording planes *inside* the medium of investigation:

$$P(z_{i+1}) = W^{-1}(z_{i+1},z_i)P(z_i)W^{-1}(z_i,z_{i+1}) \quad (2)$$

for $i = 1,2,\ldots,N$.

The focussed image at depth level z_{i+1} is obtained by selecting from the unfocussed image $P(z_{i+1})$ the data around zero travel time.

It is shown that the proposed method is most suitable for media with variable propagation velocities and variable absorption.

The practical aspects and an illustrative example of this focussing technique will be given in part II of this paper.

INTRODUCTION

The objective of most echo-techniques is the collection of information on the *internal* structure of a medium without destructive penetration. Experience has taught that raw echo recordings need extensive processing before reliable information can be extracted from it. The most fundamental processing technique for echo recordings is generally referred to as '*image reconstruction*'. Image reconstruction is applied in ultrasonic echo techniques (Johnson *et al.*, 1979), tomography (Brooks and Dichiro, 1976), radio astronomy (Fomalont, 1979) and particularly in seismology (Claerbout, 1976; Berkhout, 1980). Advanced acoustic image reconstruction techniques combine different echo measurements (multi-channel processing) such that *diffractions* are 'collapsed', the distortion (in position and amplitude) of *reflections* is corrected for and *noise* is attenuated. In the different disciplines different terminologies are used:

- *Focussing* (ultra-sonic acoustics)
- *Synthetic aperture* (radio astronomy)
- *Migration* (seismic exploration)

In the following we will introduce *synthetic focussing* as an *acoustic* image reconstruction technique based on the scalar wave equation. Application involves two steps (Figure 1):

1. *Wave field extrapolation*
   Using the scalar wave equation, the recorded data is transformed into a series of new recordings which represent simulated registrations at new positions of the recording plane.

2. *Imaging.*
   The imaging principle formulates that the upper parts (i.e. the data around *zero* travel time) of the simulated recordings, related to recording planes *inside* the medium of investigation, form together the *focussed* result.

The wave field extrapolator is derived from the Kirchhoff integral for plane surfaces (summation approach) or directly from a simplified version of the wave equation (finite-difference approach). For the wavenumber-frequency approach the wave field extrapolator can either be derived from the Kirchhoff integral or directly from the Fourier-transformed wave equation. Wave field extrapolation can be applied *recursively,* i.e. the extrapolated output is used as input for the next extrapolation step (Figure 2). Recursive extrapolation has the significant advantage that 'local' velocities can be used and, therefore, spatial velocity variations can be properly handled. It is important to realise that in complicated situations a reliable solution can only be obtained if, in addition to *recursive* extrapolation, the velocity input and the focussed image are defined in the *depth* domain. As in complicated situations the initial velocity depth model is not accurately known, an *iterative* application is called for.

Modeling and synthetic focussing are techniques which are closely related; they are based on the same theory: modeling involves *forward* extrapolation and simulates the effects of wave field propagation; synthetic focussing involves *inverse* extrapolation and removes the effects of wave field propagation. In the space-frequency domain

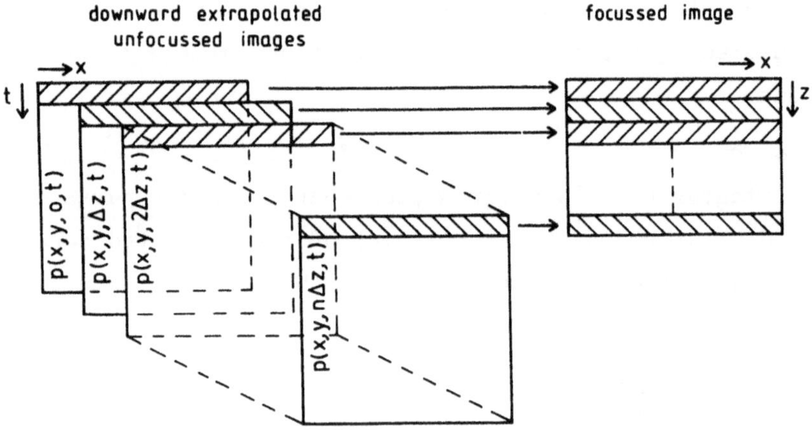

Fig.1. A focussed image may be synthesized by a number of data strips, each strip being defined by the upperpart (i.e. the data around *zero* travel time) of an *un*focussed image which is simulated for a recording plane *below* the surface.

it can be very easily seen that for each frequency component forward extrapolation is realised by *convolution* along the spatial axes; inverse extrapolation is realised by *deconvolution* along the spatial axes, the deconvolution operator being derived from the forward model. This is a very important observation as fundamental problems such as *stability* of the inverse extrapolator, optimum spatial *bandwidth* and optimum spatial *resolution* can be fruitfully approached from the theory of inverse filtering.

In figure 3 the influence of *forward* extrapolation on the upward travelling wave field of a buried point source (dipole at depth level z_o) is schematically shown. Note that after each extrapolation step two changes can be seen:

a. The acoustic energy shifts away from the apex of the response

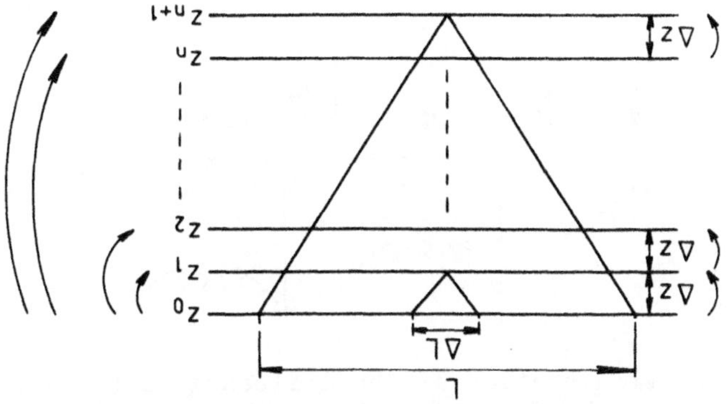

Fig.2. In recursive extrapolation the output of the previous extrapolation step is used as input for the next step; in *non*recursive applications the starting data is used as input for all extrapolation steps.

curve.

b. The apex is shifted towards larger travel times ($\Delta z/c$ for each step).

In Figure 4 the influence of *inverse* extrapolation on the upward travelling wave field of a buried point source (dipole) is schematically shown. Once again two changes can be seen after each extrapolation step:

a. The acoustic energy shifts towards the apex of the response curve;

b. The apex is shifted towards smaller travel times ($\Delta z/c$ for each step).

If the timeshift $\Delta z/c$ is not applied, a time-delayed result is obtained which leaves the position of the apex unchanged. The latter has practical advantages.

Note that in Figure 4 the focussed image of the buried point source

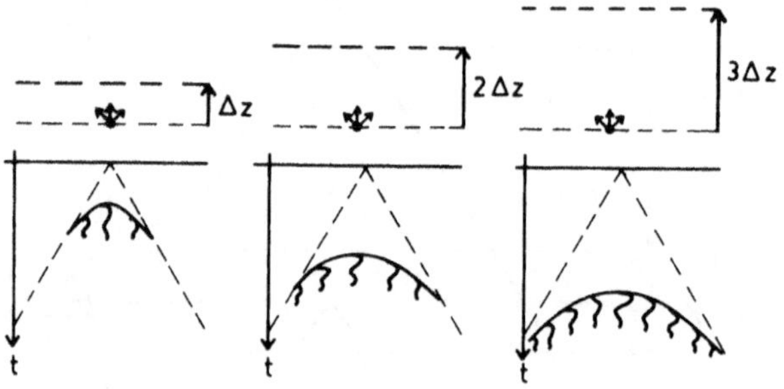

Fig.3. Schematic illustration of the influence of forward extrapolation. For *constant-velocity* media the response curve is defined by a hyperboloid, the lateral position of the apex being given by the lateral position of the point source.

can be obtained from the unfocussed data, related to the recording plane $z=z_o$, at travel time zero or at travel time $3\Delta z/c$ for the time-delayed version. We will see that inverse extrapolation involves a summation procedure along a point-diffractor response curve with a complex-valued frequency-dependent apodization.

Modeling and synthetic focussing are based on the wave equation for compressional waves (P-waves):

$$\nabla^2 P + k^2(1 - j\eta)^2 P = \nabla \ln\rho . \nabla P, \tag{1}$$

where

$$\nabla = \vec{i}_x \frac{\partial}{\partial x} + \vec{i}_y \frac{\partial}{\partial y} + \vec{i}_z \frac{\partial}{\partial z} \tag{2}$$

$$\nabla^2 = \frac{\partial^2}{\partial x^2} + \frac{\partial^2}{\partial y^2} + \frac{\partial^2}{\partial z^2} \tag{3}$$

and

$P = P(x,y,z,\omega)$ represents *pressure* (in fluids) or a scaled version of the average of the principal *stresses* (in solids).

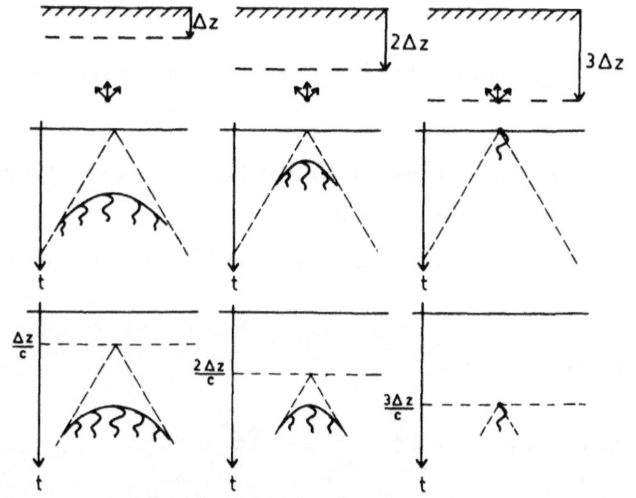

Fig.4. Schematic illustration of the influence of inverse extrapolation. The upper part shows p(x,z,t) for four different recording planes. The lower part shows the *time-delayed* versions: $p(x,z,t-n\Delta z/c)$.

$k = k(x,y,z,\omega) = \omega/c(x,y,z,\omega)$.

$c = c(x,y,z,\omega)$ represents the frequency-dependent phase velocity; if $\eta = 0$ (no absorption) then $c = c(x,y,z)$ represents the propagation velocity.

In this paper transmission coefficients are ignored but angle-dependent reflection coefficients are included with the aid of the *scattering matrix*. Moreover, multiply reflected and/or multiply diffracted energy will be ignored (Borne's approximation).
In the following we will derive expressions for the *forward* extrapolator first. From these expressions the *inverse* extrapolator will be defined.

FORWARD EXTRAPOLATION

In echo-techniques forward travelling P-waves occur in two ways:

a. Waves travelling *downward* from the surface to the inhomogeneities of interest;

b. Waves travelling *upward* from the inhomogeneities back to the surface.

In the forward problem both type of waves have to be extrapolated. We will review the basic principles now.
Forward wave field extrapolation can be quantified by two basic equations:

1. *The Taylor series*

$$P(z_o \pm \Delta z) = P(z_o) \pm \frac{\Delta z}{1!}\frac{\partial P}{\partial z_o} + \frac{\Delta z^2}{2!}\frac{\partial^2 P}{\partial z_o^2} \pm \ldots\ldots \tag{4a}$$

Hence, given the data itself *and* the derivatives towards z in the data-plane $z = z_o$ then the data in the plane $z = z_o \pm \Delta z$ can be computed with the aid of (4a).
In (4a) the positive sign must be used for downward travelling waves and the negative sign must be used for upward travelling waves (note that "downward" means increasing z).

2. *The wave equation*

$$\nabla^2 P + k^2(1 - j\eta)^2 P = 0. \tag{4b}$$

Standard acquisition techniques do not measure derivatives and, therefore, the derivatives towards z as needed in the Taylor series must be computed from the data itself. We will see that the *wave equation* facilities this computation:

$$\frac{\partial P}{\partial z_o} = \pm\, jkH(x,y,\omega) * P(x,y,z_o,\omega); \tag{5a}$$

$$\frac{\partial^2 P}{\partial z_o^2} = (\pm jk)^2 H(x,y,\omega) * H(x,y,\omega) * P(x,y,z_o,\omega); \tag{5b}$$

etc.

In expressions (5a) and (5b) the negative sign relates to downward travelling waves and the positive sign relates to upward travelling waves; * denotes convolution.

If we combine expressions (4a) and (5) the basic expression for *forward* extrapolation is obtained:

$$P(z_o \pm \Delta z) = P(z_o) + \frac{(-jk\Delta z)}{1!} H * P(z_o) + \frac{(-jk\Delta z)^2}{2!} H*H*P(z_o) + \ldots.$$

or

$$P(z_o \pm \Delta z) = \left[\delta(x)\delta(y) + \frac{(-jk\Delta z)}{1!}H + \frac{(jk\Delta z)^2}{2!} H * H + \ldots\right] * P(z_o)$$

or

$$P(z_o \pm \Delta z) = \left[\delta(x)\ \delta(y) + G(x,y,\Delta z,\omega)\right] * P(z_o) \tag{6a}$$

or

$$P(x,y,z_1,\omega) = W(x,y,\Delta z,\omega) * P(x,y,z_o,\omega), \tag{6b}$$

where $W(x,y,\Delta z,\omega)$ represents the forward extrapolator for the extrapolation distance $\Delta z = |z_1 - z_o|$. Note that $G(x,y,\Delta z,\omega)$ plays the role of a spatial *prediction-error* filter.

If we assume a pressure point source in the plane $z = z_o$, i.e. a dipole

$$P(x,y,z_o,\omega) = \delta(x)\delta(y), \tag{7a}$$

then the pressure in the plane $z = z_1$ is given by

$$P(x,y,z_1,\omega) = W(x,y,\Delta z,\omega). \tag{7b}$$

Therefore $W(x,y,\Delta z,\omega)$ may be considered as a *spatial impulse response*. We will see that in the summation approach W is given by the Kirchhoff integral for plane surfaces, i.e. the Rayleigh integral. Hence, in the summation approach summation of *all* terms of the Taylor series has already been carried out, resulting in an analytical expression for W. In 'finite-difference techniques' a *truncated* version of the Taylor series expansion of W is used (Claerbout, 1976).

To illustrate the foregoing let us consider expressions (5), (6) and (7) for loss-free, homogeneous media. For this simplified situation the wave equation may be written as

$$\frac{\partial^2 P}{\partial x^2} + \frac{\partial^2 P}{\partial y^2} + \frac{\partial^2 P}{\partial z^2} + k^2 P = 0, \tag{8a}$$

where k is independent of the spatial coordinates.
Fourier transformation of (8a) towards x and y yields

$$\frac{\partial^2 \tilde{P}}{\partial z^2} + (k^2 - k_x^2 - k_y^2)\, \tilde{P} = 0, \tag{8b}$$

where $\tilde{P} = \tilde{P}(k_x, k_y, z, \omega)$.
Equation (8b) can also be written as

$$\frac{\partial P}{\partial z} = \pm j \sqrt{k^2 - k_x^2 - k_y^2}\, \tilde{P}. \tag{8c}$$

Bearing in mind that the expression of a plane wave travelling in the positive z-direction is given by

$$A\, e^{-j(\omega/c_z)z}$$

and the expression of a plane wave travelling in the negative z-direction is given by

$$A\, e^{+j(\omega/c_z)z},$$

c_z being the phase velocity along the z-axis ($c_z = c/\cos\alpha$), then it can easily be seen that in (8c) we must choose a negative sign when we are dealing with *downward* travelling waves and we must chose the positive sign when we are dealing with *upward* travelling waves. Hence, if we compare (8c) with (5a) we may conclude

$$\tilde{H}(k_x,k_y,\omega) = \sqrt{1 - (k_x^2 + k_y^2)/k^2} \tag{9a}$$

or, in terms of the tilt angle α,

$$\tilde{H}(k_x,k_y,\omega) = \cos\alpha \tag{9b}$$

or, using Berkhout (1980, appendix E),

$$H(x,y,\omega) = \frac{k}{2\pi} j_1(kr_o)/r_o, \tag{9c}$$

where $r_o = \sqrt{x^2 + y^2}$ and j_1 represents the first-order spherical Bessel function. Now if we consider the Taylor series in the (k_x,k_y,z,ω) domain and use (8c) then we may write:

$$\tilde{P}(z_1) = \left[1 + \left(-j\sqrt{k^2 - k_x^2 - k_y^2}\right)\frac{\Delta z}{1!} + \left(-j\sqrt{k^2-k_x^2-k_y^2}\right)^2\frac{\Delta z^2}{2!} + \ldots\right]\tilde{P}(z_o) \tag{10a}$$

or, bearing in mind the series expansion of the exponential function,

$$\tilde{P}(z_1) = e^{-j\sqrt{k^2-(k_x^2+k_y^2)}\Delta z}\,\tilde{P}(z_o). \tag{10b}$$

For the special situation that $k_x^2 + k_y^2 > k^2$ it follows from (10b)

$$\tilde{P}(z_1) = e^{-\sqrt{(k_x^2 + k_y^2)-k^2}\,\Delta z}\,\tilde{P}(z_o). \tag{10c}$$

To ensure finite pressure values the *negative* exponential has to be chosen in (10c).

If we compare (10b) and (10c) with (6b) then we obtain

$$\tilde{W}(k_x,k_y,\Delta z,\omega) = e^{-j\sqrt{k^2-(k_x^2+k_y^2)}\Delta z} \quad \text{for } k_x^2 + k_y^2 \leq k^2; \tag{11a}$$

$$\tilde{W}(k_x,k_y,\Delta z,\omega) = e^{-\sqrt{(k_x^2+k_y^2)-k^2}\,\Delta z} \quad \text{for } k_x^2 + k_y^2 > k^2. \tag{11b}$$

The part at $k_x^2 + k_y^2 > k^2$ is called the *evanescent* wave field of spatial impulse response W. Note that the evanescent wave field attenuates exponentially with distance.

Application of the inverse Fourier transformation to (11), as e.g. derived by Berkhout (1980, appendix D), yields

$$W(x,y,\Delta z,\omega) = \frac{\Delta z}{2\pi} \frac{1 + jkr}{r^3} e^{-jkr}, \qquad (11c)$$

where $r = \sqrt{x^2 + y^2 + \Delta z^2}$.

The time domain version of (11c) was used by French (1975) in a discussion on seismic migration of oblique profiles.

INVERSE EXTRAPOLATION

In the foregoing we have derived that forward extrapolation can be described by spatial convolution for each temporal frequency component (Figure 5 a):

$$P(z_1) = W(z_1,z_o) * P(z_o) + N(z_1), \qquad (12)$$

where $W(z_1,z_o)$ is a symbolic notation for $W(x,y,\Delta z,\omega)$ with $\Delta z = |z_1 - z_o|$. Equation (12) quantifies the effect of wave propagation from depth level z_o to depth level z_1. The term $N(z_1)$ specifies the noise present on depth level z_1. Note that $N(z_1)$ may be spatially coherent.

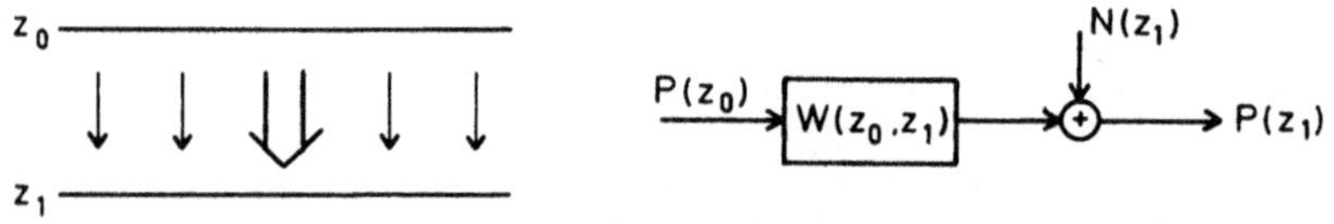

Fig.5(a). The physical model of wave propagation between depth levels z_o and z_1 together with additive noise. The convolution operator $W(z_o,z_1)$ quantifies the forward propagation effects in layer (z_o,z_1).

Now if we want to compensate the propagation effect in layer (z_o,z_1), we need an operator such that (Figure 5b):

$$<P(z_o)> = F(z_o,z_1) * P(z_1), \tag{12b}$$

$<P(z_o)>$ representing an estimate of the exact pressure at z_o. We will call F the *inverse* extrapolator. If we substitute (12a) in (12b) we obtain

$$<P(z_o)> = F(z_o,z_1) * W(z_1,z_o) * P(z_o) + N'(z_o), \tag{13}$$

where $N'(z_o)$ represents the filtered noise.

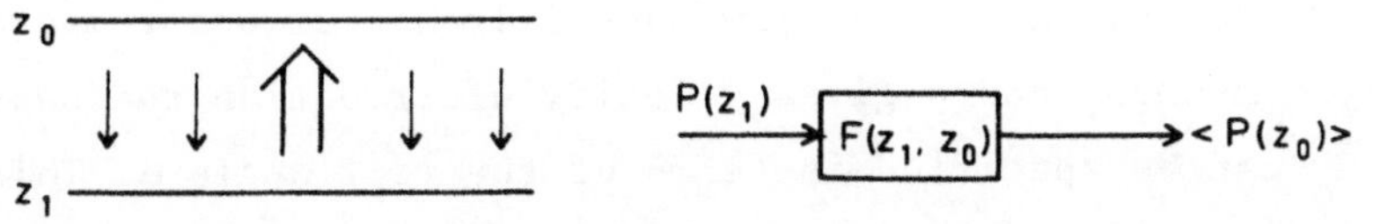

Fig.5(b). Inverse extrapolation can be realised by spatial *de*-convolution. Application of deconvolution operator $F(z_1,z_o)$ compensates for the forward propagation effects in layer (z_o,z_1).

From (13) it can be seen that $<P(z_o)>= P(z_o)$ if $N'(z_o) = 0$ (noise free situation) and if

$$F(x,y,\Delta z,\omega) * W(x,y,\Delta z,\omega) = \delta(x)\delta(y) \tag{14a}$$

or, after Fourier tranformation

$$\tilde{F}(k_x,k_y,\Delta z,\omega) = 1 / \tilde{W}(k_x,k_y,\Delta z,\omega). \tag{14b}$$

Operator (14) is unstable and cannot be used as such. This important fact can be easily understood if we consider the loss-free homogeneous situation.

Substitution of (11) into (14) yields

$$\tilde{F}(k_x,k_y,\Delta z,\omega) = e^{+j\sqrt{k^2-(k_x^2+k_y^2)}\,\Delta z} \qquad \text{for } k_x^2 + k_y^2 \leqslant k^2, \tag{15a}$$

$$\tilde{F}(k_x,k_y,\Delta z,\omega) = e^{+\sqrt{(k_x^2+k_y^2)-k^2}\,\Delta z} \qquad \text{for } k_x^2 + k_y^2 > k^2. \tag{15b}$$

Expression (15b) defines an exponentially increasing operator which is unacceptable in practical situations. This also means that result (15a) can never be realised. It can be shown that in situations with *absorption* the unstability is even more pronounced.
In practice several alternatives are of interest:

1. *Band-limited inversion*

$$\tilde{F} = \tilde{W}_o/\tilde{W}, \tag{16a}$$

where $\tilde{W}_o$ represents a spatial low-pass filter. For one temporal frequency component and a given velocity distribution the pass-band of $\tilde{W}_o$ can be specified in terms of the tilt angle α. This is a valuable option, particular in finite-difference techniques, where W has been properly described upto a maximum tilt angle only (α_m).

2. *Least-squares inversion*

$$\tilde{F} = \frac{\tilde{W}^*}{|\tilde{W}|^2 + |\tilde{N}|^2}. \tag{16b}$$

Note the interesting similarity with two-sided least-squares temporal deconvolution, where the *time* wavelet and the spatially-*incoherent* noise spectrum should be specified (see e.g. Berkhout, 1977). Here we are dealing with the *spatial* wavelet and the spatially-*coherent* noise spectrum; temporal frequency plays the role of a *parameter*.
Hence, if we carry out inverse extrapolation with the aid of (16b) then shot-generated noise can be optimally taken into account. Note that spatial wavelet estimation means velocity analysis.

3. *Matched filtering*

$$\tilde{F} = \tilde{W}^*. \tag{16c}$$

It is interesting to realise that if we do not consider absorption and evenescent waves then (16c) defines a pure spatially zero-phasing procedure.

Note that the well-known inverse summation operator (Schneider, 1978),

$$F = \frac{\Delta z}{2\pi} \frac{1 - jkr}{r^3} e^{+jkr}, \tag{17}$$

is obtained by applying the matched-filter concept to the summation operator for loss-free homogeneous media (11b).

SUMMATION APPROACH

The summation approach to wave field extrapolation is based on the Kirchhoff-integral for plane surfaces, i.e. the Rayleigh integral. The derivation of the Rayleigh integral assumes homogeneous media and can be carried out in two ways:

1. Starting with the theorem of Gauss, Green's theorems are derived. If we chose for the two scalar fields in Green's second theorem:

   a. the pressure field we are interested in (P)

   b. the pressure field of a single monopole (G_1)

   then, by substituting the wave equation for both wave fields in Green's second theorem, the Kirchhoff integral is obtained:

$$P_A = \frac{1}{4\pi} \oiint_S \left[P \frac{\partial G_1}{\partial n} - \frac{\partial P}{\partial n} G_1 \right] dS, \tag{18}$$

where

P = pressure field of interest;

$\frac{\partial P}{\partial n}$ = component of the pressure field along the normal $\vec{n}$ of closed surface S ($\vec{n}$ pointing *inward*);

G_1 = pressure field of one monopole, situated in a point A *inside* S.

If we chose for S a plane surface and we modify G_1 to G_2 such that $G_2 = 0$ *on* S then the Kirchhoff integral can be rewritten in terms of the Rayleigh integral:

$$P_A = \frac{1}{4\pi}\iint_S P \frac{\partial G_2}{\partial n} dS$$

or, substituting the expression for G_2 and using for S the horizontal plane $z = z_o$,

$$P_A = \frac{|z_A - z_o|}{2\pi}\iint_S P \frac{1 + jk_1\Delta r}{\Delta r^3} e^{-jk_1\Delta r} dxdy, \tag{19}$$

where $k_1 = k(1 - j\eta)$, $P = P(x,y,z_o,\omega)$ and $\Delta r = \sqrt{(x_A-x)^2+(y_A-y)^2+(z_A-z_o)^2}$. Note that if $p_A = p(x_A,y_A,z_A,t)$ need be computed upto a finite recording time only, integration area S becomes finite as well (finite aperture area).

2. Starting with the wave equation,

$$\frac{\partial^2 P}{\partial x^2} + \frac{\partial^2 P}{\partial y^2} + \frac{\partial^2 P}{\partial z^2} + k^2(1 - j\eta)^2 P = 0, \tag{20a}$$

a Fourier transformation towards the spatial variables x and y is carried out:

$$\frac{\partial^2 \tilde{P}}{\partial z^2} + (k_1^2 - k_x^2 - k_y^2)\,\tilde{P} = 0, \tag{20b}$$

where $k_1 = k(1 - j\eta)$.

Equation (20b) formulates the one-dimensional Helmholtz equation along the z-axis; its forward solution is well-known:

$$\tilde{P}(k_x,k_y,z,\omega) = \tilde{P}(k_x,k_y,z_o,\omega)\; e^{-j\sqrt{k_1^2-(k_x^2+k_y^2)}\,|z-z_o|}. \tag{21}$$

On physical grounds the imaginary part of the square root in (21) must always be chosen negative. Now if we take the inverse Fourier transform of (21) with respect to the spatial Fourier variables k_x and k_y then Rayleigh integral (19) is obtained again.

For a given $\Delta z = |z_A - z_o|$, i.e. A lies in the plane $z_1 = z_o \pm \Delta z$, we may write

$$W(x_A-x, y_A-y, \Delta z, \omega) = \frac{\Delta z}{2\pi} \frac{1 + jk_1\Delta r}{\Delta r^3} e^{-jk_1\Delta r} \tag{22a}$$

and Rayleigh integral (19) can be reformulated as

$$P(x_A, y_A, z_1, \omega) = \iint_S W(x_A-x, y_A-y, \Delta z, \omega)\, P(x, y, z_o, \omega)\, dxdy$$

or

$$P(z_1) = W(z_1, z_o) * P(z_o), \tag{22b}$$

* denoting convolution along the spatial axes. Please note the similarity with equations (6a) and (6b).

As we shall see, formulation of wave field extrapolation in terms of *spatial convolution* in the temporal frequency domain is extremely advantageous in bringing out all essential aspects of forward and inverse extrapolation in a simple way.

To obtain an expression for W in the two-dimensional situation, we have to integrate with respect to y:

$$\begin{aligned} W(x, \Delta z, \omega) &= \int_{-\infty}^{+\infty} W(x, y, \Delta z, \omega)\, dy \\ &= \frac{1}{2\pi} \int_{-\infty}^{+\infty} \frac{1 + jk_1 r}{r^2} \cos\phi\, e^{-jk_1 r}\, dy \\ &= \frac{-jk_1}{2} \cos\phi\, H_1^{(2)}(k_1 r), \end{aligned} \tag{23}$$

where $\cos\phi = \Delta z/r$ and $H_1^{(2)}$ represents the first-order Hankel function of the second kind.

In the far field (kr >> 1) the expressions for W become somewhat simpler:

$$W(x,y,\Delta z,\omega) \approx \frac{jk_1}{2\pi} \cos\phi \frac{e^{-jk_1 r}}{r} \tag{24a}$$

$$W(x,\Delta z,\omega) \approx \sqrt{\frac{jk_1}{2\pi}} \cos\phi \frac{e^{-jk_1 r}}{\sqrt{r}}, \tag{24b}$$

where $\sqrt{j} = e^{j\pi/4}$.

For an extensive discussion on the derivation of 2-dimensional and 3-dimensional summation operators the reader is referred to Berkhout (1980, chapters 5 and 6).

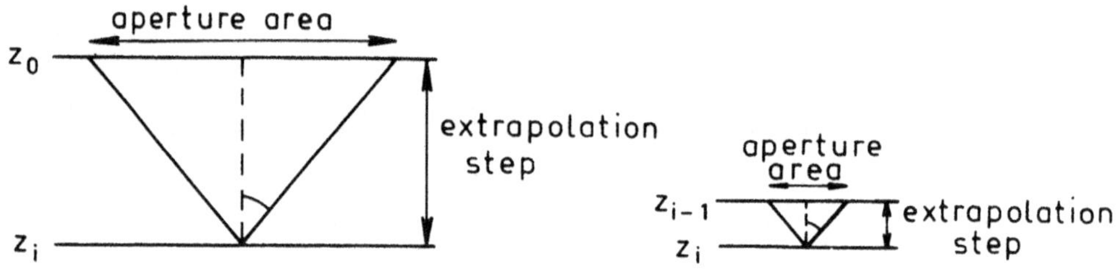

Fig.6. For small extrapolation steps small volumes are involved and, therefore, *local* propagation velocities can be used.

Wave field extrapolation with summation operators can be applied both recursively and nonrecursively. In recursive applications Δz is small ($\Delta z < \lambda/2$). Consequently, for a given aperture angle the conical volume for which one velocity value has to be specified to compute one extrapolated pressure value will be small (Figure 6). This means that a *local* velocity value can be used, which may be taken differently for each extrapolated pressure value. Hence we may conclude that, although the Rayleigh integral was derived for homogeneous situations, vertical and lateral velocity variations can be properly handled by summation operators if the wave field extrapolation procedure is carried out *recursively*. Note that for situations with *lateral* velocity variations (22) defines a *space-variant* convolution. In nonrecursive applications the final extrapolation result is computed by one step only. Hence, for deep data very large extrapolation steps occur and, therefore, very large volumes are involved for which one velocity value has to be specified. Of course, this procedure will never lead to satisfactory results in situations with significant *lateral* velocity variations. Still, nonrecursive summation extrapolation is often applied in the *time* domain as it is a fast wide-angle procedure (seismic applications):

a. In modeling applications the extrapolated data (*forward* extrapolation) is directly computed for the desired reference plane ($z = 0$);

b. In focussing applications the extrapolated data (*inverse* extrapolation) need be computed for $t = 0$ only.

In the foregoing we have seen that in situations with laterally changing velocities forward extrapolation can be formulated in terms of space-variant convolution. From systems theory (e.g. consider the formulation of Kalman filtering) we know that the relation between input and output of nonstationary systems can be beautifully described by means of matrices. For the *two-dimensional* situation the matrix formulation of (22) becomes:

a. In terms of *column* vectors

$$\vec{P}(z_1) = W(z_1, z_o)\, \vec{P}(z_o) \tag{25a}$$

b. In terms of *row* vectors

$$\vec{P}^I(z_o)\, W(z_o, z_1) = \vec{P}^I(z_1). \tag{25b}$$

In expressions (25) we have defined:

$$\vec{P}^I(z_o) = \left(P(o,z_o,\omega),\ P(\Delta x,z_o,\omega),\ ..\ P(n\Delta x,z_o,\omega),\ ..\ P(N\Delta x,z_o,\omega)\right)$$

= data vector (row vector) containing the pressure values in the data plane $z = z_o$. If z_o defines the surface ($z_o = 0$) then $\vec{P}^I(z_o)$ defines a *source array* (of course for one frequency component only).

$W(z_o,z_1)$ = forward propagation matrix between the depth levels z_o and z_1. If all elements of $\vec{P}^I(z_o)$ are zero except at $x = n\Delta x$, where $P(n\Delta x,\Delta z,\omega) = 1$, then $\vec{P}^I(z_1)$ represents the response of a spatial impulse (=dipole) at $(n\Delta x,z_o)$, i.e. $\vec{P}^I(z_1 = \vec{W}^I(n\Delta x,\Delta z,\omega)$. Hence, symbolically,

$$W(z_o,z_1) = \begin{bmatrix} \vec{W}^I(o,\Delta z,\omega) & & & \\ & \vec{W}^I(\Delta x,\Delta z,\omega) & & \\ & & \ddots \; \vec{W}^I(n\Delta x,\Delta z,\omega) & \\ & & & \ddots \; \vec{W}^I(N\Delta x,\Delta z,\omega) \end{bmatrix}$$

Note that $\vec{W}(n\Delta x,\Delta z,\omega)$ is defined by (22a) with the local velocity of layer (z_o,z_1) around $x = n\Delta x$. Note also that without lateral velocity variations W represents a *Toeplitz* matrix.

The symbol 'I' means interchanging rows and columns. Hence, it follows from (25a) and (25b) that $W(z_o,z_1) = W^I(z_1,z_o)$. In the following we have chosen for the row vector presentation; thus the data response of one source measured on an *horizontal* plane at some given depth level, is defined by *row* vectors.

In Figure 7 the total data flow is shown for a response from depth level $z = z_m$ due to one source or one source array at the surface $z = z_o$. If we use the symbol S for the *down*going pressure field and

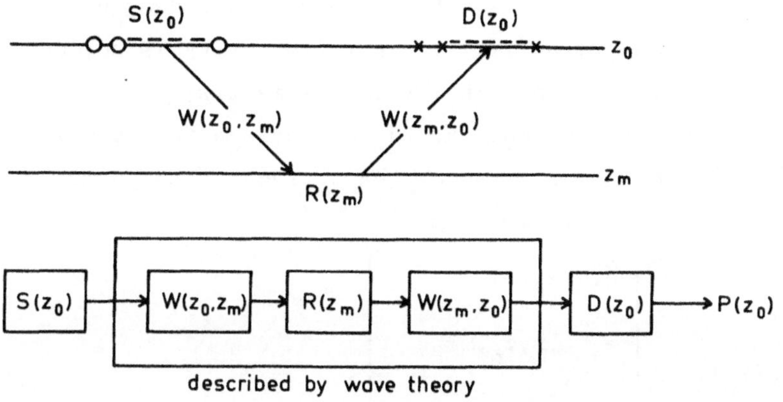

Fig.7. Physical model for the simulation of acoustic echo-data. Here the response from an acoustic discontinuity at depth level z_m is considered.

P for the *up*going pressure field then we may write in terms of vectors and matrices:

$$\vec{P}^I(n\Delta x, z_o, \omega) = \vec{S}^I(n\Delta x, z_o, \omega)\left[W(z_o,z_m)R(z_m)W(z_m,z_o)\right] D(z_o), \quad (26)$$

where

$\vec{S}^I(n\Delta x, z_o, \omega)$ = row vector defining the source array on the surface around $x = n\Delta x$;

$W(z_o,z_m)$ = forward propagation matrix between the depth levels z_o and z_m;

$W(z_m,z_o)$ = forward propagation matrix between the depth levels z_m and z_o; assuming that the principle of reciprocity holds, then $W(z_m,z_o)=W^I(z_o,z_m)$;

$R(z_m)$ = scattering matrix at depth level z_m; if there exists a reflector at z_m then the nth row defines the angle-dependent reflection coefficient at $x = n\Delta x$; if the reflectivity is not angle-dependent (locally-reacting assumption) then R is a diagonal matrix and, finally, if the reflectivity is not depending on x then R is a *Toeplitz* matrix;

$D(z_o)$ = detector matrix, the nth column representing the detector array on the surface at $x = n\Delta x$: $\vec{D}(n\Delta x, z_o, \omega)$. Hence,

$$D(z_o) = \begin{bmatrix} \vec{D}(o,z_o,\omega) & & & \\ & \vec{D}(\Delta x,z_o,\omega) & & \\ & & \vec{D}(n\Delta x,z_o,\omega) & \\ & & & \vec{D}(N\Delta x,z_o,\omega) \end{bmatrix};$$

$\vec{P}^I(n\Delta x, z_o, \omega)$ = row vector defining the pressure values as recorded by the detector arrays on the surface and generated by one source array on the surface at $n\Delta x$ (i.e. one echo recording).

Now if we consider separately many responses from different sources ($n = 1, 2, \ldots, N$) and combine all the row vectors $\vec{S}^I(n\Delta x, z_o, \omega)$ in a source matrix $S(z_o)$ and all the row vectors $\vec{P}^I(n\Delta x, z_o, \omega)$ in a data matrix $P(z_o)$ then we obtain the matrix formulation of a multi-record data set:

$$P(z_o) = S(z_o)\left[W(z_o,z_m)R(z_m)W(z_m,z_o)\right] D(z_o) \qquad (27a)$$

or, taking the responses from all depth levels of interest,

$$P(z_o) = S(z_o)\left[\sum_{m=1}^{M} W(z_o,z_m)\, R(z_m)\, W(z_m,z_o)\right] D(z_o). \qquad (27b)$$

Note that for *recursive* applications

$$W(z_o,z_N) = W(z_o,z_1)\,W(z_1,z_2)\,\ldots\ldots\,W(z_{M-1},z_M). \qquad (28)$$

From (27b) and (28) it follows that multi-record data sets can also be formulated as

$$P(z_m) = W(z_m,z_{m+1})\,P(z_{m+1})\,W(z_{m+1},z_m) + R(z_m) \qquad m = 1,2,\ldots \qquad (29a)$$

and

$$P(z_o) = S(z_o)\left[W(z_o,z_1)P(z_1)W(z_1,z_o)\right]D(z_o). \qquad (29b)$$

In the above expressions

$$S(z_o) = \begin{bmatrix} \vec{S}^I(o,z_o,\omega) & & & \\ & \vec{S}^I(\Delta x,z_o,\omega) & & \\ & & \vec{S}^I(n\Delta x,z_o,\omega) & \\ & & & \vec{S}^I(N\Delta x,z_o,\omega) \end{bmatrix}$$

and

$$P(z_o) = \begin{bmatrix} \vec{P}^I(o,z_o,\omega) & & & \\ & \vec{P}^I(\Delta x,z_o,\omega) & & \\ & & \vec{P}^I(n\Delta x,z_o,\omega) & \\ & & & \vec{P}^I(N\Delta x,z_o,\omega) \end{bmatrix}$$

Expressions (29) define the basic equations for modeling. Here we have defined W with the aid of the Rayleigh integral.

One of the most difficult tasks in the forward problem is the specification of the scattering matrices $R(z_m)$. In the practice of modeling $R(z_m)$ is generally specified as a *diagonal* matrix (locally-reacting assumption). This actually means that with most current modeling software the *propagation* effects are evaluated only.

Finally, it follows from (27b) that *zero-offset* data (source position = detector position) is given by

$$P(n\Delta x, z_o, \omega) = \vec{S}^I(n\Delta x, z_o, \omega)\left[\sum_{m=1}^{M} W(z_o, z_m)R(z_m)W(z_m, z_o)\right]\vec{D}(n\Delta x, z_o, \omega), \tag{30}$$

n=1, 2, ..., N.

Now let us use the basic forward model, as defined by expressions (29), to formulate the *inverse* problem. First we will rewrite (29a):

$$P(z_o) = W(z_o, z_m)\left[S(z_o)R(z_m)D(z_o)\right] W(z_m, z_o)$$

or

$$P(z_o) = W(z_o, z_m)\ \bar{R}(z_m)\ W(z_m, z_o). \tag{31}$$

The matrices W, S and W, D may be interchanged if we assume that the spatial impulse response vector $\vec{W}$ does not change within the length of a source and detector pattern. Note that S and D have an averaging influence on R unless we are dealing with single sources and single detectors. The objective of inverse extrapolation is compensation for downward propagation matrix $W(z_o, z_m)$ and upward propagation matrix $W(z_m, z_o)$ such that an estimate of $\bar{R}$ is obtained.

Hence, this type of extrapolation involves *matrix invrsion:*

$$\begin{aligned} <\bar{R}(z_m)> &= F(z_m, z_o)P(z_o)F(z_o, z_m) \\ &= \left[F(z_m, z_o)W(z_o, z_m)\right]\bar{R}(z_m)\ \left[W(z_m, z_o)F(z_o, z_m)\right] \\ &= W_o(z_m)\ \bar{R}(z_m)W_o(z_m). \end{aligned} \tag{32}$$

If $W_o(z_m)$ represents an unity matrix we obtain after inversion

$<\bar{R}(z_m)> = \bar{R}(z_m)$. However, due to the band limitation of the spatial wavelet W, this result will never be reached (see also expressions (16)).

From the foregoing we may conclude that

1. The basic inverse problem is given by

$$F(z_m,z_o)\, W(z_o,z_m) = W_o(z_m), \tag{33a}$$

where each row of W_o contains the samples of a spatial band-pass filter $\vec{W}_o(n\Delta x, z_m, \omega)$:

$$W_o(z_m) = \begin{bmatrix} \vec{W}_o^I(o,z_m,\omega) & & & \\ & \vec{W}_o^I(\Delta x,z_m,\omega) & & \\ & \ddots & \vec{W}_o^I(n\Delta x,z_m,\omega) & \\ & & \ddots & \vec{W}_o^I(N\Delta x,z_m,\omega) \end{bmatrix}$$

Note that in the situation of matched filtering $F = W^*$.

2. Inverse extrapolation involves two matrix multiplications with inverse F:

   a. To compensate for the propagation effect of *upward* travelling waves

$$Q(z_o) = P(z_o)F(z_o,z_m); \tag{33b}$$

   b. To compensate for the propagation effect of *downward* travelling waves

$$<\bar{R}(z_m)> = F(z_m,z_o)Q(z_o). \tag{33c}$$

It can be easily verified that (33b) defines inverse extrapolation of *source gathers* (rows of P) and (33c) defines inverse extrapolation of *detector gathers* (columns of Q).

Finally, one problem remains to be discussed: In order to obtain $<\bar{R}(z_m)>$ the influence of

$$\sum_{i=m+1}^{M} W(z_m,z_i)\ \bar{R}(z_i)W(z_i,z_m)$$

has to be eliminated from the inverse extrapolation result. Bearing in mind that $W(z_m,z_i)$ involves travel times of at least $i\Delta z/c$, $<\bar{R}(z_m)>$ can be obtained (after inverse extrapolation upto depth level z_m) in the *time* domain at *zero* travel time (imaging principle):

$$<\bar{r}(z_m)> = \frac{1}{2\pi}\int_{\omega_{min}}^{\omega_{max}} P(z_m)\ d\omega \quad \text{(multi-channel imaging).}$$

Generally, only zero-offset traces are considered for imaging (Claerbout, 1976). Hence, in the situation of zero-offset imaging the diagonal elements of matrix $P(z_m)$, i.e. $P_{n,n}(z_m)$, are used only:

$$<\bar{r}_{n,n}(z_m)> = \frac{1}{2\pi}\int_{\omega_{min}}^{\omega_{max}} P_{n,n}(z_m)\ d\omega \quad \text{(single-channel imaging)}$$

As was mentioned before, in seismic applications the summation approach is mostly used *non*recursively in the *time* domain for two- and three dimensional data, the absorption being neglected. Hence, expressions (24a) and (24b) are used with the assumption $\eta = 0$:

$$W(x,y,\Delta z,t) \approx \frac{\cos\phi}{2\pi r}\, d_1\left(t - \frac{r}{c}\right), \tag{34a}$$

where $r = \sqrt{x^2+y^2+\Delta z^2}$;

$$W(x,\Delta z,t) \approx \frac{\cos\phi}{\sqrt{2\pi r}}\, d_{\frac{1}{2}}\left(t - \frac{r}{c}\right), \tag{34b}$$

where $r = \sqrt{x^2+\Delta z^2}$.

In expression (34a) $d_1(t)$ represents the inverse Fourier transform of a band-limited version of $j\omega/c$ and $d_{\frac{1}{2}}(t)$ represents the inverse Fourier transform of a band-limited version of $\sqrt{j\omega/c}$.

Using for the inverse problem a spatial matched filter, it follows from (32a) and (32b) that inverse extrapolation in the summation approach means convolution in the space-time domain with

$$f(x,y,\Delta z,t) = \frac{\cos\phi}{2\pi r} d_{-1}\left(t + \frac{r}{c}\right), \tag{35a}$$

$$f(x,\Delta z,t) = \frac{\cos\phi}{\sqrt{2\pi r}} d_{-\frac{1}{2}}\left(t + \frac{r}{c}\right), \tag{35b}$$

where $d_{-1}(t)$ is related to $-j\omega/c$ and $d_{-\frac{1}{2}}(t)$ is related to $\sqrt{-j\omega/c}$. It is interesting to realise that conventional focussing involves a nonrecursive summation operator in the time domain with time-delayed pressure output, without apodization and without time differentiation:

$$f(x,z,t) = \delta\left(t + \frac{r - z}{c}\right) \quad \text{for} \quad r \leqslant r_{max}$$

$$= 0 \qquad \text{for} \quad r > r_{max}.$$

Following this paper we will present a paper in which the practical aspects of the focussing technique given are envisaged. The relationship with conventional focussing will be discussed in more detail and some interesting examples will be given.

CONCLUSIONS

1. It has been shown that acoustic focussing can be described by an inverse extrapolation procedure which simulates unfocussed images for recording planes *inside* the medium of investigation. The focussed result is automatically obtained by selecting the data around zero travel time after each extrapolation step.

2. The recursive version of the proposed method is most suitable for media with spatial variations in velocity and absorption.

3. From the theory it follows that the proposed inversion process compensates for the propagation effects in the medium. In its most general form it involves the inversion of the upward and downward propagation matrix.

4. Conventional focussing can be considered as a *simplified* inversion process, assuming loss-free homogeneous media and neglecting all amplitude aspects.
   In conventional focussing *vertical* travel times are not compensated (floating time reference).

5. The properties of the transducers and the method of data collection have been included in the theory.

REFERENCES

French, W.S., 1975, *Computer migration of oblique seismic reflection profiles:* Geophysics, vol. 40, p. 961 - 980.

Claerbout, J.F., 1976, *Fundamentals of geophysical data processing:* New York, McGraw-Hill.

Berkhout, A.J., 1977, *Least-squares inverse filtering and wavelet deconvolution:* Geophysics, vol. 42, p. 1369 - 1383.

Schneider, W.A., 1978, *Integral formulation in two and three dimensions:* Geophysics, vol. 43, p. 49 - 76.

Berkhout, A.J., 1980., *Seismic migration - imaging of acoustic energy by wave field extrapolation-:* Amsterdam/New York, Elsevier North Holland Publ. Comp.

ACOUSTIC IMAGING BY WAVE FIELD EXTRAPOLATION

PART II: PRACTICAL ASPECTS

J. Ridder, A.J. Berkhout, L.F. v.d. Wal

Delft University of Technology
Department of Applied Physics
Group of Acoustics
P.O. Box 5046, 2600 GA Delft, The Netherlands

ABSTRACT

In this paper the practical aspects of acoustic imaging will be envisaged, based on the theoretical considerations given in the foregoing paper (Part I).
Three different methods of data acquisition, particularly suitable in performing *synthetic focussing*, will be discussed, i.e.

. Plane wave method;

. Zero offset method;

. Complete dataset method.

It will be shown that the complete dataset allows the most general imaging technique and therefore may be used in inhomogeneous media. Both the plane wave and the zero offset method can be derived from the complete dataset and may be considered as simplified versions. However, some restrictions have to be made. The plane wave method is computationally simpler but it assumes that a plane wave stays a plane wave during propagation. From a data acquisition point of view the zero offset method is least intricate, but due to motion artefacts serious problems may be encountered during imaging.

The discussion of the three techniques will be based on the following criteria:

- Instrumentation
- Resolution
- Imaging
- Signal to noise ratio

Finally an example will be given, showing the result obtainable with a zero offset data acquisition technique using a tissue equivalent ultrasound phantom.

INTRODUCTION

In the foregoing paper the *theoretical* considerations of wave field extrapolation have received ample attention. Applying the theory in practice, e.g. medical diagnosis, several interesting problems are encountered.

Far the most delicate problem is the aspect of data acquisition. If the medium under investigation is time-invariant (seismic applications, non destructive testing) no restrictions have to be made concerning the data acquisition time. In medical diagnosis, however, the time interval during which all data samples should be acquired may be very small, due to the movement of both organs and tissue. The simple reason for this is that in synthetic focussing the *r.f. signal* is recorded over a large aperture and to avoid *motion artefacts* of the focussed image, the medium should not move more than a fraction of a wavelength during data acquisition. Such problems can only be coped with, using very fast data acquisition techniques.

A second problem may be encountered in realizing the source- and detector array's, i.e. the transducer. As stated above, echo-data are collected over a large aperture, while in general both the *sample aperture* and the *sample interval* should be less than a wavelength. Therefore application of synthetic focussing in medical

imaging, using a frequency range of 1 - 10 MHz, implicates a high standard of transducer technology.
Further problems may arise in regarding signal to noise ratio, and the special purpose hardware, needed to acquire the data *and* perform the actual focussing.

In the following the problems mentioned above will be discussed, while comparing different data acquisition techniques, which may be used in synthetic focussing.

DATA ACQUISITION TECHNIQUES

In this section we will discuss three different data acquisition techniques, i.e.

- . Plane wave method;
- . Zero offset method;
- . Complete dataset method.

1. Plane wave method.

The basic principle of the plane wave method (PW) is schematically shown in Figure 1.

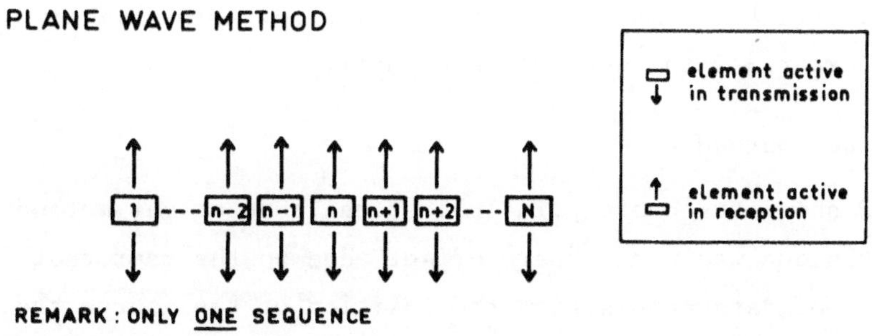

Fig.1.: Schematic illustration of the basic principle of the plane wave method(PW).

During transmission all individual elements of a "large aperture" linear array are activated simultaneously, which will result in a plane wave propagating into the medium.
Note that satisfying results will be obtained only, if the plane wave stays a plane wave during propagation to the maximum depth level.

In the reception mode the responses of all elements are recorded *individually*. Each element should be able to record echo's over a large angle, i.e. should be highly omnidirectional. It is emphasized here, that *one transmission cycle* will suffice to obtain all necessary data samples. Therefore the PW method is most suitable for synthetic focussing of moving structures.

Referring to the foregoing paper, we described the general form of the *forward model* in the *frequency domain* (Equation (27b)):

$$P(z_o) = S(z_o)\left[\sum_{m=1}^{M} W(z_o,z_m)R(z_m)W(z_m,z_o)\right]D(z_o). \tag{1}$$

Using PW data acquisition, source matrix $S(z_o)$ will reduce to a single row vector $\vec{S}^I(z_o)$, since we are dealing with *one* large source array only and thus $P(z_o)$ will reduce to a simple row vector $\vec{P}^I(z_o)$ as well. In general all individual elements of $\vec{S}^I(z_o)$ will be identical.
Note also that for small elements (i.e. point detectors) the detector matrix $D(z_o)$ may be described by a *scalar* matrix (i.e. a diagonal matrix with identical elements).

2. Zero offset method.

Figure 2 shows the basic principle of the zero offset method (ZO). This technique was named "Zero offset" due to the fact that in this type of data acquisition the position of the source and detector coincide (i.e. there is no *spatial offset* between source and detector.

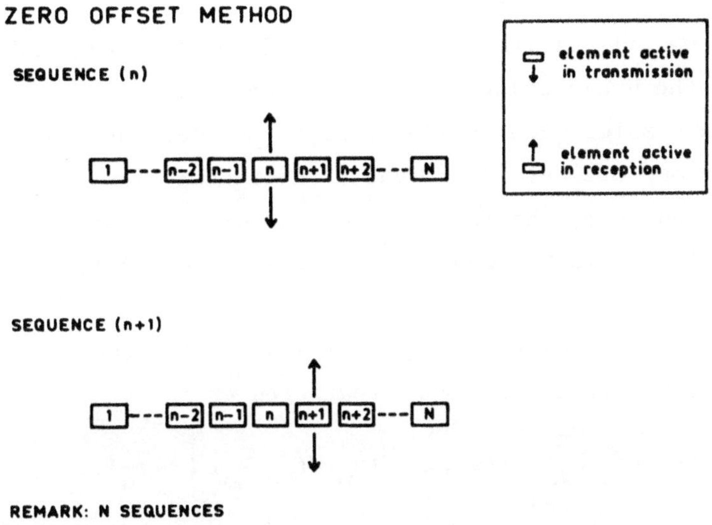

Fig.2.: Schematic illustration of the basic principle of the zero offset method(ZO).

Only one element is activated in transmission *and* reception. Therefore N sequences (N = number of sample points within the aperture) will be needed to obtain all necessary data samples. In the foregoing paper we described *one* ZO sequence as (Equation (30))

$$P(n\Delta x, z_o) = \vec{S}^I(n\Delta x, z_o)\left[\sum_{m=1}^{M} W(z_o, z_m)R(z_m)W(z_m, z_o)\right]\vec{D}(n\Delta x, z_o), \tag{2}$$

where row vector $\vec{S}^I(n\Delta x, z_o)$ denotes the source, and column vector $\vec{D}(n\Delta x, z_o)$ the detector at position $n\Delta x$ within the aperture. For small sources and detectors, both $\vec{S}^I(n\Delta x, z_o)$ and $\vec{D}(n\Delta x, z_o)$ will contain one non-zero element only. In general the N sequences, given by Equation (2) for n = 1,2, ...N, are denoted as a simple row vector $\vec{P}^I(z_o)$.

3. Complete dataset.

Finally we will discuss the complete dataset method (CD). Figure 3 shows the basic principle of this technique.
The transmission mode is identical to the one described above for ZO; only one element is activated in transmission. During reception, however, the responses of *all* elements are recorded individually. Again N sequences are needed to collect all the data samples.
Since in this situation both $S(z_o)$ and $D(z_o)$ reduce to scalar matrices, equation (1) may be rewritten as

$$P(z_o) = S(z_o)\left[\sum_{m=1}^{M} W(z_o,z_m)R(z_m)W(z_m,z_o)\right] D(z_o), \tag{3}$$

where $S(z_o)$ and $D(z_o)$ are *scalar* functions dependent of frequency only.

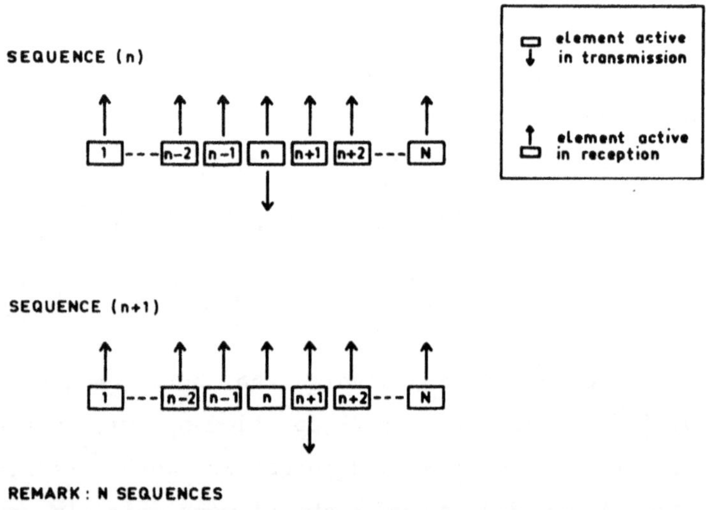

Fig.3.: Schematic illustration of the basic principle of the complete dataset method(CD).

Note that contrary to the PW and ZO method $P(z_o)$ need be represented by a *matrix* and not by a simple row vector.

It is easily understood that the complete dataset is the most general form of data-acquisition, since all the acoustic raypaths between the individual sources and detectors are taken into account seperately.

Both the PW and ZO method can be derived from the CD method in an illustrative way. In PW acquisition all sources are activated simultaneously and therefore one detector response will contain contributions from all source positions. In the case of a CD data matrix $P(z_o)$, the individual elements of a column vector each represent the contribution of *one* individual source to a given detector. PW data is therefore simply obtained from $P(z_o)$ by summing the column elements.

In the special case of ZO data-acquisition selected acoustic raypaths are regarded only: the raypath from a source to a detector located at the same position. In other words, from CD data matrix $P(z_o)$ only the diagonal elements are taken into account.
Note that this result *cannot* be neatly represented by a *single* matrix equation as e.g. Equation (3), since all non-diagonal elements have to be obliterated from the CD data matrix *afterwards*.

INSTRUMENTATION

An important aspect in realizing a synthetic focussing system is formed by the complexity of the data acquisition system. This complexity is strongly influenced by three factors:

- the amount of data samples obtained during data-acquisition;
- the number of elements within the aperture;
- the required number of parallel data acquisition channels.

Transducer

	PW	ZO	CD
# Elements	N	1→N	N
Elementspacing (d)	1/2λ	1/4λ	1/4λ

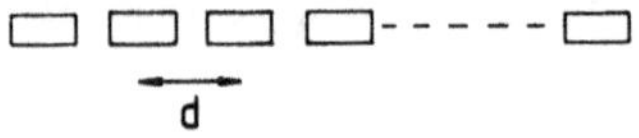

Fig.4.: Amount of data to be processed for PW,ZO and CD data acquisition.

In Figure 4 the total amount of data points is given for each of the three data acquisition techniques. The number of elements within a "large aperture" transducer is denoted as N. The number of samples per element (M) is determined by data acquisition time *and* sample frequency.

For PW *and* ZO the total amount of data points is given by: N x M. If, for instance, we use a 256-element transducer, a data acquisition time of 200 μsec per sequence and a sample frequency of 10 MHz, we will obtain not less than 2^{19} data samples.

Using the CD method we will collect this number for *each* individual sequence. Here the total amount of data is given by: N x N x M, i.e. 2^{27} in the given example. Although this seems an enormous quantity, we now have characterized each possible acoustic raypath from a single source to a single detector individually. Therefore the complete dataset offers the best prospects of handling correctly both spatial velocity variations *and* absorption during image reconstruction.

Using PW data acquisition, the medium is "illuminated" with a plane sound wave, and during transmission no lateral phase differences are introduced over the transducer surface. To avoid *spatial aliasing* during imaging, the sample interval d should be less than half a wavelength: $d \leqslant \lambda/2$, as shown in Figure 5.
The ZO and the CD method, however, introduce lateral phase differences along the transducer surface in transmission *as well as* in re-

Amount of data to be processed

	PW	ZO	CD
# Datapoints	N×M	N×M	N×N×M

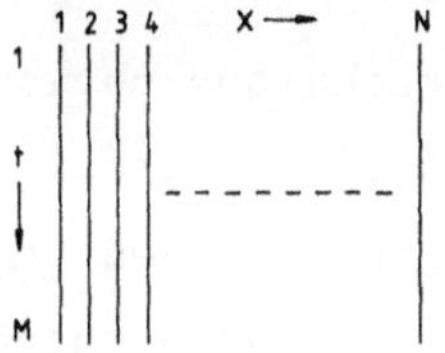

Fig.5.: Demands on elementspacing (d) for PW,ZO and CD data acquisition.

ception. For these methods the sample interval should therefore not exceed a quarter of a wavelength: $d \leqslant \lambda/4$ (see Figure 5).
In the high frequency range (1 - 10 MHz) this demand may impose severe practical problems, because very small inter-element distances have to be realized.
Note also that the number of elements within a given aperture increases by a factor of two compared to the number required in the PW method.

As stated in the previous section, using the PW method only one sequence is needed to collect all necessary data samples. On the other hand the data acquisition system itself becomes rather complex since we need *one* data acquisition channel for *every* individual detector.

This large number of parallel data channels cannot be avoided however, if, in imaging moving structures, a fast imaging technique proves to be the only solution.

In general data acquisition using the ZO or CD method will take N times as long, since N sequences are needed before all necessary data are collected. Both methods therefore tend to be rather sensitive for motion artefacts. Using once more a 256-element transducer, a centre frequency of 3 MHz and a data acquisition time of 200 µsec per sequence, structures moving over 2 mm/sec will cause a perceptible "blurring" in the imaged result, i.e. motion artefacts.

The main advantage of the ZO technique obviously consists of the fact that a single data acquisition channel will suffice. After each sequence the data channel may be switched to the next detector. Note that increasing the number of data channels will not reduce data-acquisition time.

This does not apply to the CD method, where, like in PW data acquisition one channel is needed for every individual detector.

Figure 6 summarizes the results on parallel data acquisition channels and the corresponding minimum and maximum acquisition times for all three data acquisition techniques respectively.

From this section we may draw the following conclusions:

1. The plane wave method is least sensitive for motion artefacts;

2. The zero offset method needs a single data acquisition channel only, while the data acquisition systems for PW and CD generally consists of N parallel channels;

	PW	ZO	CD
# DA-Channels			
max	N	N	N
min	1	1	1
Total DA-Time			
min	T	N×T	N×T
max	N×T	N×T	N×N×T

Fig.6.: Summary on the number of data acquisition channels and corresponding acquisition time for PW,ZO and CD data acquisition.

3. The complete dataset offers the best prospects for image reconstruction of inhomogeneous media. However, the amount of data to be processed is enormous.

RESOLUTION

The *lateral* resolution of all pulse-echo imaging systems is among others influenced by:

- the inability of any imaging system to reconstruct the *evanescent* wave field (also known as the *super near field*);
- the finite dimensions of the transducer aperture area.

Referring to the foregoing paper, the evanescent wave field was defined as that part of the spatial impulse response which attenuates exponentially with distance. The inability to reconstruct this specific part of the wave field determines an ultimate limit on the lateral resolution (Berkhout, 1980, chapter 12), i.e. the best result we may ever hope to achieve using *any* acoustical image reconstruction technique.

In the upper part of Figure 7 this ultimate limit (expressed as spatial impulse response W(ρ,f)), is shown for the two dimensional situation.

Note that in the super near field W(ρ,f) is *dependent on frequency only*.

Super near-field:

$$W(\rho,f) = \frac{k}{\pi}\,\frac{\sin(kx)}{kx}$$

Finite aperture:

$$W(\rho,f) = \underset{PW}{\frac{\sin(k\rho)}{k\rho}} \quad \underset{ZO}{\frac{\sin(2k\rho)}{2k\rho}} \quad \underset{CD}{\frac{\sin^2(k\rho)}{(k\rho)^2}}$$

where

$$\rho = \frac{lx}{2\sqrt{x^2+z^2}}$$

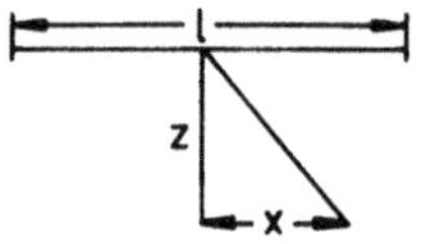

Fig.7: Spatial impulse responses due to the super near field (top) and finite transducer aperture (bottom).

The lower part of Figure 7 shows the spatial impulse responses for PW,ZO and CD data acquisition, as determined by the finite aperture area(1).

For PW and ZO data acquisition the given results will be illustrated with an example. Let us consider six diffractors located in a homogeneous medium as shown in Figure 8.

Four diffractors are located on the main axis of the transducer at depth levels of 10,30,70 and 150 mm respectively.

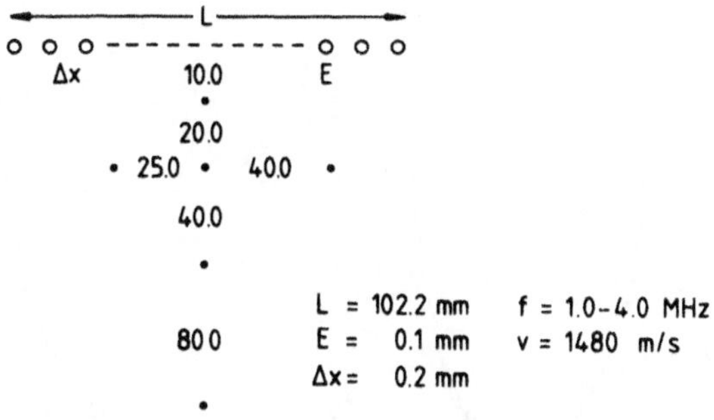

Fig.8: Schematic illustration of the model, used in comparing the lateral resolution achievable with PW and ZO data acquisition.

At a depth level of 30 mm two more diffractors are located at 25 mm to the left and 40 mm to the right of the transducer main axis. The transducer consists of 512 elements, 0.1 mm in width and at 0.2 mm intervals, thus covering a total aperture area of 102.2 mm. The -6 dB bandwidth of the transmitted signal ranges from 1 upto 4 MHz.

Both for PW and ZO all 512 responses were simulated and focussed using the *matched filter* version of the focussing operator, as mentioned in Equation (16c) in the foregoing paper:

$$\tilde{F} = \tilde{W}^* \tag{4}$$

For the PW method the focussed result at the 30 mm depth level is shown in Figure 9.

Figure 9 clearly shows that the influence of the lateral position of the diffractors on the focussed result is *not* very pronounced. Only at levels under -20 dB the imaged diffractors begin to show perceptible differences.

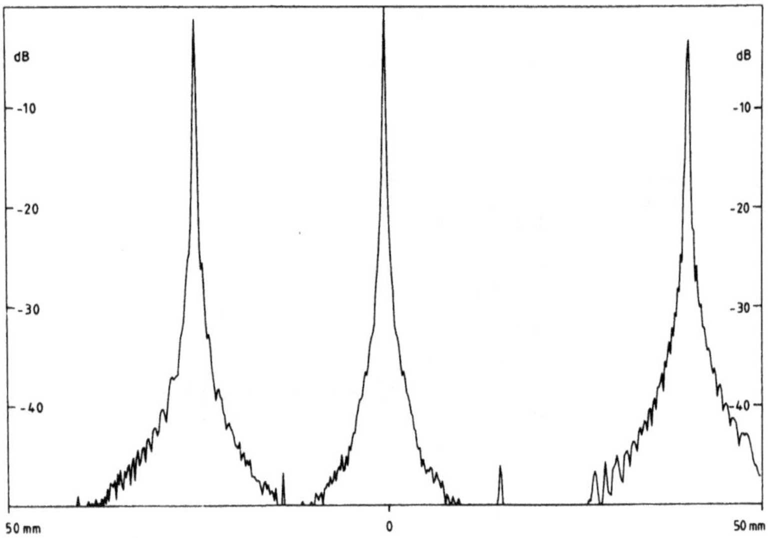

Fig.9.: Focussed result for the PW method at a depth level of 30 mm (see Figure 8).

The table in Figure 10 summarizes the beamwidths of the four imaged diffractors on the main axis for PW and ZO data acquisition at three different levels (i.e. -6 dB, -10 dB and -20 dB respectively).
The third column denotes the beamwidths, which may be obtained using a conventional 2D-imaging system with a comparable aperture of 104.0 mm. The given numbers match those achievable with a dynamically focussed linear array system as described by Ridder (1975). This system uses an *active* transducer area of 24.5 mm per sequence and operates at a centre frequency of 3.1 MHz.
Note that at a depth level of 30 mm the PW and ZO method still yield idéntical values for lateral resolution, which is caused by the fact that during image reconstruction the evanescent wave field is not taken into account. In the given example minimum lateral resolution is entirely determined by this phenomenon upto a depth level of ca. 50 mm.
Only at larger depth levels the influence of the finite transducer aperture becomes apparent. As to be expected, in this area

	Beamwidth (mm)								
	PW			ZO			Conv.		
Depth (mm)	-6	-10	-20	-6	-10	-20	-6	-10	-20 dB
10	0.3	0.4	0.7	0.3	0.4	0.7	2.5	3.2	--
30	0.3	0.5	1.0	0.3	0.4	1.0	2.3	2.9	5.7
70	0.5	0.7	1.6	0.4	0.5	1.0	2.3	3.7	4.8
150	1.0	1 2	30	04	0.7	1.5	3.0	4.0	5.0

Fig. 10.: Table summarizing the beamwidths at four different depth levels, as shown in Figure 8, using the PW method, the ZO method and conventional focussing.

significant differences occur between the PW and the ZO method.

Summarizing we may draw the following conclusions from this section on resolution:

1. In terms of achievable lateral resolution the zero offset method offers the best prospects.
2. The lateral resolution obtained with plane wave data acquisition is less than the resolution offered by both the ZO and CD method.
3. The inability to reconstruct the evanescent wave field during focussing, causes the maximum achievable lateral resolution to be limited. This ultimate limit depends on frequency only.
4. A diagnostic imaging system, based on synthetic focussing, using either one of the discussed data acquisition techniques, will by far surpass the performance of conventional imaging systems.

IMAGING

In this section on imaging two aspects will be discussed. First we will consider the concept of *maximum dipping angle*, where by dipping angle we understand the angle between a plane interface inside the medium (i.e. a reflector) and the transducer surface. The *maximum* dipping angle indicates the maximum slope (or dip) of such an interface by which its responses may still be recorded, *using* a given data acquisition technique.
From mere geometrical considerations it follows that for the plane wave method the maximum dipping angle is two times as small as for ZO or CD data acquisition. This is illustrated by Figure 11, where the *boundary* rays are schematically shown.

The second aspect of imaging we will consider is the *angle-dependent reflection coefficient*.
In general a reflector is described as a plane interface between two media, which have different acoustic impedances, i.e. they differ in density ($\rho_1 \neq \rho_2$), in propagation velocity ($c_1 \neq c_2$) or in both.

Geometrical considerations

	PW	ZO&CD
α (max)	1/2 arctg(l/z)	arctg(l/z)

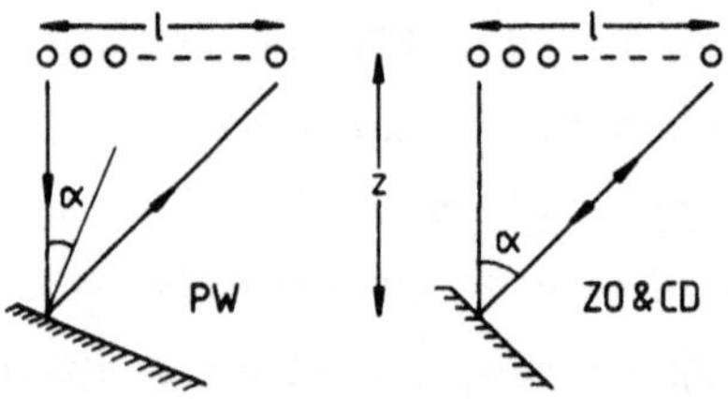

Fig.11.: Schematic illustration of the concept of maximum dipping angle, for PW,ZO and CD data acquisition.

It can be shown that, due to a difference in *propagation velocity*, the reflection coefficient (r) for an incoming plane wave becomes dependent on its angle of incidence (α):

$$r(\alpha) = \frac{\rho_2 c_2 \cos\alpha - \rho_1 \sqrt{c_1^2 - c_2^2 \sin^2\alpha}}{\rho_2 c_2 \cos\alpha + \rho_1 \sqrt{c_1^2 - c_2^2 \sin^2\alpha}} \qquad (5)$$

To illustrate this Figure 12 shows angle-dependent reflection curves for a water/skin interface (1) and a blood/liver interface (2 & 3). The values for ρ and c are derived from literature (Kak (1976) and Wells (1977)):

1. water: $\rho = 1010\ kg/m^3$ $c = 1500\ m/s$
   skin : $\rho = 836\ kg/m^3$ $c = 1950\ m/s$
2. blood: $\rho = 1060\ kg/m^3$ $c = 1560\ m/s$
   liver: $\rho = 1000\ kg/m^3$ $c = 1570\ m/s$
3. blood: $\rho = 1060\ kg/m^3$ $c = 1560\ m/s$
   liver: $\rho = 1000\ kg/m^3$ $c = 1560\ m/s$

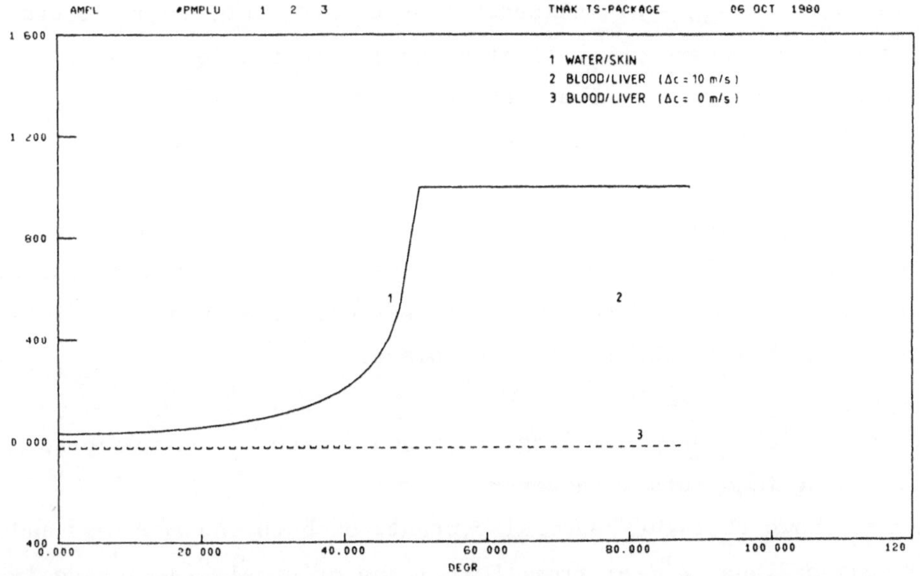

Fig. 12.: Angle-dependent reflection curves for three different interfaces inside the human body.

Note that in the case of a water/skin interface *total reflection* occurs at an angle of incidence of ca.45°. It may therefore be expected that medical imaging systems using a waterbath-coupling, will not be able to transmit *any* acoustical energy into the patients body at angles exceding 45°.
Note also that even a slight variation in propagation velocity may change the path of the reflection curve rather strongly. While at curve 3 for instance the reflection coefficient is independent of the angle of incidence, since Δc (liver/blood) = 0, at curve 2, where Δc(liver/blood) = 10 m/s, the reflection coefficient becomes *zero* for $\alpha \approx 60^{o}$, and above this angle changes its sign.

From the above we conclude that, when using apertures, which cover several hundreds of wavelengths, it becomes rather important in which way an interface is irradiated.

As shown in Figure 13, when using the PW method, only *one angle* is taken into account (i.e. the dipping angle of the reflector). For media which are strongly inhomogeneous, this angle may either coincide with an angle of incidence, where total reflection occurs ($r(\alpha) = \pm 1$), or where the reflection coefficient becomes zero.
For the ZO method we are always concerned with normal incidence, i.e. $r(0)$, since source and detector are located at the same position.
Finally, for CD data acquisition several angles of incidence occur, and therefore the CD method may be used to determine $r(\alpha)$, or rather the values for ρ and c of the two different media, provided the interface between them may be regarded as plane.

There is one last aspect of imaging we would like to mention here, without going into details however.
It may be shown that the spectral contents of both the *recorded* and the *focussed* signal differ from that of the original signal used in transmission. This change in spectral behaviour not only depends

$$r(\alpha) = \frac{\rho_2 c_2 \cos\alpha - \rho_1\sqrt{c_1^2 - c_2^2 \sin^2\alpha}}{\rho_2 c_2 \cos\alpha + \rho_1\sqrt{c_1^2 - c_2^2 \sin^2\alpha}}$$

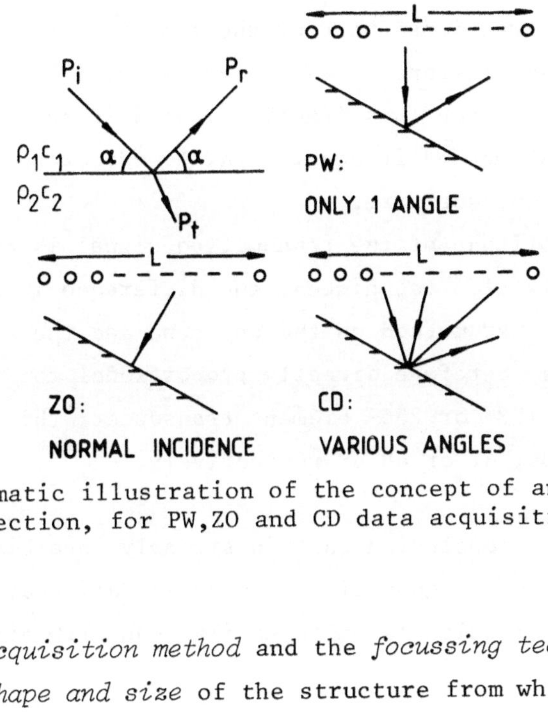

Fig.13.: Schematic illustration of the concept of angle-dependent reflection, for PW,ZO and CD data acquisition.

on the *data acquisition method* and the *focussing technique*, but also on the *shape and size* of the structure from which the recorded responses originate. These phenomena in itself will be discussed in a seperate publication, which will appear in the near future.

To summarize this section on imaging we conclude the following:

1. The maximum dipping angle, using the PW method is two times as small as for ZO and CD data acquisition.
2. In media, which are strongly inhomogeneous, the PW method may cause errors in the imaging of reflectors, since only one angle of incidence is considered.

3. The CD method is most suited to characterize inhomogeneous media, since angle-dependent reflection coefficients may be determined and both density (ρ) and propagation velocity (c) of sublayers may be estimated.

SIGNAL TO NOISE RATIO

In synthetic focussing the signal to noise ratio is mainly influenced by the fractional part of the total aperture, which is activated in transmission.
For PW data acquisition this fractional area is equal to 1, while for the ZO and CD method it equals 1/N, N indicating the number of elements within the aperture.
Provided the amplitude of the transmitted signal is equal for all three data acquisition techniques, the difference in signal to noise ratio between the PW method on the one hand and the ZO and CD method on the other, is therefore directly proportional to N. Using, for example, a 64-, 128- or 256- element transducer, this difference will amount to 36, 42 or 48 dB respectively.

This leads to the conclusion that in strongly absorbing media, like e.g. the human body, a good signal to noise ratio may be achieved with PW data acquisition far more easily than with either ZO or CD acquisition.

SYNTHETIC FOCUSSING -AN EXAMPLE-

To give an impression of the results obtainable by using a synthetic focussing technique, as described in Part I and II of this publication, this last section will be dedicated to an example.

For this example we selected a *tissue equivalent ultrasound phantom* (type 412), available from RMI (Radiation Measurements Inc., Middleton, Wisconsin, U.S.A.).
This phantom is filled with a tissue equivalent mixture of gel

and graphite particles, which has an over-all attenuation of ca. 0.07 dB/mm/MHz. A cross-section of the phantom is shown in Figure 14.

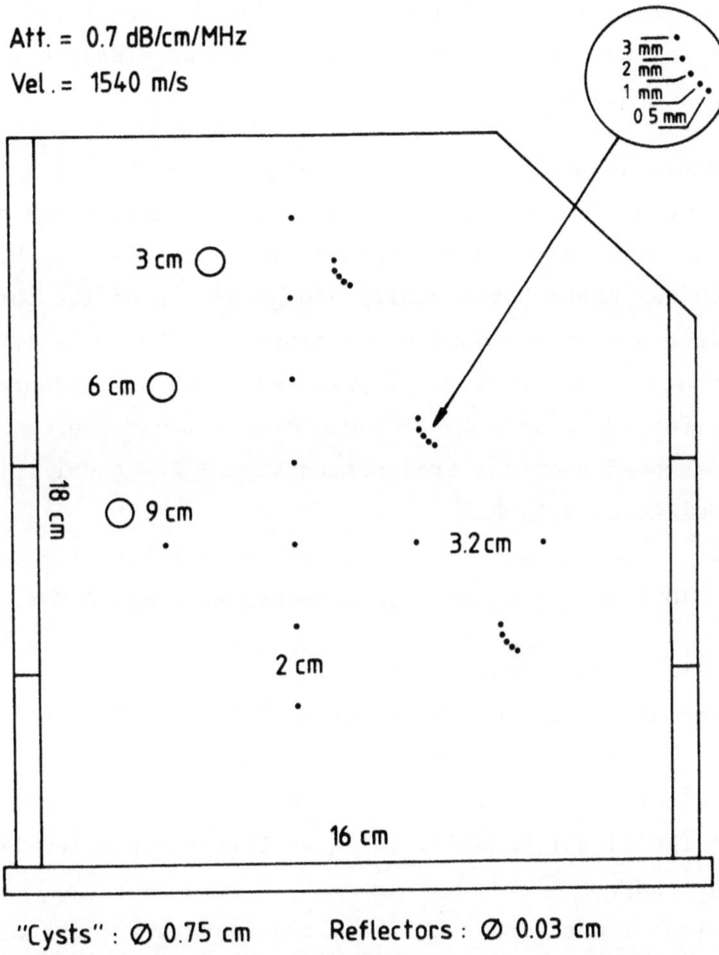

Fig. 14.: Cross section of a tissue equivalent ultrasound phantom (type 412, available from RMI).

In addition to the gel this object also contains three "cysts" (ϕ 7.5 mm) and a number of nylon filaments (ϕ 0.3 mm). Part of these threads are located at regular axial and lateral intervals of 20 and 32 mm respectively, while the other filaments are grouped in three "bowl-shaped" clusters. One of these clusters is shown in detail in the top right corner of Figure 14. The lateral distance between the threads equals 1 mm, while the axial distance varies from 3, 2, 1 to 0.5 mm.

In this example we performed the data acquisition by using the ZO method, but we *did not* use a "large aperture" linear array however. Instead of an array we used a single element transducer, 300 μm in width, which was moved horizontally along the top of the phantom. By taking into acount 512 sample positions, at 200 μm intervals, we actually simulated an aperture of 102.5 mm. Such a procedure is generally referred to as a *synthetic aperture* technique.
The centre frequency of the transmitted signal was equal to 3 MHz, with a bandwidth of 1.5 MHz.
The received responses were digitized using an 8 bit A/D-convertor (Biomation 8100) operating at a sample-frequency of 10 MHz.

Part of the focussed result is shown in Figure 15.
Note that the appearance of the lower cyst is more "noise-like" than that of the upper one. Referring to the fact that we used a very small element (300 μm) both in transmission and reception, a decrease of the signal to noise ratio at larger depth levels is hardly surprising.
It should also be noted that the recorded responses cover a wide dynamic range, since there is a large difference between the reflection coefficient of a gel/nylon interface (filaments) and that of a gel/liquid interface (cysts).
To obtain a satisfactory result in displaying the final image (Figure 15), the focussed responses of the nylon filaments were

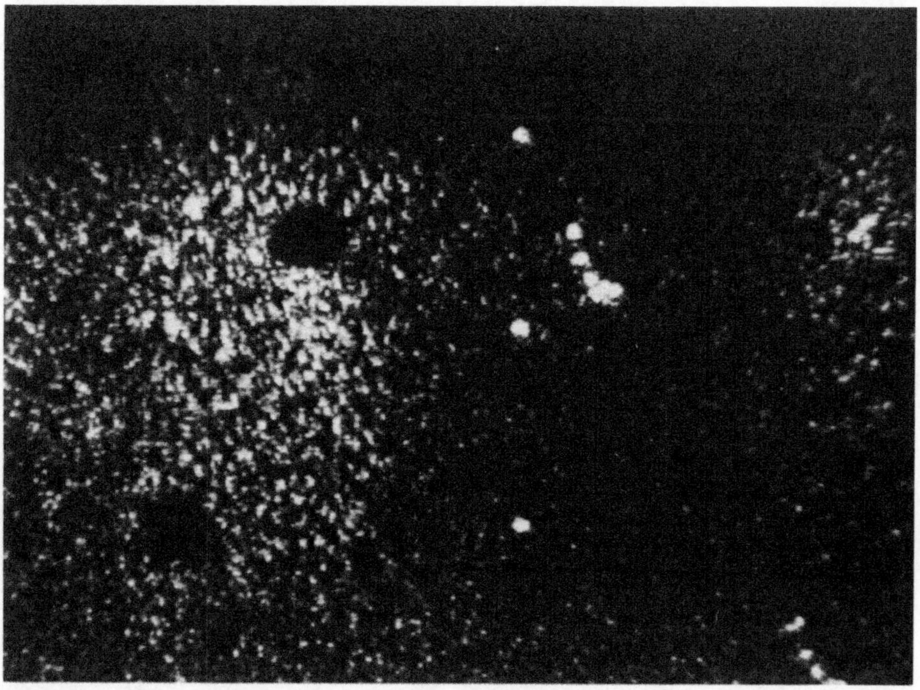

Fig.15.: Focussed result of the upper part of the RMI ultrasound phantom, using ZO data acquisition.

clipped 30 dB beneath their maximum value.
In spite of this gross clipping the five threads in the upper cluster may still be distinguished individually. A small artefact however, which is probably a sidelobe effect, may be seen in between the fourth and fifth filament.

CONCLUSIONS

In this second paper on synthetic focussing a number of practical aspects of PW,ZO and CD data acquisition has been discussed. Each of the three techniques showed to have its strong and weak sides. Summarizing we conclude the following:

PW method:

Advantages
- data acquisition time is small and therefore (fast) moving structures may be imaged without motion artefacts;
- good signal to noise ratio.

Disadvantages
- lateral resolution is less than the ZO and CD method;
- less suitable for imaging of (strongly) inhomogeneous media and reflectors at large dipping angles.

ZO method:

Advantages
- only one data acquisition channel is needed;
- high lateral resolution.

Disadvantages
- sensitive for motion artefacts;
- less suitable for imaging of (strongly) inhomogeneous media;
- problems may arise in achieving a good signal to noise ratio in absorbing media.

CD method:

Advantages
- good lateral resolution;
- very suitable for imaging (strongly) inhomogeneous media.

Disadvantages - number of data samples to be manipulated is enormous;
- sensitive for motion artefacts;
- problems may arise in achieving a good signal to noise ratio in absorbing media.

In general we may state that the choice of a data acquisition technique *strongly* depends on the properties of the medium to be imaged, which may be either time-invariant or not, inhomogeneous or not, etc., etc.

ACKNOWLEDGEMENT

We would like to thank our colleagues from the Echocardiographic Laboratory at the Thorax Centre of the Erasmus University in Rotterdam for their essential support in performing the actual data acquisition on the tissue equivalent ultrasound phantom and in displaying the focussed result as shown in Figure 15 of this paper.

REFERENCES

Ridder, J., 1975, *Development of a real-time dynamically focussed ultra-sonic system for two dimensional imaging* (in Dutch): Delft, doctoral thesis.

Kak, A.L., Fry, F.J., 1976, *Acoustic impedance profiling: an analytical and physical model study;* Proceedings of Ultrasonic Tissue Characterization Seminar, NBS Special Publication 453.

Wells, P.N.T., 1977, *Biomedical Ultrasonics* ; Academic Press.

Berkhout, A.J., 1980 , *Seismic migration - imaging of acoustic energy by wave field extrapolation-:* Amsterdam/New York, Elsevier North Holland Publ. Comp.

A FAST ULTRASONIC IMAGING SYSTEM FOR MEASURING UNSTEADY VELOCITY FIELDS IN AIR

Bernd O. Trebitz

Graduate Aeronautical Laboratories
California Institute of Technology
Pasadena, CA 91125

INTRODUCTION

Advances in ultrasonic imaging in medicine have prompted the development of ultrasonic imaging systems to be used as diagnostic tools in fluid mechanics. The fact that a sound wave propagating through a medium is convected with the flow velocity can be utilized to probe the velocity field. Such a measurement technique, being nonintrusive, does not require any scattering particles and may be applied in optically opaque media, as pointed out by Johnson et al.[1] In contrast with their work, in which individual time of flight measurements are made, in the present work the entire object wave is measured in both amplitude and phase along a linear transducer array. Hence, Fresnel transforming techniques may be applied to reconstruct the flow field in depth. Furthermore, fast acquisition of the acoustical data offers the possibility of tracking unsteady flow fields. The working medium is air because of its smaller sound speed compared to water, which results in a larger sensitivity to velocity disturbances.

SYSTEM CHARACTERISTICS

The principle of transmission imaging is briefly reviewed in Fig. 1. A large area piezoelectric transducer serves as the transmitter of ultrasound, which passes through a test section containing the object. In this case, the object can be described as a spatially and temporally changing acoustic index of refraction $n(\underset{\sim}{x},t) = c_o/c$, where c_o is the speed of sound in the undisturbed fluid and $c = c_o + \Delta c$. Δc is the change of c due to the fluid flow, which can be due to a temperature or a velocity disturbance in the sound path. For velocity disturbances alone, Δc is just the flow velocity.

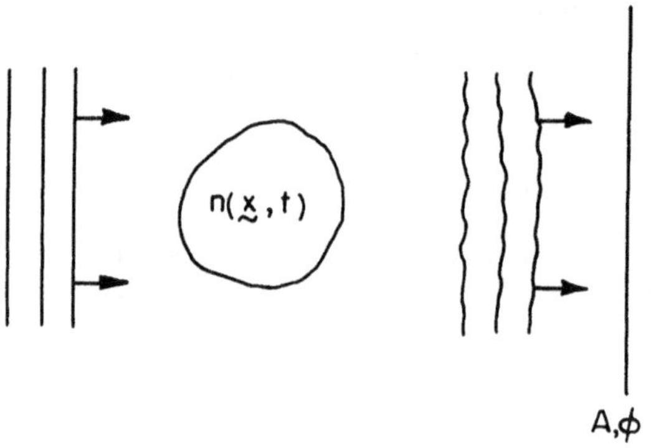

Fig. 1. Transmission Imaging

In other words, the sound wave is convected with the local flow velocity and thus amplitude and phase of the transmitted wave are modified.

These modifications are detected by a 100 element linear piezoelectric transducer array, which is 3.5" long and 0.5" wide. Amplitude and phase distribution along the array are measured directly at a rate of 100 kHz. After A/D conversion, the digital information is recorded on magnetic tape resulting in a scanning rate of 500 Hz.

Actual flow measurements are taken relative to the undisturbed "no flow" condition. This is accomplished by first propagating ultrasound through quiescent air in order to establish the response of every single array element. Then the flow is turned on and measurements are taken under the same driving conditions. Amplitude ratios and phase shifts with respect to the reference values are computed, which eliminate varying channel responses and entirely determine the object wave. Since all data are stored on magnetic tape, various digital processing techniques can be applied to the same image. Furthermore, the fact that amplitude and phase information is available allows one to apply Fresnel transforming techniques for reconstruction of the entire field in the transmitter-receiver plane.

SIGNAL PROCESSING

The essential element in the signal processing chain is a phase sensitive detector circuit, shown in Fig. 2. An analog switch, which is activated by the reference signal E_R, passes either the amplified signal E_s or no signal at all. If E_R and E_s are of the same frequency, one obtains, after low pass filtering E_{LP}, a DC output signal E_o which is proportional to the amplitude of E_s and to the phase of E_s relative to E_R. To demonstrate the phase dependence, Fig. 2 also shows three examples of phase relationships between E_R and E_s. If the input signal E_s is sinusoidal with amplitude A and in phase with the reference signal E_R, the output signal E_o is proportional to the area under the first half cycle of the sinusoid and this turns out to be A/π. Similarly, if the same input signal lags the reference signal by 90°, the resulting output signal is zero. For a phase lag of 180°, one obtains $-A/\pi$. Amplitude and phase of E_s can be determined uniquely by mixing E_s with two reference signals, one is in phase and the other one is leading by 90° with respect to the driving signal of the transmitter. The former is proportional to $A\sin\phi$, the latter one to $A\cos\phi$, where ϕ is, up to multiples of 2π, the phase angle between the transmitted and the received signal.

The block diagram of the entire system is shown in Fig. 3. A function generator supplies a sinusoidal signal, which is amplified to drive the transmitting transducer continuously. The receiver array samples the acoustic wave along a line and passes the signals after preamplification to the phase sensitive detectors. The two reference signals at quadrature, designated "sin" and "cos", are supplied by a phase-locked-loop circuit, which is driven by the same function generator that drives the ultrasonic transmitter. The output signals of the phase sensitive detectors are converted into an 8-bit word with an accuracy of ± .8%. Overall noise and crosstalk between adjacent channels are held below this limit. The digital output is then recorded on magnetic tape to be processed on a computer. Alternatively, the system output can be viewed in real time on a CRT display. The address generator controls the data flow to tape recorder and CRT display.

The scanning rate of 500 Hz is achieved by parallel processing the output signals of all 96 receiver elements in use. Every element has its own preamplifier and two phase sensitive detector circuits. The preamplifiers are located close to the array on 12 printed circuit boards. Every four preamplified signals are transferred to one processing printed circuit board, as shown on Fig. 4. Each of

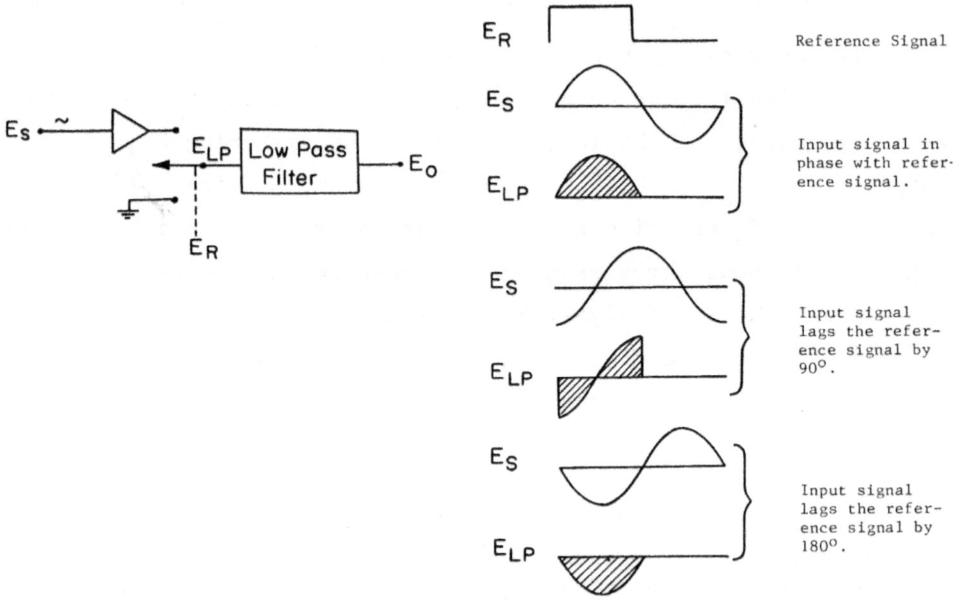

Fig. 2. Phase Sensitive Detector Circuit

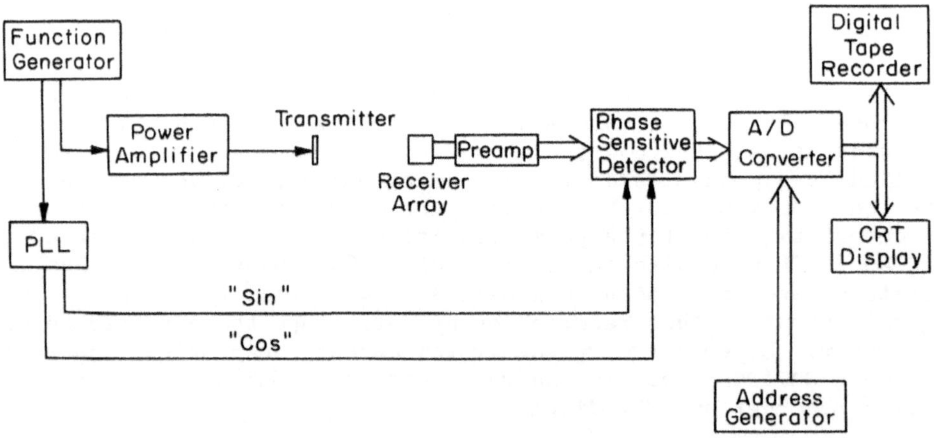

Fig. 3. System Block Diagram

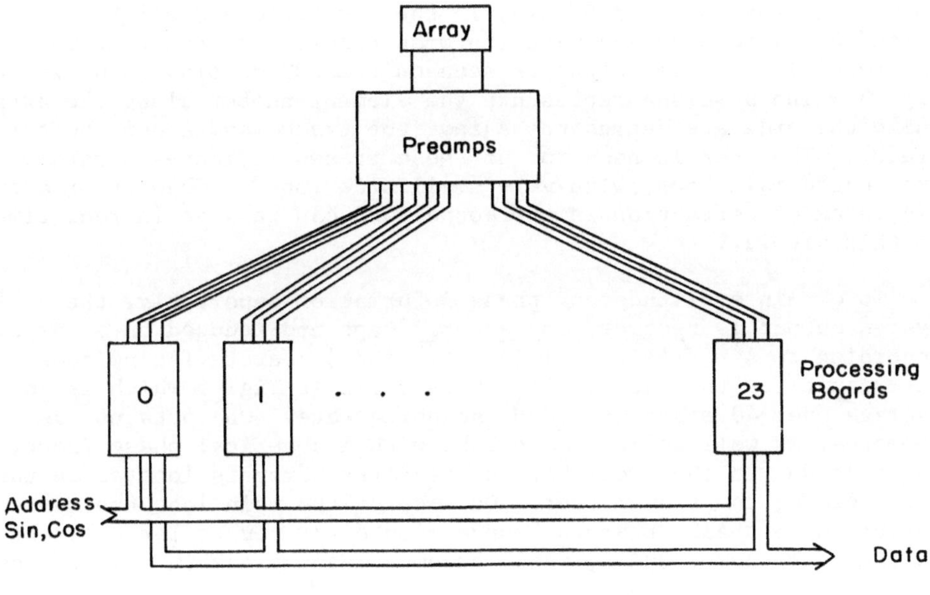

Fig. 4. Structure of Processing Electronics

the 24 processing boards contains eight phase sensitive detectors and one 8-bit A/D converter. The 8-bit address word activates the output of one processing board at a time, which is then transferred through a common data bus to the digital tape recorder. While the next processing board is activated, the previous one has already started the conversion of its next channel output, such that the conversion is completed when this particular board is addressed again. Parallel processing combined with the described scanning technique amount to the scanning rate of 500 Hz.

PRELIMINARY RESULTS

A piezoelectric transducer measuring one inch in diameter was used at 118.7 kHz as the transmitter. The receiver array was mounted approximately 5.5" away in the far field and slightly tilted to avoid standing waves between transmitter and receiver. The resulting system output as seen on the CRT display is shown in Fig. 5. The abscissa represents the element number along the array, while the ordinate represents $A\sin\phi$ (top trace) and $A\cos\phi$ (bottom trace). The overall behavior of these traces indicates a rather pronounced main lobe, with very small side lobes. Changes of acoustic index of refraction in the sound path can be seen in real time on this display.

To obtain amplitude and phase information separately, the system output is recorded on magnetic tape and reduced on a computer according to $A = \sqrt{(A\sin\phi)^2 + (A\cos\phi)^2}$ and $\phi = \text{arctg}\ (A\sin\phi/A\cos\phi)$. The result for the data of Fig. 5 is shown in Fig. 6 which is an average over 40 scans at 500 Hz scanning rate. One obtains, as expected, as well defined main lobe with a spherical phase front, which is due to the fact that the receiver array is located in the far field of the transmitter. Outside of the main lobe the system output is so small, that the conversion error causes large errors in the phase computation, which becomes meaningless in this region.

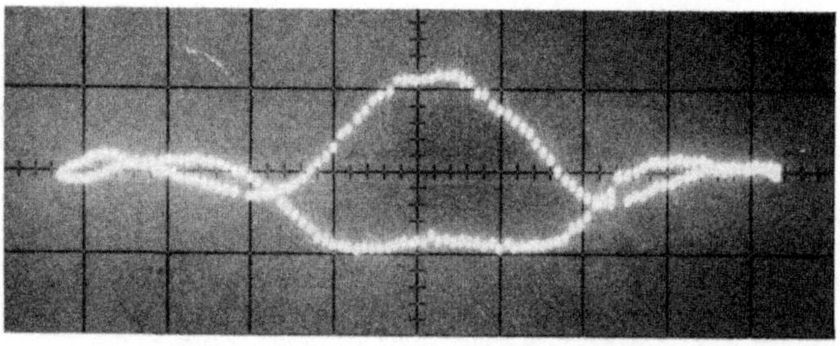

Fig. 5. CRT Display of System Output Without Flow
Top Trace, $A\sin\phi$; Bottom Brace, $A\cos\phi$

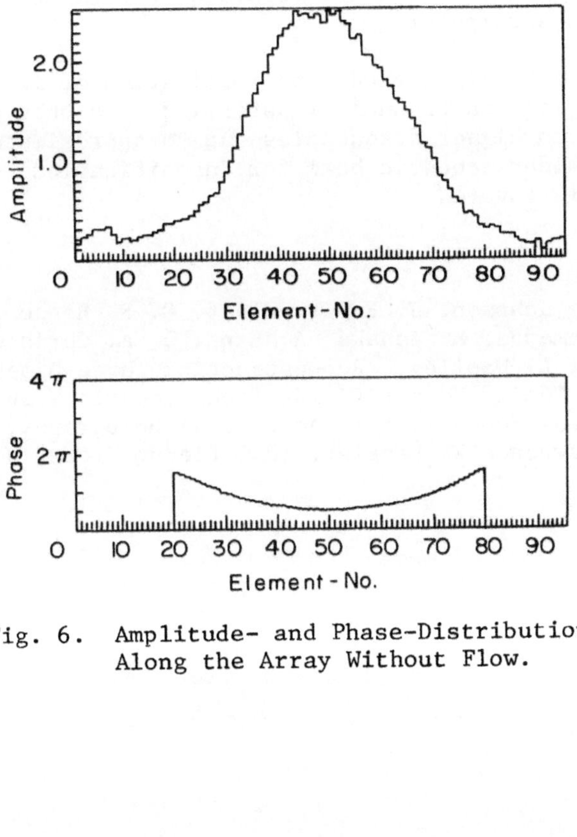

Fig. 6. Amplitude- and Phase-Distribution Along the Array Without Flow.

Measurements will now be taken with a flow turned on, but under the same driving conditions, such that any change in amplitude and phase of the sound wave are entirely due to the flow. Preliminary results from a swirling jet flow have been obtained but are not yet ready for publication.

CONCLUSIONS

An ultrasonic imaging system capable of measuring amplitude and phase of a transmitted sound wave in air along a 100 element piezoelectric transducer array was designed and fabricated. The scanning rate of 500 Hz allows one to track unsteady flow fields up to that frequency. Data are recorded on a digital tape recorder and reduced on a computer.

Further work will include the application of image processing techniques as well as Fresnel transforming. In this case, it is planned to go to higher frequencies, up to approximately 300 kHz, and to use a wider acoustic beam for insonification, such that the entire array is covered.

REFERENCE

1. S. A. Johnson, J. F. Greenleaf, C. R. Hansen, W. F. Samayoa, M. Tanaka, A. Lent, D. A. Christensen, and R. L. Woolley, Reconstructing Three-Dimensional Fluid Velocity Vector Fields from Acoustic Transmission Measurements, in: "Acoustical Holography," Vol. 7, Lawrence W. Kessler, ed., Plenum Press, New York (1977).

METHODS FOR EFFICIENT COMPUTATION OF THE IMAGE FIELD OF HOLOGRAPHIC LENSES FOR SOUND WAVES

Jakob J. Stamnes and Tore Gravelsæter

Central Institute for Industrial Research
P.B. 350 Blindern
Oslo 3, Norway

ABSTRACT

Methods for efficient computation of the image field of acoustical holographic lenses are presented. The methods are valid for large field angles and large relative apertures and apply at short as well as at long wavelengths. Some of the techniques are based on the Kirchhoff diffraction integral, others on the boundary-diffraction-wave integral. For rapid computation of these integrals we employ three different techniques for different positions of the object and observation points: In the case in which the integrand has small variations we employ the technique of phase and amplitude approximations; in shadow boundary regions we use a uniform asymptotic technique; and in other regions we use a second order asymptotic method.

1. INTRODUCTION

Over the past few years efforts have been expended to construct holographic or zone-plate lenses for sound waves . The working principle of such lenses as well as their advantages over conventional lenses, are discussed elsewhere[1-4]. In this paper we address the problem of computing the image field of holographic lenses.

By a holographic lens we mean a zone-plate-like structure, as shown in Fig. 1. It consists of a plane parallel plate with a concentric pattern of circular grooves, which give the desired focusing effect.

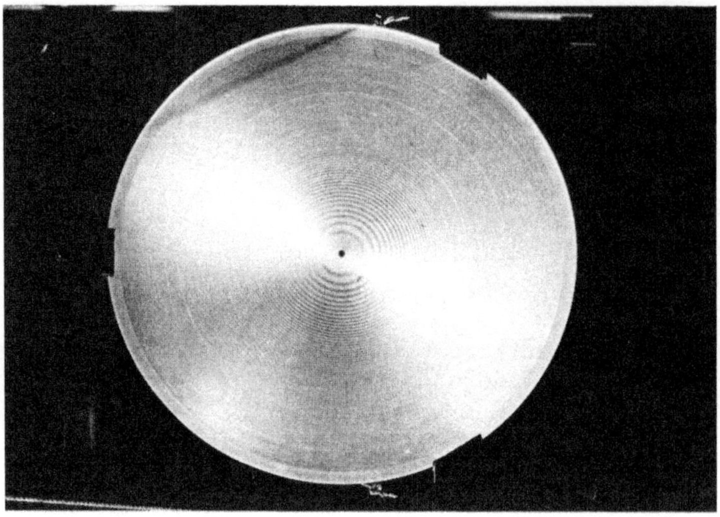

Fig. 1. Holographic lens for sound waves.

To make the computational problem tractable mathematically we assume that the only effect of the grooves is to retard that part of the incident wave front that passes over them relative to the part that passes between them. Thus, we neglect multiple reflections within the grooves, and also multiple scattering from the edges of the grooves. Also, we assume that the plane parallel plate in which the grooves are engraved, is sufficiently thin that the acoustical wave is not aberrated significantly on passage through it. Finally, we neglect multiple reflections within the plane parallel plate, and we neglect the presence of shear waves within the lens so that a scalar theory is adequate.

Under the assumptions above, our physical model of the problem is that of diffraction by an array of concentric annular apertures, each annulus having a particular value of phase and amplitude associated with it. This diffraction problem has been treated earlier[5], but only in the paraxial approximation for the case of a plane wave normally incident on the aperture. The paraxial approximation limits the validity of method so that the transverse extent of the source, the aperture, and the observation area must be small compared to the distance from the aperture to the source or to the observation point. Therefore, we shall not make use of this approximation.

When it comes to the formulation of our problem within the frame work of scalar diffraction theory, the question arises of whether to use either of the two Rayleigh-Sommerfeld diffraction integrals or the Kirchhoff diffraction integral. Although our methods would work equally well with any of these three diffraction

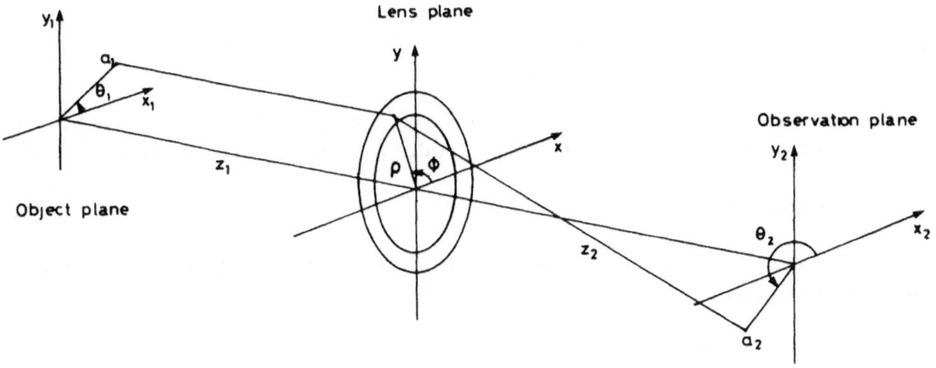

Fig. 2. Geometry of diffraction problem.

integrals, we have chosen to use the Kirchhoff integral, simply because there are reasons to believe that it corresponds better to the physical phenomenon of diffraction[6].

The geometry of our diffraction problem is shown in Fig. 2, where for simplicity only one annular aperture is drawn. The z axis of the coordinate system passes through the center of the aperture, and the object plane, the aperture plane, and the observation plane are at $z=-z_1$, $z=0$, and $z=z_2$, respectively. Polar coordinates (a_1,θ_1), (ρ,ϕ), and (a_2,θ_2) are used to specify respectively the object point, a typical point in the aperture, and the observation point.

A sound source at a point $P_1(a_1,\theta_1,z_1)$ emits a spherical, monochromatic wave which is diffracted by an annular aperture of inner radius ρ^- and outer radius ρ^+. The diffracted sound field is observed at some point $P_2(a_2,\theta_2,z_2)$. The total field at P_2 due to a concentric array of annular apertures is the sum of the fields due to each separate annulus.

According to the Kirchhoff diffraction integral, the field at P_2 is given by[7]

$$u(P_2) = \frac{-iC}{2\lambda}\int_0^{2\pi}\int_{\rho^-}^{\rho^+} G(\rho,\phi)\exp[iF(\rho,\phi)]d\rho d\phi, \tag{1}$$

where $k = 2\pi/\lambda,\lambda$ being the wavelength, and

$$G(\rho,\phi) = \frac{\rho}{r_1 r_2}\left[\frac{z_1}{r_1}\left(1+\frac{i}{kr_1}\right) + \frac{z_2}{r_2}\left(1+\frac{i}{kr_2}\right)\right], \tag{2}$$

$$F(\rho,\phi) = k(r_1+r_2), \tag{3}$$

with

$$r_\ell = [z_\ell^2 + a_\ell^2 + \rho^2 - 2\rho a_\ell \cos(\phi-\theta_\ell)]^{\frac{1}{2}} \ ; \ \ell = 1,2 \ . \tag{4}$$

The complex factor C in (1) allows for the possibility that the aperture has a phase ψ and an amplitude A associated with it, i.e.,

$$C = A\exp(i\psi) \ . \tag{5}$$

Alternatively, we may use the boundary-diffraction-wave (BDW) theory to rewrite (1) as[8]

$$u(P_2) = u^{(g)}(P_2) + u^{(d)}(P_2) \ , \tag{6}$$

where $u^{(g)}$ is equal to the incident field in the geometrically insonified region, and is equal to zero in the geometrical-shadow region, and

$$u^{(d)}(P_2) = C\ [I(\rho^+) - I(\rho^-)] \ , \tag{7}$$

where

$$I(\rho) = \int_0^{2\pi} g(\phi)\exp[if(\phi)]d\phi \ , \tag{8}$$

with

$$f(\phi) = k(r_1+r_2) \ , \tag{9}$$

$$g(\phi) = \rho g_1(\phi)/4\pi r_1 r_2 g_2(\phi) \ , \tag{10}$$

$$g_1(\phi) = -z_2[\rho-a_1\cos(\phi-\theta_1)]-z_1[\rho-a_2\cos(\phi-\theta_2)] \ , \tag{11}$$

$$g_2(\phi) = r_1 r_2 + [\rho-a_1\cos(\phi-\theta_1)][\rho-a_2\cos(\phi-\theta_2)] + a_1 a_2 \sin(\phi-\theta_1)\sin(\phi-\theta_2) - z_1 z_2 \ . \tag{12}$$

From a computational point of view, the advantage of the BDW theory is that it reduces the original double integral in (1) to a sum of two single integrals, as given in (7). How large computational savings this reduction will lead to, depends of course on the width $(\rho^+-\rho^-)$ of the annulus. The disadvantage of the BDW integral is that its integrand blows up as the observation point approaches the shadow boundary. Thus, if it is to be used in that area, special care must be taken to avoid this difficulty.

Our computational task is to use either (1) or (8) to compute the field diffracted by each annular aperture and add these fields to obtain the total field. There are three reasons why this task is not as straightforward as it may appear at first glance.

First, if the wavelength is small compared to the size of the aperture, the integrals in (1) or in (8) will undergo so many oscillations over the integration domain that a direct numerical integration becomes very time consuming, even with today's large computers. Second, the field around an image point has a complicated, rapidly varying structure, which makes it necessary to evaluate the field at a large number of observation points to get sufficient information about the image. Third, one would like the computational scheme to be fast enough to make the method valuable as a practical design tool.

Realizing that standard numerical integration routines are too slow for our purposes, we set out to develop alternate methods to evaluate the integrals in (1) and (8). Two possible methods, one based on local phase and amplitude approximations, the other on asymptotic techniques, are discussed in this paper.

The organization of the paper is as follows: In Sec. 2 we discuss the method based on phase and amplitude approximations applied to the BDW integral and to the Kirchhoff integral. Then we discuss in Sec. 3 the method of stationary phase for double integrals applied to the Kirchhoff integral, with particular emphasis on uniform asymptotic approximations. Finally, in Sec. 4 we compare the methods, point out their advantages and difficulties, and discuss in which situations to use what method.

2. METHOD OF LOCAL PHASE AND AMPLITUDE APPROXIMATIONS

We begin by considering the one-dimensional integration problem, represented by the boundary-diffraction-wave (BDW) integral in (8). For notational convenience we omit from now on the explicit ρ dependence in (8). In preparation to implement the method of local phase and amplitude approximations we divide the integration interval in N subintervals so that we may write (8) in the form

$$I = \sum_{n=1}^{N} I_n \ , \tag{13}$$

with

$$I_n = \int_{\phi_{L,n}}^{\phi_{U,n}} g(\phi)\exp[if(\phi)]d\phi \ ; \ \phi_{L,n}=\phi_{M,n}-\tfrac{1}{2}\Delta_n ; \ \phi_{U,n}=\phi_{M,n}+\tfrac{1}{2}\Delta_n , \tag{14}$$

where Δ_n and $\phi_{M,n}$ are respectively the width and the midpoint of the n-th subinterval. In each subinterval $\phi_{M,n}-\tfrac{1}{2}\Delta_n < \phi < \phi_{M,n}+\tfrac{1}{2}\Delta_n$ we approximate the amplitude function and the phase function by second degree polynomials, i.e., we write

$$g(\phi) \cong g_{LIN}(\phi) = g_0 + g_1\phi + \tfrac{1}{2}g_2\phi^2 , \tag{15}$$

$$f(\phi) \cong f_{LIN}(\phi) = f_0 + f_1\phi + \tfrac{1}{2}f_2\phi^2 . \tag{16}$$

Next, we substitute (15) and (16) into (14) and linearize the argument of the exponential function by expanding $\exp(\frac{1}{2}if_2\phi^2)$ in a Taylor series and retaining only the first two terms. Then we obtain

$$I_n \cong I_n^{(APP)} = \int_{\phi_{L,n}}^{\phi_{U,n}} (g_0+g_1\phi+\tfrac{1}{2}g_2\phi^2)(1+\tfrac{1}{2}if_2\phi^2)\exp[i(f_0+f_1\phi)]d\phi . \tag{17}$$

As a result of our approximations we have obtained an integral that may readily be evaluated analytically. A straightforward calculation yields

$$I_n^{(APP)} = A_n + iB_n , \tag{18}$$

where [with $F_0 = f_0 + f_1\phi_{M,n}$; $F_1 = \frac{1}{2}\Delta_n f_1$]

$$A_n = \Delta_n[X\cos F_0 - Y\sin F_0] , \tag{19}$$

$$B_n = \Delta_n[X\sin F_0 + Y\cos F_0] , \tag{20}$$

$$X = g_0h_0(F_1) - \tfrac{1}{2}\, g_2h_2(F_1) - g_1f_2h_3(F_1) , \tag{21}$$

$$Y = -g_1h_1(F_1) - \tfrac{1}{2}g_0f_2h_2(F_1) + \frac{1}{4}\, g_2f_2h_4(F_1) , \tag{22}$$

with

$$h_k(x) = \frac{d^k}{dx^k}\, h_0(x) \; ; \; k = 1,2,3,4 \; ; \; h_0(x) = \sin(x)/x . \tag{23}$$

Explicit expressions for $h_k(x)$ [k = 1,2,3,4] in (23) are given in Ref. 9. Also, in that reference explicit expressions are given for the coefficients g_k and f_k[k = 0,1,2] that minimize the difference between I_n in (14) and $I_n^{(APP)}$ in (17).

Next, consider the two-dimensional integration problem, represented by the Kirchhoff integral in (1). As in the previous case we divide the angular interval in N subintervals so that we may write (1) in the form

$$u = \frac{-iC}{2\lambda} \sum_{n=1} u_n , \tag{24}$$

where

$$u_n = \int_{\phi_{L,n}}^{\phi_{U,n}} \int_{\rho^-}^{\rho^+} G(\rho,\phi)\exp[iF(\rho,\phi)]d\rho d\phi \ . \tag{25}$$

Since the spatial extent of the integration area usually is less in the radial direction than in the angular direction for an annulus of a holographic lens, we prefer to perform the radial integration first. For that purpose one could use the one-dimensional integration technique presented above for the BDW integral. However, since the formulas after the first integration then would be somewhat complicated, we use a different procedure.

To obtain a simple formula after the first integration we linearize the phase function in the radial direction, i.e., in each subdomain $\phi_{M,n}-\frac{1}{2}\Delta_n < \phi < \phi_{M,n}+\frac{1}{2}\Delta_n$, $\rho^- < \rho < \rho^+$ we write

$$F(\rho,\phi) \cong F_{LIN}(\rho,\phi) = F_0(\phi) + F_1(\phi)\rho \ . \tag{26}$$

Also, we assume that the amplitude function $G(\rho,\phi)$ varies slowly over the integration interval $\rho^- < \rho < \rho^+$, so that it may be replaced by its value at the midpoint. Then we obtain from (25)

$$u_n = 2 \int_{\phi_{L,n}}^{\phi_{U,n}} G(\rho_M,\phi)\text{sinc}(F_1(\phi)\Delta\rho)\exp[iF_{LIN}(\rho_M,\phi)]d\phi \ , \tag{27}$$

where

$\text{sinc}(x) = \sin(x)/x$, and

$$\Delta\rho = \tfrac{1}{2}(\rho^+-\rho^-) \ ; \ \rho_M = \tfrac{1}{2}(\rho^++\rho^-) \ . \tag{28}$$

It has been assumed that $\Delta\rho$ is small enough to make the linearization procedure presented above sufficiently accurate. If $\Delta\rho$ is too big for that, the ρ-interval is to be divided in subintervals, and the linearization procedure is to be applied to the ρ-integration over each subinterval.

Next, we consider the angular integration in (27). If the amplitude function $G(\rho_M,\phi)\text{sinc}(F_1(\phi)\Delta\rho)$ varies slowly compared to the exponential function $\exp[iF_{LIN}(\rho_M,\phi)]$, then we apply the same procedure to the integral in (27) as we did previously to the BDW integral in (8). If the amplitude function in (27) does not vary slowly compared to the exponential function, we split up the sinc-function in its two exponential parts. As a result the integral in (27) becomes a sum of two integrals, both with the same amplitude function $[2iF_1(\phi)\Delta\rho]^{-1} G(\rho_M,\phi)$, and with exponential functions

$\exp[iF_{LIN}(\rho_M,\phi) \pm F_1(\phi)\Delta\rho]$. To each of these integrals we again apply the same technique as we applied in the preceding case to the BDW integral.

A somewhat simpler integration scheme is proposed in Ref. 10. That scheme involves a linear approximation to both the amplitude function and the phase function as compared to the parabolic approximation used in (15)-(16).

3. METHOD OF STATIONARY PHASE

As mentioned in the introduction, one of the reasons diffraction integrals are hard to evaluate numerically, is that their integrands often are rapidly oscillating functions, particularly at short wavelengths. The oscillating nature of the integrand is associated with the phase function given in (3). For a sufficiently large value of k, relatively small variations in ρ or ϕ may cause the phase $F = k(r_1+r_2)$ to vary over many 2π-intervals. As a result the exponential function $\exp(iF)$ in (1) will oscillate so rapidly that the Kirchhoff diffraction integral may be evaluated by the method of stationary phase for double integrals.

According to that method[11], only neigborhoods of critical points of various kind contribute to the integral in (1) as k gets large. Critical points of the first kind are points (ρ_1,ϕ_1) inside the domain of integration or at its boundary at which the phase is stationary, i.e., $F_\rho(\rho_1,\phi_1) = F_\phi(\rho_1,\phi_1) = 0$. Critical points of the second kind are points (ρ_2,ϕ_2) at the boundary of the domain of integration at which the tangential derivative of the phase function vanishes, i.e., $\rho_2 = \rho^+$ or $\rho_2 = \rho^-$, and ϕ_2 is the solution of $F_\phi(\rho_2,\phi) = 0$. Since the curve bounding the aperture is smooth, there are no critical points of the third kind, i.e., corner points. The value of the integral in (1) is the sum of the various contributions described above.

It may be shown[12] that the contribution $u_1(P_2)$ from an interior stationary point corresponds to that of geometrical optics. Thus, if the observation point P_2 lies in the shadow region, then $u_1(P_2) = 0$, and if P_2 lies in the geometrically insonified region then

$$u_1(P_2) = \exp(ikR)/R \, , \tag{29}$$

where R is the geometrical distance from the object point to the observation point.

Also, it may be shown[12] that there are either 2 or 4 critical points of the second kind on each of the two boundaries of the annulus, and that each of these critical points gives a lowest order contribution of the form

$$u_2^{(1)}(P_2) = \pi^{\frac{1}{2}} k^{-3/2} G_{0,0} \exp[ikF_{0,0}+i\alpha\pi/2]/|F_{0,2}|^{\frac{1}{2}}F_{1,0} , \qquad (30)$$

where

$$H_{m,n} = \frac{1}{m!n!} \frac{\partial^{m+n}}{\partial\rho^m\partial\phi^n} H(\rho,\phi) \Bigg|_{\rho=\rho_2,\phi=\phi_2} , \qquad (31)$$

with H standing for either F or G. The next term in the asymptotic series is given by

$$u_2^{(2)}(P_2) = u_2^{(1)}(P_2)\frac{i}{k} Q , \qquad (32)$$

where Q is a somewhat lengthy algebraic expression which involves the derivatives of G up to the second order and of F up to the fourth order. Explicit formulas for Q and all the derivatives appearing in (30) and (32) are given in Ref. 12.

Evidently, the formulas (30) and (32) break down if $F_{1,0}$ or $F_{0,2}$ tends to zero. A uniform asymptotic expansion which is valid as $F_{1,0} \to 0$ is developed in Ref. 12, where it is also shown that the vanishing of $F_{1,0}$ takes place as the observation point crosses the shadow boundary. The uniform expansion involves the Fresnel integrals.

The second order derivative $F_{0,2}$ approaches zero for observation points in the region in which we have a transition from 2 to 4 critical points on any of the two boundaries of the annulus. Thus, we have a situation in which there is a coalescence of three critical points into one, which means that not only the second order derivative $F_{0,2}$ vanishes at the point of coalescence, but also the third order derivative, $F_{0,3}$. In this case a uniform asymptotic approximation will involve the parabolic cylinder function[13].

4. COMPARISON OF METHODS

A thorough evaluation and detailed comparison of the methods outlined in this paper have been carried out[12], from which we now quote some pertinent results.

As far as the computational speed is concerned, we have found that the method of local phase and amplitude approximations (MPA) is about 10-15 times faster than a direct numerical integration performed by the Simpson integration routine. Also, we have found that if the annulus is so wide that a radial subdivision is necessary in the Kirchhoff integral to provide sufficient accuracy, then the BDW integral gives a substantial reduction in computation time compared to that for the Kirchhoff integral. Thus, as a general rule one should use the BDW integral rather than the Kirchhoff integral, whenever the former is applicable. For both integrals, however, the computation time increases rapidly as the wavelength decreases.

The computational speed of the method of stationary phase (MSP) is independent of the size of the wavelength. Thus, for any given geometry the MSP will become substantially faster than the MPA when the wavelength gets sufficiently short. As a general rule one may say that the MSP always is preferable to the MPA, as far as speed is concerned.

However, as we already have pointed out the BDW breaks down when the observation point gets close to the shadow boundary. Also, the MSP breaks down if $F_{1,0} \to 0$ or $F_{0,2} \to 0$. In the former case the MSP has been repaired through the development of a uniform asymptotic approximation, which remains valid as $F_{1,0} \to 0$.

In Ref. 12 very simple expressions are given which can be used to decide in which situations the BDW, the MSP and the uniform MSP are not valid. By the help of these expressions one can decide when to choose the MPA instead of the other methods listed above. Thus, one can minimize the computer time by always selecting the fastest of the methods that are valid for the geometry and wavelength in question.

Generally speaking, the uniform MSP is to be used for observation points near the shadow boundary, the BDW with the MPA is to be used for observation points in regions where $F_{0,2} \to 0$, and the second order MSP is to be used in other regions.

Finally, we point out that our asymptotic methods bear close resemblance to the geometrical theory of diffraction (GTD) of Keller[14]. However, whereas the GTD breaks down at the shadow boundary, our uniform MSP remains valid in that region.

In a separate paper[3] the techniques described above are applied to study the imaging properties of holographic lenses for sound waves.

REFERENCES

1. H. Heier and J.J. Stamnes, Acoustical Cameras for Underwater Surveillance and Inspection, Proc. Underwater Technology Conference, Bergen, Pergamon Press (1980).
2. H. Heier, A Wide-Angle Diffraction Limited Holographic Lens System for Acoustical Imaging, in A.F. Metherell (ed.), Acoustical Imaging, Vol. 10, Plenum Press.

3. J.J. Stamnes and T. Gravelsæter, Image Quality and Diffraction Efficiency of Holographic Lenses for Sound Waves, in A.F. Metherell (ed.), Acoustical Imaging, Vol 10, Plenum Press.
4. O. Weiss and C. Scherg, An Ultrasonic Fresnel Lens Imaging System with High Lateral Resolution, Proc. Ultrasonics International, 534-539 (1979).
5. A. Boivin, On the Theory of Diffraction by Concentric Arrays of Ring-Shaped Apertures, J. Opt. Soc. Am., 42:60-64 (1952).
6. E.W. Marchand and E. Wolf, Consistent Formulation of Kirchhoff's Diffraction Theory, J. Opt. Soc. Am. 56:1712-1722 (1966).
7. M. Born and E. Wolf, Principles of Optics, 4th ed., Pergamon, New York, Sec. 8.3.2. (1970).
8. M. Born and E. Wolf, Principles of Optics, 4th ed., Pergamon, New York, Sec. 8.9 (1970).
9. H.M. Pedersen and J.J. Stamnes, Evaluation of diffraction integrals using local phase and amplitude approximations (in preparation).
10. J.J. Stamnes, New Methods and Results in Focusing, Proc. ICO Conference on Optics in Four Dimensions, Ensenada, Mexico, American Institute of Physics, (1980).
11. J. Focke, Asymptotische Entwicklungen Mittels der Methode der Stationären Phase, Ber. Verh. Saechs. Akad. Wiss., Leipzig, 101:1-48 (1954); G. Braun, Zur Methode der Stationären Phase, Acta Phys. Austriaca 10:8-33 (1956); D.S. Jones and M. Kline, Asymptotic Expansion of Multiple Integrals and the Method of Stationary Phase, J. Math. Phys. 37:1-28 (1958); N. Chako, Asymptotic Expansions of Double and Multiple Integrals Occurring in Diffraction Theory, J. Inst. Maths Applics 1:372-422 (1965).
12. J.J. Stamnes and T. Gravelsæter (in preparation).
13. L.B. Felsen and N. Marcuwitz, Radiation and Scattering of Waves, Chap. 4, Prentice-Hall, Inc., New Jersey (1973).
14. J.B. Keller, Geometrical Theory of Diffraction, J. Opt. Soc. Am. 52:116-130 (1962).

IMAGE QUALITY AND DIFFRACTION EFFICIENCY OF A HOLOGRAPHIC LENS FOR SOUND WAVES

Jakob J. Stamnes, Tore Gravelsæter and Otto Bentsen

Central Institute for Industrial Research
P.B. 350 Blindern
Oslo 3, Norway

ABSTRACT

A holographic lens for sound waves has been constructed and its imaging properties investigated theoretically and experimentally. First we discuss the advantages of holographic lenses over both conventional lenses and phased arrays. Then we use the techniques described in a separate paper for a theoretical study of the image quality and diffraction efficiency of the holographic lens. Finally, we compare the theoretical results with those obtained in a water tank experiment.

1. INTRODUCTION

Holographic or zone-plate lenses offer many advantages over conventional lenses in acoustical imaging[1-3], the prime one being that one can increase the diameter of the holographic lens without increasing the thickness. Thus, in making lenses of large diameters to obtain systems that combine long range with high resolution, one would avoid problems of absorption, weight, and material costs, if holographic lenses were used instead of conventional ones.

Another advantage, which has to do with image sharpness, is that one without additional difficulty can produce holographic lenses which are equivalent to aspheric conventional lenses[3]. Thereby one can decrease the number of elements needed to obtain a desired degree of aberration correction.

A question that naturally arises in a discussion of acoustical imaging is whether one should use lenses or phased acoustical arrays. We feel there are several reasons why lenses are preferable in many situations. The use of a phased array, which commonly is applied both as transmitter and receiver, requires sophisticated signal processing to provide an image. Also, since the image usually is built up point by point by using the phased array to focus sound on one object point at a time, the imaging process is slow at long working distances. In contrast a lens images all points in the object simultaneously. It does all the required signal processing in parallel in real time, so that live images may be obtained without difficulty.

On the other hand, if a high image repetition rate is not required, one needs only use a one-dimensional (rotating or scanning) detector array in combination with a lens to obtain a true three-dimensional picture. In general, a true three-dimensional picture cannot be obtained by a linear phased array.

Another advantage is that the acoustical array detector for the camera system can be made much simpler than for the phased array system, since in the former case it is not necessary to detect the phase of the signal, but only the power. Here it should be admitted that the phase information may be important if image processing is to be performed. However, a lens used in combination with a phase sensitive detector array would still be preferable to the phased array, since all the time consuming image transformations are done in real time by the lens.

Thus, in summary, the camera system would provide images at a higher rate and to a lower expense than would the phased array system.

In a separate paper[4] a diffraction model for holographic lenses is presented, and methods based on this model are developed for computing the image field of holographic lenses. We apply in this paper the methods in Ref. 4 to study the imaging properties of a holographic lens for sound waves.

The objective of the study is twofold. First of all, we want to provide theoretical results based on the diffraction model in Ref. 4 for comparison with measured results, as far as the image field of an actual lens is concerned. To establish agreement between theory and experiment is important, as we want to use the methods based on the diffraction model as a practical tool in design and evaluation of holographic lenses.

The second objective of the paper is to apply the methods in Ref. 4 to study the image quality and the diffraction efficiency of a holographic lens for sound waves.

The paper is organized as follows: In Section 2 the working principle of a holographic lens is explained, as well as how it is related to the Fresnel lens and the Fresnel zone plate. Also, the parameters of an actual test lens are given. Section 3 contains a comparison of the imaging properties of a holographic lens with those of a Fresnel zone plate. A thorough study of the image quality of the holographic test lens is presented in Section 4, and the diffraction efficiency is investigated in Section 5. Section 6 is devoted to a comparison of the theoretical results with those obtained in a water tank experiment. Finally, a summary of the main results of the paper is given in Section 7.

2. THE HOLOGRAPHIC LENS

The working principle of a holographic lens is explained in Fig. 1. Suppose for simplicity that a plane wave with wave front parallel to the plane surface of the lens in Fig. 1(a) is incident from below. Then only the hatched areas in the figure will have any influence on the transmitted wave front, except for a phase factor of $2n\pi$, where n is an integer. Thus, the lens shown in Fig. 1(b) produces the same transmitted wave front as the one in Fig. 1(a), except that there is a discontinuous phase jump of $2n\pi$ between two different zones in the former case.

We refer to the lens in Fig. 1(b) as a holographic Fresnel lens. In contrast to the holographic Fresnel lens, an ordinary Fresnel lens has a random phase jump from one zone to another, so that the wave fields that are due to different zones do not interfere constructively in the focal region. As a consequence, the diffraction-limited resolution of an ordinary Fresnel lens is determined by the width of the zones, whereas the resolution of a holographic Fresnel lens is determined by the diameter of the lens.

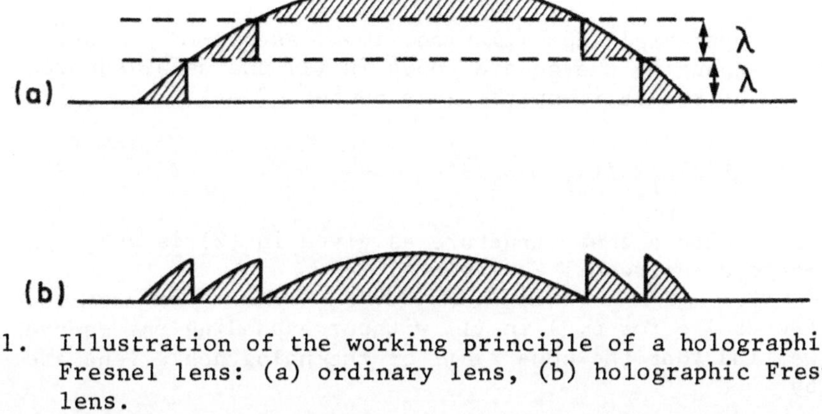

Fig. 1. Illustration of the working principle of a holographic Fresnel lens: (a) ordinary lens, (b) holographic Fresnel lens.

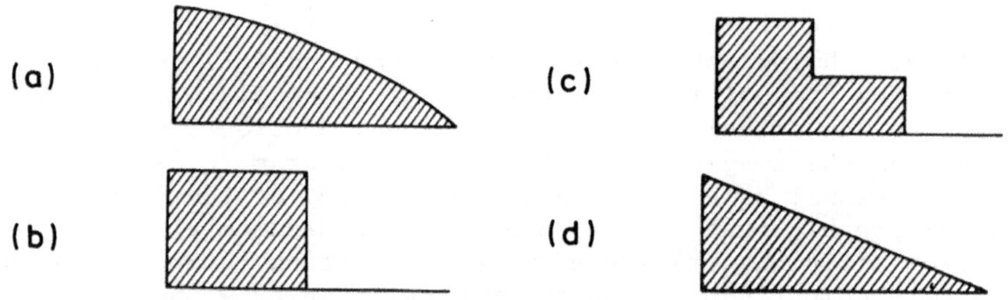

Fig. 2. Figure (a) shows the profile of one of the zones in Fig. 1(b). Figures (b), (c), and (d) illustrate various ways of approximating or quantizing the phase function in (a).

The actual profile of one of the zones in Fig. 1(b) is shown in Fig. 2(a). In practice it may be easier, however, to make the staircase profiles shown in Figs. 2(b) and (c). The number of steps in the staircase we refer to as the number of quantization levels of the phase function. Obviously, when the number of quantization levels becomes infinitely large we obtain the saw-tooth or straight-line profile in Fig. 2(d). In Section 5 we study how the diffraction efficiency varies with the number of quantization levels.

A holographic lens has been constructed that gives an optimum image of a point on the axis. The principle of construction is that the difference between the acoustical path length from the object point to the image point via each zone of the lens and the direct acoustical path length between the object and the image point is to be an integer number of wavelengths. Thus, if $\rho(n)$ is the radius of the n-th zone, λ is the wavelength, z_1 is the object distance, and z_2 is the image distance, we have

$$(\rho^2(n) + z_1^2)^{\frac{1}{2}} + (\rho^2(n) + z_2^2)^{\frac{1}{2}} = n\lambda + z_1 + z_2 \quad . \tag{1}$$

In the paraxial approximation, i.e., as $\rho(n) << z_1, z_2$, we obtain by expanding the square roots in (1) the usual square root dependence of the radii on the zone number, i.e.,

$$\rho(n) \simeq [2n\lambda z_1 z_2/(z_1 + z_2)]^{\frac{1}{2}} \quad . \tag{2}$$

A lens with a zone structure as given in (2) is usually referred to as a Fresnel zone plate.

If we solve for $\rho(n)$ in (1) without expanding the square roots we find that the zone radii of the holographic lens are given by

$$\rho(n) = \frac{\lambda^2}{n\lambda+z_1+z_2}[n^4/4 + n^3(z_1+z_2)/\lambda+n^2((z_1+z_2)^2+z_1z_2)/\lambda^2$$

$$+ 2nz_1z_2(z_1+z_2)/\lambda^3]^{\frac{1}{2}} \quad , \qquad (3)$$

which reduces to the expression used in Ref. 1, in the case that $z_1 = \infty$, $z_2 = f$.

The parameters of the lens that was constructed for our test purposes are $z_1 = z_2 = 2000$ mm and $\lambda = 0.7$ mm. The lens has 100 zones, and the phase of each zone is quantized in two levels. A picture of the lens is shown in Fig. 1 of Ref. 4. The lens is made of a plane parallel plate of perspex into which a concentric pattern of circular grooves is engraved. The radii of the grooves are given in (3), and the profile of each groove is as shown in Fig. 2(b). From (3) it follows that the diameter of the lens is $2\rho(100) = 0.75$ m.

3. COMPARISON BETWEEN THE HOLOGRAPHIC LENS AND THE FRESNEL ZONE PLATE

The difference between the zone radii in (3) of the holographic lens and those in (2) of the corresponding Fresnel zone plate is displayed in Fig. 3 for the case in which $z_1 = z_2 = 2000$ mm, $\lambda = 0.7$ mm. It follows from Fig. 3 that the Fresnel zone plate produces a wave front that, as the zone radius increases, becomes advanced compared to the wave front of the holographic lens. Therefore the Fresnel zone plate will exhibit positive spherical aberration so that the focal point will be shifted toward the lens.

The focal shift is clearly demonstrated in Fig. 4, in which the intensity distributions along the axis are displayed for the zone plate and for the holographic lens. In addition to the shift in focal position of 36 mm, the maximum intensity is reduced by 16.3% in the case of the Fresnel zone plate.

The intensity distributions in the focal plane of the two lenses are displayed in Fig. 5. We see from Fig. 5 that the secondary lobes of the intensity pattern that is due to the Fresnel zone plate are more prominent, as is to be expected because of the spherical aberration.

In many practical situations the encircled energy is a better measure of image quality than the intensity distributions in Fig. 5. The encircled energy is defined as the fraction of the total energy that is contained within a circular area centered at

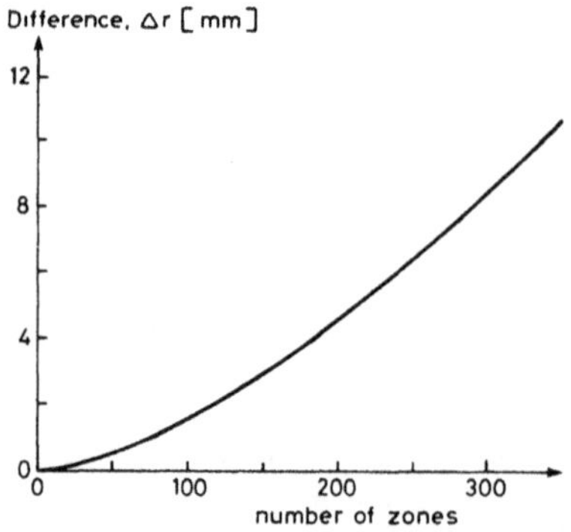

Fig. 3. Plot of the difference between the zone radii of the holographic lens [equation (3)] and those of the Fresnel zone plate [equation (2)] as a function of zone number.

the image point[5]. Plots of the encircled energy for the two lenses are presented in Fig. 6, from which it follows that the holographic lens is better than the Fresnel zone plate as far as encircled energy within the main lobe is concerned.

From Fig. 3 it is clear that the spherical aberration may be reduced by stopping down the lens. To illustrate this aberration reduction we computed the intensity distribution along the axis for the two lenses, paying attention only to contributions from the first 50 zones. The spherical aberration was then reduced so that the shift in focal position was only 16 mm and the intensity reduction at the focal point only 1.3%.

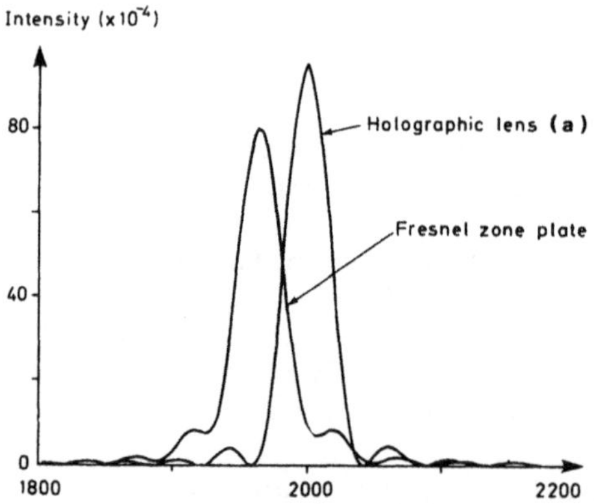

Fig. 4. Plots of the axial intensity distributions in the neighborhood of the image points of a holographic lens and the corresponding Fresnel zone plate.

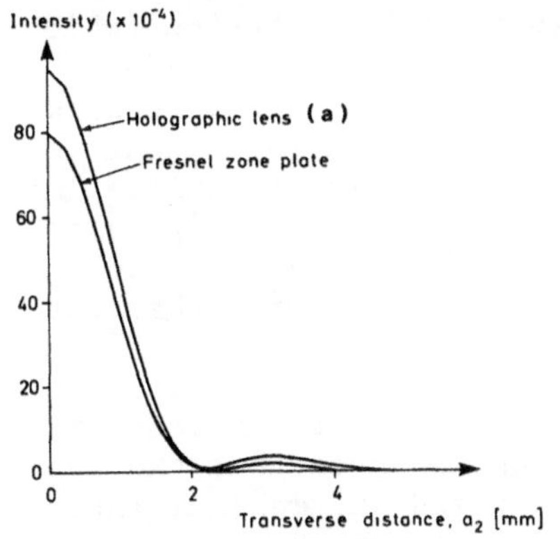

Fig. 5. Plots of the intensity distributions for a holographic lens and for the corresponding Fresnel zone plate in their image planes.

4. IMAGE QUALITY OF THE HOLOGRAPHIC LENS

To illustrate the image quality of the holographic lens specified in Section 2, we now present computed results for the image field.

Ideal Imaging

We start by considering the case of ideal imaging, in which the object point is on the axis at the position $z_1 = 2000$ mm, which is the object position for which the lens was constructed.

The intensity distribution along the axis has been presented already in Fig. 4(a). It is readily seen to have close resemblance

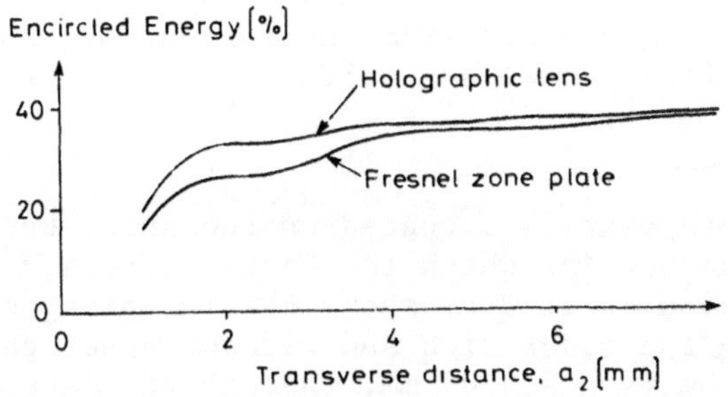

Fig. 6. Encircled energy curves for the holographic lens and for the corresponding Fresnel zone plate in their image planes.

with the $[\sin(u/4)/(u/4)]^2$-distribution, which an ordinary aberration free lens produces[6]. Since $u = \pi(D/z_2)^2\Delta z/2\lambda$, where D is the lens diameter and Δz is the distance from the focal point, we find that the width of the main lobe along the axis is $16\lambda(z_2/D)^2 = 79.6$ mm, in the case of an aberration free lens comparable to our holographic lens. From the curve in Fig. 4(a), in which the sampling distance is 4 mm, the width of the main lobe is found to be 80 mm, a result which agrees very well with that for the aberration free lens.

As another measure of the resemblance between the two on-axis intensity distributions, we note that the ratio between the intensity at the focal point and the maximum intensity in the first side lobe is 22 in the case of the aberration free lens and 20 in the case of the holographic lens.

The intensity distribution in the focal plane, presented in Fig. 5(a), is to be compared with the $(J_1(v)/v)^2$-distribution[6] produced by an ordinary aberration free lens. Since $v = \pi(D/z_2)r/\lambda$, where r is the radial distance from the focal point, we find that the width of the main lobe is $2.44(z_2/D)\lambda = 4.55$ mm. From Fig. 5(a), in which the sampling distance is 0.25 mm, the width of the main lobe is found to be 4.50 mm. The ratio between the intensity at the focal point and the intensity in the first side lobe, is 57 in the case of an aberration free lens[5] and 52 in Fig. 5(a). Thus, again reasonably good agreement is found between the results for the holographic lens and those for an ordinary aberration free lens.

Fig. 7 shows the intensity distribution in some selected planes parallel to the image plane. Fig. 7(a) corresponds to the image plane, and Figs. 7(b), (c), and (d) correspond to observation planes that are situated respectively 10 mm, 20 mm, and 30 mm in front of the image plane. We see from Fig. 7 that the depth of field is somewhere between 10 mm and 20 mm, depending on how much reduction in the maximum intensity one allows. These curves agree fairly well with the corresponding curves for an aberration free lens. The latter curves may be obtained from the contour map constructed by Linfoot and Wolf from Lommel's data[6].

Spherical Aberration

If the object point is situated on the axis, but at a position different from the one for which the lens was constructed, we get spherical aberration. Fig. 8(a) shows the intensity distributions along the axis in the cases with and without spherical aberration. The object position is a = 4000 mm, whereas the lens was constructed for an object position of $z_1 = 2000$ mm. The intensity distributions in the image planes are displayed in Fig. 8(b), On comparison of Figs. 8(a) and (b) with Figs. 4 and 5, respectively, we recognize the typical results of spherical aberrations.

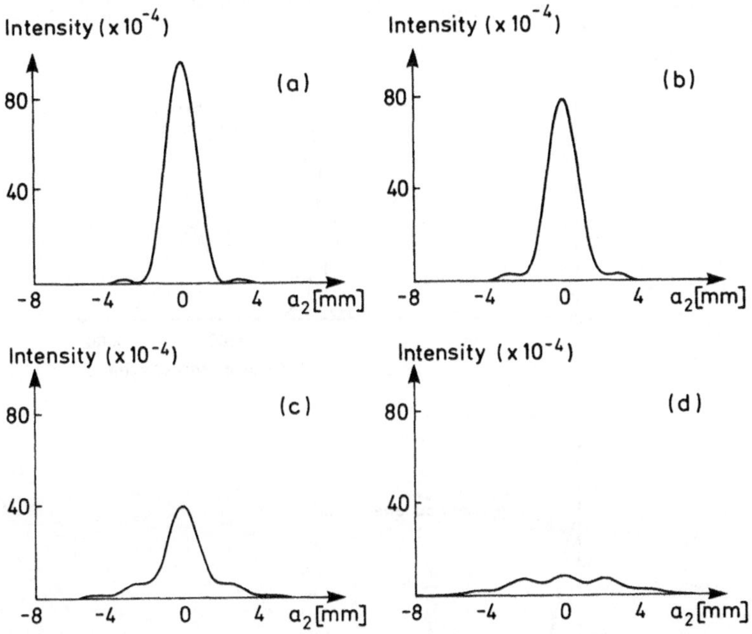

Fig. 7. Intensity distributions for the holographic lens in some selected planes parallel to the image plane. (a) corresponds to the image plane; (b), (c), and (d) correspond to planes that lie respectively 10 mm, 20 mm, and 30 mm in front of the image plane. a_2 is the transverse distance from the axis to the observation point.

The aberration free distributions in Figs. 8(a) and (b) were obtained by computing the image field of a lens with construction data $z_1 = a = 4000$ mm and $z_2 = b = af/(a-f)$, where the focal length $f = 1000$ mm, as before. Thus, we have $z_2 = 1333$ mm. The diameter of this lens was chosen to be the same as for the other lens, i.e., 750 mm.

Chromatic Aberration

One of the main drawbacks of holographic lenses is their large chromatic aberrations. To study the influence of such aberrations we computed the image field that results when a point source on the axis at a distance $z_1 = 2000$ mm from the lens radiates with a wavelength of 0.649 mm (a change of about 7%). The resulting intensity distribution along the axis is shown in Fig. 9, from which it follows that the intensity at the focal point is reduced by 15%, and that the width of the main lobe has increased from 80 mm [cf. Fig. 4(a)] to 104 mm. This change in width agrees well with the

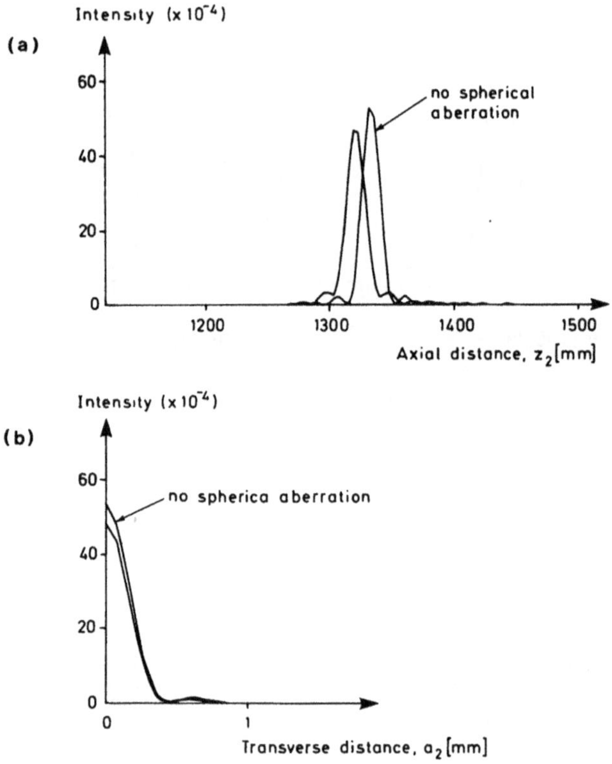

Fig. 8. Illustration of spherical aberration. The object point is displaced along the axis to a position a = 4000 mm. The object position for which the lens is constructed, is z_1 = 2000 mm. (a) Axial intensity distributions in the neighborhood of the image points of a holographic lens with and without spherical aberrations. (b) Intensity distributions in the image plane of a holographic lens with and without spherical aberration.

changes in f-number and wavelength: in the case of an aberration free lens the width is $16\ \lambda'(b/D)^2 = 16 \cdot 0.649(2348/750)^2 = 102$ mm.

However, the shape of the intensity pattern is similar to that of the ideal pattern in Fig. 4(a). Also, the ratio between the heights of the main lobe and the first side lobe is about 20 in both cases.

Fig. 9 shows that there is a shift in the image position from 2000 mm to 2348 mm behind the lens. This shift agrees fairly well with the lens formula $b = af'/(a-f')$, where $f' = f\lambda/\lambda'$. With $a = z_1 = 2000$ mm, $f = 1000$ mm, $\lambda = 0.7$ mm and $\lambda' = 0.649$ mm, we obtain $b = 2341$ mm.

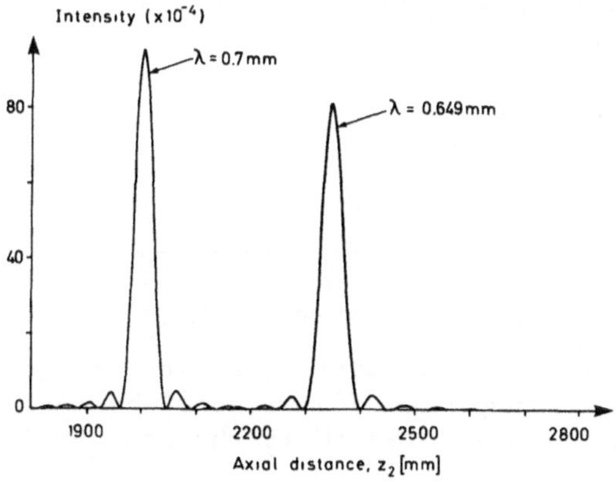

Fig. 9. Illustration of chromatic aberration: Intensity distribution along the axis with and without chromatic aberration. The change in wavelength is about 7%.

The intensity distribution in the image plane is shown in Fig. 10, from which we see that the main lobe is slightly wider than in the case of an aberration free lens. This change in width is mainly due to the changes in f-number and wavelength, as the aberration free curve was computed for an image distance of z_2 = 2000 mm, and a wavelength of λ = 0.7 mm. Fig. 11 shows that the chromatic aberration has little influence on the encircled energy. The energy content within each lobe is the same as in the aberration free case, but, as mentioned above, the width of the main lobe is slightly larger in the case of aberrations, due to the changes in f-number and wavelength.

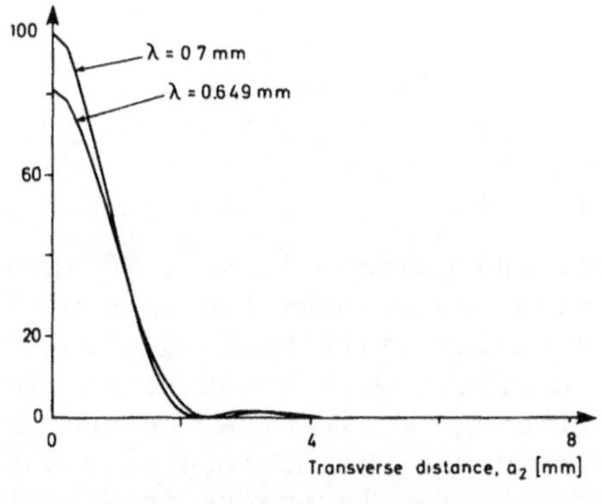

Fig. 10. Illustration of chromatic aberration: Intensity distribution in the image plane with and without chromatic aberration. The change in wavelength is the same as in Fig. 9.

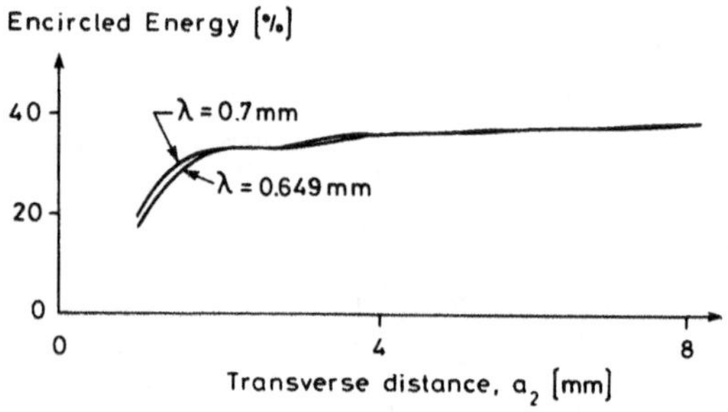

Fig. 11. Illustration of chromatic aberration: Encircled energy in the image plane with and without chromatic aberration. The change in wavelength is the same as in Fig. 9.

Thus, we may conclude that a change in wavelength of 7% does not degrade the image seriously, as long as the insonification is fairly monochromatic. However, a polychromatic insonification would lead to serious image degradation because of the shift in the image position. If Δb is the allowable amount of defocusing, then the allowable change in wavelength follows from $b' = b \pm \Delta b$, where $b' = af'/(a-f')$ with $f' = f\lambda/\lambda'$. With $\lambda = 0.7$ mm, $b = 2000$ mm, $f = 1000$ mm, and $\Delta b = \pm 15$ mm [cf. Fig. 7], we find that $\Delta\lambda/\lambda \simeq 0.4\%$.

Off-axis Aberrations

Since our holographic lens is constructed for imaging at a magnification of unity ($z_1' = z_2 = 2f$), it exhibits no coma[7-8]. The field of view therefore, is limited by astigmatism.

To illustrate the astigmatism we computed the image field that is due to a point source lying 5° off-axis in the object plane. A plot of the intensity distribution in image space along the axis, that goes through the object point and the center of the lens, is presented in Fig. 12, from which it follows that the maximum intensity along this axis is obtained at a point which lies at a distance of 1972 mm behind the lens. Next, we computed the intensity in the image plane, which is parallel to the lens at a distance of 1972 mm behind it, along lines going through the point of maximum intensity found above, and making 0°, 45°, 90° and 135° with the meridian plane (the plane containing the axis of the lens and the object point). The intensity distributions along lines making an angle of 90° with each other, were found to be similar as shown in Fig. 13. From these findings it is clear on comparison with Fig. 9.9 of Ref. 5 that we have an astigmatic intensity pattern, typical for the central plane. The maximum intensity in Fig. 13 is only about 40% of the value that one finds in the absence of aberrations.

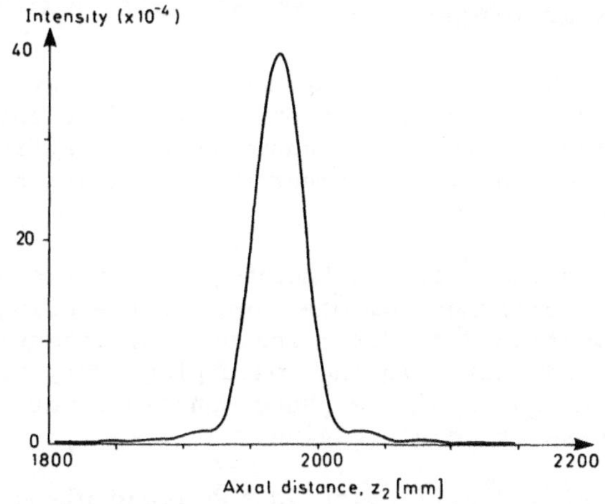

Fig. 12. Object point 5° off-axis: plot of the intensity distribution in image space along an axis which passes through the object point and the center of the lens.

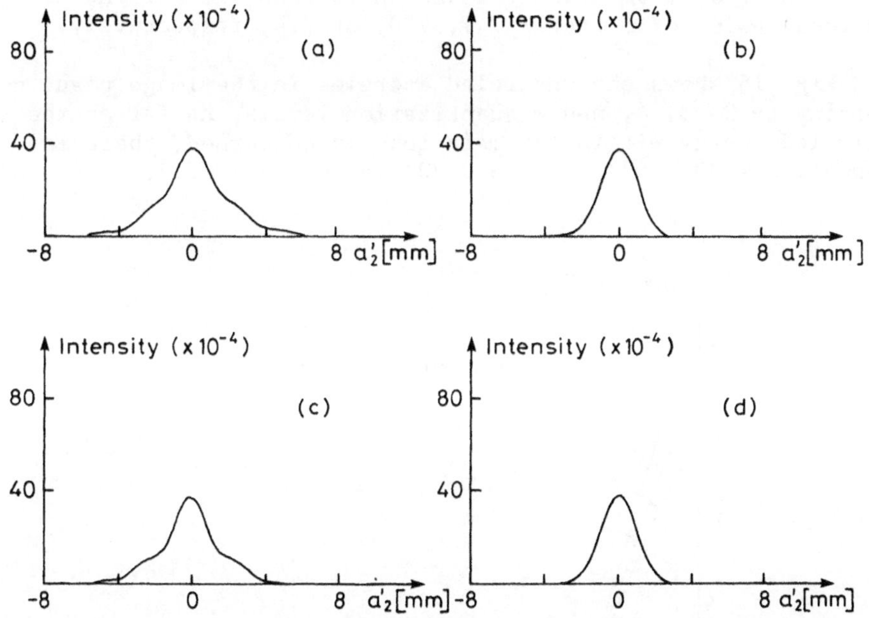

Fig. 13. Object point 5° off-axis: plots of the intensity distributions along different lines in an observation plane that is parallel to the lens and passes through the point of maximum intensity in Fig. 12. The lines (a), (b), (c), and (d) make 0°, 45°, 90°, and 135°, respectively, with the meridian plane.

5. DIFFRACTION EFFICIENCY

It is well known that holographic lenses have higher order focal points, so that only part of the incident energy goes into the desired image. Also, it is known that the efficiency may be increased by increasing the number of quantization levels of the phase function.

To obtain a quantitative knowledge about the diffraction efficiency we have performed computations for the holographic lens specified in Section 2 to determine how the intensity distribution and the encircled energy in the image plane vary with the number of quantization levels of the phase function. Our computations are based on the case of ideal imaging.

The intensity distribution in the image plane corresponding to 2, 3, and 4 quantization levels are shown in Fig. 14, together with the result corresponding to an infinite number of quantization levels (the straight-line approximation). It follows from Fig. 14 that an increase in the number of quantization levels from 2 to 3, from 2 to 4, or from 2 to ∞, gives an increase in the intensity at the focal point of a factor 1.7, 2.0, or 2.4, respectively.

Fig. 15 shows the encircled energies in the image plane corresponding to 2, 3, 4, and ∞ quantization levels. As far as the encircled energy within the main lobe is concerned, there is respectively 33%, 57%, 67%, and 80% in the case of 2, 3, 4, and

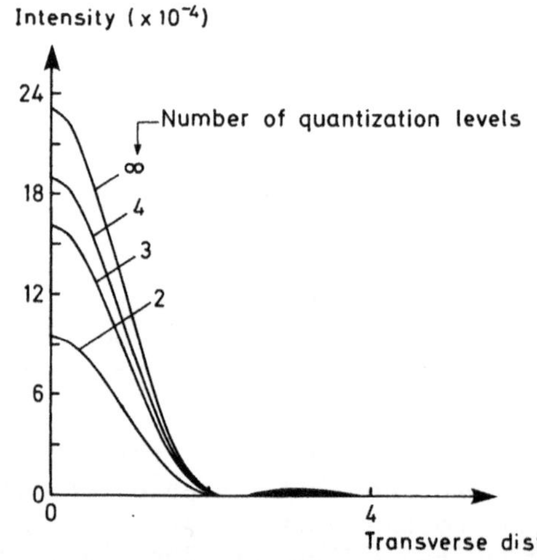

Fig. 14. Diffraction efficiency: plots of the intensity distributions in the image plane corresponding to different numbers of quantization levels.

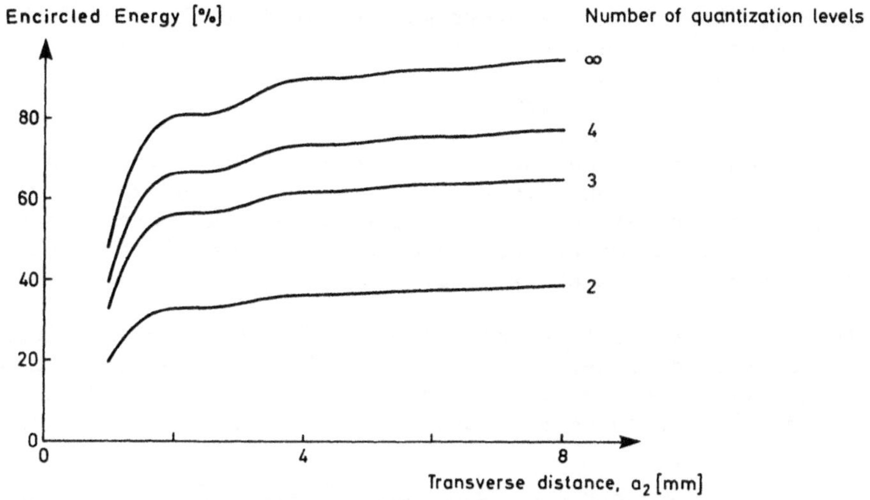

Fig. 15. Diffraction efficiency: encircled energy curves corresponding to different numbers of quantization levels.

∞ quantization levels. Thus, in going from a staircase approximation of the phase function with 4 levels to a straight-line approximation, there is a gain of 13% in diffraction efficiency, as far as the energy contained within the main lobe is concerned.

Since the encircled energy contained within the main lobe of an aberration free lens is 83.8%, there is a loss in diffraction efficiency, as far as the energy contained within the main lobe is concerned, of about 4% by using the straight-line approximation to the phase function.

6. EXPERIMENTAL RESULTS

As mentioned in Section 2, our test lens is made out of a plane parallel plate of perspex, into which a concentric pattern is engraved of one hundred circular grooves, the phase of each groove being quantized in two levels. The thickness of the plate is 30 mm, and the sound speed within the plate is 1.826 times that in water. The construction data [cf. equation (3)] are $z_1 = z_2 =$ 2000 mm, and $\lambda = \lambda_o = 0.645$ mm. The wavelength λ_o corresponds to a frequency of 2.5 MHz in water at 15 ^oC.

The experiments were performed in a water tank, which is 7 m by 4 m in area and 2 m in depth. No precautions were taken to reduce reflections from the water surface or from the concrete walls and bottom of the tank. Still, the reflections did not cause any serious problems, as the intensity of the side lobes was higher than the background intensity.

As a sound source we used a spherical PZT disc, which was mounted so as to give an effective source diameter of approximately 1 mm. The detector was made in the same way as the source. An aperture of 1.5 mm in diameter was placed at the center of the sphere, defined by the PZT transducer disc.

The image was built up point by point by scanning the detector in a predetermined manner over the image plane by the use of computer controlled step motors. The computer also controlled the measurements of the sound pressure, which was recorded on magnetic tape.

The resonance frequency of the PZT transducers was 2.0 MHz. Since the transducers had low sensitivity at the construction frequency of the lens (2.5 MHz), the measurements were performed at a frequency of 2.123 MHz, corresponding to a wavelength of λ_1 = 0.691 MHz. The change in wavelength is about 7%. We know from Section 4 that the main effect of the chromatic aberration introduced by this change in wavelength, is to give a shift in the axial image position, and thereby also in f-number.

In Fig. 16(a) the intensity distribution over an area in the image plane of 20 mm by 20 mm is displayed as a 7 level gray scale picture on a Tektronix 4027 terminal. The sampling distance was 0.75 mm in each direction, and the original picture was in colours. Fig. 16(b) shows the theoretical distribution, corresponding to the case of ideal imaging, displayed in the same way. A comparison between the two figures reveals that the first side lobe of the measured pattern is severely degraded. The energy in about two thirds of the area of the first side lobe is missing, and the remaining part of the lobe contains more energy than would be expected. This asymmetric defect is caused by the source: by rotating the source a certain angle around the axis of the lens, the image pattern with its asymmetric defect rotates the same angle.

Fig. 17 shows a comparison of the intensity distributions in Figs. 16(a) and (b) along a vertical line through the center in Fig. 16(a). Except for the asymmetry due to the above mentioned defect of the source, the agreement is fairly good. It should be noted, however, that in the theoretical evaluation, one has not taken into account the influence of spherical aberration, which arises because of the finite thickness of the lens and because of the change in wavelength (spherochromatism).

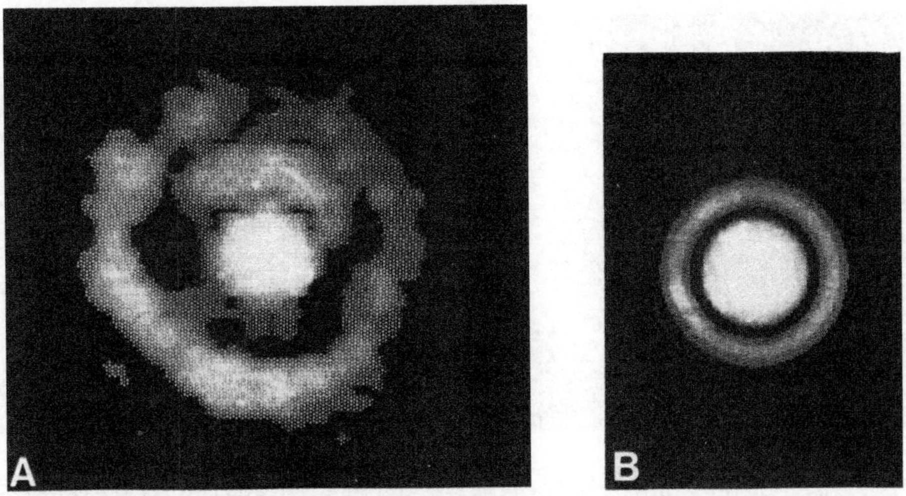

Fig. 16. Intensity distributions in the image plane: (a) measured, (b) theoretical.

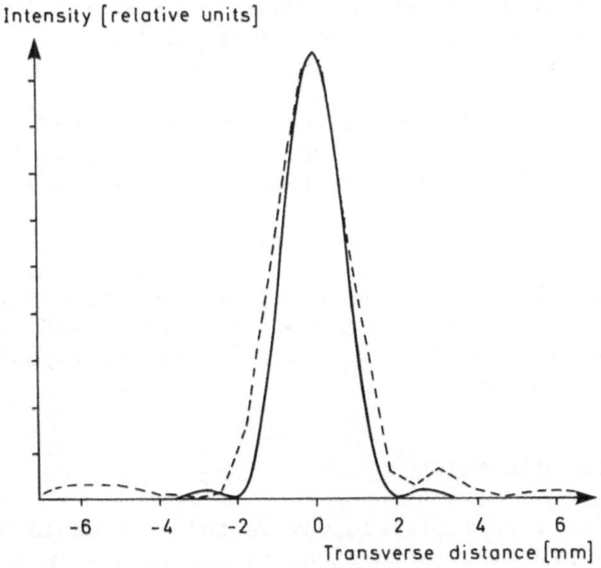

Fig. 17. Theoretical (solid line) and measured (dashed line) intensity distributions along one selected line in the image plane.

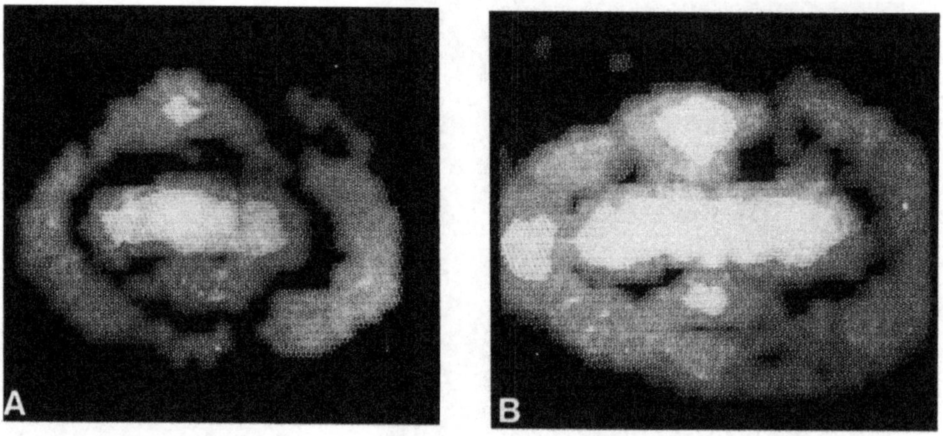

Fig. 18. Object point off-axis: measured intensity distributions. (a): 5° off-axis; (b): 10° off-axis.

The experimental width of the main lobe is 5.3 mm, and the theoretical width is $2.44\ (b'/D)\lambda_1 = 3.9$ mm. The difference between the two widths is caused mainly by the finite width of the detector aperture. By convolution with a detector aperture of 1.5 mm in diameter the theoretical lobe of 3.9 mm in diameter would become 5.4 mm in diameter.

To obtain intensity distributions that are due to object points that lie off-axis, the lens was tilted about its nodal point (i.e., its center), and the intensity was measured in the same receiving plane as in the case when the lens was not tilted. Figs. 18(a) and (b) show the resulting intensity distributions for tilt angles of 5° and 10°, respectively. The effects of astigmatism are clearly visible. It is, however, not possible to make a direct comparison between these measured distributions and the theoretical ones displayed in Fig. 13, since the latter corresponds to a different image plane.

7. SUMMARY AND DISCUSSION

In Section 1 we explained why we think it would be advantageous in many practical situations to use lenses in acoustical imaging. We also gave some reasons why holographic lenses would be preferable over conventional ones.

In Section 3 we compared a holographic lens with its Fresnel-zone-plate counterpart, and showed that the former is superior, as it is not afflicted with spherical aberration in the case of ideal imaging.

Using methods based on diffraction theory we have analyzed in detail the imaging properties of a holographic lens, and shown how its image is deteriorated by various aberrations. It was shown that one can tolerate relatively large displacements of the object point along the axis without running into problems with spherical aberration. Also, when the holographic lens was used with monochromatic radiation a change in wavelength of 7% lead to no serious problems. However, when used with polychromatic radiation only a wavelength band of 0.4%, or maybe even less, would be permissible. The lens under investigation does not suffer from coma, but astigmatism limits its field of view to less than $\pm 5^{\circ}$.

We also have determined how the diffraction efficiency of holographic lenses varies with the number of quantization levels of the phase function. In particular, we have shown that the use of a straight-line approximation to the phase within each zone, yields an energy content within the main lobe that is only about 4% less than for a conventional lens.

Measured results for the image field of an actual lens agree fairly well with the theoretical results. However, a detailed comparison would require a better knowledge of the sound field that is incident on the lens, and would also require that one includes in the theoretical analysis the effects of spherical aberration due to the finite thickness of the lens, and due to the change in wavelength.

We point out that before lenses can be used with success in acoustical imaging, a suitable detector array needs to be developed. As mentioned in Ref. 2, such a detector array should have high sensitivity, resolution compatible with that of the lens, fast read out, and low production costs. An idea of how to make such a detector array is presented in Ref. 2.

Finally, we mention that a holographic doublet, which offers diffraction limited imaging within a field of view of $\pm 20^{\circ}$ is described in Ref. 3.

REFERENCES

1. O. Wess and C. Scherg, An Ultrasonic Fresnel Lens Imaging System with High Lateral Resolution, Ultrasonics International, 534-539 (1979).

2. H. Heier and J.J. Stamnes, Acoustical Cameras for Underwater . Surveillance and Inspection, Proc. Underwater Technology Conference, Bergen, Pergamon Press (1980).

3. H. Heier, A Wide-Angle Diffraction Limited Holographic Lens System for Acoustical Imaging, in A.F. Metherell (ed.), Acoustical Imaging, Vol. 10, Plenum Press.

4. J.J. Stamnes and T. Gravelsæter, Methods for Efficient Computation of the Image Field of Holographic Lenses for Sound Waves, in A.F. Metherell (ed.), Acoustical Imaging, Vol. 10, Plenum Press.

5. M. Born and E. Wolf, Principles of Optics, 4th ed., Pergamon, New York, Sec. 8.5.2 (1970).

6. M. Born and E. Wolf, Principles of Optics, 4th ed., Pergamon, New York, Sec. 8.8 (1970).

7. W.C. Sweatt, Describing holographic optical elements as lenses, J. Opt. Soc. Am. 67:803-808 (1977).

8. W.T. Welford, Aberrations of the Symmetrical Optical System, Academic Press, London (1974).

AN UNDERWATER FOCUSED ACOUSTIC IMAGING SYSTEM*

P. Maguer*, J.F. Gelly+, C. Maerfeld+, and G. Grall+

*GESMA/DCAN - 29240 BREST NAVAL
+Thomson-CSF ASM Division
06802 Dagnes-sur-Mer
France

INTRODUCTION

We describe an acoustical focusing imaging device specially designed for the study of diffused back scattering in the case of objects with very low roughness in comparison to the wavelength (such as machined metallic immersed objects).

Acoustical imaging of objects with high impedance mismatch with the propagating medium is characterized by a drastic contrast between specular echoes and diffused scattering by the roughness.

Imaged objects published elsewhere[1,2,3] show a bright point distribution. In our case of object with low variations of the radius of curvature, there are too few number of bright points in order to ensure a perfect pattern recognition, and supplementary information must be extracted from the diffused scattering as in optics. The apparatus is mainly composed of an acoustical thin lens and an oscillating mirror enabling a one-dimensional array to scan the real frontal image. The immersed part also includes array amplifiers and mechanical control electronic unit.

The immersed part is connected with a cable to the exploitation console in which the 64 channels are multiplexed to form the video signal. A photograph of the system is shown in Fig. 1. Time trigger is given by a mirror position and the image is optically displayed on a variable persistence CRT.

---

*Work supported by DCAN/GESMA France

Insonification of the object can be obtained by the array itself (internal transmission) or by an auxilliary projector (external transmission).

GENERAL FEATURES

The observation of low level diffused echoes required very good sensitivity abilities. Transmission losses in the focusing device are minimized by the use of a well-corrected and unique lens. Few centimeter detail resolution at a range of 10 meters leads to a 3 dB directivity of about 0.2 degree. The choice of high centre frequency of 2 MHz allows reasonable dimensions and enhances diffusing effects. Furthermore, propagation losses are prohibiting the use of higher frequencies.

Mechanical scanning was suggested by the extreme difficulty to make and process a high quality two-dimensional array.

Focusing in the 2 - 10 m range is mechanically performed by the displacement of the lens. Fig. 2 is a functional diagram of the immersed part. In the external transmission mode where the array transducers are only receivers, the frame rate is limited by the mirror angular rotation speed.

In the internal transmission mode where array transducers are alternatively transmitting and receiving, the rotation speed cannot be higher than one angular resolution cell by round trip of the pulse. Fig. 3 is a schematic of the signal processing, control and display units.

Imaging part

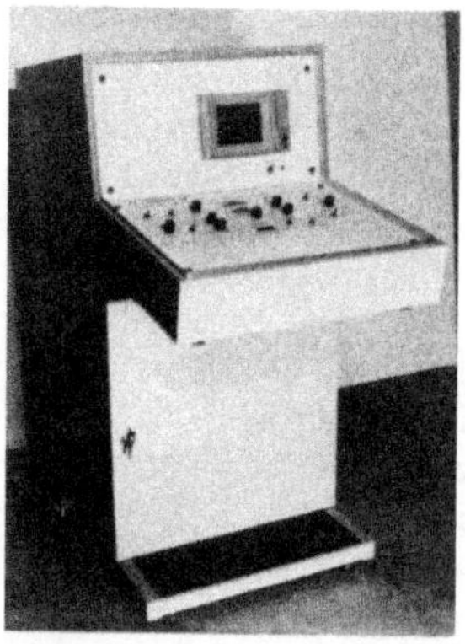

Exploitation console

Fig. 1 : Underwater imaging focusing system

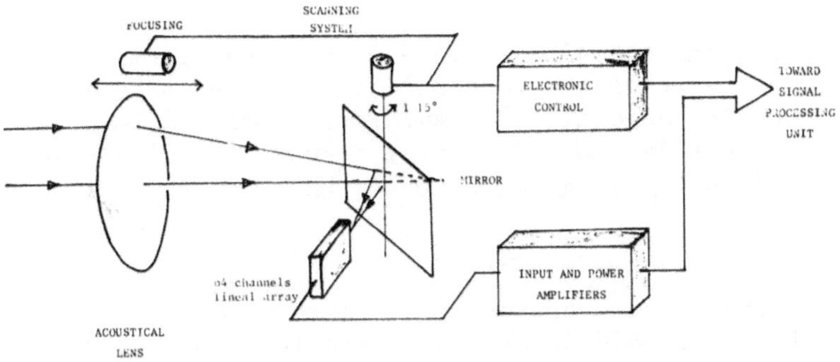

Fig. 2 : Functional diagram of immersed part

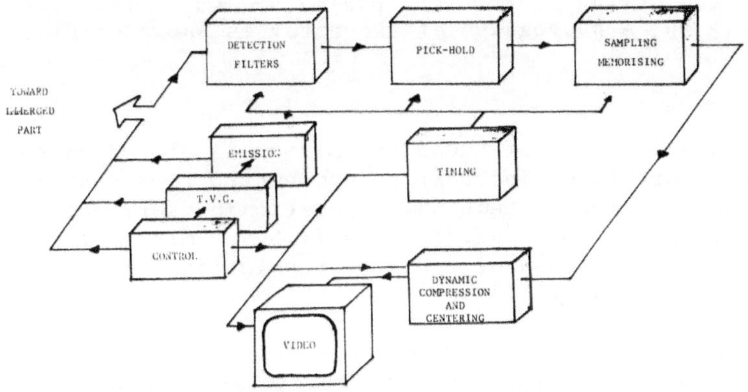

Fig.3 : Schematic of the signal processing

LENS

The focusing element of the system is a thin biconcave plexiglass acoustical lens. The high degree of precision of surface machining has been performed using carefully stabilization techniques. The lens is corrected for spherical aberation and coma free for the distance of 5 m. A removable apodizing screen has also been realized. It is made of a specially designed polyurethan which has an acoustic impedance matched to that of water and shows a strong absorption coefficient to allow for gaussian weighting by thickness variation.

Focal length is 0.497 m (0.477 with screen). Fig. 4 shows directivity patterns of the lens and demonstrates side lobe level suppression by the apodizing screen. The radius of the field curvature is about 145 mm and one-way transmission losses are 6.5 dB (15.5 dB with the screen).

ARRAY

The 64 transducers of the linear array are 1.5 mm spaced and each element is subdivided into four parallelepipeds for mode optimization. Water matching and backing isolation are provided by two similar quarter wave sections.[4]

3 dB-bandwidth is 270 kHz, centred on 2 MHz. The key difficulty in this kind of device is to keep a sufficient angular acceptance compatible with the lens apertures : 30° acceptance is achieved by taking special care to avoid inter-element coupling (mode selection, and design of the wear plate). One-way conversion losses are about 13 dB. A photograph of the array is shown on Fig. 5.

MIRROR

The rotation axis of the plane mirror is placed on the center of the field curvature, in order to compensate this distorsion along one of the image dimension. A positioning servo-control permits the observation of 100 resolved lines ($\pm$15° of mirror rotation). The total image is formed of 100 x 64 resolution cells (about 20° x 12.8° operating field).

In the external C.W. transmission mode the frame rate is depending on the maximum speed of mirror rotation : in order to minimize inertial momentum and prevent unsuitable acoustic surface wave propagation, the mirror is made of "soft" acoustical material ; reflexion losses are about 1 dB, and the upper scanning rate is 200 resolution lines/second.

SIGNAL PROCESSING

The input dynamic range is 46 dB. After amplification, linear detection integration, peak detection and analog storage, the 64 channels are sampled at 64 kHz. The dynamic range of the CRT display is 15 dB. The 15 dB corresponds either to the total compressed input dynamic or to an uncompressed portion of it. With this arrangement, a separation of specular and diffused echoes is possible to some extent.

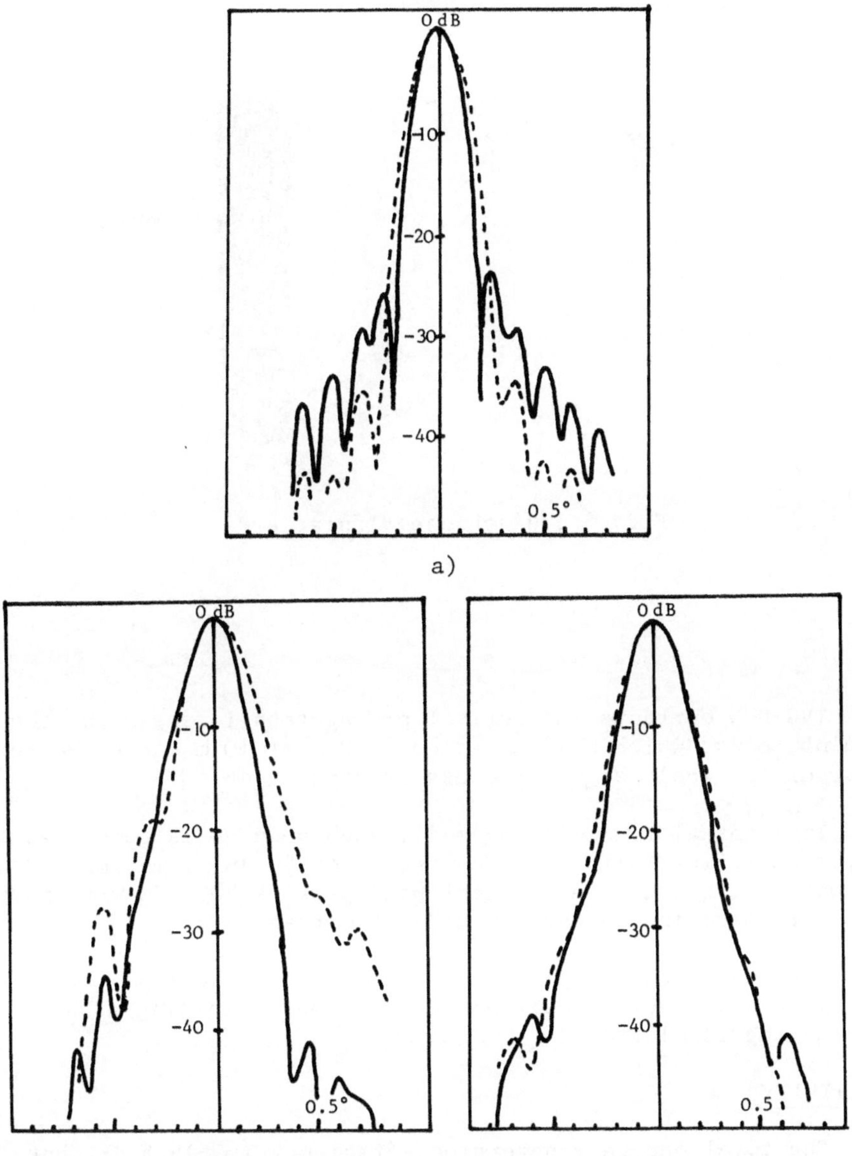

Fig. 4 : Directivity patterns of the lens (dotted line, with apodizing screen)
a) Axial directivity pattern
b) 7.5° off-axis directivity pattern (meridional plane)
c) 7.5° off-axis directivity pattern (sagital plane)

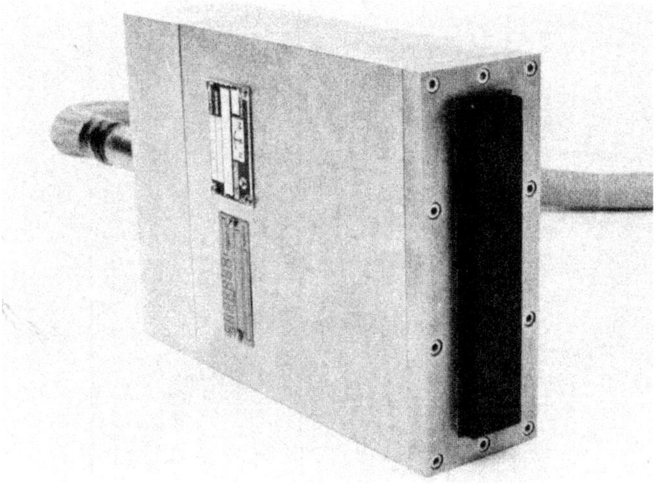

Fig. 5 : 64 channel linear array

TVG-DVG corrects theoretical propagation losses along the depth of focus, which maximum value is 2 m at 10 m range. Noise factor of the whole signal process is about 9 dB.

In internal transmitting mode, each channel is connected on a power amplifier (with power control over 50 dB) ; maximum power is 1 Watt per channel. Acoustical band-pass is 200 kHz with possibility of operating in pure or modulated frequency.

A key particularity of the processing is the capability of forming an adjustable receiving temporal gate yielding a sonar-like ranging resolution.

SENSITIVITY

The total one-way conversion efficiency is -19.5 dB, but losses could be easily reduced by 10 dB by the optimization of the different sections. Noise power in 400 kHz is about -116 dB/W. Consequently the minimum detectable acoustic intensity is $2.5\ 10^{-9}$ W/cm^2 on the array or 4.10^{-13} W/cm^2 on the lens.

IMAGING OF ROUGH OBJECTS

Consider a given plane surface, plane wave insonified, with low roughness in comparison to the wavelength. It can be shown that image intensity is separable in two parts called "coherent" and "uncoherent" part respectively identified with specular and diffused field.

With Kirchoff approximation[5], and if the object is in the nearfield of the imaging system (Fresnel distance of the imaging pupil) we obtain the following relations for the averaged image intensity :

$$\langle I_{coh}\rangle = G^{-2} \cdot I_o \cdot P^2(K\bar{a}, K\bar{b}) \cdot F \qquad (1)$$

$$\langle I_{inc}\rangle = G^{-2} \cdot I_o \cdot K^2 \cdot \bar{c}^{-2} \cdot S(K\bar{a}, K\bar{b}) * P^2(K\bar{a}, K\bar{b}) \cdot F \qquad (2)$$

I_o : incident plane wave intensity

G : magnification

$P(K\bar{a},K\bar{b})$: pupil function of the lens

S : spectral density of the surface roughness

$\bar{a},\bar{b},\bar{c}$: sum of direction cosines of incident and reflexion direction

K : wave number

Where: $F = R^2\, e^{-K^2\bar{c}^2\, 2\sigma_z^2}$ with :

σ_s : RMS height of the surface

R : plane wave reflexion coefficient.

* indicates the covolution product

For a surface which has a correlation function which is gaussian and stationnary, we obtain :

$$S(K\bar{a}\ ,\ K\bar{b}) = 2\pi\, \sigma_z^2\, L^2\, e^{-K^2L^2(\bar{a}^2 + \bar{b}^2)}$$

If the correlation distance L is large in comparison to the resolution of the imaging system, $S(K\bar{a}, K\bar{b})$ acts as a Dirac for the convolution product. The diffused echo cannot be distinguished from the specular reflexions, so it is theoretically impossible to separate specular and diffused field of objects where surface correlation distances are larger than one resolution cell, e.g. if $L^2 \gg \lambda^2\, d^2/S_L$, where λ is the wavelength, d the object distance and S_L is the surface of the entrance pupil of the imaging system. If d is 5 m, λ = 0.75 mm, S_L = 700 cm^2, the diffused field will only be imaged if the correlation distance is less than 1.4 cm. In the opposite case, if $L^2 \ll \lambda^2\, d^2/S_L$, Eq. (2) reduces to a spatial filtering with a bandpass given by the entrance pupil :

$$\langle I_{inc}\rangle = \frac{F}{2\pi} \cdot \frac{S_L}{d^2}\, K^4\, \bar{c}^{-2}\, \sigma_z^2\, L^2\, e^{-K^2L^2(\bar{a}^2 + \bar{b}^2)} \cdot I_o$$

And the diffused echo can be detected out of the specular spot which is contained in the $P(K\bar{a}, K\bar{b})$ non-zero value domain.

According to the effective sensitivity and the propagation losses we expect to observe diffused scattering with 10 m range objects insonified with 4 W/m^2, provided that $\sigma_z L > 10^{-8}$ i.e. assuming σ_z and L are of the same order, objects of 0.1 mm roughness amplitude. This intensity can be achieved on the 100 x 64 resolution cells with a transmitting power of about 10 W. It is therefore theoretically possible to observe diffused echoes in standard manner on relatively smooth surfaces.

For the imaging system objects cannot be considered as a plane when :

$$R_c \lambda << \frac{d^2 \lambda^2}{S_L}$$

where R_c is the local radius of curvature. This case makes easier the diffused echo observation provided $L^2 < R_c \lambda$, since the reduction of dynamic range from specular to diffused echo is : $(d^2\lambda/R_c S_L)^2$, and the extend of the specular spot is also reduced. However the diffused scattering expression is unchanged and observable incidences are still dependent on the spectral density S(Ka,Kb). The lower the ratio L/λ, the lower the directivity of the scattered field, and for certain type of surfaces grey scale should be observable and lead to optical-like imaging.

We have formed an acoustical imaging of a plate with RMS roughness of about 1/10 mm (corundum grains bounded on stainless steel plate). As we can see in Fig. 6, this plate is visible over a large range of incidences, and the specular spots do not "pollute" the whole picture. At 70° of incidence the SNR at the output of

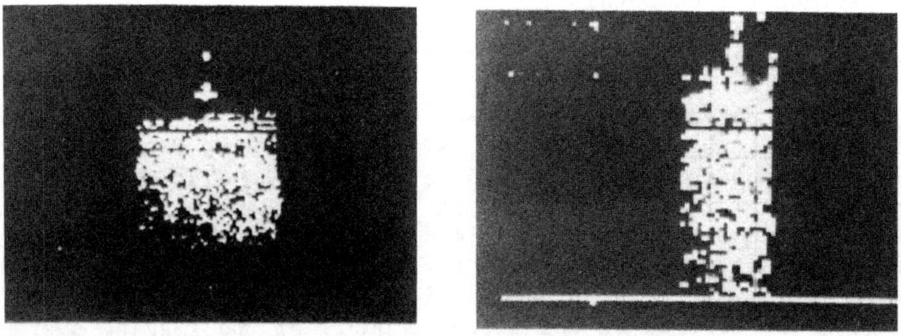

a) b)

Fig. 6 : Rough plate
a) Normal incidence b) 70° incidence

preamplifiers is still about 12 dB, with an insonification power of 0.2 W per array transducer (internal transmission mode), in agreement with the evaluation of Eq. (2).

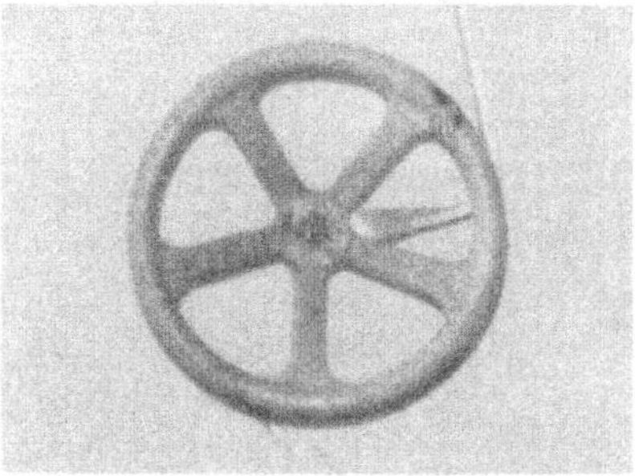

a) Optic image

b) Distance 3 m
Normal incidence

c) Distance 3 m
45° incidence

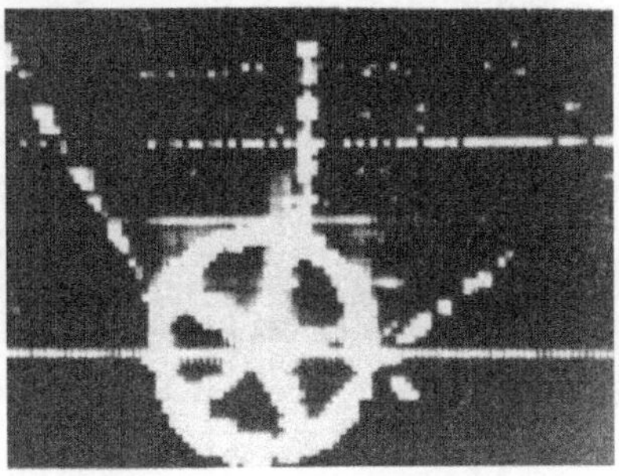

d) Distance 2.2 m
normal incidence

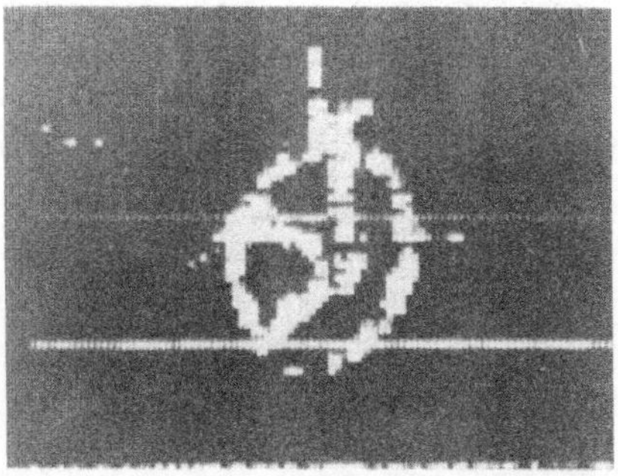

e) Distance 2 m, 45° incidence

Fig. 7 : Imaging of 0.2 m diameter wheel

The pictures in Fig. 7 are acoustical images of a painted steel wheel, the roughness which has not been evaluated. An important part of the object is visible over many incidence angles, although the major part of the information is obviously due to bright points of specular origin.

The effect of the side lobe level appears to be less restrictive than that of the SNR, as it is proved by comparing the images with and without apodizing screen. Near-in side lobes 20 dB down from the main lobe performed a sufficiently scatter isolation, provided farther outside lobes are rapidly decreasing (case of the lens used without its apodizing screen).

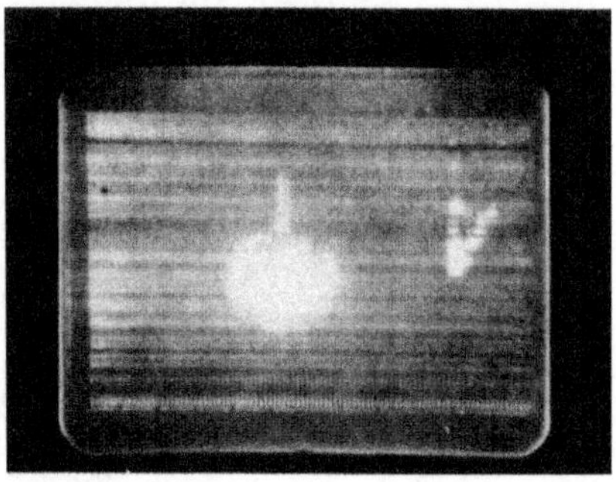

Fig. 8
0.15 m diameter smooth sphere - distance 2.5 m

Fig. 8 shows an obtained image of a 15 cm diameter solid aluminium sphere with a very smooth surface (RMS roughness less than 0.01 mm). Gradual decreasing of intensity from the centre to the edge of the image is observed and the object is easily recognized. However this good result is rather surprising since Eq. 2 predicts a diffused level lower than the thermal noise of the system. A possible explanation of this unexpected high quality imaging is the effect of surface wave re-radiation of the whole surface of the sphere.

Further analysis of the back-scattering of such smooth surfaces should be performed to give quantitative evaluation. In most cases, the quality of the images has not been improved by using the 10 % F.M capability of the apparatus. Keeping in mind the practical constraints of the field-operation of this sort of imaging system (sources rather grouped and far from the objects), it seems difficult to perform a strong decorrelation of the insonification.
The choice of the transmission modes (external or internal) has a minor influence on picture aspect (excepted of course when positioning a lot of projectors around the object).

Finally, considerable improvement in object pattern recognition is provided by the use of depth resolution. This "sonar like" processing, which has no equivalent in optics, is an important compensation of the difficulties encountered in this type of front plane imaging system.

CONCLUSION

Initial results demonstrate the possibility of obtaining diffused images of objects with roughness very much lower than the wavelength and for low insonification power (a few watts). This is due to the high sensitivity of the imaging device.

Image interpretation is made easier by the capability of compressing or centring the dynamic range and by that of a sonar-like depth resolution.

The sensitivity could be 10 dB increased, mainly using recently developed higher efficiency array.

The principal constraint of the system is the use of a mechanical scanning (oscillating mirror), which limits the frame rate to about 1 image/second. Electronically scanned two-dimensional arrays are now in study in order to avoid these difficulties. Other experiments are now intended with the aim to determine limits of smooth hard surface imaging.

ACKNOWLEDGMENT

The authors thank Mr. Lagier for the design and realization of the lens, Mr. Defranould who is at the origin of array design and Mr. Guthmann for the electronic processing realization.

REFERENCES

[1] N.L. Warner "An ultrasonic Imaging sonar" EASCON 74 - IEEE (18 D 173)

[2] C.M. Jones, G.A. Gilmore "Sonic cameras", JASA, Vol. 59, number 1, pages 72-85, (Janv 1967)

[3] A. Mano and K. Nitadori "An experimental underwater viewing systems using acoustical holography" 1977 Ultrasonic Symposium Proceeding IEEE

[4] Thomson-CSF Report DASM 80/06/146 PhD "Etude des réseaux de transducteurs piézoélectriques"

[5] P. Beckman, P. Spizzichino "The scattering of electromagnetic waves from tough surfaces (Chap. 3)" Pergamon Press (1963)

ACOUSTICAL IMAGING VIA COHERENT RECEPTION OF SPATIALLY COLOURED TRANSMISSION

P. Tournois

Thomson-CSF, ASM Division
06802 Cagnes-sur-Mer
France

INTRODUCTION.

If the transducers of an array transmit different signals simultaneously, the insonified space is "coloured" since a particular resulting signal corresponds to each direction. At the receiver's end, classical beamforming is performed as if insonification were isotropic. Next, each formed beam output is fed into the filter matched to the signal transmitted into that particular direction ; a gain in angular resolution results, which is equal to that of the product of the transmit-receive antenna patterns although no directional transmission was required.

1. COLOURED TRANSMISSION.

Let us assume N transducers E_j., j = 1 to N, located at points T_j, in a 3-dimensional space, and simultaneously transmitting N different signals $C_j(t)$, non intercorrelated (Fig. 1).

This set of signals covers a frequency band B around a centre frequency f_o.

Each point P_i in the space receives a signal $S(P_i)$ which is the superposition of all the signals C_j at this point, i.e. :

$$S(P_i) = \sum_{j=1}^{N} a_{ij}\, C_j\,(t - \tau_{ij}) \qquad (1)$$

where a_{ij} is the amplitude of signal C_j at point P_i and τ_{ij} the time delay between E_i and P_i.

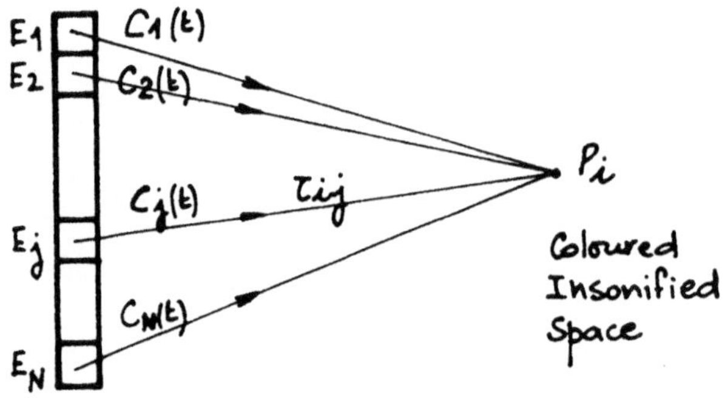

Fig. 1
Principle of coloured transmission

If for simplification sake, we further assume that all the signals are transmitted at the same amplitude level, and that propagation loss is independent of C_j at point P_i, it follows that the $S(P_i)$ is a strong function of :

- the nature and the number of the transmitted codes C_j
- the geometry of the transmitter array
- the co-ordinates of P_i.

Insonification of the space is therefore strongly anisotropic and it can be said that the space is "coloured" by the set of transmitters E_j.

If θ,φ,d are polar co-ordinates of point P_i w.r.t. origin O, $S(P_i)$ will be :

- a function of θ and φ in the far-field

$$S(P_i) = S_i(\theta,\varphi) \quad (2)$$

- a function of θ,φ and distance $d = OP_i$ in the near-field

$$S(P_i) = S_i(\theta,\varphi,d) \quad (3)$$

2. THE SIGNALS TO BE TRANSMITTED.

The major types of signals C_j, which can be transmitted simultaneously, while covering the bandwidth B, are (Fig. 2) :

- N CW pulses with different frequencies and same duration T, thus each covering a bandwidth $b = 1/T$, such that $B \geqslant Nb$.

- N orthogonal codes, with same duration T, each covering the same bandwidth B (e.g. PSK, PN ...) and such that $BT >> N$

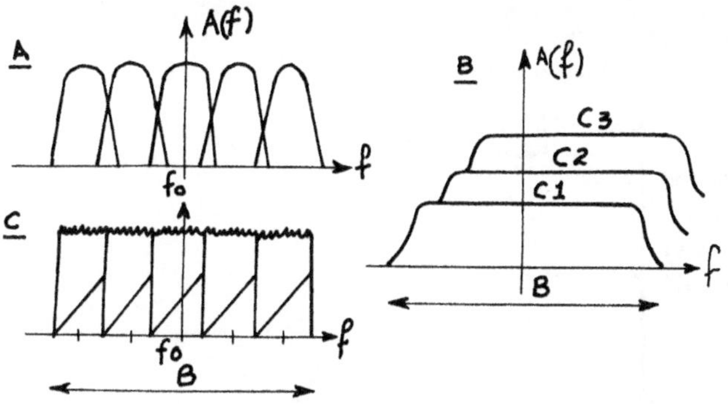

Fig. 2
Types of signals transmitted

- N linear FM ramps with identical slopes and duration hence bandwidth $b \leqslant B/N$ and centred around various frequencies such that $BT \gg N$

3. ECHO RECEPTION BY MEANS OF AN OMNIDIRECTIONAL RECEIVER.

The signal $S(P_i)$ received by P_i is back-scattered by this point and received by an omnidirectional receiver R.

A) To discriminate between the echoes due to the various transmitters E_j, N filters $C_j^*(t)$ matched to the transmitted signals $C_j(t)$, are placed in parallel at the output of R (see Fig. 3A).

For each point P_i, the delay times τ_{ij} are known and the delay time differences introduced by the geometry of transmitters E_j w.r.t. to P_i, can be compensated by means of a delay line network placed at the filter outputs.

In particular, when P_i is in the far-field (see Eq. 2), this delay line network forms the angle resolving beams which could have been formed at the transmitter's end by feeding the same signal into all transmitters E_j.

When P_i is closer to the antenna, the delays can be programmed to account for distance and thus ensure dynamic focusing.

Finally, weights W_j can be easily introduced at the matched filter outputs to reduce the side lobe levels of the angular directivity pattern.

B) In the far-field case, an alternate method to obtain directive beams consists in using filters $S_i^*(\theta,\varphi)$ matched to signals $S(P_i) \equiv S_i(\theta,\varphi)$ which are independent and thus distinguishable through correlation (see Fig. 3B).

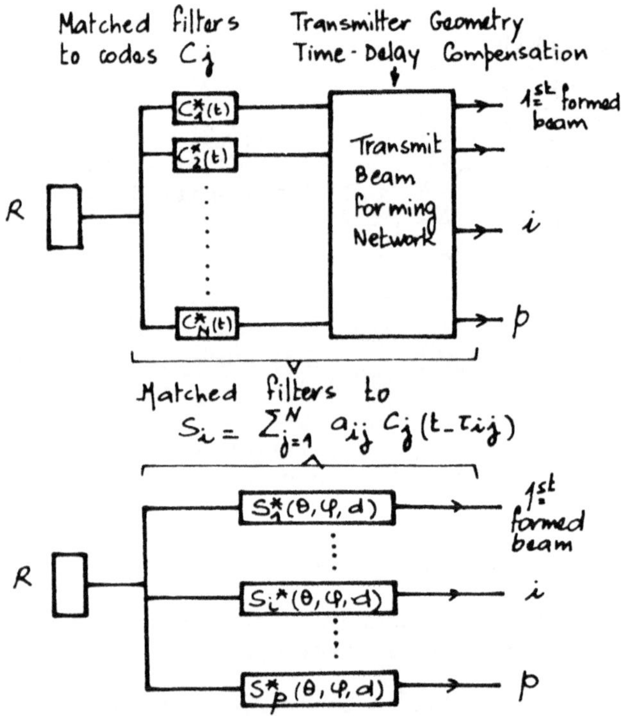

Fig. 3
Reception schemes in the case of M-transmitter one receiver

Dynamic focusing can be performed provided these filters are fast programmable transversal filters with a large enough number of coefficients to track the function of distance $S_i^*(\theta,\varphi,d)$ as time evolves.

In addition, these filters can include a weighting function for side-lobe reduction.

The two methods just described yield rigorously identical results. The delay line network of the former clearly shows the angular directivity function performed, which depends on the transmitting antenna geometry ; whereas the latter method demonstrates that the matched filters compress the back scattered signals $S(P_i)$ with bandwidth B, and as a consequence, the range accuracy is proportional to 1/B.

4. ECHO RECEPTION BY MEANS OF AN M-ELEMENT RECEIVING ARRAY.

When insonification is isotropic -as in the case of single omnidirectional transmitter- and when operating in the far-field, it is well-known and classical to use an M-element receiving array to perform beamforming, and weighting can be introduced to reduce the side lobes of the resulting directivity pattern. The beam directivity then depends on the geometry of the receiving antenna.

In a near-field operation, dynamic focusing may be achieved by using programmable delay lines in the beamforming network.

This beamforming technique, classical in the case of a spatially isotropic transmission, can also be used in our case of coloured transmission : each thus formed beam output, $S(P_i)$ is therefore known since the points P_i must then belong to a particular direction in space.

Therefore, placing the matched filter $S^*(P_i)$ at each of the classical formed beam outputs(see Fig. 4A) yields the same results as those described in the previous paragraph : a range resolution proportional to 1/B because of signal compression, and an angular directivity pattern related to the transmitting array geometry. This latter directivity is identical to that obtained with the block diagram of Fig. 3A or 4B, which is nothing else than a beamforming at the transmitter end.

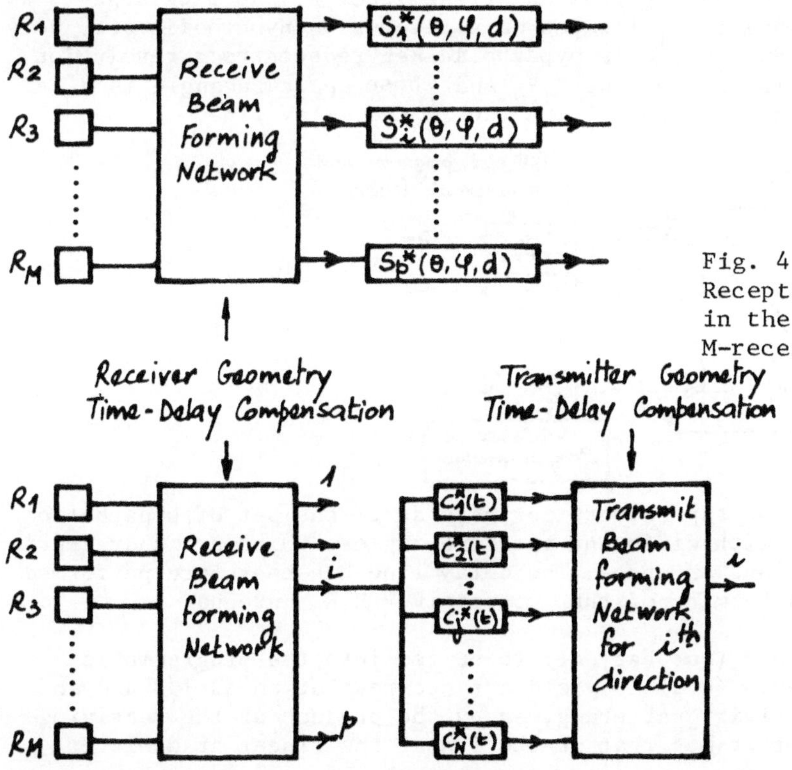

Fig. 4
Reception schemes in the case of M-receivers

With respect to angular resolution, the net result of the association of classical reception beamforming and matched filters $S^*(P_i) = S_i^*(\theta,\varphi,d)$ in the case of coloured transmission, is a product of directivity patterns: the reception angular directivity pattern of the receive antennas times the transmission directivity pattern of the transmit antennas. And this is achieved with no directive transmission required thus allowing for very fast image acquisition.

Three simple examples of application of the coloured transmission principle are now described.

5. TWO-RECEIVER ECHO RECEPTION.

Two receivers R_1 and R_2 separated by a distance L, receive echoes due to the coloured transmission of an N-transducer linear array of length L, whose axis is parallel to R_1R_2.

The output signals from receivers R_1 and R_2 are respectively fed into the end transducers T_1 and T_2 of a surface wave tapped delay line. The two contradirective waves thus generated are tapped at each of the Q taps (Fig. 5). Each tap performs the in-phase summation of echoes coming from any transmitter E_j and back-scattered by the points P_i located on a revolution hyperbolic net. Very far from the antennae this hyperbolic net reduces to a revolution cone whose axis is the line R_1R_2 and whose aperture angle is a function of the location of the tap on the delay line.

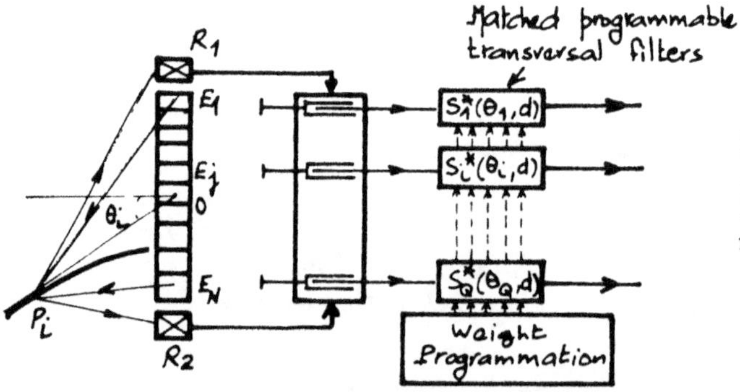

Fig. 5
Application to a 2-receiver, N-transmitter system

The set of tap outputs corresponds to the set of hyperbolic nets or cones with different aperture angles, which are characteristic of directions in space. The delay line has therefore performed classical beamforming of the two-receiver R_1R_2 antenna.

Each tap output has only to be fed into the programmable matched filter $S^*(\theta,d)$ to yield a range resolution as 1/B and an angular directivity pattern given by the product of the receiving antenna pattern times that of the linear the linear transmitter array.

In the present particular case the final directivity function obtained is of the form :

$$\sin \left[\frac{2\pi L}{\lambda} \sin\theta \right] / \left[\frac{2\pi L}{\lambda} \sin\theta\right] \qquad (4)$$

i.e. identical to that of an unweighted receiving antenna with length L.

6. ECHO RECEPTION ON A LINEAR N-RECEIVER ANTENNA WITH LENGTH 2L.

a) Two-transmitter case :

Two transmitters E_1, E_2, located at the ends of the receiving antenna, send two simultaneous linear FM ramps with identical duration T and bandwidth B ; an up-chirp C_1 is sent by E_1 and a down-chirp C_2 by E_2.

Each receiver output R_k is fed into two filters matched to C_1 and C_2 (see Fig. 6) ; the 2N resulting signals are processed in a beamforming network to form beams with a directivity function given by Eq. 4 when no weighting is employed.

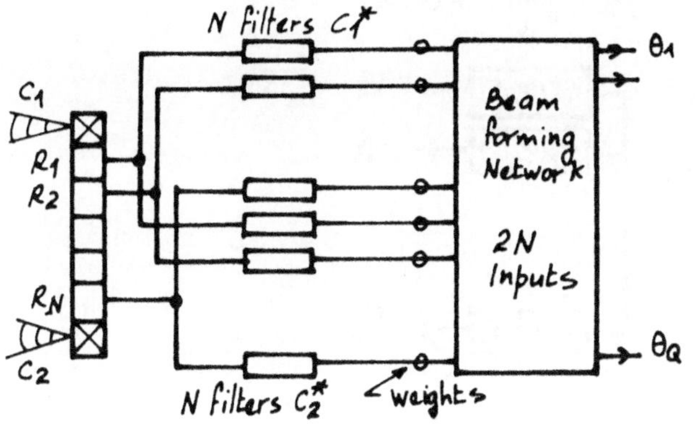

Fig. 6
Application to a N-receiver - 2 transmitter system

To improve side lobe reduction, 2N weight $W_1^{C_1} \dots W_N^{C_1}$, $W_1^{C_2} \dots W_N^{C_2}$ can be inserted on the C_1^* and C_2^* filter outputs.

b) N-transmitter case :

Assuming the N transducers of the receiving antenna are also used to transmit N codes C_i, the block diagram of the full receiver is identical to that shown in Fig. 4A. And the resulting directivity function is of the form (no weight case) :

$$\sin^2\left(\frac{\pi L}{\lambda} \sin\theta\right) / \left(\frac{\pi L}{\lambda} \sin\theta\right)^2$$

since the transmit and receive directivity patterns are identical.

7. THREE DIMENSIONAL IMAGING.

Let us now consider a vertical linear N transmitter array with length L simultaneously transmitting N orthogonal signals ; and a similar receiving array placed horizontally as shown in Fig. 7.

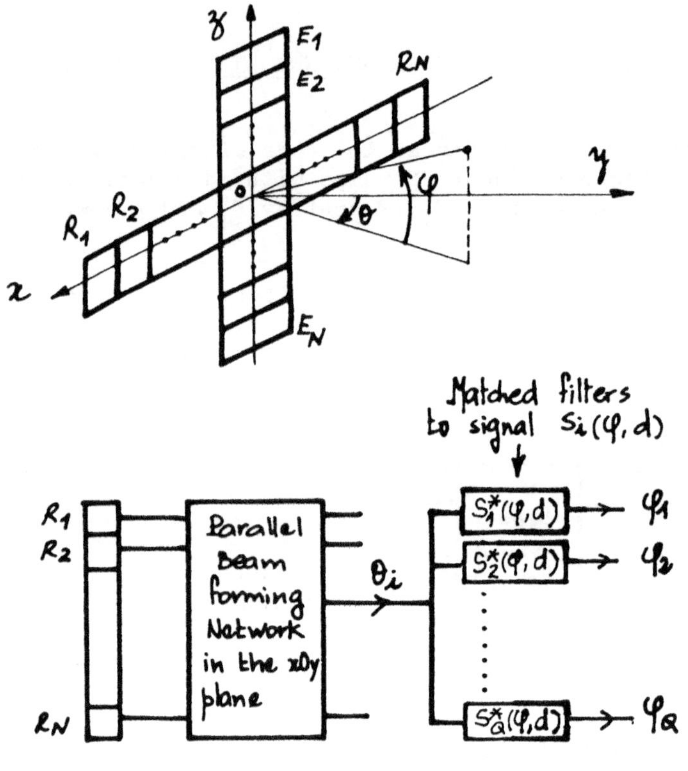

Fig. 7
Application to 3-dimensional imaging

The N- receiver outputs are processed in a first classical beamforming network to form Q beams with high θ-directivity and poor ϕ-directivity (vertically fanned-beams).

The outputs of each of these horizontal beams are fed into a set of Q parallel filters matched to the signals $S(P_i) = S(\varphi, d)$. These signals are indeed very much φ-dependent and θ_i-invariant because the transmit antenna is vertical. Q directions are thus formed in the vertical plane for one bearing direction.

· The net result is again the multiplication of the reception (horizontal plane) and transmission (vertical plane) directivity patterns. But it has not been necessary to transmit a directional beam and change its orientation from line to line. The whole image is obtained in a single shot.

CONCLUSION.

The principle of coloured transmission allows for the multiplication of the directivity patterns of the receive antenna and of the transmit antenna without having to achieve directional transmissions.

When using directional transmissions, it is generally necessary to have them scan the space in order to cover the field of observation, and this considerably reduces the frame rate. A much higher rate is made possible with the spatially coloured transmission scheme.

If classical non-directional transmissions are employed in order to cover the whole field of observation instantaneously, and if one uses directional reception composed of many parallel beams, shifted in angle, the angular resolution only depends on the receive antenna. With the coloured transmission scheme, the final angular resolution is a function of the transmit and receive antennae. When both functions are performed by the same antenna, the angular resolution is improved in nearly a 2:1 ratio and the spurious side lobe level is much decreased because of the multiplication of the transmit and receive directivity patterns.

FLAW DETECTION AND IMAGING BY HIGH-RESOLUTION SYNTHETIC PULSE PULSE HOLOGRAPHY

Helmut Ermert and James O. Schaefer

Institut für Hochfrequenztechnik
Universität Erlangen-Nürnberg
D-8520 Erlangen, West-Germany

INTRODUCTION

Many applications in the field of non-destructive testing are concerned with flaw detection and/or imaging. In such applications conventional holography using acoustic waves has been applied with some success. This technique has some disadvantages which limit the applicability of conventional (monofrequency) holography /1/. Conventional holography has very poor axial resolution, and with optical reconstruction significant image distortion as well as a combination of acoustical and optical speckle noise are seen. Good axial resolution can be obtained using wide-band holography and with computer reconstruction the problem of laser generated optical speckle in optical reconstruction is eliminated.

A synthetic-pulse holography technique is proposed here which offers several advantages over conventional acoustical holography. Good lateral resolution is obtained by using a bandwidth with a large center frequency as well as a large synthetic aperture /2/. Good axial resolution is obtained by synthesizing a pulse of short duration /3/. Results are given of images made of artificial defects in steel.

PRINCIPLES OF THE IMAGING CONCEPT

Referring to figure 1 we have an aperture from which an image area is to be viewed. The larger this aperture is the better the lateral resolution is (for a given range). A transmitter and receiver are moved across this aperture. The transmitter illuminates the image area with a quasi-CW signal (70 - 100 milliseconds in duration). The receiver,which has very low electrical and acoustical cross-coupling is used to pick up all reflected echos. In the following discussion we will consider the special cases of the transmitter coincident with the receiver (negligible spatial seperation) and cross-sectional imaging only.

Let us represent the transmitted and received signals as follows:

$F_{TR}(x_A, \omega) \longleftrightarrow f_{TR}(x_A, t)$: transmitted signal (1)

$F_R (x_R, \omega) \longleftrightarrow f_R (x_A, t)$: received signal (2)

$f_{TR}(x_A, t)$ and $f_R (x_A, t)$ represent the complex time domain Fourier transformation of F_{TR} and F_R

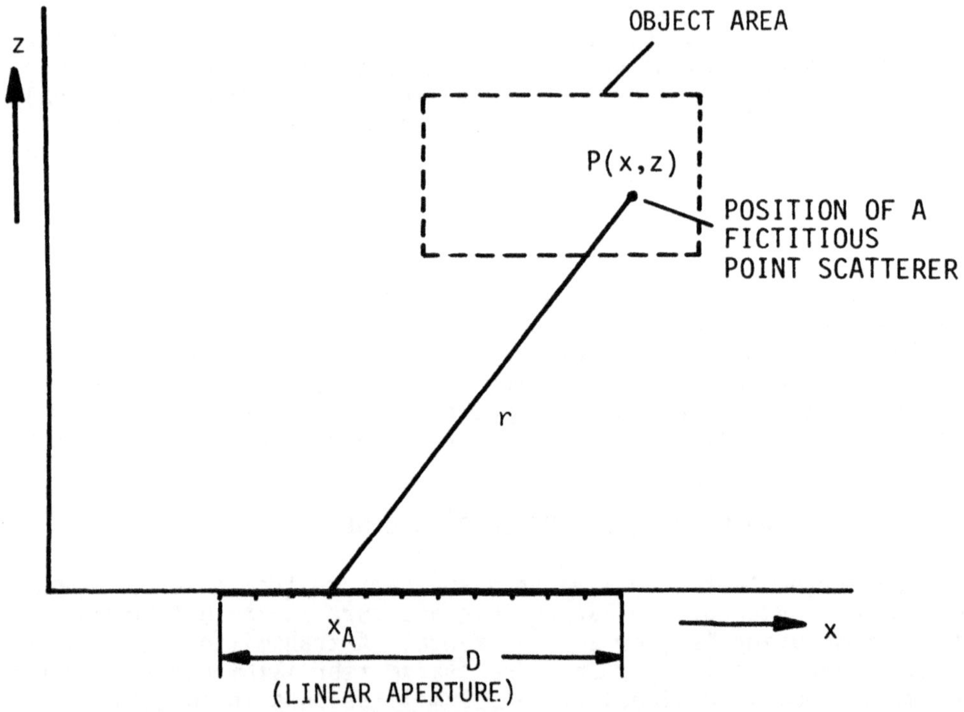

FIGURE 1: Geometry of reconstruction

The backscattered signal f_F from a stationary ficticious scattering point in the image area will be an exact replica of the time domain representation of the transmitted signal. It, however, will be delayed by the roundtrip time of flight to and from the point P(x,z)

$$f_F(x_A, t, x, z) = f_{TR}(x_A, t-\tau) \tag{3}$$

where $$\tau = \frac{2r}{c} = \frac{2}{c}\sqrt{(x-x_A)^2 + z^2} \tag{4}$$

A factor of $1/r^2$ is neglected as well as factors from dispersive and attenuating media. The transducer directivity can also be included into this term. These factors can be all easily added for a more exact solution. These extra terms will only complicate the analysis, and the fundamental principle will be less obvious. So for the sake of simplicity, we will assume non-dispersive and non-attenuating media. We will also assume that the image area is far enough away from the aperture so that the $1/r^2$-term is essentially constant and the transducer illuminates the image area over the whole aperture.

It is now possible to construct an image function from f_R and f_F. The image intensity at a point (x,z) is given by:

$$I(x,z) = \left| \int_{x_A}\int_{-\infty}^{+\infty} f_R(x_A, t) \cdot f_F^*(x_A, t, x, z)\, dt \cdot dx_A \right| \tag{5}$$

$$I(x,z) = \left| \int_{x_A}\int_{-\infty}^{+\infty} f_R(x_A, t) \cdot f_{TR}^*(x_A, t-\tau)\, dt \cdot dx_A \right| \tag{6}$$

This technique of image construction is similar to the cross-correlation technique used in other quasi-holographic reconstructions /5/. It should be noted that in this analysis all constant factors have been suppressed, since they represent nothing more than an intensity scaling factor in the final image.

The above equation can be transformed back into the frequency domain as follows:

$$F_F(x_A, \omega, x, z) = e^{-j\omega\tau} F_{TR}(x_A, \omega) \tag{7}$$

this follows since f_F is nothing more than a time shifted version of f_{TR} and thus using Parseval's theorem /4/ we can transform equation (6) into:

$$I(x,z) = \left| \int_{x_A} \int_{-\infty}^{\infty} F_R(x_A,\omega) \cdot F_{TR}^{*}(x_A,\omega) e^{j\omega\tau} d\omega dx_A \right| \qquad (8)$$

This is the frequency domain version of the reconstruction.

MULTIFREQUENCY HOLOGRAPHY (SYNTHETIC PULSE CONCEPT)

In the case of bandlimited signals it is necessary only to integral over the band of transmitted signals. Since the transmitted signals were discrete frequencies and were not varied continously, the intergral over frequency becomes a discrete summation from ω_1 to ω_{NF}. Thus

$$I(x,z) = \left| \int_{x_A} \sum_{n=1}^{NF} F_R(x_A, \omega_n) \cdot F_{TR}^{*}(x_A,\omega_n) e^{j\omega_n\tau} dx_A \right| \qquad (9)$$

At this point we still have the choice of which frequencies and amplitudes to transmit. The decision about the frequencies and their corresponding weighting is what determines the synthetic pulse. We may thus represent the synthetic pulse (in the time domain) as a discrete Fourier transform of the transmitted set of frequencies.

$$f_{TR}(x_A, t) = \sum_{n=1}^{NF} W_n(x_A) e^{j\omega_n t} \longrightarrow F_{TR}(x_A, \omega) \qquad (10)$$

where W_n is a complex weighting function over the transmitted frequency spectrum, which is used for pulse shaping.

Similary the received signal can be represented as:

$$f_R(x_A,t) = \sum_{n=1}^{NF} A_n(x_A) e^{j\omega_n t} \longrightarrow F_R(x_A,\omega) \qquad (11)$$

where A is a complex value representing the phase and amplitude of the measured signal.

The image function (9) can be written as:

$$I(x,z) = \left| \int_{x_A} \sum_{n=1}^{NF} A_n(x_A)\cdot W_n^*(x_A) e^{j\omega_n \tau} \, dx_A \right| \tag{12}$$

Since the aperture is not continuously sampled but rather discretely sampled, the integral over x_A become a summation.

Equation (16) can then be rewritten as:

$$I(x,z) = \left| \sum_{\ell=1}^{NXA} \sum_{n=1}^{NF} A_{n\ell} \cdot W_{n\ell}^* e^{j\omega_n \tau_\ell} \right| \tag{13}$$

where NXA is the total number of aperture points sampled. The time delay becomes

$$\tau_\ell = \frac{2}{c} \cdot \sqrt{(x-x_{A\ell})^2 + z^2} \tag{4a}$$

The index ℓ denotes the discrete aperture position.

We can also view equation (13) as a discrete Fourier transform of f_R weighted by the function W. So we can rewrite (13) as:

$$I(x,z) = \left| \sum_{\ell=1}^{NXA} f_R' \left(x_{A_\ell}, t = \tau_\ell\right) \right| \tag{14}$$

in the case where the transmitted spectrum is a rectangle function (that is all transmitted amplitudes are equal), then the function f_R' is equal to f_R and the image is nothing more than the summation of the complex valued A-scans evaluated at τ_ℓ roundtrip time of flight. This is equivalent to image reconstruction in the time domain.

MONOFREQUENCY HOLOGRAPHY

In monofrequency holography only one frequency is transmitted at each position. This results in an image which has very good lateral resolution but very poor axial (longitudinal) resolution /6/, /7/. The mathematics of monofrequency reconstruction is as follows:

$$f_{TR}(x_{A_\ell}, t) = e^{j\omega_1 t}$$ transmitted (single frequency) signal (15)

$$f_R(x_{A_\ell}, t) = A_{1_\ell}\, e^{j\omega_1 t}$$ received signal (16)

from equation (13) we have:

$$I(x,z) = \left|\sum_{\ell=1}^{NXA} A_{1_\ell}\, e^{j\omega_1\, \tau_\ell(x,z)}\right| \tag{17}$$

This reconstruction results in very poor axial resolution due to the fact that only one frequency is used.

RANDOM FREQUENCY HOLOGRAPHY

A considerable saving in computing time can be realized if we transmit from each aperture point one and only one frequency. However, if we change the frequency from position to position in a random fashion we will be able to maintain reasonable axial resolution characteristic of a broadband system.

The analysis goes as follows. If we transmit a single frequency from aperture point x_A then the signal can be represented as

$$f_{TR}(x_{A\ell}, t) = W_{n(\ell)}\, e^{j\omega_{n(\ell)} t} \tag{18}$$

The received signal is thus

$$f_R(x_{A\ell}, t) = A_{n(\ell)}\, e^{j\omega_{n(\ell)} t} \tag{19}$$

where $n = n(\ell)$ is a random number

ℓ is the aperture position index (1 to NxA)

$\omega_{n(\ell)}$ is a randomly chosen frequency within the transducer bandwidth.

Using the same approach as with the multifrequency evenly spaced frequency case, we have:

$$I(x,z) = \left| \sum_{\ell=1}^{NXA} A_{n(\ell)} \, C^{*}_{n(\ell)} \, e^{j\omega n(\ell)\, \tau_{\ell}(x,z)} \right| \tag{20}$$

Notice that this last image function is a sum over only the aperture points since only one frequency was used at each aperture position. This technique does not incorporate nearly the amount redundancy as in the case of multifrequency evenly spaced frequency so as a result the background noise is somewhat higher. However the reduction in computation time by a factor of NF (the number of frequency) must be considered when deciding which algorithm to use.

RECONSTRUCTION ALGORITHM

The reconstruction algorithm goes as follows. A set of complex valued A-scans is created by fast Fourier transforming the received data at each aperture point. This set of complex A-scans can be thought of as a time matrix, each column corresponding to a particular aperture point and each row corresponding to a time of flight. This time matrix will contain values only at times corresponding to the time of flight from the aperture point to a reflecting object and back. The time of flight to a point in the object area is calculated from each aperture point. This time of flight corresponds to a complex amplitude in the time matrix. These complex values are added together (one value for each aperture point). The total image is formed by performing this process over a grid of values which samples the object area. The values are then plotted in a two dimension grid with the grey scale indicating the image intensity.

The calculation of the times of flight involve the computation of square roots. This is a fairly time consuming calculation and in order to minimize reconstruction time a modified lookup table concept is used.

"FLEXIBLE" LOOKUP TABLE

In order to do the image reconstruction in a reasonable amount of time it is necessary to minimize the number of calculations required. One way of doing this is to calculate one time a lookup table of all values used in the program and instead of calculating them each time, they are read into memory in large groups. This concept can reduce reconstruction time by nearly two orders of magnitude. The problem is that a versatile lookup table requires a huge amount of memory capacity (several megabytes).

We have therefore settled on a compromise between calculating all values each time for reconstruction and calculating none and using only a lookup table. If the sample spacing in the object area is an integer multiple of aperture sample spacing, then it will be necessary to calculate considerably fewer square roots. This results in a computation time reduction by a factor of (1/NX + 1/NXA) where NX is the total number of points per line in the object area; NXA is the number of measurement points in the aperture. This flexible lookup table gives a line by line reconstruction which offers a considerable savings in time over straight forward calculation and a considerable saving in memory capacity required for a stored lookup table.

SAMPLING OF DATA

Before the data is taken, the question of how much data to be recorded in order to reconstruct a reasonable image must be addressed. The image quality is a function, of course, of how much data is taken. A compromise is often necessary then to decide what the minimum amount of data is, which will still result in an acceptable image.

FREQUENCY DOMAIN SAMPLING

The measured data, consists of a complex spectrum in the frequency domain. This spectrum is sampled (in the frequency domain) and is Fourier transformed to form a discrete time sequence. The total bandwidth of the spectrum ΔF determines the transformed pulse duration (from a point scatterer). This bandwidth determines the axial resolution of the system. If the sampled frequencies are equally spaced apart by Δf, then this spacing Δf determines the degree of ambiguity in the final image. This ambiguity arises from the periodicity characteristic of a discrete Fourier transformation. The ambiguities appear as "ghosts" in the image. In order to have an image area free of ghosts the Δf should be chosen as:

$$\Delta f \leqq \frac{2}{TMAX - TMIN} \tag{21}$$

where TMAX and TMIN are the maximum and minimum times of flight from the aperture to the object area.

In the case where random frequencies are chosen within the bandwidth of the transducer, the problem of ambiguities is less pronounced, because the ghosts are also randomly distributed throughout the image area and do not superimpose each other. They do however add a component to the background noise level of the image. This addition background noise must be considered in the light of considerably reduced computation time due to spare frequency sampling.

TIME DOMAIN SAMPLING

The final image is a complex superposition of a sequence by Fourier transformation of a bandlimited signal. This transformation is at discrete time points. If these time points are to be on accurate representation of that actual sample without aliasing, then samples in the time domain must be separated by no more than:

$$\Delta t = \frac{1}{2\, f_{max}}$$

where f_{max} is the highest frequency component in the transmitted signal.

APERTURE AND OBJECT AREA SAMPLING

It is necessary to chose the size of the aperture and the number of measurement points within it. The aperture size is determined by the directivity pattern of the transducer. It makes no sense to have a large aperture and not be able to illuminate the object area from every point in the aperture. Some criterion must be used to limit the aperture size, such as the 3 db beamwidth of the transducer.

The separation between measurement points in the aperture is determined by the highest spatial frequency in the aperture. This is determined by the velocity of sound in the material, the frequency of operation, and the angular separation of the aperture from the object area. If we sample the aperture at twice the highest spatial frequency we should (by sampling theory) be able to reproduce the distribution in the aperture. Thus the separation between aperture samples is given by:

$$\Delta xA = \frac{1}{2}\, \frac{C}{f_{max} \cdot \cos\theta_{min}} \qquad (22)$$

where c is the sound velocity, f_{max} is the maximum frequency transmitted, θ_{min} is the minimum viewing angle from the aperture to any object point.

Sampling in the object area is related to the time domain sample interval. The time domain signal samples are separated by Δt. This corresponds to a sample spacing in the object area of

$$\Delta t \cdot c \cdot \cos\theta_{min} = \Delta x \text{ (object sample separation in the x direction)} \qquad (23)$$

$$\Delta t \cdot c \cdot \sin\theta_{min} = \Delta z \text{ (object sample separation in the z direction)} \qquad (24)$$

EXPERIMENTAL SETUP

In this set up (see Fig. 2) the controller used for data acquisition was a HP 9825 A. It controlled the frequency and amplitude on an HP 8165 programmable oscillator. The transducer phase and amplitude were continuously monitored on an HP 3575 phase-gain meter. The output of the phase gain meter was digitized and stored on an floppy disc system. The data was then processed on Telefunken TR 440 computer and output was either a printed output or a 16 grey scale TV monitor output.

The transducers used were Krautkrämer 2 MHz 45° shear wave transducers. They were operated from 1.5 MHz to 3.5 MHz, and oil coupled to the surface of the steel. The transducer position was varied along the steel under computer control using a digital positioner. The orientation and relative positions are shown in Fig. 3.

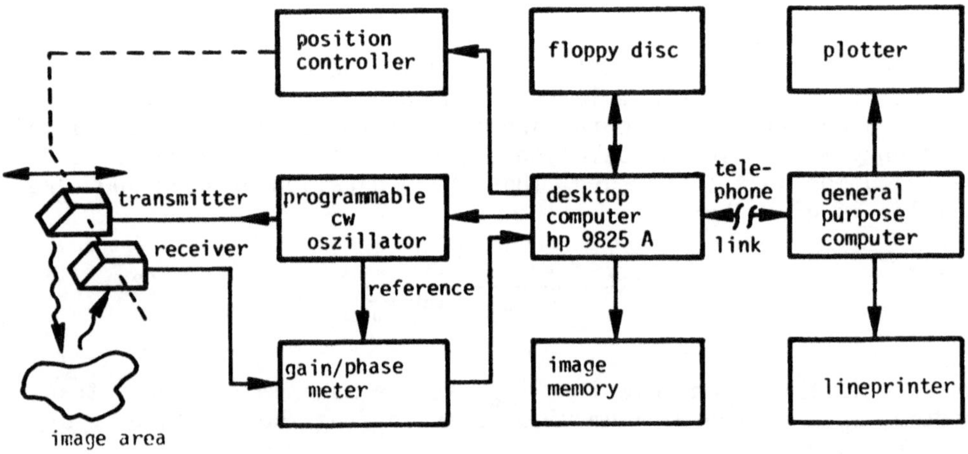

FIGURE 2: System block diagram

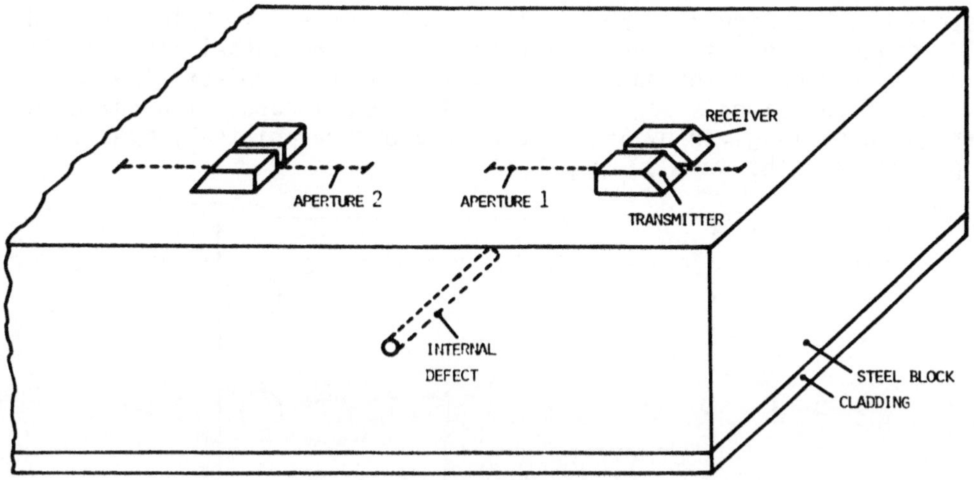

FIGURE 3: Transducer and aperture orientation

RESOLUTION

An analysis of resolution of multifrequency holography system was done in /6/. The results will be shown but not derived. The lateral resolution is given by:

$$\delta_L \simeq 0.82 \ (z \cdot \lambda) \ /D \qquad (25)$$

where D = aperture size

z = depth of object area (center)

λ = wavelength of center frequency signal

The axial resolution is given by:

$$\delta_A \simeq 0.42 \cdot c/\Delta F \qquad (26)$$

where: c = speed of sound in the medium

ΔF = bandwidth of the transmitted signals

RESULTS

The following figures show the images of artificially generated defects in steel. The defects were holes (or triangles) bored through steel blocks. Figure 4 shows a sketch of four - 2 mm holes drilled 50 mm deep into a section of a reactor pressure vessel. This scale of this sketch is the same as all images of this object. The bandwidth of the signals was 1.5 to 3.5 MHz in all cases. The number of frequencies is given by NF, number of aperture points by NXA. The aperture lengths were 30 mm.

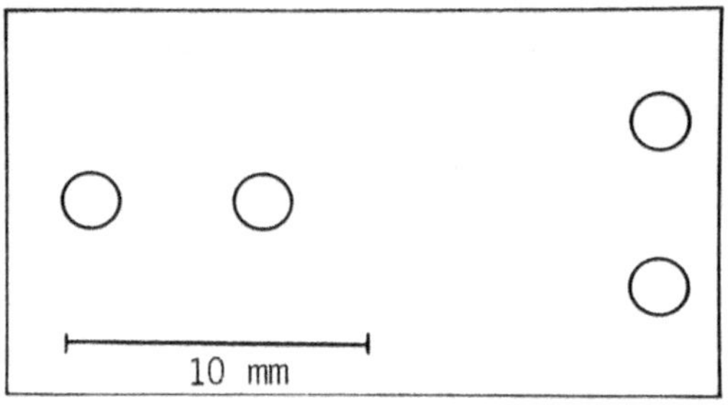

Fig. 4 Sketch of 4 - 2 mm dia holes drilled into a steel block, figures 5 - 8 are images of this object

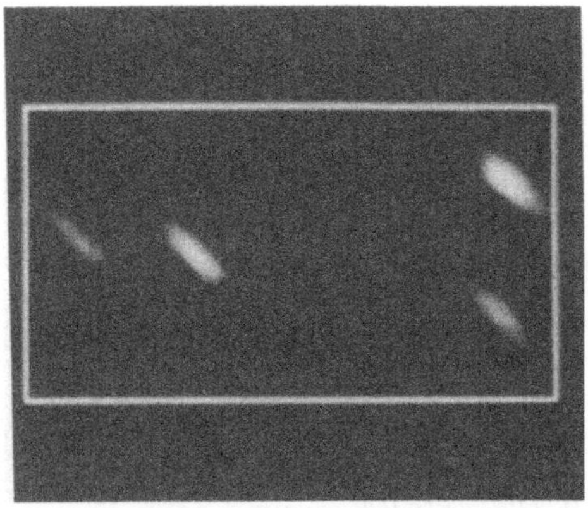

FIG. 5: Images made with NF = 64, aperture 1 (NXA = 53)

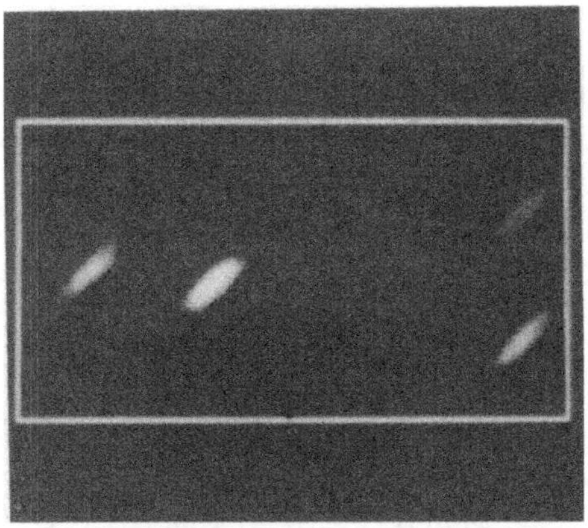

FIG.6: Image made with NF = 64, aperture 2 (NXA = 53)

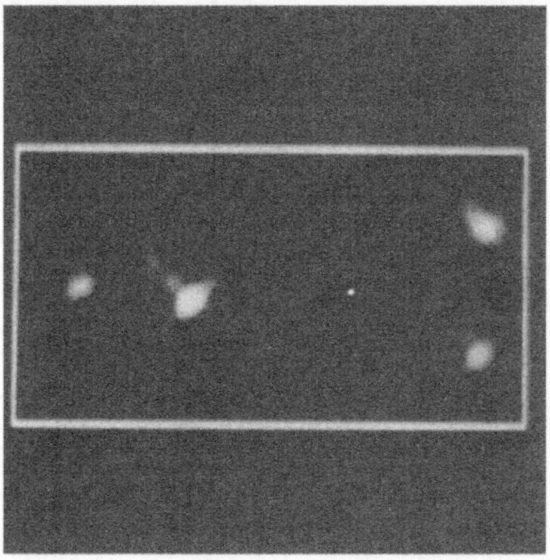

FIG. 7: Image made with NF = 64, aperture 1 (NXA = 53) and aperture 2 (NXA = 53)

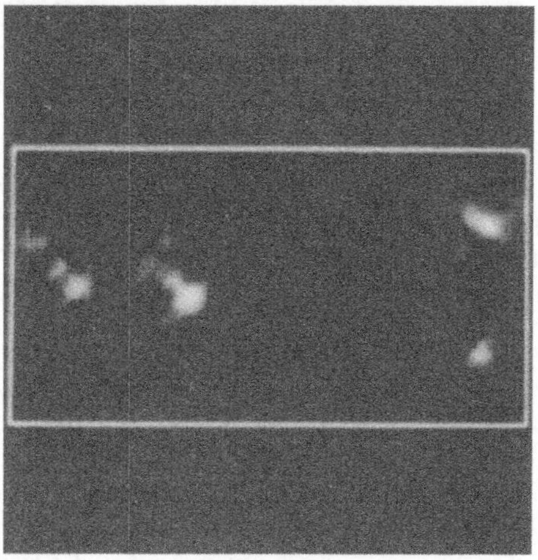

FIG. 8: Image made with NF = 1, frequencies chosen randomly in ΔF aperture 1 (NXA = 371) and aperture 2 (NXA = 371)

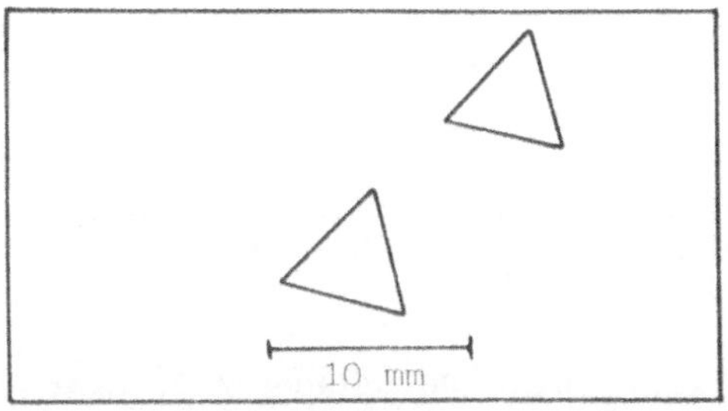

FIG.9: Sketch of triangles bored in steel

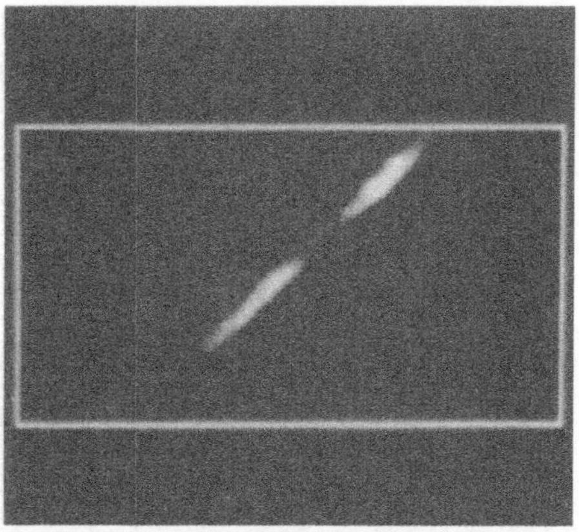

FIG. 10: Image of fig. 9 with NF = 64, aperture 2, NXA = 53.

CONCLUSION

A theoretical derivation was given describing the principle of operation of a multifrequency holography system. Using this concept synthetic pulses can be formed by altering the spectrum of a transmitted set of signals. A flexible lookup table concept was used to speed up the reconstruction. Using these algorithms images were generated. This technique is a quasi-CW technique and no adverse effect were seen due to standing waves or back wall reflections.

This work was supported by the Gesellschaft für Reaktorsicherheit, D-5000 Köln, Federal Republic of Germany (BMFT-Project 150 422).

BIBLIOGRAPHY

/1/ Ahmed, M., Wang, K., Metherell, A., Holography and its Application to Acoustic Imaging. Proceedings of IEEE, 67:4, pp. 466 - 483, (April 1979).

/2/ Frederick, J., Sydel, J., Fairchild, R., Improved Ultrasonic Nondestructiv Testing of Pressure Vessels, Yearly Progress Report, Aug 1, 1974, Nuclear Regulatory Commission NUREG 0007-1, Washington, D.C. (1974).

/3/ Sato, T., Ikeda, O., Ohshima, H., Fujikura, H., A Few Effective Pre-processings in Synthetic Aperture Sonar System, Acoustical Holography ed., Kessler, vol. 7, pp 569 - 582 (1977).

/4/ Papoulis, A., Signal Analysis, McGraw Hill, New York (1977).

/5/ Leith, E., Quasi-Holographic Techniques in the Microwave Region, Proceedings of the IEEE, 59:9, pp 1305 - 1318 (Sept 1971).

/6/ Ermert, H., Karg, R., Multifrequency Acoustical Holography, IEEE Transactions on Sonics and Ultrasonics, SU - 26:4, pp 279 - 286 (July 1979).

/7/ Erhard, A., Wüstenberg, H., Kutzner, J., The accuracy of Flaw Size Determination by Ultrasonics, British Journal of Non-Destructive Testing, pp 39 - 43 (Jan 1979).

/8/ Dick, M., Dick, D., McLeod, F., Kindig, N., Ultrasonic Synthetic Aperture Imaging, Acoustical Holography, ed. Kessler, Vol. 7, pp 327 - 346 (1977).

/9/ Duck, F., Johnson, S., Greenleaf, J., Samayoa, W., Digital Image Focussing in the Near Field of a sampled Acoustic Aperture, Ultrasonics, p 83 - 88 (March 1977).

A THREE-DIMENSIONAL SYNTHETIC FOCUS SYSTEM

K. Liang, B. T. Khuri-Yakub, C-H Chou, and G. S. Kino

Edward L. Ginzton Laboratory

Stanford University, Stanford California 94305

ABSTRACT

Synthetic focus imaging techniques suitable for reconstructing 3-D acoustic images of flaws inside silicon nitride are described. A 50 MHz imaging system consisting of a precision scanner, a microcomputer controller, and a minicomputer image processor has been developed for this purpose. A square synthetic aperture is used to image flaws in flat disc samples and a cylindrical synthetic aperture is used in the cylindrical rod case. We have developed the theory to predict the imaging performance of the two aperture geometries. The respective Point Spread Functions are simulated and agree well with theoretical results. Special attention is given to reconstructing images of specular reflectors. Computer simulations based on theoretical flaw models have been carried out.

INTRODUCTION

The principle and operation of two-dimensional synthetic focus systems have been well demonstrated.[1-5] Three-dimensional imaging is a logical extension of the two-dimensional schemes to attain good definitions in all three directions. Whereas the two-dimensional synthetic focus methods can be likened to the action of a physical cylindrical lens, the three-dimension approach is analogous to the operation of a physical spherical lens.

In our present application, we are interested in reconstructing three-dimensional acoustic images of flaws inside silicon nitride samples. Silicon nitride is a ceramic which has become increasingly important for use as a structural material in recent years. It is light yet extremely tough and can withstand very high temperatures. It has a wide range of applications including turbine blades, rotors, ball bearings, etc. However, a basic problem is that ceramics are brittle. Relatively small flaws can critically affect the mechanical strength of the materials. Some common flaws found in silicon nitride are carbon, silicon, iron, boron and nitride inclusions, and voids. They affect the fracture characteristics of the host material in different ways. Some are apparently innocuous, but most are detrimental. Thus, it is necessary to determine the presence, size, and nature of single internal flaws as small as 100 μm .[6] Three-dimensional synthetic focus imaging techniques are employed to accomplish the task. The reconstructed image essentially gives the reflectivity distribution of the sample examined, from which one can determine the locations and sizes of the defects. The image also contains artifacts characteristic of the nature of the defect being imaged.

For obvious reasons, the geometry of the synthetic aperture has to be tailored to that of the sample. Since the choice of aperture geometry governs the characteristics of an imaging system, silicon nitride samples of different geometric shapes have to be studied individually. We have confined our attention to two specific geometries: (a) flat disc samples 2.5 cm diameter by 0.5 cm thick; and (b) cylindrical rod samples 0.635 cm diameter and 10 cm long. Samples with the latter geometry are of special interest because they can be mounted in a mechanical testing system and stressed to fracture so as to verify predictions based on our NDE measurements. The longitudinal wave velocity in ceramic materials is on the order of 10 km/sec , so it is necessary to operate in the 50 to 100 MHz frequency range to obtain the resolution required. First, a sample is immersed in a water bath and scanned in C-scan reflection mode to locate the transverse positions and depths of the flaw sites. Then a sampled aperture is synthesized over a relatively small volume of interest around the flaw, as depicted in Figs. 1a and 1. This operation limits the number of reconstruction points required to a reasonable value. In our imaging system, a single 50 MHz focused transducer operated in pulse-echo mode is used. The focal point of the incident beam is located at the water/silicon nitride interface to approximate a point source and also, because of the large velocity mismatch at the interface, to create a wide angle transmitted beam insonifying the interior of the sample. For the flat disc geometry, pulse-echo data is collected over a square 8 x 8 element aperture with sampling point spacing of two wavelengths. For the cylindrical case, an open-ended sampled cylindrical aperture is used. It consists of a stack of sixteen rings, two wavelengths apart, each ring having 32 evenly distributed sampling points. Image reconstruction is accomplished by back-projecting the pulse-echo data on a minicomputer. The back projection procedure is illustrated in Figs. 2a and 2b.

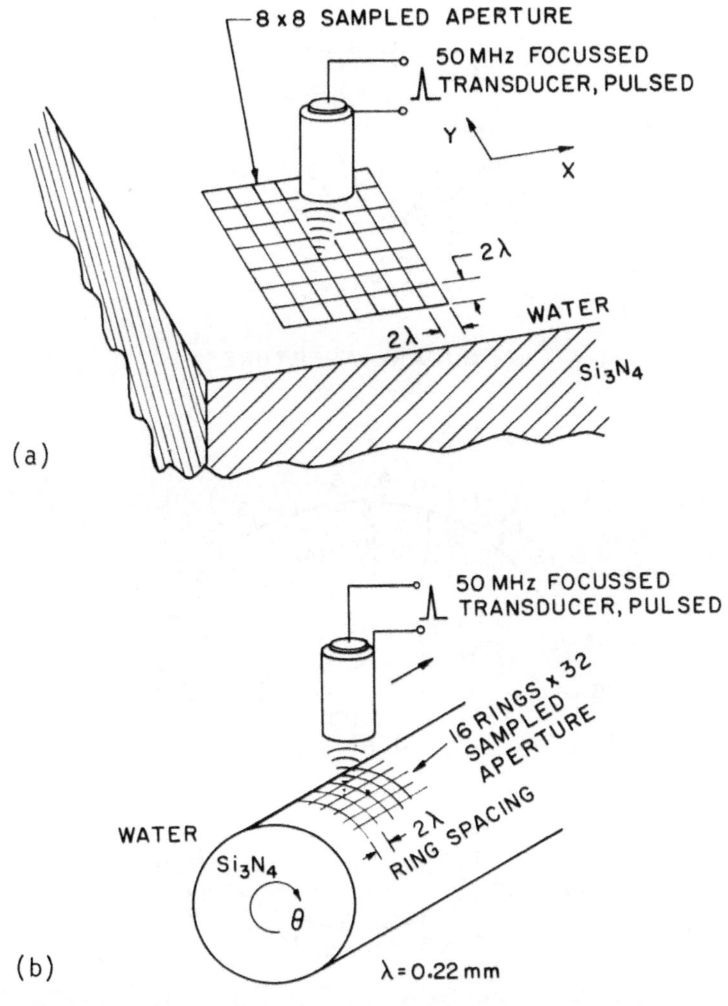

Fig. 1. Imaging flaws inside silicon nitride samples using synthetic aperture techniques: (a) flat disc samples; and (b) cylindrical rod samples.

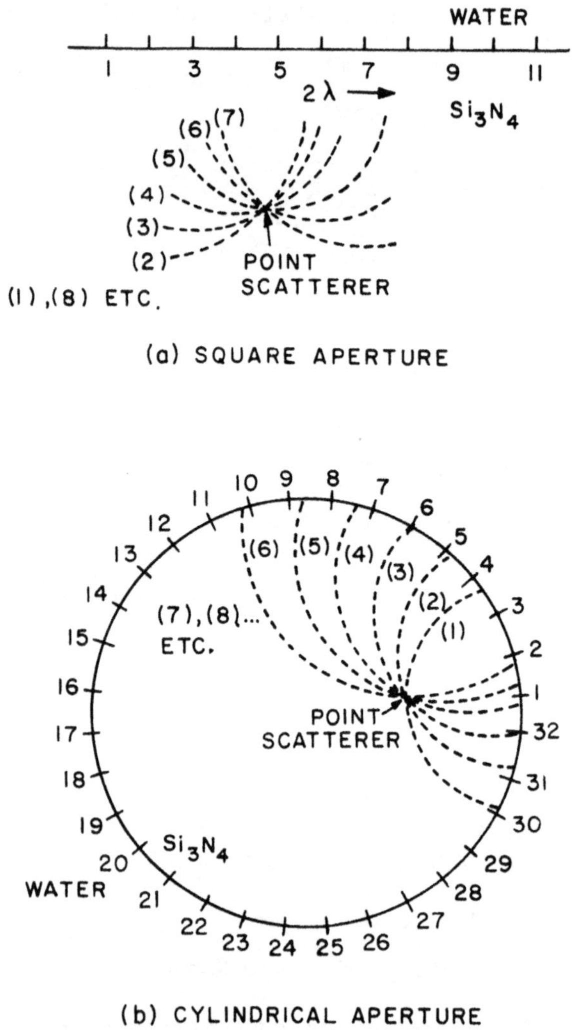

Fig. 2. Image reconstruction by back-projection.

We have carried out theoretical analyses and computer simulations to gauge the performance of the two aperture geometries. Much effort was devoted to establishing the respective Point Spread Functions (PSF), the conventional way of specifying the resolving power of an imaging system. Inherent in the PSF concept is the assumption that a finite size reflector is regarded as a collection of point scatterers; thus, every point in the aperture responds to every point in the reflector. This assumption is invalid for strong specular reflectors, as is the case in imaging finite size flaws inside silicon nitride samples. This fact is illustrated in Fig. 3 where we have a two-dimensional fully enclosed aperture imaging a finite size circular defect. Based on the geometric optics argument, a sampling point on the aperture tends to respond only to a point on the defect boundary directly facing it. In reconstructing the image point by back-projection, because of the finite width of the transmitted pulse (typically a few RF cycles), a limited number of sampling points in the vicinity of the actually excited aperture element will contribute constructively. Outside of this neighborhood, the contributions due to the other sampling points are either negative or random. Hence 100% coherence summing at an image point is not possible, giving rise to undesirably high sidelobe levels. We concluded that the connection between the PSF and the "resolution" of specular reflector images is not clear.

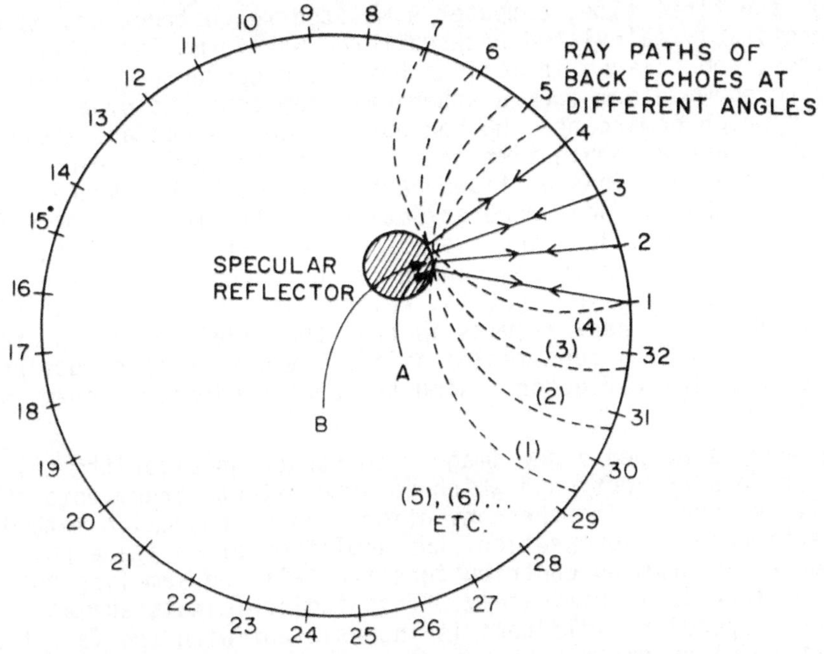

Fig. 3. Imaging specular reflector by synthetic focus technique.

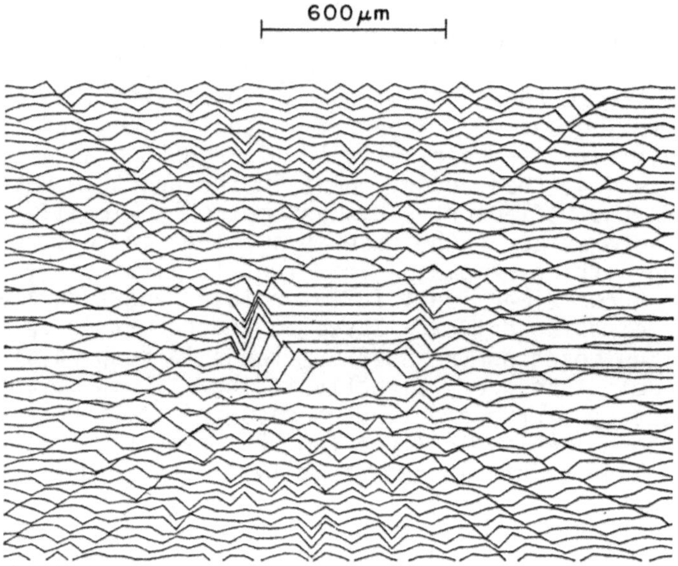

(a) VOID

For the first time, computer simulated reconstructions based on theoretically calculated backscattered waveforms for spherical void, iron, and silicon inclusions have been conducted. A typical backscattered waveform from a spherical flaw consists of a front-face echo which demarcates the boundary of the defect and trailing echoes which are a combination of back-face echo and mode-converted waves.[7] Different kinds of flaws have distinct backscattered signals. Synthetic focus reconstruction results in an image which shows the defect boundary and the associated "ring" artifact, corresponding to later echoes, which is characteristic of the nature of the flaw, as shown in Figs. 4a, 4b, and 4c. However, a significant drawback to these results is that the sidelobe levels are very high. This is a fundamental difficulty with reconstructing images of specular reflectors using the back-projection scheme we have described.

We have developed a new image reconstruction algorithm called "Selective Back-Projection," which to some extent circumvents the difficulties inherent in the conventional back-projection method. In conventional back-projection, the amplitude of an image point is evaluated by summing contributions from all the sampling points in the aperture indiscriminately. This indiscriminate summation in the case of specular reflectors is inconsistent with the fact that an object point on the defect boundary is apparently "visible" to only a localized group of sampling points in the aperture. Thus,

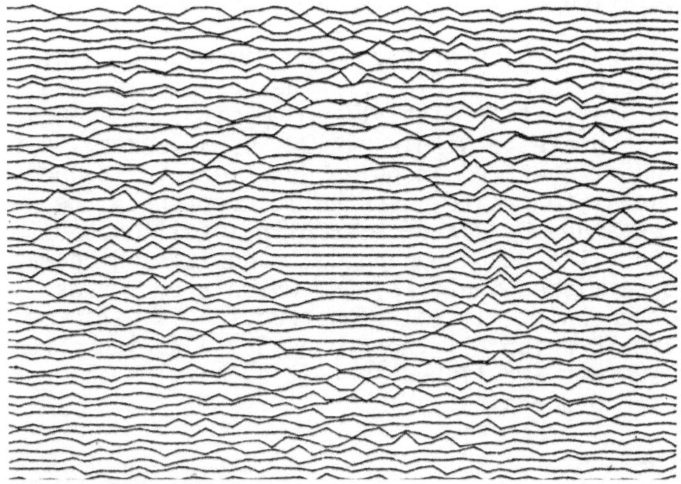

(b) SILICON INCLUSION

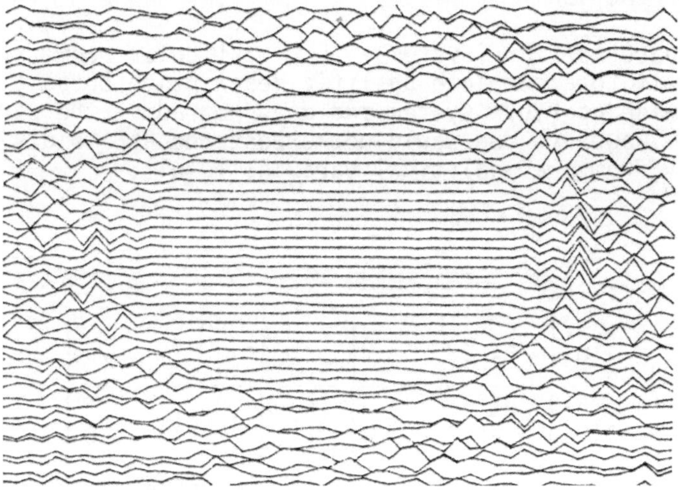

(c) IRON INCLUSION

Fig. 4. Computer simulation images of different types of circular defects inside silicon nitride: (a) void; (b) silicon inclusion; and (c) iron inclusion.

the obvious strategy is to only sum sampling points that contribute constructively and deliberately leave out sampling points that do not matter. There is also the necessity to distinguish between random and meaningful contributions, which is discussed in detail below. Based on computer simulation results, there is a dramatic improvement in the images generated by selective back-projection over those by conventional back-projection.

HARDWARE DESCRIPTION

The functional blocks of the three-dimensional synthetic focus imaging system are shown in Fig. 5. The scanning mechanism consists of a precision X-Y stage with ±7 μm positional accuracy and a rotator mount accurate to within 1/100 of a degree. A single 50 MHz transducer operated in pulse-echo mode is scanned over the sample of interest to create a synthetic aperture. The pulse-echo data is time expanded by use of a sampling oscilloscope to facilitate data collection. The scanning and data acquisition operations are coordinated by a microcomputer system which also serves as an off-line data storage unit. Image reconstruction is performed on a mini-computer system in which all the processing algorithms reside. The two computer systems are linked so that data files can readily be retrieved for processing.

THEORY OF OPERATION

Point Spread Function

(1) Square Aperture System. This particular geometry does not lend itself to a complete analytic description because it is basically a wide aperture system and does not admit the paraxial approximation. However, one would not expect the PSF to deviate significantly from the analytical paraxial result:[8]

$$H(x,y) = \frac{\sin\left[\pi \frac{2D}{\lambda z}(x - x_0)\right] \sin\left[\pi \frac{2D}{\lambda z}(y - y_0)\right]}{\sin\left[\pi \frac{2\ell}{\lambda z}(x - x_0)\right] \sin\left[\pi \frac{2\ell}{\lambda z}(y - y_0)\right]} \qquad (1)$$

where D is the width of the square aperture, ℓ is the element spacing in x and y, and (x_0, y_0, z) is the location of the point target. The 4 dB lateral single point definition is therefore

$$d_x = d_y = \frac{\lambda z}{2D} \qquad (2)$$

Note that the resolution is twice as good as a receiver array focused on a point insonified by a separate source because in a synthetic aperture system the signal path is twice the distance from an array element to an object point.

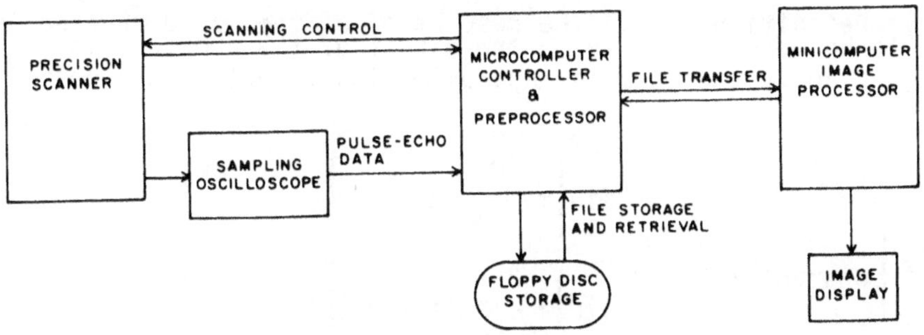

Fig. 5. Three-dimensional synthetic focus imaging system.

Range resolution is primarily determined by the pulse width. The typical impulse response of the transducer used in the imaging system approximates a Gaussian envelope RF pulse four and a half cycles long.

Since the aperture is undersampled by a factor of 4 , grating lobes are expected to occur where

$$\Delta x = n(N-1)\frac{z\lambda}{2D} \tag{3}$$

and

$$\Delta y = m(N-1)\frac{z\lambda}{2D} \tag{4}$$

$n, m = 1, 2, 3, \ldots$, where $N \times N$ is the number of elements in the sampled square aperture. However, because short pulses are used, there is no longer coherent summing at locations far removed from the point target. Thus, the grating lobe levels are reduced by approximately M/N, where M is the number of RF cycles in the pulse.

(2) Cylindrical Aperture System. Again we do not attempt to derive the exact expression for the PSF of the system. Rather, we divide the difficult overall problem into two much simpler, mathematically tractable, albeit non-exact, parts. Through this exercise, we can estimate the resolution of the system. The validity of the approximation can easily be checked by computer simulation. We will consider the resolution in the z direction and that in the (r,θ) plane separately.

(a) Definition in z. The resolution of the z direction is assumed to be dependent only on the extent of the aperture in z. Further, we apply the paraxial approximation to obtain the definition in z

$$H(z)\Big|_{\text{constant } r,\theta} = \frac{\sin\left[\pi \frac{2D_z}{\lambda R}(z - z_0)\right]}{\sin\left[\pi \frac{2\ell_z}{\lambda R}(z - z_0)\right]} \tag{5}$$

where D_z is the width of the aperture in z, ℓ_z is the element spacing in z, and R is the distance of the point target from the aperture. The 4 dB single point definition in z is therefore

$$d_z = \frac{R\lambda}{2D_z} \tag{6}$$

Also, since the system is undersampled in z by a factor of 4, grating lobes are present. However, the use of short pulses suppresses the grating lobe levels. The aberration in the grating lobes due to the non-paraxial nature of the focusing at the grating lobe locations also serves to reduce their amplitudes.

(b). Definition in the (r,θ) plane. The PSF in the constant z plane is assumed to depend only on angular (θ) distribution of the aperture and the derivation is given as follows. Consider imaging a point scatterer located at (r_0,θ_0) using a continuous aperture of radius a, as shown in Fig. 6. The image obtained by back-projection can be represented mathematically by

$$H(\vec{r}) = K \int_0^{2\pi} \frac{F\left[\frac{2(R_0 - R)}{v}\right] \exp\left[j\omega \frac{2(R_0 - R)}{v}\right]}{(RR_0)^{1/2}} \, d\phi \tag{7}$$

$$R = \left[r^2 + a^2 - 2\,ar\cdot\cos(\phi - \theta)\right]^{1/2} \tag{8}$$

$$R_0 = \left[r_0^2 + a^2 - 2\,ar_0\cos(\phi - \theta_0)\right]^{1/2} \tag{9}$$

where $F(t) \exp(j\omega t)$ is the transmitted pulse, a is the radius of the aperture, ϕ is the angular position of an infinitesimal array element subtending an angle $d\phi$ at the center of the aperture, K is the amplitude scale factor, and v is the velocity of the medium. Assuming that $F(\mu) = 1$ at $\mu = 0$ and $F(t)$ is several cycles long, we can write

$$H(\vec{r}) = K \int_0^{2\pi} \frac{\exp\left[j\omega \dfrac{2(R_0 - R)}{v}\right]}{(RR_0)^{1/2}} d\phi \tag{10}$$

In order to evaluate Eq. (7), we make use of the relation valid for $a > r$ and any value of r,[9]

$$H_0^{(2)}(2kR) = J_0(2kr)\, H_0^{(2)}(2ka) + 2\sum_{n=1}^{\infty} J_n(2kr)\, H_n^{(2)}(2ka) \cos n(\phi - \theta) \tag{11}$$

where $k = \omega/v$. Assuming $ka >> 1$, the $H_n^{(2)}$'s can be approximated by their asymptotic forms. Therefore, we obtain the following relations

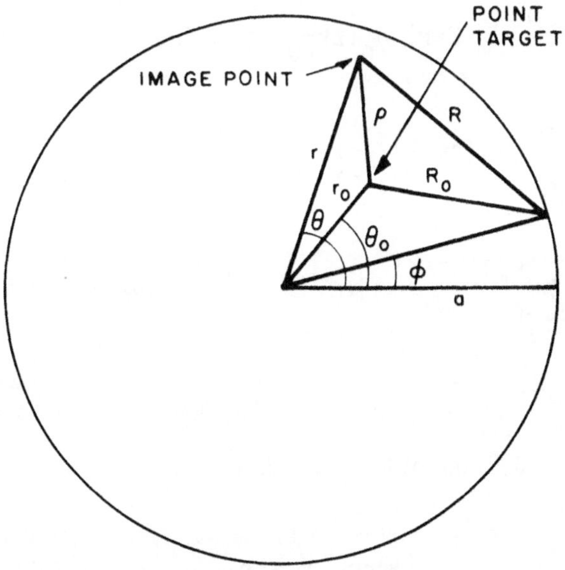

Fig. 6. Imaging a point target using a circular aperture.

$$\frac{a^{1/2}\, e^{-2jkR}}{R^{1/2}} \simeq J_0(2kr)\, e^{-2jka} + 2\sum_{n=1}^{\infty} e^{j(n\pi/2)} J_n(2kr)\, e^{-2jka} \cos n(\phi - \theta) \tag{12}$$

and

$$\frac{a^{1/2}\, e^{2jkR_0}}{R_0^{1/2}} \simeq J_0(2kr_0)\, e^{2jka} + 2\sum_{n=1}^{\infty} e^{-j(n\pi/2)} J_n(2kr_0)\, e^{2jka} \cos n(\phi - \theta_0) \tag{13}$$

It follows that

$$\begin{aligned} \frac{a e^{2jk(R_0 - R)}}{(RR_0)^{1/2}} = {} & J_0(2kr)\, J_0(2kr_0) + 4\sum_{m=1}^{\infty}\sum_{n=1}^{\infty} e^{j\left[(m - n)\pi/2\right]} J_m(2kr)\, J_n(2kr_0) \\ & + 2J_0(2kr)\sum_{n=1}^{\infty} e^{-j(n\pi/2)}\, J_n(2kr_0)\, e^{2jka} \cos n(\phi - \theta_0) \\ & + 2J_0(2kr_0)\sum_{n=1}^{\infty} e^{j(n\pi/2)}\, J_m(2kr)\, e^{-2jka} \cos m(\phi - \theta) \end{aligned} \tag{14}$$

Since

$$\int_0^{2\pi} \cos m(\phi - \theta) \cos n(\phi - \theta_0)\, d\phi = \begin{cases} 0 & \text{when } m \neq n \\ \pi \cos n(\theta - \theta_0) & \text{when } m = n \end{cases} \tag{15}$$

Therefore

$$\int_0^{2\pi} \frac{e^{2jk(R_0 - R)}}{(RR_0)^{1/2}}\, d\phi \simeq \frac{2\pi}{a} J_0(2kr_0)\, J_0(2kr) + 2\sum_{n=1}^{\infty} J_n(2kr_0)\, J_n(2kr) \cos n(\theta - \theta_0) \tag{16}$$

It is known that[9]

$$J_0(2k\rho) = J_0(2kr_0)\, J_0(2kr) + 2\sum_{n=1}^{\infty} J_n(2kr_0)\, J_n(2kr) \cos n(\theta - \theta_0) \tag{17}$$

where ρ is the distance between $\vec{r}$ and $\vec{r}_0$, i.e.,

$$\rho = \left[r^2 + r_0^2 - 2rr_0 \cos(\theta - \theta_0)\right]^{1/2} \tag{18}$$

Thus, the PSF of a continuous cylindrical aperture at the constant z focal plane reduces to the simple result

$$H(r,\theta) = \frac{2\pi}{a} KJ_0(2k\rho) \tag{19}$$

The Rayleigh definition d_R is therefore given by

$$d_R \simeq \frac{0.6\lambda}{\pi} \simeq 0.2\lambda \tag{20}$$

Hence, in principle, a remarkable resolution of 0.2λ can be attained. The reasons for this surprisingly good theoretical definition are two-fold. Firstly, a cylindrical system instead of a spherical one is being considered. Therefore, the first zero of the J_0 rather than the J_1 Bessel function dictates the Rayleigh definition. This gives an improvement by a factor of 1.6 but with higher sidelobe levels and thus a relatively poorer two-point definition. Secondly, because in a synthetic aperture system any ray suffers twice the phase shift that it would in a single lens system, there is a factor of 2 improvement in definition.

Also note that in the expansion of the $H_n^{(2)}$'s in Eqs. (12) and (13) we only use the large ka assumption. Therefore, the PSF expression given by Eq. (19) is valid for any general point target location (r_0, θ_0).

(3) Effect of Sampling and Grating Lobes. To consider the effect of sampling the aperture, assume we have N evenly spaced sampling points. We introduce the aperture function

$$A(\phi) = \sum_{n=0}^{N-1} \delta(\phi - \phi_n) \tag{21}$$

where

$$\phi_n = 2n\pi/N \tag{22}$$

Note that the period of the function is $2\pi/N$. Therefore, Eq. (21) can be expressed as a Fourier series.

$$A(\phi) = \frac{1}{2\pi} \sum_{m=0}^{\infty} \exp(jmN\phi) \tag{23}$$

With the effect of sampling taken into account, Eq. (6) can be rewritten in the form

$$H(\vec{r}) = \frac{K}{2\pi} \int_0^{2\pi} \frac{F\left[\frac{2(R_0 - R)}{v}\right] \exp\left[j\omega \frac{2(R_0 - R)}{v}\right]}{(RR_0)^{1/2}} \sum_{m=0}^{N-1} e^{jmN\phi} \, d\phi \tag{24}$$

Following similar arguments to those already given with $F(u) = 1$, the PSF of a sampled cylindrical aperture in the constant z focal plane is given by the relation

$$H(r,\theta) = \sum_{m=0}^{\infty} J_{mN}(2k\rho) e^{-j\psi_m} \tag{25}$$

where ψ_m is a phase factor. The maximum of the leading term, J_0, gives the main lobe. The other terms account for the presence of grating lobes due to sampling. The nearest grating lobe is approximately given by the first maximum of $J_N(2k\rho)$ for large N. We have found empirically from numerical tabulation of Bessel functions that this occurs where

$$\rho \approx \frac{N + 2}{4\pi} \lambda \tag{26}$$

Equivalently, the first grating lobes are located $.4(N + 2)dR$ from the main lobe. Compared with a rectilinear imaging system with the same Rayleigh definition, the grating lobes are closer to the main lobe but have lower amplitudes. Specifically, for a 32-element cw system, the first maximum of J_{32} is where

$$\rho = 2.71\lambda \tag{27}$$

and has a value 14.4 dB below the main lobe.

IMAGING SPECULAR REFLECTORS

Consider, for instance, a spherical defect of radius b, at the center of an array of radius a, $a >> b$. Suppose we are trying to obtain an image of a point P on the defect boundary, a distance R from the synthetic aperture array, as shown in Fig. 7. When a transducer is opposite the point P at the point A on the aperture, the response is maximum. Similarly, when the transducer is moved to point B, a point Q on the radial line passing through B and the center Q of the circle has maximum response. If the system is focused on the point P, the signal arriving at B from Q has the wrong time delay. The error in time delay is

$$\Delta T = \frac{2(BP - BQ)}{v} \tag{28}$$

where v is the velocity in the medium. Let angle POQ be θ. Since a or $R >> b$, it can be shown that

$$\Delta T \simeq \frac{4a}{v} \sin^2 \frac{\theta}{2} \tag{29}$$

If we regard the sampling points as essentially continuous, the total contribution to the received signal is

$$\Phi = \int_0^{2\pi} \exp \left[j\omega \frac{4a}{v} \sin^2 \frac{\theta}{2} \right] d\theta \tag{30}$$

where we have ignored the amplitude variation due to the change in signal path length. For θ small, Eq. (8) becomes a Fresnel integral. It can be shown that the main contribution to this integral is approximately from the region where

$$4ka \sin^2 \frac{\theta}{2} < \frac{\pi}{2} \tag{31}$$

where $k = \omega/v$. This corresponds to contributions having the same sign. The elements making the main contribution to the image point P are within the angular range

$$-\frac{1}{2}\sqrt{\frac{\pi}{2ka}} < \sin\frac{\theta}{2} < \frac{1}{2}\sqrt{\frac{\pi}{2ka}} \tag{32}$$

Points outside this range may give in-phase or out-of-phase contributions.

As an example, in silicon nitride, at a frequency of 50 MHz , $\lambda = 220\ \mu m$. A flaw 600 µm in diameter corresponds to $kb = 8.6$. Thus, the angular range 2θ over which all the contributions to received signals are positive, is approximately 50° . For 32 elements evenly distributed around the aperture, this result implies that only five of them give cumulative contributions to the image of the point P and the remaining elements essentially contribute randomly.

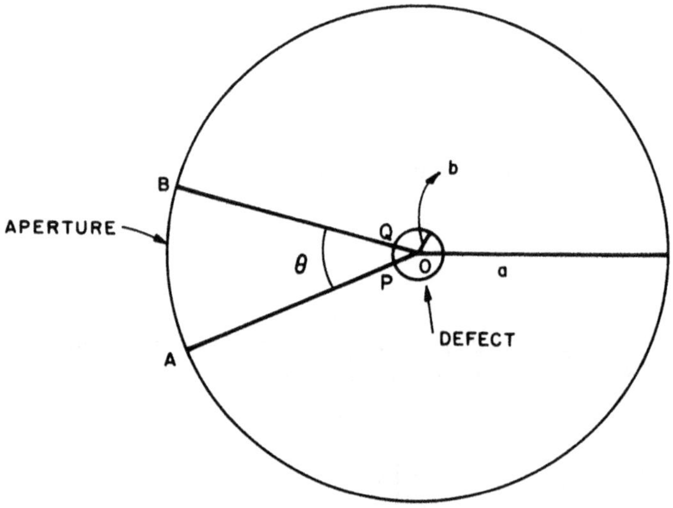

Fig. 7. Imaging a finite size circular specular reflector using a circular aperture.

SELECTIVE BACK-PROJECTION

Figure 8 diagrams the entire synthetic focus procedure. Without loss of generality, a two-dimensional circular aperture is chosen as a specific example. The aperture synthesis step is mathematically a mapping from the object field $f(x,y)$ into the data field which is a set of time series. For the purpose of illustration, the time series are ordered in the form of a matrix with each row corresponding to the complete pulse-echo data record at a sampling point. The columns are the progressive time entries of the pulse-echo data records. The synthetic focus step is simply the evaluation of the amplitude of each pixel in the image plane with the aid of a focus map generated independently based on geometric considerations. The focus map maps out a meandering path in the data field matrix, along which the contribution from each sampling point should be picked up for a particular image pixel.

As explained before, conventional back-projection indiscriminately sums up all the contributions. The selective back-projection scheme sums up only the subset of contributions from adjacent elements that are of the same sign and give maximum total magnitude. The basis for this strategy is shown in Fig. 3. Although point A on the defect boundary is only physically "visible" to array element 1, in the reconstruction process, the sidelobe due to point B, which physically only excites array element 2 can influence the amplitude at A because of the use of a finite width transmitted pulse. If the sidelobe contribution from B, or equivalently the back-projected contribution from array element 2, is in phase with the main lobe contribution from array element 1, then the main level is boosted. For a smooth specular reflector, constructive contributions should only come from a connected neighborhood of the aperture. Since the data is automatically searched to find the packet of numbers which gives maximum contribution, one does not have to make any a priori assumptions about the orientation and radius of curvature of the surface of the specular reflector.

It is important to point out that even though the main lobe of an image point is boosted by sidelobe contributions from neighboring points, the improvement in overall amplitude is meager because only a small number of array elements are involved. A sidelobe from an image point tends to leave streaks in the image. These streaks come from lone contributions in the selective summing step. Therefore, to reduce the undesirably high sidelobe levels, lone contributions are discarded, and the corresponding pixel amplitudes are set to zero. In other words, one can set up a criterion that an object point has to be "seen" by at least m array elements to discriminate against unwanted sidelobes.

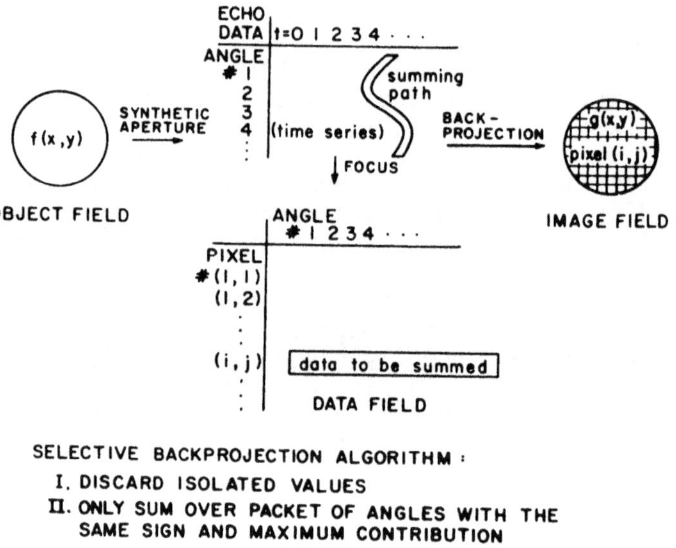

Fig. 8. Selective back-projection algorithm.

COMPUTER SIMULATIONS

Point Spread Function

(1) Square Aperture System. The PSF for an 8 x 8 element aperture at the focal plane parallel to the aperture is shown in Fig. 9a. The simulated point target is located at $z = 2.5$ mm. A 4-1/2 cycle Gaussian envelope transmit pulse is used. Since the aperture has 3.09 mm sides and the definition is improved by a factor of 2 in synthetic aperture imaging, the configuration is equivalent to a f/0.4 lens system. The 4 dB definition given by the computer simulation result is very close to the theoretically predicted 0.4λ according to Eq. (2). The nearest grating lobe is about 3.5λ away and is 14 dB down from the main lobe level. However, the paraxial result given by Eq. (3) shows that the first grating lobes should be 2.84λ from the main lobe. Moreover, the PSF is not space-invariant because the image field is non-paraxial.

Nonlinear processing has been shown to be very effective in suppressing sidelobe and grating lobe levels.[5] Fig. 9b shows the PSF of the system after nonlinear processing. The square root of the magnitude of the pulse-echo data is taken with the sign

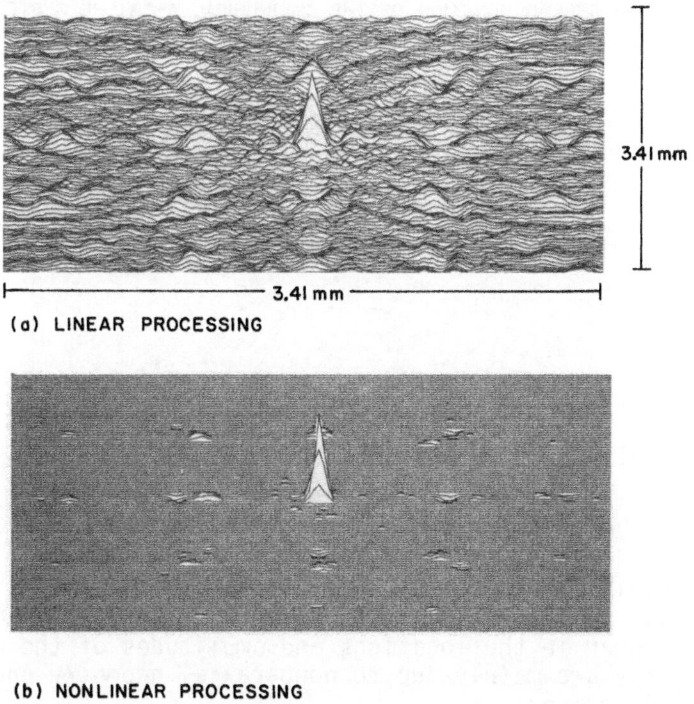

Fig. 9. Point spread function of the square aperture system.

preserved before back-projection, and the image amplitude is squared after back-projection. The reduction in sidelobe level is remarkable, and the nearest grating lobe is now 28 dB below the main lobe, an improvement of a factor of 2 as predicted by simple theory.

(2) Cylindrical Aperture System. Figure 10 shows the PSF of a sampled cylindrical aperture having 16 stacks x 32 elements . The simulated point target is located on the z axis at z = 3.3 mm.. The Rayleigh definition of the system in the (r,θ) plane is very close to the theoretical value of 0.2λ . The first grating lobes occur about 2.8 from the main lobe, in good agreement with the theoretical value of 2.71 (Eq. (27)). The amplitude of the grating lobe is roughly 20.4 dB down from the main lobe. This result is 6 dB better than that for the cw case, as can be expected because of the use of a short transmit pulse.

The definition in the z direction should be comparable to that of a f/.24 lens system. The 4 dB definition in z is verified to be very close to theory. The first grating lobe is found about 4λ from the main lobe as opposed to the theoretical value of 3.6 .

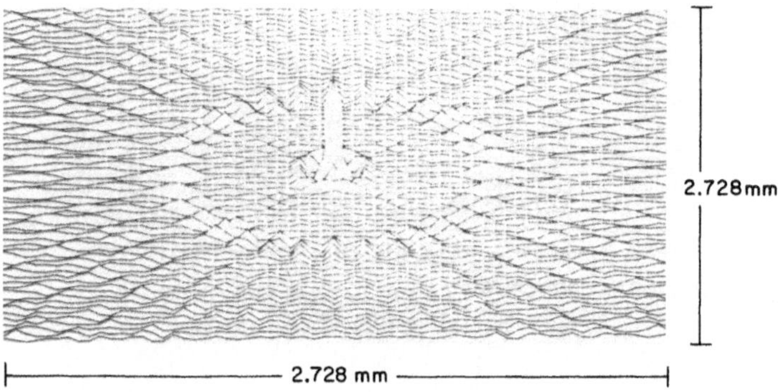

Fig. 10. Point spread function of the cylindrical aperture system.

To summarize, theoretical predictions are generally in good agreement with computer simulation results. Significant discrepancies do appear in the locations and amplitudes of the grating lobes, but they are mainly due to nonparaxial geometry and use of short transmit pulses.

RECONSTRUCTION OF SPECULAR REFLECTORS

Because the selective back-projection algorithm can easily be implemented on a two-dimensional system without excessive computation time, we have, for the purposes of illustration, reconstructed images of a void and of silicon and iron inclusions for a two-dimensional system on the computer. 32 evenly distributed array elements are used. To simulate the pulse-echo data, theoretically derived backscattered waveforms for voids, silicon, and iron inclusions are employed. The calculation is based on the work of Ying and Truell[10,11] on scattering from spherical objects. The pulse-echo data for each array element is generated based on geometric optics considerations. The signal path is assumed to be along the line segment between the array element and the point on the defect boundary directly opposite the element, as shown in Fig. 3. Selective back-projection is then applied to reconstruct the images of the various single defects. The amplitude of each image point is evaluated by inspecting the contributions from all 32 array elements. Only the packet of contributions which gives maximum magnitude is kept. The number of entries in the packet (corresponding to the number of array elements involved) is checked, and if it is more than two, the amplitude of the image point is set to equal the packet sum. Otherwise, the image point value is forced to zero. Note that not all the array elements in the packet contribute equally. The element nearest to the normal of the defect surface at the image point will contribute the most. The neighboring contributions will fall off approximately cosinusoidally. Thus, the selective back-projection algorithm, when used to image specular reflectors with dimensions greater than one wavelength, is equivalent to a limited angular aperture imaging system with cosinusoidal apodization.

Figures 11a and 11b show the images of a void 600 μm in diameter obtained by conventional and selective back-projection methods respectively. There is decidedly great improvement in the clarity of the defect boundary in Fig. 11b. The defect boundary amplitude is boosted by 3 to 4 dB compared to that obtained by conventional back-projection, and the far-out sidelobe is almost completely annihilated by the selective back-projection process.

Similar improvement can readily be observed in Figs. 12 and 13 for the cases of silicon and iron inclusions.

Even greater improvement in the image quality can be achieved when selective back-projection is combined with nonlinear processing. This is clearly evident in Fig. 11c for the case of a void.

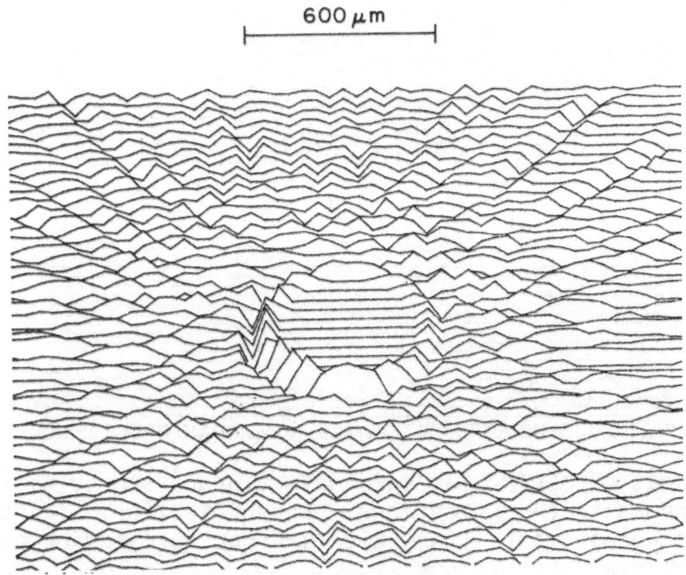

(a) CONVENTIONAL BACKPROJECTION

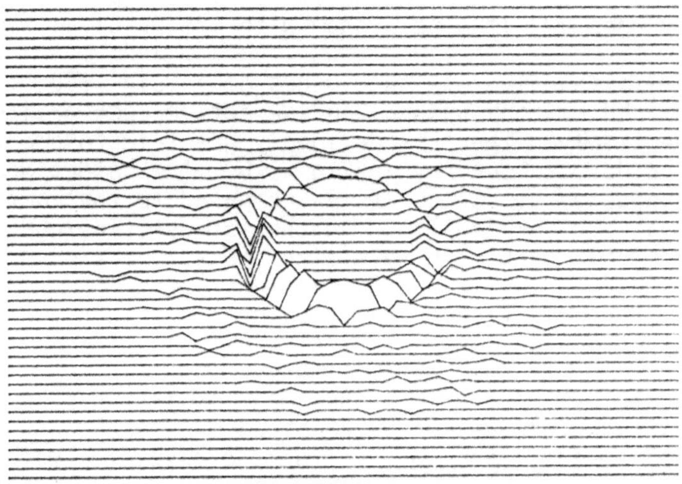

(b) SELECTIVE BACKPROJECTION

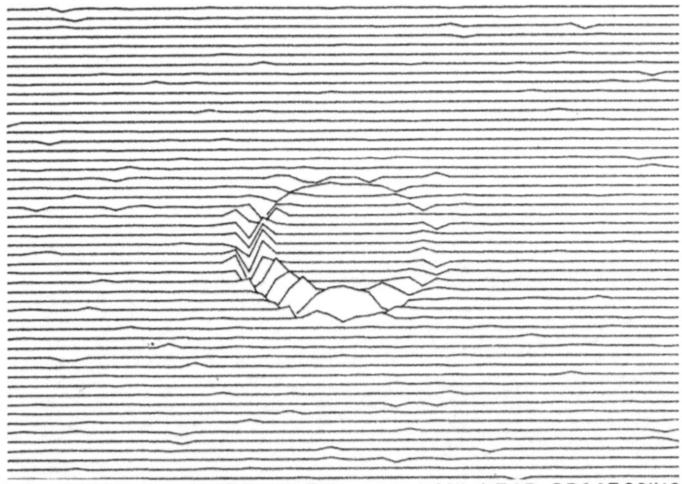

(c) SELECTIVE BACKPROJECTION AND NONLINEAR PROCESSING

Fig. 11. Images of a circular void defect processed by different schemes: (a) conventional back-projection; (b) selective back-projection; (c) selective back-projection and nonlinear processing.

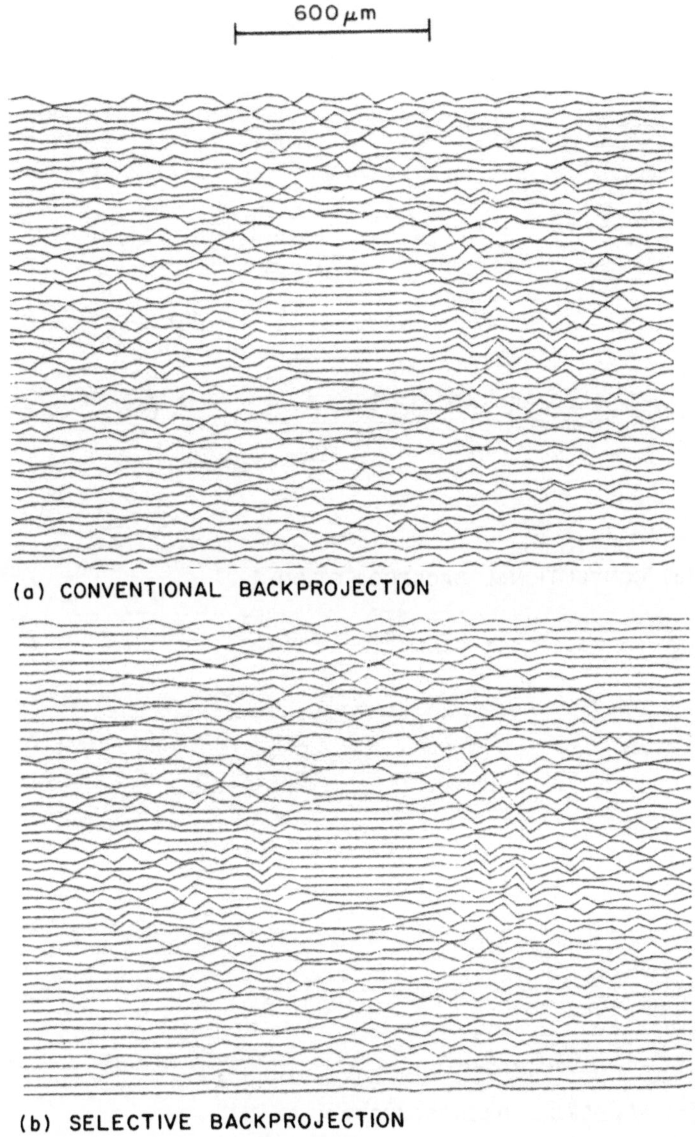

Fig. 12. Images of a silicon inclusion defect processed by different schemes: (a) conventional back-projection; and (b) selective back-projection.

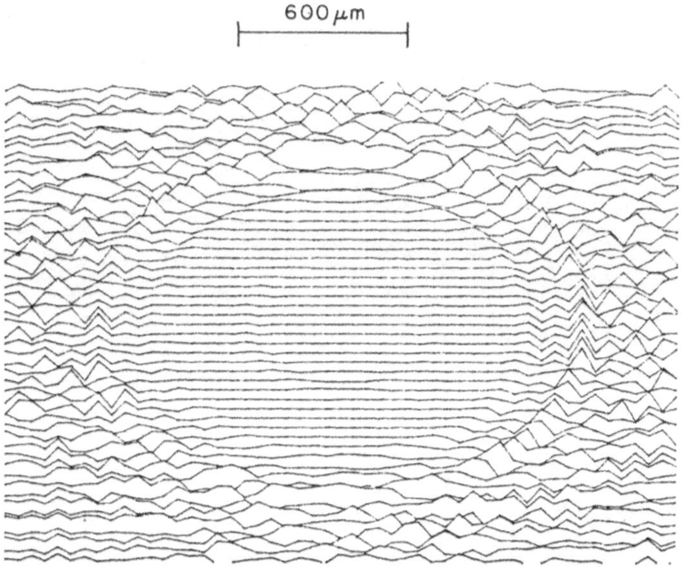

(a) CONVENTIONAL BACKPROJECTION

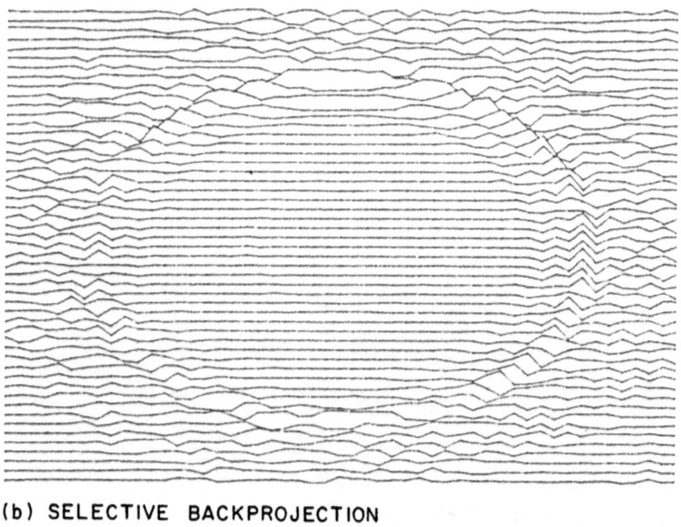

(b) SELECTIVE BACKPROJECTION

Fig. 13. Images of an iron inclusion defect processed by different schemes: (a) conventional back-projection; and (b) selective back-projection.

CONCLUSION

We have developed the theories to characterize the imaging performance of square and cylindrical synthetic apertures. Most of the theoretical predictions are confirmed by computer simulation results. We have established that the three-dimensional synthetic focus technique is capable of extremely good single point resolution in all three directions. The result on the PSF for the cylindrical aperture configuration is of particular importance because of its generality. The theoretical analysis was carried out without making any assumption about the position of the image point relative to the aperture. This result may also apply to tomographic systems because of similarity in aperture geometry. In addition, we have analyzed the effect of sampling the aperture. Sampling introduces grating lobes whose positions are nearer to the main lobe than in a rectilinear system but whose amplitudes are far weaker. Short pulse operation tends to reduce the grating lobe levels.

Specular reflectors present special difficulties for synthetic aperture imaging. Computer simulations show that the conventional back-projection technique is seriously inadequate in reconstructing images of specular reflectors because of the resulting high sidelobe levels. A selective back-projection technique has shown great promise in enhancing the images by suppressing sidelobe amplitudes.

We hope to obtain experimental results on our three-dimensional imaging system in the near future to substantiate the simulation results.

ACKNOWLEDGMENT

This work was supported by the Advanced Research Projects Agency under Rockwell International Subcontract No. RI-1246.

REFERENCES

Chou, C-H, Khuri-Yakub, B. T., Kino, G. S., and Evans, A. G., Defect Characterization in the Short Wavelength Regime, accepted for publication, J. NDE, August, 1980.[7]

Corl, P. D., Grant, P. M., and Kino, G. S., A Digital Synthetic Focus Acoustic Imaging System for NDE, Proc. Ultrasonic Symposium, Cherry Hill, New Jersey, September, 1978.[3]

Corl, P. D., and Kino, G. S., A Real-Time Synthetic Aperture Imaging System, Acoustic Holography, Vol. 9, K. Wang, ed., 1979.[5]

Corl, D., Kino, G. S., Behar, D., Olaisen, H., and Titchener, P, Digital Synthetic Aperture Acoustic Imaging System, Proc. ARPA/AFML Review of Progress in Quantitative NDE, La Jolla, California, July, 1979.[4]

Corl, P. D., Kino, G. S., DeSilets, C. S., and Grant, P. M., A Digital Synthetic Focus Acoustic Imaging System, Acoustic Holography, Vol. 8, A. F. Metherell, ed., 1978.[1]

Erdelyi, Magnus, Oberttinger, and Triconni, Higher Transcendental Functions, Bateman Manuscript Project, Vol. 2, California Institute of Technology, McGraw Hill.[9]

Evans, A. G., Tittmann, B. R., Kino, G. S., and Khuri-Yakub, B. T., Ultrasonic Flaw Detection in Ceramics, Proc. ARPA/AFML Review of Progress in Quantitative NDE, Asilomar, California, June, 1976.[6]

Johnson, G., and Truell, R., Numerical Computation of Elastic Scattering Cross-Section, J. Appl. Phys., Vol. 36, p. 3466, 1965.[11]

Kino, G. S., Corl, D., Bennett, S., Peterson, K., Real-Time Synthetic Aperture Imaging System, Proc. Ultrasonic Symposium, Boston, Mass., November, 1980.[8]

Kino, G. S., Grant, P. M., Corl, P. D., and DeSilets, C. S., Digital Synthetic Aperture Acoustic Imaging for NDE, Proc. ARPA/AFML Review of Progress in Quantiative NDE, La Jolla, California, July, 1978.[2]

Ying, C. F., and Truell, R, Scattering of a Plane Longitudinal Wave by a Spherical Obstacle in an Isotropically Elastic Solid, J. Appl. Phys., Vol. 27, p. 1086, 1956.[10]

A REAL-TIME SYNTHETIC APERTURE DIGITAL ACOUSTIC IMAGING SYSTEM

S. Bennett, D. K. Peterson, D. Corl, and G. S. Kino

Edward L. Ginzton Laboratory
Stanford University
Stanford, California 94305

INTRODUCTION

We present recent results obtained with a 32-element real-time synthetic aperture acoustic imaging system. The system, which was described at last year's Acoustic Imaging Conference, has been developed in two main directions. First, a comprehensive testing program has enabled us to diagnose and correct a number of sources of electronic error, and this, coupled with the addition of high quality grey scale display systems, has resulted in greatly improved image quality. Second, in recognizing the need for quantitative as well as qualitative interpretation of NDE images, we have developed a communication channel between the imaging system and a mini-computer. This facility allows us not only to process experimental data collected from the imaging system with greater sophistication than is possible using the real-time hardware alone but also enables us to obtain images synthesized from theoretical models generated in the computer and rapidly reconstructed in the imaging system. This is an extremely useful facility since it is possible to compare the experimental performance of the system directly with what would be expected theoretically.

The operation of the real-time imaging system, illustrated in Fig. 1, can be summarized as follows. A multiplexer is used to address, in turn, each element of a piezoelectric array. A short rf pulse of about 3.3 MHz center frequency, typically 2.5 cycles long between 20 dB points, is used to excite one element of the array through the multiplexer. Signals reflected from an object are received at the same transducer and passed through the multiplexer to an amplifier, into an A-to-D converter, and then into a 1024 x 8 bit RAM. This process is repeated in turn for each of the

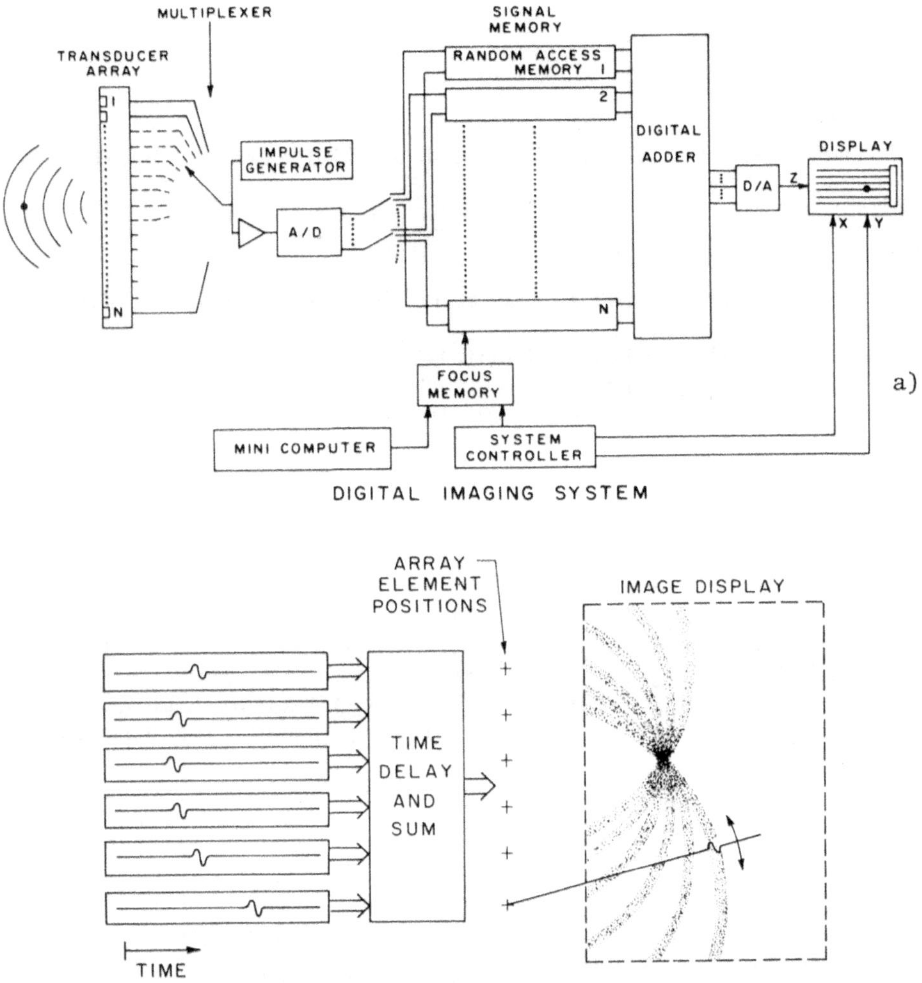

Fig. 1. The digital acoustic imaging system. (a) Block diagram of the hardware components; (b) An illustration of the time delay and sum reconstruction algorithm.

32 elements of the array, thus storing information from the object plane in 32 RAMs, called the signal memory. The stored signals are then available for further processing to reconstruct an image. Because the information is built up element by element, the device behaves as a synthetic aperture imaging system. It has the advantage that, as only one element is used at a time, only one amplifier and A-to-D converter is required. On the other hand, because a signal is transmitted from only one element at a time, the transducer array is required to have high sensitivity and efficiency to obtain good images.

A focused image is formed by applying the appropriate time delays to each of the digitized echoes, then summing the contribution from each array element. The sum signal is envelope-detected by taking its modulus and passing the output through a D-to-A converter and low-pass filter to an intensity display. The time delay information is stored as a look-up table in another set of RAMs called the focus memory.

New data is collected, focused, and then displayed at a rate of approximately 30 frames per second using a digital clock rate of the order of 10 MHz .

THEORY OF IMAGING

We have discussed the theory of imaging in previous papers. Here, we will emphasize the lateral point spread function, two point resolution, errors due to time quantization, and non-linear processing.

Single Point Definition

In a theory given in earlier papers, we treated the array as a periodic system.[1,9] Here, for the purpose of the present discussion, we are only interested in points near the object. Consequently, it is possible to treat the array as a continuous uniform aperture so that mathematically the sums involved become integrals. As can be seen from our earlier treatments, the effect of having a finite number of elements in a short pulse system of this kind is to introduce grating lobes that are relatively weak.

Consider an aperture of width D in which an element of length dx' at the position $(x',0)$ is excited by a pulse of the form $\mathrm{Re}\, F(t)e^{j\omega t}$, as shown in Fig. 2. The signal reflected from a point (x_0,z_0) to the differential element dx' will therefore be of the form

$$V(t)dx' = \mathrm{Re}\, F(t - 2R_{x',x_0}/v)\, \exp[j\omega(t - 2R_{x',x_0}/v)]dx' \quad (1)$$

where

$$R_{x',x_0} = \sqrt{(x_0 - x')^2 + z_0^2} \quad (2)$$

Suppose the system is focused on the point (x,z_0) . We shall evaluate the signal arriving at x' at the time $t' = t + 2R_{x',x}/v$, which is equivalent to assuming that the system is focused on the

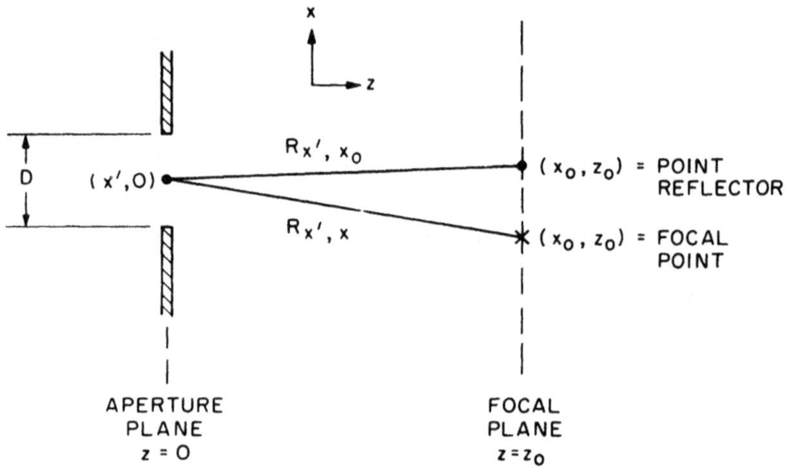

Fig. 2. Geometry and coordinates used in the discussion of a continuous aperture imaging system.

point (x, z_0) . In this case, the lateral point spread function (PSF) will be of the form

$$H(x) = \int_{-D/2}^{D/2} F(t' - 2\Delta R/v)\, e^{j\omega(t' - 2\Delta R/v)}\, dx' \tag{3}$$

where $\Delta R = R_{x',x_0} - R_{x',x}$ and D is the length of the array.

By using the paraxial approximation and, in addition, carrying out the observation at a time $t' = (x_0^2 - x^2)/z_0 v$ of a point on the focal plane $z = z_0$, Eq. (3) can be simplified to yield the point spread function in the form

$$H(x) = \int_{-D/2}^{D/2} F[2(x_0 - x)x'/vz_0]\, e^{4j\pi(x_0 - x)x'/z_0\lambda}\, dx' \tag{4}$$

If the pulse is long, then for points in the neighborhood of the object, we can take $F[2(x - x_0)x'/vz_0)] = 1$, so that

$$H(x) = D\,\frac{\sin 2\pi(x - x_0)D/z_0\lambda}{2\pi(x - x_0)D/z_0\lambda} \tag{5}$$

The amplitude of this function drops by 4 dB from its maximum value at two points a distance d_x apart, where

$$d_x = \frac{z_0\lambda}{2D} \tag{6}$$

Thus d_x is the single point definition. This definition is a factor of two better than for a single lens with the same aperture. This is because in the synthetic aperture system, signals travel from a single element to a point on the object and back, twice the distance that a ray travels from the point on the object to the lens in a single lens system.

Before proceeding further, it is convenient to normalize the lateral coordinate and write $X = x/d_x$. All amplitude plots in this paper use this normalized argument. Equation (5) now becomes

$$H(X) = D\ \mathrm{sinc}(X) \tag{7}$$

where $\mathrm{sinc}X = \sin\pi X/\pi X$. This is plotted in Fig. 3 ($\sigma = \infty$; see below for an explanation of the symbol σ).

There is a coherent addition of the two amplitudes and therefore, at least for nearby points, the system behaves like a coherent imaging system. For points far from each other, this is not the case because the short pulse used in the synthetic aperture system results in incoherence, basically because there cannot be phase additions between the return echoes from the two points if the difference in the time delays of the signals received from them is more than the length of the pulse. For a pulse M cycles long, this implies that the sidelobe structure due to phase additions tends to disappear at a distance $\sim Md_x$ from the focal point.[9] Figure 3 shows the lateral PSF for a Gaussian envelope pulse where $F(t) = \exp[-(\omega t/\sigma)^2]$. When $\sigma = \infty$, the signal is CW; when $\sigma = 2\pi$, the pulse is about 2.5 cycles in duration.

It can be shown that for a periodic array, the grating lobe amplitude will be reduced by a factor of M/N where N is the number of elements in the array.[9] Thus, the system behaves as a coherent imaging system for nearby points and as an incoherent imaging system for points sufficiently far apart. This is of great advantage because it makes it possible to image relatively large objects without major interference from neighboring points on the object, as would occur with most coherent imaging systems.

Two Point Definition

Suppose we now try to image two equal amplitude point reflectors at the positions $(x_0 \pm \delta/2, z_0)$. The response function to

$$H(X) = \int_{-1/2}^{1/2} \exp\left[-\left(\frac{2\pi(X - X_0)x'}{\sigma}\right)^2\right] \cos\left(2\pi(X - X_0)x'\right) dx'$$

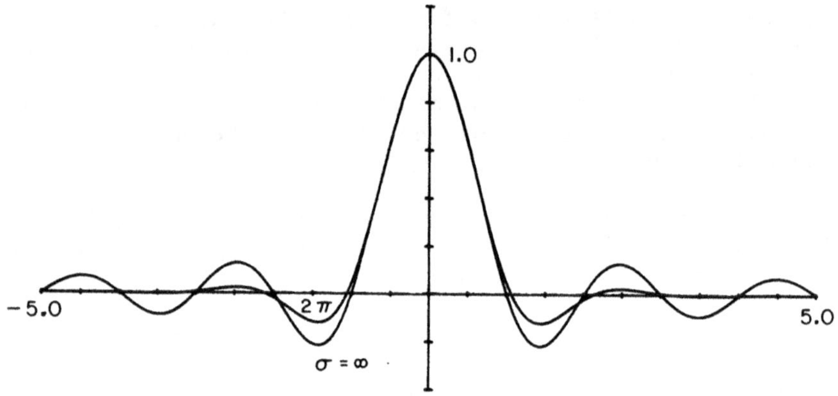

LATERAL P.S.F. WITH GAUSSIAN MODULATED PULSES

Fig. 3. The lateral PSF for the imaging system shown in Fig. 2, using CW excitation ($\sigma = \infty$) and 2.5 cycle Gaussian weighted pulses ($\sigma = 2\pi$).

the two reflectors will be

$$P(X) = H(X + \Delta/2) + H(X - \Delta/2) \tag{8}$$

where $\Delta = \delta/d_x$.

It is convenient to make use of the Sparrow criterion rather than the Rayleigh criterion for a two-dimensional system of this kind. This suggests that the two points can be distinguished from each other if there is no perceptible dip in amplitude when the imaging system is focused at a point midway between them. On this basis, the two point definition h_s corresponds to $\Delta = 1.32$ or

$$h_s = 1.32 d_x \tag{9}$$

Figure 4 shows $P(x)$ for three different values of Δ .

$$H(X) = \int_{-1/2}^{1/2} \left[\cos\left(2\pi(X - \Delta/2)x'\right) + \cos\left(2\pi(X + \Delta/2)x'\right) \right] dx'$$

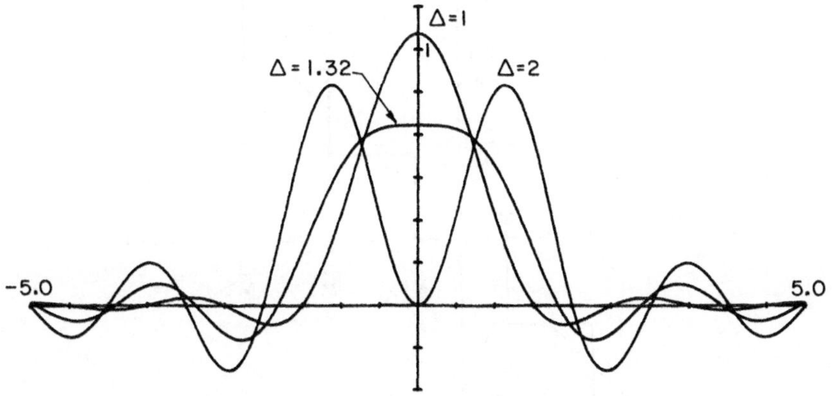

TWO POINT LATERAL RESOLUTION USING C.W. INSONIFICATION

Fig. 4. Image of two points laterally separated by $\Delta = 1.0$, 1.32, and 2.0.

Errors Due to Time Quantization

One difficulty that might arise with a digital imaging system is the generation of sidelobes due to phase or timing errors.[11] If we suppose that there are μ samples per rf cycle, the maximum possible timing errors are $\pm\tau/2\mu$ where $\tau = 2\pi/\omega$. Thus, the digital system time delay will vary in a series of steps in the manner shown in Fig. 5a. Therefore, the timing error is ε where

$$\begin{aligned} \varepsilon &= T_{(digital)} - T_{(required)} \\ &= 2[R_{(digital)} - R_{(required)}]/v \qquad (10) \end{aligned}$$

This error ε varies in the manner shown in Fig. 5b. We see that the error is a sawtooth function of $T_{(required)}$.

The point spread function given by Eq. (3) now becomes

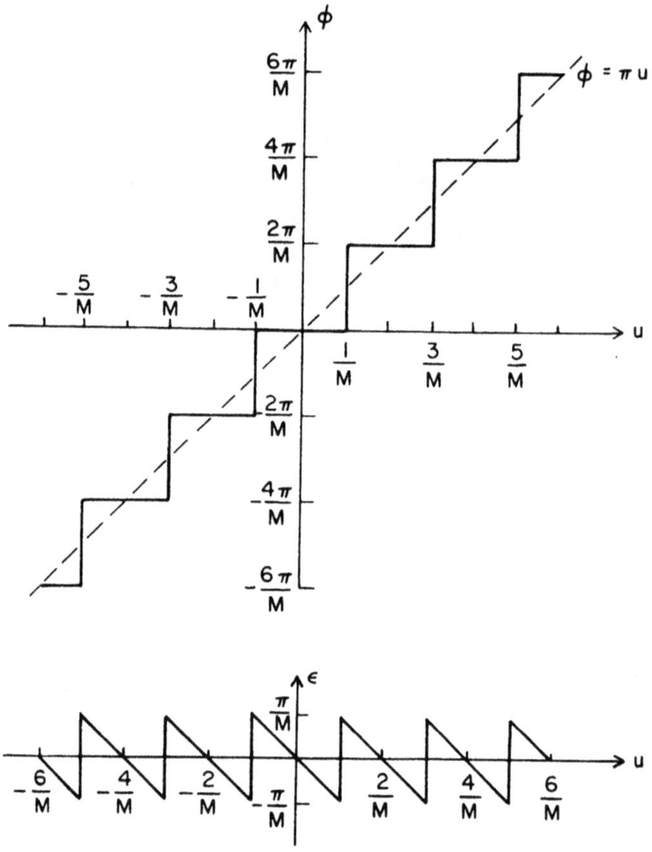

Fig. 5. Quantization of the phase (time) variable. (a) The broken line shows the continuous relationship desired. The solid line shows the quantized approximation to the dashed line; (b) A plot of the error due to phase quantization.

$$H(x) = \int_{-D/2}^{D/2} F(t' - 2\Delta R/v)\ e^{j\omega(t' - 2\,\Delta R/v)}\ e^{j\omega\varepsilon}\ dx' \tag{11}$$

As ε is a periodic function of time and of x'^2, we can write

$$e^{j\omega\varepsilon} = \sum_m A_m e^{2jm\mu\pi R_0/v} \tag{12}$$

where

$$A_m = (-1)^m \frac{\sin \pi/\mu}{\pi(m + 1/\mu)} \tag{13}$$

The effect is to lower the mainlobe amplitude corresponding to $m = 0$ by $\mathrm{sinc}(1/\mu)$. The largest harmonic term in Eq. (12) corresponds to $m = -1$. Using the paraxial approximation once more and putting $t' = (x_0^2 - x^2)/z_0 v$, we can determine the contribution of the mth term to the integral. For CW signals, this integral, which contains square law terms in x' in the argument of the exponential, can be evaluated by the method of minimum phase. The effective sidelobe level, i.e. the ratio of the sidelobe level to the mainlobe level, is

$$S_m = \frac{A_m}{A_0}\sqrt{\frac{z_0\lambda}{D^2 m\mu}} = \left|\frac{1}{mM + 1}\right|\sqrt{\left|\frac{z_0\lambda}{D^2 m\mu}\right|} \tag{14}$$

for the mth harmonic term. For example, for $x = 80$ mm , $\lambda = 0.45$ mm , $m = -1$ $D = 16$ mm ,and $\mu = 3$, the estimated sidelobe level due to digital phase errors is -20.5 dB . At the same time, the mainlobe is reduced in amplitude due to digital phase errors by 1.7 dB . In fact, because there may be additions due to several harmonic terms, the sidelobe levels may be one or two dBs worse than this figure.

Finally, it should be noted that this is precisely the result we would expect physically if we regard the beam arriving at the array of width D as passing through a subsidiary m^{th} focal point at $z = z_0/(1+mM)$ just as occurs with a Fresnel lens. This beam would occupy a width DmM at the focal plane. Thus, it is as if there were a uniform background signal due to the subsidiary focus. In fact, the effect of phase quantization errors is only important near the mainlobe because with a pulsed system, F(t) is non-zero for only a limited region around the focal point.

Nonlinear Processing

In earlier papers we have described the use of non-linear processing techniques to improve the sidelobe level.[1,2,3] We have since modified our original arguments to take account of interactions between signals reflected from more than one point. We consider an array of N elements focused on a point A, as illustrated in Fig. 6. A signal reflected from a point reflector at A is assumed to have an amplitude a at each transducer element. Thus, the total output, after suitably delaying and summing the signals from each element, is Na. An individual transducer element can receive a signal at the same time as A from a point B, which lies on a circle of the same radius as that passing through A and with its center at the transducer. Let the signal from B have an amplitude b at the transducer element of interest. When imaging with short pulses, the echoes from A and B will coincide in time for only one element of the array; all other array elements will "see" A and B as two distinct echoes. If V_{AB} is the image intensity at A with both points present and V_B is the image intensity at the same point with only point B present, then we will say that we can just observe A when $V_{AB} = 2V_B$. Thus, with linear processing we can observe point A if $a > 2b/N$.

Now suppose we use n^{th} root non-linear compression at each transducer before image reconstruction and n^{th} power expansion after image reconstruction. If the signal at the mth element is V_m, we take $|V_m|^{1/n} \operatorname{sgn} V_m$, delay and add the resultant signals and then take the nth power of the summed output. In this case, with the system focused on A, the resultant output is

$$V_{AB} = \left[(N - 1)a^{1/n} + (a + b)^{1/n}\right]^n \tag{15}$$

We have supposed for simplicity that b and a have the same sign. We are interested in the case when $b \gg a$. Thus, we can write

$$V_{AB} \approx \left[(N - 1)a^{1/n} + b^{1/n}\right]^n \tag{16}$$

Now if $a = 0$, the background signal is V_B where

$$V_B = b \tag{17}$$

The point A is observable if $V_{AB} = 2V_B$, i.e. if

$$(N - 1)a^{1/n} + b^{1/n} > 2^{1/n}b^{1/n} \tag{18}$$

or

$$a > \frac{(2^{1/n} - 1)^n}{(N - 1)^n} b \tag{19}$$

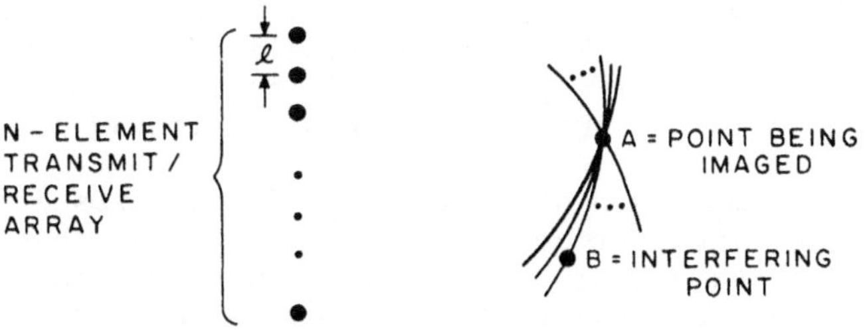

Fig. 6. Imaging the point A in the presence of an interfering point B .

For $n = 2$, this result implies that $a > 0.17b/(N - 1)^2$. Thus, we would expect to be able to observe far weaker reflectors in the presence of interference if we use non-linear instead of linear processing. We do not use an indefinitely large value of n because of noise problems and because of artifacts generated by interactions between several sources.

In the past we have always confined our discussions of non-linear processing to point reflectors which were widely separated. Below, we consider the effects of non-linear processing for points which may be quite close to each other so that their echoes overlap in time. For linear processing we may rely on the superposition principle, but for non-linear processing no such intuitive aid is available. We should not be surprised when non-linear processing gives results which are very different from those obtained with linear processing.

If we make the same assumptions which lead to Eq. (4) (paraxial, etc.), then the signal received by a differential element dx' is

$$V(x) = F[2(x_0 - x)x'/vz_0] \exp[j4\pi(x_0 - x)x'/\lambda z_0] \qquad (20)$$

With non-linear processing, the total output image amplitude is

$$H(x) = \left\{ \int_{-D/2}^{D/2} |V(x)|^{1/n} \mathrm{sgn}[V(x)]\,dx' \right\}^n \qquad (21)$$

This non-linear lateral PSF is plotted in Fig. 7 for $n = 1, 2$, and 3. The reduction in sidelobe level with non-linear processing is quite apparent. Also, the mainlobe width is slightly decreased.

Figure 8 shows the two-point lateral response for the Sparrow separation, $\Delta = 1.32$. Here again, non-linear processing improves the image quality; the two peaks are distinct for $n = 2$ and 3 (non-linear processing) but not for $n = 1$ (linear processing).

A careful comparison of Figs. 7 and 8 indicates a problem peculiar to non-linear processing. Figure 7 shows that the peak response for all values of n is unity. However, in Fig. 8, the two-point response decreases in amplitude as n increases. This is due to an interaction between the echoes of the two points. This interaction is shown more explicitly in Fig. 9a. Here we have two points separated by $\Delta = 8$. The amplitude of the point at $-\Delta/2$ is varied from 0 to 10, while the amplitude of the point at $+\Delta/2$ is held constant at unity. Notice that as the amplitude of the left hand point increases, it suppresses the observed amplitude of the right hand point.

$$H(X) = \left\{ \int_{-1/2}^{1/2} \cos^{1/n}\left(2\pi(X - X_0)x'\right) dx' \right\}^n$$

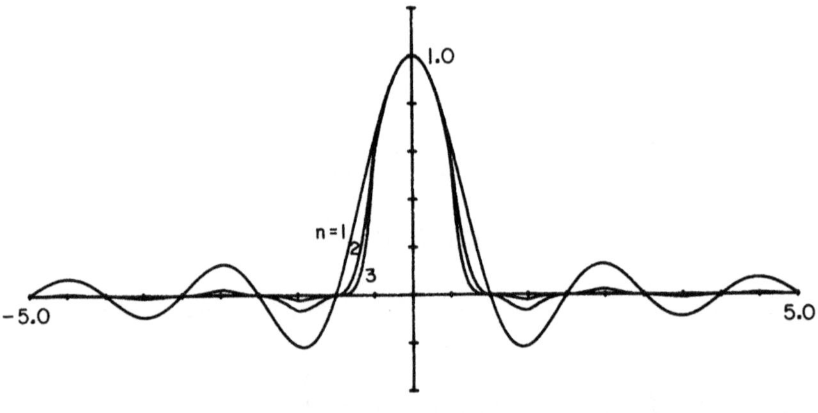

Fig. 7. Lateral PSF with linear $(n = 1)$, square root/square non-linear $(n = 2)$, and cube root/cube non-linear $(n = 3)$ Processing.

TWO POINT LATERAL RESOLUTION WITH
NON-LINEAR PROCESSING AND C.W. INSONIFICATION

$$H(X) = \left\{ \int_{-1/2}^{1/2} \left[\cos\left(2\pi(X-\Delta/2)x'\right) + \cos\left(2\pi(X+\Delta/2)x'\right) \right]^{1/n} dx' \right\}^{n}$$

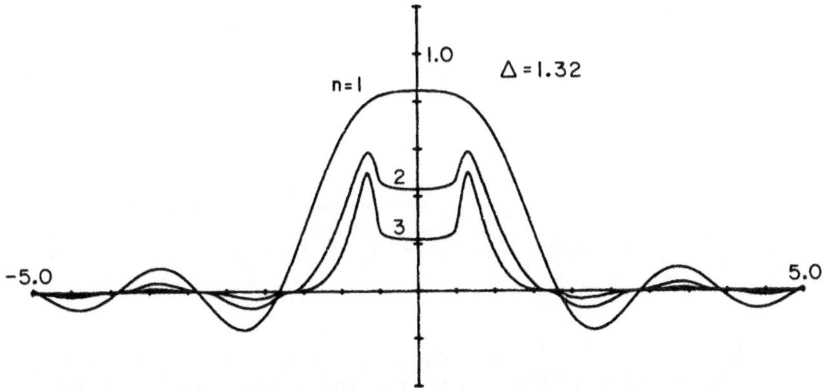

Fig. 8. Image of two points separated by $\Delta = 1.32$, with linear ($n = 1$) and non-linear ($n = 2,3$) processing.

The interaction between points can be decreased by using pulses instead of CW imaging signals. While CW echoes from the two points overlap for all times, pulse echoes only partially overlap in time, thus decreasing the interaction between points.

Figure 9b shows the results obtained with non-linear processing applied to a pulse imaging system. The pulses are Gaussian modulated cosine bursts of approximately 2.5 cycles in duration. Now the suppression of the right hand point is not as severe, although still noticeable.

Thus, the improvement to be obtained by non-linear processing depends on the ratio of the amplitudes of the interfering sources and their distance apart. In many cases, a considerable improvement can be obtained, especially with short pulse systems where there is a limited amount of interference between most of the points within the field of view, as there often is in NDE applications. In medical applications, the interference is much more severe because

$$H(X) = \left\{ \int_{-1/2}^{1/2} \left[1 \cdot \cos\left(2\pi(X-\Delta/2)x'\right) + b \cdot \cos\left(2\pi(X+\Delta/2)x'\right) \right]^{1/n} dx' \right\}^{n}$$

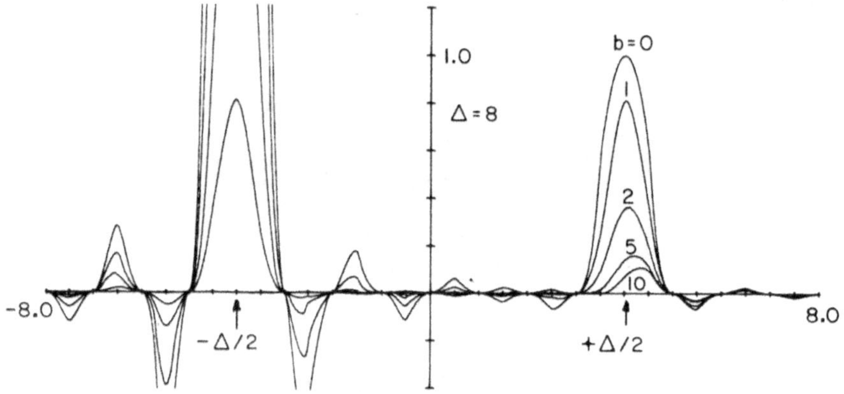

Fig. 9a. With non-linear processing and CW insonification, the observed amplitude of the unity amplitude reflector at $+\Delta/2$ diminishes as the intensity of the reflector at $-\Delta/2$ increases $(b = 0. \rightarrow 10.)$.

of the relatively dense field of reflectors that are present so that non-linear processing may not be of much help.

IMAGING PERFORMANCE

In this section we compare real-time images with their simulated analogues. Two types of objects are considered here: single point reflectors and two closely spaced point reflectors.

The simulation of point reflectors employs a very simple model. The point reflector is assumed to be a perfect reflector with an omni-directional reflection pattern. Diffraction losses are ignored. In essence, the acoustic pulse received is just a replica of the transmitted acoustic pulse with a time delay equivalent to the transit time from the transducer element to the point reflector and back.

$$H(X) = \left\{ \int_{-1/2}^{1/2} \left[1 \cdot e^{-(2\pi(X-\Delta/2)x'/\sigma)^2} \cos\left(2\pi(X-\Delta/2)x'\right) + B \cdot e^{-(2\pi(X+\Delta/2)x'/\sigma)^2} \cos\left(2\pi(X+\Delta/2)x'\right) \right]^{1/n} dx' \right\}^n$$

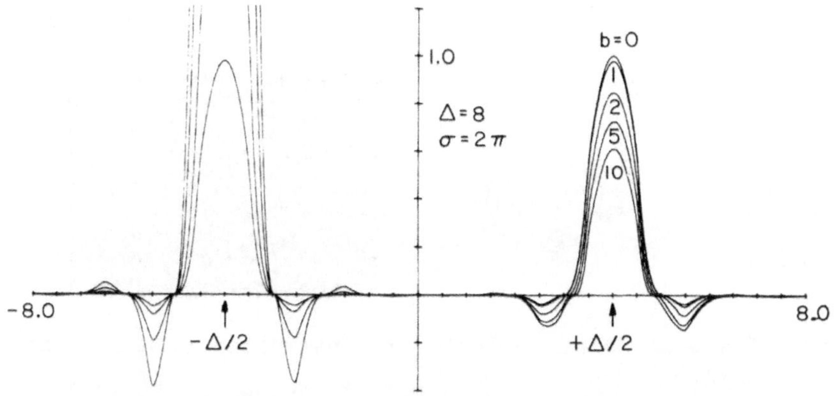

Fig. 9b. The use of Gaussian pulse (2.5 cycles duration) insonification reduces the interaction between the two points.

For the purposes of simulation, the 32-element transducer array is modeled as 32 equally spaced elements (.5 mm center-to-center) with identical cosine-carrier Gaussian-envelope pulses (3.3 MHz center frequency, 2.5 cycles between 20 dB points).

The first test object chosen for comparison was a point reflector, on axis, at 80 mm range in water. To obtain experimental data for this object, we used a .25 mm (.5λ, where λ is the acoustic wavelength of the center frequency of the pulse) diameter wire. The image obtained from the simulated data is shown in Fig. 10. The corresponding image obtained from real data is shown in Fig. 11. The images are presented in four different formats:

(1) a photograph of the display screen. "Tic" marks on all photographs represent 10 mm ;
(2) a "3-dimensional" plot of a selected portion of the image 20λ x 20λ (9.1 mm x 9.1 mm) ;

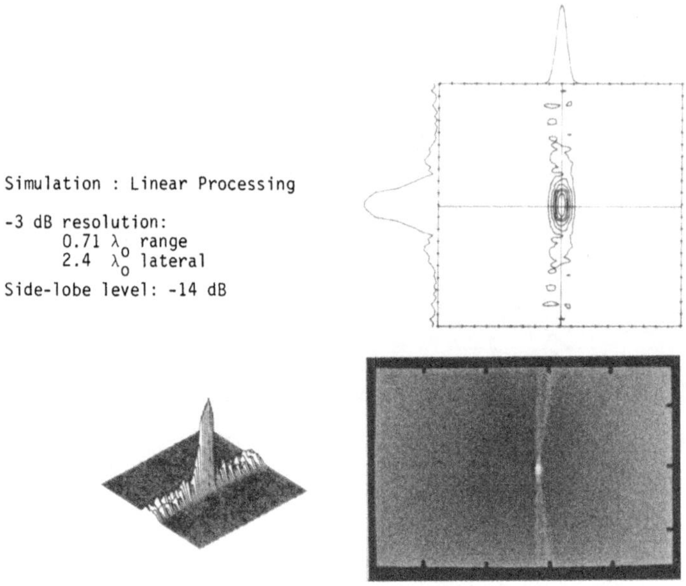

Fig. 10. Image of point reflector reconstructed from simulated echo data.

(3) a contour plot of the same portion shown in the "3-D" plot. "Tic" marks on the axes represent $\lambda = .45$ mm . Contour levels are at .85 , .70 , .55 , .40 , .25 , and .10 of the maximum height;

(4) horizontal and vertical cross-sections through the contour plot. The plane of intersection for the horizontal (or vertical) line in the contour plot. The corresponding cross-sections are plotted along the horizontal (or vertical) axis.

For ease of comparison, all graphs were normalized before plotting. Image data was envelope detected (absolute value followed by low-pass filter) before display.

For this imaging experiment, the range of the target was 80 mm, and the imaging aperture was 16 mm . For standard optics, this would be an f/5 system. However, since we use round trip delays, instead of one way delays for focusing, we effectively halve the f/number of the imaging system. Thus, the above experiments use an imaging system which has an effective f-number of F/2.5 , where F = f/2 is the <u>effective</u> f/number. To first approximation, the range resolution is determined by the pulse length and the lateral resolution is determined by the imaging wavelength and f/number.

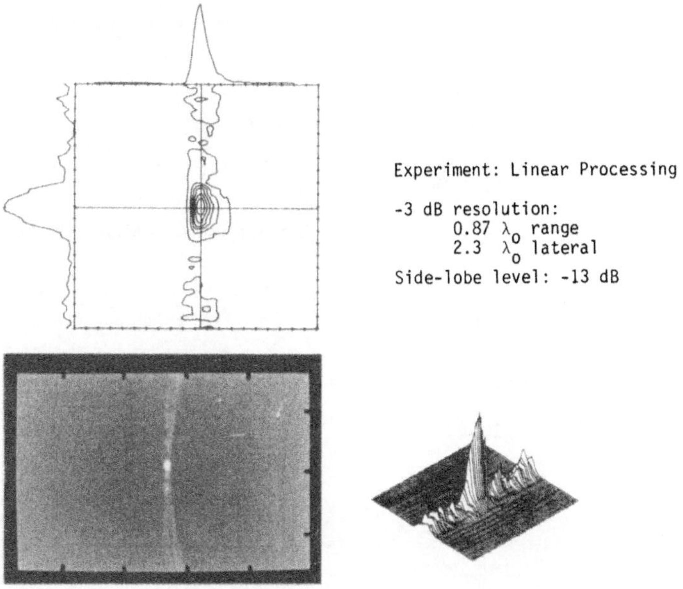

Fig. 11. Image of a thin wire (.25 mm diameter) at 80 mm range.

Paraxial, CW optics predicts a lateral resolution of about Fλ . The range resolutions indicated in Figs. 10 and 11 are comparable to λ (.45 mm) . The lateral resolutions indicated in these same figures are all slightly larger than Fλ (1.1 mm) . The transducer array used for real data acquisition had two defective elements. These missing elements would be expected to lead to slightly higher sidelobe levels and asymmetrical distribution of sidelobes around the mainlobe.[10]

Figure 12 shows the simulated data reconstructed using non-linear processing (square root compression/square expansion). Figure 13 shows real data reconstructed using the same non-linear algorithm. The advantage of non-linear processing is apparent here. For the simulated data, the first sidelobes ("near-in" sidelobes) have dropped from -14 dB for linear processing down to -25 dB for non-linear processing. A similar, but not quite so dramatic, reduction in sidelobe levels is obtained using real data from the wire target. At present, we sample the echoes at a 10.5 MHz rate. This gives only 3 samples per wavelength. The preceding discussion of phase errors would predict that the mainlobe amplitude will be reduced by sinc (1/3) = -1.7 dB .

The range and lateral resolutions of real images are comparable to those of the simulated images, being approximately λ and Fλ ,

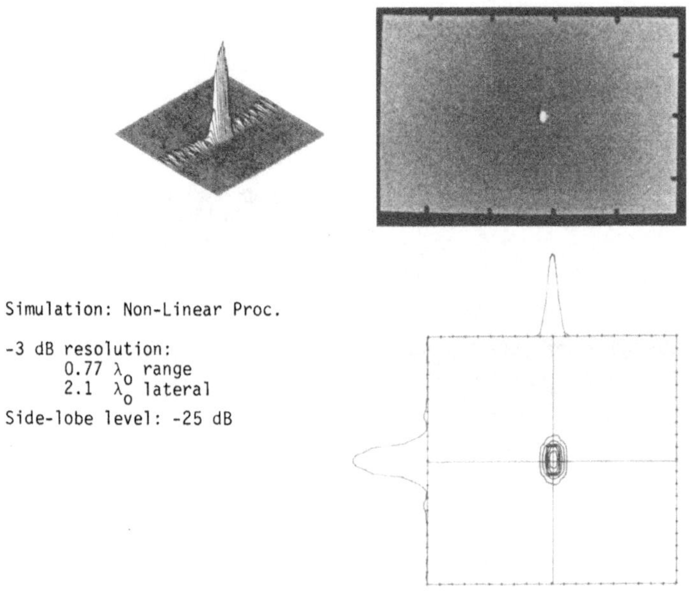

Fig. 12. Simulated echo data (same as used in Fig. 10) reconstructed using square root/square non-linear processing.

respectively. Non-linear processing provides a marked decrease in sidelobe level without any deterioration of the single point resolution.

Figure 14 shows an image reconstructed from a simulation of two point targets at a range of 80 mm, laterally separated by 2 mm. Figure 15 shows an image of two small wires (.25 mm diameter) separated by 2 mm. The two objects are resolvable for both simulated and real echo data. The image of real data has a small bump on axis, behind the two wires. We believe this is due to a mode converted echo which returns later than the front face reflection. Such delayed echoes are also observed for single, thicker (e.g. 1.25 mm diameter) wire targets. Alternatively, this delayed echo may be due to a scattering interaction between the two wires.

REAL-TIME IMAGING APPLICATIONS

In this section we present a series of images demonstrating the performance of the imaging system in a variety of practical situations. Images made with three different types of arrays are shown: longitudinal waves in water,[6,7] surface waves on aluminum

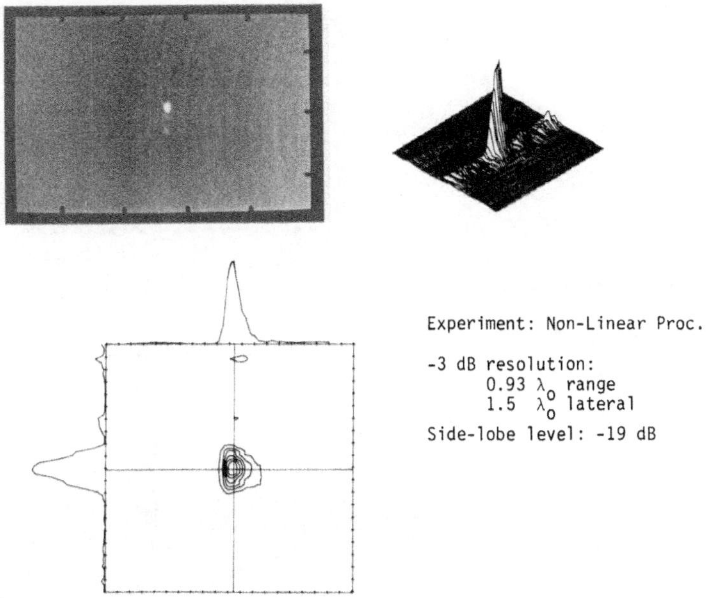

Fig. 13. Experimental echo data (same as used in Fig. 11) reconstructed using square root/square non-linear processing.

using an edge-bonded array,[8] and longitudinal waves in aluminum using a new direct contact array. The latter array represents an important extension to the imaging system for NDE applications, since it is often either inconvenient or impractical to provide a water buffer between the transducer and the test specimen. A similar contacting shear wave transducer array is currently being developed.

Figure 16 demonstrates the ability of the system to distinguish small targets close to a large specular reflector. The benefits of non-linear processing are clear.

The next images (Fig. 17) were obtained using an edge-bonded surface wave array. The surface waves are coupled to the test specimen by means of a polyethylene strip and a thin smear of liquid couplant (Sonotrac). The acoustic images clearly show the three saw cuts, the corner of the block, and ghost images of the saw cuts reflected in the corner of the block. It is also just possible to discern two short bars, one in the upper left corner and the other along the bottom edge of the image. These correspond to the two edges of the block, indicating that the angular acceptance of the array is at least $\pm 45°$. The value of non-linear processing in reducing sidelobe levels is apparent in the non-linear image of Fig. 17.

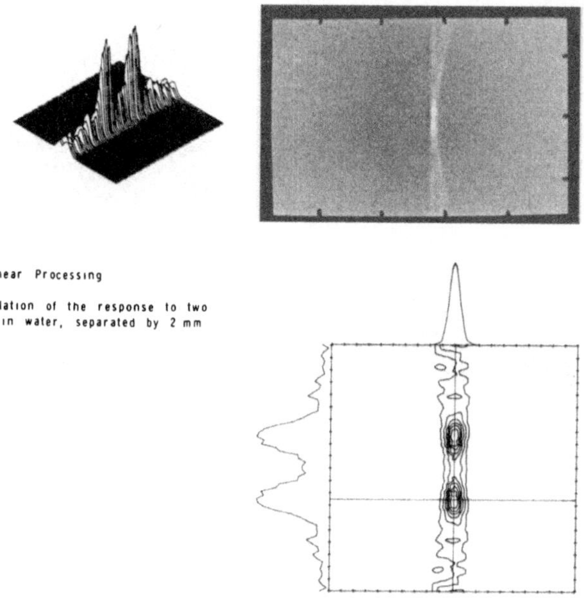

Fig. 14. Image of two point reflectors, separated by 2 mm , reconstructed from simulated echo data.

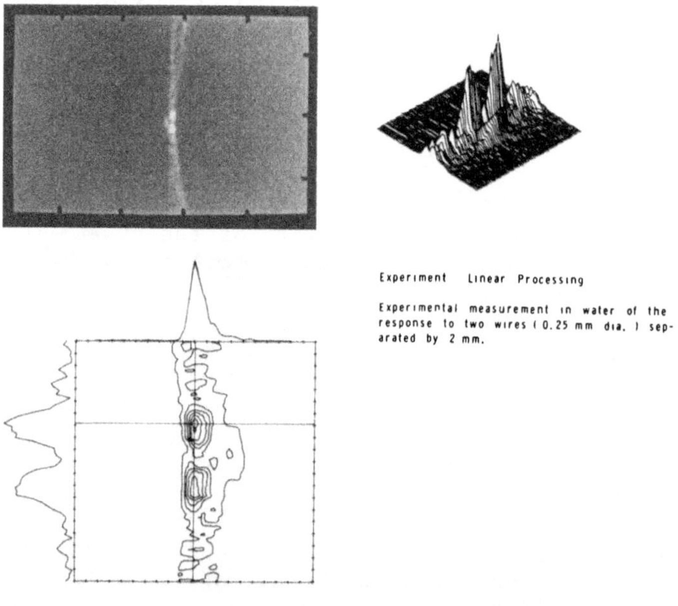

Fig. 15. Image of two thin wires (.25 mm diameter), laterally separated by 2 mm , at 80 mm range.

LONGITUDINAL WAVES IN WATER

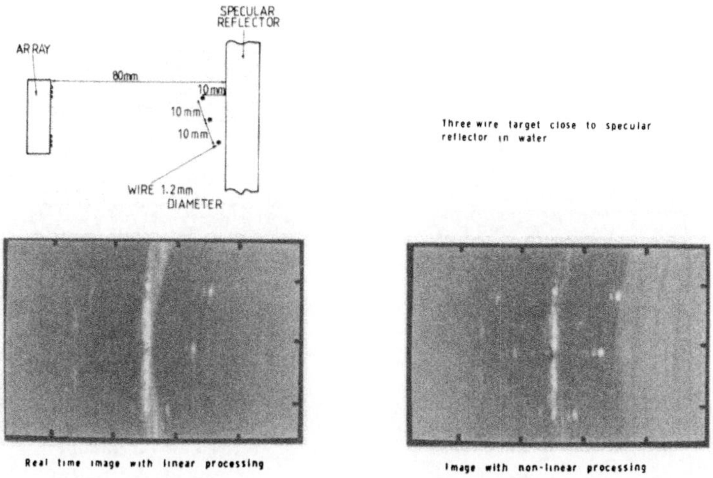

Fig. 16. Image of three wires (1.25 mm diameter) in front of a specular reflector in water.

SURFACE WAVES: TEST SPECIMEN

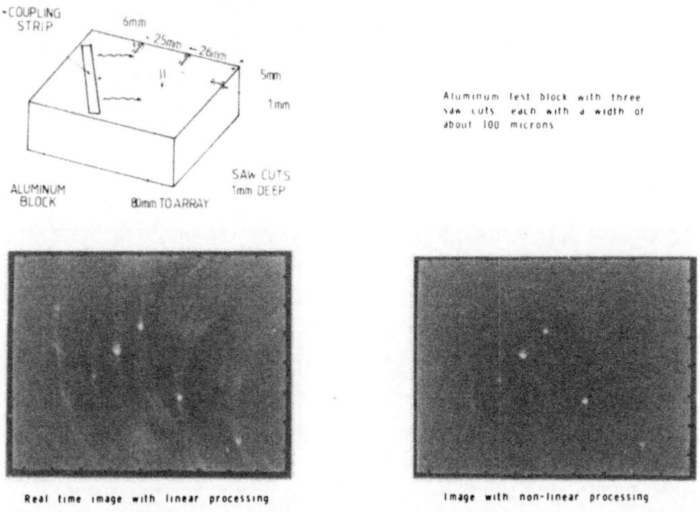

Fig. 17. Surface acoustic wave image of a test block with three saw cuts along its edges.

SURFACE WAVES: FATIGUE CRACK

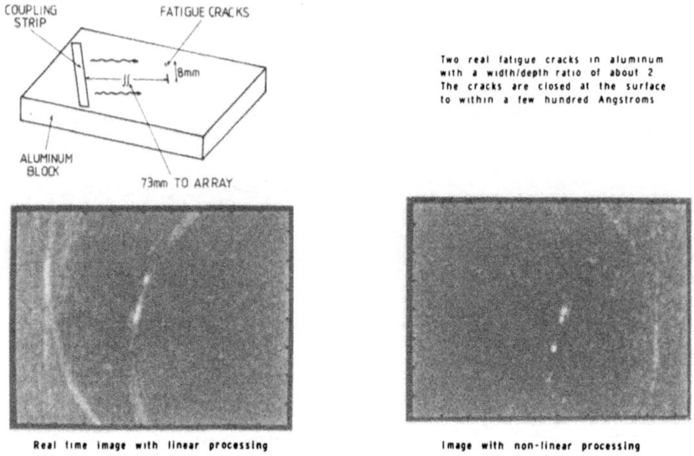

Fig. 18. Surface acoustic wave image of fatigue cracks in aluminum sample.

DIRECT CONTACT LONGITUDINAL WAVES IN METAL

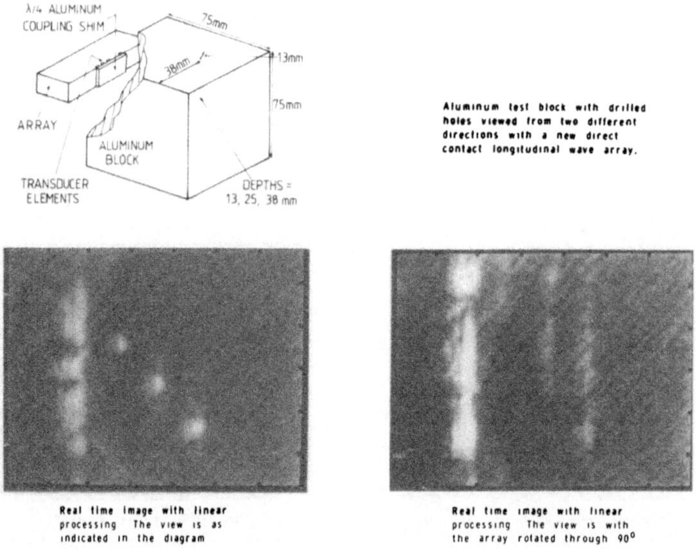

Fig. 19. Image of three 1/16 inch drill holes in an aluminum block, taken with a new direct contact longitudinal wave array.

A further example of surface wave imaging is shown in Fig. 18. Here the test specimen was an aluminum plate with two fatigue cracks, measuring 3 mm and 1 mm wide, with a width to depth ratio of about 2 . The cracks are closed at the surface to within a few hundred Angstroms.

The images in Fig. 19 were taken with the new direct contact longitudinal wave array. The test specimen was an aluminum block with 1/16 inch holes drilled in it. The placement of the array and the block for the two viewing directions is indicated in Fig. 19. The two linear real-time images are also shown in the figure. It should be emphasized that these images are the first that we have taken using this new type of array and therefore represent only preliminary results.

CONCLUSIONS

We have discussed the theoretical performance of a digital synthetic aperture imaging system and given examples of acoustic images obtained in real-time with such a system.

The theoretical approach is similar to physical optics theory but differs in some important respects. First, the transmit/receive nature of our imaging aperture results in a lateral resolution twice as good as an equivalent single lens system. Second, we may image with pulsed rather than CW excitation. This gives reduced grating lobe and sidelobe levels and improves the range resolution. Finally, we have explored the effects of non-linear processing on the synthetic aperture reconstruction technique. Non-linear processing gives slightly better resolution and substantially lower sidelobe levels but has the disadvantage that adjacent object points interact with each other. The use of pulsed excitation can decrease this interaction.

We have presented a series of simulated and experimental images to demonstrate the single point and two points response of the real-time imaging system, showing that the system achieves diffraction limited resolution images at a rate of about 30 frames per second.

Finally, several images of test samples were presented, indicating the utility of the imaging system for nondestructive evaluation. These include imaging in water, with surface waves on aluminum, and with bulk longitudinal waves in aluminum.

ACKNOWLEDGEMENTS

We wish to acknowledge the work of Alan Selfridge and Pierre Khuri-Yakub in developing the arrays used in this imaging system.

This work was supported by the Air Force Office of Scientific Research under Contract No. F49620-79-C-0217.

REFERENCES

1. P.D. Corl, G.S. Kino, C.S. DeSilets, and P.M. Grant, "A Digital Synthetic Focus Acoustic Imaging System," Acoustic Imaging, vol. 8, A.F. Metherell (ed.), Plenum Press, New York, 1980.
2. G.S. Kino, P.M. Grant, P.D. Corl, and C.S. DeSilets, "Digital Synthetic Aperture Acoustic Imaging for NDE," Proc. ARPA/AFML Review of Progress in Quantitative NDE, La Jolla, California, July 1978.
3. P.D. Corl, P.M. Grant, and G.S. Kino, "A Digital Synthetic Focus Acoustic Imaging System for NDE," Proc. Ultrasonics Symposium, Cherry Hill, New Jersey, September 1978.
4. D. Corl, G.S. Kino, D. Behar, H. Olaisen, and P. Titchener, "Digital Synthetic Aperture Acoustic Imaging System," Proc. ARPA/AFML Reivew of Progress in Quantitative NDE, La Jolla, California, July 1979.
5. P.D. Corl and G.S. Kino, "A Real-Time Synthetic-Aperture Imaging System," Acoustic Imaging, vol. 9, K. Wang (ed.), Plenum Press, New York, 1980.
6. G.S. Kino and C.S. DeSilets, "Design of Slotted Transducer Arrays with Matched Backings," Ultrasonic Imaging, vol. 1, pp. 189-209, 1979.
7. G.S. Kino, B.T. Khuri-Yakub, A. Selfridge, and H. Tuan, "Development of Transducers for NDE," Proc. ARPA/AFML Review of Progress in Quantitative NDE, La Jolla, California, July 1979.
8. H.C. Tuan, A.R. Selfridge, J. Bowers, B.T. Khuri-Yakub, and G.S. Kino, "An Edge-Bonded Surface Acoustic Wave Transducer Array," Proc. Ultrasonics Symposium, New Orleans, Louisiana, September 1979.
9. G.S. Kino, "Acoustic Imaging for Nondestructive Evaluation," Proc. IEEE, vol. 67, pp. 510-523, May 1979.
10. J. Havlice, G. Kino, W. Leung, H. Shaw, K. Toda, T. Waugh, D. Winslow, and L. Zitelli, "An Electronically Focused Two-Dimensional Acoustic Imaging System," Acoustic Holography, vol. 6, N. Booth (ed.), Plenum Press, New York, 1975.
11. K. Bates, private communication, 1980.

EFFECT OF THE PHOTOCONDUCTIVE LAYER ON THE RESOLUTION OF OPTO-ACOUSTIC TRANSDUCERS

Behzad Noorbehseht,* Glen Wade and Carl Schueler

Department of Electrical & Computer Engineering
University of California
Santa Barbara, CA 93106

ABSTRACT

Resolution of opto-acoustic transducers (OAT's) may be determined using linear systems theory. An OAT is treated as a linear spatial system. Each of its constituent layers is regarded as a subsystem characterized by a unique spatial transfer function. In a previous paper we have shown that the resolution of OATs is determined, to a very good approximation, by the transfer function of the piezoelectric layer alone. In this paper we show that the transfer function of the piezoelectric layer is affected by the presence of the photoconductive backing layer in such a way as to lower the resonant frequency of the OAT. We show also that the normalized resolution, the resolution measured in wavelengths at the operating frequency of the device, is affected if the OAT is operated off resonance. The normalized resolution increases for frequencies below the resonant frequency and decreases for frequencies above resonance.

INTRODUCTION

In previous papers[1,2] we have shown how the resolution characteristics of an Opto-Acoustic transducer may be studied using linear systems theory. We have shown that the OAT may be

---

* Now with the Electrical Engineering Department, University of Houston, Houston, Texas 77004

regarded as a linear spatial system with an optical input and an acoustical output. Thus it may be characterized by a unique transfer function called the Opto-Acoustic Spatial Transfer Function (OASTF). The OASTF consists of two parts: one representing the diffraction effects, called the diffraction-limited spatial transfer function; the other accounting for the aberrations, called the aberration-limited spatial transfer function. An OAT-based imaging system, such as that depicted in Fig. 1(a), may therefore be represented by the block diagram of Fig. 1(b), where H_D represents the diffraction-limited transfer function, and H_A the aberration-limited transfer function.

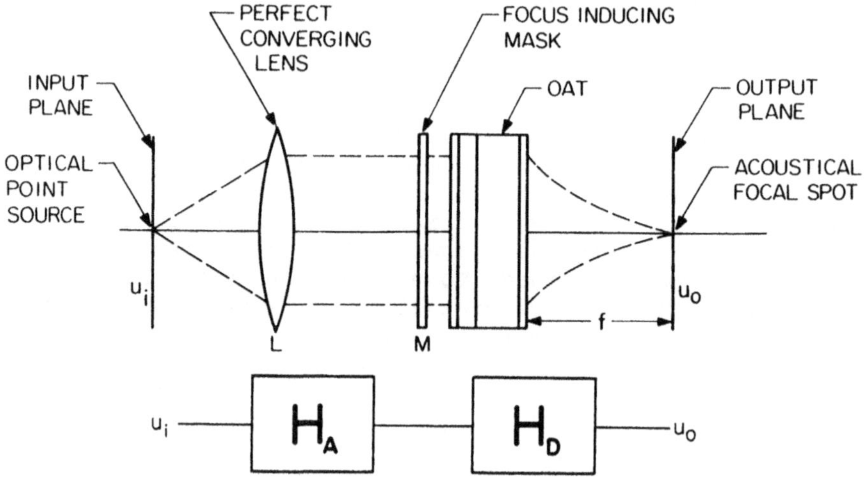

Figure 1 (a) A simplified scheme to achieve a focused sound beam using the OAT.

(b) A block diagram representation of the system shown in (a).

Our subject of discussion in this paper will be H_A. We have previously shown it to be almost identical to the transfer function of the piezoelectric layer, but the piezoelectric layer spatial response is affected by the presence of the photoconductive backing layer. We show that this effect causes the OAT resonant frequency to be lower than the resonant frequency of the piezoelectric layer alone. We also show how the resolution of the OAT may be changed by varying the operating frequency of the device.

ABERRATION-LIMITED TRANSFER FUNCTION

a. No Backing Layer

Because of their small thickness, the photoconductive and electrode layers have much broader spatial transfer functions than the piezoelectric layer does. Therefore, only the piezoelectric layer need be considered to derive H_A, the aberration-limited transfer function of the OAT[1].

To find the one-dimensional piezoelectric transfer function, we obtain the Fourier transform of the output due to a line electrode input pattern. First we relate the input electrode pattern to the elastic wave excitation in the transducer. Then we decompose the elastic waves into their spatial frequency components and find the transmitted phase and amplitude of each component. (Fig. 2).

Following the above procedure yields[2,3]

$$H_A(f_x) = \frac{T_{12}(\alpha)}{jk_1\cos\alpha} \quad \frac{[1-\exp(-jk_1d_1\cos\alpha)][1-\Gamma_{10}(\alpha)\exp(-jk_1d_1\cos\alpha)]}{1-\Gamma_{12}(\alpha)\Gamma_{10}(\alpha)\exp(-jk_1 2d_1\cos\alpha)} \tag{1}$$

where $k_1 = 2\pi/\lambda_{pl}$ and λ_{pl} = wavelength of the compressional wave in the piezoelectric material. T_{12} and Γ_{12} are the transmission

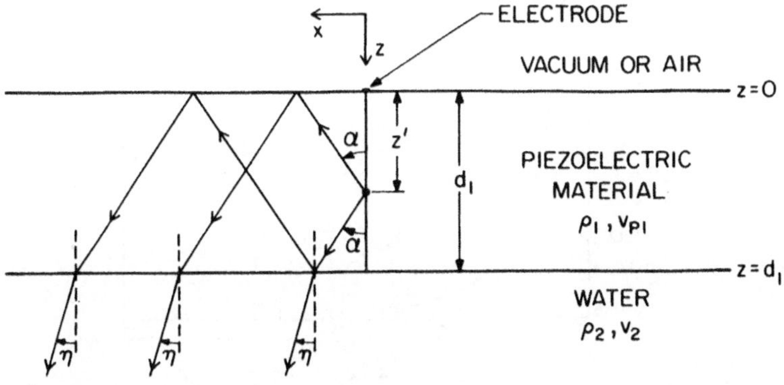

Figure 2 Generation and transmission of a plane-wave inside the piezoelectric layer due to a line electrode excitation.

and reflection coefficients at the piezoelectric-water interface, and Γ_{10} is the reflection coefficient at the piezoelectric-vacuum interface. f_x, α, and η are related by

$$f_x = \frac{\sin \alpha}{\lambda_{p1}} = \frac{\sin \eta}{\lambda_2} \qquad (2)$$

where f_x is the spatial frequency variable. α and η are the angles of the incident and transmitted waves, respectively. Numerical calculation of (1) was carried out for the case of a PZT-5 transducer, having a thickness d_1 equal to half the compressional wavelength at an operating frequency of 3 MHz. A list of material properties of PZT-5 is given in Table I.

Figure 3 shows a plot of (1), where the horizontal axis is the normalized spatial frequency ν_x given by

$$\nu_x = f_x \lambda_2 \qquad (3)$$

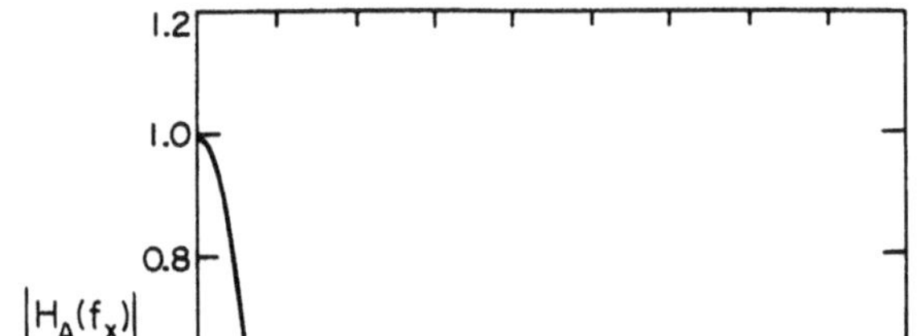

Figure 3 Magnitude of the aberration-limited transfer function versus normalized spatial frequency $f_x\lambda_2$.

Table I Material Properties of PZT-5 and CdS

	Density (10^3 Kg/m^3)	Elastic Constants (10^{10} N/m^2)	P-Wave Velocities (10^3 m/s)	SV-Wave Velocities (10^3 m/s)
PZT-5	$\rho_1 = 7.75$;	$c_{11}^E = 12.1$;	$V_{p1} = (c_{11}^E/\rho_1)^{1/2} = 3.951$;	$V_{sv1} = (c_{44}^E/\rho_1)^{1/2} = 1.650$
		$c_{44}^E = 2.11$		
CdS	$\rho_3 = 4.82$;	$c'^E_{11} = 9.07$;	$V_{p3} = (c'^E_{11}/\rho_3)^{1/2} = 4.338$;	$V_{sv3} = (c'^E_{44}/\rho_3)^{1/2} = 1.766$
		$c'^E_{44} = 1.504$		

where λ_2 is the acoustic wavelength in water. As can be seen in fig. 3, the transfer function goes to zero at $\nu_x = \nu_{xcl}$ which corresponds to $\alpha = 90°$. For $\nu_x > \nu_{xcl}$, the corresponding plane waves are evanescent[4], and we have ignored evanescent waves.

The shape of the transfer function, which determines the resolution of the device, is itself determined by the various parameters in (1). In particular it is affected by the applied temporal frequency f and the reflection coefficient Γ_{10} at the back side of the piezoelectric layer. Both of these parameters are affected by the presence of the photoconductive layer. These are studied in the next section.

b. Effect of the Photoconductive Backing Layer

In this section we include the effect of the photoconductive backing layer, because in actual OAT structures the piezoelectric layer is typically covered on one side by a photoconductive layer. The effect of this layer is to modify the expression for the reflection coefficient Γ_{10} in (1), without any change in the form of the equation. This results in a change of the resonant frequency of the OAT.

The modified reflection coefficient, denoted here by Γ_{eff}, can be determined by solving the problem of reflection of a plane elastic wave at the boundaries of a solid layer separating a solid and a liquid (in this case vacuum) medium. The complete solution to this problem is too cumbersome to use in practical problems. Therefore, we have solved the problem by invoking some reasonable simplifying assumptions. Our resulting expression for Γ_{eff} takes the form[2]

$$\Gamma_{eff} = \Gamma_{13} + \frac{T_{13}\, T_{31}\, \Gamma_{30}\, \exp(-j2k_3 d_3 \cos\beta)}{1 - \Gamma_{31}\, \Gamma_{30}\, \exp(-j2k_3 d_3 \cos\beta)} \tag{4}$$

where β is the angle of the elastic wave relative to the z-axis and k_3 is its wave number in the photoconductive layer. d_3 is the thickness of that layer. The other symbols are graphically defined in fig. 4. In general, $\Gamma_{eff} \neq \Gamma_{10}$. While Γ_{10} is real, Γ_{eff} is complex and has nonzero phase. However, we have found that for thin photoconductors such as those encountered in typical OAT structures (i.e. thicknesses from 10 to 200 μm), $|\Gamma_{eff}| \cong |\Gamma_{10}|$ ($|\;|$ ≡ absolute magnitude).

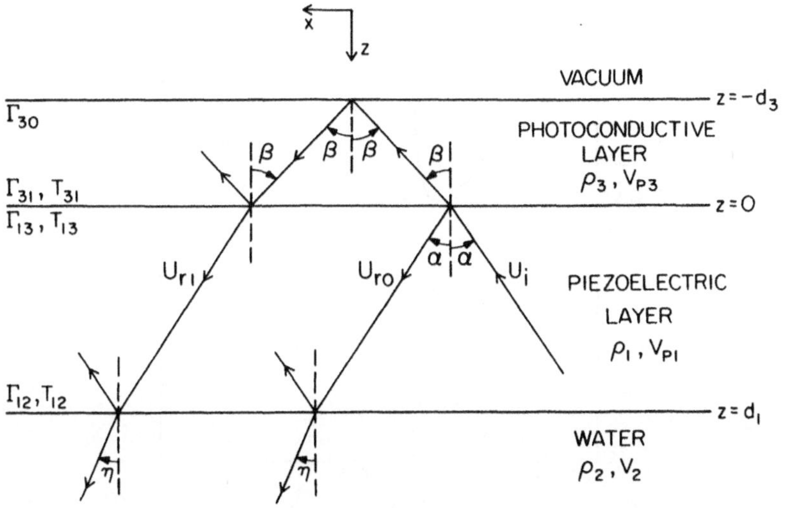

Figure 4 View of the OAT along the y-axis showing reflection and transmission of an incident wave U_i, at the photoconductive layer and water boundaries of the piezoelectric layer. Electrode layers are not shown.

Our numerical calculations are based on the use of PZT-5 as the piezoelectric material, and Cadmium Sulphide (CdS) as the photoconductive material. Relevant material properties of CdS are given in Table I. Figure 5 shows $|\Gamma_{eff}|$ for arbitrary values of $d_3 << \lambda_{p3}$. Note that $|\Gamma_{eff}|$ is essentially independent of d_3. Figure 6 shows the phase of Γ_{eff} versus ν_x for four values of $d_3 << \lambda_{p3}$. For $d_3 = 0$, $\Gamma_{eff} = \Gamma_{10}$.

Because Γ_{eff} is complex, the resonant frequency f_o' of the OAT is different from the resonant frequency f_o of the air-backed transducer. If the OAT is operated at f_o' instead of f_o, the shape of H_A is unaffected by the photoconductive layer and is the same as that shown in fig. 3. f_o' can be determined numerically. Table II lists f_o' for the values of d_3 given in

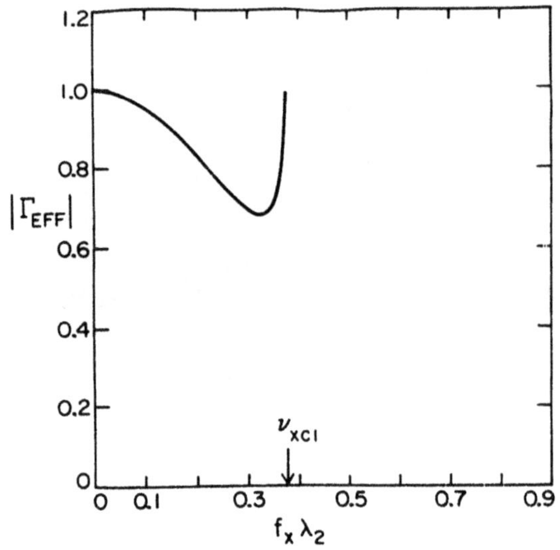

Figure 5 Variations of the magnitude of Γ_{eff} versus normalized spatial frequency $f_x\lambda_2$.

Table II Resonance Frequencies of the OAT for Different Thicknesses of the Photoconductive Layer

CdS Layer Thickness $(d_3/\lambda_{p3}) \times 10^2$	0.0	2.075	4.149	6.224	8.229
Normalized Resonance Frequency (f'_o/f_o)	1.0	0.972	0.949	0.920	0.894

fig. 6. The values of d_3 are normalized by the compressional wavelength λ_{p3} in the photoconductor at f_o. These correspond to 30, 60, 90, and 120 μm at 3 MHz.

Figure 7 is a plot of the normalized resonant frequency f_o'/f_o as a function of normalized photoconductor thickness d_3/λ_{p3}. This plot may be used to obtain f_o' for an arbitrary f_o,

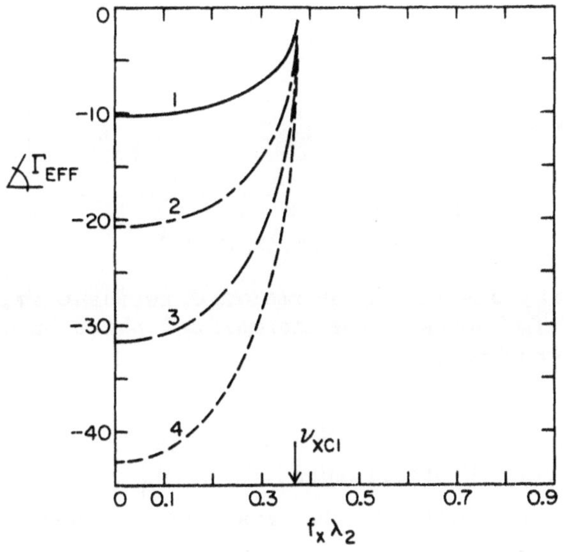

Figure 6 Variations of the phase of the reflection coefficient Γ_{eff} with normalized spatial frequency $f_x\lambda_2$ for small values of d_3.

1. $d_3 = 0.02075\ \lambda_{p3}$ (30 μm at 3 MHz)
2. $d_3 = 0.04149\ \lambda_{p3}$ (60 μm at 3 MHz)
3. $d_3 = 0.06224\ \lambda_{p3}$ (90 μm at 3 MHz)
4. $d_3 = 0.08229\ \lambda_{p3}$ (120 μm at 3 MHz)

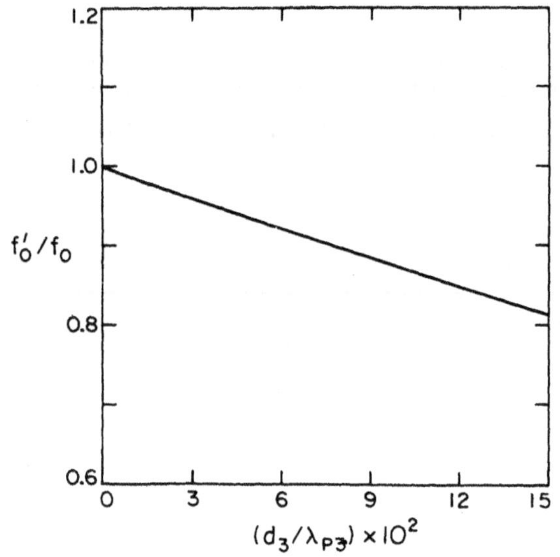

Figure 7 Variations of the normalized resonant frequencies of the OAT versus the normalized photoconductive layer thicknesses.

provided $d_3/\lambda_{p3} \leq 0.15$ (corresponding to d_3 = 216 μm at 3 MHz). If we operate the OAT at the resonant frequency f_o of the air-backed transducer, (in our case f_o = 3 MHz) then the shape of $|H_A|$ is changed. This change in $|H_A|$ depends on d_3. Figure 8 shows plots of $|H_A|$ for the four values of d_3 given in Table II. The peaks in $|H_A|$ occur at values of ν_x given by

$$\nu_x = (v_2/v_{p1})(1 - (f_o'/f)^2)^{\frac{1}{2}} \quad , \qquad f \geq f_o' \qquad (5)$$

where f_o' is the resonant frequency of the OAT and f is the operating frequency of the OAT (in this case f = f_o, the resonant frequency of the air-backed transducer), v_2 and v_{p1} are the compressional wave velocities in water and in the piezoelectric material, respectively.

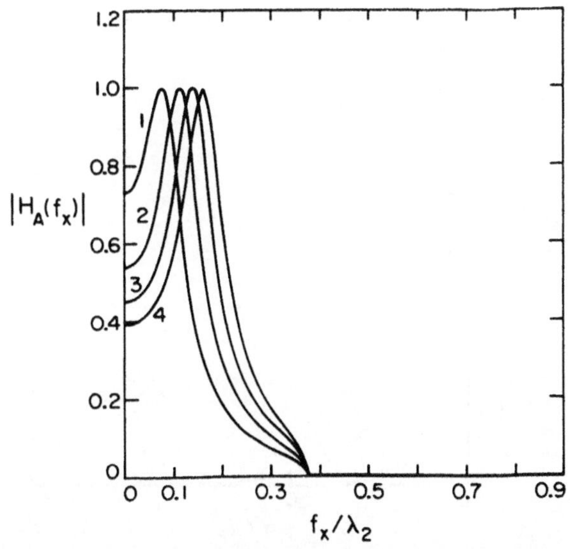

Figure 8 Variations of the magnitude $|H_A|$ versus normalized spatial frequency $f_x\lambda_2$. Each curve was obtained at the resonant frequency of the piezoelectric layer.

1. $d_3 = 0.02075\ \lambda_{p3}$ (30 μm at 3 MHz).
2. $d_3 = 0.04149\ \lambda_{p3}$ (60 μm at 3 MHz)
3. d_3 $0.06224\ \lambda_{p3}$ (90 μm at 3 MHz)
4. $d_3 = 0.08299\ \lambda_{p3}$ (120 μm at 3 MHz)

The transfer functions in fig. 8 correspond to wider line-spread functions than the one in fig. 2. As an example, fig. 9 shows the line-spread functions for an OAT with the same specifications as above, with d_3 = 30 μm, for two values of f. The wider curve corresponds to $f = f_o$ while the narrow one corresponds to $f = f'_o$.

To further illustrate this effect, plots of $|H_A|$ and the corresponding line-spread functions for values of f close to the

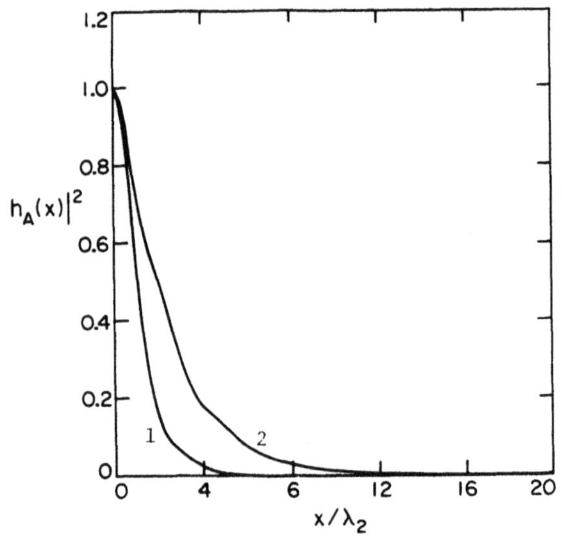

Figure 9 Variations of the magnitude squared of $h_A(x)$ versus x/λ_2 with d_3 = 30 μm, f_o = 3 MHz.

1. $f = f'_o$

2. $f = f_o$

OAT resonant frequency f'_o, for an OAT with d_3 = 30 μm, are shown in figs. 10 and 11. It can be seen in fig. 10 that when $f > f'_o$, there is a reduction in $|H_A|$ at lower spatial frequencies, compared to when $f = f'_o$. (The highest spatial frequency is still limited to ν_{xcl}.) This results in widening the corresponding line-spread function, as shown in fig. 11. A decrease of f below f'_o results in widening $|H_A|$, relative to the $f = f'_o$ case, because the OAT is no longer resonant for any of the spatial frequencies in the range $0 \leq \nu_x \leq \nu_{xcl}$. This in turn results in a narrower line-spread function, as shown for two cases in fig. 11, and as explained below.

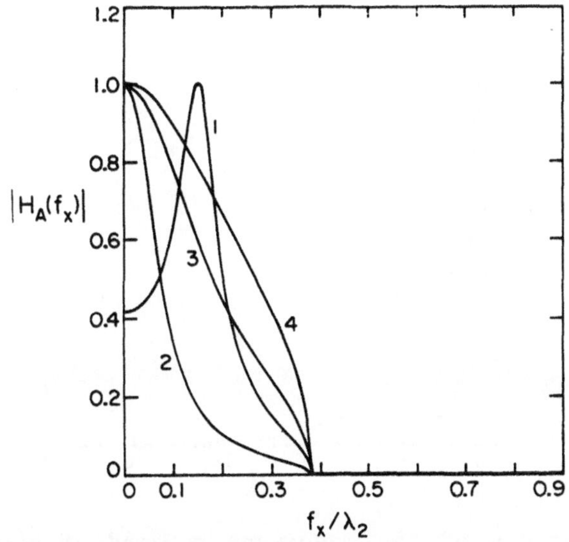

Figure 10 Variations of the magnitude of $H_A(f_x)$ versus f_x/λ_2, for d_3 = 30 μm and f_o = 3 MHz.

1. $f = 1.1\ f_o'$

2. $f = f_o'$

3. $f = 0.9\ f_o'$

4. $f = 0.8\ f_o'$

DISCUSSION

Although the transfer function of the photoconductive layer has no direct effect on the aberration-limited transfer function H_A, the photoconductive layer does affect the resolution of the OAT by loading the piezoelectric layer, so that the OAT resonant frequency is reduced. As long as the OAT is operated at its resonant frequency, the aberration-limited transfer function is unaffected by the presence or absence of the photoconductive layer.

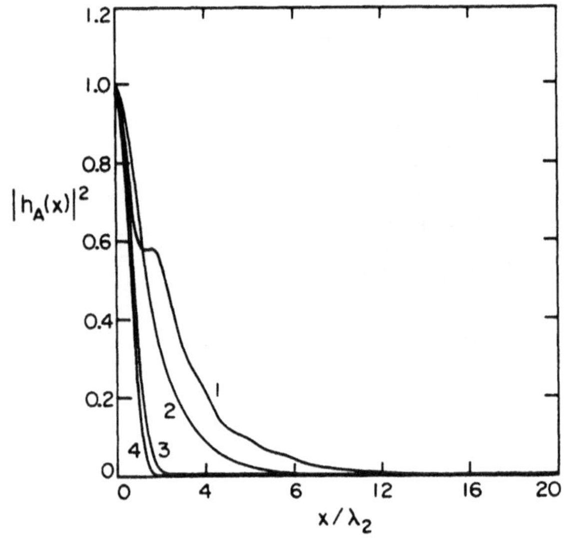

Figure 11 Variations of the magnitude squared of $h_A(x)$ versus x/λ_2 for d_3 = 30 μm and f_o = 3 MHz.

1. $f = 1.1\ f'_o$

2. $f = f'_o$

3. $f = 0.9\ f'_o$

4. $f = 0.8\ f'_o$

If the OAT is operated below resonance, then normalized resolution is enhanced. If the OAT is operated above resonance, then normalized resolution is worsened. This can be explained with reference to fig. 12. When the OAT is operated on resonance, the plane wave corresponding to zero spatial frequency, when $\alpha = 0$, is resonant and therefore suffers the least attenuation before it is transmitted into water. All the other plane waves, corresponding to the other spatial frequencies, are attenuated more. This is because their effective wavelengths λ_{eff} in the thickness direction exceed λ_{pl}, as illustrated in fig. 12. This can be seen by noting that λ_{eff} and λ_{pl} are related by

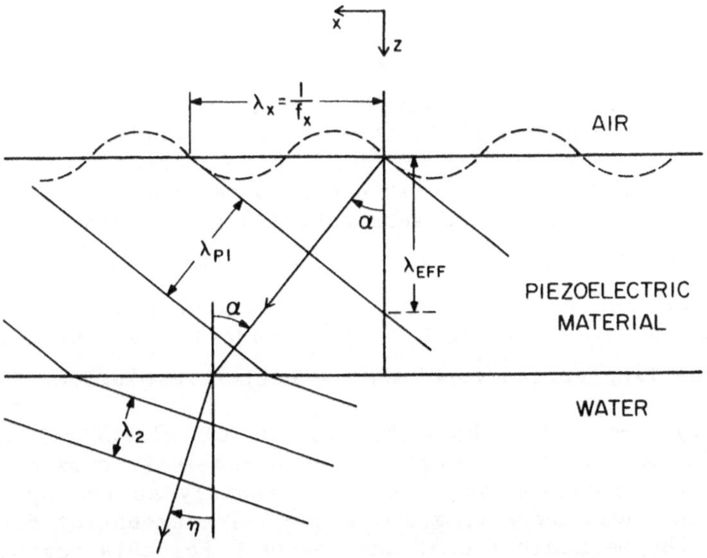

Figure 12 One of the spatial frequencies generated by the electrode pattern in fig. 2, together with its corresponding plane-wave in the transducer. This plane-wave is eventually transmitted into water as shown.

$$\lambda_{eff} = \lambda_{pl}/\cos\alpha \tag{6}$$

When the OAT is operated at a frequency higher than its resonant frequency, the zero spatial frequency is no longer resonant because its corresponding wavelength λ_{pl} is shorter than λ'_o, the required wavelength for resonance. However, depending on the applied frequency, a spatial frequency could exist for which $\lambda_{eff} = \lambda'_o$. Consequently, at that spatial frequency, $|H_A|$ exhibits a peak, as shown in fig. 8, and by curve 1 of fig. 10. This results in worse resolution, because we measure resolution at the 40% intensity of the line-spread function.[5] Notice, in fig. 11, that the top of the line-spread function is narrower than for the $f = f'_o$ case. The reduction in low spatial frequency content of H_A results in a broadening at the base of the line-spread function, where resolution is defined.

For operating frequencies below resonance, none of the spatial frequencies in the range $0 \leq \nu_x \leq \nu_{xcl}$ are resonant because the lowest effective wavelength λ_{eff} in the thickness direction is too large to satisfy the resonance condition. If the OAT has reasonably high Q, then a small reduction in applied frequency Δf below resonance, will more rapidly reduce the amplitude of spatial frequencies close to resonance than it will spatial frequencies that are substantially off resonance. $|H_A|$ will be wider than at resonance since the peak of $|H_A|$ at the origin is lowered more rapidly than at higher spatial frequencies. This increase in width of $|H_A|$ causes a decrease in the width of the line-spread function and better resolution.

Finally, we note that the amount of electrical energy converted to mechanical energy in the piezoelectric material is a maximum at resonance and decreases rapidly as the operating frequency is moved away from resonance. The amount of decrease depends on the mechanical Q of the device. For this reason, the device can only be operated at frequencies which are not too far off resonance, and this is why we have considered only cases where $\Delta f \ll f'_o$.

ACKNOWLEDGEMENT

The authors wish to thank the Automation, Bioengineering, and Sensing Systems Section of the National Science Foundation's Engineering Division for its generous support of this work, and Tracy Hamilton of the Electrical and Computer Engineering Department at UCSB for preparing the manuscript.

REFERENCES

1. B. Noorbehesht and G. Wade, "Spatial Frequency Characteristics of Opto-Acoustic Transducers," Acoustical Imaging, Vol. 9, K. Wang Ed., Plenum Press, New York, 1980.

2. B. Noorbehesht and G. Wade, "Resolution of Opto-Acoustic Transducers," IEEE Trans. Sonics and Ultrasonics, 1980. (Submitted).

3. B. Noorbehesht, G. Flesher, and G. Wade," Spatial Response of Arbitrarily Electroded Piezoelectric Plates by Plane-Wave Decomposition," Ultrasonic Imaging, Vol. 2, No. 2, 1980.

4. J.W. Goodman, Introduction to Fourier Optics, McGraw-Hill, New York, pp. 50-51, 1968.

5. M. Ahmed, "The Response of Piezoelectric Face Plates Used in Ultrasonic Imaging Systems," IEEE Trans. Sonics and Ultrasonics, SU-25, No. 6, pp. 330-339, November 1978.

INVESTIGATION OF A LIQUID CRYSTAL ACOUSTO-OPTIC CONVERSION CELL

W. Hamidzada, S. Letcher and S. Candau

Department of Physics
Univ. of Rhode Island
Kingston, RI 02881 USA

Univ. Louis Pasteur
4, Rue Blaise Pascal
67070 Strasbourg, France

An acousto-optic conversion cell using nematic liquid crystals provides a low-cost method for planar detection of an ultrasonic radiation field. Although the information received is somewhat rudimentary, in some cases it could be considered as a replacement for an array of transducers. Nematic liquid crystals are fluids that, because of the long-range order of the orientation of their long molecules, are optically uniaxial. Proper surface treatment of a substrate will cause the molecules--and the optic axis--to align normal to the surface, and a thin layer of material between two treated glass plates will have uniform alignment throughout the volume with the optic axis normal to the plates. This is the usual configuration of an acousto-optic conversion cell. When placed between crossed polarizers, light at normal incidence is blocked, but mechanical (or electrical) perturbations will cause the optic axis to tip and light to be transmitted. Because of the weakness of the elastic restoring forces for orientational distortions, very weak perturbations are observable.

It has been known for some time that an acoustic field provides sufficient power to perturb the orientation,[1-4] but only in the last five years have thorough studies been undertaken. Although a number of models have been proposed for the coupling of a sound wave to the molecular orientation, it has now been established that the coupling is primarily via acoustic streaming.[5-14] Conoscopic studies and irradiation of a segment of a cell composed of an annular channel[13] have clearly demonstrated the existence of streaming in the thin layer of liquid crystal. Streaming would not be one's first choice for the mechanism of an acousto-optic device. In addition to being a square-law detector responsive only to the integrated acoustic intensity, the response is slow and the resolution is poor due to the large-scale flow. Nothing can be done about the square-law nature of the detector, but the latter two conditions can be ameliorated. An electric field applied across a nematic with negative dielectric anisotropy ($\varepsilon_{\perp} > \varepsilon_{\parallel}$) will increase the sensitivity and shorten the rise time and when applied to a sample with positive anisotropy will reorient the optic axis and shorten the decay time.[4,12] Some mixtures of nematics that have a frequency-dependent anisotropy exhibit both effects.[12] The resolution can be improved by introducing a grid network between the glass plates that restricts the liquid flow to dimensions on the order of a wavelength.[9,10,12] The grid has the added advantage of supporting the glass plates, which can thus be made very thin.

For oblique incidence of a cw sound beam, the flow is mainly parallel to the plane of the layer and the transmitted optical intensity is predicted[13] to obey

$$J = J_o + AI_t^n, \tag{1}$$

where J is the transmitted optical intensity, I_t is the acoustic intensity, and n = 4. In cw experiments on cells with grids, Eq. (1) is satisfied with n between 3.5 and 4.[10] No model exists, however, to describe the flow in a cell with a grid. One would expect a complicated pattern including, perhaps, microstreaming generated at the grid edges.[15]

Even in a cell without a grid, quantitative deviations from the streaming model are not unexpected. The theory is linear in the sense that only small angles of molecular tilt are considered, but a finite-amplitude sound wave can create finite tilt angles and can also generate defects, in the form of disclinations, that interfere with the flow.

In order to test further the nature of the streaming mechanism and to determine to what extent this type of cell can be used in a pulsed mode, we have studied the response of the cell to pulsed ultrasound. The apparatus is shown schematically in Fig. 1.

The rf pulses applied to the transducer are characterized by a peak-to-peak voltage, V, pulse width, t, and repetition period, T. The instantaneous acoustic power at the cell is proportional to V^2 and the integrated acoustic power goes as V^2t/T. All of the measurements reported here were made at 4.5 MHz.

Figure 2 shows the transmitted optical intensity, J, in arbitrary units, for a range of values for t and T when V = 263 volts. Generally, the optical signal increases with increasing total acoustic power, but for very low power there is no response ant at high power the optical signal saturates. The optical signal as a function of duty cycle, t/T, for constant pulse height is shown in Fig. 3. Most of the points fall on a universal curve, ranging from no response when $t/T < 2 \times 10^{-3}$ to saturation when $t/T > 2 \times 10^{-2}$.

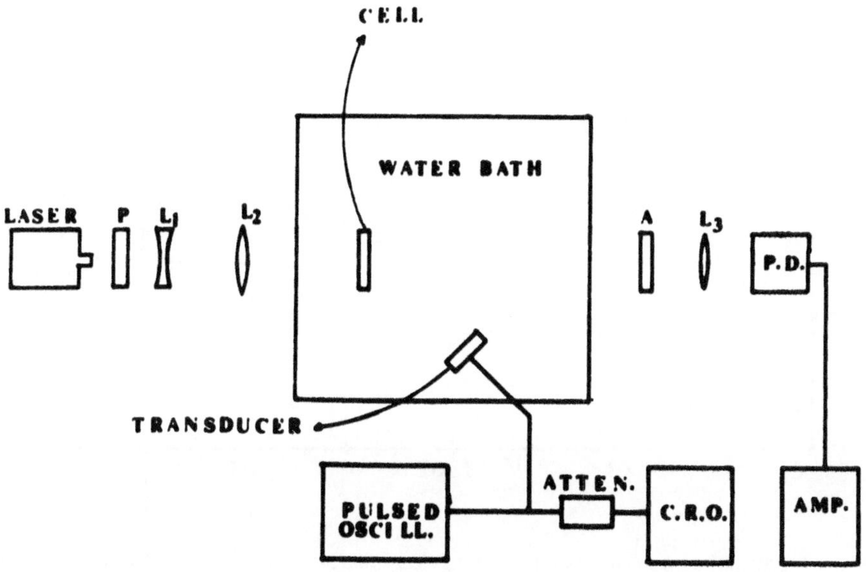

Fig. 1. Apparatus for study of liquid crystal cell.

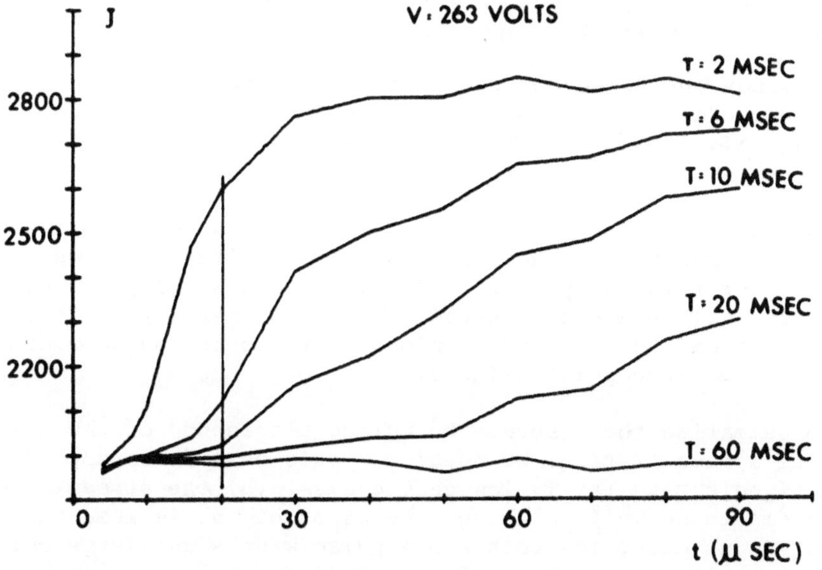

Fig. 2. Transmitted optical intensity, J (arbitrary units) vs. pulse width for constant pulse amplitude.

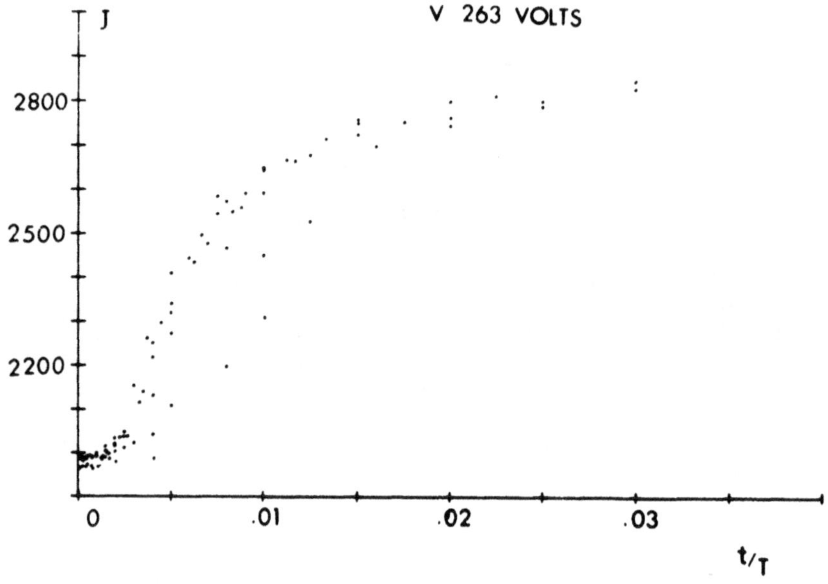

Fig. 3. Transmitted optical intensity vs. acoustic duty cycle for constant pulse amplitude.

The points that do not fall on this curve represent very short pulse widths, $t < 20$ μsec. Except for very short pulses, then, the pulsed mode can generate dc streaming whose effect depends on the fraction of time the sound is on.

To study the effect of the total integrated power, including the pulse amplitude, on the response of the cell, a log-log plot was made of $J - J_o$ vs. V^2t/T This gave an exponent, n, between 2 and 3, not in very good agreement with the model or with the cw results.

In Fig. 4, we have plotted T vs. V^2t for various combinations of the three parameters that give the same, fairly weak, optical intensity. For relatively large duty cycles and small pulse heights, the points fall on a single straight line, satisfying Eq. (1). The departures from the line occur primarily for short pulse widths and corresponding large pulse heights.

To summarize the response to pulsed ultrasound of the nematic cell with spacer grating, we note: (a) this cell does respond to pulsed signals with fairly low duty cycles, (b) the streaming mechanism seems to be valid although the exponent, n, is less than in the continuous case, (c) both short pulse widths and large pulse amplitudes result in departures from Eq. (1).

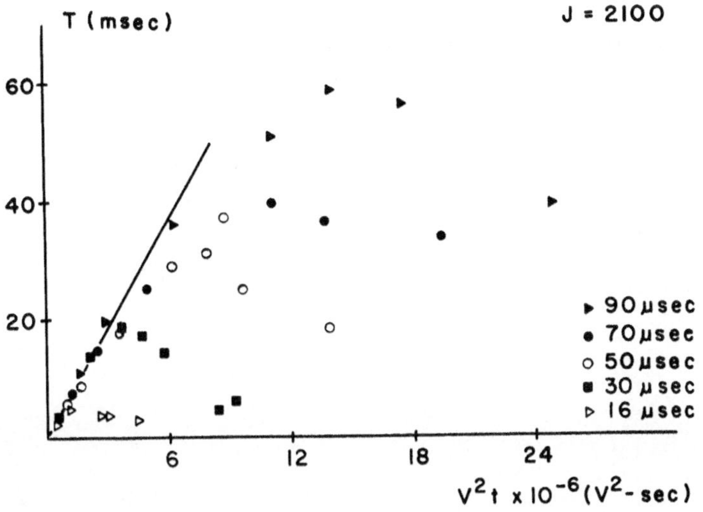

Fig. 4. T vs. V^2t for constant J. The straight line behavior represents Eq. (1).

REFERENCES

1. V. Zolina, Trudy Lomonosov Inst. Akad. Nauk. SSSR 8, 11 (1936).
2. L.W. Kessler and S.T. Sawyer, Appl. Phys. Lett. 17, 440(1970).
3. H. Mailer, K.L. Likins, T.R. Taylor and J.L. Fergason, Appl. Phys. Lett. 18, 105 (1971).
4. P. Greguss, Acustica 29, 52 (1973).
5. K. Miyano and Y.R. Shen, Appl. Phys. Lett. 28, 473 (1976).
6. S. Nagai, A. Peters, and S. Candau, Rev. Phys. Appl. (Paris) 12, 21 (1977).
7. C. Sripaipan, C.F. Hayes, and G.T. Fang, Phys. Rev. 15A, 1297 (1977).
8. S. Letcher, J. Lebrun, and S. Candau, J. Acoust. Soc. Am. 63, 55 (1978).
9. S. Nagai and K. Iizuka, Japan J. Appl. Phys. 17, 723 (1978); Mol. Cryst. Liq. Cryst. 45, 83 (1978).
10. J. Lebrun, S. Candau, and S. Letcher, J. Phys. (Paris), Colloq. C3, 40, C3-298 (1979).
11. C.F. Hayes, Mol. Cryst. Liq. Cryst. 59, 317 (1980).
12. J.N. Perbet, M. Hareng, and S. Leberre, Rev. Phys. Appl. (Paris) 14, 569 (1979).

13. S. Candau, A. Ferre, A. Peters, G. Waton, and P. Pieranski, Mol. Cryst. Liq. Cryst., to be published.
14. J.L. Dion, J. Appl. Phys. 50, 2965 (1979).
15. W. Nyborg, in "Physical Acoustics", vol. IIB, W.P. Mason, ed., Academic Press, New York (1968).

PERFORMANCE OF ULTRASOUND TRANSDUCER AND MATERIAL CONSTANTS OF PIEZOELECTRIC CERAMICS

Jun-ichi Sato, Masami Kawabuchi, and Akira Fukumoto

Matsushita Research Institute Tokyo, Inc.

4896 Ikuta Tama-ku Kawasaki, 214 Japan

1. INTRODUCTION

In recent years, real-time ultrasound diagnostic systems have proved to be very effective for a wide range of clinical usages. These real-time systems use piezoelectric transducers arrayed side by side and composed of small plank-shaped piezoelectric ceramics. These transducers must be efficient and give high resolution images in order to provide accurate diagnosis.

At present materials used for ultrasound transducers are mostly ceramics of $BaTiO_3$, $PbNb_2O_6$, $Pb(Zr, Ti)O_3$ or $Pb(Mg_{1/3} Nb_{2/3}, Ti, Zr)O_3$. The characteristics of ultrasound transducers are closely related to the material constants of these ceramics. At the same time, the electrical impedance characteristics of the transducer should be matched to its driving and receiving electrical impedance characteristics. At higher ultrasound frequencies this matching becomes more important, since greater efficiency of the transducer is required due to an increased acoustic loss in human tissue and to various limitations on the selection of electronic components which become more severe at higher ultrasound frequencies.

The article reports on a method to select piezoelectric materials for optimum operation of the diagnostic system. After deriving this method, this principle is used in trying to derive a new piezoelectric material which gives better matching to the electronics and gives better efficiency and resolution than conventional piezoelectric materials.

2. PULSE RESPONSE OF ULTRASOUND TRANSDUCER

In this section the equivalent circuit as shown in Fig. 1 is used to calculate the pulse response of an ultrasound transducer for diagnostic imaging systems. In this equivalent circuit the voltage transfer ratio is given by

$$H(j\omega) = \frac{\dot{E}_\ell}{\dot{E}_s} = \frac{2R_\ell Z_{ot}}{\{\dot{A}Z_{ot}+\dot{B}+\dot{Z}_s(\dot{C}Z_{ot}+\dot{D})\}\{\dot{A}Z_{ot}+\dot{B}+\dot{Z}_\ell(\dot{C}Z_{ot}+\dot{D})\}} \quad (1)$$

where $\dot{Z}_s=R_s+jX_s$, $\dot{Z}_\ell=R_\ell+jX_\ell$, $\dot{E}_s$ and $\dot{E}_\ell$ are the input and output signal amplitudes respectively, R_s and R_ℓ are the source and load resistances respectively, X_s and X_ℓ are the source and load reactances respectively, Z_{ot} is the acoustic impedance of the load medium, $\dot{A}$, $\dot{B}$, $\dot{C}$ and $\dot{D}$ are four terminal constants of the ultrasound transducer and ω is the angular frequency.[1] In the case of an ultrasound transducer which has a two-layer acoustic transformer and is air-backed, four terminal constants are dependent on the parameters shown in Table 1. When the ultrasound transducer is driven by the driving signal f(t), the received signal g(t) across the load resistance R_ℓ is given by

$$g(t) = \frac{1}{\sqrt{2\pi}}\int_{-\infty}^{\infty} H(j\omega)F(j\omega)e^{j\omega t}d\omega \quad (2)$$

where $F(j\omega)$ is the Fourier transform of the driving signal f(t)

$$F(j\omega) = \frac{1}{\sqrt{2\pi}}\int_{-\infty}^{\infty} f(t)e^{-j\omega t}dt \quad (3)$$

and t is the time.

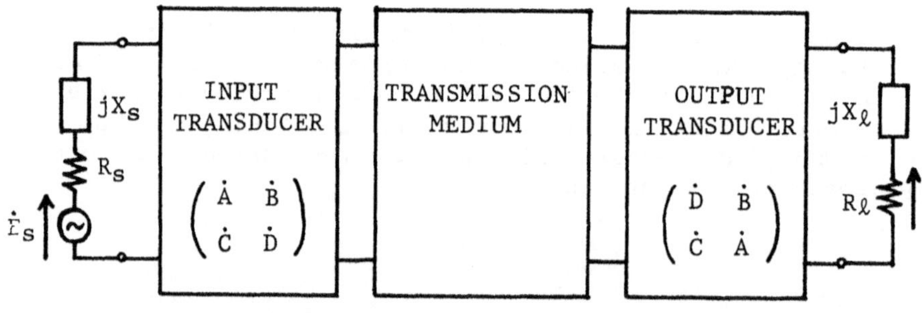

Fig. 1. Equivalent circuit of ultrasound transducer system

Table 1. Parameters affecting characteristics of ultrasound transducer

Components	Parameters
Piezoelectric vibrator	1. Electromechanical coupling coefficient 2. Dielectric constant 3. Sound velocity 4. Density 5. Physical size (thickness, width and length)
Acoustic transformers	1. Sound velocity 2. Density 3. Thickness

3. DEFINITION OF PARAMETERS TO EVALUATE AN ULTRASOUND TRANSDUCER

In an ultrasound transducer system as shown in Fig. 1, three parameters are defined to evaluate the performance of the ultrasound transducer. These parameters are α, the efficiency, DR, the dynamic range and $D_n(A)$, the normalized axial resolution.

The efficiency α is given by

$$\alpha = 20\log(V_\ell/V_s) \quad (4)$$

where V_s and V_ℓ are the maximum voltages of the driving signal f(t) and the received signal g(t) respectively as shown in Fig. 2. This value corresponds to the maximum voltage received at the ultrasound transducer, when it is driven by V_s and the transmitted wave into the loss-free propagation medium is reflected by a perfect reflector.

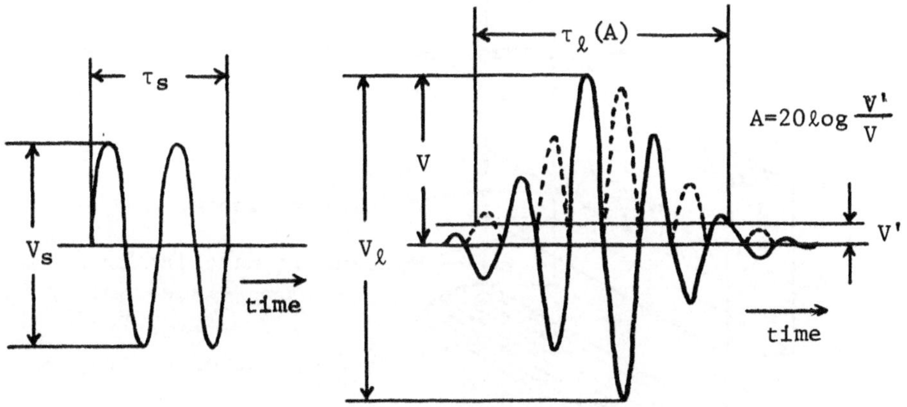

(a) Driving signal f(t) (b) Output signal g(t)

Fig. 2. Wave forms of driving signal and output signal

The dynamic range DR is given by

$$DR = 20\log(V_\ell/2\sqrt{2}V_n) \qquad (5)$$

where V_n is the effective noise voltage generated thermally in the source resistance R_s and the load resistance R_ℓ. The noise voltage V_n is given by

$$V_n = V_{sn}\cdot\frac{V_\ell}{V_s} + V_{\ell n} \qquad (6)$$

where $V_{sn} = \sqrt{4kTBR_s}$, $V_{\ell n} = \sqrt{4kTBR_\ell}$, k is the Boltzman's constant, T the absolute temperature and B the band width. This value corresponds to the signal voltage range which can be detected by the ultrasound transducer.

The normalized axial resolution $D_n(A)$ is given by

$$D_n(A) = \frac{\tau_\ell(A)}{\tau_s} \qquad (7)$$

where $\tau_\ell(A)$ is the A[dB] ring down period of the received signal g(t), and τ_s the duration period of the driving signal f(t).

The optimum combination of these parameters of any ultrasound transducer is the highest efficiency, the highest dynamic range and the smallest normalized axial resolution.

4. CALCULATION

In this section RF pulse response of an ultrasound array transducer with the construction as shown in Fig. 3 is calculated

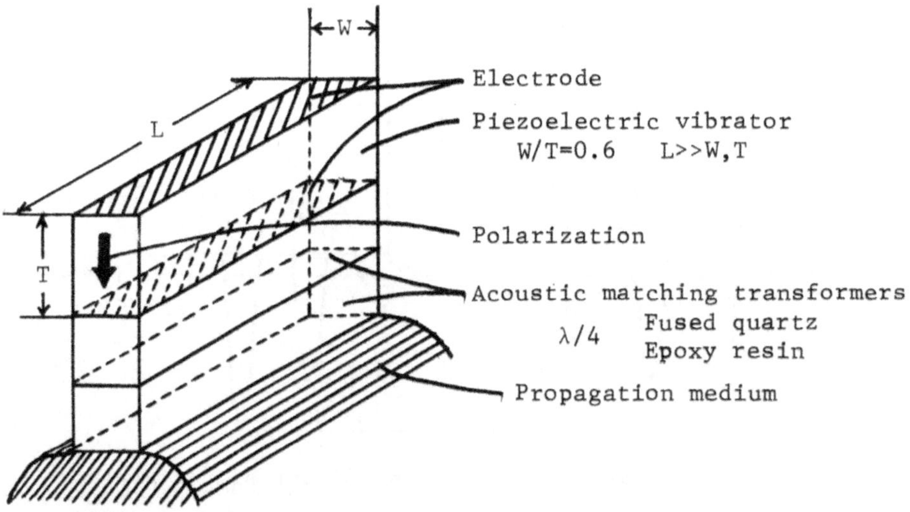

Fig. 3. Construction of ultrasound transducer used for present analysis

by Eq.(2) where the values of R_ℓ and $X_s (=X_\ell)$ are chosen in such a way that the maximum DR value is obtained. Various combinations of acoustic transformers are tried.[2,3] The combination used in the present article was a quarter wave length thick fused quartz with the acoustic impedance of 13.1×10^6 [kg/m^{2}s] and a quarter wave length thick epoxy resin with the acoustic impedance of 3.1×10^6 [kg/m^{2}s]. The quarter wave length is defined at RF frequency f_s of the driving signal f(t). It is assumed in the present calculation that

$$
\begin{aligned}
&f_s \cdot L = 30 \ [\mathrm{MHz \cdot mm}],\\
&W/T = 0.6,\\
&f_s/f_0 = 0.9,\\
&n = 24,\\
&\text{and } R_s = 50 \ [\Omega]
\end{aligned}
\qquad (8)
$$

where L is the length, W the width, T the thickness, f_0 the antiresonance frequency of the piezoelectric vibrator and n the number of elements to make a co-phase aperture.

Fig. 4 shows the dependence of the dynamic range DR on the normalized source resistance $R_s\omega_0 C_0$ and the electromechanical coupling coefficient k_e of the piezoelectric vibrator. Values used for this calculation are T=300[K], B=4[MHz] and V_s=30[V]. Fig. 4 shows that the range of $R_s\omega_0 C_0$ satisfying DR$\geqq$120[dB] becomes wider as k_e increases. Because of a large acoustic propagation loss in human tissue, an ultrasound diagnostic imaging system should have a large DR of the received signal.

Fig. 5 shows the dependence of the efficiency α and Fig. 6 shows the dependence of the normalized axial resolution $D_n(A)$, A=-20[dB] and -40[dB] on the normalized source resistance $R_s\omega_0 C_0$ and the electromechanical coupling coefficient k_e of the piezoelectric vibrator. As shown in Fig. 5, α increases as k_e increases or $R_s\omega_0 C_0$ decreases. As shown in Fig. 6, $D_n(-20)$ and $D_n(-40)$ tend to decrease as k_e increases but there is an optimum range of $R_s\omega_0 C_0$ for them.

Fig. 7 shows calculation results of the dependence of α, $D_n(-20)$ and $D_n(-40)$ on $R_s\omega_0 C_0$ and k_e. The line shaded portion indicates the region where the following relations hold:

$$
\begin{aligned}
&\alpha \geqq -12 \ [\mathrm{dB}]\\
&\mathrm{DR} \geqq 120 \ [\mathrm{dB}]\\
&D_n(-20) \leqq 1.50\\
&D_n(-40) \leqq 6.25.
\end{aligned}
\qquad (9)
$$

By using Eq.(8), the normalized source resistance $R_s\omega_0 C_0$ can be converted into the value of $\varepsilon_e^S/\varepsilon_0$ where ε_e^S is the dielectric constant at constant strain of the piezoelectric vibrator and ε_0 is the dielectric constant of vacuum.

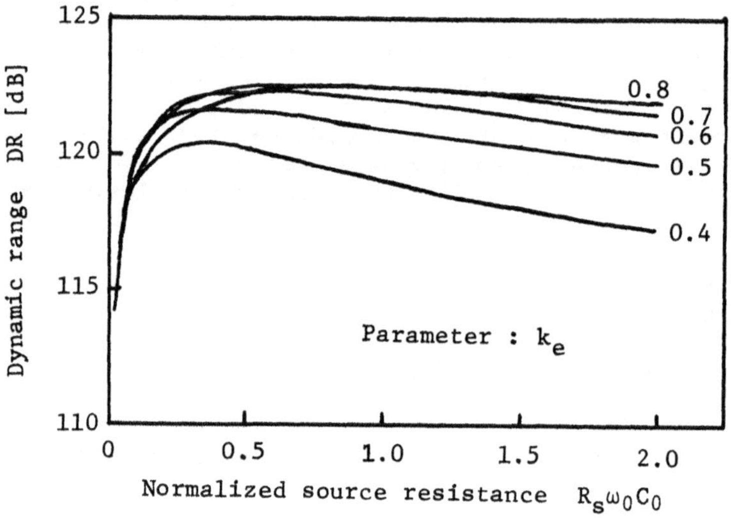

Fig. 4. Dependence of dynamic range DR on normalized source resistance $R_s\omega_0 C_0$ and electromechanical coupling coefficient k_e

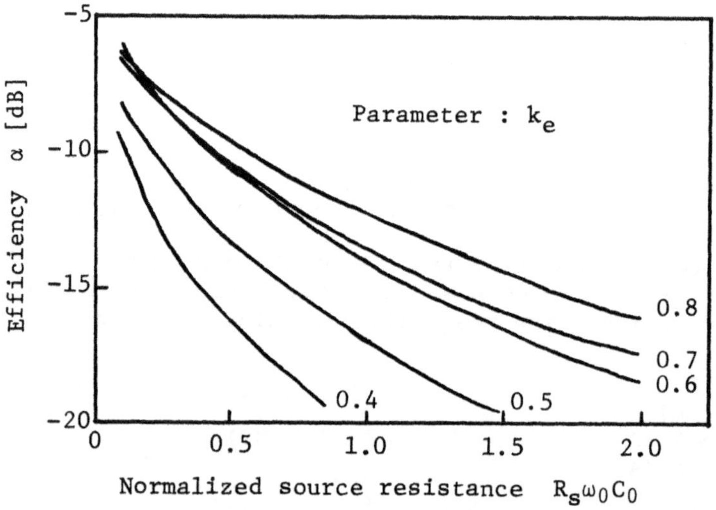

Fig. 5. Dependence of efficiency α on normalized source resistance $R_s\omega_0 C_0$ and electromechanical coupling coefficient k_e

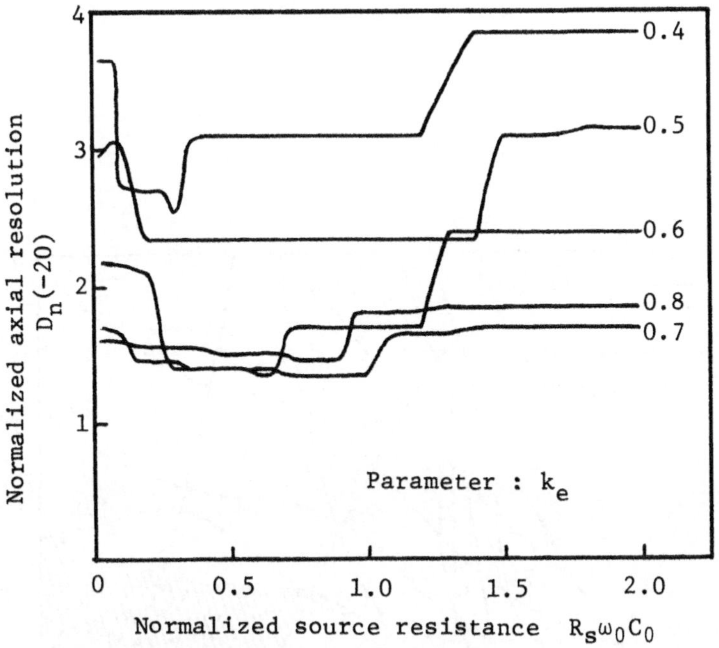

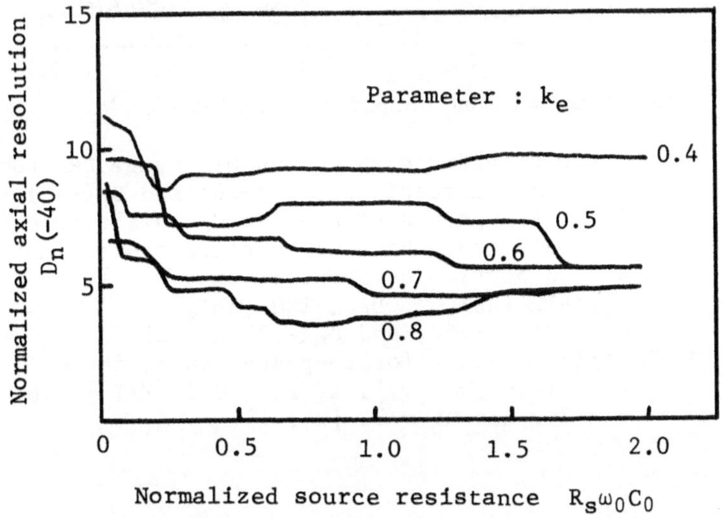

Fig. 6. Dependence of normalized axial resolution $D_n(-20)$ and $D_n(-40)$ on normalized source resistance $R_s\omega_0C_0$ and electromechanical coupling coefficient k_e

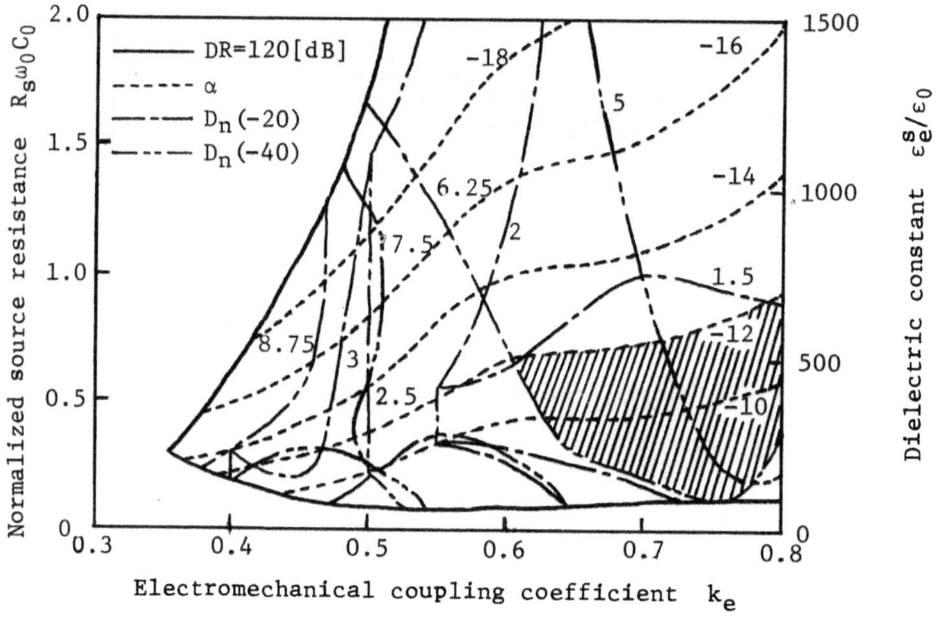

Fig. 7. Dependence of efficiency α, dynamic range DR and normalized axial resolution $D_n(-20)$ and $D_n(-40)$ on normalized source resistance $R_s\omega_0C_0$, dielectric constant $\varepsilon_e^s/\varepsilon_0$ and electromechanical coupling coefficient k_e. Line shaded portion corresponds to

$\alpha \geq -12$ [dB], DR $\geq$ 120 [dB],

$D_n(-20) \leq 1.50$, and $D_n(-40) \leq 6.25$.

Dielectric constant for co-phase transducer at

$f_s \cdot L = 30$ [MHz·mm], $f_s/f_0 = 0.9$, W/T = 0.6,

n = 24, and $R_s = 50$ [Ω].

5. COMPARISON OF MATERIAL SUITABLE FOR HIGH FREQUENCY CO-PHASE ARRAY TRANSDUCERS

In our previous article, it is reported that the electro-methanical coupling coefficient k_e depends on the width to thickness ratio W/T of the piezoelectric vibrator.[4] When W/T is sufficiently small, k_e and $\varepsilon^S_e/\varepsilon_0$ are close in values to k'_{33} and $\varepsilon^{S'}_{33}/\varepsilon_0$, respectively.[5] Fig. 8 shows values of the electromechanical coupling coefficient k'_{33} and the dielectric constant $\varepsilon^{S'}_{33}/\varepsilon_0$ for typical ceramic materials. The region for good high frequency co-phase transducers is given by the line shaded area. As seen in Fig. 8, there are few conventional materials in this region. The cross-hatch shaded region indicates the new material PCMUS-series developed by the present research. PCMUS-1(R) is $Pb(Mg_{1/3}Nb_{2/3}, Ti, Zr)O_3$ ceramics [6,7] and PCMUS-2(R) and PCMUS-3(R) are $PbSr(Ti, Zr)O_3$ ceramics. Table 2 lists the material constants on two examples of these materials PCMUS-1(R) and PCMUS-2(R) and several conventional materials frequently used for ultrasound transducers.[8]

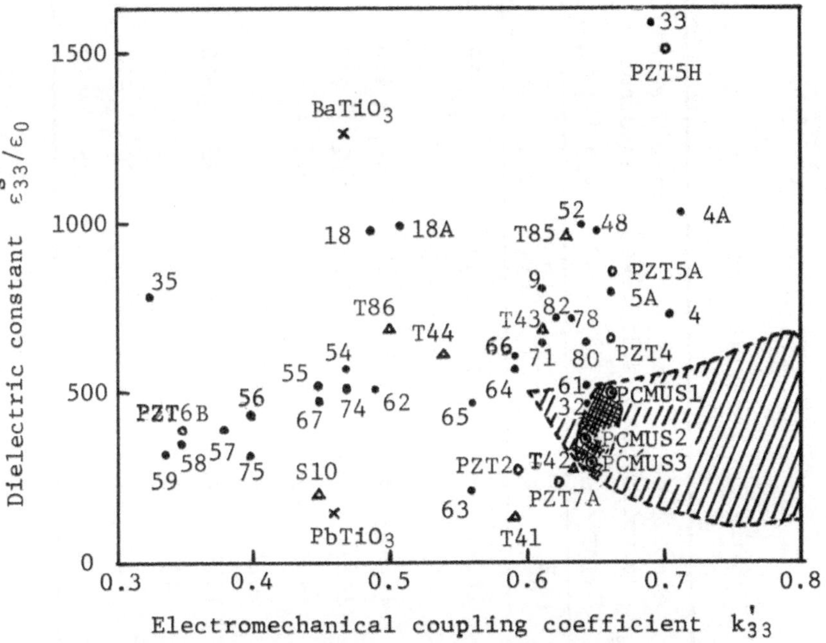

Fig. 8. Electromechanical coupling coefficient k'_{33} and dielectric constant $\varepsilon^{S'}_{33}/\varepsilon_0$ for typical piesoelectric ceramic materials (Calculated by authors)

Numbers : Matsushita
PZT : Vernitron
T and S : Toshiba

Table 2. Material constants of typical piezoelectric ceramics

Materials	Elastic compliance $\times 10^{-12}[m^2/N]$		Elastic stiffness $\times 10^{10}[N/m^2]$		Piezoelectric constant $\times 10^{-12}[C/N]$		Dielectric constant		Electromechanical coupling coefficient		
	s_{33}^E	s_{33}^D	c_{33}^E	c_{33}^D	d_{33}	d_{31}	$\varepsilon_{33}^T/\varepsilon_0$	$\varepsilon_{33}^S/\varepsilon_0$	k_t	k_{33}	k'_{33}
PCMUS-1	12.9	8.5	13.1	15.3	176	-88	785	484	0.54	0.70	0.66
PCMUS-2	14.5	7.6	11.8	16.0	211	-93	734	369	0.55	0.69	0.64
PCM-5A	17.6	8.7	11.6	14.9	367	-186	1710	784	0.50	0.71	0.66
PCM-33	19.3	8.7	11.5	15.5	575	-262	3530	1518	0.50	0.74	0.69
PZT-5A	18.8	9.46	11.1	14.7	374	-171	1700	830	0.49	0.705	0.66
PZT-5H	20.7	8.99	11.7	15.7	593	-274	3400	1470	0.505	0.75	0.70
PZT-7A	13.9	7.85	13.1	17.5	150	-60	425	235	0.50	0.66	0.62
$PbTiO_3$	8.0	6.3	13.2	16.8	51	-4	170	133	0.46	0.46	0.46
$BaTiO_3$	9.5	7.1	14.6	17.1	190	-78	1700	1260	0.38	0.50	0.47
$PbNb_2O_6$	25.4	21.8	—	—	85	$\sim$-9	225	—	0.37	0.38	—

6. FABRICATION OF HIGH FREQUENCY ARRAY TRANSDUCERS AND EXPERIMENT

According to the present calculation a high frequency array transducer was fabricated. The piezoelectric ceramic material used for the ultrasound transducer was PCMUS-2$_{(R)}$. The design conditions are f_s=5[MHz], f_s/f_0=0.9, W/T=0.6, L=6[mm], n=24 and R_s=50[Ω]. Fig. 9 shows the received signal g(t) of this transducer obtained by the pulse echo method and the calculated result of the RF pulse response.

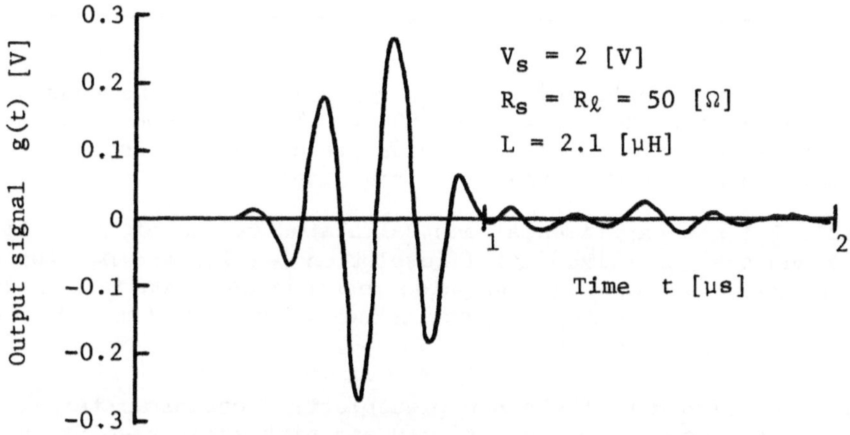

(a) Calculation

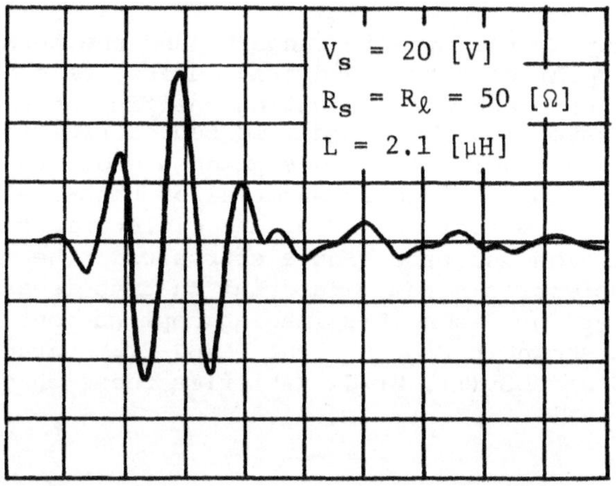

(b) Experiment x = 0.2 [μs/div]
y = 1 [V/div]

Fig. 9. RF pulse response of ultrasound transducer fabricated by present development. (Calculation and Experiment)

Table 3. Comparison of calculated and experimental values of α, $D_n(-20)$ and $D_n(-40)$ for transducers using PCMUS-2$_{(R)}$ ceramics

Experiment			Calculation		
α [dB]	$D_n(-20)$	$D_n(-40)$	α [dB]	$D_n(-20)$	$D_n(-40)$
-11.5	1.31	6.25	-10.1	1.33	6.00

The experiment was performed using a two-cycle gated sine pulse generated by a power source with R_s=50[Ω]. The ultrasound waves were reflected by an Aℓ block reflector placed 15[mm] from the transducer and received by the same transducer.

Table 3 lists experimental and calculated values of the efficiency α and the normalized axial resolution $D_n(-20)$ and $D_n(-40)$ of the transducers obtained by the pulse response as shown in Fig. 9. As shown in Fig. 9 and Table 3, the agreement between the calculation and the experiment is good.

It is concluded that the new piezoelectric ceramic material PCMUS-2$_{(R)}$ provides a high sensitivity and high axial resolution co-phase ultrasound transducer for diagnostic applications.

7. CONCLUSION

The efficiency α, the dynamic range DR and the normalized axial resolution $D_n(A)$ of ultrasound transducers are closely related to the electromechanical coupling coefficient k_e of the piezoelectric vibrator and the normalized source resistance $R_s\omega_0C_0$. It is shown that one condition for designing a good transducer is the selection of the piezoelectric material of higher electromechanical coupling coefficient and of the dielectric constant matched to the driving and receiving electronics. The characteristics of the piezoelectric ceramic material that achieves a good co-phase transducer for diagnostic imaging applications are given in the diagram as shown in Fig. 8. The PCMUS-2$_{(R)}$ ceramics, a new material composed of $PbSr(Ti, Zr)O_3$ satisties these characteristics.

ACKNOWLEDGEMENTS

We would like to thank Drs. H. Ouchi, I. Ueda and Mr. M. Nishida, Materials Research Laboratory, Matsushita Electric Industrial Co., Ltd., for the development of PCMUS$_{(R)}$ piezoelectric materials.

REFERENCES

1. E. K. Sittig, Effects of bonding and electrode layers on the transmission parameters of piezoelectric transducers used in ultrasonic digital delay lines, IEEE Trans. Sonics and Ultrasonics, SU-16, 2 (1969).
2. T. Noguchi, H. Fukukita, and A. Fukumoto, Suggestion of a design procedure of piezoelectric transducers for ultrasonic imaging devices, Trans. Inst. Electronics and Communication Engr. of Japan, 60, 781 (1977).
3. J. H. Goll, The design of broad-band fluid-loaded ultrasonic transducers, IEEE Trans. Sonics and Ultrasonics, SU-26, 385 (1979).
4. J. Sato, M. Kawabuchi, and A. Fukumoto, Dependence of the electromechanical coupling coefficient on the width-to-thickness ratio of plank-shaped piezoelectric transducers used for electronically scanned ultrasound diagnostic systems, J. Acoust. Soc. Am., 67, 1609 (1979).
5. G. S. Kino and C. S. Desilets, Design of slotted transducer arrays with matched backings, Ultrasonic Imaging, 1, 189 (1979).
6. J. Sato, M. Kawabuchi, and A. Fukumoto, Ultrasound transducers using new piezoelectric materials, Proc. 2nd Mtg. Ferroelectric Materials and Their Application, 67 (1979).
7. J. Sato, M. Kawabuchi, and A. Fukumoto, Design of ultrasound transducers using new piezoelectric ceramics, Abst. 2nd Mtg. World Fed. Ultrasound Medicine and Biology, 401 (1979).
8. H. Jaffe and D. A. Berlincourt, Piezoelectric transducer materials, Proc. IEEE, 53, 1372 (1965).

ANALOG ELECTRICAL SIMULATION OF THE TRANSIENT BEHAVIOUR OF PIEZOELECTRIC TRANSDUCERS

P. Alais, Z. Houchangnia and M.Th. Larmande

Laboratoire de Mécanique Physique
Université Pierre et Marie Curie
ERA CNRS 537, PARIS, France

INTRODUCTION

In non destructive testing and more recently in medical applications ultrasonic transducers have been used for emitting or receiving brief signals. It is specially important that these signals must decrease rapidly enough to avoid masking of weak echoes behind strong ones. For this reason diffraction effects on focused transient radiation have been the object of numerous studies in the past years. Although the impulse diffraction theory seems to be mastered now, the characteristics of the transducers remain very important and constitute very often the main limiting features of the spatial axial or longitudinal resolution achieved from either single transducers or arrays.

In the quasi harmonic case which interests submarine sonar devices, modern finite element methods have permitted to attain the radiation of complex shaped transducers [1,2] . These methods have been used as a first step in looking for the behaviour of transducers included in arrays used for medical applications [3,4] . It is important effectively not only to know the decay of the response of the transducer in association with the different damping effects due to absorption or emission but also to attain the radiation diagram of each elementary transducer which may be very important in the reduction of side lobes in the synthetized radiation. Nevertheless the only theoretical study of the transient behaviour of a transducer which may be achieved with reasonable complexity by using numerical computers is limited to the one dimensional case i.e a piston like transducer. Many papers have been published for evaluating the important effect of the damping obtained with an adequate rear medium or with an adaptative front plate [5,6,7].

The object of this paper is to present an electrical analog simulation technique which should permit to go further and to attain the transient behaviour of 2- dimensional transducers as the elementary transducer of a linear array may be reasonably approached. The general principle of the used analogy will be developed and illustrated in the case of the one dimensional propagation. A first network corresponding to this simple case has been tested and has permitted to obtain different results which may be compared with other previous studies [5]. The principle of the simulation of 2 dimensional propagation in piezo and non piezoelectric solids will be given as well as one program of research in the very next future.

A GENERAL TECHNIQUE OF ELECTRICAL ANALOG SIMULATION OF LINEAR OSCILLATING SYSTEMS.

Numerous possibilities of electrical analogies are classically well known; a long time ago, we had the occasion to present a systematical technique [8] for simulating discrete or continuous linear systems in forced harmonic oscillations. The harmonic oscillation of the system must be correctly described by a set of n variables $q_i = Q_i e^{j\omega t}$ i.e n complex amplitudes Q_i. That means for a continuous system to operate the splitting of the system in elementary volumes, the physical state of each one being described by a reduced number of Q_i. Rather than trying to build the simulation on finite difference equations approaching the involved continuous field equations, it is in general much easier to approach the different powers exchanged during the oscillation at the level of the elementary volume. If, considering the ℓth type of involved energy, the corresponding power affording this ℓ- type of energy in an elementary volume may be approached according to

$$\Pi^\ell = \sum_i \dot{q}_i p_i^\ell$$

where the summation interest the q_i involved in the description of the elementary volume and p_i^ℓ are conjugate variables associated with the ℓ-type of energy and the volume considered. One may say in the harmonic case that :

$$p_i^\ell = P_i^\ell e^{j\omega t} \quad \text{and} \quad P_i^\ell = K_{ij}^\ell V_j, \quad V_j = j\omega Q_j .$$

The complex amplitudes of conjugate variables are linearly related to the V_i through the generalized matrix impedances

K_{ij}^{ℓ} . If a correct description of the whole energy exchange of any type involved in the oscillation of the system is given by a good choice of the descriptive variables q_i and the knowledge of the associated impredances $K_{ij}(\omega)$, it is possible to simulate the oscillation of the real system at the pulsation ω by the oscillation of an electrical network at the pulsation $\omega = \omega/\nu$ obtained by association of elementary networks. To each variable q_i corresponds an electrical node submitted to an electrical potential $\varphi_i = \Phi_i e^{j\omega' t}$ and the ℓ- power exchanged at the level of the elementary volume is represented by the exchanged power interesting a corresponding elementary network :

$$\Pi^{\ell}_{analog} = \sum \varphi_i i_i^{\ell}, \quad i_\ell^{\ell} = I_i^{\ell} e^{j\omega' t}, \quad I_i^{\ell} = \Gamma_{ij}^{\ell} \Phi_j ,$$

where the I_i^{ℓ} are the complex amplitudes of intensities feeding the elementary network from outside, and $\Gamma_{ij}^{\ell}(\omega')$ the complex matricial conductance of the network.

The analog correspondance laws which permit to define the network are :

$$(1) \quad \begin{cases} \Phi_i = \lambda_i V_i , & \omega' = \omega/\nu \\ [\Phi_i, I_i^{\ell}] = \eta\, [V_i, P_i^{\ell}] , & \end{cases}$$

where $[A,B] = A^* B = |A||B| e^{j\varphi}, \varphi = \mathrm{Arg}\ B/A$, which means that the energy exchanges are transposed in the analog electrical oscillation at pulsation ω' , respecting relative phases, and intensities with the real factor η .

In these conditions we must use electrical conductances $\Gamma_{ij}(\omega')$:

$$(1') \quad \begin{cases} [\Phi_i, \Gamma_{ij}^{\ell} \Phi_j] = \eta\, [V_i, K_{ij}^{\ell} V_j] , \\ I_i^{\ell} = \frac{\eta}{\lambda_i} P_i^{\ell} , \quad \Gamma_{ij}^{\ell}(\omega') = \frac{\eta}{\lambda_i \lambda_j} K_{ij}^{\ell}(\omega) . \end{cases}$$

The arbitrary choice of the λ_i is limited for convenience to very few different values depending of the physical nature of the corresponding V_i . Of course the excitation of the system from outside is also described by conjugate variables P_i simulated by external intensities I_i obeying the same correspondance laws.

In general, any linear physical system may be simulated in

the harmonic case for a given value of the pulsation ω i.e it is always possible to find electrical networks satisfying the correspondance laws 1 and 1' [8] , for a given pulsation ω But in some cases, the electrical simulation is much stronger in that sense that the relations

$$(2) \qquad \Gamma_{ij}(\omega') = \frac{\eta}{\lambda_i \lambda_j} K_{ij}(\omega) \ , \qquad \omega' = \omega/\nu \ ,$$

remain valid not only for a given value of ω but for a whole band-width interesting the real system and the electrical analog network. In that very interesting cases the simulation is valid for any excitation belonging to the said bandwidth and may be used to study the transient behaviour of the system which will be used here.

UNIDIMENSIONAL STUDY OF THE THICKNESS MODE OF A TRANSDUCER

In this case the transducer of area A and thickness e may be split in elementary volumes $A\,\Delta x$ where the elementary thickness Δx remains much smaller than the smallest wave length to be considered in the oscillation. If the faces of the transducer are the planes $x = 0$ and $x = e$, and $e = n\,\Delta x$, the entire kinematics of the oscillation is correctly known from the knowledge of the $n+1$ displacements $q_i(t)$ at the level of the

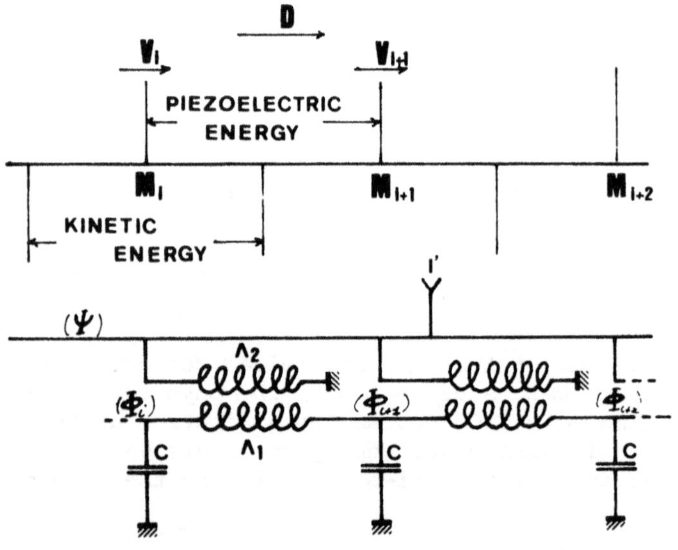

Fig. 1 - Electrical analog simulation of the thickness mode of a transducer

planes $x_i = i\Delta x$, i = 0,....., n . In the thermodynamical point of view, the q_i are extensive variables and if we want to attain the piezoelectric energy through symmetrical matrices we have to retain the electrical induction $\vec{D}$ as the complementary electrical variable. In this special case of the uni dimensional thickness mode, we know that due to the equation $\operatorname{div}\vec{D}=0$ $\vec{D}$ is a spatially uniform field $\vec{D} = D(t)\,\vec{x}$. If we want to attain the analog network through the harmonic case as we did previously, the correct set of variables could be the $n+2$ (Q_i and D^*) complex amplitudes

$$q_i = Q_i^* e^{j\omega t} , \quad D = D^* e^{j\omega t}.$$

For convenience we shall choose as a final set

$$V_i = j\omega Q_i \quad , \quad I = j\omega A D^*$$

i.e complex amplitudes of the velocities v_i and the intensity i delivered to the transducer, respectively, (Fig. 1).

In these conditions the piezoelectric energy may be attained through the expressions of the stress T and the electrical field E associated to the mechanical strain $S = dq/dx$

$$(3)\quad \begin{cases} T = \mu S - hD = \mu \dfrac{\Delta q}{\Delta x} - hD , \\ E = -hS + \beta D = -h\dfrac{\Delta q}{\Delta x} + \beta D . \end{cases}$$

For the elementary volume $A\Delta x$, the power associated to the variation of piezoelectrical energy will be

$$\Pi = AT\Delta v + Ei\Delta x$$

The pure part of mechanical energy $(D=0)$ will be characterized by

$$\Pi_m = \left[\Delta V, \mu A \frac{\Delta q}{\Delta x}\right] = \left[\Delta V, \frac{1}{j\omega}\frac{\mu A}{\Delta x}\Delta V\right],$$

while the pure part of dielectric energy $(q=0)$

by $$\Pi_e = \left[I, \beta\Delta x D\right] = \left[I, \frac{1}{j\omega}\frac{\beta\Delta x}{A} I\right].$$

If we choose to associate to our set of variables (V_i, I) the potentials of the $n+2$ nodes of an electrical network through:

$$(4)\quad \Phi_i = \lambda_1 V_i \quad , \quad \Psi = \lambda_2 I ,$$

it is easy to check that Π_m and Π_e will be simulated by the powers exchanged by inductances Λ_1 and Λ_2 at the pulsation $\omega' = \omega / \nu$ according to :

(5)
$$\begin{cases} \Pi'_m = \eta \left[\Delta V, \frac{1}{j\omega} \frac{\mu A}{\Delta x} \Delta V\right] = \left[\Delta\Phi, \frac{1}{j\omega'\Lambda_1} \Delta\Phi\right] \\ \qquad \Rightarrow \Lambda_1 = \nu \lambda_1^2 \Delta x / \eta \mu A \\ \Pi'_e = \eta \left[I, \frac{\beta \Delta x}{j\omega A} I \right] = \left[\Psi, \frac{1}{j\omega'\Lambda_2} \Psi \right] \\ \qquad \Rightarrow \Lambda_2 = \nu \lambda_2^2 A / \eta \beta \Delta x \end{cases}$$

Then, the total piezoelectric energy will be simulated by a transformer connected as indicated on the Fig 1 the matricial conductance of which may be written :

(6)
$$\begin{vmatrix} I_1 \\ I_2 \end{vmatrix} = \frac{1}{j\omega'} \begin{vmatrix} \frac{1}{\Lambda_1} & \frac{1}{M} \\ \frac{1}{M} & \frac{1}{\Lambda_2} \end{vmatrix} \begin{vmatrix} \Delta\Phi \\ \Psi \end{vmatrix}$$

where the coupling factor $K = (\Lambda_1 \Lambda_2 / M^2)^{1/2}$ must be equal to the piezoelectric coupling factor $(h^2/\mu\beta)^{1/2}$. It is easy to check that the other energy involved in the oscillation i.e the kinetic energy is characterised for an elementary volume $A \Delta x$ centered at the level of a plane $x = x_i$ by the power :

$$A \Delta x \rho v_i \dot{v}_i \Rightarrow \Pi_c = \left[V_i, j\omega\rho A \Delta x V_i \right],$$

which may be simulated in the electrical analogy by the power Π'_c exchanged at the pulsation ω' by the capacitance:

(7)
$$\begin{cases} \Pi'_c = \eta \left[V_i, j\omega\rho A \Delta x V_i \right] = \left[\Phi_i, j\omega' C \Phi_i \right], \\ C = \frac{\eta \nu}{\lambda_1^2} \rho A \Delta x \end{cases}$$

It should be remarked that the splitting in volumes associated to the kinetic energy is interlaced with the one associated to the piezoelectric energy so that for the extrem nodes at the level of interfaces $x = 0$ and $x = e$ we must retain $C/2$ instead of C in the network. It is obvious that the simulation of a non piezoelectric solid for example an adaptative front plate will reduce to a classical delay line of inductances and capacitances,(Fig 2). The flux of energy radiated in and infinite medium of acoustical impedance ρc at the rear or the front face of the transducer adapted or not by a front plate which may be written $\Pi_R = A T v = A \rho c v^2$, will be simulated by the power dissipated by a resistance

$$(8) \quad \begin{cases} \Pi_R = \eta\,[V, A\rho c V] = [\Phi, \frac{1}{R}\Phi] \\ R = \lambda_1^2/\eta A\rho c \end{cases}$$

It should be noted for simulating the electrical excitation of the transducer that the analog potential Ψ is associated to the intensity delivered to the transducer so that the electrical analog conductances are related according to (2), to the electrical impedances Z involved in the problem i.e the electrical impedance of the transducer itself or the impedance of adaptative electrical components,

$$(2') \quad \Gamma(\omega') = \frac{\eta}{\lambda_2^2} Z(\omega)$$

In the same way a serial connection of an inductance in the real process is simulated by the parallel connection of the analog corresponding capacitance in the analog network.

At the end, le potential V imposed to the transducer is transposed in the analogy through the intensity $I' = (\eta/\lambda_2)V$. It may be checked that the whole network is realized with electrical components satisfying the relations (2) not only for one given value of ω but for a reasonable bandwith limited essentially by the technology used for building transformers. Such a network may be used to study not only the well known harmonic response of a transducer but also its transient behaviour.

EXPERIMENTAL RESULTS OBTAINED FROM THE UNIDIMENSIONAL ANALOGY

We have built a network simulating unidimensional transducers built with a TiZnPb ceramic presenting a thickness mode coupling factor k = 0,45, imposing the mutual coupling of the transformers. The whole thickness of the transducer was simula-

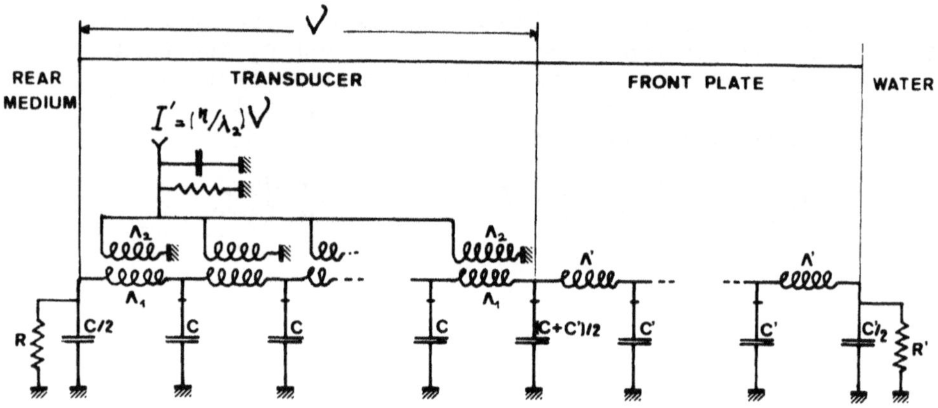

Fig.2 - Network simulating a transducer and its front plate with both faces radiating.

ted by twelve cells (Fig.2). The central frequency of the analog oscillation corresponds obviously to the period $T = 24 (\Lambda_1 C)^{1/2}$ and may be adapted by an adequate choice of C. Due to the relations (2), the resistances R, R' simulating the rear and front radiations and the iterative resistance $(\Lambda'/C')^{1/2}$ of the line simulating the front plate should be related to the iterative resistance $(\Lambda_1/C)^{1/2}$ according to the ratios of the corresponding acoustical conductances $(\rho c)^{-1}$ to the acoustical conductance $(\mu \rho_0)^{-1/2}$ of the piezoelectric ceramic. The retained values of Λ', C' were chosen to simulate a $\lambda/4$ front plate with six cells, at the central frequency.

The figures 3 and 4 show transfer functions of the velocity of the front face related to the electrical potential amplitude imposed to the transducer in the frequency interval $(0, 2\omega_0)$ for the particular cases of perfect mechanical matching of the transducer at both faces (Fig.3) or only at the front face, the rear face remaining free (Fig.4). The retained electrical tuning of the analog network obtained from a capacitor and a resistance shunting the Λ_2 coils was simulating a serie RL circuit resonant at ω_0 driving the transducer with a Q of about 2. These results have been compared with those

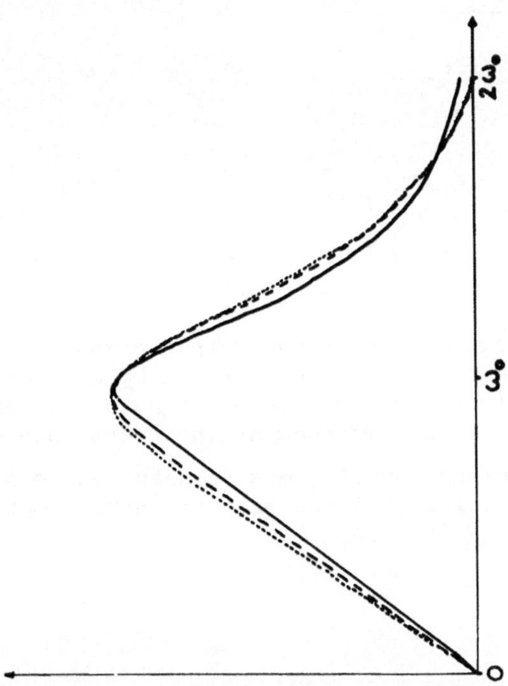

Fig. 3 - Both rear and front faces perfectly matched

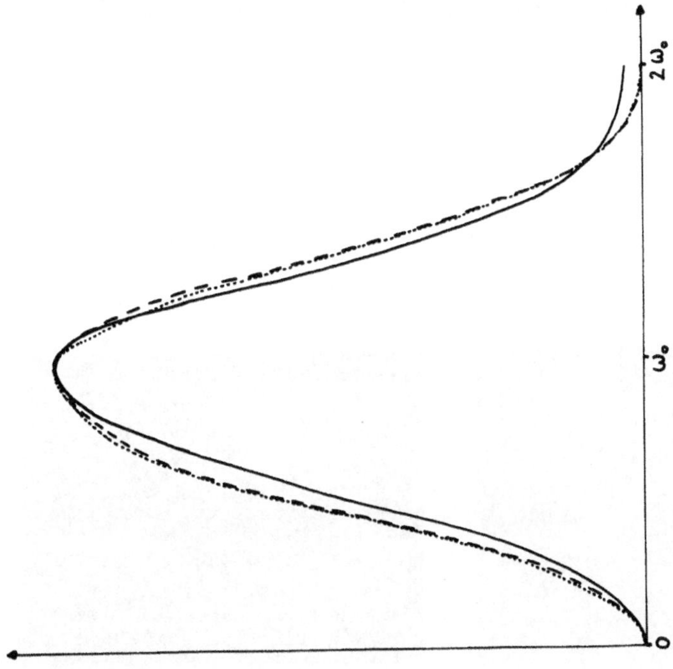

Fig. 4 - Front face matched rear face free

Output pressure frequency response for a transducer driven by a tuned RL circuit having electrical Ω of 2

.... Analytically computed;
------- Ideal network;
——— Real network

a b

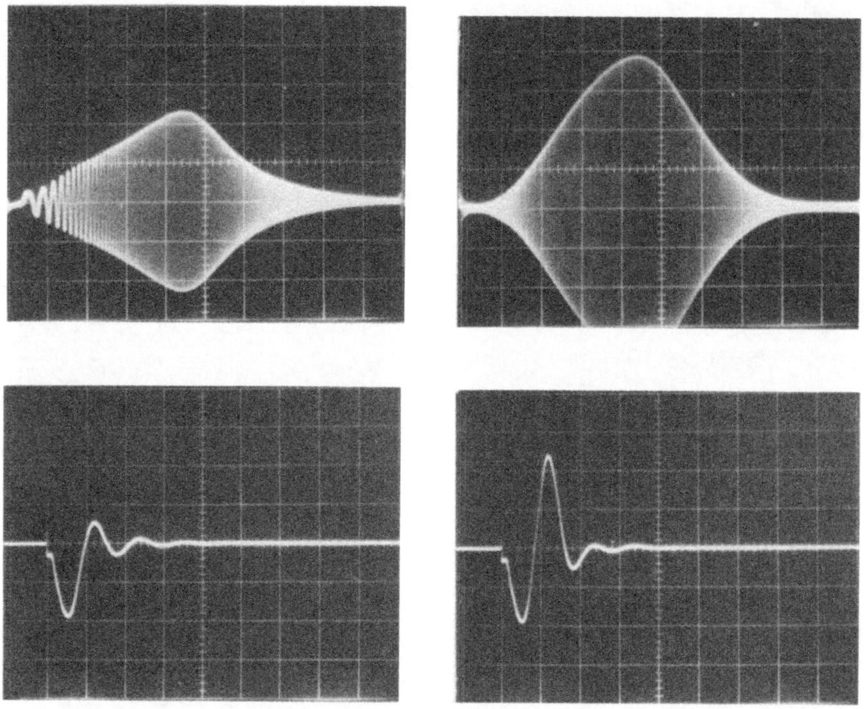

c d

Fig.5 a-b Analog results of the output pressure frequency response of a transducer driven by a tuned RL circuit (Q=2), perfectly matched at both rear and front faces (Fig 5a) and matched at the front face only (fig 5b).

c-d Transient output pressure response to a step exciting voltage of the same transducer matched respectively as in a and b.

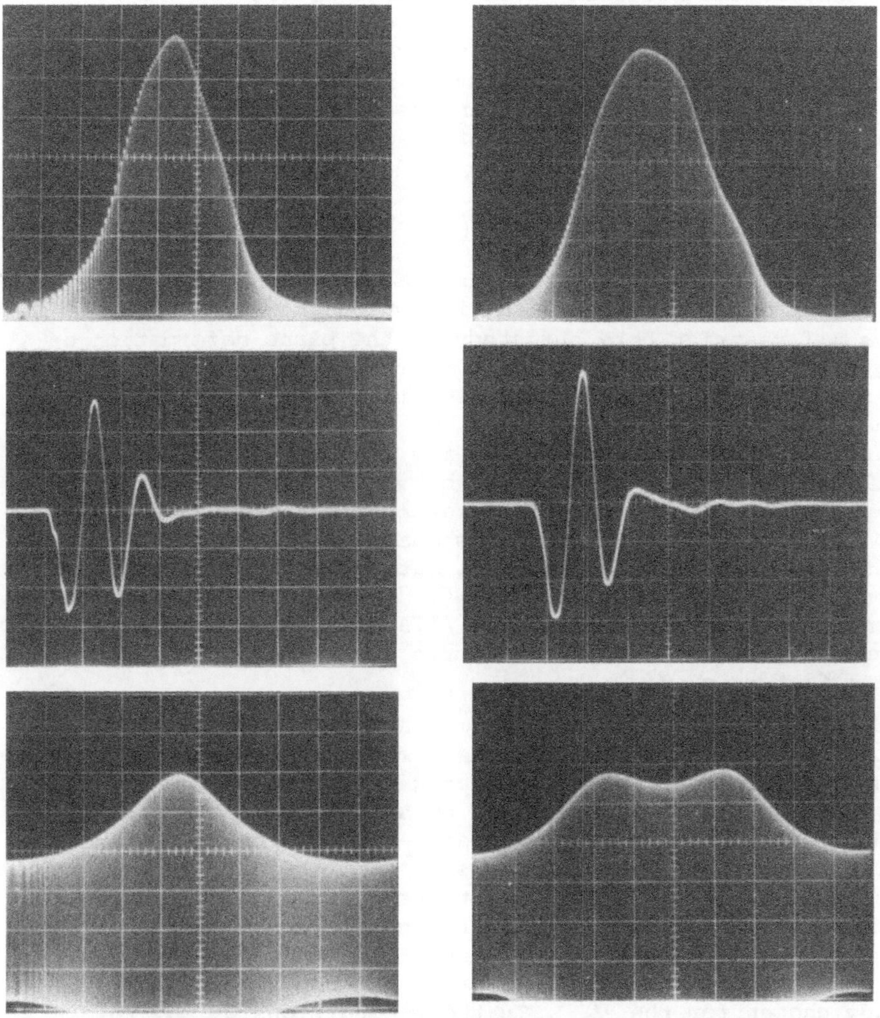

e f g h i j

Fig.5 e-f Analog results of the output pressure frequency response of a transducer driven by a tuned RL circuit (Q=2), partially adapted at the rear face ($Z_o/3{,}5$, radiating in water ($Z_o/20$) through one matching ($\lambda/4$) layer (Fig 5e) and through two matching layers (Fig.5f)

g-h Transient output pressure response to a stop exciting voltage of the same transducer matched respectively as in e and f.

i-j Transfer functions pressure versus pressure of the matching layers used in e and f respectively and radiating in water.

ones obtained by numerical computing of the analytic solution and also of the solution given by the equations of the ideal network. The good agreement justifies both the principle of the analogy and the technique used for realizing the network. Figures 3 and 4 are also very similar to the results presented in [5] . The fig.5 show different results obtained in terms of the same transfer function or of the transient response to a step function.

SIMULATION OF BIDIMENSIONAL TRANSDUCERS

By bidimensional transducers we mean transducers the oscillation of which may be described through two space variables. We shall restrict here our study to the plane deformation of piezp-electric cylindrical bars of rectangular section and of the elementary transducers used in linear arrays. As a reasonable approach of the problem, we shall assume, taking in account the geometry given by the fig. 6, that the deformation is exclusively plane and described by x_1 x_3 variables and also that the electrical induction exercised through the plane electrodes orthogonal to x_3 remains quasi parallel to x_3, $\vec{D} = D_3 \vec{x_3}$ That means that in the classical matricial piezoelectric relations we may admit : $S_2 = S_4 = S_6 = 0$, $D_1 = D_2 = 0$ which reduces them as indicated on Fig.6. The (1,3,3') relation gives account of the involved piezoelectric energy and of the 3 coupling factors α, β, γ.

The (5) relation concerns pure mechanical shear energy. Generalizing the simulation of a one dimensional transducer it is easy to understand that a correct description of the transducer oscillation may be attained by splitting in elementary volumes $L \Delta x_1 \Delta x_3$ and retaining as a correct set of (extensive) variables the complex amplitudes V_1 and V_3 of the velocity according to the x_1, x_3 directions at the level of a finite number of points associated to the splitting as indicated on Fig.7, and the complex amplitudes of the temporal derivative of the electrical induction, $j\omega D_3$ at the same points already chosen for the V_3 . In these conditions, we may retain the analog correspondance laws for the electrical potentials Φ of the nodes of the analog network :

$$(9) \quad \begin{cases} \Phi_1 = \lambda_1 V_1 \\ \Phi_3 = \lambda_3 V_3 \\ \Delta_1 \Phi_0 = j\omega \lambda_0 L \Delta x_1 D_3 \end{cases} \qquad \begin{matrix} \Pi'_{analog} = \eta \, \Pi_{real} \\ \omega' = \omega / \nu \end{matrix}$$

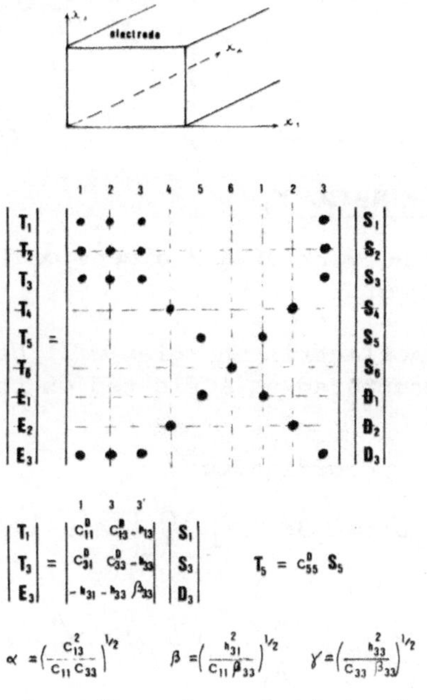

Fig.6 Simplified piezoelectric relations for a 2 dimensional transducer. α is a purely mechanical coupling coefficient. γ and β are electromechanical coefficients for the thickness and lateral modes respectively.

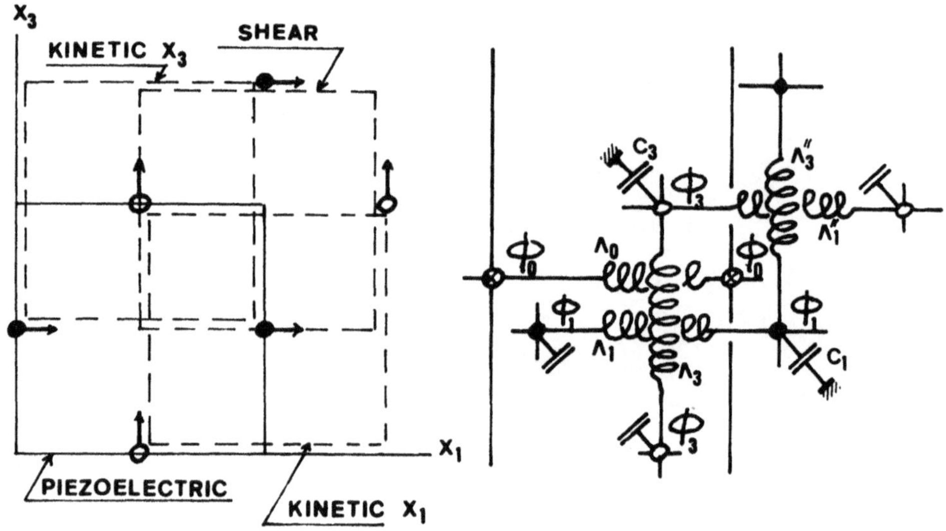

Fig.7 - Analog network of a 2 dimensional piezoelectric transducer

The kinetic and piezoelectric energies will be then approached on elementary volumes centered as indicated on the Fig.7 and simulated as follows :

- The kinetic energy according to

$$(10)\quad \begin{cases} \eta\,[V_1, j\omega\rho L \Delta x_1 \Delta x_3 V_1] = [\Phi_1, j\omega' C_1 \Phi_1] \\ C_1 = \dfrac{\eta\nu}{\lambda_1^2}\,\rho L \Delta x_1 \Delta x_3 , \end{cases}$$

through the capacitances C_1.

- The kinetic energy according to x_3 , through the capacitances

$$(10')\quad C_3 = \frac{\eta\nu}{\lambda_3^2}\,\rho L \Delta x_1 \Delta x_3 .$$

- The piezoelectric energy associated to the (1,3,3') matrix will be represented by a 3 coil-transformer with the coupling coefficients α, β, γ imposed by the piezoelectric material the matricial conductance of which may be written :

$$\begin{vmatrix} I_1 \\ I_3 \\ I_0 \end{vmatrix} = \frac{1}{j\omega'} \begin{vmatrix} \frac{1}{\Lambda_1} & + & + \\ + & \frac{1}{\Lambda_3} & - \\ + & - & \frac{1}{\Lambda_0} \end{vmatrix} \begin{vmatrix} \Delta_1 \Phi_1 \\ \Delta_3 \Phi_3 \\ \Delta_1 \Phi_0 \end{vmatrix} \tag{11}$$

where the non diagonal components may be attained through the coupling coefficients α, β, γ and the signs imposed classically by usual ceramic piezoelectric materials. Taking in account that reversing the (arbitrary) sign of one potential reverse two of the 3 non diagonal coefficients it appears that there are 2 classes of 3- matrices : positive ones with 3 or 1 positive non diagonal coefficient and negative ones with 2 positive. It is easy to check the matricial impedance, inverse of (11) is of the positive type and must be simulated by a transformer with a serial magnetic circuit wherever a negative one should be represented by a parallel circuit. The diagonal coefficients of (11) are obtained through immediate analog relations

$$(12) \quad \begin{cases} \eta \left[\frac{\Delta_1 V_1}{\Delta x_1}, \; L \Delta x_1 \Delta x_3 \frac{C_{11}}{j\omega} \frac{\Delta_1 V_1}{\Delta x_1} \right] = \left[\Delta_1 \Phi_1, \; \frac{1}{j\omega' \Lambda_1} \Delta_1 \Phi_1 \right] \\ \Lambda_1 = \frac{\nu \lambda_1^2}{\eta} \frac{\Delta x_1}{L C_{11} \Delta x_3}, \end{cases}$$

in the same way

$$\Lambda_3 = \frac{\nu \lambda_3^2}{\eta} \frac{\Delta x_3}{L C_{33} \Delta x_1}, \tag{12'}$$

$$(13) \quad \begin{cases} \eta \left[j\omega D, \; L \Delta x_1 \Delta x_3 \beta D \right] = \left[\Delta_1 \Phi_0, \; \frac{1}{j\omega' \Lambda_0} \Delta_1 \Phi_0 \right] \\ \Lambda_0 = \frac{\nu \lambda_0^2}{\eta} \frac{L \Delta x_1}{\beta \Delta x_3}. \end{cases}$$

- the shear energy associated to the relation $T_5 = C_{55} S_5$ is not obviously represented from the fact that the shear strain S_5 must be approached by

$$j\omega S_5 \simeq \frac{1}{2} \left\{ \frac{\Delta_3 V_1}{\Delta x_3} + \frac{\Delta_1 V_3}{\Delta x_1} \right\},$$

and the shear energy simulated by a 2 coil-transformer of coupling coefficient 1, the matricial conductance of which may be written :

$$(14)\qquad \begin{vmatrix} I_1 \\ I_3 \end{vmatrix} = \frac{1}{j\omega'} \begin{vmatrix} \frac{1}{\Lambda''_1} & + \\ + & \frac{1}{\Lambda''_3} \end{vmatrix} \begin{vmatrix} \Delta_1 \Phi_3 \\ \Delta_3 \Phi_1 \end{vmatrix}$$

where

$$(15)\qquad \begin{cases} \eta\left[\frac{\Delta_1 V_3}{2j\omega\Delta x_1},\ L\,\Delta x_1 \Delta x_3\, C_{55}\frac{\Delta_1 V_3}{2\Delta x_1}\right] = \left[\Delta_1\Phi_3,\ \frac{\Delta_1\Phi_3}{j\omega'\Lambda''_1}\right], \\ \Lambda''_1 = \frac{4v\lambda_3^2}{\eta}\ \frac{\Delta x_1}{C_{55}\,L\,\Delta x_3}, \\ \Lambda''_3 = \frac{4v\lambda_1^2}{\eta}\ \frac{\Delta x_3}{C_{55}\,L\,\Delta x_1}. \end{cases}$$

The fact that the electrical induction D_3 depends only of x_1 due to the equation $\operatorname{div}\vec{D} = 0$ is transposed in the network by connecting the nodes N_0 associated to the same x_1 . So on one column the "electrical" coils of the 3-coil transformers are fed in parallel by a global intensity I which is the analog transposition of the voltage V imposed to the electrodes according to the relation :

$$(16)\qquad I = \frac{\eta}{\lambda_0} V$$

As this voltage is imposed constant on the surface of metallic electrodes it will be simulated by the same intensity I fed to the different columns of transformers serially connected as indicated on the Fig. 8 . As for the one dimensional transducer the electrical transducer impedance Z is transposed according to the electrical conductance Γ of the network and the potentiel Φ of excitation of the analog network is related to the intensity J feeding the transducer :

$$(17)\qquad \begin{cases} \Gamma(\omega') = \frac{\eta}{\lambda_0^2} Z(\omega) \\ \text{with}\quad I = \Gamma\Phi,\quad V = ZJ,\quad \Phi = \lambda_0 J. \end{cases}$$

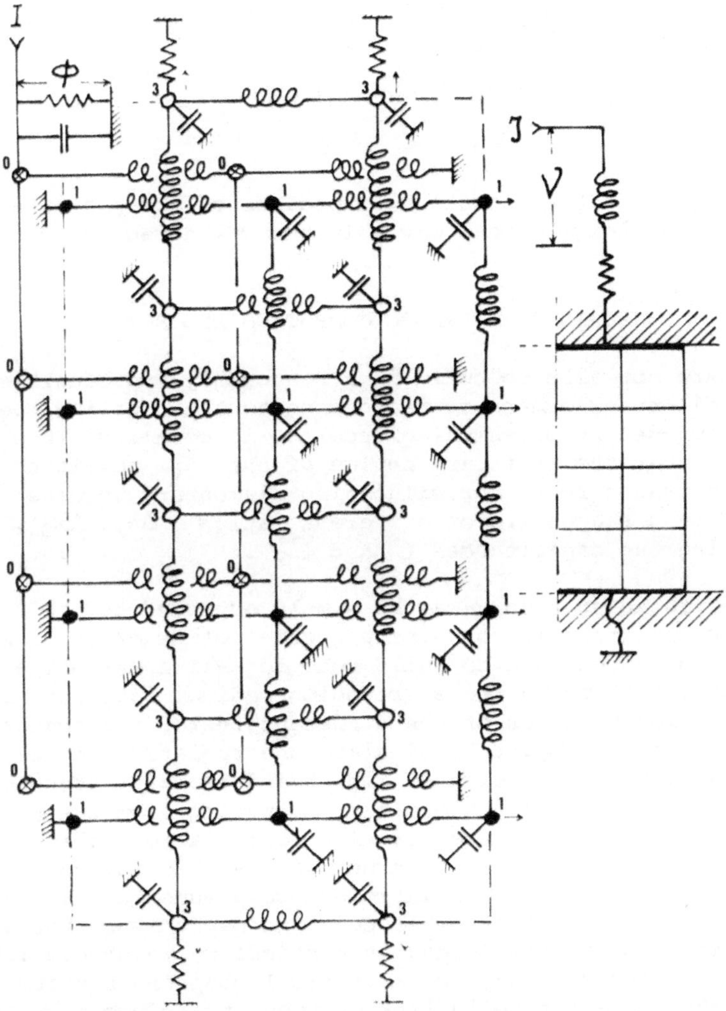

Fig.8 Analog network simulating the symmetrical oscillation of a piezoelectric bar of rectangular section radiating at the electrode faces, the other faces remaing free.

As an example the complete network simulating the symmetrical oscillation of a bar radiating in the x_3 direction and not acoustically charged in the x, direction, is represented in the Fig. 8 . It is easy to check that the resistance simulating radiation in a medium of acoustical impedance ρc are related to the iterative impedance of the delay lines Λ_3, C_3 included in the network according to :

$$R = \left(\frac{\Lambda_3}{C_3}\right)^{1/2} \cdot \frac{\rho_0 c_3}{\rho c} , \tag{18}$$

Where $\rho_0 c_3$ is the acoustical impedance of the piezoelectric material for longitudinal waves in the x_3 direction.

EXPERIMENTAL ANALOG COMPUTING PROJECT

We are now able to build in a reasonably reproducible way the 3 coil-transformers corresponding to the coupling coefficients imposed by classical piezoelectric ceramics. It should be noted that the arbitrary choice of the λ_i permit to use the same transformers for simulating elementary volumes $L\,\Delta x_1\,\Delta x_3$ of different shapes i.e of different ratios $\Delta x_1/\Delta x_3$, just by varying the capacitances C_1 and C_3 .

The preceding theory may be simply extended to the 2 dimensional propagation in non piezoelectric solids or liquids, so that it is conceivable to build once for all a network simulating the radiation in the surrounding medium, water, to attain the radiation diagrams of the different tested transducers. But it seems rather simpler to simulate the radiation in the infinite medium like in the case of Fig. 8 and to have access to the radiation diagram through ordinary computing from the experimental measurements of the transient oscillations observed of the radiating surface. It is true that is this way we neglect the radiation coupling mechanism but we assume that it should remain small in front of other coupling mechanisms that we want to attain through this study. In particular, we should like in a very next future study the mechanical coupling effects related to the different solid links between transducers by simulating at least one transducer, the two adjacent ones in the array, the front adaptating front plate and the rear medium. The dielectric coupling effect which is not implicitly taken in account due to our approach of the induction field may be taken in account through a correction of the external excitation obtained from a classical electrostatic study.

CONCLUSION

The possibility has been shown to attain through an electrical analog simulation the transient behaviour of relatively complex shaped piezoelectric transducers.

A first experimental approach of this work has been done by obtaining through a first network results already given by numerical computing in the simple case of the one dimensional transducer.

We plan to build in the very next future a network which should permit to attain the mechanical damping and the mechanical coupling involved between transducers in a linear array and to evaluate their effects on the transient oscillation and the radiation diagram.

REFERENCES

[1] R.R. SMITH, J.T. HUNT, and D. BARACH - " Finite element analysis of acoustically radiating structures with applications to sonar transducers, J.A.S.A., 54, 5 pp 1277-1288 1973.

[2] D. BOUCHER - " Calcul des Modes de Vibration de transducteurs piezoelectriques par une methode elements finis - perturbation.
Thèse presentée à l'Université du Maine, France 1979.

[3] J. SATO, M. KAWABUCHI and A. FUKUMOTO " Dependence of the electro-mechanical coupling coefficient on the width-to-thickness ratio of plank-shaped piezoelectric transducers used for electronically scanned ultrasound diagnostic systems " J.A.S.A., 66, 5 Nov. 1979.

[4] J. SATO, H. FUKUKITA, M. KAWABUCHI and A. FUKUMOTO " Farfield angular radiation pattern generated from arrayed piezoelectric transducers ", J.A.S.A., 67, 1, pp. 333-335, Jan 1980.

[5] G.K. LEWIS - " A matrix technique for analyzing the performance of multilayered front matched and backed piezoelectric ceramic transducers ". Acoustical Imaging Vol 8 pp 395-416.

[6] T.R. MEEKER - " Thickness mode piezoelectric transducers " Ultrasonics, Janv 1972, pp 26-36.

[7] R.H. COURSANT - " Les transducteurs ultrasonores ". Acta Electronica, 22, 2, 1979, pp 129-141.

[8] P. ALAIS - " Méthode générale de calcul de systèmes linéaires en oscillation forcée à partir de considérations énergétiques " Annales de l'Ass. Int. pour le Calcul Analogique, n°4, Oct 67.

OPTIMIZATION CRITERIONS FOR THE PIEZOELECTRIC TRANSDUCERS USED IN ACOUSTICAL IMAGING

J. Hue, M.G. Ghazaleh, C. Bruneel, and R. Torguet

LABORATOIRE O.A.E. - E.R.A. CNRS N°593

UNIVERSITE DE VALENCIENNES 59326 VALENCIENNES CEDEX
FRANCE

INTRODUCTION.

In the past few years, new piezoelectric transducer materials (polymers like PVC, PVF, PVF_2,...) appeared. They seem more interesting than ceramics for acoustic applications, particularly for medical imaging, because they are compliant and their acoustical impedance is closer to that of water than ceramic impedance.

In this paper, our aim is not to define new ideas but only to summarize the main characteristics needed to optimize the performance of a transducer. It's necessary to consider the whole problem, since some authors draw erroneous conclusions by taking only a particular property into account. For example, with the material PVF_2, very large bandwidth may certainly be obtained, but at the expense of the efficiency which becomes too weak for medical applications.

A large bandwidth and a good efficiency constitute the two main characteristics for the transducers used in the medical area.

After a detailed study of those characteristics, the performances of a polymer material (PVF_2) and a ceramic material (PZT5) will be compared.

A) PARAMETERS CARACTERIZING A TRANSDUCER

A transducer is good if three criterions are satified that is to say, the efficiency, the bandwidth and the sensitivity which must be greater as possible.

1) The efficiency defined.

- for an emitting transducer as the ratio of the acoustical energy which is transmitted into the propagation medium to the maximal electrical energy which can be delivered by the generator;

- for a receiver it is defined as the ratio of the available electrical energy to the incident acoustical energy.

2) The bandwidth is usually defined as a 3 dB bandwidth normalized to maximal acoustical power.

3) The sensitivity is defined as the acoustical energy with a unity signal to noise ratio for the electrical energy received. The sensitivity is related to the efficiency, so that the efficiency and the bandwith are only taken account.

The transducer can be considered as a reciprocal component.

B) PHYSICAL PARAMETERS OF A TRANSDUCER

1) A Lossless transducer is well defined using the following parameters :

- The acoustical impedances
  of the transducer Z_p
  of the front propagation medium Z_{mF}
  of the back loading medium Z_{mb}
- The electromechanical coupling factor K
- The relative permittivity ε_r

According to the classical formula |1| , |2|

$$Z_{el} = -j\,X_s + \frac{K^2\,X_s}{x\,\pi}\left|\frac{2Z_p\,(\cos\theta - 1) + j(Z_{mF} + Z_{mb})\sin\theta}{(Z_{mF} + Z_{mb})\cos\theta + j\left(Z_p + \frac{Z_{mF}\,Z_{mb}}{Z_p}\right)\sin\theta}\right|$$

where X_s is the reactance of the static capacitance C_s of the transducer defined as :

$$C_s = \frac{\varepsilon_r \varepsilon_o A}{L}$$

with L the sample thickness and A the area and where x is the reduced frequency $\frac{f}{f_o}$ and θ is equal to $\pi . x$.

A lossless transducer may be then represented by the following electrical circuit (fig. 1)

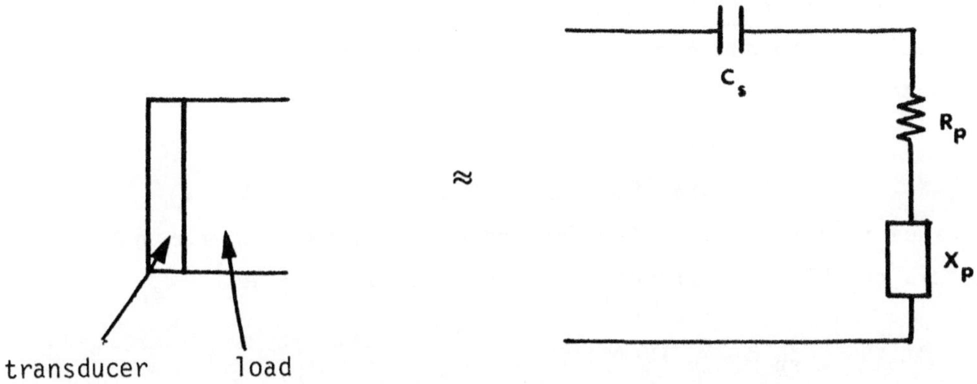

Figure 1. Equivalent electrical circuit for a lossless transducer where R_p and X_p are the piezoelectric radiation resistance and reactance, respectively.

In order, to obtain a good flat efficiency over a wide band width, the transducer must be matched to the generator which two conditions.

- R_p quite constant on the widest possible bandwidth.
- R_p as high as possible for providiny a small electrical quality factor defined as :

$$Q_E = \frac{X_p + X_s}{R_p}$$

An acoustical matching of the transducer to the propagation medium leads to slow variations of R_p over a wide band and a high electromechanical coupling factor provides a small electrical quality factor.

2) Effect of the internal losses.

Two kinds of losses must be distinguished by their effects on the efficiency of the transducer.

. The acoustic losses reduce the signal amplitude without increasing the noise.

. The dielectric losses decrease the efficiency and increase the noise, strongly reducing the signal to noise ratio.

In a first approximation, to include dielectric losses, it is sufficient to add a resistance, shunting the static condenser, whose value is

$$R_{11} = \frac{1}{\omega \, C_s \, tg \, \delta}$$

Since $tg^2\delta << 1$, in the most practical cases, one can substitute this parallel resistance by a serie one defined as

$$R_{sp} \simeq X_s \, tg \, \delta$$

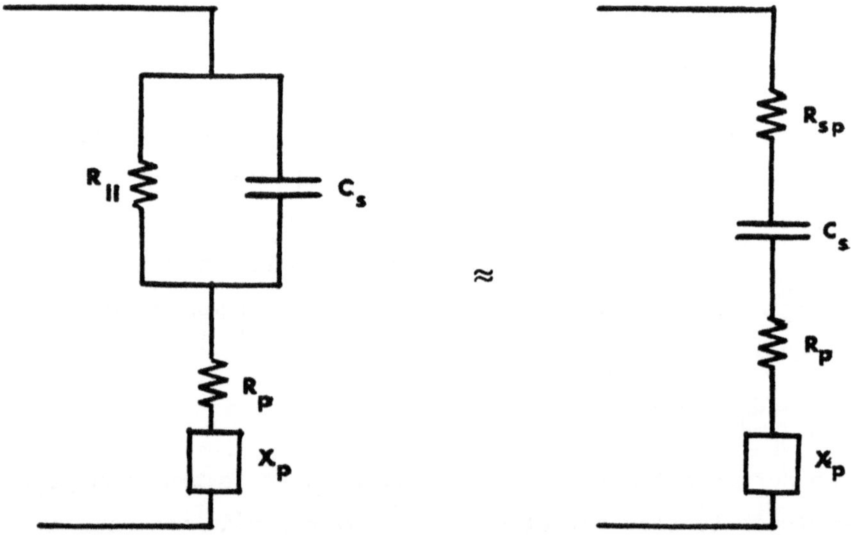

Figure 2. Equivalent electrical circuit of a transducer with dielectric losses.

which viewed from the active part of the transducer has to be added to the generator internal resistance.

$$R_{geq} = R_g + R_{sp}$$

C) PRACTICAL REALIZATION OF THE OPTIMIZED CONDITIONS

The ideal case is realized when the acoustical matching condition is satisfied. The electrical matching has then only to be performed.

Otherwise, the acoustical matching may be obtained using quarter wavelength multilayers |3| .

On the figure 3, a practical example is given for medical application where glass and lucite are used to match the PZT ceramic to the biological tissue.

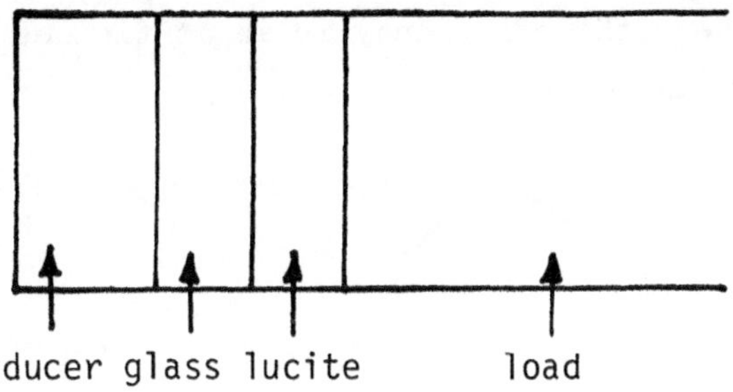

Figure 3. Practical configuration of a transducer used in medical application.

This method give good results that is to say a bandwidth reading one octave and conversion losses lower than 3dB (classical values retained).

D) EXPERIMENTAL SET UP OF NON OPTIMIZED CONDITIONS CURRENTLY USED.

In practical realizations, the large bandwith condition is generally obtained at the expense of the efficiency. Two methods give this result :

- the backing solution increases the bandwidth but severaly decrease the efficiency. The major part of the energy is then dissipated inside the damper which has an acoustical impedance of the same order than that of the transducer.

The conversion losses are generally of the order of 10 dB .

- another compromise can be achieved when using an electrical method. A theoretical study has shown that without an acoustical matching it is however possible to increase the bandwidth |4| . This method consists on balancing, in a first step, the capacitance C_s with an inductance at the resonnant frequency f_o . In a second step, the piezoelectric radiation resistance at the resonnant frequency R_{po} is chosen larger than the generator internal resistance.

In the case of a ceramic radiating into water, the curves of efficiency and bandwidth versus ratio R_{po} over R_g is plotted in figure 4 where the efficiency is in continuous line and bandwith is in dotted line.

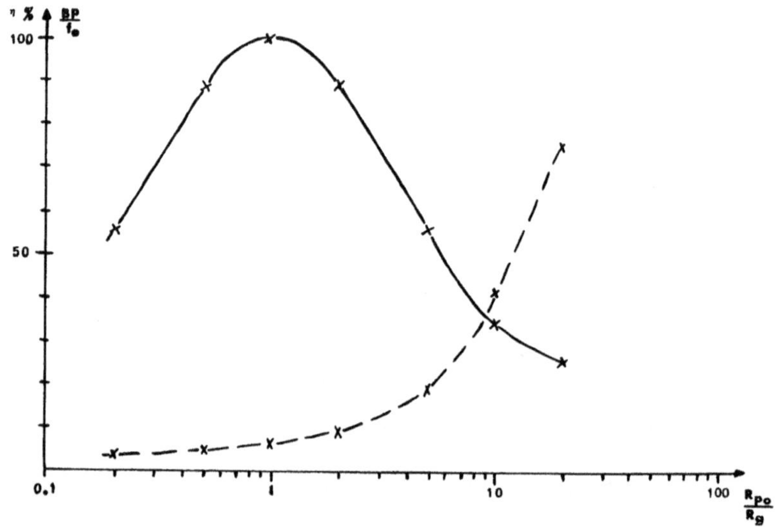

Figure 4. Evolution of the efficiency and the bandwidth versus R_{po}/R_g ratio.

For a ratio of the order of 20, a relative bandwith of 80 percent is obtained with an efficiency of 25 percent, that is to say a conversion loss of 6 dB .

The figure 5 shows the evolution of efficiency versus frequency corresponding to the precedent situation .

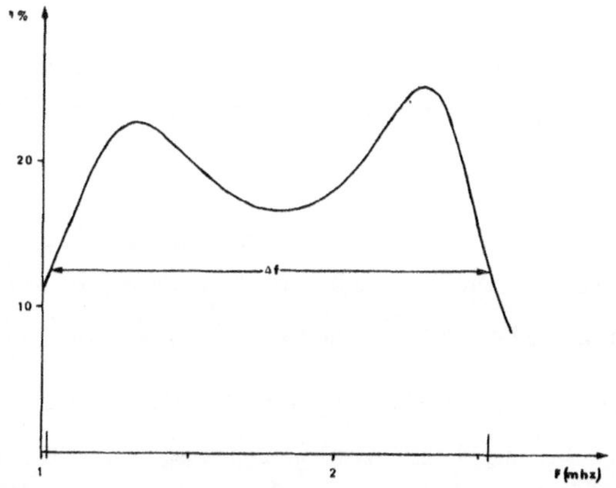

Figure 5. Evolution of the efficiency versus frequency for a R_{po}/R_g ratio equal to 20.

In pratical lossy case, the ratio $R_{po}/R_{g\,eq}$ has an upper limit

$$\frac{R_{po}}{X_{so}\,tg\ \delta} .$$

This may cause a limitation to the increase of the bandwidth by the electrical mismatching method described above.

Experiments are actually under progress to show the bandwidth increase for low dielectric loss transducer materials (eg Li Nb O_3) together with the limitation is this increase for high loss angle transducer materials.

E) THE PVF_2 BEHAVIOR.

When taking into account its mechanical properties, that is to say :

- its compliance
- its acoustical impedance close to the water one.

The PVF_2 or PVDF appears as a good material for medical applications.

However, two essential parameters do not allow to obtain good results.

First, the low value of the electromechanical coupling factor K^2, which equals only 2 percent, to be compared with the 50 percent of the ceramic.

Secondly, the dielectric loss factor (tg δ) equal to .2 at 2 MHz, which leads to the dissipation inside the transducer of an energy ten times greater than the acoustical energy radiated.

On other hand it is actually difficult, owing to the frequencies used in the medical applications, to obtain thick well polarized samples, to be used under resonnant conditions.

Experimental results are in good agreement with the theoretical predictions. The experiments have been performed using a 26 μm thickness sample with back reflector. The resonnant frequency is then of about 21,3 MHz and the working frequency of about 2,5 MHz. The measured conversion loss amounts to 28 dB instead the theorical value of 24 dB .
The theorical calculations in the resonnant conditions for a sample with a thickness of 200μm give conversion loss of about 13 dB.

F) CONCLUSION.

It appears that the most critical parameters of piezoelectric material is the electromechanical coupling factor K .

A value of K^2 of about 50 percent is needed in order to obtain a large bandwith together with a good efficiency. That explains the large use of materials like PZT ceramics or lithium niobate.

The low value of K^2 of PVF_2 cancels strongly the other advantages of this material for medical applications.

At a first time, the new piezoelectric polymers must have a high electromechanical coupling factor and at a second time the dielectric loss factor (tg δ) must be decreased.

F) REFERENCES.

|1| E. DIEULESAINT and D. ROYER
Ondes élastiques dans les solides, Masson and Cie Paris, 1974.

|2| B.A. AULD
Acoustic field and waves in solids, Vol. 1, John Wiley and Sons, New York, 1973.

|3| J.H. GOLL and B.A. AULD
IEEE Transactions on Sonics and Ultrasonics. Vol. SU 22 n°1, January 1975.

|4| M.G. GHAZALEH and all
Unpublished.

COMPARISON OF DIFFERENT PIEZOELECTRIC TRANSDUCER MATERIALS FOR OPTICALLY SCANNED ACOUSTIC IMAGING

C. W. Turner and S. O. Ishrak

Department of Electronic & Electrical Engineering
King's College
London, England

INTRODUCTION

Recent work has shown that optically-scanned piezoelectric transducers may be employed for transmission imaging of objects with a high spatial frequency content, in the 1-10 MHz frequency range[1]. The optically scanned transducer shown in Fig. 1, consists of a monolithic piezoelectric receiving transducer with its front face electrode removed and replaced by a semiconductor layer which has a semi-transparent metal film evaporated on it. A scanning light beam samples the local amplitude and phase of an incident acoustic wavefront sequentially, by photoconductively switching the semiconductor. In practical devices, however, an unwanted background reference signal from the dark regions of the transducer is present, largely because of the semiconductor capacitance. Brightness modulation of the scanning light beam overcomes this problem by producing sidebands originating only from the switched (or illuminated) part of the transducer. Typically a modulation frequency of 50kHz is used with an acoustic wave frequency of 3MHz. An acoustic image obtained with this device is shown in Fig. 2.

In this paper, we first discuss the limitations on sensitivity and spatial resolution of the piezoelectric layer in the optically-scanned transducer. The physical properties of PVF_2, $LiNbO_3$ and PZT will then be deduced by comparing electrical impedance measurements with those predicted by a modified Mason model. The expected sensitivity and performance of each piezoelectric material in an optically-scanned transducer will be computed, using equivalent circuits for the imaging device. Finally, experimental results obtained in optically-scanned transducers composed of PVF_2/Si, PZT/Si and $LiNbO_3$/Si sandwiches are presented.

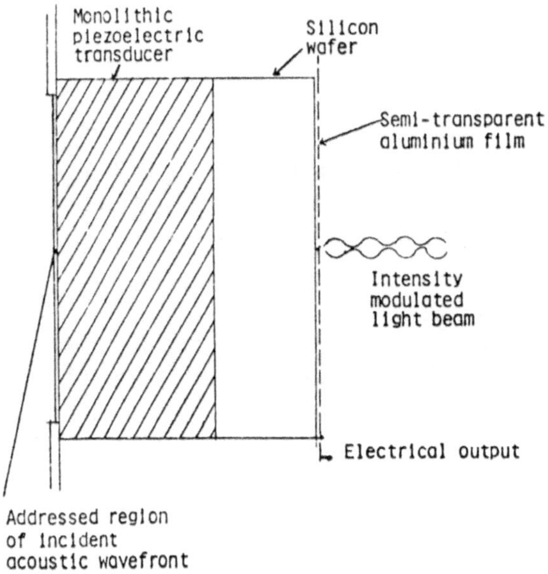

Fig. 1 The optically scanned transducer

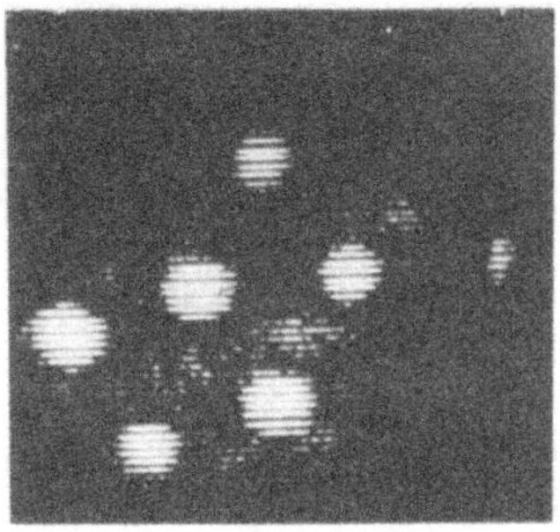

Fig. 2 An optically scanned transmission image of 2 mm holes in rubber spaced 2 mm apart (64 line image at a frame rate of 10Hz)

FACTORS AFFECTING THE SENSITIVITY AND RESOLUTION

a. Resolution Considerations

In general, the two dimensional set of current generators at the sideband frequency does not faithfully represent the acoustic wavefront. Three main factors influence the final image:

1. Size of the addressing light spot;
2. Spatial response of the photoconductive layer;
3. Spatial response of the piezoelectric layer.

Only the last effect listed above will be considered here. The piezoelectric layer is assumed to be mechanically air-backed, since the adjacent silicon layer is much less than $\lambda/_{10}$ thick.

The acoustic image on the receiving faceplate may be represented by a characteristic angular spectrum of plane waves (i.e. a superposition of plane waves propagating in different directions). Each plane wave component is refracted at the piezoelectric/water interface resulting in a transformed angular spectrum (and hence distorted acoustic image) in the piezoelectric transducer. For angles of incidence less than the critical angle (θ_c) and assuming an unbounded piezoelectric medium, spatial frequencies corresponding to the transformed (refracted) spectrum in the piezoelectric are identical to those in the water. For angles greater than θ_c, however, the refracted plane wave components are evanescent in the piezoelectric (as long as $\lambda_{water} < \lambda_{piezoelectric}$, as is usually the case). The largest permitted angle of incidence is thus θ_c. Consequently, the highest spatial frequency, f_{max}, that can be resolved by an imaging transducer is given by,

$$f_{max} = \frac{\text{Sin}\theta_c}{\lambda_w} = \frac{C_w}{\lambda_w C_t} = \frac{1}{\lambda_t}$$

where C_w, C_t and λ_w, λ_t are the velocities and wavelengths of sound in water and in the transducer.

The resolution of the piezoelectric transducer is therefore determined by the wavelength of sound in the transducer material and <u>not</u> by the wavelength in the load medium (water).

In addition, the finite thickness of the piezoelectric layer causes oblique standing waves to be set up, as shown in Fig. 3a, which result in further spatial filtering of the input angular spectrum. The extent of this lateral spreading is a function of the

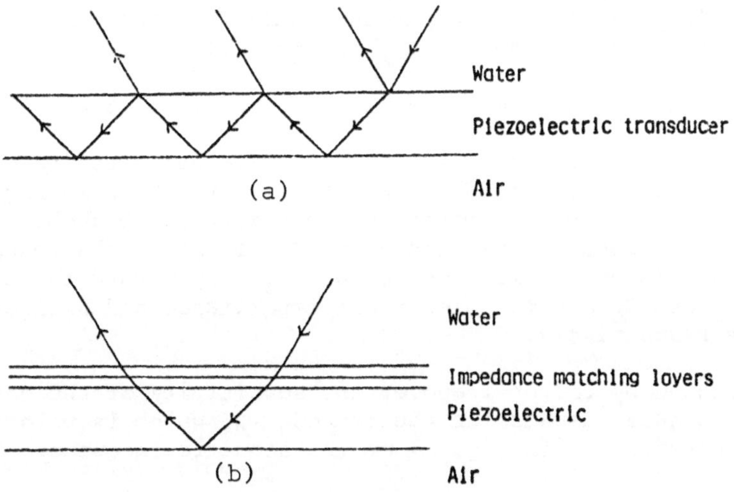

Fig. 3 Lateral spreading effects in piezoelectric transducers

mechanical loss, the reflection coefficient at the water/piezoelectric boundary and the thickness of the imaging plate. These effects have been investigated in detail by several workers[2,3]. It has been shown that the spreading may be considerably reduced by the addition of impedance matching layers (see Fig. 3b).

In this paper, the mechanical loss and the radiation resistance into water for PVF_2, $LiNbO_3$ and PZT-5A half-wave resonant transducers will be determined and related to their experimental resolutions.

b. Sensitivity Considerations

Consider the optically-scanned transducer equivalent circuits at the acoustic frequency (Fig. 4a) and at the sum frequency (Fig. 4b) Each branch represents one resolvable acoustic element whose size is determined by the extent of lateral cross-coupling through the effects discussed above. The sensitivity of the transducer depends on five factors listed below:

1. Magnitude of current through R_s' (in Fig. 4a) at the acoustic frequency (i.e. I_r' in Fig. 4a).

2. Efficiency of sideband generation at the sum frequency per unit I_{ap} (in Fig. 4a) for a given optical illumination level.

3. Intensity of the illuminating light beam.

4. Photoconductive switching efficiency.

5. Available power at the output terminals from the current generator at the sum frequency (I_s in Fig. 4b).

We will be mainly concerned here with factors 1 and 5 above.

Each acoustic element (Z_T) in the equivalent circuits of Fig. 4 may be represented by a modified Mason circuit, shown in Fig. 5, consisting of an acoustic branch and an electrical branch. R_a, L_a, and C_a are functions of the mechanical Q and K_t^2 of the transducer, while R_r depends on the load medium and I_a represents the acoustic source. C_o and R_o are the electrical capacitance and dielectric loss resistance respectively.

As stated by factor 1 above, the sensitivity of the optically scanned transducer depends on the current I_r', which is related to the current generator I_a (in Fig. 5) at the acoustic frequency.

The magnitude of this current is influenced by the mechanical absorption loss, the electrical shunting effects of C_o and R_o, the reflected power at the piezoelectric/water boundary and the

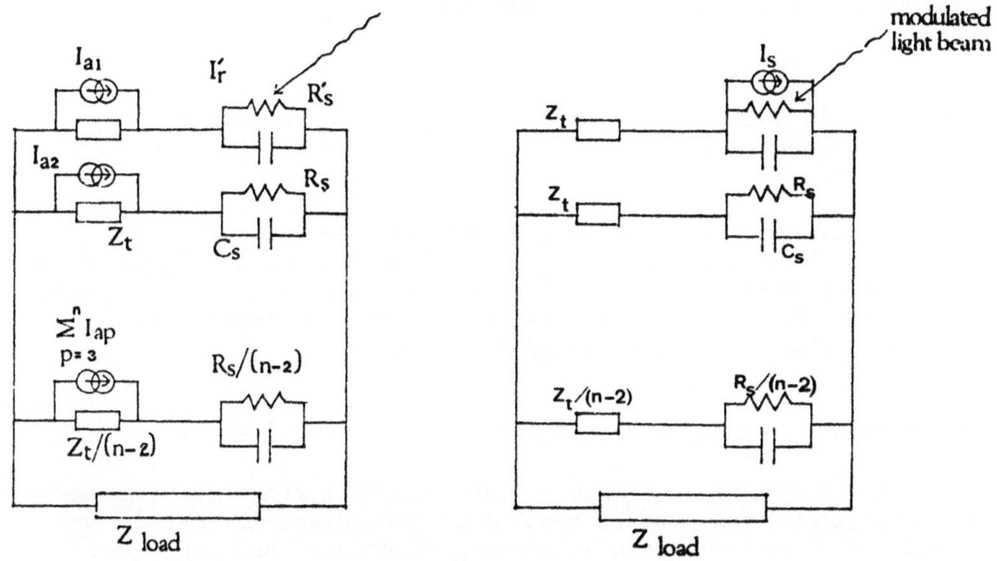

Fig. 4 Optically-scanned transducer equivalent circuits at the acoustic frequency (a) and at the sum frequency (b)

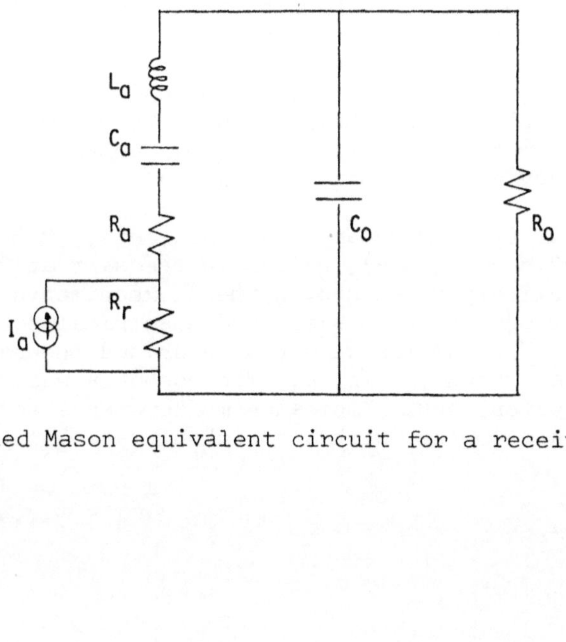

Fig. 5 Modified Mason equivalent circuit for a receiving piezoelectric transducer

electromechanical coupling constant K_t.

In addition, the power available at the output from the current generator I_s at the sideband frequency (i.e. condition 5 stated previously) is determined by the signal loss in the series impedance presented by Z_T and the shunting effect of the (n - 1) dark elements.

The above parameters vary strongly between PZT, PVF_2 and $LiNbO_3$. In the following sections, the elements of the Mason model will be determined and used in the equivalent circuits of Fig. 4 to compute the relative sensitivities of the three materials when employed in the optically-scanned transducer.

EVALUATION OF PHYSICAL PROPERTIES FOR THE PIEZOELECTRIC TRANSDUCER

It is clear from the above discussion that the mechanical and electrical properties of the monolithic piezoelectric transducer determine its usefulness as an imaging faceplate, both in terms of resolution and sensitivity. In particular, the mechanical Q, the electromechanical coupling constant, K_t, the dielectric constant ε_r and the loss tangent, $\tan\delta$, have been deduced for PVF_2, $LiNbO_3$ and PZT transducers, through the evaluation of the Mason equivalent circuit.

Electrical input impedance measurements around resonance were first made for each sample, when mechanically loaded by air on both sides. The equivalent circuit parameters R_a, L_a, C_a and C_o were then varied in a computer program to obtain the best theoretical fit to the experimental data points. The mechanical Q and K_t^2 were deduced from the computed values of the circuit elements. (This technique for estimation of piezoelectric parameters follows closely that employed by several other workers[4,5,6] in studying thin PVF_2 films.) The procedure was then repeated with the sample loaded by water on one side, for calculation of the radiation resistance. (Although the properties of PZT-5A and $LiNbO_3$ are well known, we have repeated measurements on them for the purposes of this comparison, and for testing the accuracy of our computer program).

A PZT-5A plate (2.5 cm square), a $LiNbO_3$ disc (2.5 cm diameter) and a PVF_2 film (2.5 cm square), half-wave resonant at 2MHz, 2.95MHz and 3.00MHz respectively were used in these experiments. All the samples were electroded on both sides and electrical contact was made with short wires. The samples were bonded around the edges to an aluminium plate, as shown in Fig. 6. The assembly was then mounted on the wall of a water tank. Impedance measurements were made with a HP Vector Impedance meter, with and without water in the tank.

Fig. 7 is an example of the matched theoretical and experimental curves for the PVF_2 transducer when loaded by air on both sides (except around the edges). The following points regarding the measured PVF_2 parameters should be noted:

(1) The unloaded mechanical Q (≈6) for PVF_2 is less than those reported previously for thinner specimens. The discrepancy may be partly due to the different mechanical conditions, or to the particular origin of our specimen. The values for K_t^2 (≈.024), ε_r (≈7) and tan δ (0.25) for PVF_2, however, agree reasonably with other reported figures. Our investigations have been limited to only one sample at present.

(2) The PVF_2 has a loaded Q of less than 3, which follows from its known good match to water and its large mechanical loss. This extremely low Q limits the accuracy in the measurement of the radiation resistance.

Tables 1, 2 and 3 summarize the measured mechanical and electrical properties of the three materials. Their relative significance in determining the ultimate sensitivity and resolution of the optically-scanned transducer will be discussed in the following section.

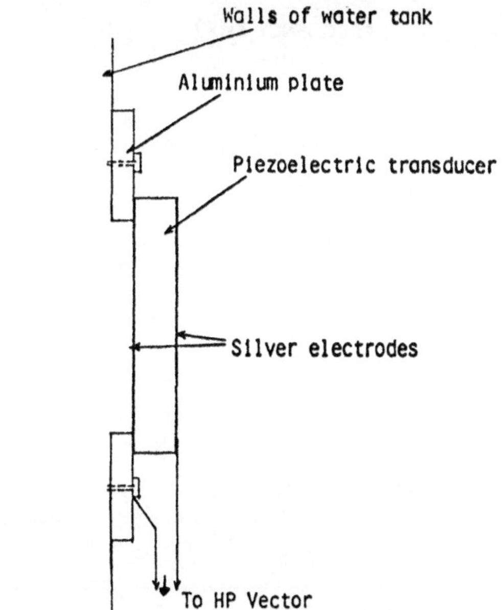

Fig. 6 Experimental conditions for the impedance measurement

thickness = 0.35 mm
$Q = 6 \pm 0.5$
$K_t^2 = 0.024 \pm 0.001$
$\varepsilon_r \approx 7$
$\tan\delta = 0.25$

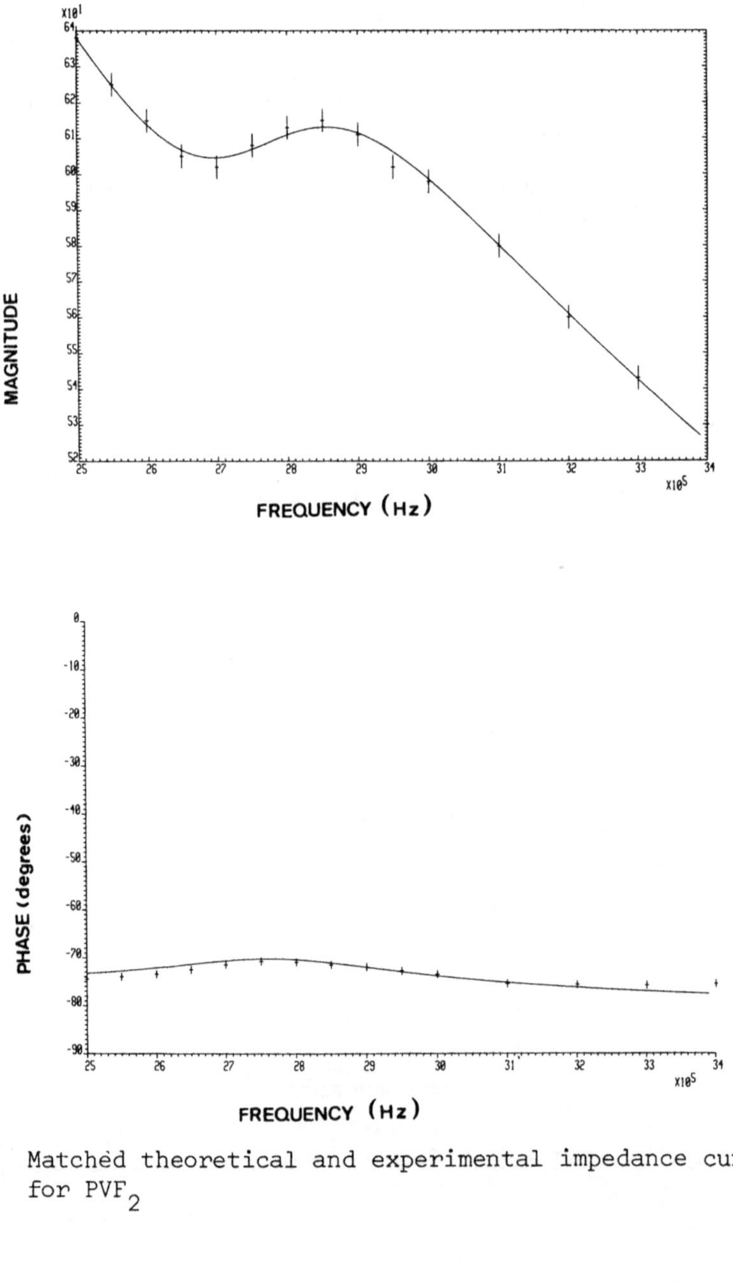

Fig. 7 Matched theoretical and experimental impedance curves for PVF_2

Table 1

Sample	Thickness	Unloaded Q	K_t^2	ε_r	Tan δ
PZT-5A	1.1 mm	45 ± 2	0.4 ± .02	850	<.08
$LiNbO_3$	1.25 mm	142 ± 7	0.56 ± .03	39	<.03
PVF_2	0.35 mm	6 ± .5	0.024 ± .001	≃7	.25

Table 2

Sample	Loaded Q	Theoretical Q (assuming no mechanical losses)	R_r (Radiation Resistance) (Ω)
PZT-5A	25 ± 2	31	3Ω ± 1
$LiNbO_3$	32 ± 2	36.6	29Ω ± 2
PVF_2	2.6 ± 0.3	4.02	3K ± .5K

Table 3

Sample	Impedance Ratio to Water Zt/Zw	Wavelength Ratio to Water λt/λw	Effective Mechanical Loss Resistance (Ω) (includes radiation into aluminium collar)
PZT-5A	20	2.91	1.6 ± .5
$LiNbO_3$	23.2	4.94	8 ± 1
PVF_2	2.55	1.44	5K ± 0.7K

COMPUTER SIMULATION RESULTS

A comparison of the physical parameters listed in Tables 1, 2 and 3 suggests that $LiNbO_3$, with its high electromechanical coupling constant (K_t), relatively low dielectric constant (ε_r) and negligible loss tangent has a significantly higher sensitivity than PZT-5A. PVF_2 has an even lower dielectric constant than $LiNbO_3$ and is additionally a good impedance match to water, but its low K_t and its electrical and mechanical losses degrade its sensitivity considerably. The relative performance of the three piezoelectric materials in the optically-scanned transducer has been studied quantitatively through a computer model based on the equivalent circuits shown in Fig. 4. Z_T (in Fig. 4) was substituted by the modified Mason equivalent circuit discussed earlier. Silicon of 13KΩ-cm resistivity and 150 μm thickness has been assumed for the results to be presented here.

Fig. 8 compares plots (obtained from the simulation), of the maximum available output power as a function of acoustic frequency, for each of the three transducers.(i.e. PZT/Si, $LiNbO_3$/Si and PVF_2/Si). It is clear that the qualitative arguments presented above are consistent with these results.

As has been mentioned earlier, the transducer permittivity plays an important role in the final sensitivity of the device. For piezoelectric materials with low ε_r, capacitive shunting of the current generators at the acoustic frequency is small. However, the associated increase in transducer impedance (Z_T in Fig. 4) results in greater signal loss at the sideband frequency. The results for the sensitivity of $LiNbO_3$ and PVF_2 point to an optimum value for the dielectric constant, roughly equal to that of the semiconductor (≃12 for silicon).

In practical optically-scanned imaging transducers, the equivalent circuits of Fig. 4 have to be modified to include a series capacitance representing the piezoelectric/semiconductor interface region. PVF_2 and $LiNbO_3$ are relatively unaffected by this series capacitance while PZT-5A, with its large ε_r, is seriously compromised and is least tolerant to non-uniformities in the mechanical clamping of the semiconductor to the piezoelectric.

EXPERIMENTAL RESULTS

a. Experimental Conditions

For the purposes of this comparison, a single 13KΩ-cm silicon wafer, 2.5 cm in diameter and 150 μm thick, coated with a 300Å semi-transparent aluminium film, was employed in turn with the PZT-5A, $LiNbO_3$ and PVF_2 transducers. The silicon was mechanically clamped

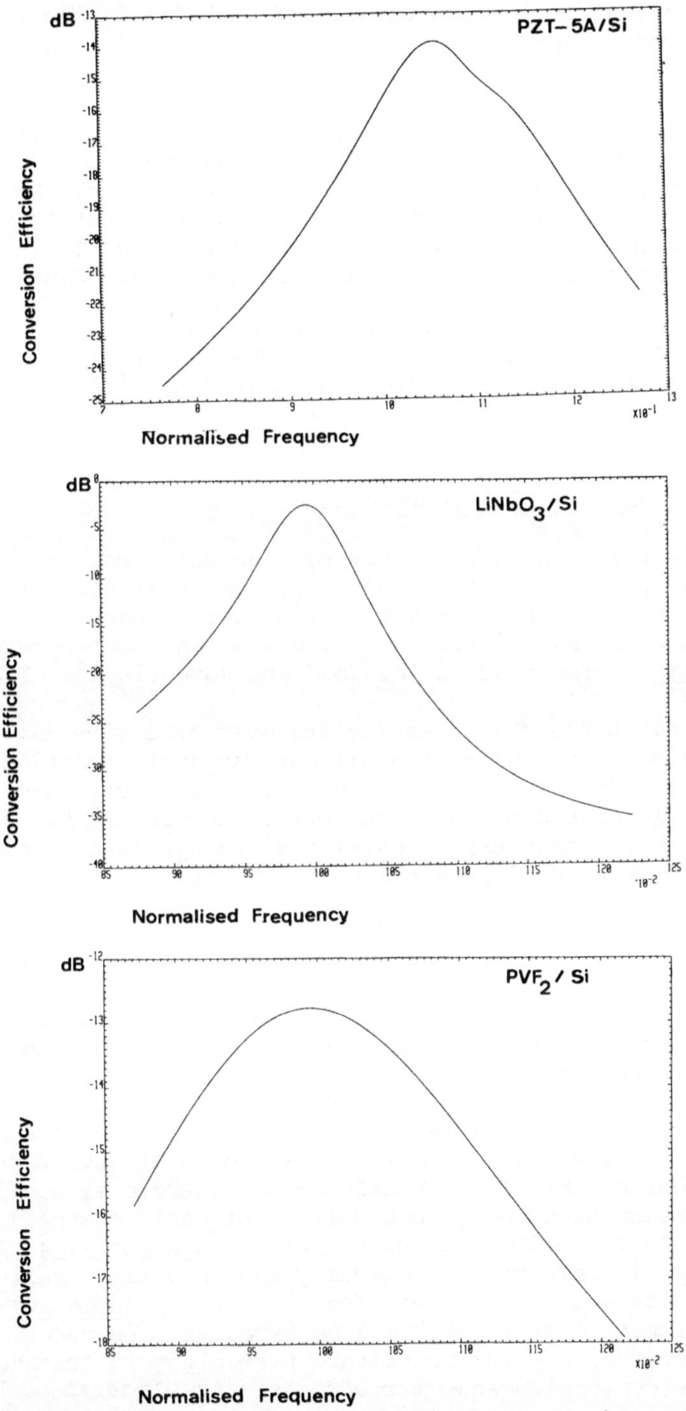

Fig. 8 Computed plots of conversion efficiency against normalised frequency

to the piezoelectric for experimental flexibility. The scanning optical source was provided by an oscilloscope CRT, Z-modulated at 50kHz.

A 2.5 cm square PZT-5A transmitting transducer, half-wave resonant at 2.3MHz and operated close to the loaded resonant frequency of the receiving transducer (1.97MHz for PZT-5A, 2.85MHz for $LiNbO_3$ and 2.75MHz for PVF_2), served as the acoustic source. Transmission acoustic images of test objects placed against the receiving transducer were obtained through optical scanning.

The electrical output at the sideband frequency from each of the three composite transducers (PZT/Si, $LiNbO_3$ and PVF_2/Si) was observed on a Spectrum Analyser, and simultaneously displayed on a CRT, as in Fig. 9.

b. Relative Sensitivity Measurements

Each optically scanned receiving transducer was insonified with a roughly uniform beam and the best obtainable sideband levels were compared at the Spectrum Analyser. The outputs were corrected for their respective transmitter insertion loss and post receiver amplifier gain. The results obtained are summarised in Table 4.

Note that $LiNbO_3$/Si is easily the most sensitive device, followed by PVF_2/Si. Since the silicon was just mechanically clamped to the transducers, the interface capacitance effects discussed earlier, could have diminished the efficiency of the PZT/Si transducer. However, these experimental results agree reasonably well with the computer model predictions.

c. Resolution

Quantitative figures of merit for the resolution of the imaging faceplates were obtained through (i) acceptance angle and (ii) spatial response measurements.

(i) Acceptance angle measurements: In general, the size of a receiving (or radiating) element is related to its polar response through a Fourier transform relationship. Therefore, acceptance angle measurements, with the composite (optically scanned) transducer illuminated by a single stationary spot ($< \lambda_{piezoelectric}$ in diameter), lead to valuable information regarding the faceplate resolution. For an element size equal to the wavelength in the load medium an angular acceptance range close to $\pm 90^o$ may be expected. (However, in practice this is never the case for monolithic piezoelectric transducers, because mode conversion andother effects take place at much smaller angles.)

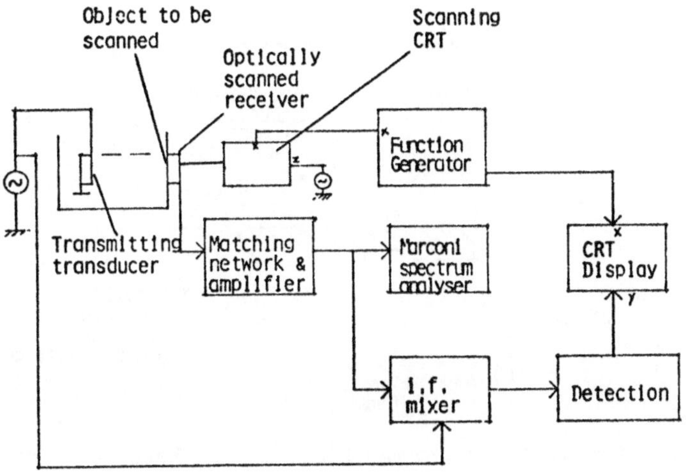

Fig. 9 Experimental arrangement for one-dimensional transmission imaging

The relative polar responses for the three transducers (PVF_2/Si, PZT/Si and $LiNbO_3$/Si) were experimentally evaluated, using the arrangement shown in Fig. 10. To obtain a reasonable S/N ratio the transmitting point transducer had to be 1.5 mm in diameter, which meant that the output was influenced by the transmitter's polar response for angles greater than $\pm 25^o$. Also, the S/N ratio was always poor, because the effective power density at the receiver plane was extremely small. The above reasons, and the relatively crude mechanical scanning mechanism limited the accuracy of the measurement. However, reasonable estimates could still be made.

Fig. 11 compares the polar diagrams of the three transducers. Note that PVF_2 has a 3dB halfwidth much greater than 20°, compared to 6° for PZT-5A and 4° for $LiNbO_3$. (In fact, the PVF_2 polar response was limited by the transmitter). The large angular acceptance range for PVF_2 follows from its good impedance and velocity match to water (see Table 1). In addition, the large acoustic loss ensures effective damping of any internal lateral spreading effects. PZT and $LiNbO_3$, on the other hand, are both poorly matched to water, which points to a moderate resolution. This is confirmed by their small acceptance angles. The slightly better velocity match to water in PZT accounts for its marginally wider polar response.

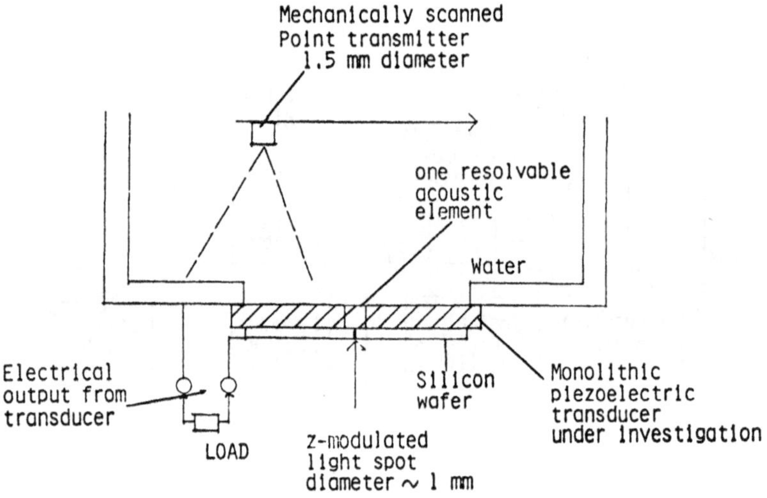

Fig. 10 Experimental arrangement for Polar Response Measurement

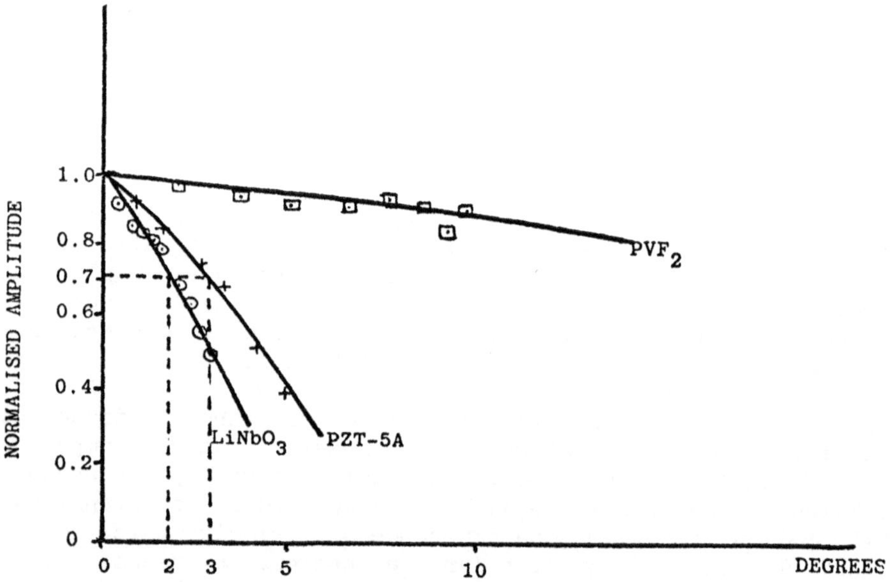

Fig. 11 Polar Response

Table 4

Transducer	Transmitter insertion loss (relative to that for PZT)	Post-receiver gain	Signal level at Spectrum Analyser	Net relative signal level
PZT-5A/Si	0dB	0dB	0dB	0dB
$LiNbO_3$/Si	- 3dB	- 4dB	+ 25dB	+ 32dB
PVF_2/Si	- 2dB	- 5dB	- 2dB	+ 5db

Table 5

Transducer	Operating frequency	λ_W	λ_t	Slit size in wavelengths		Measured 3dB spatial response in wave-lengths
PZT-5A	1.97MHz	767 μm	2.2 mm	1.3 λ_w	.45λ_t	3λ_w
$LiNbO_3$	2.85MHz	525 μm	2.6 mm	1.9 λ_w	.38λ_t	4λ_w
PVF_2	2.75MHz	544 μm	780 μm	1.84λ_w	1.28λ_t	1.4λ_w

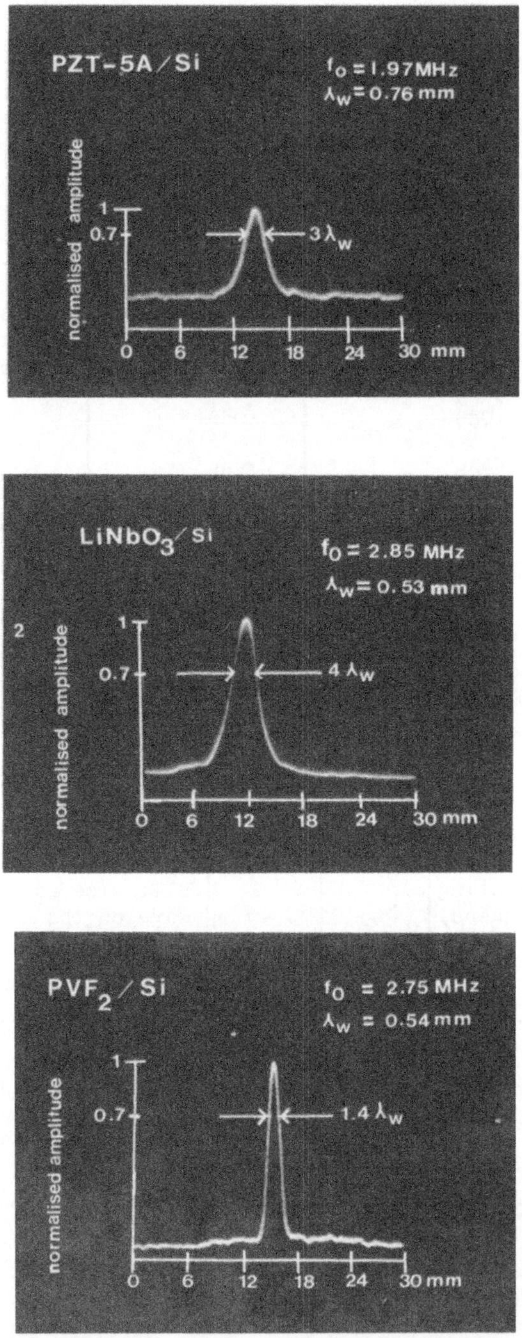

Fig. 12 Spatial responses of PZT-5A/Si, $LiNbO_3$/Si and PVF_2/Si transducers to a 1 mm slit

The above arguments, supported by the acceptance angle measurements, therefore indicate a significantly superior resolution for the PVF_2 monolithic transducer.

(ii) Spatial response measurements: The relative resolutions for the three transducers were then evaluated through optically-scanned acoustic imaging of test objects. A 1 mm slit in acoustically absorbent material (rubber) was placed against the receiver transducer, which was then insonified by a roughly uniform acoustic beam. Line scans were recorded for the three transducers in turn and the results obtained are presented here.

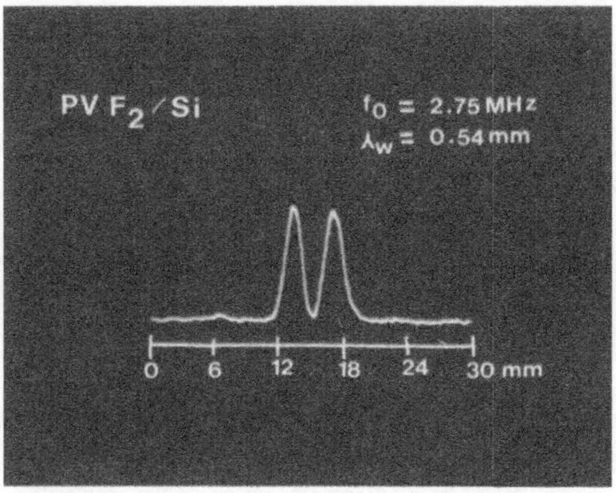

Fig. 13 Spatial response of PVF_2/Si transducer to two 1 mm slits, 1 mm apart

The normalised widths of the 1 mm slit for the different operating frequencies are listed in Table 5. Fig. 12 compares the experimentally obtained line scans. Note that the 1 mm slit is well within the resolving capability of the PVF_2 transducer. With the existing experimental arrangement the resolution was limited by the light spot size. An alternative test object consisting of two 1 mm slits, 1 mm apart was imaged successfully using the PVF_2/Si transducer (Fig. 13).

The superior resolution of PZT compared to $LiNbO_3$, predicted by the shorter acoustic wavelength (and wider acceptance angle) is also confirmed in Fig. 12.

CONCLUSIONS

The comparative performance of $LiNbO_3$, PZT and PVF_2 as optically-scanned acoustic imaging transducers can be summarised as follows:

(1) $LiNbO_3$ has by far the largest sensitivity, but it suffers from a poor impedance and velocity match to water.

(2) The large dielectric constant of PZT-5A means that its sensitivity is worse than $LiNbO_3$. Its acoustic impedance and velocity are only marginally better matched to water.

(3) PVF_2 has a very good resolution mainly because of its impedance and wavelength match to water. Although its sensitivity is reduced by mechanical and electrical losses, its overall performance in optically-scanned imaging transducers is superior to that of other available piezoelectric materials.

ACKNOWLEDGEMENTS

We gratefully acknowledge the support of Thorn EMI Central Research Laboratories, who also supplied the sample of PVF_2.

REFERENCES

1. Turner, C. W., Ishrak, S.O., Fox, D. R., "Optically Scanned Amplitude and Phase probing of acoustic fields", paper presented at the 1979 Ultrasonics Symposium (to be published in SU).

2. Auld, B. A., Drake, M. E., Roberts, C. G., Applied Physics Letters, Vol. 25, No. 9 (1974)

3. Noorbehesht, B., Flesher, G. and Wade, G., Ultrasonic Imaging, Vol. 2, No. 2 (1980)

4. Bui, L., Shaw, H. J., and Zitelli, L., IEEE Trans. Sonics and Ultrasonics, Vol. SU-24 No. 5 (1977)

5. Ohigashi, H., Journal of Applied Physics, Vol. 43, No. 3 (1976)

6. Callerame, J., Tancrell, R. H., and Wilson, D. T., 1978 Ultrasonics Symposium Proceedings.

DARK FIELD ACOUSTIC MICROSCOPE

I.R. Smith and D.A. Sinclair

Dept. of Electronic and Electrical Engineering
University College London,
Torrington Place, London WC1E 7JE

ABSTRACT

In a scanning acoustic microscope dark field imaging can be achieved by departing from collinearity in the arrangement of receiving and transmitting lenses. We have recently shown that one can also obtain dark field images in a collinear arrangement by replacing one lens with a plane wave transducer and by placing a stop in the centre of one of the lenses. The technique is however difficult to apply to high frequency instruments. In the present paper we report on an alternative configuration in which the transducer is formed on the front surface of the lens. An annular electrode pattern is used to excite the transducer. Disconnecting the central region achieves dark field conditions. More generally, by prescribing the amplitude of the signals applied to each distinct electrode in a multielement annular transducer, it is possible to vary the spatial frequency response of the instrument over wide limits. This enables one to adapt the performance of the instrument to the requirements set by specific objects. These techniques will be illustrated with reference to human tissue characterisation and the detection of cracks in metals.

INTRODUCTION

The Scanning Acoustic Microscope {1}, first described by Quate and Lemons {2}, is proving a useful and flexible tool for non destructive, high resolution imaging of a wide range of biological and solid state objects. Several different imaging modes have been demonstrated including confocal transmission and reflection arrangements {3}, plane transducer-spherical transducer geometries {4} and an off-axis dark field configuration {5}. In this paper, we present results from a new dark field arrangement {6} suitable for both transmission and reflection imaging. Dark field systems are particularly appropriate for the examination of highly transmissive, weakly diffractive objects. For this class of object most of the illumination energy passes straight through without being diffracted and consequently contains no useful contrast information. Dark field systems are insensitive to this energy and only detect the diffracted field from the object. In this way a large DC term in the received signal is removed which enables the sensitivity of the microscope to be significantly increased.

In a previous dark field acoustic microscope system {5}, one lens of a transmission scanning acoustic microscope (SAM) was tilted to an off axis position so that the beam from the transmit lens did not quite fall within the acceptance angle of the receiving lens. An object, placed in the focal plane, diffracted some energy onto the receiving lens, and so increased the detected signal. A dark field image could therefore be produced. To bring the system into alignment required that the focal regions of the lenses were coincident in space. Alignment of this off-axis dark field configuration is difficult for high resolution systems since the focal spot size approaches 1 μm. Also, because of the assymetry of the lens arrangement, only a portion of the high spatial frequency diffraction components are detected and hence the system is mainly sensitive to edges.

These problems may be overcome by adopting an on-axis dark field geometry as shown in Figure 1. The object is illuminated by a normally incident plane wave produced by a large aperture piezoelectric transducer. A conventional SAM lens is used to collimate and detect the energy transmitted through the object. The undiffracted wave that passes through the object will hit the lens surface normally and produce a broad spectrum of plane waves within the lens rod. However, because of the high velocity ratio across the lens surface, (necessary to minimise spherical aberration), only the central portion of the lens will actually contribute to the field in the lens rod. Since the transducer on the back of the lens rod has a finite angular acceptance only the lower of the spatial frequencies in the lens rod will be detected and contribute to the output voltage. By placing an absorbent or reflective stop on the front face of the lens, it is possible to prevent these spatial frequencies reaching the transducer and hence the system is dark field.

When an object is placed in the focal plane of the system some of the incident energy will be diffracted or scattered. These waves will be collimated by the lens onto the transducer and produce an output. The system is only sensitive to the diffracted energy coming from the lens focal region. Conventional mechanical scanning is then used to build up the picture point by point.

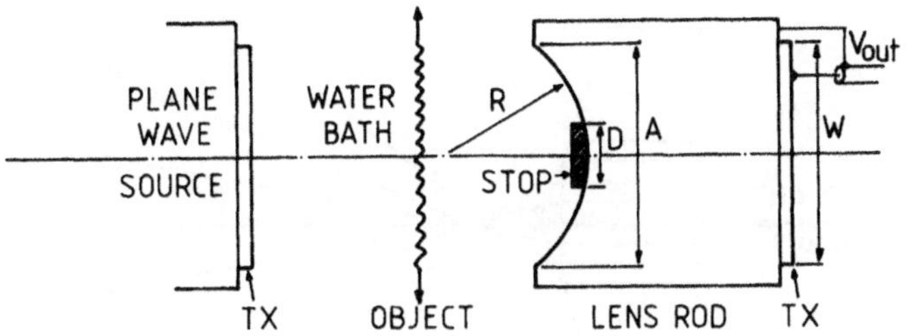

Figure 1. Dark field SAM geometry for 11 MHz operation. R = 16mm, A = 25mm, W = 25mm, n = 4.3 and D = 4mm.

LENS STOP THEORY

The imaging equation of such a system is easily obtained from that of the conventional confocal two lens SAM. Thus the image spatial frequency spectrum, V(f), can be written

$$V(f) = H\{t(r)\}.H\{U_1(r)\} \tag{1}$$

where, in cylindrical coordinates, f is the spatial frequency, r is the radial coordinate, $U_1(r)$ is the lens sensitivity function and t(r) is the object transmittance. H{X} denotes the zero order Hankel transform of X. In the space domain this is equivalent to convolving the lens sensitivity function with the transmittance of the object. An approximate expression for the lens sensitivity function is

$$U_1(r) = \frac{AJ_1(\pi Ar/\lambda F)}{(\pi Ar/\lambda F)} - \frac{DJ_1(\pi Dr/\lambda F)}{(\pi Dr/\lambda F)} \tag{2}$$

where J_1 is the Bessel function of the first order, F is the focal length of the lens, A is the lens aperture, D is the stop diameter and λ is the wavelength. Figure 3 shows the calculated lens sensitivity function for the geometry of Figure 1. The effect of the aperture stop is to narrow the central lobe of the lens sensitivity and to increase the sidelobe level.

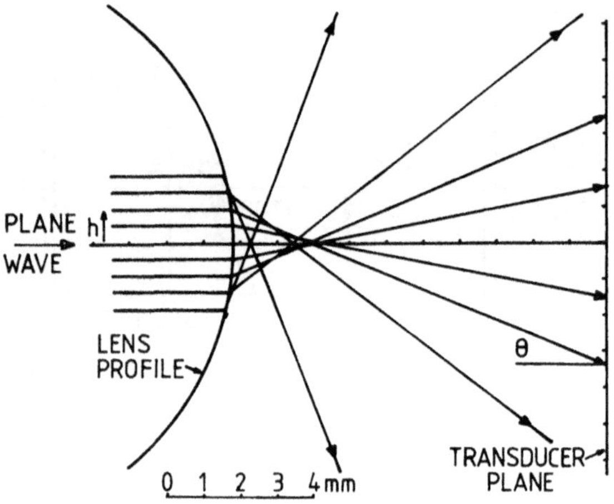

Figure 2. Ray tracing for dark field geometry of Figure 1.

Selection of the stop size is a compromise between retaining acceptable side lobe levels and achieving a satisfactory level of suppression of the undiffracted energy. The zero order spatial frequency component incident upon the lens produces a broad range of spatial frequencies within the lens rod. Since the receiving transducer has a finite width, W, it will give an appreciable output for the range of incident spatial frequencies f_o for which $|f_o| < 2/W$. It is necessary to make the stop size sufficiently large to block all of these detectable components. A criterion for the minimum size of stop may be obtained from the ray tracing model of Figure 2. A ray striking the lens at a height h will cross the transducer at an angle $\theta = (n-1)(h/R)$, where n is the velocity ratio at the lens-water interface and R is the lens radius. The criterion for minimum stop size is that a ray striking the periphery of the lens stop should correspond to a spatial frequency of f_o hitting the lens. In this case the stop diameter must be

$$D_{min} = \frac{4R\lambda}{(n-1)W} \tag{3}$$

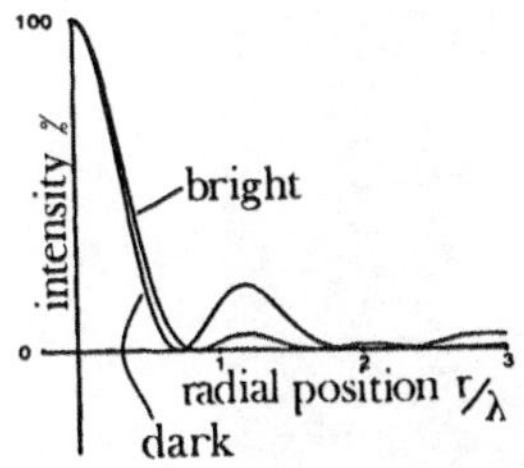

Figure 3. Computed lens sensitivity function with and without stop for dark field geometry of Figure 1.

At the other extreme, rays parallel to the axis and incident upon the lens beyond the critical angle suffer total internal reflection and do not reach the receiving transducer. This sets an upper limit on the stop size of

$$D_{max} = \frac{2R}{N} \tag{4}$$

It should be noted that an acceptably small aperture stop diameter depends on the phase sensitive nature of the receiving transducer. If the transducer was sensitive to intensity, the stop would have to block all the energy incident upon the lens that might reach the transducer. It would, therefore, have to be of much larger diameter than when used with a phase sensitive transducer since, in that case, some of the rays normally incident upon the lens may hit the transducer provided they do not fall within its angular acceptance. This shows that the technique may not be directly transposed into an incoherent optical system without using an excessively large stop.

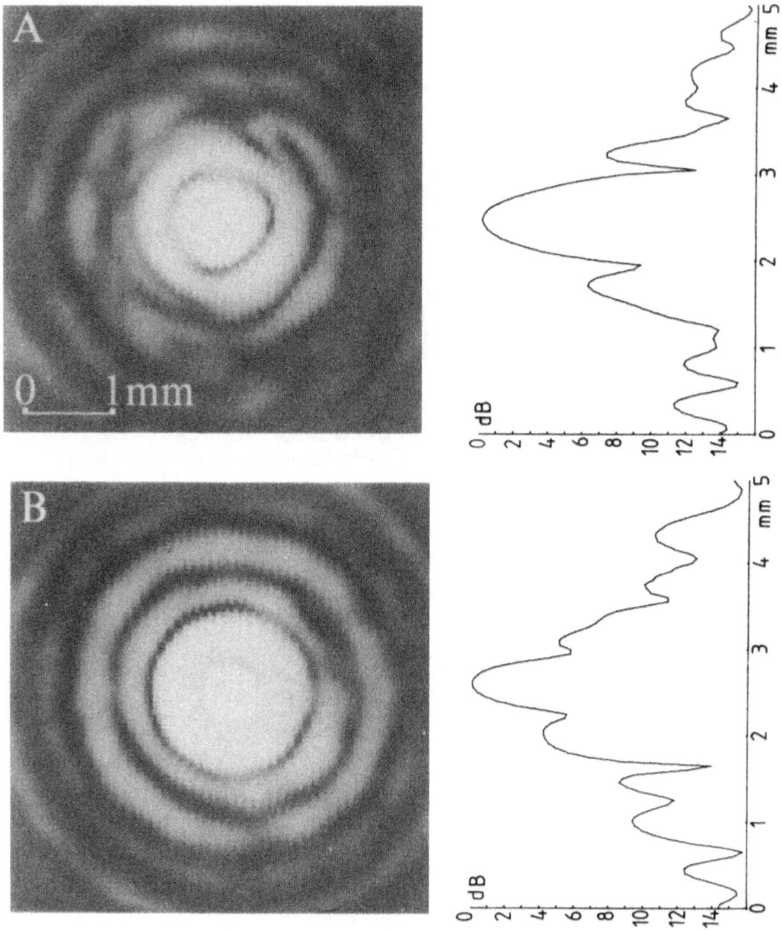

Figure 4. Measured focal plane distribution of lenses, showing line scans through the focal point
a) bright field b) dark field

BEAM PROBING OF LENS WITH STOP

The spatial sensitivity, G(r), of the quartz lens used in the experiments described below was measured using a high resolution acoustic probe {8}. The lens and stop dimensions are given in Figure 1. Line scans through the focal plane with and without the aperture stop are shown in Figures 4a) and 4b). It can be seen that the effect of the stop is to decrease the width of the lens main lobe width by a factor of 1.2 and to increase the sidelobe level by 3dB. While this latter effect will not seriously degrade the system performance, it does explain the increased prominence of the sidelobe response in the images below.

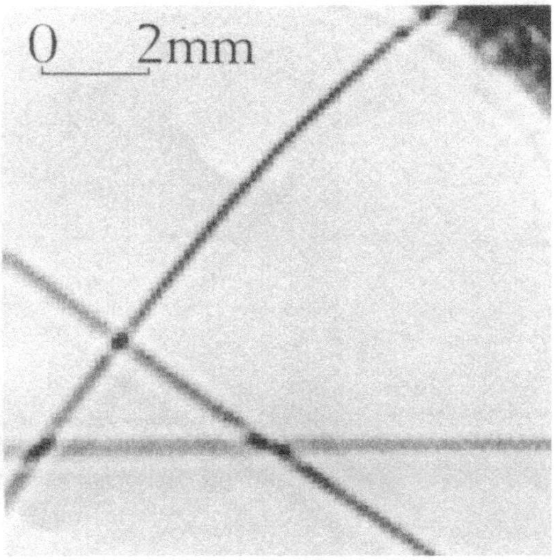

a) bright field

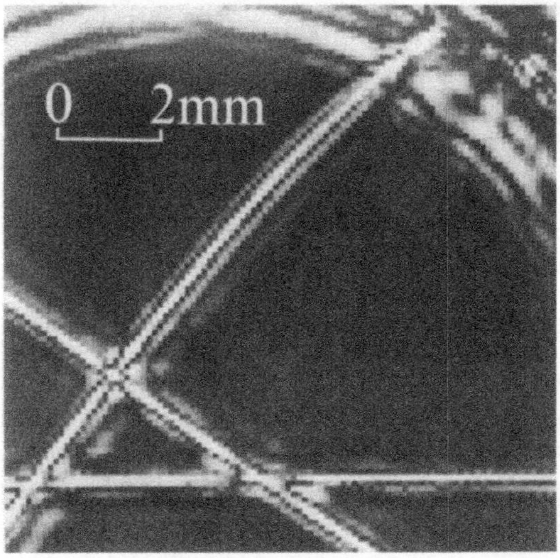

b) dark field

Figure 5. Acoustic images of wire array. The wire diameter is one water wavelength.

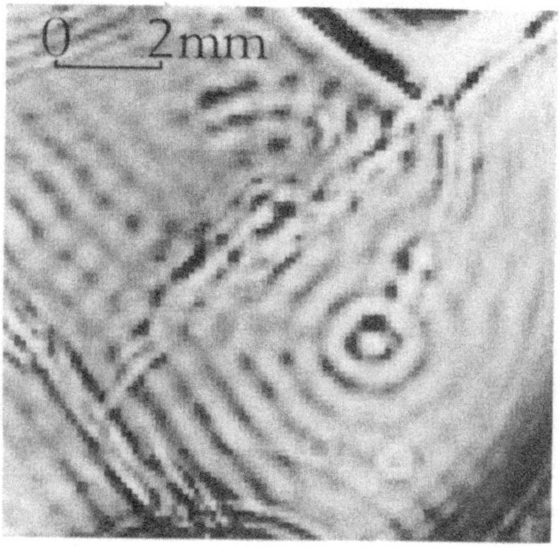

(a)

(b)

Figure 6. Dark field image of thin (less than tenth wavelength) crack in glass cover slip:
a) dark field image
b) schematic of crack.

EXPERIMENTAL RESULTS FROM LENS-STOP SYSTEM

A dark field SAM was constructed as shown in Figure 1. The stop produced a zero order suppression of approximately 20 dB. Figures 5a) and 5b) show bright and dark field images of a one wavelength diameter wire array. The wires are clearly resolved and there is also some indication of the increased sidelobe level in the dark field image. Figure 6 shows an application of the system in which a very thin ($\lambda/10$ across by 2 λ deep) crack was made in a glass cover slip. The branches of the crack are visible in the image, as is an air bubble. The ripple like structure in the image is due to standing waves running transversely across the object. These are present because the acoustic pulse length was substantially greater than the thickness of the cover slip.

SPHERICAL TRANSDUCER GEOMETRY FOR DARK FIELD IMAGING

We have developed a new lens structure for dark field imaging, shown in Figure 7, which offers some advantages. In this system, we use a spherical transducer as a focussed receiver. This type of transducer has been demonstrated at high frequency {4}, and is efficient for high resolution microscopy {9}. The transducer has an annular electrode pattern, the inner electrode being analogous to the field stop in the previous system. With a plane wave incident, the phase of the detected signal varies very rapidly across the transducer. Only the central portion integrates to give a substantial signal. We can electrically isolate the inner electrode so that the signal from the outer electrode yields a dark field image. If we include the inner electrode, then a bright field image is formed. Thus, by the use of an electrical switch, dark or bright field images may be selected without disturbing the acoustic components of the system. We have built a microscope using this spherical transducer configuration, and now present some results demonstrating its performance and applications.

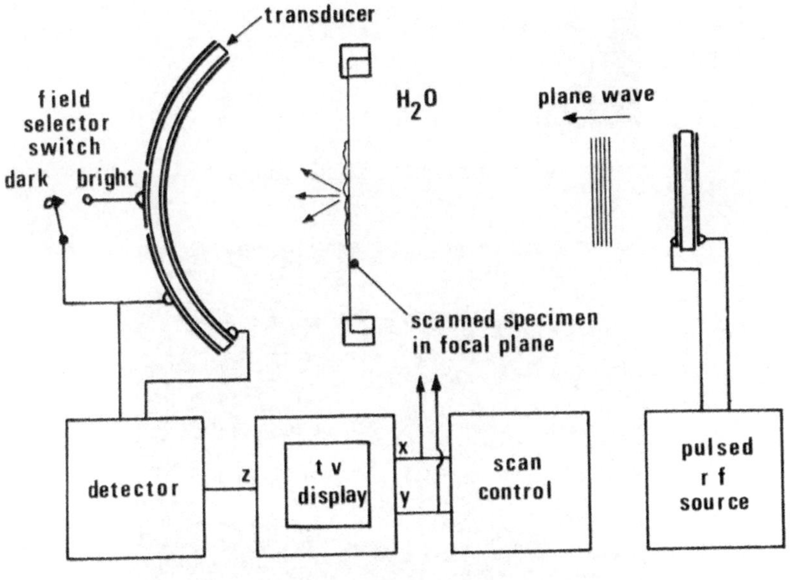

Figure 7. Dark field spherical transducer geometry.

EXPERIMENTAL RESULTS WITH SPHERICAL TRANSDUCER SYSTEM.

In our new dark field microscope, operating at 2.5 MHz, the spherical transducer was made from PZT-4a with R = 25mm, A = 25mm and D = 8mm.

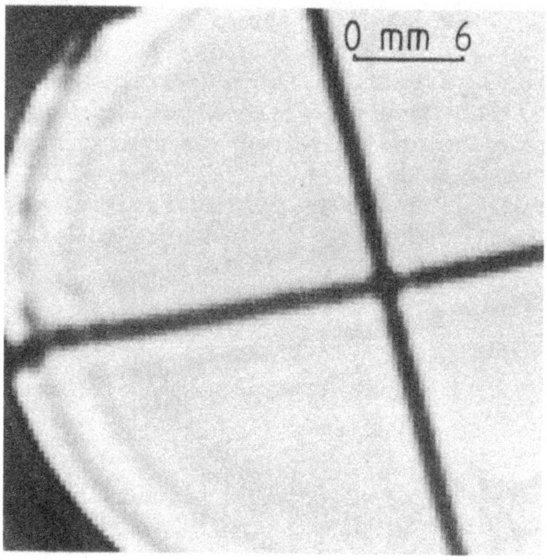

a) bright field image

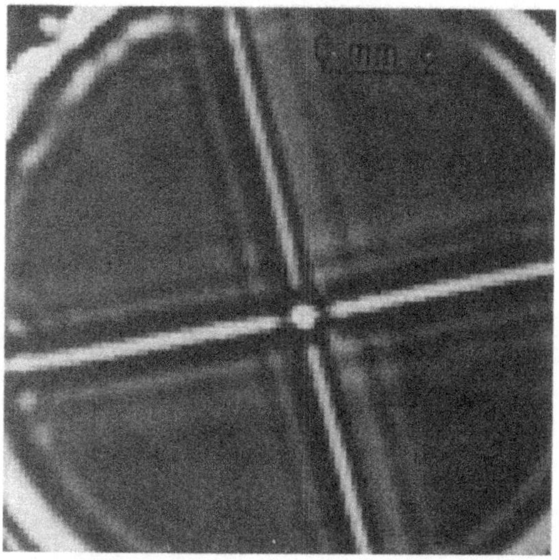

b) dark field image

Figure 8. Acoustic images of two $4/3\lambda$ wires.

Figure 8 compares bright and dark field images of a wire (thickness $4/3\lambda$), which demonstrates the expected response of the system, showing a large amount of energy diffracted by the wire. We note that the resolution of the dark field image is not seriously degraded. Figure 9 shows a line scan taken through a section of the wire, and bright and dark field responses may be compared. The large degree of zero order suppression achieved may be seen, together with the slight increase in sidelobe level. We have been investigating non-destructive testing applications for dark field microscopy, and Figure 10 shows one example. It is an image of two sheets of steel, 4mm thick, which have been argon arc butt welded. The heating caused by the welding process has induced local elastic variations in the steel, and so the weld is visible, running across the image. Because of the increased detection sensitivity achieved by the use of the dark field scheme, we can see a region of additional diffraction running along the centre of the weld (arrowed), which must correspond to some non-uniformity in the weld.

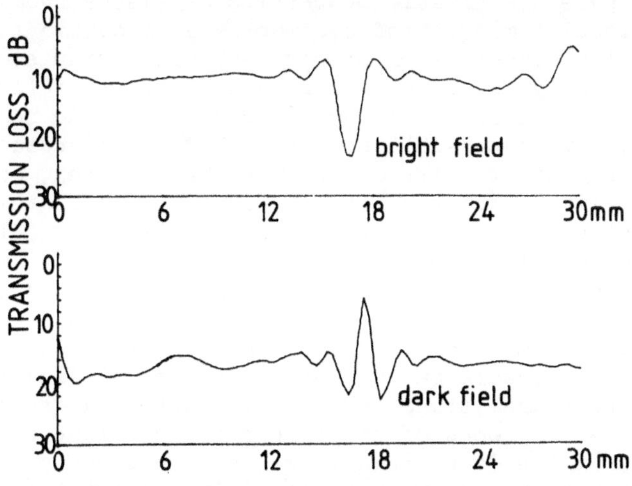

Figure 9. Line scans from Figure 8:
a) bright field
b) dark field

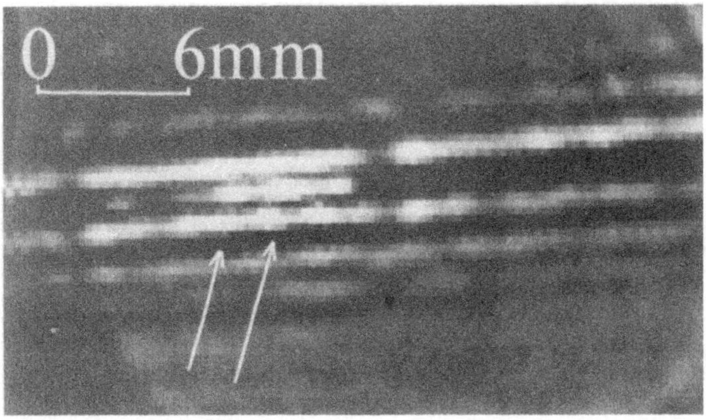

Figure 10. Acoustic image of a butt weld showing region of non-uniformity (arrowed).

In another application of acoustic microscopy, we are interested in computing the elastic properties of human tissue from SAM images. It is intended to compare the elastic properties of normal and diseased tissue, with the aim of enhancing the diagnoses that can be made using clinical B-scan instruments. In order to compute the elastic properties of tissue from acoustic images, we model a microtomed section as an acoustic resonator and analyse its transmission properties. This simple model enables remarkably accurate measurements to be made {10}, provided that there is no strong diffraction in the region of measurement. It is possible to use dark field microscopy to measure the amount of diffraction over the specimen, and so indicate the likely accuracy of the elasticity measurements. Figure 11 shows typical results in this application. The specimen is a section of pig muscle, with a band of fatty tissue running through the middle. In the bright field image, the muscle shows a characteristic pattern, and the fat has a varying transmittance. Comparing this with the dark field image, it can be seen that the muscle has a similar pattern, indicating that the changes of transmittance are due in part to diffraction. However, there are zones of negligable diffraction, showing up as dark regions in the dark field image, and in these places greater confidence may be placed in the elasticity computation. Similarly, within the fat, the majority is only very weakly diffractive and is suitable for analysis. There is an elastic region in the very centre of the band which is clearly diffracting, as seen in the dark field image, and which would generate spurious information in an elasticity determination.

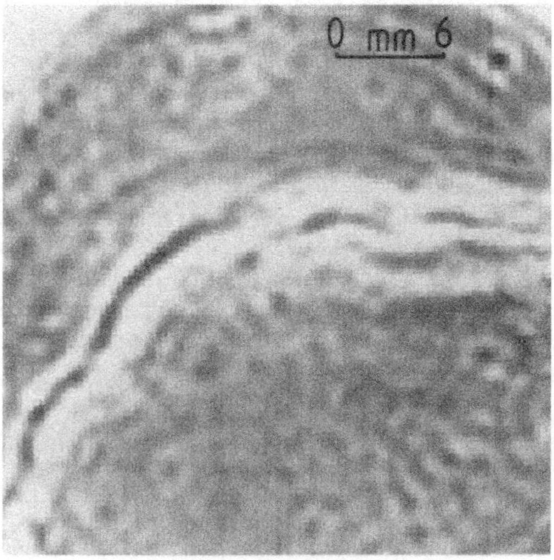

a) bright field

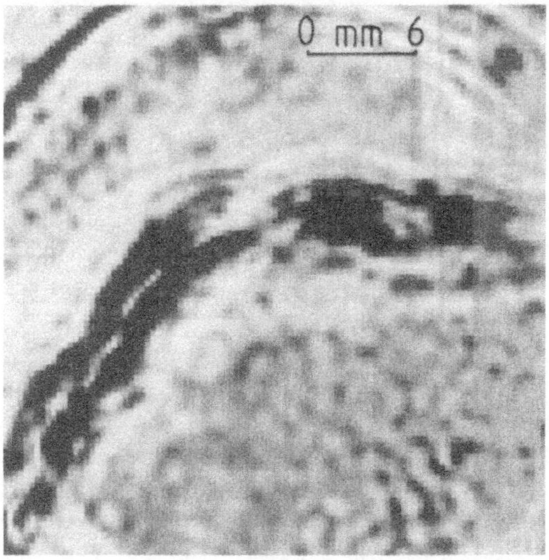

b) dark field

Figure 11. Acoustic images of pig tissue

DARK FIELD MICROSCOPY IN REFLECTION

For the very highest resolution in acoustic microscopy, it is normal to use a reflective system, because this reduces the problem of system alignment. It is thus natural to attempt the extension of dark field work to a reflection scheme. Figure 12 shows how this might be achieved {11}. A pulse is transmitted from the central portion of the transducer, and is incident upon the specimen in the focal plane of the source. Because of the small size of the central portion, the field over the focal plane is a diverging spherical wave. However, in the narrow focal region, this is effectively a plane wavefront. The reflected energy is detected on the outer electrode, which is sensitive only to diffracted energy, as we have already seen. Thus the "unused" part of the transducer provides a convenient method for achieving reflection dark field microscopy.

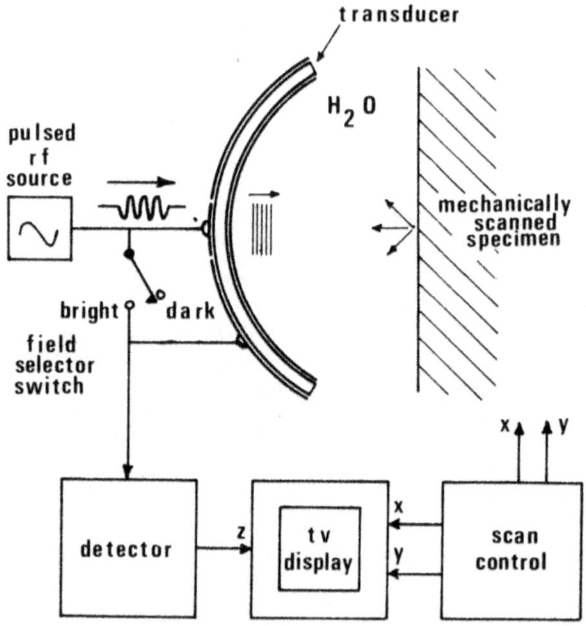

Figure 12. Reflection dark field geometry

CORRELATION MICROSCOPE

We note that in designing our dark field microscope with a spherical transducer, we are presented with a simple method for altering the lens pupil function. It is possible to prescribe any electrode pattern on the transducer surface, and to weight the signals from the electrodes using amplifiers and phase shifters. With several electrodes it will be possible to discriminate between particular scattering components in the image. However, by weighting the individual components and then integrating them, we can realise a generalised pupil function, P(x,y):

$$G(x,y) = \mathcal{F}\{P(x,y)\} \exp\left\{\frac{jk(x^2+y^2)}{2R}\right\} \tag{5}$$

where $\mathcal{F}\{P(x,y)\}$ is the Fourier transform of P(x,y), evaluated at $f_x = x/R\lambda$ and $f_y = y/R\lambda$, and k is the (generally complex) wavenumber of the sound waves in the water. Hence, the imaging equation becomes {7}:

$$V(x_s,y_s) = \iint \mathcal{F}\{P(x,y)\} \exp \frac{jk(x^2+y^2)}{2R} U_2(x,y) \times t(x-x_s,y-y_s)\, dx\, dy \tag{6}$$

where x_s and y_s are the scan coordinates, and $U_2(x,y)$ is the plane wave source distribution, which is a constant. If we arrange for the lens sensitivity function to be the complex conjugate of the same desired object feature,

$$\tau^*_{obj}(x,y)$$

then the imaging equation becomes:

$$V(x_s,y_s) = \iint \tau^*_{obj}(x,y).\ t(x-x_s,\ y-y_s)dx\, dy \tag{7}$$

and the image is the correlation of that sensitivity function with the transmittance of the specimen. By a suitable choice of pupil function, found using (5), setting $\tau^*_{obj} = G(x,y)$, the detection sensitivity of the microscope is enhanced to the required feature. The microscope has become an analogue pattern recognition unit, and uniquely, it is correlating the mechanical properties of the specimen.

It will be possible to implement this scheme in low frequency microscopes, using amplifiers and phase shifters with several electrodes on the transducer. Alternatively, we note that in high resolution microscopes there is a substantial reflection loss between the lens and the imaging medium. In gaseous argon {9} and in liquid helium {12} the reflection loss is approximately 55 dB. These losses are conventionally reduced to negligable proportions by the use of a quarter wavelength matching layer. By varying the thickness of the matching layer, it will be possible to alter the spatial frequency response of these instruments over this wide dynamic range. This might be achieved by the use of ion beam etching.

ALTERNATIVE DARK FIELD ARRANGEMENTS

Since the SAM is a linear system, it would, in principle, be possible to make a dark field system by suppressing the constant unwanted signal with a filter in the time domain, rather than a stop in the space domain. This would require the introduction of a high pass filter after the RF detector in the system. However, this scheme has two main disadvantages. Firstly, the constant signal is only removed after it has been amplified and detected at RF. This means that a large portion of the dynamic range available in the RF components is taken up with this unchanging signal and so the effective system dynamic range is reduced. Secondly, because the image is raster scanned and the orthogonal scan rates will differ by a factor equal to the number of lines per picture, typically 100 - 250, the system will not have the desired uniform spatial frequency response. If the high pass filter cut off frequency is set correctly with respect to the fast scan, then spatial frequencies in the slow scan direction will see a filter with cut off frequency 100 to 250 times too high. An alternative method would be to digitise the image data and perform spectral modifications in a computer {13}. This does not remove the problem of the limited dynamic range mentioned above however. Clearly a system employing the correct original spatial frequency response as discussed above is the best solution.

CONCLUSION

We have demonstrated a new method of dark field acoustic microscopy which overcomes the difficulties of previous systems. Since the sound source used to illuminate the object is unfocussed, the constraints on alignment are considerably relaxed.

A spherical geometry has been described which may be used as the basis for a high resolution system employing photolithographically masked, sputtered zinc oxide transducers. Bright or dark field operation may be conveniently selected with this system by the operation of a single switch. The spatial frequency response of the system is symmetrical and, for a given aperture, the lens collects as much as possible of the diffracted energy. Although the system does suffer a small loss in resolution over the confocal geometry, we have shown with beam probing results that this is less than a factor of root 2. Applications for dark field microscopy have been discussed, in both non-destructive testing, where the enhanced sensitivity may be helpful, and in the analysis of tissue samples, where the new information is an aid to understanding conventional acoustic micrographs.

The spherical lens geometry described can be used to form a correlation microscope. This would be a pattern recognition type of instrument, identifying regions of mechanical similarity on the specimen.

ACKNOWLEDGEMENTS

The authors are grateful to the Medical Research Council for supporting the research project and to the Science Research Council for the award of studentships. The authors also wish to thank Prof. E.A. Ash and Dr. H.K. Wickramasinghe for helpful discussions and encouragement during the work and Prof. G.S. Kino and Dr. S.D. Bennett for suggesting the reflection geometry. W. Duerr kindly provided the acoustic beam probing results.

REFERENCES

{1} C.F. Quate, A. Atalar, H.K. Wickramasinghe, "Acoustic Microscopy with Mechanical Scanning - A Review", Proc.IEEE, 67 8. pp. 1092 - 1114 (1979)

{2} R.A. Lemons and C.F. Quate, "Acoustic Microscope - Scanning Version", Appl. Phys. Lett. 24, (4), p. 163 (1974).

{3} V.B. Jipson, "Acoustic Microscopy of Interior Planes", Appl. Phys. Lett 35 (5), pp. 385-387 (1979)

{4} N. Chubachi, J. Kushibiki, T. Sannomiya and T. Iyama, "Performance of Scanning Acoustic Microscope Employing Concave Transducers", 1979 Ultrasonic Symp. Proc. IEEE New Orleans, La, U.S.A. pp. 415-418 (1979)

{5} W.L. Bond, C.C. Cutler, R.A. Lemons and C.F. Quate, "Dark Field and Stereo Viewing with the Acoustic Microscope", Appl. Phys. Lett., 27, pp. 270 - 272 (1975).

{6} D.A. Sinclair and I.R. Smith, "Dark Field Acoustic Microscopy", El. Lett. 16, 16, pp. 627 - 629 (1980)

{7} H.K. Wickramasinghe,"Contrast and Imaging Performance in the Scanning Acoustic Microscope", J. Appl. Phys. 50, pp. 664 - 672 (1979)

{8} W. Duerr, D.A. Sinclair and E.A. Ash, "A High Resolution Acoustic Probe", El. Lett., to be published (1980)

{9} H.K. Wickramasinghe and C.R. Petts, "Gas Medium Acoustic Microscope", Proc. of the Rank Conference on Scanned Image Microscopy, Royal Society, London, Ed. E.A. Ash, Academic Press, to be published (1980).

{10} S.D. Bennett, "Coherent Techniques in Acoustic Microscopy", Ph.D. Thesis, University of London, pp. 102-146 (1979).

{11} G.S. Kino and S.D. Bennett - Private communication.

{12} J. Heiserman, D. Ruger and C.F. Quate, "Cryogenic Acoustic Microscopy", J. Acoust. Soc. Am. 67 (5), pp. 1629-1637 (1980).

{13} V.B. Jipson, "Acoustic Microscopy at Optical Wavelengths" Ph.D. Thesis, Stanford University, Ca., U.S.A. (1979)

ACOUSTIC MICROSCOPY IN NON-DESTRUCTIVE TESTING

B. Nongaillard, and J.M. Rouvaen

Laboratoire d'Optoacoustoélectronique ERA 593 CNRS

Université de Valenciennes 59326 VALENCIENNES CEDEX FRANCE

In the past few years, scanning acoustic microscopy has been recognized as a valuable method for observing microstructures which were hardly resolved using standard optical techniques. Indeed, it is the case of transparent biological samples which needs time consuming coloration techniques to be observed using optical microscopy. It is also the case for the observation of thick or opaque samples which gives only, when using optical techniques, informations about the surface structures. At the present time the spatial resolution of acoustical methods is comparable to that obtained with optical microscopes (say 0.5 μm). Moreover, in the pulse emission mode, it is possible to focus the acoustic beam under the sample surface and to detect structures like microdefects lying under the surface. A first practical example is the observation of microdefects near the surface of electronic microcircuits, which needs a very high spatial resolution (1μm) and a very small depth of focus under the surface sample (1 to 10 μm). The utilization of mercury as a coupling medium decreases the surface echo amplitude and allows the detection of the echo on the defect [1]. Indeed the mechanical impedances of the mercury, the lens and the electronic microcircuit (say alumina) are very close to, and moreover the propagation losses in this liquid metal material are quite negligible in this high frequency domain of ultra sound. A second example, which is the particular subject of our study, is the detection and the observation of defects lying at a depth of several millimeters inside dense materials like alumina samples. The non negligible acoustic attenuation suffered from inside these materials calls for the reduction of the operating frequency in the neighbourhood of 100 MHz. The spatial resolution which may be expected for such a frequency in alumina amounts to 10 μm on the surface and 80 to 100 μm inside the material. So that defects embedded in dense and thick samples, which are hardly detectable by destructive methods of metallurgical microscopy, may be detected by acoustic methods. For this subject, it is necessary to determine precisely the attainable spatial ·esolution with respect to the depth of the focalization and the geometrical parameters of the lens.

Only two parameters are very important in this study : the velocity ratio of the coupling medium and of the sample and the depth of the focalization plane inside the material. Actually, what is well known is the behaviour of acoustic microscope used to visualize details inside a sample with an acoustic velocity very close to that of the coupling medium (say biological tissues and water), the spatial resolution is then limited only by diffraction effects and the spherical aberration is negligible (1,2). Moreover the depth of the focal plane is not changed and the spatial resolution in this case is given by the geometrical optics formula. Now when studying media with acoustic velocities differing wore and more from that of the coupling medium the results are not the same. In the non destructive testing applications, the velocity ratio is often greater than 4 (4 for metallurgical applications, 7 for alumina samples). The subject has been studied in several articles (3,4) but more work is required since the results given apply to velocity ratios lower or equal to 2 (3) or use only a geometrical optics approximation and neglect diffraction effects(4). Using geometrical optics approximation, the theoretical results show that the spatial resolution inside the material is strongly reduced by the goemetrical aberrations introduced by the surface of the sample. The experimental results are not in good agreement with these predictions and very good images have been obtained inside electronic-microcircuits (1). This can be explained by the fact that the geometrical theory doesn't take into account the repartition of the energy in the elementary acoustic beams which have been deflected by the surface. This phenomenon depends also on the shape of the sample surface. In the microelectronic domain only flat surfaces are studied but our problem is to visualize defects inside flat or round samples. At first, only plane samples have been considered since this appears in fact as the poorer resolution case if we consider a sample surface and a lens with radius of curvature of the same sign, the incidence angle of the acoustic beam is strongly reduced, so that the aberrations decrease.

The acoustic microscope, that is to say the lens, coupling medium and sample surface has been modelized and the propagation equation has been numerically solved. Let as consider the geometry depicted in fig. 1.

The solution of the propagation equation for the velocity potential may be written as

$$\Phi(M) = \frac{R}{\sqrt{\lambda_2}} \int_{-\Theta_m}^{\Theta_m} A(P) \frac{\exp j\left[k_1 R(1-\cos\Theta) + \vec{k}_2.\vec{PM}\right]}{\sqrt{PM}} \cos(\vec{n},\vec{PM})\, d\Theta$$

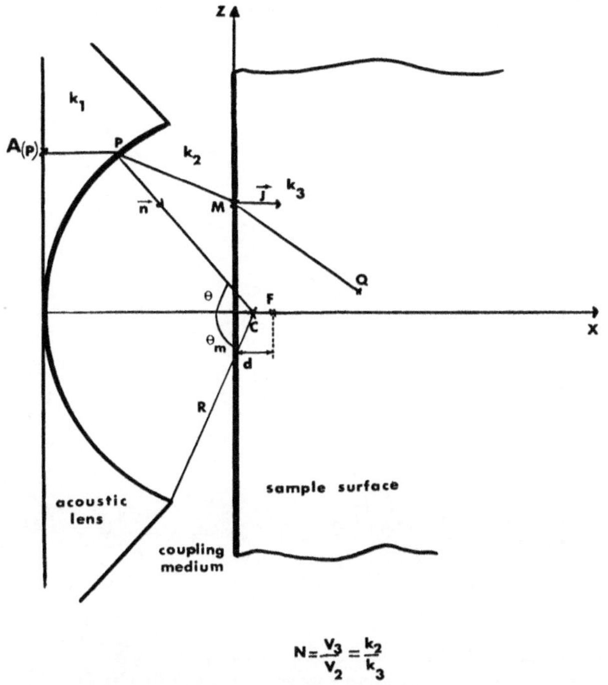

Fig. 1 - Notation used in the computer calculation.

It may be seen that these calculations have been performed for a cylindrical lens, that is a monodimensionnel problem. The diffraction spectrum of a rectangular slit is a sinc function whose argument is a function of the aperture dimensions and the results are easily extrapoled to a spherical lens by replacing the previous function by the classical $\frac{J_1(X)}{X}$. This approximation decreases the time computation and increases the precision of the results. The next approximation in this formula is to neglect the coupling medium attenuation, but this phenomena doesn't affect the shape of the acoustic field repartition in the focal plane.

Below the sample surface, the velocity potential is then given by :

$$\Phi(Q) = \frac{1}{\sqrt{\lambda_3}} \int_{-ndz}^{+ndz} \Phi(M) \frac{\exp j\left[\vec{k}_3 . \vec{MQ}\right]}{\sqrt{MQ}} \cos(\vec{j},\vec{MQ}) \, dz$$

where 2n + 1 is the number of spatial samples on the surface where the acoustic velocity potential has been computed in the first stage. The longitudinal variations along X axis for samples with velocity ratios varying from 2 to 7 are shown in Fig. 2 and the transverse variations along Z axis are given in Fig. 3.

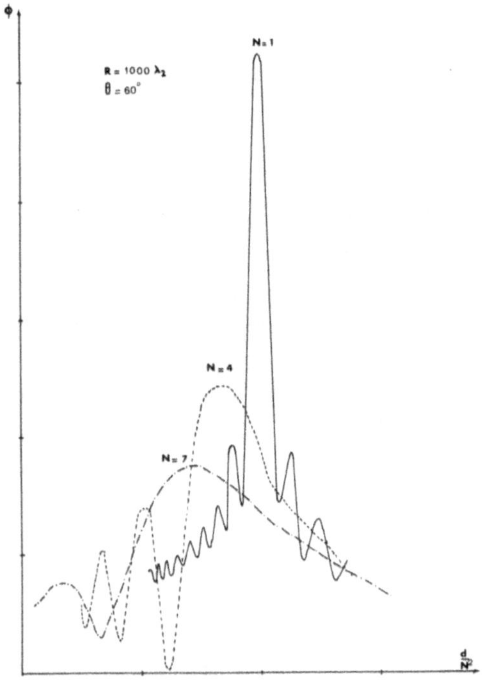

Fig. 2 - Longitudinal variation for the acoustic velocity potential versus index N.

These curves have been drawn for a distance d between the sample surface and the focal plane in the coupling medium, which has been kept constant. The value of this parameter has been choosen as R/6 in a very first time. Indeed, in the case of high values of the distance d (R/2 for example), a great number of spatial samples has to be stored in the first stage of the computation and the second calculation step is then more and more time consuming. Several conclusions may be drawn from these curves in the caso of a flat sample, which is much pessimistic as that of round samples. A point always exists on the longitudinal X axis where the amplitude is a maximum but its abscissa doesn't correspond longer to that of the paraxial focus for the high velocity ratios. The relative departure may be as high as a 30 percent decrease with respect to the paraxial focus position with respect to the surface. This feature may be explained since, for high velocity ratios, the acoustic energy is strongly deflected by the surface and is not convergent in the focal plane. The examination of the transverse variations leads to the conclusion that high secondary lobes exist, even in the focal plane, but that the width of the principal maximum remains of the order of the acoustic wavelength in the observed medium. It may also be seen on the axial distribution curves that the depth of field increases together with the velocity ration, if the angular apeture of the

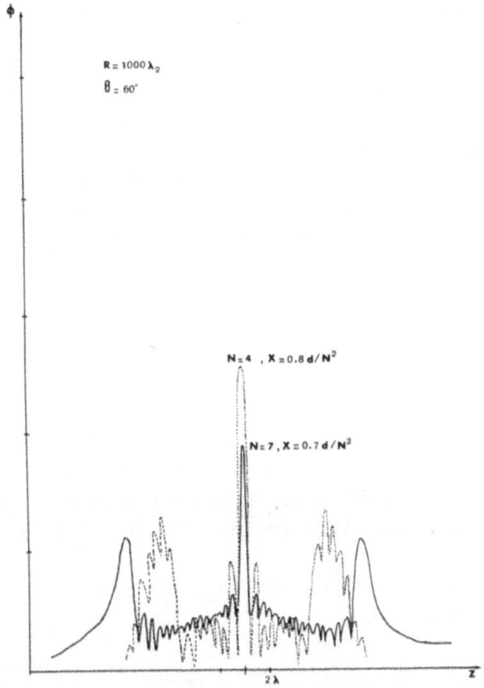

Fig. 3 - Transverse variation of the acoustic velocity potential versus index N.

lens is kept constant. Such a feature may be explained since the critical reflection angle for the incident acoustic beam on the sample surface decreases when the acoustic velocity of the explored medium becomes higher. Our results show that when the distance d is kept constant the geometrical aberrations lead effectively to a reduction in the effective apertureof the lens, the spatial resolution remaining of one acoustic wavelength in the examined sample. The results which have been shown, have been computed for a lens radius of curvature 1000 λ_2 large which is the geometry used in the experimental set up to detect defects inside alumina samples at a 100 MHz operating frequency.

Now, if we want to move the surface sample and to detect defects deeper and deeper, the spatial resolution will decrease as the ratio $\frac{(2n+1)dz}{d}$ for the geometry depicted in fig. 1, since the beam size on the surface sample is kept constant. The case of round sample surfaces may be shortly explained using geometrical optics approximations since,we have seen that no geometrical aberrations appear. This surface acts as a divergent lens and the acoustic beam is then focused deeper inside the material than in the case of a flat surface sample. This phenomenon doesn't affect the resolution but must be taken into account for the choice of the lens radius.

Finally, images of sections lying at a ten to twenty wavelengths depth inside rocky samples have been obtained and presented in an earlier publication. Carbon fibers casted in polymer resin shells have also been imaged in order to detect cracks and other defects. High frequency applications like study of integrated circuits are also experimented by others workers [(1)] and their results show also the feasibility of imaging defects buried under the surface of the sample using acoustic microscopy.

References :

1) ATTAL and G. GAMBON(1978) "Signal Processing in the reflection acoustic microscope" Electron Lett., 14,472-473.
2) RA LEMONS and CF QUATE (1974) Appl . Phys. Lett. 25,251.
3) B. NONGAILLARD, JM ROUVAEN et al (1978) Visualization of thick specimens using a reflection acoustic microscope, J. Appl. Phys. 50, 1245-1249.
4) VB JIPSON (1979) Acoustic microscopy of interior planes, Appl. Phys. Lett., 35, 385-387.

IMAGING INTERIOR PLANES BY ACOUSTIC MICROSCOPY

J. Attal, G. Cambon and J.M. Saurel

Centre d'Etudes d'Electronique des Solides, CNRS

USTL, 34060 - MONTPELLIER-CEDEX, FRANCE

ABSTRACT

It has been recently demonstrated that the acoustic microscope can visualize image planes relatively deep beneath the surface with a lateral resolution approaching one acoustic wavelength in the solid. This has been achieved using mercury or gallium as the medium connecting the acoustic lens to the object. These liquids with their solid-like impedance better improve the acoustic power penetrating in the object and minimize the aberration effects caused by refraction at the liquid-solid interface. In this paper we discuss on the suitability of both liquids devoting a large place to imaging with mercury. We present some unique features of this kind of microscopy such as test of bonding of two silicon wafers with indium when temperature is changed or studies of electrical breakdown of layers of indium and gold through a wafer of glass or silicon.

1. BACKGROUND

The acoustic microscope has been first developed by C.F. Quate and his collaborators (1) at Stanford University in 1974. It has now been advanced to the point where it is useful for viewing microscopic objects sized larger than 0.2 to 0.5 µm. For several years biological specimens have been the widest scope of applications and only recently the interest of this instrument turned to solid state applications. Indeed most of solid state devices are silicon made, material which is completely opaque to visible light but can be readily penetrated by an acoustic wave. Application can be extended to other semiconductors and most of metals. In that field, observation of structures buried in the bulk can bring useful informations for a non destructive evaluation.

Figure 1 gives a schematic configuration of the microscope operating in reflection mode (2). A collimated acoustic wave is generated with a zinc oxide thin film transducer at a driving frequency lying between 50 MHz and 1 GHz. The acoustic beam is focused down to the sample through a spherical lens constituted by a sapphire-liquid interface. This liquid which was formerly water also serves to acoustically connect the lens to the object. The size of the acoustic beam at the focal point is of the order of one acoustic wavelength. The wave reflected by the sample is collected back to the lens which recollimates the beam before detection by the transducer. To get rid of spurious reflection inside the sapphire and to prevent the reflected signal from mixing to the incident wave we are operating in pulse mode. An original technique of detection gives the amplitude and the phase of the reflected wave with a fairly good signal to noise ratio. The object is mechanically scanned along two orthogonal directions in the focal plane of the lens in such a way that the acoustic reflection coming from the sample is recorded point by point. The fast scan (30 to 100 Hz) is obtained from the movement of a vibration which gives a peak to peak amplitude of 5 millimeters. The slow scan (0.1 to 1 Hz) is assured by a micromotor which lifts up and down the total assembly. Inductive pick-up coils are used to display the line scans on the imaging scope in synchronism with the object, by transducing the X and Y positions.

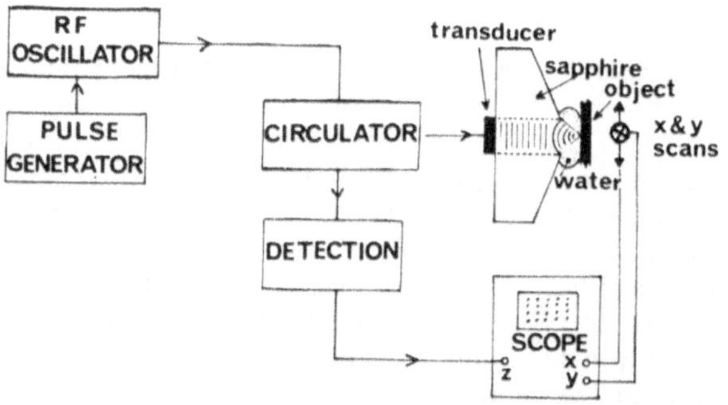

Fig. 1. Schematic configuration of the acoustic microscope operating in reflection mode.

2. RESOLUTION AND PENETRATION

The actual resolution of the microscope is sensitive to the four following parameters :
(i) frequency of the acoustic waves primarily limited by the absorption of the liquid. Higher the frequency, better the resolution will be.
(ii) position of the object with respect to the lens. Figure 2 shows the situation of a convergent beam penetrating in a solid object. We see a large spreading of the acoustic distribution from the paraxial focus (for small incidence angles) to the surface of the object. Calculations can be done by applying Snell's law (3) and show that the aberrations drastically increase with the incident angle θ_i or when the object is moved towards the lens.
(iii) the sound velocity in the object. Shear waves and longitudinal waves are generally present after conversion at the interface and both of them can be focused in the object (fig. 3). However shear waves with a shorter wavelength improve the resolution.
(iv) the acoustic index of refraction which is the ratio between the velocity in the object (transverse or longitudinal) to the velocity of the liquid. Best values of resolution will be obtained when this ratio will be close to 1.

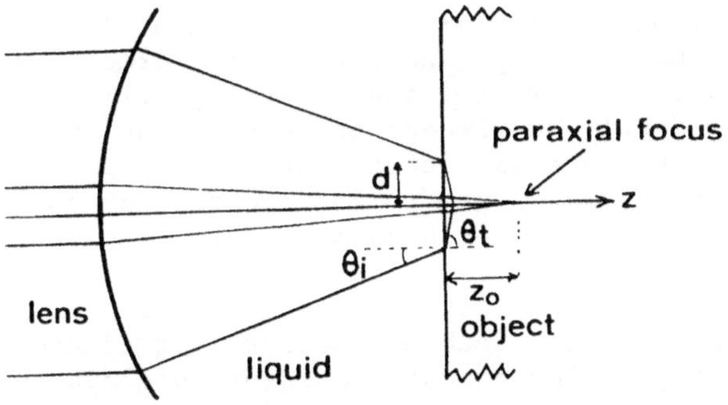

Fig. 2. Geometrical ray tracing for the internal focusing.

3. NECESSITY OF IMAGING SOLID STATE OBJECTS WITH METALLIC LIQUIDS

Since its introduction no extensive work has been really undertaken in the way of imaging in the bulk. Only recently some resear-

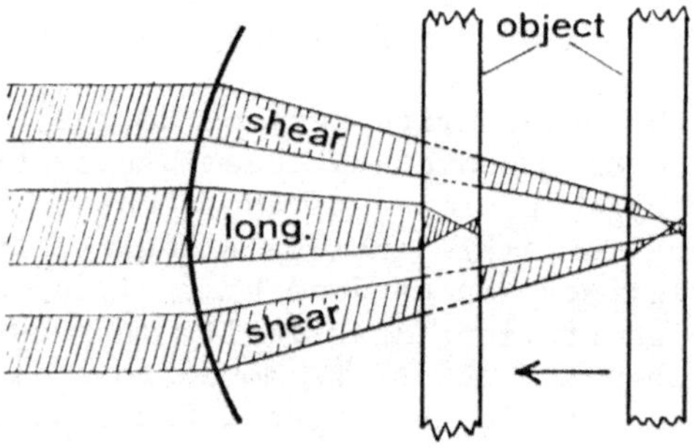

Fig. 3. Focusing of shear modes and longitudinal modes by moving the object towards the acoustic lenses. The shaded areas represent the efficient parts of the beams.

chers have been interested in this problem (4). In fact, if most solid state devices are transparent to ultrasounds, we are to face to the drastic problem of the reflection at the interface due to the impedance mismatch liquid-solid. Unfortunately most common liquids have a mechanical impedance one to two orders of magnitude less than solids. It turns out that a strong reflection occurs at the interface and only 10 percent or less of the energy is used to "illuminate" the internal structures of the object. We cannot match the impedances as we use to do in optics by coating the object with a multilayer system because we are operating with a convergent beam. If we search for liquids with a solid-like impedance, only metallic liquids are suitable. Mercury and gallium are in fact the best candidates since both of them are liquid at room temperature and exhibit very low acoustic absorption compared to other liquids metallic or not.

4. ACOUSTIC AND WETTING PROPERTIES OF MERCURY AND GALLIUM

Several problems have to be considered when we use metallic liquids :

- First we have to know the intrinsic acoustic parameters such as velocity, impedance, absorption loss which usually follows a square law dependance with frequency. Table 1 gives these values compared to water. Velocities and densities are quite different for both metallic liquids but their impedances are similar. At the first glance, gallium should be prefered to mercury for its solid-like

TABLE 1

Intrinsic acoustic parameters of Hg and Ga compared to water.

Liq.	Temp. (°C)	Velocity $x10^{-5}$ (cm sec^{-1})	Density (g cm^{-3})	Impedance $x10^{-5}$ (g sec^{-1} cm^{-2})	Absorption α/f^2 $x10^{17}$ (sec^2 cm^{-1})	Coeff. of Merit M*
Hg	23.8	1.450	13.6	19.7	5.8	1.89
Ga	∼ 30	2.87	6.1	17.5	1.58	1.82
H_2O	20	1.5	1	1.5	25	1

* defined in Ref. 6.

velocity which approachs most of shear velocities of solids and its lower absorption.
- In fact the crucial problems with liquid metals are the acoustic contacts between lens and object. This becomes worse when we mechanically scan the object. In terms of absorption and velocity we have demonstrated that gallium is better than mercury, unfortunately, these parameters does not guarantee a good acoustic image since it depends also of the behaviour of the liquid in contact with the object. Gallium produces an amalgam with most solids and metals in particular, these impurities travel in the liquid and scatter the ultrasound or give ghost structures in the images. By evaporating a protective layer of SiO_2 we have succeeded in the way of imaging with gallium but this technique needs improvements. In contrast, mercury does not spontaneously wet most of solids (except gold tin, lead and a few other ones) but still ensures a good acoustic contact with a scanning up to five millimeters, its strong surface tension makes it balling up and imaging just requires a droplet filling the lens cavity squeezed by the plane surface of the object.

5. IMAGE RESOLUTION MEASUREMENTS

Resolution has been measured according to Kino's two spots definition (5) as a function of penetration of the acoustic beam with three different objects mostly used in integrated circuits : silicon, fused silica and aluminium. The liquid used is mercury, the samples to image are wafers 100 to 300 microns thick polished on both sides. By moving on the object towards the lens, we have been able to focuse both longitudinal and shear waves in the object. The resolution is of course better with shear waves and all the images have been recorded with this type of wave. Table 2 gives a survey of the resolution achievable as a function of penetration at 135 MHz.

TABLE 2

Resolution in shear mode operation as a function of thickness and material

Surface	Fused glass		Si		Al
(in mercury)	145 μm	320 μm	100 μm	200 μm	230 μm
12 μm	15 μm	30 μm	20 μm	30 μm	20 μm

6. APPLICATION OF IMAGING WITH MERCURY

a) Test of bonding of two silicon wafers with indium layers.

This application of acoustic microscope is unique and there is no non destructive way to do this experiment. We first deposit on one face of two silicon wafers a layer of 2000 Å of indium through a microscopic grid as set up in figure 4. Then we do the bonding by pressing the two wafers, the indium evaporations facing each other. At room temperature, if the bonding is poor we can only see the grid pattern evaporated on the silicon wafer # 1 in contact with the liquid.

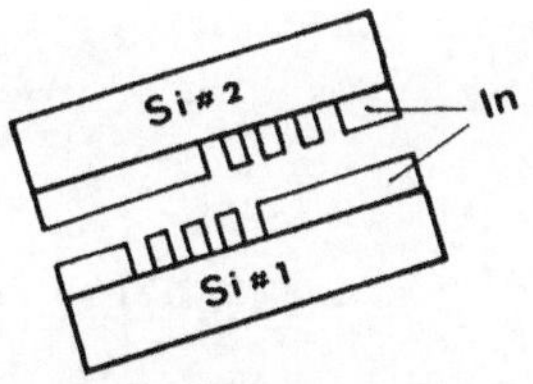

Fig. 4. Set-up of the two silicon wafers with evaporated indium patterns ready for bonding.

Figures 5 (a), (b), (c), show the evolution of the bonding after diffusion. The images of the grid pattern deposited on the silicon wafer # 2 are somewhat an indication of the quality of the bonding. The first photograph (a) is obtained after two hours of diffusion at 135° C, the second (b) after one more hour at 190° C, the last (c) after leaving the sample one month at room temperature. Indubitably there is a large change in the quality of the bonding. We clearly understand the improvement when we pass the melting point of indium but we assume that a further change of bonding back to room temperature could be due to a further diffusion of indium.

b) Studies on the breakdown of strip film through the substrate

Two simple experiments of electric breakdown and ion migration have been attempted on metallic strip films deposited on glass substrate. The thickness of the glass wafers was around 150 microns and two sorts of materials for deposition has been selected : gold and indium. The experiments consist in increasing the current flow

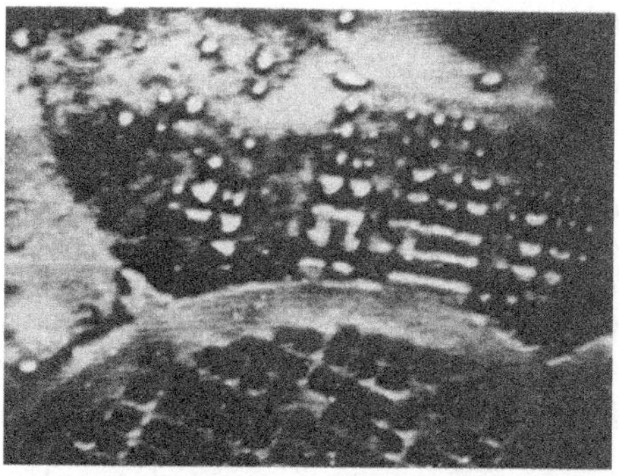

a) After 2 hours of diffusion at 135° C.

b) After one more hour at 190°.

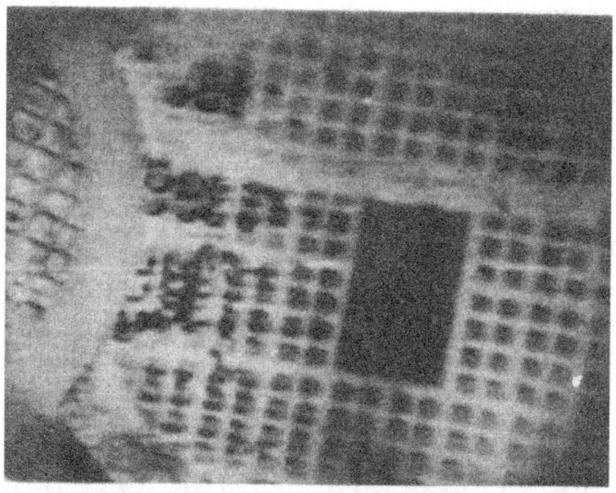

c) After leaving the sample one month at room temperature.

Fig. 5. Images of the bonding of two silicon wafers with evaporated patterns. Frequency is 135 MHz.

in the strip until breakdown. Images are taken at 135 MHz in sequence each time current is increased. The direction of the current is from right to left. Figure 6 show the recordings of a gold strip of 0.5 millimeter width, and 2800 Å thick when the current density varies from 0 to about 1.1×10^4 A/mm^2. No significant change over a few minutes has been noticed when the current density was below 0.7×10^4 A/mm^2. For higher currents there is non reversing changes, small white dots appear in the image, the number of them increasing with the current. Finally the breakdown occurs for values around 1.1×10^4 A/mm^2 followed by the thermal breakdown of the glass. Brightness and contrast have been kept at the same level (except for the last photograph) and we notice a large variation of these parameters from one image to the other. As a proof we have recorded the reflective signal corresponding to one line scan (fig. 7) and we can see a decrease by about a factor five of the average intensity. This change is related to the scattering of the ultrasound by the new created structure of the gold film.

Figure 8 shows the acoustic images of an indium strip of 0.5 millimeter width and 4500 Å thick. As previously no significant change has been noticed with current densities below 1×10^3 A/mm^2. In contrast with gold strip there is only a slight increase of the average brightness but a large local change in the left part of the image. We observe very clearly domains formation characterized by white and dark spaces, both of them moving towards the cathode when electric field is increasing. There is a ion migration in the indium strip along the direction of the applied electric field. The final breakdown occurs in the dark region which presummably is high resistivity. The change in contrast can be related to a variation of thickness of the layer and quantitative measurements will be done after calibration in a next future.

CONCLUSION

These experiments show the high capability of the acoustic microscope to bring out informations about change of structures buried in the bulk. It is the unique way to control the quality of the bonding of two opaque materials and to understand the evolution of structures of strip films deposited on substrate when electric current is applied. The latest results is of the utmost importance since we can monitor ion migration occuring in electronic devices.

AKNOWLEDGMENTS

The authors are indebted to Professor C.F. Quate of Stanford University for helpful discussions and encouragements. This work was partially support by the DGRST and NATO research grant No. 1885.

i = 0

i = 0.76×10^4 A/mm^2

i = 0.83×10^4 A/mm^2

i = 0.89×10^4 A/mm^2

i = 0.96×10^4 A/mm^2

i = 1.05×10^4 A/mm^2

i = 1.1×10^4 A/mm^2
breakdown

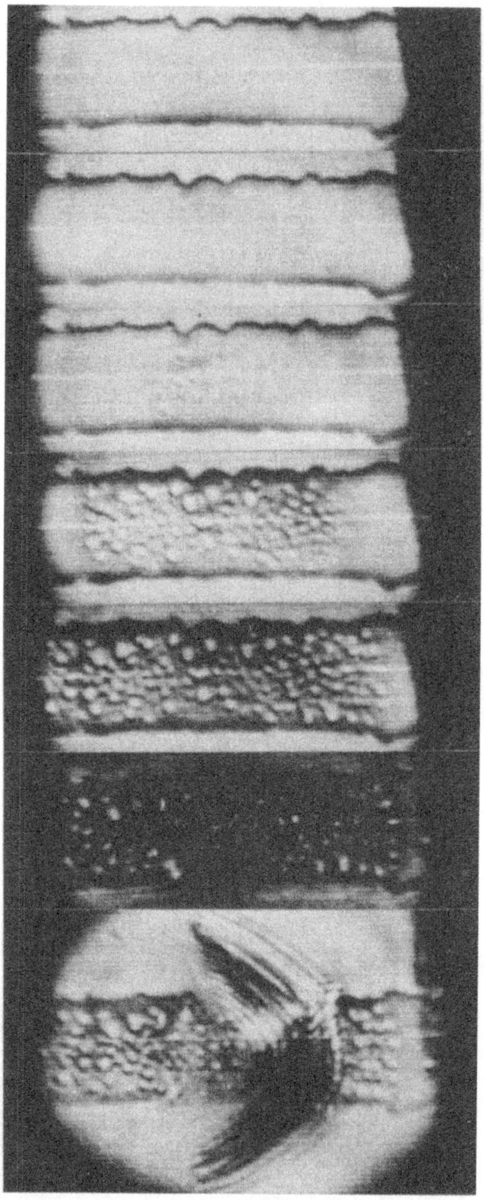

Fig. 6. Acoustic images of a gold strip through the glass substrate when current density is changed. The brightness of the bottom picture has been increased. Frequency is 135 MHz, width of the strip is 0.5 mm.

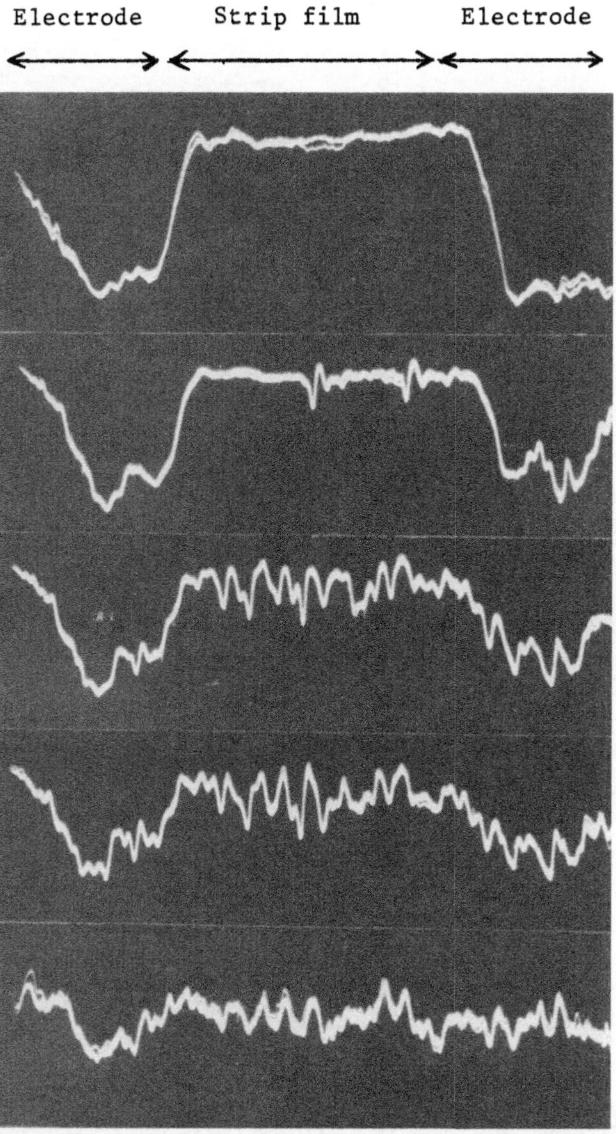

Fig. 7. Reflected acoustic signal corresponding to one line scan of the gold strip. From top to bottom current densities are respectively : (in A/mm^2) 0, 0.79 x 10^4, 0.86 x 10^4, 0.93 x 10^4, 1.x10^4. Frequency is 135 MHz.

$i = 0$

$i = 1\text{x}10^3$ A/mm^2

$i = 1.5\text{x}10^3$ A/mm^2

$i = 2\text{x}10^3$ A/mm^2

$i = 2.5\text{x}10^3$ A/mm^2

$i = 3\text{x}10^3$ A/mm^2
breakdown

Fig. 8. Acoustic images of an indium strip through the glass substrate when current is increased. Frequency is 135 MHz, width of the strip is 0.5 mm.

REFERENCES

1. Lemons R.A. and Quate C.F. Acoustic microscope-scanning version Appl. Phys. Lett. 24, 163-165, 1974.

2. Attal J. and Cambon G. Non destructive testing of electronic devices by acoustic microscopy. Revue de Physique Appliquée, 13, 815-819, 1978.

3. Attal J. Acoustic microscopy : imaging microelectronic circuits with liquid metals. International Conference on "Scanned Image Microscopy" London 22-24 Sep. 1980, (o be published).

4. Sinclair D.A. and Ash E.A. : Bond integrity evaluation using transmission scanning acoustic microscopy. Elect. Lett. (in press).

5. Kino G.S. Fundamentals of scanning systems. Internat Symposium on Scanned Image Microscopy 22 to 24 Sept. 1980 (to be published).

6. Attal J. and Quate C.F. Investigation of some low ultrasonic absorption liquids, J. Acoust. Soc. Am. 59, p. 69-73, 1976.

IMAGING TECHNIQUES FOR ACOUSTIC MICROSCOPY OF MICROELECTRONIC CIRCUITS

R. L. Hollis and R. Hammer

IBM Thomas J. Watson Research Center,
P. O. Box 218, Yorktown Heights, NY 10598 USA

ABSTRACT

Following the work of Lemons and Quate,[1] an 800 MHz acoustic microscope has been constructed, operating in the reflection mode for examination of microelectronic materials and devices. To aid in the interpretation of the acoustic images, a method has been developed which produces false color acoustic micrographs of high quality. This technique, along with derivative and logarithmic processing is applied to the acoustic examination of plated structures on organic/ceramic substrates. A standard optical microscope is operated in tandem with the acoustic microscope to allow convenient and absolute comparison between the optical and acoustic images.

INTRODUCTION

In recent years, the pulsed-reflection acoustic microscope has shown promise as a tool for visualizing subsurface defects in microelectronic devices and also as a method for characterizing materials. Figure 1 is a schematic representation of our 800 MHz instrument, showing the principal electronic elements and typical signals.[2] The approximate resolution at optimum focus is 1.5 μm. In our system, the image is stored in an analog electron beam-addressable memory (scan converter) before it is displayed at TV line rates on a CRT monitor.

From the outset, interpretation of the images obtained by acoustic microscopy has been difficult due in part to the phase-sensitive nature of the acoustic detector and also the need to separate purely topographical features from acoustic features which depend on the density and elastic properties. Instruments of the type illustrated in Fig. 1 often exhibit good signal-to-noise characteristics, approaching perhaps 30 dB. Surprisingly high contrast ratios are obtained in many cases from samples which show poor optical contrasts. Indeed, the nature of this contrast mechanism has been the subject of a number of recent studies.[3-6] It is found that the transducer voltage $V(z)$

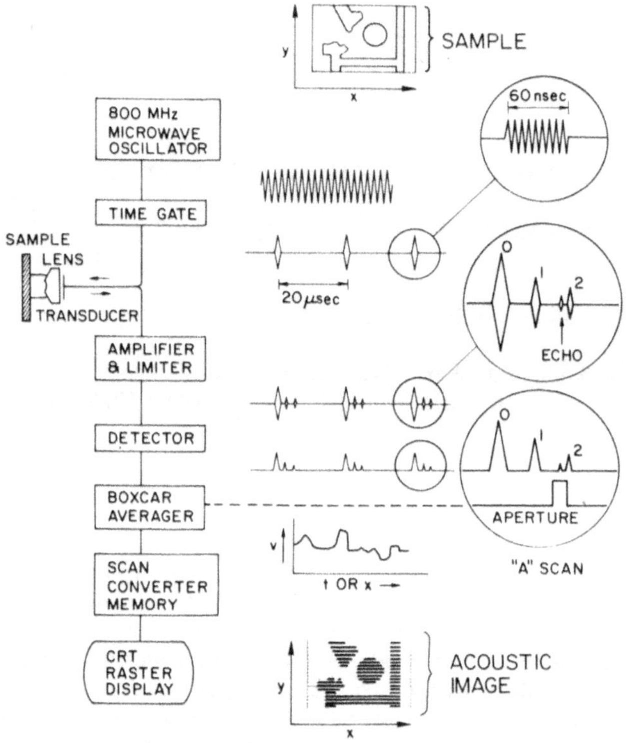

Fig. 1. Schematic diagram of 800 MHz pulsed-reflection acoustic microscope.

shows a series of peaks and valleys as a function of the spacing z, where z is the distance measured normal to the sample surface from the optimum focal position. Curves of $V(z)$ are characteristic of the elastic material properties. There is a great deal of information present in the signal reflected back from the sample, and it is natural to devise techniques for processing and displaying it in meaningful ways. Techniques such as nonlinear harmonic imaging,[7] differentiation,[8] phase imaging,[9,10] dark-field acoustic microscopy,[11] and stereo viewing[11] have been discussed. More recently, another useful technique has emerged: that of displaying an acoustic image in which one dimension represents lateral position on the sample (x) and the other dimension is the focal position (z) relative to the sample surface, with the transducer output $V(x,z)$ determining gray scale. Using this technique, Jipson[12] was able to distinguish subtle differences in the acoustic images of high-leakage and normal FET gate devices. The utility of two-dimensional Fourier transform processing for spatial filtering and image enhancement in acoustic microscopy has also been described.[12]

In our own recent work in acoustic microscopy of microelectronic devices, we have found several simple imaging techniques which either aid in the interpretation of the acoustic information or enhance its display. In this paper we describe the utility of composite A-scan, edge-enhancement, logarithmic processing, and false-color methods. These are all simple methods which do not require computer processing.

In order to demonstrate these techniques, we have taken as our sample a test structure consisting of plated metal lines on an organic film (polyimide) which is on a metallized ceramic wafer. The polyimide layer is 6 μm to 8 μm in thickness and the metallurgy is Cr-Cu-Ni, 4 to 5 μm in thickness. The lines are 12 μm in width connected to pads which are approximately 50 μm by 100 μm. A schematic cross-section of this test structure is shown in Fig. 2.

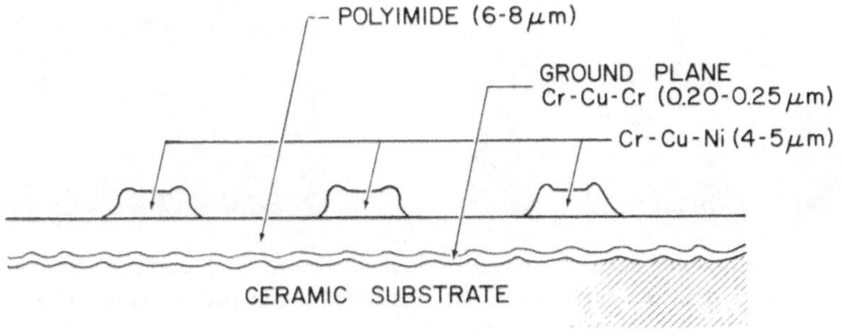

Fig. 2. Schematic cross-sectional view of plated test structure.

COMPOSITE A-SCAN

Normal gray-scale imaging [see Fig. 3(a)] is useful for defect detection and comparison with optical images, but does not provide quantitative data. A composite A-scan mode of imaging is shown in Fig. 3(b). The composite A-scan is extremely easy to produce. The block diagram of Fig. 4 shows the basic method. The acoustic signal V is summed with the sensed y-position to form the y-deflection on the CRT. A portion of the sensed y-position is summed with the sensed x-position to form the x-deflection. A constant voltage determines the beam brightness. With the composite A-scan, the transducer voltage can simply be read off the image at any point for quantitative comparisons between different features. Spatial resolution in the fast-scan direction is retained, but must be sacrificed in the slow-scan direction to avoid a confusing display. Thus the image of Fig. 3 is restricted to 64 scan lines. In our microscope, the acoustic lens and the sample are independently driven by voice-coils from digitally-synthesized waveforms, making it easy to interchange fast- and slow-axes, thereby allowing us to choose which axis (x or y) is to be the high resolution axis.

DERIVATIVE AND LOGARITHMIC COMPRESSION

We have found that derivative edge-enhancement can materially aid in the interpretation of acoustic images. Such techniques, of course, are not new and are

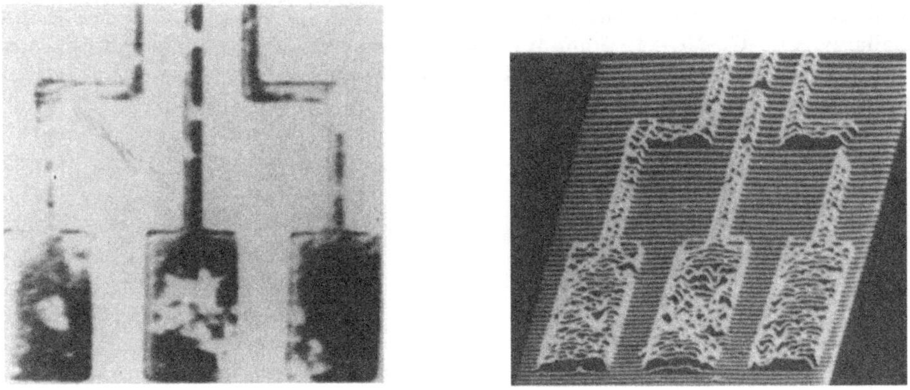

Fig. 3. (a) Normal acoustic image of test structure, (b) composite A-scan acoustic image in which the transducer voltage is displayed *vs.* *xy*-position.

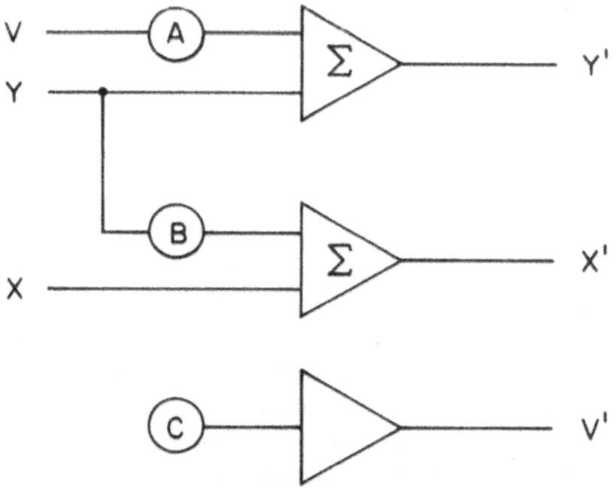

Fig. 4. Block diagram of circuit to produce the composite A-scan image. A, B, and C are gain adjustments.

employed extensively, for example, in scanning electron microscope (SEM) imaging. In acoustic microscopy, Wickramasinghe and Heiserman[8] have demonstrated that normal acoustic images can be combined with first and second derivative images to provide enhancement of edges and weak features. In the simplest case, which we employ, the averaged acoustic return signal is differentiated with respect to time, which produces a spatial derivative with respect to the x (fast) scan direction. Figure 5 shows normal and derivative images at two focal positions separated by approximately 4 μm.

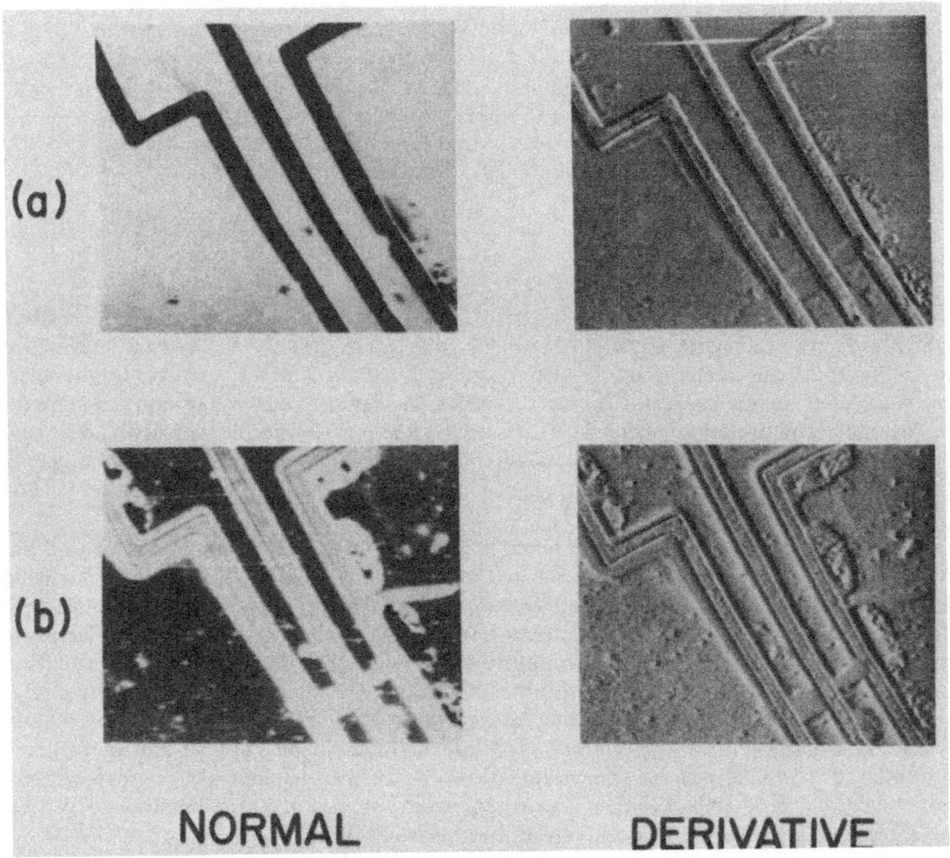

Fig. 5. (a) Normal and derivative acoustic images. (b) Same as (a) but with lens position 4 μm closer to the sample.

Most work in acoustic microscopy has employed lenses of high numerical aperature which give extremely shallow depths of field. Because of the $V(z)$ response, nonplanar samples are difficult to work with, and even perfectly planar samples are difficult to level properly. To correct this problem, a dynamic leveling system can be used which utilizes a PZT stack to drive the acoustic lens vertically from signals derived from the x-y scanning system.[13] However, in the absence of such a system, fringes may appear in images from non-level samples. The derivative imaging mode tends to eliminate these effects from the images, and one obtains black and white features on a neutral gray background. It must be pointed out, however, that in some cases this may be a disadvantage, since depth information may be lost. The derivative has the effect of compressing the *dynamic range* of the acoustic signal into the available gray-scale range of the video display device and photographic recording medium. As is evident from Fig. 5, many small features not readily apparent in the normal images are clearly defined in the derivative case.

Logarithmic compression is also extremely useful when it is desirable to compress the dynamic range. In this mode of operation the acoustic signal is input to a logarithmic amplifier whose output is stored in the scan-converter memory. Figure 6 shows comparative examples of acoustic images obtained by (a) normal gray-scale method, (b) logarithmic compression, where weak features (appearing white in this case) are enhanced, (c) derivative edge-enhancement previously mentioned, and (d) logarithm-of-derivative which combines the advantages of (b) and (c).

FALSE COLOR IMAGING

Schemes for producing false-color acoustic micrographs have been described which are direct acoustic analogs of optical color micrographs.[14] That is, images are to be formed from acoustic information obtained at two or three acoustic frequencies, color-coded to produce the micrograph. Unlike the optical case where marked changes in the dielectric function occur in a limited frequency range, the density and elastic constants which determine the acoustic signal are not very sensitive to small changes in the acoustic frequency.[15]

In the method we describe here, a single acoustic frequency is used to obtain false-color images. Often it is desirable to combine information from several acoustic focal planes in a single color-coded representation. Since typically the acoustic lens has high numerical aperature, the depth of focus is shallow. If a series of acoustic micrographs is taken at different depths, each micrograph will contain information primarily from that depth. By coding each of three such micrographs in the colors (*e.g.* red, yellow, and blue) a false-color composite is obtained. If such a procedure were to be carried out with an ordinary optical microscope rather than an acoustic microscope it would not be successful. For the reflection acoustic microscope, the color-coding procedure works well because of the $V(z)$ response of the lens/transducer, as will be evident from the following discussion.

In acoustic microscopy utilizing piezoelectric films for detection, the output signal is sensitive to both the phase and amplitude of the reflected acoustic wave. Typically, the reflected signal shows a series of peaks and valleys as the lens-to-object spacing is varied.[3] In a sample comprised of a multiplicity of layers or with layered surface topography such as those encountered in integrated circuits or planar packages, this property of the acoustic response gives rise to light and dark areas in the acoustic micrograph corresponding to high and low elevations on the sample (i.e. different focal planes).

A $V(z)$ curve for the sample under discussion, obtained with the acoustic lens focused on the top surface of a metal line is shown in Fig. 7(a). The response exhibits a typical series of peaks and valleys decreasing in magnitude away from the $z = 0$

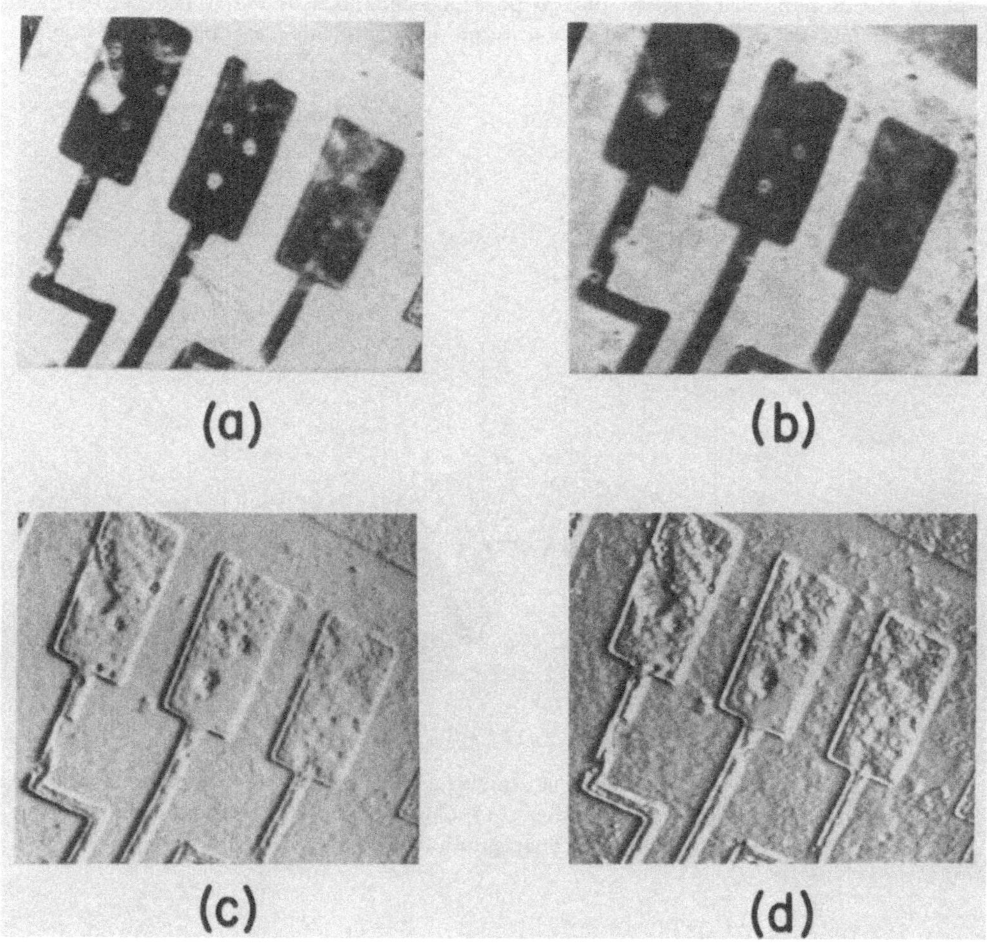

Fig. 6. (a) Normal gray-scale, (b) logarithmic, (c) derivative, (d) logarithm-of-derivative acoustic micrographs.

peak. In the region surrounding the metal lines, the $V(z)$ curve [Fig. 7(b)] shows an unusual behavior due to the presence of both the polyimide layer and the underlying ceramic substrate. In the acoustic images, regions which appear light correspond to large V. For an image taken at $z = 0$ some regions (*e.g.* the metal lines) appear light, whereas in an image taken at $z = -26$ μm, for example, those same regions appear dark (corresponding to smaller V). Thus by changing the lens-to-object spacing by the proper distance as indicated in Fig. 7, given regions of the sample which are all at the same elevation can be made to appear dark or light relative to regions at other elevations. This is sometimes referred to as a "phase-reversal effect" and is well-known in acoustic microscopy. By taking advantage of this effect, light regions at one focal position can be encoded in color, *e.g.* red, whereas those same regions appear to be black at another focal position due to phase cancellations at the transducer. These black regions will not be overlaid by a different encoding color, *e.g.* blue, and so on.

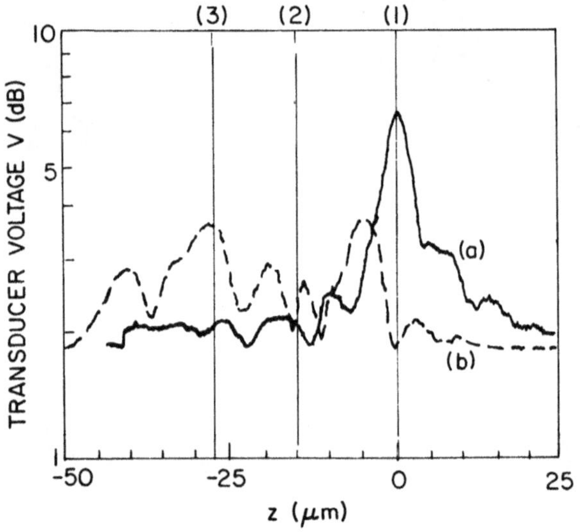

Fig. 7. *V*(*z*) response curves for test structure. Trace (a) is for the metal lines, (b) for surrounding area. Numbers (1), (2), and (3) refer to focal positions used for the false-color acoustic micrograph of Fig. 8.

Typically, each acoustic micrograph image is displayed on a black-and-white television monitor. A simple method for accomplishing the false-color overlay process is to take a multiple-exposure photograph of the monitor with color film, using a different transparent color filter for each acoustic focal position. Red, yellow, and blue filters can be used to selectively activate the red, yellow and blue dye layers in the film. Alternatively, the image information can be processed by a computer and presented on a color display monitor. To avoid image degradation, precise registration between exposures is required. In our instrument, registration reproducibility of ± 0.25 μm is achieved using a digitally-commanded servo system.[2] Figure 8 illustrates schematically the false-color process. Three images taken at the three focal positions indicated in Fig. 7 are color-coded in red, yellow, and blue and are combined into a single composite false-color micrograph containing all of the information of the separate micrographs. (Unfortunately, the expense of publishing a full-color image in the present proceedings prevents showing the actual color micrograph.)

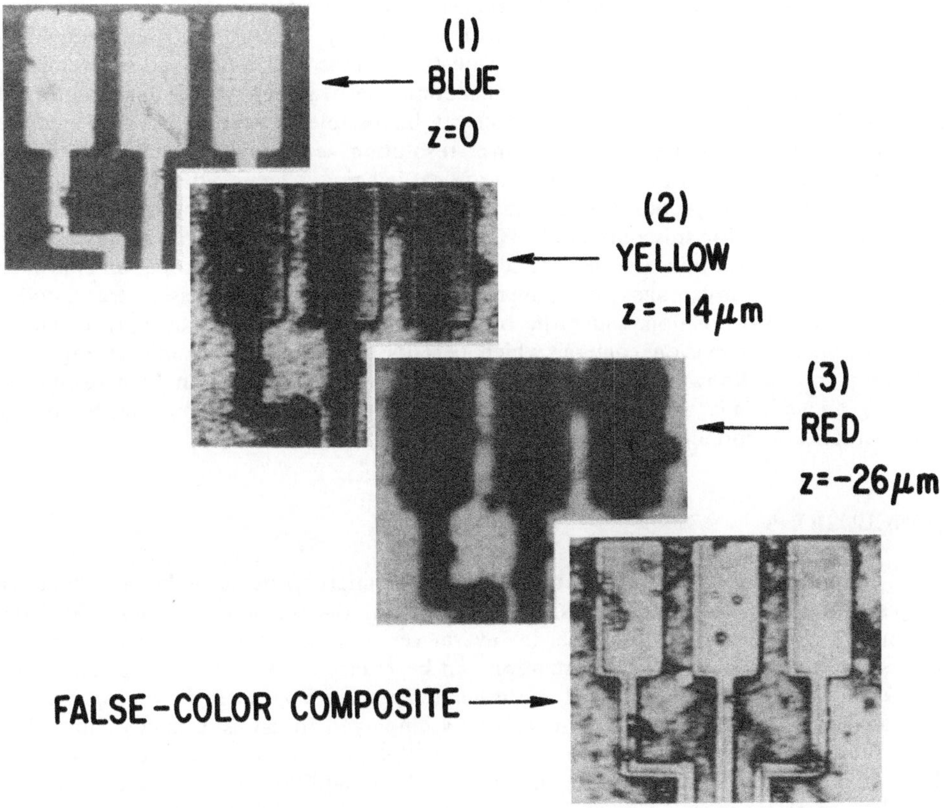

Fig. 8. Schematic illustration of acoustic micrographs taken at three different focal positions, in precise registration with each other. The separate planes are color-coded to form the composite false-color acoustic micrograph.

Although a small penalty in resolution is paid by operating at non-optimum focal positions, the ability to combine information from several focal positions (which may be at or below the sample surface) into a single false-color micrograph has proved to be a valuable asset for interpretation of the acoustic information. Note that this procedure is distinctly different from a trivial mapping of a single focal-plane gray-scale image into color.

COMPARISON OF OPTICAL AND ACOUSTICAL IMAGES

As a separate issue, we stress that a scanning acoustic microscope should include as an adjunct an optical microscope to permit fast sample alignment and absolute comparison between surface optical features and acoustic features. In previous instruments, which were built primarily for experimentation, this was attempted by first taking an optical photograph of a region of interest, and then examining approximately the same area with the acoustic microscope, usually at several focal positions. This awkward procedure was sometimes satisfactory for regularly patterned samples like integrated circuits, but was nearly impossible to accomplish with samples of unstructured appearance. In the lower-resolution scanning laser acoustic microscope (SLAM) instrument, acoustical and optical imaging is achieved simultaneously.[16] For high-resolution reflection acoustic microscopy, a combination acoustic/optical lens has been proposed for this purpose.[17] We have taken the expedient of simply mounting a standard optical microscope in tandem with the acoustic microscope, an approach which has several advantages of its own. The sample is then precisely translated a short distance into position under the optical or acoustical lens. The optical microscope includes a television camera, which permits electronic processing, storage and display. To our knowledge, this feature has not been included in high-resolution scanning acoustic microscopes elsewhere, and would be of prime importance to any future practical embodiment of the instrument.

CONCLUSIONS

Most of the published results of acoustic microscopy have been normal *xy* images whose gray-scale represents the magnitude of the transducer signal. We have shown that by processing this signal in several very simple ways, substantially greater understanding of the acoustic information can be realized. With the exception of the unique false-color imaging method, the other techniques have been in widespread use in other fields, but have not been extensively employed in acoustic microscopy. We have successfully employed these imaging techniques to acoustic microscopy of thin-film structures on silicon and gallium arsenide, in addition to the plated structures described in this paper. The present discussion has been intentionally restricted to very simple techniques which do not require the use of a digital computer. More elaborate methods can be used to good avail with digital processing; indeed the literature abounds with procedures from other fields which we feel confident can be applied with similar utility to acoustic microscopy.

ACKNOWLEDGMENTS

We express our gratitude to C. F. Quate, who introduced us to acoustic microscopy and provided lens/transducer assemblies, and to V. B. Jipson who advised us in many of the intricacies involved. We thank D. A. Thompson for his foresight in suggesting the project, and T. H. DiStefano for his interest, help, and enthusiasm. We thank R. J. Lang for his careful work in fabricating the mechanical portion of the instrument, and D. R. Vigliotti for his assistance with the microwave equipment. The test sample used in this study was fabricated by M. Blakeslee.

REFERENCES

[1]R. A. Lemons and C. F. Quate, *Appl. Phys. Lett.* **25**, 251 (1974).

[2]For a brief description of our acoustic microscope, particularly the microwave and digital voice-coil scanning electronics, see R. L. Hollis and R. Hammer, in Proceedings of the International Symposium on Scanned Image Microscopy, The Royal Society, London, September 1980 , Academic Press, (to be published).

[3]R. G. Wilson, R. D. Weglein, and D. M. Bonnell, *in* Semiconductor Silicon 1977, *Proc.* **77**(2), ed. by H. R. Huff and E. Sirtl, Electrochemical Society, Princeton, 431 (1977).

[4]R. D. Weglein and R. G. Wilson, *Electronics Lett.* **14**(12), 352 (1978).

[5]A. Atalar, *J. Appl. Phys.* **49**(10), 5130 (1978).

[6]W. Parmon and H. L. Bertoni, *Electronics Lett.* **15**(21), 684 (1979).

[7]R. A. Lemons and C. F. Quate, *Science* **188**, 905 (1975).

[8]H. K. Wickramasinghe and J. Heiserman, *Electronics Lett.* **13**(25), 776 (1977).

[9]H. K. Wickramasinghe and M. Hall, *Electronics Lett.* **12**(24), 637 (1976).

[10]J. Attal and G. Cambon, *Electronics Lett.* **14**(15), 472 (1978).

[11]W. L. Bond, C. C. Cutler, R. A. Lemons, and C. F. Quate, *Appl. Phys. Lett.* **27**(5), 270 (1975).

[12]V. B. Jipson, Ph.D. dissertation, Stanford University, 1979.

[13]A dynamic leveling system due to D. Rugers was implemented by V. B. Jipson [see Ref. 11].

[14]U. S. Patent No. 4,028,933, R. A. Lemons and C. F. Quate, (1977).

[15]Nevertheless, it should be possible using a multiple-frequency technique to obtain color micrographs containing depth information derived from the frequency dependence of the $V(z)$ function (V. B. Jipson, private communication).

[16]In addition to side-by-side acoustical and optical presentations, superposition of acoustical and optical micrographs on a color TV monitor in which each image is color-coded is briefly discussed by L. W. Kessler, *J. Acoust. Soc. Am.* **55**(5), 909 (1974).

[17]R. C. Addison (unpublished).

LIST OF PARTICIPANTS

M. F. Adams
University of Sheffield
177 Industry St.
Sheffield, U.K.

P. Alais
Universite' Paris VI
Lab. de Mécanique Physique
2, Place de la Gare de Ceinture
78210 St-Cyr l'Ecole - FRANCE

R. Ankri
C.G.R. Ultrasonic
9, Chaussée de Paris
77102 Meaux - FRANCE

J. Attal
Centre d'Etudes Electroniques des Solides
U.S.T.L., Place E. Bataillon
34000 Montpellier - FRANCE

M. Auphan
L.E.P.
3, Avenue Descartes
94450 Limeil-Brévannes - FRANCE

V.M. Baborovsky
Tube Investments Ltd.
TIRL, Hinxton Hall
Saffron Walden - U.K.

S.L. Bailey
"Ultrasonics" Journal
IPC, P.O. Box 63, Bury Street
Guildford, Surrey - U.K.

J.C. Bamber
Institute of Cancer Research
Clifton Avenue
Sutton, Surrey - U.K.

P. Bardouillet
FRAMATOME
Tour Fiat, Tour de la Coupole
92084 Paris La Défense - FRANCE

J.F. de Belleval
Universite de Compiegne
B.P. 233
60206 Compiegne - FRANCE

O. Bentsen
Central Institute for Industrial Research
Askergt 1
Oslo 4 - NORWAY

G. Berger
Laboratoire de Biophysique
CHU Cochin Port-Royal
24, rue du Fb. St-Jacques
75674 Paris - FRANCE

A. J. Berkout
Delft University of Technology
Depart. of Applied Physics
2600 GA Delft - NETHERLANDS

M. Bernard
E.N.S.T.
46, rue Barrault
75013 Paris - FRANCE

G. Blanc
E.T.C.A.
16 bis, Av. Prieur de la Côte
d'Or, 94114 Arcueil - FRANCE

G.P. Blanschong
D.R.E.T.
26, Bd. Victor
75996 Paris Armée - FRANCE

M. Boynard
U.E.R. Biomedicale Service de
Biophysique
45, rue des Saints-Pères
75006 Paris - FRANCE

H. Brettel
Gesellschaft f. Strahlen u.
Umweltforschung, m.b.H.
Abteilung
Angewandte Optik/R
Ingolstädter Landstraffe 1
D-8042 Neuherberg - GERMANY

P.H. Brown
Central Research Lab. of
Thorn
EMJ LTD
Shoenberg House, Trevor Road
Hayes, Middlesex - U.K.

C.D. Bruneel
Université de Valenciennes
Université
59326 Valenciennes - FRANCE

C.B. Burckhardt
Hoffmann-Laroche and Co. AG
4000 Basel - SWITZERLAND

D.S. Cairns
Diagnostic Sonar LTD
5 Young Square
Livingston EH54 9BX - U.K.

M. Certo
Société C.T.S.E.
B.P. 12081
Milan - ITALY

M. Clement
Lab. de Biophysique CHU
Cochin
Port Royal
24, rue du Fb. St-Jacques
75674 Paris - FRANCE

R. M. Condomines
Faculté de Médecine - Dép.
Biophysique
Université Autonome de
Barcelone - SPAIN

Courtel
Service de Biophysique CHU
Cochin
24, rue du Fb. St-Jacques
75674 Paris - FRANCE

Ph. Defranould
Thomson C.S.F.
06802 Cagnes s/Mer - FRANCE

R.J. Dickinson
G.E.C. Hirst Research Centre
Wembley, Middlesex - U.K.

H. Diepers
Siemens AG
Veit Stoff Str. 44
D-8552 Höchstadt - GERMANY

J.L. Dion
Université du Quebec
C.P. 500
Trois Rivières, Québec,
CANADA

J. Donjon
L.E.P.
3, Av. Descartes
94550 - Limeil-Brevannes, FRANCE

A. Dverinckx
Rohe Scientific Corporation
Philips Medical Systems
Santa Ana, CA - U.S.A.

J.C. Dufay
Labo Régional des Ponts et Chaussées
11, rue Laplace
41000 Blois - FRANCE

Ph. Durouchoux
D.R.E.T.
26, Bd Victor
75996 Paris Armée - FRANCE

H. Ermert
Institut für Hochfrequenztechnik
der Universität Erlangen-Nürnberg
Cauerstr. 9
8520 Erlanger - GERMANY

G. Faber
Delft University of Technology
Depart. of Applied Physics
2600 GA Delft - NETHERLANDS

M. Fink
Université Paris VI
Lab. de Mécanique Physique
2, Place de la Gare de Ceinture
78210 St-Cyr l'Ecole - FRANCE

P.J. Finsterwald
A.D.R. Ultrasound
224 S. Priest
Tempe, Arizona - U.S.A.

R. Fox
King's College University of London
University of London
London - U.K.

J. Frohly
Université de Valenciennes
Université
59326 Valenciennes - FRANCE

C. Gazanhes
Laboratoire de Mécanique et Acoustique
31, chemin J. Aiguier
13009 Marseille - FRANCE

J.F. Gelly
Thomson C.S.F.
06802 Cagnes s/Mer - FRANCE

G. Grabner
Siemens AG
Veit Stoff Str. 44
D-8552 Höchstadt - GERMANY

G. Grall
Thomson CSF - DASM
2, rue du Mal Juin
29217 Le Conquet - FRANCE

G. Grau
I.F.P.
1 et 4 Avenue de Bois Préau
92506 - Rueil-Malmaison, FRANCE

A. Gravelle
FRAMATOME
39, rue Amédé Usséglio
92350 Le Plessis Robinson-FRANCE

M.A. Greenfield
Dept. of Radiology
Univ. of Calif.
Los Angeles, CA. 90024 - U.S.A.

P. Hartemann
Thomson CSF (LCR)
Domaine de Corbeville
91401 Orsay - FRANCE

P.Y. Hennion
Université Paris VI
Lab. de Mécanique Physique
2, Place de la Gare de Ceinture
78210 St-Cyr L'Ecole - FRANCE

A. Hermanne
Free University of Brussels
Pleinlaan, 2
1050 Brussels - BELGIUM

A. Herment
INSERM
ERA CNRS 785, Hôpital Broussais
96, rue Didot
75674 Paris - FRANCE

C.R. Hill
Institute of Cancer Research
Royal Marsden Hospital
Sutton, Surrey - U.K.

D. Hiller
Instut für Hochfrequenztechnik
der Universität Erlanger - Nürnberg
Cauerstr. 9
8520 Erlangen - GERMANY

K. Hogmoen
Det Norske Veritas
P.O. Box 300
N-1322 Hovik - NORWAY

R.L. Hollis
I.B.M. Research
P.O. Box 218
Yorktown Hts, NY 10598 - U.S.A.

P. Holst
Brüel & Kjaer Industri A/S/
Naerum Hovedgade 18
2850 Naerum - DENMARK

B. Hosten
Lab. de Mécanique Physique
Université de Bordeaux I
33405 Talence - FRANCE

Z. Houchangnia
Université Paris VI
Lab. de Mécanique Physique
2, Place de la Gare de Ceinture
78210 St-Cyr L'Ecole - FRANCE

H. Hue
Université de Valenciennes
Université
59326 Valenciennes - FRANCE

L. Hutchins
Depart. of Medical Physics
Royal Postgraduate Medic. School
Hammersmith Hospital
London - U.K.

S.O. Ishrak
Dept. of Electronic and Electrical Engineering
Kings' College
London - U.K.

E. Jacobsen
Norwegian Underwater Institute
Post Box 6
N-5034 Laksevag - NORWAY

G.J. Kioz
Varian Assoc.
Varian 611 Hansen Way
Palo Alto, Ca. 94303 - U.S.A.

F. Jensen
KAS in Gentofte
Plantagevej 12
DK-3460 Birkeroed - DENMARK

J.P. Jones
University of California Irvine
Dept. RAD. SCI.
Irvine, California - U.S.A.

C. Kammoun
Universite Paris VI
Lab. de Mécanique Physique
2, Place de la Gare de Ceinture
78210 St-Cyr L'Ecole - FRANCE

E. Karrer
Hewlett-Packard
1501 Page Mill Rd.
Palo Alto, CA. 94304 - U.S.A.

K. Katakura
Hitachi Ltd.
1-280 Higashi-koigakubo
Kokubunji-shi, Tokyo - JAPAN

B.T. Khuri-Yakub
Stanford University
Edward L. Ginzton Lab. Rm. 10
Stanford, CA. - U.S.A.

A. Kristensen
Electronics Research Lab.
7034 Trondehim
NTH - NORWAY

F.A. Kuypers
Philips Medical Systems
Philips Medical Systems
NETHERLANDS

M. La Greve
Université Paris VI
Lab. de Mécanique Physique
2, Place de la Gare de Ceinture
78210 St-Cyr L'Ecole - FRANCE

J.L. Lamarque
Laboratoire de Radio Diagnostic
Centre Hospitalo Universitaire
St-Eloi
34000 Montpellier - FRANCE

C.T. Lancee
Erasmus University
P.O. Box 1738
3000 DR Rotterdam - NETHERLAND

M.T. Larmande
Université Paris VI
Lab. de Mécanique Physique
2, Place de la Gare de Ceinture
78210 St-Cyr L'Ecole - FRANCE

J. Larson
Hewlett-Packard
1501 Page Mill Road
Palo Alto, CA. 94304 - U.S.A.

H. Lasota
Institute of Tele-communication
Technical University of Gdansk
Gdansk - POLAND

J. Lefebvre
Université de Valenciennes
Université
59326 Valenciennes - FRANCE

J.P. Lefebvre
CNRS - Lab. de Mécanique et d'Acoustique
Chemin J. Aiguier
Marseille - FRANCE

S.V. Letcher
University of Rhode Island
Dept. of Physics
Kingston, RI 02881 - U.S.A.

M. Lutkemeyer-Hohmann
Krupp Atlas-Electronik
Sebaldsbrücker Heerstr.235
Postfach 448545
D 2800 BREMEN 44

G. Maderlechner
Siemens FL FKS 12
Otto-Mahn-Ring 6
D 8000 München 83 - GERMANY

M. Mahfuz Ahmed
University of California, Irvine
8401 Middle town Lane
Westminster, CA. 92683 - U.S.A.

P. Maguer
French Navy - DGA
GESMA - DCAN
29240 Brest Naval - FRANCE

J. Marini
FRAMATOME
Chemin du Bois Martin
78160 Marly Le Roi - FRANCE

J.A. McKnight
R.N.P.D.L. UKA.E.A.
Risley Warrington
Cheschire - U.K.

A.F. Metherell
South Bay Hospital
514 North Prospect
Redondo Beach, CA. 90277 - U.S.A.

D. Michaux
C.H.U. Cochin
24, rue du Fb. St-Jacques
75014 Paris - FRANCE

S. Miller-Jones
A.D.R. Ultrasound
2224 St. Priest
Tempe, Arizona - U.S.A.

B. Nongaillard
Université de Valenciennes
Université
59326 Valenciennes - FRANCE

J. Ophir
University of Texas, Medical School
307 Gershwin
Houston, TX - U.S.A.

P. Peronneau
INSERM
ERA CNRS 785, Hôpital Broussais
96, rue Didot
75674 Paris - FRANCE

J. Perrin
C.H.U. Cochin, Port Royal
24, rue Fb. St-Jacques
75674 Paris - FRANCE

E.J. Pisa
Rohe/Philips
28131 Casitas Ct.
Laguna Niguel, CA. - U.S.A.

B. Poiree
D.R.E.T.
26, Bd. Victor
75996 Paris Armée - FRANCE

L. Pourcelot
Biophysique Medicale
U.E.R. de Médecine
2 bis, Bd Tonnellé
37032 Tours Cedex - FRANCE

L. Prud'hom
C.G.R.
9, Chaussée de Paris
Meaux - FRANCE

G. Quentin
Université, Paris VII
4 Place Jussieu
75005 - Paris, FRANCE

M. Rauch
VALEO
6, rue Gambetta
93406 St.-Ouen - FRANCE

R.J. Redding
Design Automation Lmt.
Berkshire
London - U.K.

W. Rey
Philips Research Laboratory
2, Av. van Becelaere, Ble 8
B-1170 Bruxelles - BELGIUM

J. Ridder
Delft University of Technology
Dept. of Applied Physics.
2600 GA Delft - NETHERLANDS

U. Roeder
Gesellschaft f. Strahlen u.
Umweltsforschung m.b.H.
Abteilung
Angewandte Optik/R
Ingolstäder Landstraffe 1
D-8042 Neuherberg - GERMANY

J. Roux
Lab. de Mécanique Physique
Université de Bordeaux I
33405 Talence - FRANCE

C. Rusca
Institut für medizinische
Radiologie
Dienst für Radiologische Physik
Kantonspital
Basel - SWITZERLAND

A. Sales
Laboratoire de Physique Expérimentale C.H.U. Cochin
24, rue du Fb. St-Jacques
75014 Paris - FRANCE

G. Salvini
ENSP - Université III
St. Jérôme
13397 Marseille Ced. 4 - FRANCE

J.S. Sandhu
University of London, Chelsea College
22, Sundorne Road
Charlton S.E. 7 - U.K.

J. Sato
Matsushita Research Institute
Tokyo, Inc.
4896 Ikuta Tam-ku
Kawasaki - JAPAN

J.O. Schaefer
Institut für
Hochfrequenztechnik
der Universität Erlangen-Nürnberg
Cauerstr. 9
8520 Erlangen - GERMANY

C. Scherg
Gessellschaft für Strahlen-
und Umweltforschung
Abteilung für Angewandts
Optik
Neüherberg - GERMANY

D. Ph. Schmidt
Technisch Physique Dienst
TNO-TH
Stieltjesweg 1
Delft - NETHERLANDS

H. Schomberg
Philips GmbH Forschuns
Laboratorium Hamburg
Vogt Koellnstrasse 30
D 2 Hamburg 54 - GERMANY

I.R. Smith
Dept. Electrical and
Electronic Engineering
University College London
Torrington Place
London wclE 7JE - U.K.

S.W. Smith
Bureau of Radiological Health
FDA
5600 Fishers La
Rockville, MD. - U.S.A.

J. Souquet
VARIAN
84 D Escondido
Stanford, CA. 94305 - U.S.A.

J.J. Stamnes
Central Institue for Industrial Research
Vaekerövn 21
Oslo 2 - NORWAY

C. Susini
IROE - CNR
Via Panciatichi 64
FIRENZE - ITALY

P. 'T HOEN
L.E.P.
3, Av. Descartes
94450 Limeil-Brevannes - FRANCE

A. Thomas
FRAMATOME
B.P. 13
71380 Saint-Marcel - FRANCE

F. Thurstone
Biomedical Engineering Dept.
Duke University
Durham, NC. - U.S.A.

E. Tournier
C.E.A. - Centre d'Etude Nucléaire de Grenoble
1, rue Philis de la Charce
38000 Grenoble - FRANCE

P. Tournois
Thomson CSF, ASM Division
06802 Cagnes s/Mer - FRANCE

B.O. Trebitz
Graduate Aeronautical Lab.
California Institute of Tech.
204-45 Caltech
Pasadena, CA. 91125 - U.S.A.

H. Tretout
Avions Marcel Dassault - Breguet Aviation
B4, Résidence de Champhleury
78300 Carrières s/Poissy - FRANCE

W. Urbach
Lab. Biophysique - UER Cochin
Port Royal
24, rue de Fb. St-Jacques
75674 Paris - FRANCE

K. Vammen
Brüel & Kjaer
Naerum DK-2850 - DENMARK

L.F. van der Wal
Dpt. of Applied Physics
2600 GA Delft - NETHERLANDS

D.W. van Wulfften Paltke
Delft University of Technology
Labouchärelaan
2283 EK Rÿswgh - NETHERLANDS

K. Verhulst
Technisch Physiche Dienst
TNO-TH
Stieltjesweg 1
Delft - NETHERLANDS

G. Wade
University of California
Santa Barbara, CA. - U.S.A.

O. Wess
Dornier System
Friedrichshafen - GERMANY

T.A. Wittingham
Regional Medical Physics Dept.
Newcastle General Hospital
Newcastle Upon Tyne - U.K.

D.E. Yuhas
Sonoscan Inc.
Bensenville, ILL. - U.S.A.

INDEX

Batch number: 10838292

Printed by Printforce, the Netherlands